FLOWS OF ENERGY AND MATERIALS IN MARINE ECOSYSTEMS

Theory and Practice

NATO CONFERENCE SERIES

I Ecology
II Systems Science
III Human Factors
IV Marine Sciences
V Air–Sea Interactions
VI Materials Science

IV MARINE SCIENCES

Volume 1 Marine Natural Products Chemistry
edited by D. J. Faulkner and W. H. Fenical

Volume 2 Marine Organisms: Genetics, Ecology, and Evolution
edited by Bruno Battaglia and John A. Beardmore

Volume 3 Spatial Pattern in Plankton Communities
edited by John H. Steele

Volume 4 Fjord Oceanography
edited by Howard J. Freeland, David M. Farmer, and Colin D. Levings

Volume 5 Bottom-interacting Ocean Acoustics
edited by William A. Kuperman and Finn B. Jensen

Volume 6 Marine Slides and Other Mass Movements
edited by Svend Saxov and J. K. Nieuwenhuis

Volume 7 The Role of Solar Ultraviolet Radiation in Marine Ecosystems
edited by John Calkins

Volume 8 Structure and Development of the Greenland–Scotland Ridge
edited by Martin H. P. Bott, Svend Saxov, Manik Talwani, and Jörn Thiede

Volume 9 Trace Metals in Sea Water
edited by C. S. Wong, Edward Boyle, Kenneth W. Bruland, J. D. Burton, and Edward D. Goldberg

Volume 10A Coastal Upwelling: Its Sediment Record
Responses of the Sedimentary Regime to Present Coastal Upwelling
edited by Erwin Suess and Jörn Thiede

Volume 10B Coastal Upwelling: Its Sediment Record
Sedimentary Records of Ancient Coastal Upwelling
edited by Jörn Thiede and Erwin Suess

Volume 11 Coastal Oceanography
edited by Herman G. Gade, Anton Edwards, and Harald Svendsen

Volume 12 Hydrothermal Processes at Seafloor Spreading Centers
edited by Peter A. Rona, Kurt Boström, Lucien Laubier, and Kenneth L. Smith, Jr.

Volume 13 Flows of Energy and Materials in Marine Ecosystems: Theory and Practice
Edited by M. J. R. Fasham

FLOWS OF ENERGY AND MATERIALS IN MARINE ECOSYSTEMS
Theory and Practice

Edited by

M. J. R. Fasham
Institute of Oceanographic Sciences
Godalming, Surrey, United Kingdom

Published in cooperation with NATO Scientific Affairs Division

PLENUM PRESS · NEW YORK AND LONDON

Library of Congress Cataloging in Publication Data

Main entry under title:

Flows of energy and materials in marine ecosystems.

(NATO conference series. IV. Marine sciences; v. 13)
"Proceedings of a NATO Advanced Research Institute, held May 13–19, 1982, in Carcans, France"—T.p. verso.
Includes bibliographies and index.
1. Marine ecology—Congresses. 2. Biogeochemical cycles—Congresses. 3. Food chains (Ecology)—Congresses. I. Fasham, M. J. R. II. North Atlantic Treaty Organization. Scientific Affairs Division. III. Series.
QH541.5.S3F57 1984 574.5′2636 83-20322
ISBN 0-306-41519-4

Proceedings of a NATO Advanced Research Institute, held May 13–19, 1982, in Carcans, France.

A Division of Plenum Publishing Corporation
233 Spring Street, New York, N.Y. 10013

Printed in the United States of America

PREFACE

The impetus for the conference held at Bombannes, France in May, 1982 arose out of a Scientific Committee on Oceanic Research (SCOR) Working Group on "Mathematical Models in Biological Oceanography". This group was chaired by K.H. Mann and held two meetings in 1977 and 1979. At both meetings it was felt that, although reductionist modelling of marine ecosystems had achieved some successes, the future progress lay in the development of holistic ecosystem models. The members of the group (K.H. Mann, T. Platt, J.M. Colebrook, D.F. Smith, M.J.R. Fasham, J. Field, G. Radach, R.E. Ulanowicz and F. Wulff) produced a critical review of reductionist and holistic models which was published by the Unesco Press (Platt, Mann and Ulanowicz, 1981). One of the conclusions of this review was that, whether holistic or reductionist models are preferred, it is critically important to increase the scientific effort in the measurement of physiological rates for the computation of ecological fluxes. The Working Group therefore recommended that an international meeting should be organized which would attempt to bring together theoretical ecologists and biological oceanographers to assess the present and future capability for measuring ecological fluxes and incorporating these data into models.

An approach was made to the Marine Sciences Panel of the NATO Science Committee who expressed an interest in funding such a meeting. They awarded a planning grant and a planning group was formed consisting of M.J.R. Fasham, M.V. Angel, T. Platt, R.E. Ulanowicz, G. Radach, F. Wulff and P. Lasserre. The group met at the Arcachon Biological Station of the University of Bordeaux in October 1980. It was decided to organize a meeting that would be a mixture of keynote speeches on what were considered critical or neglected aspects of the marine food chain, plus working group forums on the major functional categories of energy and matter transfer within the food chain. Professor Lasserre offered to act as local organizer and search for a suitable location in the Bordeaux region.

The proposal for a NATO Advanced Research Institute on the "Flow of Energy and Materials in Marine Ecosystems: Theory and Practice" was accepted by the NATO Marine Sciences Panel and the meeting was held at the "Les Dunes" Village de Vacances at Bombannes, Carcans, France, between the 13th and 19th May, 1982. As well as the funding from NATO generous financial assistance was received from SCOR and the French National Research Council which enabled us to provide travel assistance to many of the participants. It was generally agreed that the bringing together of biological oceanographers and ecologists from different countries and with different specializations created an intellectually stimulating and exciting meeting and I hope that some of this excitement has spilled over into the book. I also hope that the Working Group reports will fulfil their intended function of surveying the achievements and problems in their field and, perhaps more importantly, stimulating future research.

Finally, I would like to thank Prof. Pierre Lasserre, his wife Bernadette and his staff for their untiring organizational efforts during the conference which did so much to make it a success.

M.J.R. Fasham

CONTENTS

PART ONE - INVITED PAPERS

PART TWO - WORKING GROUP REPORTS

PART 1

INVITED PAPERS

IMPORTANCE OF MEASURING RATES AND FLUXES IN MARINE ECOSYSTEMS

Alan Longhurst

Ocean Science and Surveys, Atlantic
Department of Fisheries and Oceans
Bedford Institute of Oceanography
Dartmouth, Nova Scotia, Canada

INTRODUCTION

In a few decades we have come from natural history to the beginnings of predictive ecology, and probably the greatest advance is that we now know what has to be done even if we are not yet very good at it. It is my purpose to examine what chiefly retards our progress towards an ecology which will be verifiable and predictive: not in a spirit of advocacy, but simply because simulation models are themselves useful models of the integrative process in quantitative ecology, I shall examine what is their principal failing at the present time to illustrate the subject of this presentation. I shall examine whether, in my opinion, integrative ecology most lacks (i) information about how animals are distributed and in what numbers, (ii) skill in mathematical techniques in the simulation of of ecological interactions, (iii) the development of holistic, ground-plan ecological concepts on which to base more clever integration, or (iv) information concerning the rates of ecological processes.

It should come as no surprise, given the title of my presentation and my own interests, that I concur with the conclusion of the SCOR Working Group on Modelling in Biological Oceanography that what we most lack for model construction, and hence, I believe, for all of integrative ecology, is quantification of ecological rates and processes. It is a measure of the extent to which this is not yet widely accepted that many activities in ecological modelling, specialist journals, reviews and workshops, are still heavily preoccupied with the mechanics of simulation as if what is chiefly lacking is an understanding of the mathematics required by ecological integration. As Pielou (1981) has implied, the ecological activities of mathematicians, statisticians and engineers in recent years has encouraged the decoupling of ecological modelling, and hence ecological integration, from investigative ecology. Somewhat to mix my metaphors, as war is too important to be left to the generals, too much hangs on the achievement of verifiable ecology for it to be left to the mathematicians. I hope to show that what is mostly required now is sustained investigation of ecological rates and energy-flow: what Pielou (1981) has called a renewal of investigative ecology. It is my view that both the holist and the reductionist (if indeed they are really different) stand equally in need of such a renewal.

Integrative ecology in general, with simulation models as a special case, has a common structural form at all levels of abstraction; a thermodynamic model of oceanic primary production, or a model of the growth of a single species in a single location, can only be successful if four elements are appropriately developed and matched. It is perhaps only in the case of dimensionless process models that this is not correct.

These four elements, of all models representing biota in their environment, must include:

- a numerically correct description of the state variables under their initial conditions, and for a sufficient period for subsequent validation of the model or concept being tested;
- a mathematically competent statement of the rate variables required to link the state variables, containing supportable constant and parameter values as required by the rate eqations;
- a coherent statement of the forcing functions to act upon the state and rate variables, including also a sufficient description of those boundary conditions which advect properties across the boundaries of the area of interest; and
- a means of formulating and coding the model so that the mathematical structure and simulation language used are competent for the task, and accessible to all the ecologists (who may not be computer specialists) engaged in the work.

These are not easy conditions to satisfy, as I hope to show. It is the second, the suite of variables describing rates and fluxes in the ecosystem, that will more usually be lacking than the remainder. Additionally, of course, the ecological integration, be it a formal model or not, can only be as good as the basic concept with which it is constructed. In my opinion, the comments which follow concerning four formal simulation models are equally true for any technique to integrate ecological analysis into a species-specific, a location-specific or a regional description of the way in which the living world functions.

SARDINE RECRUITMENT MODEL

Figures 1 and 2 describe a conceptual model with a single output, the rate of II-group recruitment in the California sardine (*Sardinops corerulea*), formulated in 1968 by a group in La Jolla (Smith, Lasker, Hunter, Lenarz and Longhurst) but not previously published. Such conceptual models describe what seem to be the processes controlling the level of a single variable, in this case the population size of two-year old fish of a single species with a restricted geographical range, and do not attempt to model a complete ecosystem. Thus they depend very heavily on an intuitive statement concerning the ecological patheays relevant to the state variable of interest, and on a general description of the ecological matrix within which the model is to function.

The basic ecological information for this model came from the CalCOFI surveys, which provided nearly two decades of ecological and environmental data from the California Current, so that it was possible to obtain a reasonable construct for what seemed to control sardine recruitment. A similar construct could also have been formulated for the northern anchovy (*Engraulis mordax*).

Although it might have been clearer if the process flow diagram for this model had been drafted in a more familiar convention, such as the energy circuit language of Odum, or Forrester's dynamic simulation symbols, it is nevertheless clear that the model contains elements of state variables, rate variables and forcing functions. Perhaps most noteworthy is the fact that pragmatic experience with the CalCOFI data, and observations such as those of Wickett (1967) on consequences for California Current biota of biological events upstream in the north-east Pacific Ocean, now confirmed by Bernal

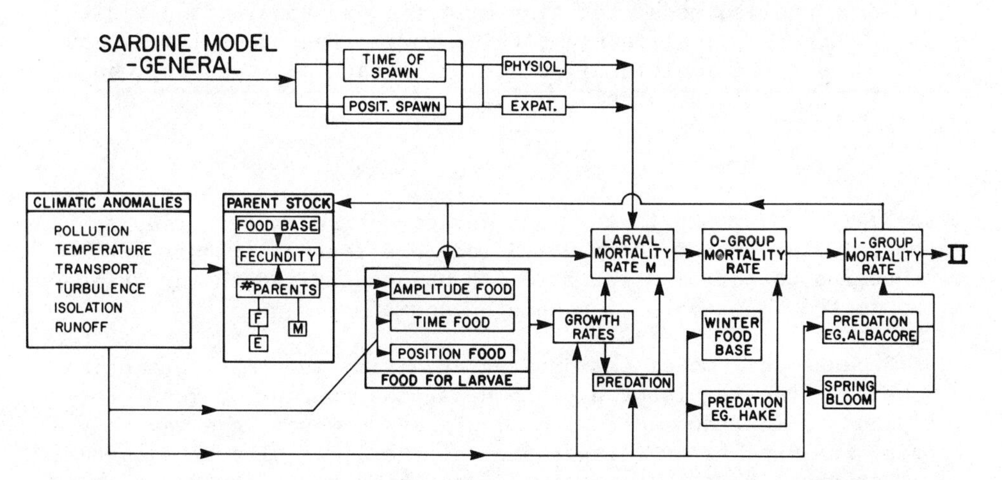

Fig. 1. General model relating climatic variability and adult stock of California sardine to II group recruitment.

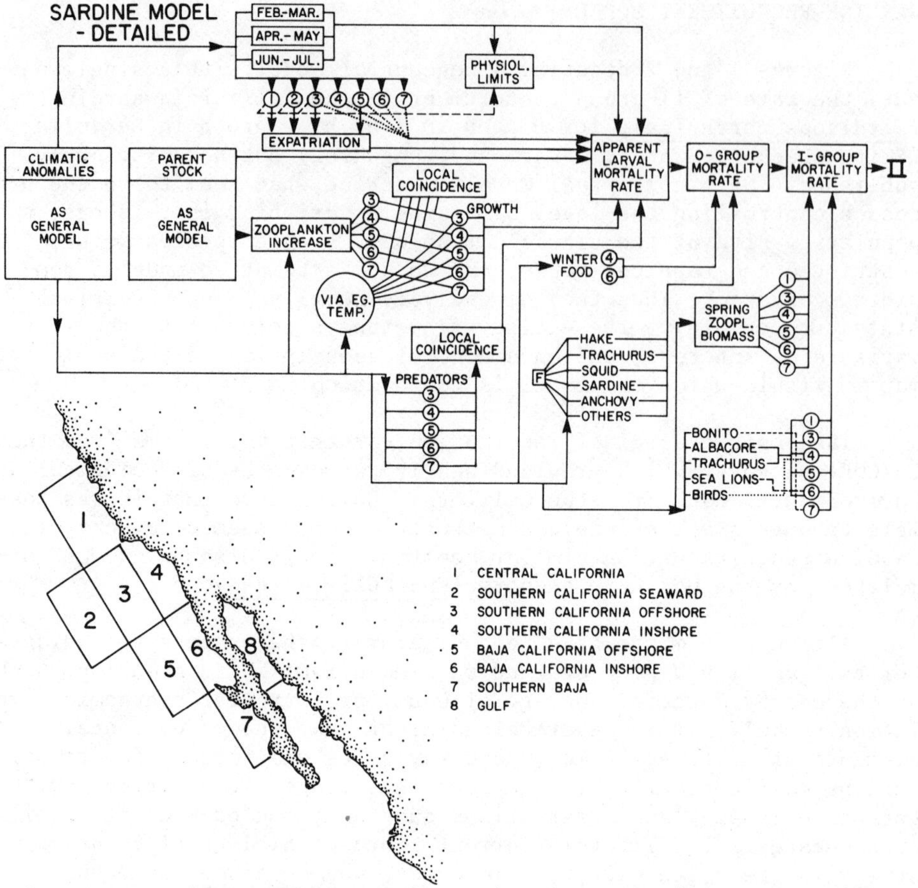

Fig. 2. Detailed model relating climatic variability and adult stock of California sardine to II group recruitment, involving individual processes in each of seven regions of the CalCOFI station grid.

(1981), led the group to consider advection and expatriation as essential elements of the model: long previous experience with time-series data from a large region had sensitized the drafting group to the importance of boundary conditions.

Because this model was drafted at a time when much attention was being given to plankton patchiness, the concept of local coincidence is introduced in several places; however, the way this concept was used is an illustration of the difficulty of even the simplest task in ecology, that of describing how biota occur in nature. Up to this period, CalCOFI, like many biological surveys even now, was designed to produce data to be expressed as numbers or biomass below a unit area of sea surface or above a unit area of substrate.

Although the general occurrence of a sub-surface chlorophyll maximum had already been introduced by Yentsch (1965), its significance was yet to be grasped. It was not for another ten years, and partly because of work done in the program which this model helped initiate, that the extent to which unit area values were misleading came to be recognized. Even if average biomass per unit area suggests that herbivores are unlikely to make a living on the plant material available, this may be illusory (e.g. Mullin and Brooks, 1976); when the detailed coincidence of biota in layers of high abundance within the water column is described (as in Figure 3) it may become much more evident how the herbivores (for example) probably do make their living.

That the sardine model could not accommodate the later demonstration by Lasker (1975) that anchovy larvae depend for their survival upon stable oceanographic conditions that allow dense layering of their food particles, and that windy or upwelling conditions effectively prevent this, illustrates the fact that models still remain very demanding of sensitive biological sampling to describe the distribution of state variables. Insensitive biological sampling, returning biomass values that are over-averaged spatially, can be totally destructive of subsequent analysis.

However, far more significant is the fact that the CalCOFI data series comprised almost entirely measurements of state variable and forcing functions; the biological data were restricted to numbers or biomass, and there was an almost total lack of information on rates of processes on relevant space and time scales. The only exception was information on the physiological rates and behaviour patterns of the larval anchovy, obtained from many years of basic research targetted onto this one life history stage of a single species. Table 1 shows the kind of information that was available when the sardine model was formulated, and which would have been capable of supporting the most critical node in an anchovy recruitment model. For sardine larvae, the parallel information base was, and still is, much less complete. An examination of the process flow diagrams for the model will reveal very many relationships based on physiological responses that could not easily be quantified even yet.

Perhaps principally for these reasons, the model was used only as a conceptual basis for the design of future research programs. Now that a further ten years' work has been completed, perhaps such a model might be feasible to code and run, to indicate our remaining areas of greatest ignorance.

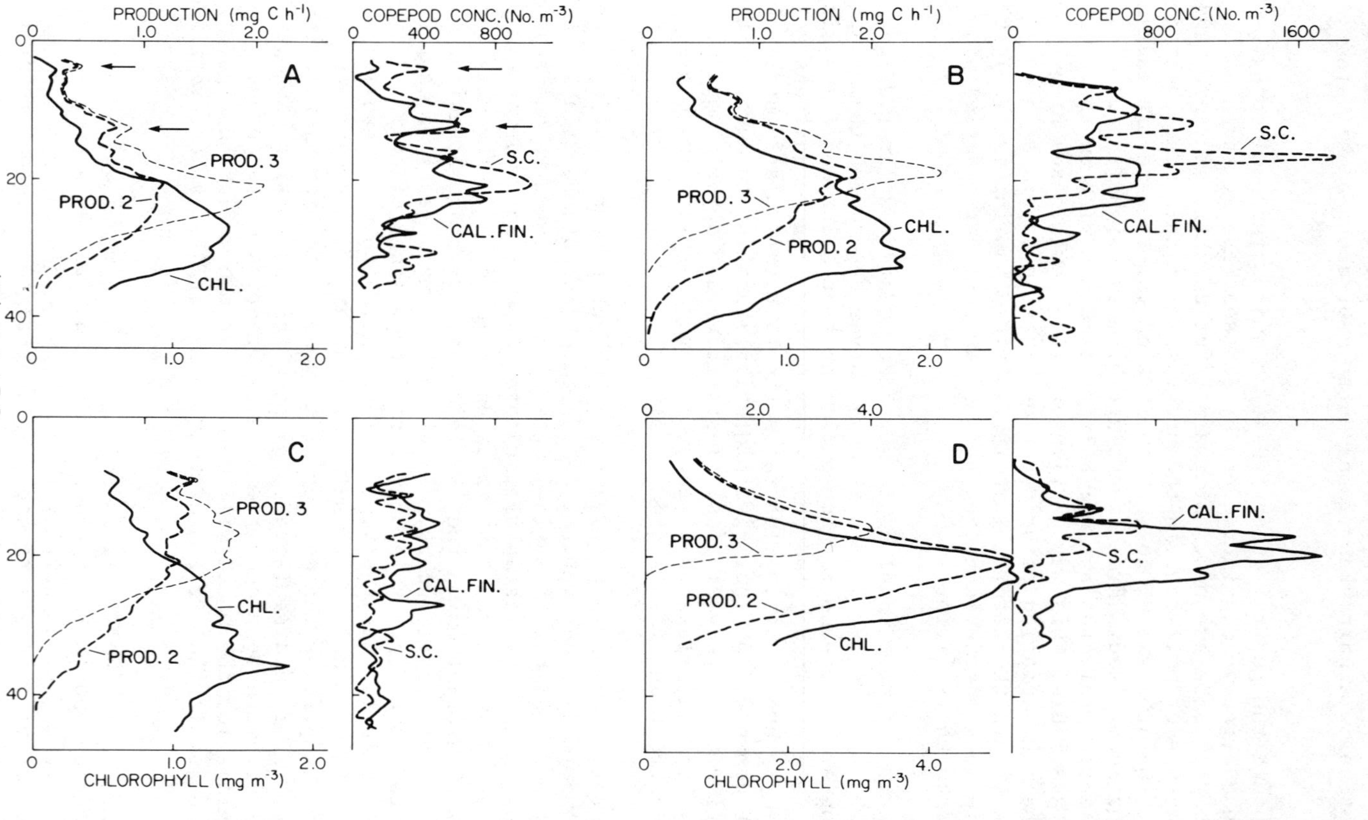

Fig. 3. Coincidence of zooplankton grazers, chlorophyll and primary production rate at four stations off Nova Scotia near the shelf-break front. Batfish data for chlorophyll and small copepods (S.C.) and _Calanus finmarchicus_ (CAL. FIN.); rate of primary production from two different models (PROD 2 & PROD 3). At three stations grazers coincide with depths of maximal production, at one station in the front, this relationship breaks down. From Herman, Sameoto and Longhurst (1981).

Table 1. Some Preliminary Estimates of Food Requirements and Feeding Behaviour for Larval Anchovy 7.75 mm TL

I. DATA

1. Age, 15 days
2. Larval length 7.75 mm
3. Larval dry weight 0.125 mg
4. Larval anchovy weight increase per day from hatching, 77×10^{-4} mg/day
5. O_2 consumption - 4.5 μl/mg/hr
6. Conversion from 1 O_2 to calories; 1 μl = 0.005 calories
7. Dry weight 1 rotifer = 1.6×10^{-4} mg
8. Approximate radius of reactive perceptive field for prey in horizontal plane
   (i) average field (includes 55% of the prey selected) r = 0.24L = 0.186 cm
   (ii) when the field includes 95% prey selected r = 0.46L = 0.356 cm
9. Swimming speed of larvae 0.92L/sec = 0.713 cm/sec
10. Feeding frequency in laboratory tanks = 59 strikes/hr
11. Percent feeding success 4.9 - 9.9 mm larvae = 79%
12. Percent time spent swimming in food search = 85%
13. Day length in laboratory, 13 hrs.

II. ASSUMPTIONS

Approximate caloric value rotifers = 5000 cal/g dry weight
Natural day length for effective feeding = 10 hrs
Approximate caloric value anchovy larvae = 5000 cal/g dry weight (5 cal/mg)
Shape of volume searched by larvae is equivalent to a cylinder or half a cylinder

III. ESTIMATES for 7.75 mm larvae

1. Basic food requirement in terms of rotifers

   Rotifers required for metabolism: 84/24 hrs
   Rotifers required for growth: 48/24 hrs
   Total rotifers required: 132/24 hrs

2. Total volume searched by larvae under various assumptions: 1.2 - 8.7L

3. An estimate of the natural prey density based on a minimum food requirement of rotifers at maximum search rates at a feeding success rate of 79%

   Maximum 139/liter
   Minimum 19.2/liter

4. Estimated number of rotifers eaten per day in laboratory = 605 rotifers/day (10,000/liter available).

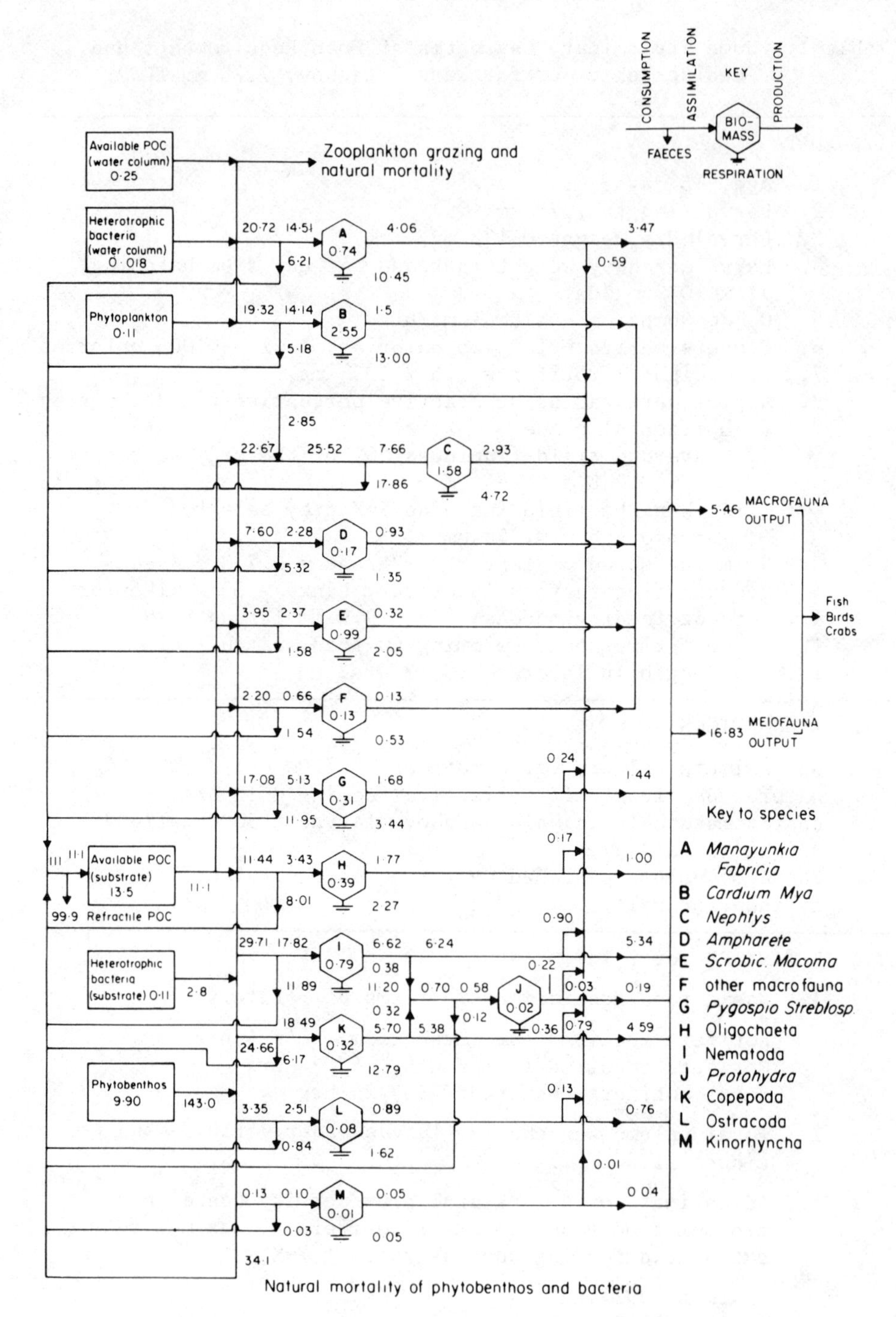

Fig. 4. Steady state energy flow diagram for the Lynher estuary mud-flat station. Standing stock biomass in $gCm^{-2}yr^{-1}$ (from Warwick, Joint and Radford, 1978).

LYNHER MUD-FLAT MODEL

This model represents energy flow through the ecosystem of a mud-flat in south-west England and includes simulations of all principal mud-flat biota, with observed abundances of vertebrate carnivores and seasonal estuarine phytoplankton considered to be forcing functions (Warwick et al., 1979). The model accommodates energy budgets for six species of benthic macrofauna, for five species of 'meiofaunal' annelids, and for five truly meiofaunal taxa: it is based on experimental data for production, respiration, assimilation and consumption for each individual taxon. The process flow diagram for this model (Figure 4) is simpler than that for the sardine model, for it includes no terms for transport, and assumes, probably correctly, that plant production in the superjacent water column, the occurrence of wading birds during winter and (less certainly) the abundance of demersal fish are not modified by the performance of the modelled benthic ecosystem.

The interest of this model is the way in which it was used to throw light on the trophic relations of a somewhat enigmatic polychaete (Nepthys hombergi). Though this was the most abundant macrofauna species (855/m^2) with a relatively high biomass, it was nevertheless surprising that its indicated energy budget required a rate of consumption about three times that of the next most abundant species, and almost as high as the total consumption of the 12.5 million nematodes of 40 species that occurred on each square meter. Though nereid and nepthyid polychaetes have traditionally been thought of as active predators because of their rapid movement and large jaws, it was known that Nereis diversicola is also capable of suspension feeding on phytoplankton by a mucous-bag technique (Harley, 1950).

Because the total net production of all small potential prey organisms for Nepthys within the community appeared to be less than

Table 2. Comparison of Simulated and Field Data for the State Variables.

	Simulation Model				Field Data			
	Max.	(Month)	Min.	(Month)	Max.	(Month)	Min.	(Month)
Phytobenthos	14.03	(Aug)	1.08	(Mar)	14.8	(May)	4.0	(Jan)
Meiofauna	1.2	(Sep)	0.9	(Jan)	2.6	(May)	0.8	(Oct)
Deposit feeders	2.5	(Jan)	2.04	(May)	2.41	(Jan	1.49	(Feb)
Nephthys	1.35	(Dec)	1.07	(Jul)	2.08	(Jun)	1.11	(Jan)

Suspension feeders are omitted because the field data are insufficient to provide monthly values, (from Warwick et al., 1978).

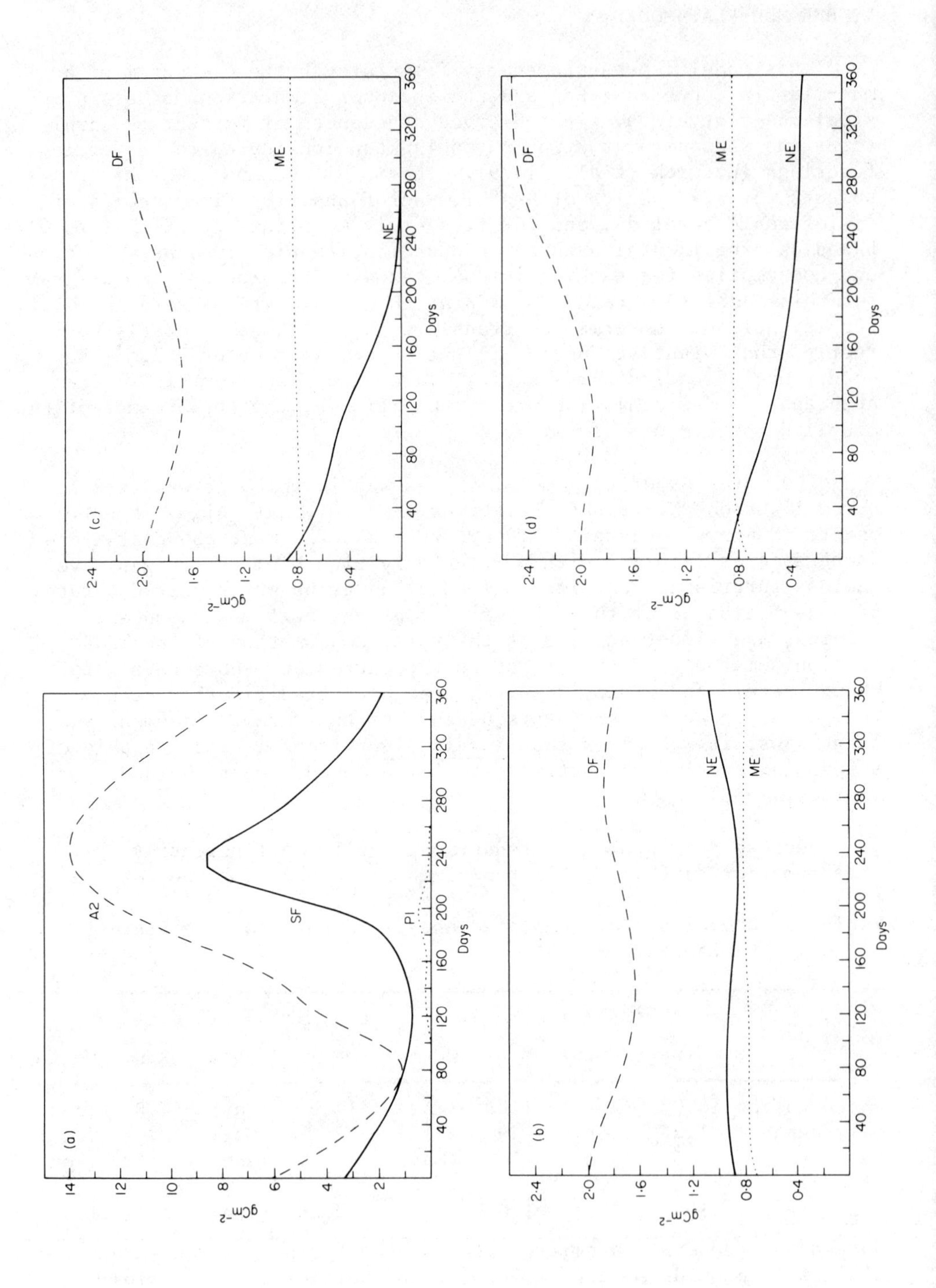
(a)
A2
SF
PI
Days
gCm⁻²
(b)
DF
NE
ME
Days
gCm⁻²
(c)
DF
ME
NE
Days
gCm⁻²
(d)
DF
ME
NE
Days
gCm⁻²

80% of the consumption demand of Nepthys alone, the model was used to investigate the consequences of different assumptions concerning the nature of the food of Nepthys. The model itself proved very stable and its output was very close to observed levels of state variables (Table 2) so some degree of confidence could be placed on its use. Three simulations, each for a one-year period, are shown in Figure 5, and these make it clear not only that Nepthys cannot maintain its observed biomass if permitted to feed only on meiofauna, but neither can it do so if permitted to feed only on phytobenthos, or mud-flat diatoms. Only if it is permitted to utilize both of these does the model allow it to follow its natural seasonal biomass cycle.

Thus, an investigation of benthic energy fluxes and their subsequent simulation in a dynamic model has revealed information (later verified by faecal examination) concerning the true trophic position of this polychaete in its community. It is also noteworthy that in this case a very thorough and painstaking quantification of all principal energy flows for this ecological unit has enabled a balanced energy budget to be constructed and tested without recourse to any unusual mathematical techniques or any new theoretical generalizations.

GEMBASE

The general ecosystem model of the Bristol Channel and Severn Estuary (GEMBASE) is an essay in regional synthesis not intended to produce a single output, as the sardine model, nor to answer a specific question, like th Lynher mud-flat model (Longhurst, 1978; Radford and Joint, 1980). In the style of the IBP biome models, it is intended to reduce a regional ecosystem to an integrated network; a description of the use of GEMBASE to simulate the ecological consequences of building a Severn tidal barrage is now available (Radford, 1982). Since GEMBASE is introduced here only to illustrate the paramount importance in ecological synthesis of adequate understanding and quantification of rates of energy flux, it is important to illustrate the gap between supply and demand for parameter values, despite the unusually high level of associated studies of ecological rates.

←

Fig. 5. Simulation of the growth of populations at the Lynher estuary mud flat station, days and standing stocks in gCm^{-2}. (b) - Nepthys is permitted to feed on phytoplankton, (c) only on meiofauna and deposit feeding benthos and (d) only on phytobenthos. A2 = phytobenthos, P1 = phytoplankton, SF = suspension feeders, DF = deposit feeders, ME = meiofauna, NE = Nepthys.

To obtain information on the system and to formulate the model a three-year survey was performed: there were 25 water chemistry and plankton surveys, and 5 benthos and sediment chemistry surveys for a total of 1200 stations, to which were added 26 helicopter surveys for a total of 650 stations to cover those parts of the system inaccessible to research ships. In order to obtain a totally independent data set for subsequent validation, a further set of surveys at a reduced level of effort was continued for five more years. Finally, a number of littoral transects were occupied on several occasions to measure macro-algae, littoral microphytes, and intertidal invertebrates.

To obtain a coherent set of forcing functions it was arranged to collect daily discharge data for all rivers entering the region, daily meteorological data for wind strength and direction, daily integrations of solar radiation, and a detailed analysis of the bed-form of the estuary to predict residual tidal currents. Data on seasonal detailed distribution of wading birds were available for the whole region from an independent survey and fishing statistics were reanalyzed to obtain information on regional fish distribution and abundances.

Simultaneously, growth and feeding rates of important biota were determined. Over a four-year period a further 39 special-purpose research cruises were devoted to this. Direct measurements were made of water column and mud-flat primary production, with some size fractionation introduced. For each of the five main benthos communities a special seasonal study was undertaken to describe the population dynamics of all species, together comprising >90% of the biomass; for intertidal communities, a review of meiofaunal respiration and production was also completed, primarily by micro-respiratory. Finally, studies of microbial heterotrophic activity and inorganic and organic nutrient dynamics were included.

Despite this very significant level of effort, which was at least equal to that put into describing the state variables, the lack of information on physiological rates and processes was still very extensive. GEMBASE was drafted as a carbon flow model between 17 biotic and 7 geographical compartments, forming a 17 x 7 interactive matrix. To describe the 43 rate variables (Table 3) the model required 150 simple differential equations tailored to the ecology of each compartment by the setting of 225 parameter values. The GEMBASE team was able to provide direct estimates for only about 10% of the required values, even after the completion of the work described above.

This situation demanded a survey of previous theoretical solutions to each of the 43 rate variables so that logical assumptions might be made for remaining constants and parameters: this required that assumptions should be made concerning a large set of widely divergent rate determining factors, for each of which it was

Table 3. List of First 24 of 43 Rate Variables Required for GEMBASE

Code	Rate variable
0103	Primary production and nutrient demand by phytoplankton
0104	Primary production and nutrient demand by plants (substrate)
0210	Intake of dissolved organic matter by planktonic heterotrophs
0211	Intake of dissolved organic matter by benthic heterotrophs
0300	Respiration of phytoplankton (water column)
0302	Organic exudation by phytoplankton
0305	Intake of phytoplankton by zooplankton
0306	Intake of phytoplankton by zoobenthos
0308	Natural mortality of phytoplankton
0400	Respiration of plants (substrate)
0402	Organic exudation by plants (substrate)
0406	Browsing rate by zoobenthos on plants (substrate)
0409	Natural mortality of plants (substrate)
0500	Respiration of zooplankton
0501	Excretion of plant nutrients by zooplankton
0502	Organic exudation by zooplankton
0505	Predation by zooplankton predators on other zooplankton
0507	Predation of zooplankton by pelagic fish
0508	Production of suspended POC by zooplankton
0600	Respiration and ecdysis of zoobenthos
0601	Excretion of plant nutrients by zoobenthos
0602	Organic exudation by zoobenthos
0607	Predation by carnivores on zoobenthos
0609	Production of deposited POC by zoobenthos

necessary to obtain estimates that were not only tailored to the relevant latitude and ecosystem but were also sensitive to the species assemblage actually known to be present in each geographical compartment; only those who have actually attempted to assemble such a series of values will appreciate how deficient the literature appeared to be for such a task.

Nevertheless, it was possible to accumulate a complete series though the assumptions that were made were in many cases rather tenuous; the model almost immediately gave output which was intuitively reasonable and was stable over periods representing more than three years, using time-steps varying from 0.1 to 1.0 days (Figures 6 and 7). It has already been noted (Longhurst, 1978) that simplistic models of BOD in estuaries are in fact capable of giving realistic output provided that forcing functions are realistic. That models based on inter-related theoretical solutions to ecological processes are capable of realism should give us encouragement about how well we understand ecological processess: it would be less remarkable if a model based on empirical relationships, such as a network of simple P/B ratios, should function successfully.

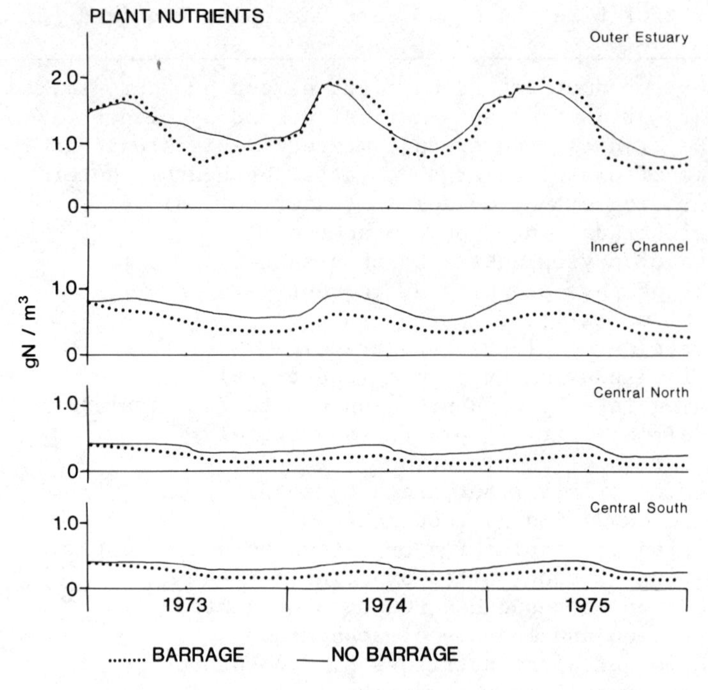
PLANT NUTRIENTS
Outer Estuary
Inner Channel
Central North
Central South
gN / m3
2.0
1.0
0
1973
1974
1975
.... BARRAGE
—— NO BARRAGE

Fig. 6.

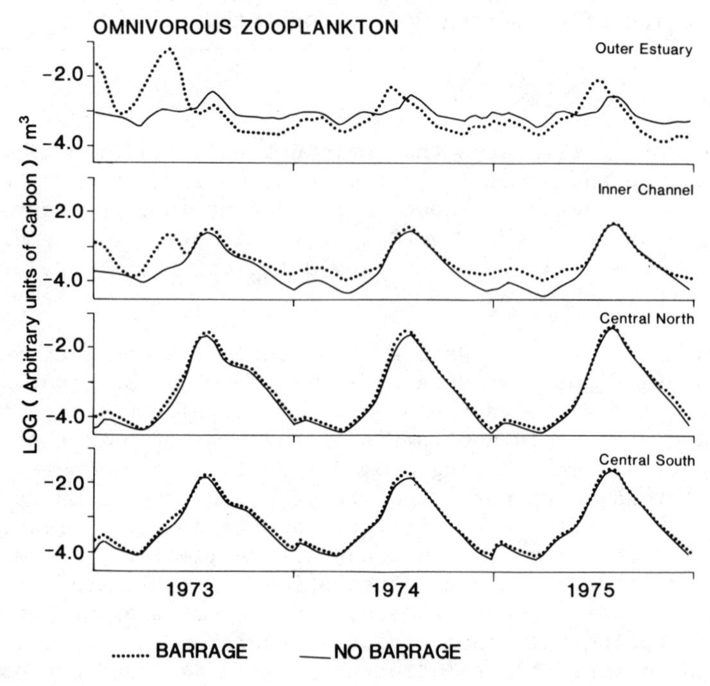
OMNIVOROUS ZOOPLANKTON
Outer Estuary
Inner Channel
Central North
Central South
LOG (Arbitrary units of Carbon) / m3
-2.0
-4.0
1973
1974
1975
.... BARRAGE
—— NO BARRAGE

Fig. 7.

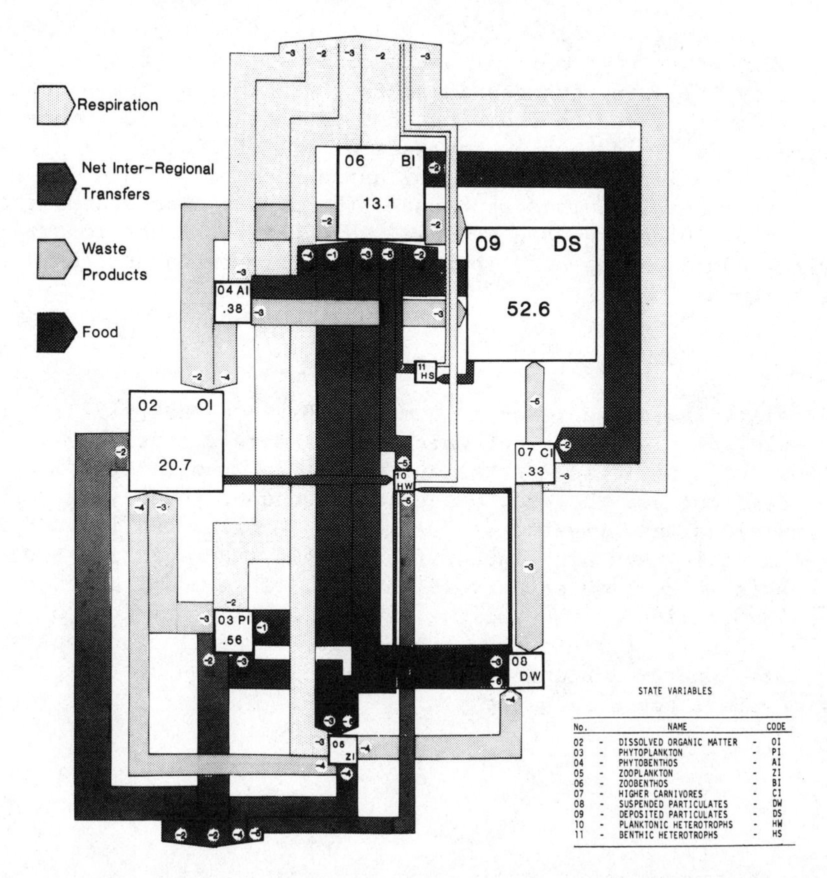

No.		NAME		CODE
02	-	DISSOLVED ORGANIC MATTER	-	OI
03	-	PHYTOPLANKTON	-	PI
04	-	PHYTOBENTHOS	-	AI
05	-	ZOOPLANKTON	-	ZI
06	-	ZOOBENTHOS	-	BI
07	-	HIGHER CARNIVORES	-	CI
08	-	SUSPENDED PARTICULATES	-	DW
09	-	DEPOSITED PARTICULATES	-	DS
10	-	PLANKTONIC HETEROTROPHS	-	HW
11	-	BENTHIC HETEROTROPHS	-	HS

Fig. 8. Mean values for 1974 assuming barrage inserted on 1 November 1972. State variables in carbon $m^{-2}d^{-2}$, rate variables the same but labels are truncated exponents only (e.g. $5.3*10^{-3}$ shown as -3). Proportions of figure are logarithmic.

Regional transfers and boundary conditions posed difficulties at least as great as those faced in the selection of the appropriate ecological construct for each individual region; such transfers were driven by a physical model (Uncles and Radford, 1980) which closely simulated salinity distribution and was itself driven by the forcing functions already described. The seasonal biological cycles in each

←

Figs. 6 and 7. Simulation of levels of nutrients and omnivorous zooplankton selected as examples of output for 17 GEMBASE state variables. Simulation assumes in-insertion of barrage on 1 November 1972 and actual forcing functions thereafter.Fig.

region were strongly influenced by transfers of biological materials generated in other regions. It was not at all always easy to distinguish, by following changes in state variables, what had been generated within a region from what had been transferred across its boundaries. For this reason, representations of biological cycles within individual regions (Figure 8) appear not to balance internally. This emphasizes the importance of a proper understanding of boundary flows of energy and material, and the paramount importance of having understanding of a spatial scale larger than the unit of interest.

MYTILUS GROWTH MODEL

Finally, I wish to refer to a model (Bayne et al., 1976) with a very simple, single output but which relies more exclusively on information on physiological rates and flows than the models discussed above. It simulates the growth over a subsequent three-year period of an individual newly-settled blue mussel (*Mytilus edulis*) under conditions of low and high intake food levels, representing two locations in an English estuary (Figure 9). This model depends not only on information on the ambient environmental conditions of temperature and salinity throughout the period of simulation, but also on the size, energy content and chemical composition of the particulate food available during the year.

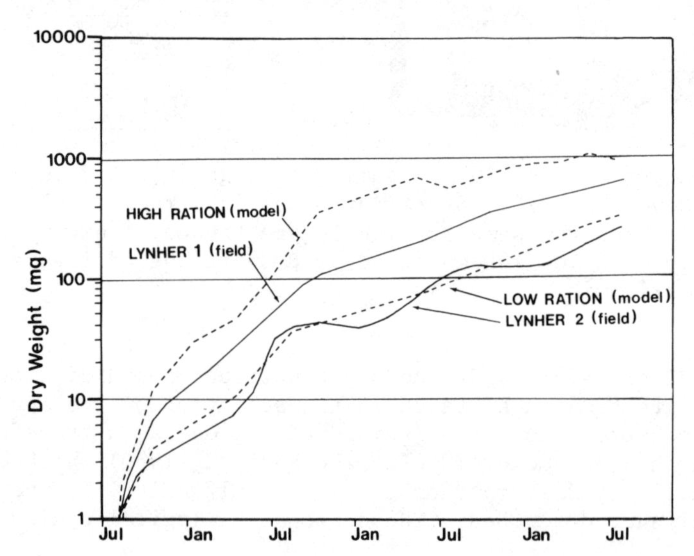

Fig. 9. Simulation of growth of a single *Mytilus edulis* over a period of three years. The discontinuous lines represent theoretical growth for high and low rations, the continuous lines represent growth in the Lynher estuary at two sites where natural conditions provide low and intermediate levels of ration.

The physiological parameters required are numerous and varied: standard and routine respiration rates as functions of temperature, salinity stress, level of ration, and metabolic body size; nitrogen intake and excretion as functions of the same set of variables; gross and net efficiency of assimilation of organic material ingested; rate of egg and gonad production; and so on. Since the population biology at each site is not modelled, virtually all inputs and variables in this model are rate variables: thus, it differs significantly from the previous models but approaches more closely to the actual experience of individual biota in the real world. Growth, reproduction and physiological performance are driven continuously by a very large spectrum of rate variables and forcing functions until terminated, usually abruptly, through interaction with a state variable representing a higher trophic level.

DISCUSSION

Since we are considering these four models only as analogues for integration in ecology we need not concern ourselves at the outset with whether or not they represent a good way of solving a particular problem. Certainly, there is a case to be made that the reductionist technique is unrealistic, and what is most lacking is a new field theory for ecology which will obviate the need for the complex interactive techniques used in these models.

More pertinent is the question to what extent each was constrained by lack of data describing the biota of interest. I believe none was seriously constrained by the quality of information available on the distribution and abundance of biota: we have sufficient understanding of variability, and sufficiently precise sampling tools that we can, if we have a mind to do it, obtain information on state variables sufficient for any task. It is more arguable, I suppose, that my examples might have been improved had they been based on a different mathematical construction, but it seems clear to me that each would have been greatly strengthened by a new level of understanding of physiological rates and processes. To take a trite example, it was neither inability to compute the added complexity, nor a lack of information on initial conditions that constrained GEMBASE to only seventeen state variables: if data for species-specific rate variables had been at hand it would have been possible to interact many more than two state variables, filter-feeders and predators, to represent the zooplankton in each region.

This argument must be balanced against the holistic contention that integrative ecology as exemplified by these four models is unrealistic, and that verifiable, predictive ecology will remain beyond our grasp until we obtain an adequate theoretical base from which to work. Such arguments recall the suggestion of Lyttleton (1977) that under some circumstances too many data may prevent deductive reasoning; would Newton, Lyttelton asked, have believed in the simple inverse-square law of planetary motion had he first of all known all the complex details describing lunar motion? On the contrary, I believe that ecology would go forward in all its aspects even more strongly than it is at present if a wider and deeper information base was available to us.

Ecology, like other sciences, proceeds on a very broad front from the empirical to the theoretical, and so it should. Because of the undoubted attraction and importance of theoretical studies there is perhaps a need occasionally to remind ourselves of the need also for systematic exploration of rates by such studies as those (e.g. Warwick et al., 1978) in which population and production dynamics were systematically described for all important taxa in each of the four communities comprising the regional benthic fauna of the Bristol Channel: this study gave a certainty to wider integrative ecology which is seldom achieved. Unfortunately, such studies are rather out of fashion: they need little innovative instrumentation, and they mostly require the application to new situations of methods no longer novel. Because the investigation of new concepts has rightly come to be so highly regarded the working ecologist may have little incentive to undertake the hard and sustained work required for the systematic investigation of rates and flows in a range of ecosystems.

It is perhaps only when a particular piece of information is sought for a particular purpose that our remaining level of ignorance is revealed. If we had, for any other group of marine organisms, the kind of information now available for the growth, mortality and fecundity for several hundreds of species of marine fish ranging over most of the world's seas and including an extremely wide range of life forms, we would be in an undoubtedly better position to make more positive statements about the functioning of marine ecosystems. A scan of recent reviews of zooplankton production rates (e.g. Mullin, 1969; Tranter, 1976; or Conover, 1979) will reveal that only a few tens of species have been systematically investigated, and that estimates for K1, K2 or P/B vary by as much as an order of magnitude under apparently similar conditions. For fish, on the other hand, Pauly (1980) has been able to tabulate estimates for the von Bertalanffy growth functions for 175 stocks of fish and to present a global data set that is internally remarkably consistent. In the absence of comparable data sets for other major componetnts of the marine biota it is much less easy to obtain the kinds of general statements concerning the dynamics of populations that can now be made for fish. For invertebrates it is necessary to proceed on the basis of far fewer data, yet the requirement for general formulations for growth rates is so imperative that attempts must be made. That they can be attempted at all is a measure of the strength of the underlying theory.

There is a very wide span of activity between the pragmatic measurement of the parameters of the von Bertalanffy or any other growth functions and the investigation of the enzymatic and molecular machinery of the photosynthetic reaction within algal chloroplasts: there is a continuum from simple population biology to comparative biochemistry that resembles a maze within which a random walk will get us nowhere. Many will spend their whole careers within one seductive alley of this maze, and will never search for its centre or seriously consider the problem of ecological integration: one of the first lessons a practical attempt in numerical ecological simulation has to teach is that just by assembling a larger pile of bricks, one may get no nearer to building a house. At some point in time one has to decide to start construction. I suggest that if the greatest need in ecology is that we obtain a more integrated understanding of rates and flows, then the most rapid move in that direction can be achieved by a greater general concern for ecological synthesis.

There is no doubt that the proportion of total effort in marine ecology which is directed towards understanding rates and flows is quite different now from what it was only twenty years ago and this conference is evidence of that. It is my intention to suggest that we still have a very long way to go.

Acknowledgements

I am grateful to my colleagues in La Jolla and Plymouth, named in the text, for permission to discuss our work in another context. Several versions of the text were discussed with Trevor Platt and Ken Mann, to whom I am also grateful. Two anonymous reviewers took me to task as an unregenerate reductionist, in which they were quite correct, but their thoughtful comments were very helpful indeed, and I am grateful to them. I also thank my present employer, the Canadian Department of Fisheries and Oceans, for permission to undertake this review.

References

Bayne, B. L., Widdows, J., and Thompson, R. J.., 1976, Physiological integrations, in: "Marine Mussels," B. L. Bayne, ed., Cambridge University Press, Cambridge.

Bernal, P. A., 1981, Advection and upwelling in the California Current, in: "Coastal Upwelling," F. A. Richards, ed., American Geophysical Union, Washington.

Conover, R. J., 1979, Secondary production as an ecological phenomenon, in: "Zoogeography and Diversity in Plankton," Van Der Spoel and Pierrot-Bults, eds., Bunge Scientific Publishers, Utrecht.

Harley, M. B., 1980, The occurrence of a filter-feeding mechanism in the polychaete Nereis diversicolor, Nature, London, 165: 734.

Lasker, R., 1975, Field criteria for survival of anchovy larvae: the relation between the inshore chlorophyll layers and successful first feeding, US Fish.Bull., 71:453.

Longhurst, A. R., 1978, Ecological models in estuarine management, Ocean Management, 4:287.

Lyttelton, R. A., 1977, The nature of knowledge, in: 'The Encyclopedia of Ignorance," R. Duncan and M. Weston-Smith, eds., Pergamon Press, Oxford.

Mann, K. H., 1976, Production on the bottom of the sea, in: "The Ecology of the Seas," Cushing and Walsh, eds., Blackwell Scientific Publication, Oxford.

Mullin, M. M., 1967, Production of zooplankton in the ocean: the present status and problems, Oceanog.Mar.Biol.Ann.Rev., 7: 293.

Mullin, M. M., and Brooks, E. R., 1976, Some consequences of distributional heterogeneity of phytoplankton and zooplankton, Limnol.Oceanog., 21:784.

Pauly, D., 1980, On the inter-relationships between natural mortality growth parameters and temperature in 175 fish stocks, J.Conseil., 39:175.

Pielou, E. C., 1981, The usefulness of ecological models: a stocktaking, Quart.Rev.Biol., 56:17.

Radford, P. J., 1982, Modelling the impact of a tidal power scheme on the Severn Estuary ecosystem, in: "Proc. Int. Symp. Energy and Ecological Modelling, Louisville, 1981," (in press).

Radford, P. J., and Joint, I. R., 1980, The application of an ecosystem model to the Bristol Channel and Severn Estuary, Water Poll.Control, 79:244.

Tranter, D. J., 1976, Herbivore production, in: "The Ecology of the Seas," Cushing and Walsh, eds., Blackwell Scientific Publications, Oxford.

Uncles, R. J., and Radford, P. J., 1980, Season and spring-neap tidal dependence of axial dispersion coefficients in the Severn - a wide, vertically-mixed estuary, J.Fluid Mech., 98:703.

Warwick, R. M., George, C. L., and Davies, J. R., 1978, Annual macrofauna production in a Venus community, Est.Coast.Mar. Sci., 7:2-5.

Warwick, R. M., Joint, I. R., and Radford, P. J., 1979, Secondary production of the benthos in an estuarine environment, in: "Ecological Processes in Coastal Environments," R. L. Jeffries and A. J. Davy, eds., Blackwell Scientific Publications, Oxford.

Wickett, W. P., 1967, Ekman transport and zooplankton concentrations in the North Pacific Ocean, J.Fish.Res.Bd.Canada, 24:581.

Yetsch, C. S., 1965, Distribution of chlorophyll and phaeopytin in the open ocean, Deep-Sea Res., 12:653.

COMMUNITY MEASURES OF MARINE FOOD NETWORKS AND THEIR POSSIBLE APPLICATIONS

Robert E. Ulanowicz

Chesapeake Biological Laboratory
Center for Environmental and Esturaine Studies
University of Maryland
Solomons, Maryland 20688-0038, USA

INTRODUCTION

Ecology is the study of the relationships of organisms with one another and with their non-living environment. The keyword to notice in this definition is "relationships". Rather than fixing attention upon the organism or population itself, we as ecologists should be primarily concerned with what transpires between populations. As we also pretend to being scientists, we aim to quantify our observations on these relationships, and this is most readily done when those interactions involve a palpable transfer of either material or energy. When we describe the species composition and densities of phytoplankton and zooplankton in an open ocean gyre, we are behaving as good quantitative biologists. Not until we attempt to balance grazing rates with rates of respiration, nutrient uptake and sinking, are we acting as quantitative ecologists.

There are those who would point out that to measure the fluxes between populations, we first need to quantify the abundances of the participants. I do not wish to deny what is a methodological necessity in most (though not all) cases. What I am proposing here, however, is the perspective that the description of the network of flow exchanges is the key to approaching the ecological problem.

The prevailing attitude among most ecosystem modellers is that one seeks to describe a flow as the result of a putative force, which may be quantified by the states of the interacting populations. For example, zooplankton grazing is usually assumed to be some multivariate function of phytoplankton density, zooplankton density, ambient temperature, relative sizes of predator and prey, etc. In writing this multivariate function, one is implicitly describing a force which causes a resultant flow - the grazing rate. This approach often works well when only one, or a very few, processes are being studied. The problem arises when we attempt to predict the course of a linked network of many flows as being the aggregate result of the simple forces. The predictions of such coupled models are usually unreliable. Furthermore, with the fixed structure inherent in this procedure one cannot treat the various "emergent surprises" which enter the development of any real ecosystem. Even if one should try to vary the structure of the model in some fashion, one is still left in need of a criterion against which to evaluate the proposed changes.

A way out of this dilemma is afforded by adopting an alternative perspective in which the community flow network is not secondary to the attributes of the constituent populations. Whence, the density, size and age structures, respiration efficiency, ethology and other properties of any member population are affected by, and in the long run are formed by, how well the state properties of the individual species contribute to some as-yet-unspecified attribute of the <u>entire</u> community network of flows. Different modes of individual behavior, physiologies, genotypes, feeding links, even species themselves, may enter and leave the community. How long they remain depends upon their contribution to some property of the whole network of exchanges.

If it sounds as though cause and effect are being confused by this emphasis upon flows, then such observation is correct. Consider, for example, the perfect causal loop (see also Hutchinson, 1948) in Figure 1. Medium flows from A to B causing further flow from B to C, which in turn engenders flow from C back to A; and the original cause (flow from A to B) becomes its own effect. Of course, if one tries to identify this loop with any real cycle (e.g. A is nitrogen in the water column, B is nitrogen in the phytoplankton and C is nitrogen in the zooplankton as in Figure 2) the picture is complicated by obligatory losses, and imports. But the causal loop remains imbedded in the real network as an element to be reckoned with.

Causal loops usually give rise to positive feedback phenomena. Any change in the status of one of the components giving rise to an

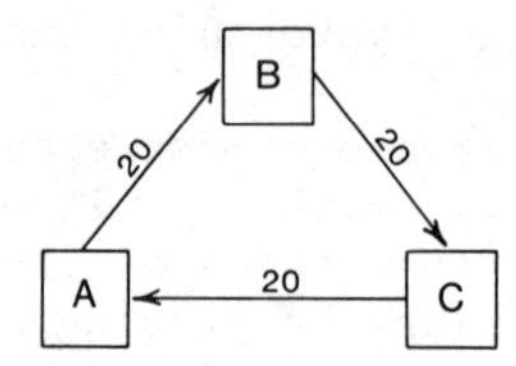

Fig. 1. An ideal flow loop (causal cycle).

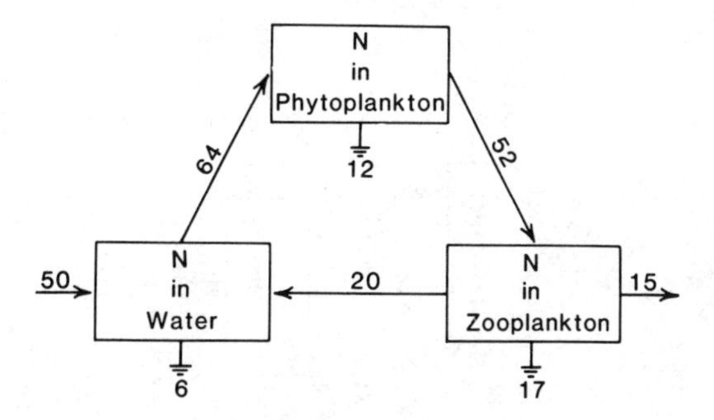

Fig. 2. A hypothetical cycle of nitrogen in a planktonic system ($gN/m^2/y$). The ground symbols represent respiratory losses.

increase in one of the flows will result in self-reward. Conversely, changes which diminish any flow in the cycle will be self-mitigated. The fluxes and the forces thereby became inextricably entwined. Now this is a bad turn of events only if one's goal is the explanation (at the subsystem level) of phenomena in a classical cause-effect fashion. If, however, one seeks (as in thermodynamics) only to quantitatively describe the eventual course of the whole system; one's task is actually made easier, because now one need measure only the perceptible flow network and ignore, for the time being, the obscure and elusive forces.

I suggest that the measurement of ecosystem flow networks is precisely the task of highest priority in ecology today. But questions remain: What constitutes an adequate description of a flow network? How can one use the flow network as a diagnostic tool in ecological research and management? And, perhaps most importantly, can one be more specific about the criterion which best describes the time evolution of an ecosystem flow network? Below I will briefly attempt to outline my opinions on these three issues with occasional examples drawn from the analysis of the carbon flow network of an estuarine marsh gut (i.e. embayment) ecosystem.

FLOW DATA REQUIRED

How much data need be collected depends largely upon decisions made at the very outset of any investigation as to what the essential components of the community will be. How the chosen aggregation affects the outcome of an analysis has always been a crucial question in ecosystems modeling (see Halfon, 1979; Cale et al., 1982; Schaffer, 1981). I will not dwell on this controversy, except to remark that my experience with the few networks I have been able to analyze leads me to believe that the conclusions drawn from global

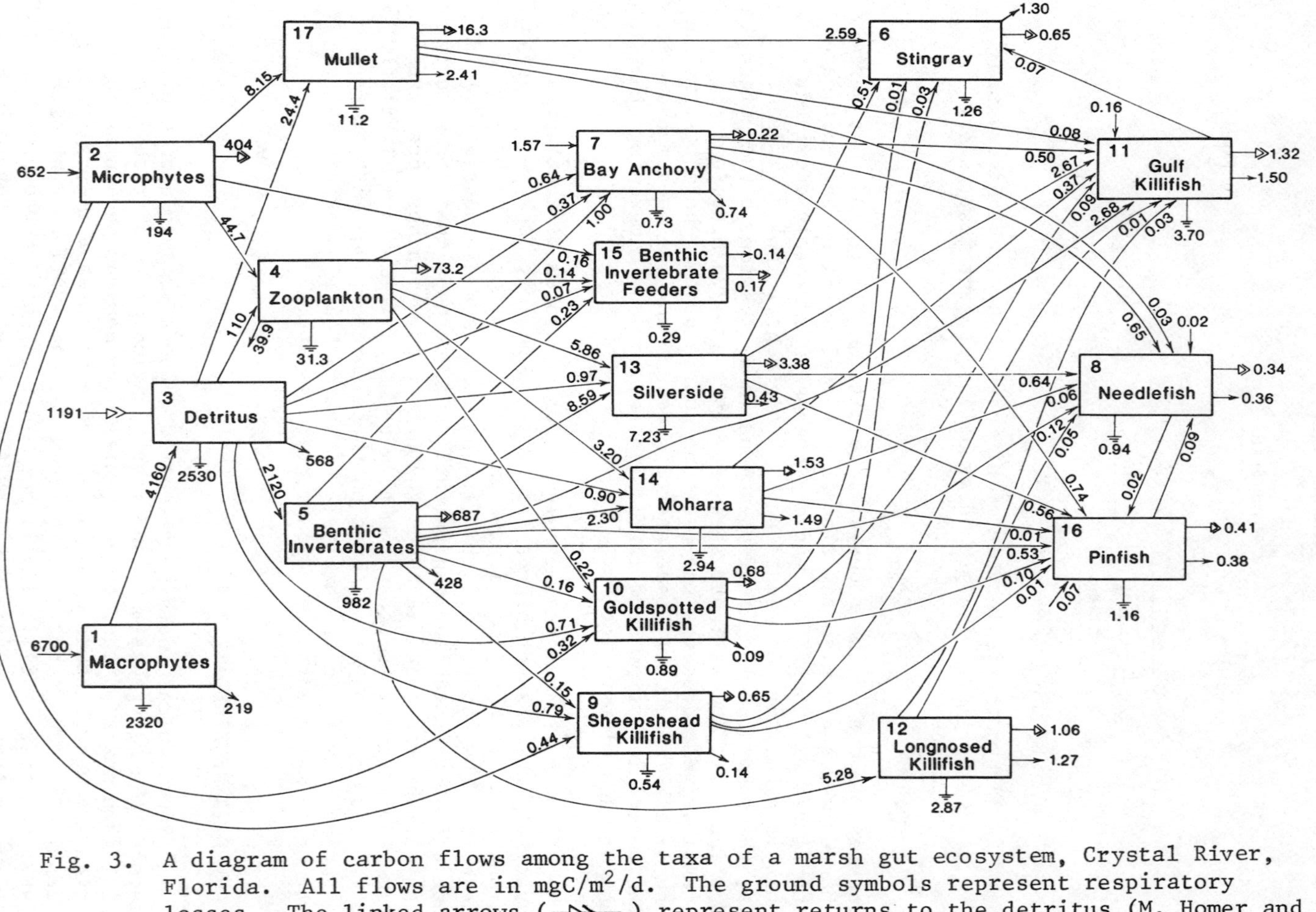

Fig. 3. A diagram of carbon flows among the taxa of a marsh gut ecosystem, Crystal River, Florida. All flows are in mgC/m^2/d. The ground symbols represent respiratory losses. The linked arrows (—▷▷—) represent returns to the detritus (M. Homer and W. M. Kemp, unpublished manuscript).

network variables seem more robust with respect to the degree of aggregation than do results issuing from deterministic modeling efforts. (Although the cycle analysis to be described later is very sensitive to the lumping of compartments).

The carbon flow network data to be used for illustration were collected by Homer and Kemp (unpublished manuscript) from a polyhaline marsh gut in the vicinity of the Crystal River estuary on the upper Gulf Coast of Florida. In the sample network 17 compartments are identified as in Figure 3. The compartments are mixed in degree of aggregation with the lower trophic level species being lumped into a few compartments, whereas the fish have been identified in several cases to the species level.

Generally speaking, the types of flows which need to be measured fall into four categories:

(1) exchanges among compartments within the system,
(2) inputs from sources outside the system,
(3) useable exports outside the system, and
(4) dissipation of medium into a form of no further use to any system.

Some investigators do not make a distinction between (3) and (4); but as will be seen later, there are compelling thermodynamic and hierarchical reasons for treating these flows separately. In our example network there are 69 internal exchanges, 6 inputs, 16 exports, and 17 respiratory flows. The flows balance around each compartment, i.e. the network is assumed to balance over the annual cycle. Steady-state is not, however, a requirement of most of the following analyses.

The given flows are in terms of carbon/area/time. Corresponding flow networks could have been measured in terms of energy, nitrogen, phosphorous, silicon, or other materials.

While the analysis of a solitary network can be performed, it is often not clear what significance to attach to the magnitudes of some of the resultant quantities. This problem can be circumvented, however, by seeking comparisons between two different systems or between the same system at different times or under different exogenous conditions. For example, we may wish to compare estuarine, coastal, gyral and abyssal marine networks. Or we may wish to contrast equitorial upwelling community networks under both light and heavy trade winds. A companion network to the example in Figure 3 has been quantified. It describes the exchanges in a nearby tidal gut under practically identical environmental conditions, save that the second community (which network is depicted in Figure 4) is continually subjected to a 6°C rise in temperature because of its proximity to thermal effluent from a nuclear power station.

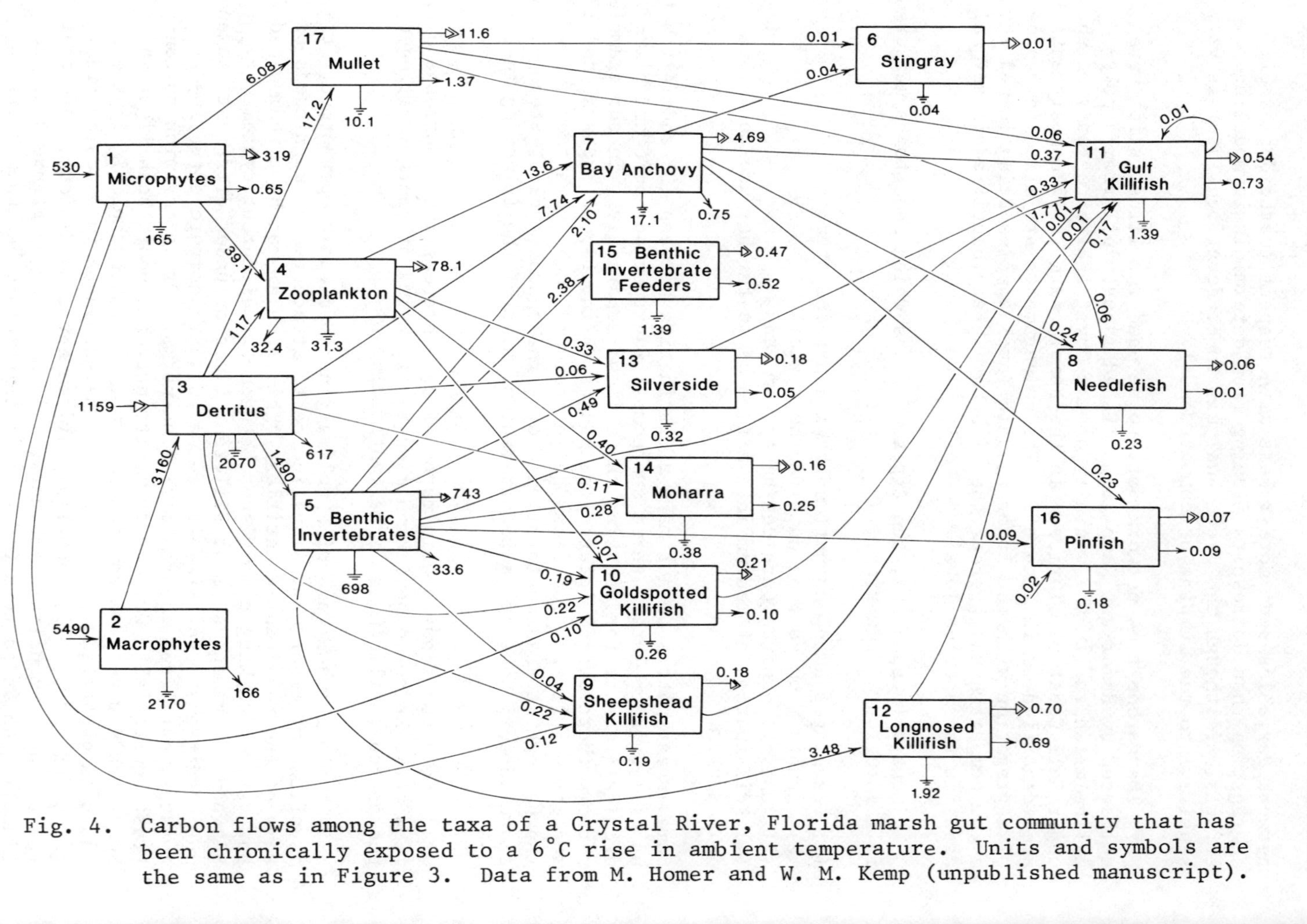

Fig. 4. Carbon flows among the taxa of a Crystal River, Florida marsh gut community that has been chronically exposed to a 6°C rise in ambient temperature. Units and symbols are the same as in Figure 3. Data from M. Homer and W. M. Kemp (unpublished manuscript).

INPUT-OUTPUT ANALYSIS

Given a flow network, one of the more detailed questions one can ask is how any one compartment is affecting (i.e. exchanging medium with) any other compartment over all pathways, direct and indirect. The direct interactions are manifest upon inspection of either the diagram of the network or of the two-dimensional matrix of internal exchanges. The indirect linkages are not always as easy to identify and enumerate. Fortunately, much information can be obtained about the indirect pathways through straightforward algebraic operations on the matrix of direct flows (see also Leontief, 1951; Hannon, 1973; Finn, 1976; Patten _et al_., 1976).

Calling P_{ij} the flow from i to j and T_i the total ouput from i, then $f_{ij} = P_{ij}/T_i$ will represent the fraction of output of i which is contributed directly to j. The F matrix for the control marsh gut appears in Table 1. If F is a matrix with components f_{ij}, it is easy to demonstrate that multiplying F by itself will yield a matrix F^2 wherein the i-jth component represents the fraction of total output from i which flows to j over all pathways of exactly two steps. Similarly, F^3 will express the fractions transferred over all pathways of length three, and so forth ad infinitum. By summing all powers of F, we obtain the so-called output structure matrix, S, wherein the i-jth component represents the fraction of output from i which flows to j over all possible pathways, i.e.

$$S = F^0 + F^1 + F^2 + F^3 + F^4 + \ldots$$

Because of the way in which F was normalized, this infinite sequence converges exactly to

$$S = (I-F)^{-1} ,$$

where I is the identity matrix (Yan, 1969), and the minus one exponent represents matrix inversion.

As an example of the use of the output structure matrix, the 1-7 component of the Crystal River marsh gut indicates that in the unperturbed web 5.46×10^{-4} of the microphyte throughput ($652\ mgC/m^2/d$), or $.356\ mg/m^2/d$, eventually reaches the Bay Anchovies (see Table 2). Curiously enough, about 15 times that amount is transferred between the same two compartments in the perturbed network.

One can also multiply S by vectors representing the inputs, exports and total throughputs of each compartment to obtain the answers to such questions as: What is the fate of each individual input to the network? Or, conversely, how much of each export can be attributed to the various inputs? For example, there are two major sources of carbon to the system, fixation of carbon by macrophytes and microphytes, with the macrophytes contribution 10.3 times greater than the planktonic production. Given that macrophytes are

Table 1. Output Fractions for the Unperturbed Crystal River Marsh Gut Ecosystem. Each Entry Represents the Fraction of the Row Throughput which Flows Directly to the Column Taxon

	Microphytes	Macrophytes	Detritus	Zooplankton	Benthic Invertebrates	Stingray	Bay Anchovy	Needlefish	Sheepshead Killifish	Gold spotted Killifish	Gulf Killifish	Longnosed Killifish	Silverside	Moharra	Benthic Invertebrate Feeders	Pinfish	Mullet
Microphytes	.0000	.0000	.6199	.0686	.0000	.0000	.0000	.0000	.0007	.0005	.0000	.0000	.0000	.0000	.0002	.0000	.0125
Macrophytes	.0000	.0000	.6210	.0000	.0000	.0000	.0000	.0000	.0000	.0000	.0000	.0000	.0000	.0000	.0000	.0000	.0000
Detritus	.0000	.0000	.0000	.0205	.3956	.0000	.0001	.0000	.0001	.0001	.0000	.0000	.0002	.0002	.0000	.0000	.0046
Zooplankton	.0000	.0000	.4738	.0000	.0000	.0000	.0041	.0000	.0000	.0014	.0000	.0000	.0379	.0207	.0009	.0000	.0000
Benthic Inv.	.0000	.0000	.3243	.0000	.0000	.0000	.0005	.0001	.0001	.0003	.0013	.0025	.0041	.0011	.0001	.0003	.0000
Stingray	.0000	.0000	.2025	.0000	.0000	.0000	.0000	.0000	.0000	.0000	.0000	.0000	.0000	.0000	.0000	.0000	.0000
Bay Anchovy	.0000	.0000	.0615	.0000	.0000	.0000	.0000	.1816	.0000	.0000	.1397	.0000	.0000	.0000	.0000	.2067	.0000
Needlefish	.0000	.0000	.2048	.0000	.0000	.0000	.0000	.0000	.0000	.0000	.0000	.0000	.0000	.0000	.0000	.0120	.0000
Sheepshd Killifish	.0000	.0000	.4710	.0000	.0000	.0217	.0000	.0000	.0000	.0000	.0072	.0000	.0000	.0000	.0000	.0072	.0000
Goldspotted Killifish	.0000	.0000	.3656	.0000	.0000	.0054	.0000	.0000	.0000	.0000	.0484	.0000	.0000	.0000	.0000	.0538	.0000
Gulf Killifish	.0000	.0000	.2003	.0000	.0000	.0106	.0000	.0000	.0000	.0000	.0000	.0000	.0000	.0000	.0000	.0000	.0000
Longnosed Killifish	.0000	.0000	.2008	.0000	.0000	.0000	.0000	.0095	.0000	.0000	.0057	.0000	.0000	.0000	.0000	.0000	.0000
Silverside	.0000	.0000	.2192	.0000	.0000	.0331	.0000	.0415	.0000	.0000	.1732	.0000	.0000	.0000	.0000	.0363	.0000
Moharra	.0000	.0000	.2391	.0000	.0000	.0000	.0000	.0094	.0000	.0000	.0578	.0000	.0000	.0000	.0000	.0016	.0000
Benthic Inv. Feeders	.0000	.0000	.2833	.0000	.0000	.0000	.0000	.0000	.0000	.0000	.0000	.0000	.0000	.0000	.0000	.0000	.0000
Pinfish	.0000	.0000	.2010	.0000	.0000	.0000	.0000	.0441	.0000	.0000	.0000	.0000	.0000	.0000	.0000	.0000	.0000
Mullet	.0000	.0000	.5005	.0000	.0000	.0796	.0000	.0009	.0000	.0000	.0025	.0000	.0000	.0000	.0000	.0000	.0000

Table 2. The Output Structure Matrix for the Unperturbed Crystal River Marsh Gut Ecosystem. Each Entry Designates the Fraction of the Row Throughput which Flows to the Column Taxon Over All Direct and Indirect Pathways

	Microphytes	Macrophytes	Detritus	Zooplankton	Benthic Invertebrates	Stingray	Bay Anchovy	Needlefish	Sheepshead Killifish	Gold spotted Killifish	Gulf Killifish	Longnosed Killifish	Silverside	Moharra	Benthic Invertebrate Feeders	Pinfish	Mullet
Microphytes	1.000	.0000	.7699	.0844	.3045	.0015	.0005	.0004	.0008	.0008	.0015	.0008	.0046	.0022	.0004	.0004	.0160
Macrophytes	.0000	1.000	.7237	.0148	.2863	.0003	.0002	.0002	.0001	.0002	.0008	.0007	.0019	.0007	.0001	.0002	.0033
Detritus	.0000	.0000	1.000	.0239	.4610	.0005	.0004	.0003	.0002	.0003	.0013	.0011	.0030	.0012	.0001	.0003	.0053
Zooplankton	.0000	.0000	.5723	1.000	.2264	.0016	.0043	.0028	.0001	.0016	.0090	.0006	.0394	.0213	.0009	.0025	.0026
Benthic Inv.	.0000	.0000	.3809	.0078	1.000	.0003	.0006	.0005	.0001	.0004	.0025	.0029	.0050	.0015	.0001	.0006	.0017
Stingray	.0000	.0000	.2360	.0048	.0934	1.000	.0001	.0001	.0000	.0001	.0003	.0002	.0006	.0002	.0000	.0001	.0011
Bay Anchovy	.0000	.0000	.1991	.0041	.0787	.0016	1.000	.1908	.0000	.0001	.1399	.0002	.0005	.0002	.0000	.2091	.0009
Needlefish	.0000	.0000	.2417	.0050	.0956	.0001	.0001	1.000	.0000	.0001	.0003	.0002	.0006	.0002	.0000	.0121	.0011
Sheepshd Killifish	.0000	.0000	.5575	.0114	.2206	.0221	.0002	.0004	1.000	.0002	.0078	.0005	.0014	.0006	.0000	.0074	.0025
Goldsptd. Killifish	.0000	.0000	.4519	.0093	.1788	.0061	.0002	.0025	.0001	1.000	.0489	.0004	.0012	.0005	.0000	.0539	.0021
Gulf Killifish	.0000	.0000	.2359	.0048	.0933	.0107	.0001	.0001	.0000	.0001	1.000	.0002	.0006	.0002	.0000	.0001	.0011
Longnosed Killifish	.0000	.0000	.2376	.0049	.0940	.0002	.0001	.0095	.0000	.0001	.0059	1.000	.0006	.0002	.0000	.0002	.0011
Silverside	.0000	.0000	.3230	.0066	.1278	.0351	.0001	.0432	.0001	.0001	.1735	.0003	1.000	.0003	.0000	.0369	.0015
Moharra	.0000	.0000	.2949	.0060	.1167	.0008	.0001	.0095	.0001	.0001	.0581	.0003	.0008	1.000	.0000	.0018	.0013
Benthic Inv. Feeders	.0000	.0000	.3302	.0068	.1306	.0002	.0001	.0001	.0001	.0001	.0004	.0003	.0008	.0003	1.000	.0001	.0015
Pinfish	.0000	.0000	.2449	.0050	.0969	.0001	.0001	.0442	.0000	.0001	.0003	.0002	.0006	.0003	.0000	1.000	.0011
Mullet	.0000	.0000	.6028	.0124	.2385	.0799	.0002	.0011	.0001	.0002	.0031	.0006	.0015	.0006	.0000	.0002	1.000

generally more refactory than microphytes, one might guess that the contributions to the exported detritus would be weighted more heavily in favor of the macrophytes. Such is not the case, however, as indirect pathways and cycling actually decrease the predominance of macrophytically-fixed carbon to microphytically-fixed carbon down to 9.7. By hindsight the reason for the decrease is apparent. Macrophytes contribute solely to the detrital pool; whereas microphytes are grazed by several species, most notably zooplankton and mullet; and are cycled in the foodweb, thereby lowering the effective respiration rate.

The matrix of specific exchanges, F, was derived from the exchange matrix, P, by normalizing according to outputs. One may also normalize according to inputs by defining $g_{ij} = P_{ij}/T_j$. The corresponding structure matrix,

$$S' = (I-G)^{-1} ,$$

contains information about the direct and indirect sources of input to each compartment. S'_{ij} describes the fraction of throughput j which is attributable to i as a source. (In the event of cycling this fraction can exceed unity).

S' possess another very useful property. It is not difficult to demonstrate that the sum of each column of S' represents the average trophic position of its respective compartment (Levine, 1980). For example, if a taxon obtains 50% of its throughput directly from outside the system, 30% along pathways one step removed from a primary input, and 20% along pathways two steps removed from original sources, then it functions 50% of a primary producer, 30% as a "herbivore" and 20% as a "carnivore". Its average trophic position is thereby 1.7 (=0.5x1 + 0.3x2 + 0.2x3). Here one assumes that all flux entering from outside the system is considered at the first trophic level; otherwise the trophic positions of the inputs would be added to the column sum.

The average trophic positions of the 17 taxa in both of the Crystal River marsh guts are displayed in Table 3. The changes in the relative trophic rankings among the taxa are rather unremarkable. Perhaps the biggest winner is the Bay Anchovy which jumped from fourth place to ninth in the trophic rankings, even though its dietary habits changed little. What is quite curious, however, is that the trophic values for practically all taxa have risen in the disturbed creek. These observations raise more questions than they answer, and to clarify the picture it is necessary to expand the scope of observation from pairwise interactions to whole cycles of medium.

CYCLE ANALYSIS

The central role which I assume cycles to play in the develop-

Table 3. Ranked Listing of the Average Trophic Levels at Which Each Taxon is Feeding. Listed on the left-hand side are the rankings of taxa in the unperturbed marsh gut. On the right side the same taxa feed at slightly different levels in an identical marsh gut which has been chronically exposed to a 6°C elevation in temperature.

	Control		ΔT Perturbed	
Rank	Taxon	Av. Trophic Value	Taxon	Av. Trophic Value
1	Microphytes	1.00	Microphytes	1.00
2	Macrophytes	1.00	Macrophytes	1.00
3	Detritus	2.34	Detritus	2.47
4	Bay Anchovy	2.70	Mullet	3.09
5	Zooplankton	2.95	Zooplankton	3.11
6	Mullet	3.00	Sheepshead Killifish	3.11
7	Sheepshead Killifish	3.02	Benthic Invertebrates	3.47
8	Benthic Invertebrates	3.34	Goldspotted Killifish	3.62
9	Goldspotted Killifish	3.51	Bay Anchovy	3.93
10	Benthic Inv. Feeders	3.51	Moharra	4.15
11	Moharra	4.00	Silverside	4.27
12	Silverside	4.13	Longnosed Killifish	4.47
13	Stingray	4.22	Benthic Inv. Feeders	4.47
14	Pinfish	4.23	Pinfish	4.58
15	Longnosed Killifish	4.34	Gulf Killifish	4.69
16	Needlefish	4.45	Stingray	4.76
17	Gulf Killifish	4.57	Needlefish	4.76

ment of an ecosystem has already been mentioned. It is desirable, therefore, to be able to identify all the cycles inherent in a network and to seek clues as to how these various cycles might interact.

The enumeration of all the cycles in a graph is not always an easy task. In tracing through the various pathways of a network it soon becomes obvious that the number of possible paths and cycles can increase geometrically (or even factorially) as the number of components in the system rises. Hence, an arbitrary search for cycles in a network of, say, 20 components has the potential for quickly saturating the capabilities of even the larger modern computers ($20! = 24 \times 10^{18}$). Fortunately, however, algorithms are now available which efficiently enumerate cycles in networks of moderate size (Read and Tarjan, 1975; Johnson, 1975; Mateti and Deo, 1976; Ulanowicz, 1982a).

One might next ask whether the members of the list of cycles fall into any natural groupings. As ecologists we are often concerned with the vulnerability of any structure. To someone familiar with a given ecosystem, it may be possible for him to trace around the links of any chosen cycle and identify the most vulnerable transfer in the loop. Once a vulnerable link has been established in each cycle, the circuits can then be grouped in such a way that all the members of any group share the same most vulnerable arc. In a real sense each vulnerable arc is a weak spot in the network, and its domain of influence upon the feedback structure is defined by the collection of cycles in which it appears, hereinafter referred to as its nexus.

To be more specific, one counts 119 distinct cycles in the graph of the control creek, as listed in Appendix A. For the sake of discussion we shall choose the smallest link in each cycle to be that loop's most vulnerable arc. The 119 cycles segregate into 41 nexuses. Several of these aggregations are rather large, one consisting of 14 separate cycles, one of 13, one of 10, three of 6, etc. The largest nexus is depicted in Figure 5 (with the vulnerable link denoted by the heavy arrow). In contrast, the heated creek foodweb contains only 46 simple cycles belonging to 30 nexuses (Appendix B). None of the large nexuses cited in the control creek survives in the perturbed system. The largest nexus in the disturbed network is a single, 4-cycle grouping.

There are further differences between the two networks when one notes the magnitudes of the vulnerable arcs. The weak links in all

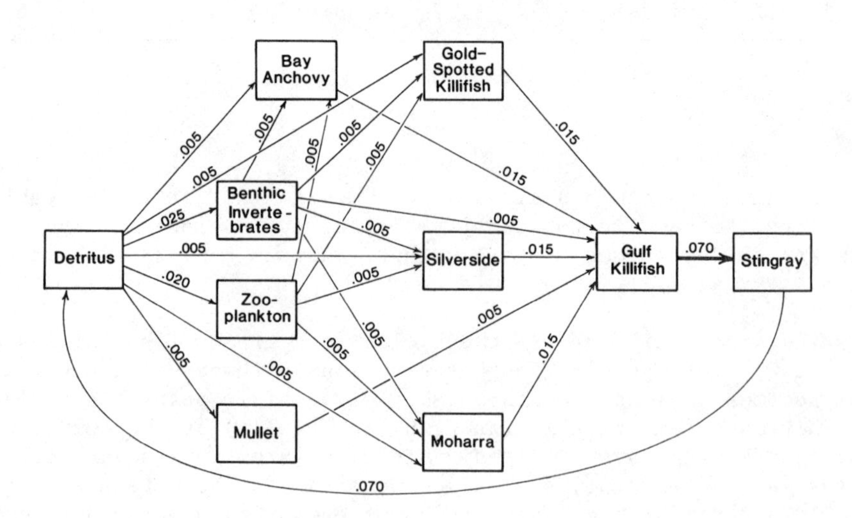

Fig. 5. The nexus of cycles associated with the cropping of Gulf Killifish (11) by Stingray (6). The flows shown (in mgC/m/d) are the amounts actually subtracted from the original flows in Figure 3 in the process of calculating the residual acyclic web.

of the many-cycled nexuses of the control system were small in magnitude, and the component cycles tended to consist of 4 or 5 links. In both systems the vulnerable arcs with the highest flows were associated with short, 2 step cycles. For example, the loop with largest flow in both webs was the immediate recycle of carbon between the detritus and the benthic invertebrates. This turnover actually increased from 687 in the control to 734 ($mgC/m^2/d$) in the perturbed system. Likewise, the turnover between zooplankton and detritus rose from 73 to 78 ($mgC/m^2/d$) in the disturbed creek. The fraction of total flow which was being cycled (see Finn, 1977) actually went up from 10.5% in the control to 14.0% in the heated system.

Some of these observations seem counterintuitive when viewed alone. For example, Finn echoed Odum's (1969) suggestion that a higher fraction of cycled flow was indicative of more mature, less disturbed networks. Nonetheless, cycle analysis reveals a coherent picture of what has happened to the heated system. It is a clear-cut instance of eutrophication. The higher-order (presumably slower) cycles have disappeared. The shorter, faster, trophically lower, turnovers now cycle more intensely. Whether the rise of the intense, short cycles caused the disappearance of the higher-order cycles (as is presumed to be the case in cultural eutrophication), or whether the increase in short cycling was occasioned by some dysfunctions in the longer loops is a moot point in the realm of cybernetics.

The trophic response of the Bay Anchovy to community stress is now clear. The diet of this species appears to consist entirely of detritus, zooplankton and benthic invertebrates - all of the players involved in the accelerated short cycles. It was in a perfect position to exploit the eutrophic changes. We also see that the general rise in trophic values after perturbation was somewhat misleading, as it was inflated by the rapid multiple passes through the shorter, but more intense, cycles.

A consistent picture of the underlying trophic dynamics in both networks can be obtained by first subtracting away all cycled flow and analyzing the residual acyclic webs. This decomposition of an arbitrary graph into cyclic and acyclic subgraphs is not a trivial problem. An approach based on the scheme of aggregation by smallest arc has been described by the author (Ulanowicz, 1982b). Suffice it to remark that in an acyclic network, any component which feeds at more than one trophic level can be unequivocally partitioned among the compartments of a finite, concatenated food chain in the sense of Lindemann (see also Ulanowicz and Kemp, 1979). In our previous example the species acting 50% as a primary producer, 30% as a herbivore and 20% as a carnivore would have its throughput, exports and respirations apportioned among the first three trophic compartments in the ratio 5:3:2. The acyclic subgraphs of the two creeks

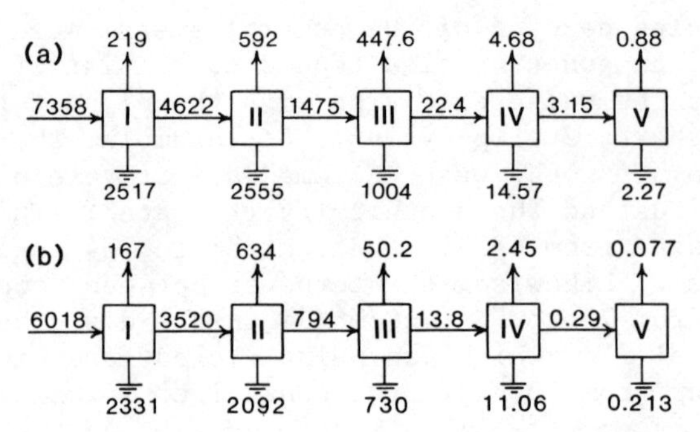

Fig. 6. Aggregated trophic flows of carbon ($mgC/m^2/d$) inherent in the web exchanges (Figures 3 and 4) after the cycled material has been removed from the network. (a) the control marsh gut; (b) the thermally stressed marsh gut.

can be mapped into straight trophic chains of length 5. The differences in the first two levels are small, but flows at the third level of the disturbed system are halved and fall off nearly tenfold at the fifth step (see Figure 6). Again, the effects of stress are more obvious at higher trophic levels.

GLOBAL COMMUNITY MEASURES

The story of network response to perturbation can be read in reverse to gain insights into the manner in which ecosystem networks develop in the relative absence of stress. Two attributes crucial to such a description are size and structure. One expects the winners in the evolutionary game to possess advantageous combinations of size and structure. (One need not think of systems competing with one another, rather one compares the actual state of the system with other putative nearby states).

A convenient measure of the size of a flow network is the total system throughput, the sum of all the individual throughputs, T_i. All other things being equal, one expects the total systems throughput to rise during the course of the development (living things grow); or, conversely, fall due to perturbation (as did the total throughput of the disturbed creek). It was also apparent how the diversity of the structure was adversely affected by perturbation and, conversely, would be expected to rise during unperturbed development. Hence, one is led to consider the diversity of the various throughputs as a measure of this complexity, i.e.

$$- \sum_i Q_i \log Q_i,$$

where Q_i ($=T_i/\sum_j T_j$) is the fraction of the total throughput particular to component i. The product of size and complexity is conveniently called the development capacity,

$$C = -T \sum_j Q_i \log Q_i,$$

where T has been taken to represent the total throughput ($\sum_j T_j$).

The inadequacy of using C as a measure of development is precisely what plagued the epochal diversity-stability arguments, namely how the compartments are related to one another does not appear in the formula. But the coefficients f_{ij} quantify how the output of one compartment is linked to the input of another. Accordingly, one may define

$$A = T \sum_i \sum_j f_{ij} Q_i \log (f_{ij}/\sum_k f_{kj} Q_k)$$

as a component of C which captures how coherently, or with what degree of definition, the elements are connected to one another (Rutledge _et al._, 1976; Ulanowicz, 1980).

One can show mathematically that C serves as an upper bound on A and that both quantities are inherently non-negative, i.e.,

$$C \geq A \geq 0$$

That portion of the capacity which does not appear as coherent structure (i.e., the difference, C-A) is termed the overhead, and may be further decomposed into three components. The first component results from the obligatory dissipative processes.

$$S = -T \sum_i r_i Q_i \log Q_i$$

where r_i is the fraction of T_i lost to dissipation, i.e. subsystem-scale processes. The second component allows for transfer to other systems.

$$E = -T \sum_i e_i Q_i \log Q_i$$

where e_i is the fraction of T_i flowing to another system. These exports make up the links in any assembled suprasystem and may be thought of as contributions to higher order structure. The residual overhead

$$R = -T \sum_i \sum_j f_{ij} Q_i \log(f_{ij} Q_i/\sum_k f_{kj} Q_k)$$

can also be shown to be non-negative and measures the average multiplicity of pathways among the components; or, alternatively, the degree of confusion as to where within the network an arbitrary output might flow. It is (like A) a property existing at the level of the system. Hence, the overhead has been partitioned along hierarchical lines.

Returning to A, the ascendency, it captures the tradeoff between system size and structure and could possibly serve as the criterion describing the time evolution of an ecosystem flow network. In fact the increases in A so paralleled the increases in other attributes of mature ecosystems (e.g. Odum, 1969) that I have elsewhere (Ulanowicz, 1980) hypothesized that self-organizing communities behave over an adequate interval of time so as to optimize their ascendencies subject to hierarchical and environmental constraints. Regardless of whether or not this hypothesis will subsequently be validated, the five quantities defined above provide a lexicon for describing community level change in flow networks.

Any data set on flows meeting the requirements specified earlier in this paper is sufficient to calculate the network ascendency and related variables. In the creek networks the ascendency has decreased from 7235 ($mgC/m^2/d$) in the control to 5349 in the disturbed creek, as one might expect from the converse of the ascendency hypothesis. The magnitude of the decrease is somewhat misleading, as it reflects mostly the decrease in systems throughput. The organizational factor in the ascendency has decreased by only 7%, indicating strong structural homeostatsis. The resistance of structure to change emphasizes the utility of the cycle analysis in dramatizing the otherwise subtle changes which have taken place in the network topology. The overall increase in the relative amount of overhead came from nearly equal increases in dissipation and redundancy. The fraction of C encumbered by tribute, E, actually fell in the disturbed network, reflecting the homeostatic response of the system in tightening the remaining cycles.

Concluding Remarks

I would like to end this brief synopsis of network flow analysis with the opinion that the approach is eminently ecological in nature. The focus is upon the interactions, rather than upon the taxa themselves. The field of vision encompasses the entire community of processes. The attitude is phenomenological, or empirical; one is concerned with a quantitative description of what actually happened, rather than with the supposition of causes, real or fictitious. This is not to denigrate the need for inference, but only to postpone suppositions until more familiarity with community level descriptions has been achieved. It is my belief that the sought-after inferences or laws will become most readily apparent at the level of the entire ecosystem.

This last probability is philosophically satisfying. Phenomena at a smaller scale may appear confusing, unexpected or contradictory. But the larger picture seems coherent and directions well-defined. Of course, this mirrors the results of the most phenomenological branch of physical science - thermodynamics. In fact ecological phenomenology, as typified by flow network analysis, may yet extend the frontiers of thermodynamics. Nevertheless, to formulate principles by searching at larger scales of observation violates the prevailing dogma of the biological sciences today — reductionism. In all likelihood the world, cybernetic as it appears, will reveal influences going in both holistic and reductionistic directions. Right now, however, such a small fraction of biological research effort is aimed at elucidating more global principles, that I believe a bottleneck impeding our deeper understanding of the natural work exists. It is to help overcome such an impediment that this conference has been called to promote the measurement of ensembles of flows occurring in marine ecosystems.

Acknowledgements

This work was supported in part by a grant from the National Science Foundation (ECS-8110035). The Computer Science Center of the University of Maryland also contributed free computer time. Dr. John T. Finn kindly reviewed the manuscript and made several very helpful comments. Contribution No. 1343, Center for Environmental and Estuarine Studies of the University of Maryland.

References

Cale, W. G., O'Neill, R. V., and Gardner, R. H., 1982, Aggregation error in nonlinear ecological models, Submitted to J.Theor. Biol.

Finn, J. T., 1976, Measures of ecosystem structure and function derived from analysis of flows, J.Theor.Biol., 56:363.

Finn, J. T., 1977, Flow Analysis: A Method for Tracing Flows through Ecosystem Models, Ph.D. dissertation, Univ. of Georgia, Athens.

Halfon, E., 1979, "Theoretical Systems Ecology," Academic Press, New York.

Hannon, B., 1973, The structure of ecosystems, J.Theor.Biol., 41:535.

Hutchinson, G. E., 1948, Circular causal systems in ecology, Ann. N.Y.Acad.Sci., 50:221.

Johnson, D. B., 1975, Finding all the elementary circuits of a directed graph, SIAM J.Comput., 4:77.

Leontief, W., 1951, "The Structure of the American Economy 1919-1939," 2nd edn., Oxford University Press, New York.

Levine, S. H., 1980, Several measures of trophic structure applicable to complex food webs, J.Theor.Biol.,83:195.

Mateti, P., and Deo, N., 1976, On algorithms for enumerating all circuits of a graph, SIAM J.Comput., 5:90.

Odum, E. P., 1969, The strategy of ecoystem development, Science, 164:262.

Pattern, B. C., Bosserman, R. W., Finn, J. T., and Cale, W. G., 1976, Propagation of cause in ecosystems, in: "Systems Analysis and Simulation in Ecology," Vol. 4., B. C. Patten, ed., Academic Press, New York.

Read, R. C., and Tarjan, R. E., 1975, Bounds on backtrack algorithms for listing cycles, paths, and spanning trees, Networks, 5:237.

Rutledge, R. W., Basore, B. L., and Mulholland, R. J., 1976, Ecological stability: an information theory viewpoint, J.Theor. Biol., 57:355.

Schaffer, W. M., 1981, Ecological abstraction: the consequences of reduced dimensionality in ecological models, Ecological Monographs, 51:383.

Ulanowicz, R. E., 1980, An hypothesis on the development of natural communities, J.Theor.Biol., 85:223.

Ulanowicz, R. E., 1982a, NETWRK: a package of computer algorithms to analyze ecological flow networks, Ref. No. 82-7 CBL of the Center for Environmental and Estuarine Studies, Solomons, Maryland.

Ulanowicz, R. E., 1982b, Identifying the structure of cycling in ecosystems, to appear in Mathematical Biosciences

Ulanowicz, R. E., and Kemp, W. M., 1979, Toward canonical trophic aggregations, Amer.Nat., 114:871.

Yan, C. S., 1969, "Introduction to input - output economics," Holt, Reinehart and Winston, New York.

APPENDIX A

Cycle Analyses (Control Creek)

3-Cycle Nexus with Weak Arc (10, 6) = .010

1. 3-10- 6- 3-
2. 3- 4-10- 6- 3-
3. 3. 5-10- 6- 3-

6-Cycle Nexus with Weak Arc (14,16) = .010

4. 3-14-16-3-
5. 3-14-16- 8- 3-
6. 3- 4-14-16- 3-
7. 3- 4-14-16- 8- 3-
8. 3- 5-14-16- 3-
9. 3- 5-14-16- 8- 3-

4-Cycle Nexus with Weak Arc (9,11) = .010

10. 3- 9-11- 3-
11. 3- 9-11- 6- 3-
12. 3- 5- 9-11- 3-
13. 3- 5- 9-11- 6- 3-

4-Cycle Nexus with Weak Arc (9,16) = .010

14. 3- 9-16- 3-
15. 3- 9-16- 8- 3-
16. 3- 5- 9-16- 3-
17. 3- 5- 9-16- 8- 3-

13-Cycle Nexus with Weak Arc (8,16) = .020

18. 3- 7- 8-16- 3-
19. 3-13- 8-16- 3-
20. 3-14- 8-16- 3-
21. 3- 4- 7- 8-16- 3-
22. 3- 4-13- 8-16- 3-
23. 3- 4-14- 8-16- 3-
24. 3- 5- 8-16- 3-
25. 3- 5- 7- 8-16- 3-
26. 3- 5-13- 8-16- 3-
27. 3- 5-14- 8-16- 3-
28. 3- 5-12- 8-16- 3-
29. 3-17- 8-16- 3-
30. 8-16- 8-

2-Cycle Nexus with Weak Arc (9, 6) = .030

31. 3- 9- 6- 3-
32. 3- 5- 9- 6- 3-

2-Cycle Nexus with Weak Arc (12,11) = .030

33. 3- 5-12-11- 3-
34. 3- 5-12-11- 6- 3-

1-Cycle Nexus with Weak Arc (17, 8) = .030

35. 3-17- 8- 3-

1-Cycle Nexus with Weak Arc (12, 8) = .050

36. 3- 5-12- 8- 3-

3-Cycle Nexus with Weak Arc (14, 8) = .060

37. 3- 14- 8- 3-
38. 3- 4-14- 8- 3-
39. 3- 5-14- 8- 3-

14-Cycle Nexus with Weak Arc (11, 6) = .070

40. 3- 7-11- 6- 3-
41. 3-10-11- 6- 3-
42. 3-13-11- 6- 3-
43. 3-14-11- 6- 3-
44. 3- 4- 7-11- 6- 3-

45. 3- 4-10-11- 6- 3-
46. 3- 4-13-11- 6- 3-
47. 3- 4-14-11- 6- 3-
48. 3- 5-11- 6- 3-
49. 3- 5- 7-11- 6- 3-
50. 3- 5-10-11- 6- 3-
51. 3- 5-13-11- 6- 3-
52. 3- 5-14-11- 6- 3-
53. 3-17-11- 6- 3-

1-Cycle Nexus with Weak Arc (3,15) = .070

54. 3-15- 3-

1-Cycle Nexus with Weak Arc (17,11) = .080

55. 3-17-11- 3-

10-Cycle Nexus with Weak Arc (16, 8) = .090

56. 3- 7-16- 8- 3-
57. 3-10-16- 8- 3-
58. 3-13-16- 8- 3-
59. 3- 4- 7-16- 8- 3-
60. 3- 4-10-16- 8- 3-
61. 3- 4-13-16- 8- 3-
62. 3- 5-16- 8- 3-
63. 3- 5- 7-16- 8- 3-
64. 3- 5-10-16- 8- 3-
65. 3- 5-13-16- 8- 3-

3-Cycle Nexus with Weak Arc (10,11) = .090

66. 3-10-11- 3-
67. 3- 4-10-11- 3-
68. 3- 5-10-11- 3-

3-Cycle Nexus with Weak Arc (10,16) = .100

69. 3-10-16- 3-
70. 3- 4-10-16- 3-
71. 3- 5-10-16- 3-

1-Cycle Nexus with Weak Arc (5, 8) = .120

72. 3- 5- 8- 3-

1-Cycle Nexus with Weak Arc (4,15) = .140

73. 3- 4-15- 3-

1-Cycle Nexus with Weak Arc (5, 9) = .150

74. 3- 5- 9- 3-

1-Cycle Nexus with Weak Arc (15, 3) = .170

75. 3- 5-15- 3-

3-Cycle Nexus with Weak Arc (7, 3) = .220

76. 3- 7- 3-
77. 3- 4- 7- 3-
78. 3- 5- 7- 3-

1-Cycle Nexus with Weak Arc (4,10) = .220

79. 3-4-10- 3-

6-Cycle Nexus with Weak Arc (8, 3) = .340

80. 3- 7- 8- 3-
81. 3-13- 8- 3-
82. 3- 4- 7- 8- 3-
83. 3- 4-13- 8- 3-
84. 3- 5- 7- 8- 3-
85. 3- 5-13- 8- 3-

2-Cycle Nexus with Weak Arc (3, 7) = .370

86. 3- 7-11- 3-
87. 3- 7-16- 3-

3-Cycle Nexus with Weak Arc (14,11) = .370

88. 3-14-11- 3-
89. 3- 4-14-11- 3-
90. 3- 5-14-11- 3-

6-Cycle Nexus with Weak Arc (16, 3) = .410

91. 3-13-16- 3-
92. 3- 4- 7-16- 3-
93. 3- 4-13-16- 3-
94. 3- 5-16- 3-
95. 3- 5- 7-16- 3-
96. 3- 5-13-16- 3-

2-Cycle Nexus with Weak Arc (7,11) = .500

97. 3- 4- 7-11- 3-
98. 3- 5- 7-11- 3-

3-Cycle Nexus with Weak Arc (13, 6) = .510

99. 3-13- 6- 3-
100. 3- 4-13- 6- 3-
101. 3- 5-13- 6- 3-

<u>1-Cycle Nexus with Weak Arc (5,10) = .610</u>

102. 3- 5-10- 3-

<u>1-Cycle Nexus with Weak Arc (9, 3) = .650</u>

103. 3- 9- 3-

<u>1-Cycle Nexus with Weak Arc (6, 3) = .650</u>

104. 3-17- 6- 3-

<u>1-Cycle Nexus with Weak Arc (10, 3) = .680</u>

105. 3-10- 3-

<u>1-Cycle Nexus with Weak Arc (3,14) = .900</u>

106. 3-14- 3-

<u>2-Cycle Nexus with Weak Arc (3,13) = .970</u>

107. 3-13- 3-
108. 3-13-11- 3-

<u>1-Cycle Nexus with Weak Arc (12, 3) 1.060</u>

109. 3- 5-12- 3-

<u>3-Cycle Nexus with Weak Arc (11, 3) = 1.320</u>

110. 3- 4-13-11- 3-
111. 3- 5-11- 3-
112. 3- 5-13-11- 3-

<u>2-Cycle Nexus with Weak Arc (14, 3) 1.530</u>

113. 3- 4-14- 3-
114. 3- 5-14- 3-

<u>2-Cycle Nexus with Weak Arc (13, 3) 3.380</u>

115. 3- 4-13- 3-
116. 3- 5-13- 3-

<u>1-Cycle Nexus with Weak Arc (17, 3) = 16.290</u>

117. 3-17- 3-

<u>1-Cycle Nexus with Weak Arc (4, 3) = 73.200</u>

118. 3- 4- 3-

<u>1-Cycle Nexus with Weak Arc (5, 3) = 686.900</u>

119. 3- 5- 3-

APPENDIX B

Cycle Analyses (Perturbed Creek)

3-Cycle Nexus with Weak Arc (6, 3) _ .010

1. 3- 7- 6- 3-
2. 3- 4- 7- 6- 3-
3. 3- 5- 7- 6- 3-

3-Cycle Nexus with Weak Arc (10,11) = .010

4. 3-10-11- 3-
5. 3- 4-10-11- 3-
6. 3- 5-10-11- 3-

2-Cycle Nexus with Weak Arc (9,11) = .010

7. 3- 9-11- 3-
8. 3- 5- 9-11- 3-

1-Cycle Nexus with Weak Arc (17, 6) = .010

9. 3-17- 6- 3-

1-Cycle Nexus with Weak Arc (11,11) = .010

10. 11-11-

1-Cycle Nexus with Weak Arc (5, 9) = .040

11. 3- 5- 9- 3-

3-Cycle Nexus with Weak Arc (8, 3) = .060

12. 3- 7- 8- 3-
13. 3- 4- 7- 8- 3-
14. 3- 5- 7- 8- 3-

2-Cycle Nexus with Weak Arc (3, 13) = .060

15. 3-13- 3-
16. 3-13-11- 3-

1-Cycle Nexus with Weak Arc (17,11) = .060

17. 3-17-11- 3-

1-Cycle Nexus with Weak Arc (17, 8) = .060

18. 3- 17- 8- 3-

4-Cycle Nexus with Weak Arc (16, 3) = .070

19. 3- 7-16- 3-
20. 3- 4- 7-16- 3-
21. 3- 5- 7-16- 3-
22. 3- 5-16- 3-

1-Cycle Nexus with Weak Arc (4,10) = .070

23. 3- 4-10- 3-

1-Cycle Nexus with Weak Arc (3,14) = .110

24. 3-14- 3-

2-Cycle Nexus with Weak Arc (14, 3) = .160

25. 3- 4-14- 3-
26. 3- 5-14- 3-

1-Cycle Nexus with Weak Arc (12,11) = .170

27. 3- 5-12-11- 3-

1-Cycle Nexus with Weak Arc (9, 3) = .180

28. 3- 9- 3-

2-Cycle Nexus with Weak Arc (13, 3) = .180

29. 3- 4-13- 3-
30. 3- 5-13- 3-

1-Cycle Nexus with Weak Arc (5,10) = .190

31. 3- 5-10- 3-

1-Cycle Nexus with Weak Arc (10, 3) = .210

32. 3-10- 3-

1-Cycle Nexus with Weak Arc (4,13) = .330

33. 3- 4-13-11- 3-

1-Cycle Nexus with Weak Arc (13,11) = .330

34. 3- 5-13-11- 3-

3-Cycle Nexus with Weak Arc (7,11) = .370

35. 3- 7-11- 3-
36. 3- 4- 7-11- 3-
37. 3- 5- 7-11- 3-

1-Cycle Nexus with Weak Arc (15, 3) = .470

38. 3- 5-15- 3-

1-Cycle Nexus with Weak Arc (11, 3) = .540

39. 3- 5-11- 3-

1-Cycle Nexus with Weak Arc (12, 3) = .700

40. 3- 5-12- 3-

1-Cycle Nexus with Weak Arc (5, 7) = 2.100

41. 3- 5- 7- 3-

2-Cycle Nexus with Weak Arc (7, 3) = 4.690

42. 3- 7- 3-
43. 3- 4- 7- 3-

1-Cycle Nexus with Weak Arc (17, 3) = 11.640

44. 3- 17- 3-

1-Cycle Nexus with Weak Arc (4, 3) = 78.100

45. 3- 4- 3-

1-Cycle Nexus with Weak Arc (5, 3) = 742.600

46. 3- 5- 3-

THERMODYNAMICS OF THE PELAGIC ECOSYSTEM: ELEMENTARY CLOSURE CONDITIONS FOR BIOLOGICAL PRODUCTION IN THE OPEN OCEAN

Trevor Platt[1], Marlon Lewis[1,2] and
Richard Geider[1,3]

[1]Marine Ecology Laboratory
Bedford Institute of Oceanography
Dartmouth, Nova Scotia B2Y 4A2, Canada
[2]Dept. of Biology
Dalhousie University
Halifax, Nova Scotia, Canada
[3]Dept. of Oceanography
Dalhousie University
Halifax, Nova Scotia, Canada

PREAMBLE

In the pelagic zone of the ocean, the primordial ecological event is the conversion by phytoplankton of radiant energy from the sun into biochemical energy. The rate at which this process proceeds is called the primary production of the ocean. In spite of its fundamental importance and its profound significance for the trophodynamics of the marine ecosystem, the absolute magnitude of primary production in the ocean is still uncertain to within a factor of ten (Eppley, 1980). More than 50 years of research effort have gone into its measurement, including 30 years with a high precision isotopic tracer technique: but instead of converging on some generally accepted figures, the estimates continue to diverge (Steemann Nielsen, 1954; Platt and Subba Rao, 1975; Eppley and Peterson, 1979; Peterson, 1980). It has become conventional, if not ritualistic, for any inconsistencies in independent estimates to be laid at the door of the ^{14}C method (Williams, 1981). The technique with the highest potential precision is therefore in danger of losing (has lost?) credibility on the grounds of accuracy. If primary production estimates up to two orders of magnitude higher than the ^{14}C figures can be given serious consideration in the literature (Sheldon and Sutcliffe, 1978; Gieskes _et al_., 1979; Johnson _et al_., 1981; Shulenberger and Reid, 1981) why not three orders or four orders higher?

In this essay we seek, by the application of elementary thermodynamic closure conditions, to set limits on the magnitudes of some important biological fluxes in the open ocean and thereby deduce the highest value of primary production that is defensible on thermodynamic grounds. As a specific example, we consider in detail the ecosystem of the oligotrophic central gyre of the north Pacific, one of the better known open ocean sites. We conclude that the vertical nitrogen flux, the radiant energy flux and the oxygen consumption due to community respiration are all consistent with a value for the net primary production that is within the range of the ^{14}C estimates.

STORAGE OF PHOTOSYNTHETICALLY USEFUL RADIATION

The thermodynamic upper limit to photoautotrophic production in the ocean is set by the rate at which photosynthetically active radiation (PAR) can be stored by organisms in the water column. This depends on the intensity of available PAR, the proportion of it that can be trapped by the autotrophic organisms (a function of the photosynthetic biomass), and the efficiency of its conversion to photosynthate.

At any depth z in the water column, the maximum rate at which radiant energy can be stored as photosynthate depends on the total optical attentuation power of the photosynthetic pigments at that depth (e.g. see Platt, 1969):

$$Q(z) = \kappa\, B(z)\, I(z) \qquad (1)$$

Here $Q(z)$ is the maximum possible energy storage rate per unit volume ($J\ m^{-3}s^{-1}$) through autotrophic processes; $B(z)$ is the photosynthetic pigment biomass ($kg\ m^{-3}$); $I(z)$ is the net downwelling irradiance ($J\ m^{-2}s^{-1}$) and κ ($m^2(kg\ Chl\ \underline{a})^{-1}$) is the attenuation coefficient of the pigments. Equation (1) is just a statement of an exponential absorption process of the type

$$\frac{\partial I(z)}{\partial z} = -f(z)I(z) \qquad (2)$$

where $f(z)$ is an attenuation coefficient$[L^{-1}]$, and we are interested only in that part of the absorption that is ascribable to autotrophic processes.

In calculating the amount of light available at depth z, however, we have to take into account all sources of attenuation, biological and physical. In this case the attenuation coefficient might be written as

$$f(z) = \zeta + \kappa B(z) \qquad (3)$$

where for simplicity we have omitted to include an explicit term for the absorption by suspended materials other than phytoplankton cells. In the open ocean we expect this to be a relatively minor contribution to absorption and in any case it is a conservative assumption for the problem under discussion. The constant $\zeta(m^{-1})$ is the linear attenuation coefficient for "pure" seawater. Then I(z) is simply

$$I(z) = I_o \exp(-\zeta z - \kappa \int_o^z B(z)dz) \qquad (4)$$

where I_o is the net downwelling irradiance just below the water surface, and the exponent is the function F(z):

$$F(z) = \zeta z + \kappa \int_o^z B(z)dz \quad . \qquad (5)$$

From (1), we can now write the upper limit to the photosynthetic energy storage rate as

$$Q(z) = I_o \kappa B(z) e^{-F(z)} \quad . \qquad (6)$$

The realised rate of energy storage as photosynthate will always be less than Q(z) because of the inevitable thermodynamic inefficiencies in the photosynthesis process. The maximum storage rate per unit of pigment biomass is

$$\frac{\partial Q(z)}{\partial B(z)} = I_o \kappa e^{-F(z)} \qquad (7)$$

The right hand side of this equation is accessible to field measurement.

We now turn to the problem of computing the photosynthesis profile. Various forms are available for the representation of the photosynthesis-light curve (Jassby and Platt, 1976; Platt _et al._, 1977; Platt and Gallegos, 1980; Gallegos and Platt, 1981): a convenient form is that used in Platt _et al._, 1980:

$$P^B(I) = P_m^B(1 - \exp(-\alpha I/P_m^B)) \quad . \qquad (8)$$

Here the parameter P_m^B that scales the amplitude of the curve is the maximum realised photosynthetic rate under prevailing environmental conditions. The superscript B indicates normalisation to the pigment biomass. We have assumed no photoinhibition, another conservative assumption for the problem. The second parameter in (8) is defined by

$$\alpha \equiv \left. \frac{\partial P^B}{\partial I} \right|_{I \to 0} \qquad (9)$$

It is the slope of the curve in the region where the efficiency of photosynthesis is maximal.

For some purposes, it is more convenient that eqn.(8) be expressed in terms of irradiation _absorbed_ per unit chlorophyll, as defined in eqn.(7), rather than irradiation _available_. To achieve this, it is easy to see that α can also be written in the form

$$\alpha(z) = \xi\kappa B(z) \tag{10}$$

where ξ is a dimensionless factor that characterises the efficiency of photosynthesis in weak light, and the normalised maximum production is expressed in energy units (J(kg Chl $\underline{a}$)$^{-1}$s^{-1}). Then (8) can be expressed as

$$\frac{P^B(I)}{P^B_m} = 1 - \exp(-\xi\kappa I/P^B_m) \quad . \tag{11}$$

Equations (6) and (11) are useful in specific cases for checking individual results at discrete depths, but we shall also need to integrate them over depth to compare the water column productivity with indirect measures of community catabolism.

Let us call the energy absorbed by autotrophs S(J m^{-2}s^{-1}) per unit area of sea surface:

$$S \equiv \int Q(z)dz \tag{12}$$

where the integral is taken over the entire water column. By definition, the non-trivial contributions to the integral will come only from the euphotic zone. The limits of integration may then be set at 0 and ∞, if it should be convenient. Clearly, the scale of S can be estimated from I_o, the PAR available just below the sea surface. S would equal I_o if seawater itself were optically transparent and if the only particulate absorbers were those containing photosynthetic pigments.

Equation (12) can be integrated without difficulty in the ideal case where the pigment biomass is distributed uniformly with depth, B(z) = B.

$$S = I_o\kappa B\int_o^\infty \exp(-(\zeta + \kappa B)z)\ dz \tag{13}$$

or

$$S = \frac{I_o(\kappa B)}{\zeta + \kappa B} \tag{14}$$

This is an instructive result: it shows that the amount of the incident energy that is absorbed by autotrophs (and is therefore available as photosynthetic potential) results from a simple partition between competing attenuation coefficients. The more abiotic sources of light absorption there are, the less light that can be stored by the phytoplankton.

At the other extreme, where all the biomass is concentrated in a narrow layer at depth z_m, i.e. $B(z) = B\delta(z-z_m)$ where δ is the Dirac delta function, we can write

$$S \leq I_o\kappa B(z)\ \exp\ (-\zeta z_m) \tag{15}$$

which states that the energy fixed in the layer cannot exceed the energy available at the depth of the layer multiplied by the effective optical cross section of the phytoplankton in the layer.

Turning to the water column productivity, $Z(\mathrm{J\ m^{-2}s^{-1}})$ we see from (11) that it takes the form

$$Z = P_m^B \int_o^\infty B(z)\left(1-\exp\left(-\frac{\xi\kappa}{P_m^B} I_o \exp\left(-\zeta z-\kappa\int_o^z B(z')dz'\right)\right)\right)dz \quad (16)$$

In making calculations based on equations (11) and (16) it is convenient, and affords the maximum generality, if we make use of the following dimensionless groups (cf. Talling, 1975): first, $P^B(z)/P_m^B$; second, $(\zeta + \kappa B)z$; and third, $\xi\kappa I_o/P_m^B$. In principle, all of these groups are accessible to measurement at sea. However, the parameters ξ, κ and P_m^B are wavelength-dependent, a feature that we have suppressed for lack of sufficient data to make an adequate treatment.

Figure 1 illustrates the application of eqn.(11) to the calculation of photosynthesis profiles for various values of the dimensionless groups, assuming a uniform distribution of chlorophyll with depth. Note that the case of non-uniform chlorophyll distribution

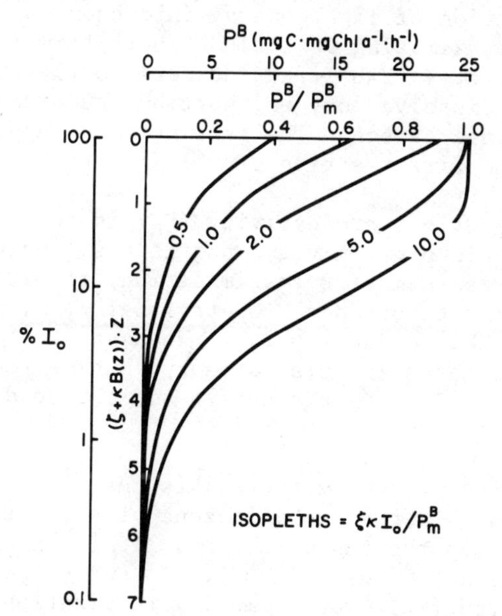

Fig. 1. Relative photosynthesis as a function of optical depth corresponding to solution of equation (11). The different curves correspond to different values of the dimensionless group $\kappa\xi I_o/P_m^B$. The upper scale is an example with P_m^B = 25 mg C·mg $\mathrm{Chl^{-1}h^{-1}}$.

can readily be recovered if the simple optical scale, $z(\zeta + \kappa B)$, is replaced by the optical path function given in eqn. (5) in which the chlorophyll profile B(z) is specified as required.

DIRECT AND INDIRECT ESTIMATES OF PRODUCTIVITY

Having considered the way in which an upper limit might be fixed for acceptable values of the primary productivity of the ocean, we now proceed to examine the estimation, direct or indirect, of productivity in real ocean situations. In comparing the results of these estimations for internal consistency, it is important to keep certain distinctions clearly in mind because the different methods do not necessarily estimate the same thing. The following classification provides a framework for discussion:

(i) P_g; Gross primary production,
(ii) P_n; Net small particle production, and
(iii) P_c; Net euphotic community production.

Gross primary production (P_g) is equal to the rate of photosynthetic conversion of light energy into chemical bond energy. Observations approximating P_g include short term (time scale $\sim$ 1 hour) carbon fixation rates; 24h light and dark bottle oxygen changes; and diel changes in dissolved oxygen, particulate organic carbon or dissolved inorganic carbon. The setting of an upper limit to P_g was considered in the first section of this paper.

Net small particle productivity (P_n) is sometimes thought of as the net autotrophic productivity, but it also includes the transformations of organic matter (assimilation, respiration and excretion) by microheterotrophs. Operationally, P_n is equivalent to the light bottle O_2 changes during 24h. We have elected to place 24h ^{14}C productivity estimates in this category, but we recognize that the interpretation of ^{14}C uptake rates is subject to discussion (Peterson, 1980).

Net community productivity (P_c) is the rate of accumulation of organic matter within the euphotic zone plus the rate of export of organic matter from the euphotic zone. It is "new production" in the sense of Dugdale and Goering (1967). As such, P_c is the excess of gross productivity (P_g) over community respiration. Direct observations of P_c include the accumulation rate of oxygen within the euphotic zone over an extended period (time scale of months), or the flux of sinking organic matter as measured by sediment traps or calculated from the respiration rates in the aphotic zone. In nutrient limited ocean provinces, P_c is constrained by the nitrogen flux to the euphotic zone, as discussed below.

In Table 2 we have collected estimates of primary production (P_g, P_n or P_c) from various stations in the oligotrophic ocean reported by different workers using a variety of techniques.

Some of the controversy surrounding recent estimates of oceanic productivity may have resulted from confusion about what the observations purport to measure and what was actually measured. A set of numbers for P_g, P_n and P_c should be internally consistent for a given piece of ocean, but we rarely (never?) have them all at the same time: we are obliged to estimate one from the other. The relationships between P_g, P_n and P_c can be examined in terms of elemental (carbon, oxygen or nitrogen) cycles or energy fluxes.

Simplified representations of the nitrogen and carbon/oxygen cycles are illustrated in Figure 2. Using the notation ρ^x_{ab} to characterise the flux of element x from compartment b to compartment a, we can define P_g, P_n and P_c in terms of these ρ^x_{ab}'s (Table 1).

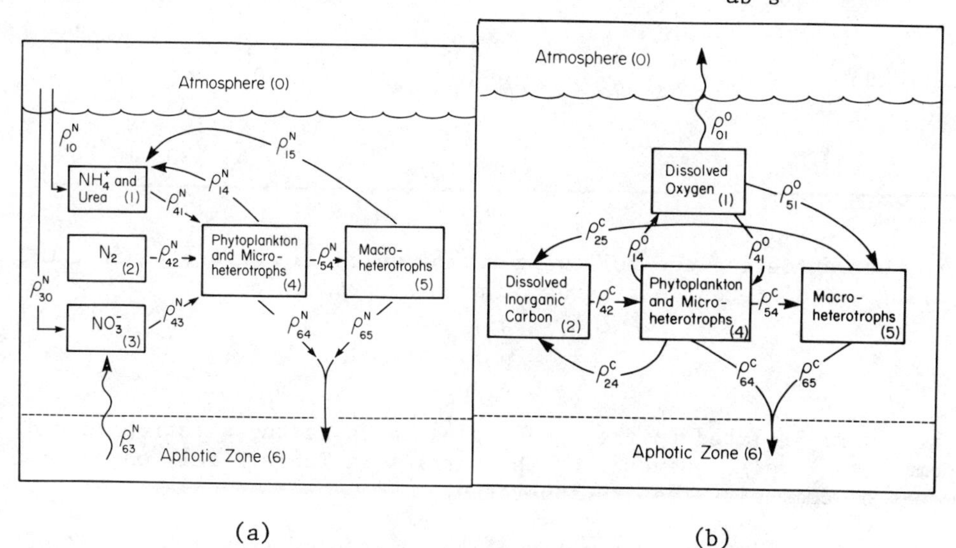

(a) (b)

Fig. 2a and 2b. (a) Simplified representation of the nitrogen cycle within the euphotic zone and interactions with the atmosphere and aphotic zone. We have classified the biological components operationally. Phytoplankton and microheterotrophs are those organisms which can be sampled effectively with a water bottle. Macroheterotrophs are those organisms which are excluded from productivity bottles. The breakdown of components is consistent with our classification of productivity measurements (Table 1, 2i, 2ii, and 2iii). (b) Simplified representation of the carbon and oxygen cycles in the euphotic zone. Explanatory comments for Figure 2a apply here as well.

Table 1. Definition of Different Measures of Primary Production in Terms of the Elemental Fluxes Shown in Figure 2a and 2b

	Gross Primary Production P_g	Net Small Particle Production P_n	Net Community Production P_c
Carbon	ρ^{c}_{42}	$\rho^{c}_{42}-\rho^{c}_{24}$	$\rho^{c}_{42}-\rho^{c}_{24}-\rho^{c}_{25}$
			$=\rho^{c}_{64}+\rho^{c}_{65}$
Oxygen	ρ^{o}_{14}	$\rho^{o}_{14}-\rho^{o}_{41}$	$\rho^{o}_{14}-\rho^{o}_{41}-\rho^{o}_{51}$
			$=\rho^{o}_{01}$
Nitrogen	$\rho^{N}_{41}+\rho^{N}_{42}+\rho^{N}_{43}$	$P^{N}_{g}-\rho^{N}_{14}$	$\rho^{N}_{64}+\rho^{N}_{65}$
			$=\rho^{N}_{63}+\rho^{N}_{30}+\rho^{N}_{10}+\rho^{N}_{42}$

We can define the following ratios which may be useful in comparing the different estimates of primary production.

$$\Pi_1 = P_n/P_g \quad \Pi_2 = P_c/P_n \quad \Pi_3 = P_c/P_g \tag{17}$$

The Π's are related to the respiration rates at the autotrophic and community levels. Several of the entries in Table 2 rely on estimates of community respiration rates.

Dark respiration rates for phytoplankton of from 10% to 50% of P_g (Raven and Beardall, 1981) imply that Π_1 has upper bounds in the interval from 0.9 to 0.5. Where significant microzooplankton and bacterial respiration occurs Π_1 would be reduced. Estimates of Π_1 based on a comparison of 6 hour and 24 hour ^{14}C productivities range from about 0.25 to 0.9 (Eppley and Sharp, 1975; Smith, 1977). Other values of Π_1 range from less than 0.1 to 0.25 based on light-dark bottle oxygen productivities (Riley, 1939; Riley <u>et al.</u>, 1949) or a comparison of ^{14}C productivity and light-dark bottle oxygen productivity (Ryther, 1954). However, the "bottle effects" associated with prolonged incubations (of three to seven days) required for determining changes in oxygen concentration may invalidate these comparisons.

The ratio Π_2 is analogous to the ratio of "new" to "total" production as constructed by Eppley and Peterson (1979). It is accessible to estimation from nitrate, ammonia and urea uptake experiments. Nitrate uptake per square metre of sea surface is presumably proportional to the vertical flux of nitrate from below the pycnocline (new production) whereas ammonia and urea uptake represent recycling of nitrogen within the euphotic zone. From the data given in Eppley and Peterson (1979) we can adopt a value of 0.1 for Π_2. This value of Π_2 will be an underestimate to the extent that nitrogen fixation and atmospheric inputs supplement the nitrate flux. A further complication that we should be prepared for in calculating Π_2 in this way is that the remineralisation rates from organic matter for carbon, nitrogen and phosphorus may not be the same.

A calculation in Conover (1978) for the Sargasso Sea, employs an upper bound on Π_2 of 0.25 based on the assumptions that:

(1) all of P_n is processed by herbivores with an assimilation efficiency of 75%, and
(2) the unassimilated organic matter leaves the euphotic zone in rapidly sinking fecal pellets.

An estimate for Π_2 of 0.1 to 0.5 can be obtained from Suess' (1980) comparison of downward particle flux with ^{14}C production for different ocean regions.

Table 3 summarises the main points of this section, showing the classification into P_g, P_n or P_c of the various ways that have been used to estimate primary production; relative magnitudes of P_g, P_n and P_c taking P_n as unity and applying the ratios Π to find P_g and P_c; and the range of values reported for P_g, P_n and P_c for oligotrophic ocean sites.

Table 2(i). Estimates of Pg

Method	mmoles C / m² d	mmoles O_2 / m² d	mmoles N / m² d	Footnote
diel O_2				
Equatorial Atlantic	74	100*	11	(1)
North Pacific near Hawaii	16	22*	2.4	(2)
Caribbean Sea	~220	~300*	~33	(3)
Tropical Atlantic (16°N, 33°W)	77	105*	12	(4)
diel POC				
Equatorial Atlantic	108*	248	28	(5)
diel ΣCO_2				
Caribbean Sea	~1300*	~1800	~200	(6)
Short term ATP Increase				
February	83	110	13	(7)
July	420	570	64	(7)
Light and dark bottle O_2				
Sargasso Sea,	140	188*	21	(8)
Calculation	46	62	69	(9)
Closure condition on energy HIGH	200	272	30	(10)
LOW	43	59	6.5	(10)

*Primary results taken from cited papers.

(1) Based on Tijssen's (1979) areal production rates for the mixed layer of stations 6 and 7 (euphotic zone = 90m and 100m respectively) increased by an amount equal to a factor of i.e. $[(Z_{eu}-Z_{mix})/Z_{mix}]0.5$ to account for the contribution to areal productivity from below the mixed layer.

(2) Gundersen et al. (1976).

(3) The volumetric oxygen evolution rate of 53 mg $Cm^{-3}d^{-1}$ (Johnson et al., 1981) was assumed to apply to a 50m deep layer.

(4) From Ivanenkov et al. (1972).

(5) Based on an areal production rate for the mixed layer at stations 6 and 7 (Postma and Rommets, 1979) increased by an amount equal to a factor of $[(Z_{eu}-Z_{mix})/Z_{mix}]0.5$ to account for the contribution to areal production from below the mixed layer, as in (1) above.

(6) The volumetric carbon fixation rate of 306 mg $Cm^{-3}d^{-1}$ (Johnson et al., 1981) was assumed to apply to a 50m deep layer, as in (3) above.

Table 2(ii). Estimates of Pn

Method	mmoles C / m^2 d	mmoles O_2 / m^2 d	mmoles N / m^2 d	Footnote
^{14}C uptake				
Offshore Atlantic	23*	30	3.5	(1)
Offshore Pacific	13*	17	2.0	
Offshore Indian	19*	25	2.9	
Central North Pacific gyre	7-13*	10-17	1-2	(2)
Pacific near Hawaii	11*	15	1.6	(3)
Net 24 hour in situ O_2 accumulation				
Pacific near Hawaii	5.8	7.9*	0.88	(4)
Light and dark bottle oxygen				
Sargasso Sea	33	45*	5	(5)
Calculation based on zooplankton grazing + export				
Sargasso Sea	13-19	18-26	2.0-2.9	(6)

*Primary results taken from the cited papers.
(1) From Table 11.4 of Platt and Subba Rao (1975).
(2) From Figure 3 of McGowan and Hayward (1978).
(3) From Gundersen *et al*. (1976).
(4) From Gundersen *et al*. (1976).
(5) From Riley's (1970) discussion, based on calculations of particle flux to aphotic zone and respiration in euphotic zone.
(6) Based on Riley (1939) light-dark bottle oxygen productivity experiments. The rate for surface waters was assumed to apply uniformly to a euphotic zone of 50m depth.

Table 2(i). (see opposite)

(7) The volumetric production rates of 9.9 mg $Cm^{-3}d^{-1}$ (February) and 50.5 mg $Cm^{-3}d^{-1}$ (July) reported by Sheldon and Sutcliffe (1978) were assumed to apply to a 100m euphotic zone.
(8) The volumetric oxygen evolution rates reported by Riley(1939) for the three stations furthest into the Sargasso Sea were averaged and assumed to apply to a 50m deep euphotic zone.
(9) From Table 1 of Eppley (1980).
(10) This study, photosynthetic storage of radiant energy.

Table 2(iii). Estimates of Pc

Method	mmoles C / m² d	mmoles O_2 / m² d	mmoles N / m² d	Footnote
Net euphotic zone O_2 accumulation				
Central North Pacific Gyre	13-36	16-49*	1.9-5.4	(1)
Pacific near Hawaii	8.5-9.2	11-12*	1.3-1.4	(2)
Tropical Atlantic	3.1	4.2*	0.46	(3)
Aphotic Zone O_2 utilization				
Sargasso Sea				
100-1700m	30	39*	4.6	
100-2000m	3.4	4.4*	0.52	(3)
200-4000m	4.2	5.4*	0.64	
Particle flux to aphotic zone				
Central North Pacific	5.7*	7.8	0.41*	(6)

*Primary results taken from the cited papers.

(1) A conservative estimate for the oxygen accumulation rate at 28°N, 155°W of 16 mmoles O_2 $m^{-2}d^{-1}$ was calculated from the excess oxygen concentrations in the water column and an assumed accumulation time of 120 days. Oxygen concentrations and relative saturation were obtained from Figure 11 of Shulenberger and Reid (1981). These were converted to excess oxygen and the total amount of excess oxygen in water column was found from a linear interpolation between observations. Another estimate for the oxygen accumulation rate of 49 mmoles O_2 $m^{-2}d^{-1}$ was calculated using the assumption that areal O_2 accumulation rate is related to the areal ^{14}C productivity by the same ratio of O_2 evolved to ^{14}C productivity for the single depth stratum examined by Shulenberger and Reid. We use a ratio of oxygen accumulation rate to ^{14}C productivity of 3.5 mmoles O_2/mmole C obtained from entries for stations A1, A2, A3 and A4 of Shulenberger and Reid's (1981) Table 1. A ^{14}C productivity of 13.3 mmoles C $m^{-2}d^{-1}$ (160 mg C $m^{-2}d^{-1}$) was obtained from Figure 3 of McGowan and Hayward (1978).

(2) Similar calculations to those described in (1) were made for the Pacific Ocean near Hawaii based on the observations of Gundersen *et al*. (1976). The relevant observations used for the calculations are:
(i) an excess of oxygen concentration of 0.37 ml ℓ^{-1} at 50m,
(ii) an accumulation time of 60 days,
(iii) a ^{14}C productivity of 0.58 mg $m^{-3}h^{-1}$ at 50m, and
(iv) a ^{14}C productivity of 128 mg C $m^{-2}d^{-1}$.

Table 3. Summary of Information from Tables 1 and 2(i), (ii) and (iii)

	P_g	P_n	P_c
Observable	Short term ^{14}C Diel C, O_2, POC ATP accumulation in short term bottle experiment (?) Light-Dark O_2	24 hour ^{14}C Light bottle O_2	O_2 accumulation Particle flux Aphotic zone O_2 consumption
Constraint (mmoles $m^{-2}d^{-1}$)	Energy flux 20-60 mmole $Cm^{-2}d^{-1}$		Mass flux for limiting nutrient 5 mmole N $m^{-2}d^{-1}$
Relative values (based on Π's)	1.1 - 4.0	1.0	0.05 - 0.3
Productivity (mmoles C $m^{-2}d^{-1}$)	16 - 1300	6 - 33	3 - 36

←

Table 2(iii). (cont.)

(3) From Ivanenkov et al. (1972).
(4) Calculated from Tables 1 and 2 of Jenkins (1980). The values differ due to the use of Worthington's (1976) estimates of ventilation rate for 100-4000m depth zone and Jenkins (1980) estimates of ventilation rate for the 100-1700m depth zone.
(5) From Table 21 of Riley (1951).
(6) From Table 1 of Knauer et al. (1979) for a single 7-day particle interceptor trap deployment. Note that a direct comparison of the carbon and nitrogen numbers implies C/N ratio of 13.9 rather than the figure of 6.7 used as a standard in this paper.

MASS BALANCE FOR NITROGEN

One of the closure calculations we shall consider is that on the nitrogen flux. Net community nitrogen demand cannot exceed the flux into the euphotic zone if productivity is to be maintained. Can nitrogen be supplied at a sufficiently high rate to support the estimated rates of primary production? In cases where productivity has been estimated by ^{14}C fixation (P_g or P_n), the question is elusive in that nitrogen could be supplied by recycling within the immediate pelagic environment at a rate whose upper limit has yet to be established. But where production is estimated from the increase in bulk oxygen concentration over an extended period, the question is much better defined. Here nitrogen must be supplied from an external source, at a rate that can be estimated.

It is generally accepted that in the oligotrophic, central ocean gyres the predominant source of nitrogen external to the euphotic zone is the nitrate pool below the pycnocline. The supply rate can be estimated by an equation of the form:

$$\text{Nitrate Flux} = K_z \frac{\partial}{\partial z} [NO_3] \quad (18)$$

where K_z is the vertical diffusivity and $\partial[NO_3]/\partial z$ is the vertical nitrate gradient. Neither our knowledge of the effective diffusion coefficient nor of the nitrate gradient is very reliable.

A number of different physical processes may contribute to vertical diffusive transport. A K_z for equation (18) could be approximated by the sum of the effective eddy diffusivities for all of these processes.

Observations of temperature microstructure yield an eddy diffusivity of about 1 to 3 x $10^{-6} m^2 s^{-1}$ (A. Gargett, 1976; Garrett, 1979) The effective eddy diffusivity for salt flux due to double diffusive processes (salt fingers) in the main thermoclines of the Sargasso Sea and the central North Pacific gyre varies from about 1 to 5 x $10^{-5} m^2 s^{-1}$ (Schmitt and Evans, 1978; Gargett, pers. comm.). Thus, the double diffusion flux dominates the strictly turbulent flux by about an order of magnitude. The highest estimates of K_z from double diffusive mechanisms approach a K_z of $10^{-4} m^2 s^{-1}$, calculated from salt and temperature distibutions within and below the main pycnocline (Munk, 1966; Garrett, 1979). To complicate matters, Jenkins (1980) argues that transport along sloping density surfaces may, in some locations (e.g. the Sargasso Sea) dominate the apparent vertical diffusion as a means of redistributing water properties.

For our estimate of the nitrate flux into the euphotic zone of the central North Pacific gyre we adopt an effective eddy diffusivity of $10^{-4} m^2 s^{-1}$, which is near the upper end of plausible values. This would include the contribution to vertical mixing by "unusual events" as reported by McGowan and Hayward (1978). We obtained a nitrate gradient of 80 μmoles m^{-4} for the 200 to 500m depth stratum, from the observations reported in the Climax II cruise report (SIO 1975). Thus, an upper bound for the nitrate supply by this mechanism is 0.8 mmoles N $m^{-2}d^{-1}$. Interestingly, a similar value can be calculated from Dugdale and Goering's (1967) nitrate uptake rate measurements for the Sargasso Sea near Bermuda. However, the nitrate uptake rates of Eppley _et al_. (1977) for the North Pacific central gyre are much lower. Although this flux provides a supply of nitrogen to the base of the euphotic zone, the problem remains of redistributing the supplied nitrogen within the euphotic zone.

Another source of unrecycled nitrogen that may be significant is nitrogen fixation. Although the rate of nitrogen fixation is usually thought to be low (Carpenter and McCarthy, 1975; McCarthy and Carpenter, 1979), the observed and estimated rates of nitrogen fixation cover 3 orders of magnitude from 1 μmole N $m^{-2}d^{-1}$ to 1 m mole N $m^{-2}d^{-1}$ (Mague _et al_., 1977; Venrick, 1974; Gundersen _et al_., 1976). Most of the research on oceanic nitrogen fixation has treated the free living cyanobacteria _Richelia_ and _Oscillatoria_, or the symbioses involving _Richelia_. Recent observations by Martinez _et al_. (1982) have demonstrated the existence of a symbiosis involving _Rhizosolenia_ (a diatom) and nitrogen-fixing bacteria. These authors also demonstrated that improper sample handling could result in low apparent fixation rates. Isolation of individual _Rhizosolenia_ mats by SCUBA techniques resulted in the highest nitrogen fixation rates. Taken together, these observations suggest that nitrogen fixation rates may equal the diffusive nitrate flux through the pycnocline.

A final source of "new" nitrogen is atmospheric dryfall and rain. Estimates of global exchange of nitrate and ammonia between the oceans and atmosphere allow the calculation of an average nitrogen loading rate of 16-54 μmoles N $m^{-2}d^{-1}$ (Fogg,1982).

Table 4 summarizes this section, with particular emphasis on the North Pacific central gyre. Also included in Table 4 are estimates of nitrogen export from the euphotic zone (Knauer _et al_., 1979) and recycling within the euphotic zone (based on the respiration rate calculations of section 4 and the observations of Eppley _et al_., 1973 and Mullin _et al_., 1975). Based on our review, we can tentatively suggest a new nitrogen supply rate of 0.5 to 1.0 mmole N $m^{-2}d^{-1}$, with nitrate transport dominating the supply in winter-spring and nitrogen fixation playing an important role in summer-autumn. Relative to these sources of new nitrogen, the contributions of rain, atmospheric dryfall and fish advection are small.

The export rate of particulate nitrogen appears to closely balance the import rate suggested above. The major component of this flux appears to be rapidly sinking particles (Knauer _et al_., 1979) with feeding by vertical migrators playing a smaller role. However, there is a need for more observations of the vertical nitrogen flux out of the euphotic zone.

Table 4. Nitrogen Sources to the Central North Pacific Gyre

	μ moles N $m^{-2}d^{-1}$	Footnote
External		
Vertical diffusion	800	(1)
N_2 fixation	1 - 1000	(2)
Rain and dry fall	16 - 54	(3)
Tuna	4.3	(4)
Recycled Nutrients		
Microzooplankton NH_4 excretion	3000 - 9000	(5)
Zooplankton NH_4 excretion	100 - 1000	(6)
Zooplankton Urea excretion	500 - 2400	(7)
Export of Nitrogen		
Mesopelagic Fish	30 - 60	(8)
Particle Flux	410	(9)

Primary results taken from the cited papers.

(1) Calculated as described in the text.

(2) Range of observed and estimated rates of N_2 fixation reported in the literature. See text for references.

(3) Fogg's (1982) estimates of atmospheric inputs of ammonia and nitrate to the oceans of 29 to 100 x 10^6 $MTyr^{-1}$ (29 - 100 x 10^{12} gyr^{-1}) were divided by the area of the world oceans (361 x 10^6 km^2) (McLellan, 1965).

(4) The potential advective flux of new nitrogen into the North Pacific central gyre associated with migrating tuna has been estimated from information provided by G.Sharp (pers. comm.). Contributions of large and small tuna were calculated separately. The biomass of large tuna was estimated at 30 kg/km^2. These tuna were assumed to metabolize 5% of their body weight per day. Assuming that dry weight = 0.2 times wet weight, and that all of the metabolized tissue is protein with a nitrogen content of 16% by weight yields a nitrogen flux of 3.4 μ moles $Nm^{-2}d^{-1}$. The biomass of small tuna was estimated at 2.5 $kgkm^{-2}$. These tuna were assumed to metabolize 15% of their body weight per day. Again, assuming that dry weight = 0.2 times wet weight, that all of the metabolized tissue is protein with a nitrogen content of 16% by weight, yields a nitrogen flux of 0.86 μ moles $Nm^{-2}d^{-1}$. These are overestimates of the nitrogen flux because the primary metabolic reserve is lipid, not protein.

Using the estimates of nitrogen flux into and out of the euphotic zone, we can suggest a P_c of about 0.75 mmoles N $m^{-2}d^{-1}$ (implying a carbon flux of 5 mmoles C $m^{-2}d^{-1}$ by using a C/N of 6.7). Since every mole of nitrogen transported vertically implies a stoichiometric amount of net oxygen evolution by phytoplankton, we can compare the nitrogen flux with the oxygen accumulation rate in the euphotic zone. Conversely, every mole of oxygen observed to accumulate in the euphotic zone implies that an equivalent amount of nitrogen had been exported.

COMMUNITY METABOLISM

Another closure condition that we can examine for the ocean ecosystem is that on the community respiration: it cannot be greater than the gross primary production plus a correction for advection. Formally,

$$\int_{\Delta z} R(z)dz \leq \int_{\Delta z} P_g(z)dz + A \tag{19}$$

where R is the metabolism of all the organisms in the community and P_g is the gross primary production. The quantity A is the algebraic sum of all the advective terms, including import of organic substrates and export by horizontal advection or sinking. A marker for the biological processes underlying (19) is the oxygen content of the water. Because oxygen is free to diffuse in the vertical we have to look at these competing processes integrated over some substantial depth interval Δz: the differential form of (19) is of limited practical application. The non-trivial contributions to $\int R(z)dz$ will extend over a much greater depth range than those to $\int P_g(z)dz$, which are confined to the euphotic zone. To the extent that (19) is not in dynamic equilibrium, oxygen will accumulate, or be depleted, in Δz.

⟵

Table 4. (Cont.)

(5) Nitrogen equivalent of estimated community respiration (this paper).
(6) Based on estimates of Mullin *et al.* (1975) and Eppley *et al.* (1973).
(7) Based on the estimates of Eppley *et al.* (1973).
(8) The advective loss of nitrogen due to feeding by mesopelagic fish was estimated as follows. The metabolic requirements of mesopelagic fish in the central gyres is about 5-10 kcal $m^{-2}yr^{-1}$ (K.H. Mann, pers. comm.). Assuming that 1 ml O_2 is consumed for every 5 cal of organic matter metabolized, applying a C/N ratio for zooplankton of 4 and a respiration quotient of 1 O_2/1 C, yields a nitrogen flux out of the equphotic zone of 30-60 moles N $m^{-2}d^{-1}$.
(9) From Knauer *et al.* (1979).

So in applying (19) to a real situation, averaging at some time scale is implied.

Metabolism of organisms is strongly size-dependent (see Platt and Silvert, 1981, and references therein). If community structure is known in terms of organism size, community metabolism may be estimated. Suppose $\beta(w)$ is a biomass density function such that $b(w) \equiv \beta(w)dw$ is the mass of organisms, per unit volume of water, with nominal weights in the interval from w to (w + dw). We call $\beta(w)$ the normalised size spectrum (Platt and Denman, 1978). Then the community metabolism (Platt and Denman, 1977) is

$$R(z) = \int M(w)\beta(w)dw \tag{20}$$

where M(w) is the weight-specific metabolism function and the integral is taken over the entire size spectrum.

Observational (Sheldon *et al.*, 1972) and theoretical (Platt and Denman, 1977; 1978) studies suggest that $\beta(w)$ may have a rather simpler form in the open ocean:

$$\beta(w) \sim w^m \quad . \tag{21}$$

The biomass spectrum b(w) can be constructed from $\beta(w)$:

$$b(w) = w\beta(w) \sim w^{m+1} \quad . \tag{22}$$

Here b(w) is the total biomass of organisms of weight w. It should be distinguished from B, the pigment biomass of the autotrophic component (but note that at the small-size end of the spectrum, b will include some B).

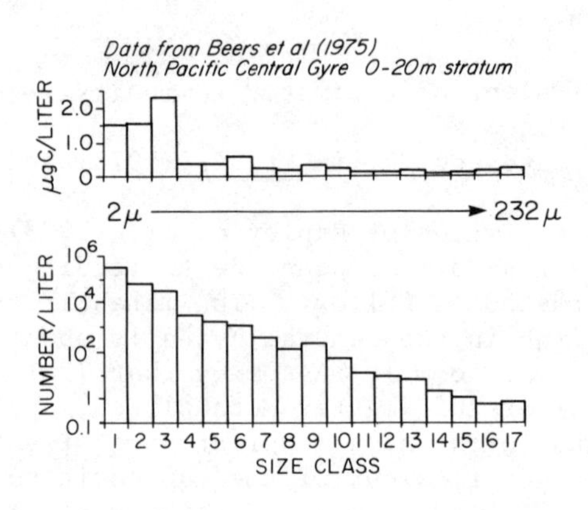

Fig. 3. Observed size distributions of biomass and numbers in the upper 20m of the North Pacific central gyre. Data from Beers *et al.* (1975).

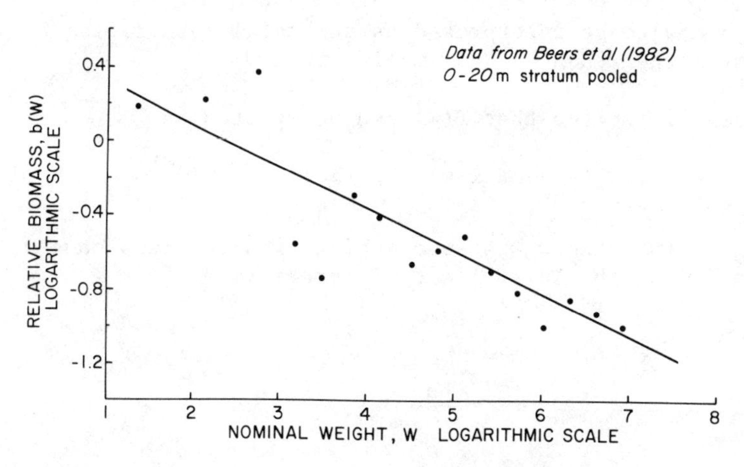

Fig. 4. Least-square fit of equation (22) to observed size distribution of biomass in the upper 20m in the North central Pacific gyre. Data from Beers *et al.* (1982).

Sheldon *et al.* (1972) considered that their biomass spectra were consistent with a numerical value of (m+1) not distinguishable from zero i.e. that they were flat. Platt and Denman (1977, 1978) derived a value of (m+1)N-0.22. For the purpose of this discussion we do not insist on any particular value for the slope of the biomass spectrum: we wish only to examine the consequences of the assumption that the biomass spectrum can be represented as a simple power function of weight. However, it is interesting to remark that the predicted slope of -0.22 (Platt and Denman, 1977), which has so far been untested, now finds some justification in the microscope-derived biomass spectra (Figure 3) published recently for the central gyre of the North Pacific by Beers *et al.* (1982). These authors note that the biomass spectrum is remarkably stable over seasons. We have fitted their spectra (Figure 4) to eqn. (22) and find a slope of -0.23±0.03 for the 0 to 20m layer and -0.20±0.02 for the 100 to 120m layer. These spectra cover the size range from 2μm to >232μm equivalent spherical diameter.

The weight-specific metabolism is in the well-known form

$$M(w) = aw^{\gamma-1} \quad . \tag{23}$$

Then eqn. (20) can be rewritten as

$$R = \int_{w_o}^{w_f} \beta(w) a w^{\gamma-1} dw \quad , \tag{24}$$

where for convenience we have suppressed the explicit dependence on depth and we have introduced the limits of integration w_o and w_f so that R is now to be interpreted as the total respiration by all organisms larger than w_o and smaller than w_f.

From (21) we see that $\beta(w)$ can be written as

$$\beta(w) = \beta_o \left(\frac{w}{w_o}\right)^m \tag{25}$$

where $\beta_o = \beta(w_o)$ and w_o is some arbitrary size class chosen to correspond with the lower limit of integration.

Then

$$R = \frac{\beta_o a}{w_o^m} \int_{w_o}^{w_f} w^{\gamma+m-1} \, dw \tag{26}$$

that is

$$R = \frac{\beta_o a}{m+\gamma} w_o^{\gamma} \left[\left[\frac{w_f}{w_o}\right]^{\gamma+m} - 1\right] \tag{27}$$

The right hand side has the dimensions $[aw^{\gamma}]$, in other words a metabolic rate like M(w), as it should. The term $(w_f/w_o)^{\gamma+m}$ is a dimensionless weight $(w^*)^{\gamma+m}$, scaled to the lowest size class w_o.

In the specific case that the slope of the size spectrum m = -1.22, we have $(\gamma+m) \sim -0.5$, since $\gamma \sim 0.72$ (Fenchel, 1974).

Then

$$R = 2\beta_o a w_o^{\gamma}(1-(w^*)^{-0.5}) \tag{28}$$

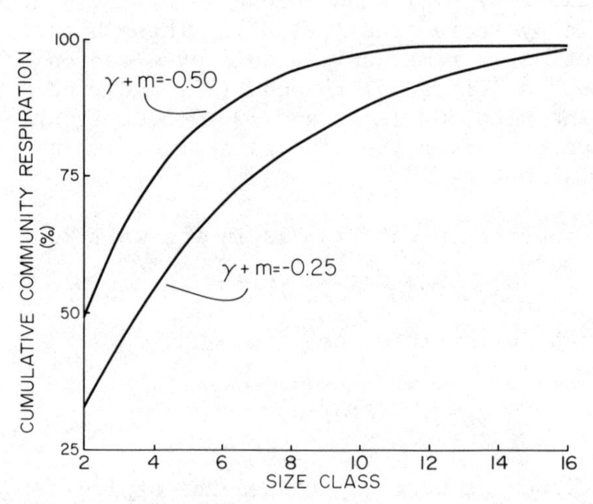

Fig. 5. Cumulative community respiration as a function of w_f, the largest size included in the integration (see equation 27).

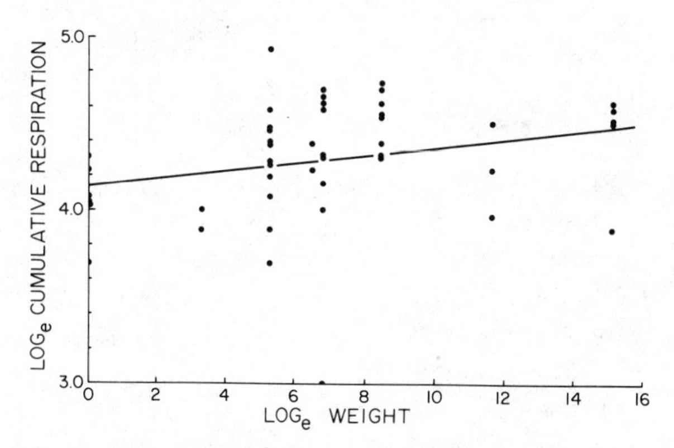

Fig. 6. Cumulative community respiration as a function of w_f, the largest size included in the integration (see equation 27). Data from Williams (1981).

The term $(w^*)^{-0.5}$ decreases rapidly with w*(dimensionless). For example when $w^* = 10^4$, $(w^*)^{-0.5}$ is reduced to 0.01, so that the term can be dropped from (27) with an error of only 1%. On an octave scale this occurs where w_f is removed by 13 or more spectral bands from w_o. Another way of saying this is that the total community respiration can be estimated with high precision by accounting for only the first thirteen octaves on the size-class scale (Figure 5). The data of Williams (1981) for coastal water (Figure 6) are consistent with this result.

We can now proceed to estimate pelagic community respiration in a particular example, using the biomass spectra reported for the north central gyre of the Pacific by Beers _et al_. (1982). These are reliable spectra constructed from microscope observations, and therefore not contaminated by detrital particles as Coulter counter spectra may be. On the other hand, they may contain underestimates of those fragile organisms that are prone to disintegrate in the fixation process.

Noting that the Beers' data encompass 17 octaves in size, we may use the following simplified form of (28) for community respiration.

$$R^{\text{community}} \sim 2\beta_o a w_o^{\gamma} \quad . \tag{29}$$

Evaluation of this expression depends on a knowledge of the abundance in the smallest size class and on the quantity a, the "intensity" of respiration. The choice of a depends on whether the organisms considered are primarily unicells or primarily metazoa. For heterothermic metazoa a ~ $2 \times 10^{-7} s^{-1}$ (Hemmingsen, 1960; Fenchel, 1974) whereas for unicellular organisms it is smaller by a factor of about 8 (Hemmingsen, 1960). From the data in Beers _et al_. (1975, Figure 9)

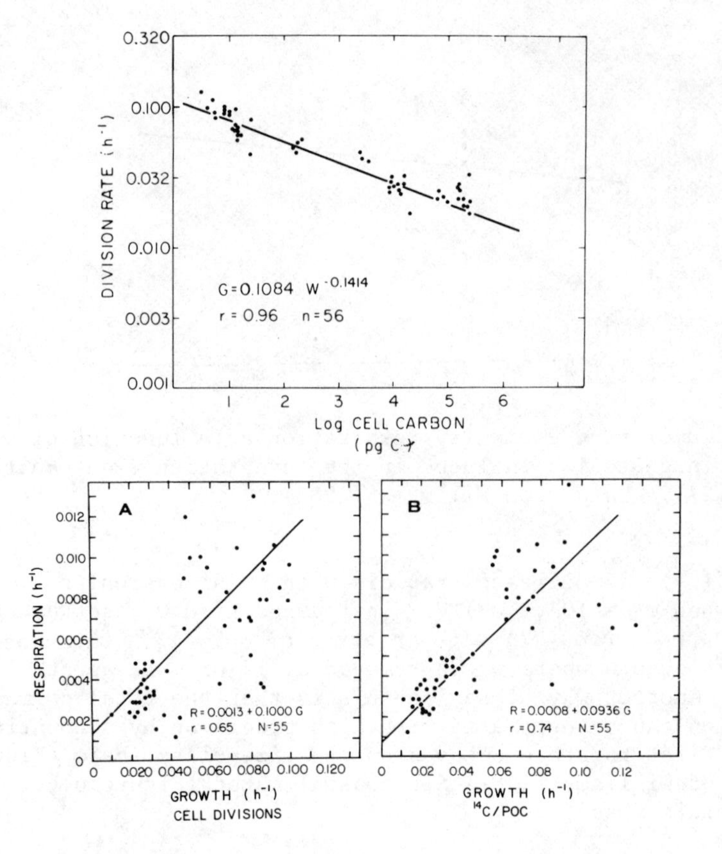

Fig. 7. (a) Division rate of light and nutrient saturated diatoms as a function of cellular carbon. From Blasco et al. (1982). (b) Respiration rates as a function of growth rates for diatoms. From Blasco et al. (1982).

we can judge that the biomass is dominated by unicells at least up to the 10th octave in size. We therefore select a value for a of $2.5 \times 10^{-8} s^{-1}$ in computing R. The recent data of Blasco et al. (1982) are useful for the calculation of allometric constants for phytoplankton metabolism (Figure 7).

As for the biomass scale, Beers et al. (1982) report their results in $\mu gC\ \ell^{-1}$. Taking 1.5 μm as the nominal radius of particles in their first octave, and assuming a specific gravity of unity, we find $w_o \sim 1.4 \times 10^{-11}$g or 9×10^{-13}g C assuming that carbon is 8% of the wet weight. The typical biomass in this octave from Beers' data is $\sim 1\ \mu g\ C\ 1^{-1}$. This is equivalent to a number density of $B_o \sim 10^6 1^{-1}$ or $10^9\ m^{-3}$. Then (29) gives

$$R = 0.08 \times 2 \times 10^9 \times 2.5 \times 10^{-8} \times (1.4 \times 10^{-11})^{0.75} g\ C\ m^{-3} s^{-1} \quad (30)$$

or

$$R = 2.4 \text{ mg C m}^{-3}\text{d}^{-1} \quad .$$

Beers et al. stress that the size spectra are very stable in time, with little difference between the 0-20m stratum and the 100-120m stratum. Let us estimate that the depth interval Δz over which this level of respiration occurs is 100m. Then the water column community respiration is 240 mg C $m^{-2}d^{-1}$, (20 mmoles $m^{-2}d^{-1}$).

Of course, this result is found assuming that eqn. (29) is an acceptable representation of the respiration of the community whose structure is described adequately by (22). Those who have no faith in the theory of the biomass spectrum may go straight to Beers data, calculate the respiration directly for the biomass in each size class and sum over all classes. They will find that the range of community respiration calculated from the eight examples of biomass spectra in Beers et al. (1982) is from 1.7 to 3.3 mg C $m^{-3}d^{-1}$ with a mean of 2.6 mg C $m^{-3}d^{-1}$!

We must now face another kind of complication in the biomass spectra. The best available data, those of Beers, contain no information on organisms less than 2 μm in size. However, the existence of autotrophic organisms smaller than this is now well-established (Waterbury et al., 1979; Johnson and Sieburth, 1979) and they may represent an appreciable proportion of the autotrophic biomass and activity (Herbland and LeBouteiller, 1981). Before calculating their contribution to the community metabolism as represented by the Beers data we first have to estimate their biomass.

We can proceed in one of two possible ways. We could assume that the biomass spectrum extrapolates beyond 2 μm with exactly the same slope (∿-0.22) as in the region of larger sizes. Examination of Beers' spectra, however, persuades us that this would lead to a serious bias in the estimation of biomass. The reason is that a consistent feature of Beers' data is a decrease in biomass with nominal size in the three smallest size classes (10μm to 2μm). We have estimated this decrease by fitting the last three spectral points in the pooled data from each depth stratum (i.e. 12 data points for each fit). We find $b(w) \sim w^{0.14}$ for the 0 to 20m stratum and $b(w) \sim w^{0.11}$ for the 100 to 120m stratum. Therefore, to estimate the contribution of the picoplankton to community respiration, we have assumed that the biomass spectrum extrapolates beyond 2μm with a slope $m = 0.14$. We take the nominal size of the smallest class to be 0.25 that in Beers' smallest class (factor of 1/64 in weight) and apply eqn. (27). We then find that including organisms less than 2 μm leads to an increase in the estimate of community respiration by about a factor of three. The revised estimate is then 720 mg C $m^{-2}d^{-1}$ (≡ 60 mmoles C $m^{-2}d^{-1}$).

THE CENTRAL GYRE OF THE NORTH PACIFIC

In this section we attempt to synthesise the material in the earlier sections by comparing the various pelagic flux estimates for the region of the central gyre of the North Pacific (28°N 155°W; see McGowan and Hayward, 1978). This is perhaps the best-studied of the oligotrophic domains of the ocean. It is also the subject of a recent paper that exposes an apparent inconsistency in two independent estimates of primary productivity (Schulenberger and Reid, 1981). The discussion will not be restricted entirely to the central gyre, however. When the data can illustrate a point, we will refer to results from other regions, as already presented in Table 2.

We begin by calculating a numerical value for the upper limit on P_g according to the procedure outlined in equations (1) to (16). For the sake of simplicity we assume that chlorophyll is distributed uniformly with depth, $B(z) = 7.5 \times 10^{-8}$ kg m^{-3}. A deep chlorophyll maximum (DCM) is invariably present in the central Pacific (data summarised in Shulenberger and Reid, 1981), but it is rather weak in the area of interest, and lies very deep. It is easy to show that, under the conservative conditions for which we calculate, its contribution to water column productivity is rather small. We take the specific absorption of the pigments to be $\kappa \sim 4 \times 10^4 m^2$ (kg Chl a)$^{-1}$, towards the high end of the values published recently by Morel and Bricaud (1981). The PAR is estimated to be 1.5×10^7 J $m^{-2}d^{-1}$ or one half the clear sky value at 20°N as given in the Smithsonian Meteorological Tables. As a first approximation we assume that this is the average flux throughout the 24h day, $\overline{I}_o = 1.7 \times 10^2$ J $m^{-2}s^{-1}$.

The efficiency of photosynthesis, in terms of absorbed energy is thought not to exceed $\xi = 0.25$ ($\lambda \sim 500$ nm; Hall and Rao, 1972). The assimilation number P_M^B is believed to have a definite upper limit, estimated on physiological grounds by Falkowski (1980) and by Gallegos and Platt (1981) to be $\sim$ 25 mg C mg $Chl^{-1}h^{-1}$ ($\equiv 2.8 \times 10^5$ J kg $Chl^{-1}s^{-1}$ in the units used here). This figure is the upper bound of large data sets published, for example, by Platt and Jassby (1976) and by Malone (1980).

These are the parameter values used in our first calculation of water column production (example A in Table 5 and Figure 8). They give a value of 2.3g C $m^{-2}d^{-1}$ or 0.2 moles C $m^{-2}d^{-1}$. The result is an upper limit in that the most generous values have been selected for the parameters and the most conservative assumptions have been made. It is an estimate that strains the credulity because of the improbability of finding all parameters at their optimal value in any given situation. Nevertheless, the estimate based on diel changes in total CO_2 in the Caribbean Sea (Johnson et al., 1981) exceeds it by a factor of six (Table 2), and it is comparable with an estimate based on diel POC changes in the equatorial north Atlantic (Postma and Rommets, 1979) and with one based on diel O_2 changes in the Caribbean

Table 5. Parameter Values for Computation of Gross Photosynthesis Corresponding to the Four Cases in Figure 2. Integrated Primary Production is also Shown

	A	B	C	D
Depth of 1% I_o (m)	100	100	100	60
I_o $(J \cdot m^{-2} \cdot s^{-1})$	$1 \cdot 7 \cdot 10^2$	$1 \cdot 4 \cdot 10^2$	$1 \cdot 2 \cdot 10^2$	$1 \cdot 2 \cdot 10^2$
κ $(Kg\ Chl\ m^{-2})^{-1}$	$4 \cdot 10^4$	$1 \cdot 6 \cdot 10^4$	$1 \cdot 6 \cdot 10^4$	$1 \cdot 6 \cdot 10^4$
P_M^B $(J\ (Kg\ Chl)^{-1} \cdot s^{-1}$	$2 \cdot 78 \cdot 10^5$	$2 \cdot 78 \cdot 10^5$	$1 \cdot 33 \cdot 10^5$	$1 \cdot 33 \cdot 10^5$
ξ (dimensionless)	0·25	0·25	0·20	0·20
B $(Kg\ Chl\ m^{-3})$	$7 \cdot 5 \cdot 10^{-8}$	$7 \cdot 5 \cdot 10^{-8}$	$5 \cdot 10^{-8}$	$5 \cdot 10^{-8}$
$\int_o^{200\ m} P\ dz$ $(mmol\ C\ m^{-2} \cdot d^{-1})$	197	108	43	30

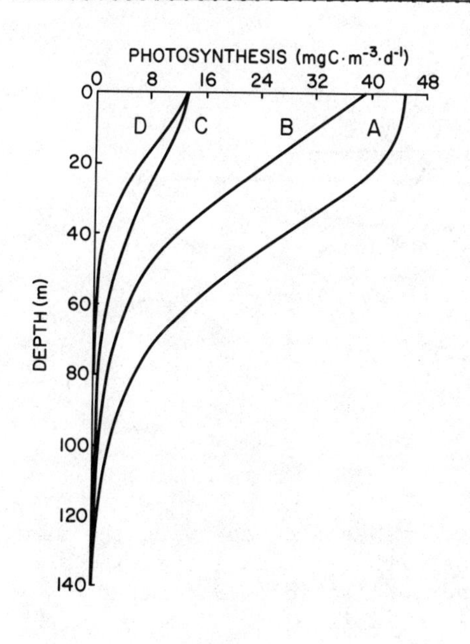

Fig. 8. Volumetric daily photosynthetic rates as a function of depth. The curves are solution to equation (II) for the four parameter sets in Table 5.

(Johnson et al., 1981). These abnormally high values cannot be justified on the grounds that the station locations have biomasses that are atypically large for oligotrophic regions.

A more believable figure for the upper limit to water column photosynthesis might be approached as follows. We take $\overline{I}_0$ to be 120 J $m^{-2}s^{-1}$ and P_M^B to be 12 mg C (mg Chl a)$^{-1}h^{-1}$, a value that would still be considered high for blue water (Platt and Subba Rao, 1975). We reduce the biomass B(z) to 5 x 10^{-8} kg Chl a m^{-3}, and take ξ to be 0.2. We take κ equal to 1.6 x 10^4 m^2 (kg Chl a)$^{-1}$ in agreement with the typical value established by Bannister (1974). The new set of parameter values corresponds to example C in Table 5 and Figure 8. It gives a magnitude of 0.51g C $m^{-2}d^{-1}$ for the water column productivity (43 mmoles C $m^{-2}d^{-1}$).

The realised water column productivity should still be lower than this figure because the non-linearity of the photosynthesis-light curve means that averaging the daily irradiance before calculating photosynthesis leads to an overestimate. As an illustration, in example B (Table 5), applying the same amount of daily solar energy according to a standard cosine curve reduces the estimate of photosynthesis by 30% compared to the figure computed for irradiance supplied at one constant rate throughout the day (Figure 9). We note also that the optical (Figure 10) data in cruise reports (SIO, 1975) are more consistent with a euphotic depth of from 60 to 85m rather than the 100m assumed above (Case D). The implication is that a

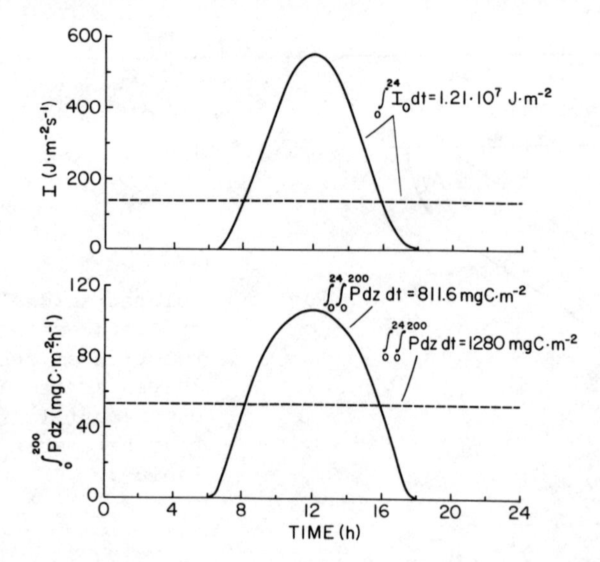

Fig. 9. The effect on integral photosynthesis of applying the same daily integral irradiance as either a constant flux (dotted) or as a cosine function (solid). Parameter values are the same as in B in Table 5.

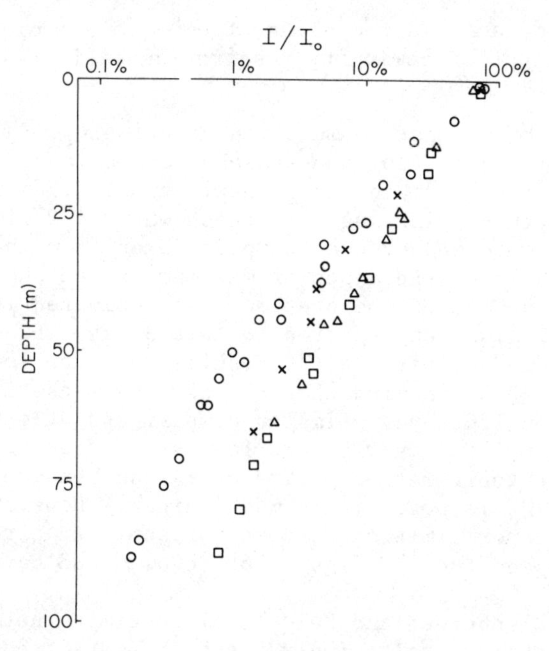

Fig. 10. Ratio of downwelling irradiance to surface irradiance for a location in the North central Pacific gyre (∿28°N, 155°W). Data from the CLIMAX II cruise (August to September 1969; SIO, 1975). Different symbols refer to different sampling dates (Δ = August 28; x = August 30; ○ = September 1; □ = September 5).

larger proportion of the available energy is being absorbed by absorbers other than pigments.

Finally, Morel and Bricaud (1981; Appendix 1) point out that the vertical attenuation coefficient used here and measured routinely in the field is an over-estimate of the true absorption coefficient. This error could bias the estimate of light absorbed by phytoplankton upwards by as much as 10%. On theoretical grounds then, we should expect the upper limit on P_g for the oligotrophic North Pacific to lie in the range from 40 to 60 mmoles C $m^{-2}d^{-1}$.

What do the measurements show? In fact, the only data that are at all applicable (recall that we have assigned the ^{14}C data to P_n) are those reported by Gundersen et al. (1976) from a station north of Hawaii. The primary production, based on diel oxygen changes, has a carbon equivalent of 16 mmoles C $m^{-2}d^{-1}$. The magnitude of P_n should be less than that of P_g. The best ^{14}C data for the central gyre is that summarised by McGowan and Hayward (1978). The values lie in the range from 7 to 13 mmoles C $m^{-2}d^{-1}$ (Table 2). Examination of Table 4 shows that this level of primary production could be supported by the calculated rate of ammonia recycling, from 3 to 9 mmoles N $m^{-2}d^{-1}$ (nitrogen equivalent of community respiration estimated earlier).

Turning now to the net community production P_c, Shulenberger and Reid (1981) estimated it from the seasonal oxygen accumulation within a particular depth stratum in the euphotic zone. They insisted that the only data worth considering were those where ^{14}C fixation and dissolved oxygen were measured on samples taken from the same bottle cast. They established the depth of maximum oxygen supersaturation (for example 66m at one of the stations) and compared it with the instantaneous ^{14}C fixation referred to that depth. From this comparison they concluded that the ^{14}C results were too low by a factor of at least two. Bear in mind that the oxygen is assumed to have accumulated during 120 days. Similar comparisons at other stations indicated that the ^{14}C data might be low by a factor of up to 5.4. Note that because their method estimates P_c and because $P_c < P_n < P_g$, the implied error of P_g is potentially much larger. However, our earlier estimate of the upper limit on P_g does not admit an error of even 5 times between P_g and the ^{14}C figures of McGowan and Hayward.

Because Schulenberger and Reid (1981) estimate net oxygen accumulation (P_c) their results imply a stoichiometric equivalent flux of particulate nitrogen out of the euphotic zone. Similarly, they imply an equivalent nitrate flux into the euphotic zone. Knauer et al. (1979) report vertical particle fluxes to the aphotic zone of 5.7 mmoles C $m^{-2}d^{-1}$. The equivalent nitrogen export due to vertical migration of mesopelagic fish is estimated to be trivial compared to this (Table 4; calculated from data compiled by K. H. Mann). The estimated maximum nitrate flux into the euphotic zone through diffusive mechanisms would support a P_c of 5.3 mmoles C $m^{-2}d^{-1}$ (Table 4). Neither the loss rate of particulate nitrogen nor the supply rate of nitrate nitrogen are consistent with "new" production (P_c) of several times the ^{14}C figures. However, the uncertain nitrogen source through nitrogen fixation could conceivably double the maximum diffusive supply rate of nitrate. This would bring the acceptable values of P_c within the range of McGowan and Haywards' ^{14}C results, but not several times higher than them.

The estimates of community oxygen utilisation, 20 mmoles C $m^{-2}d^{-1}$ based on the observed size spectra (Beers et al., 1975), or 60 mmoles C $m^{-2}d^{-1}$, extrapolated to include organisms smaller than 2 μm, are indirect measures of P_g. They are consistent with the accepted values of P_n and the range of possibilities for the ratio P_n/P_g (Table 3). They are not consistent, however, with Shulenberger and Ried's estimates for P_c and the possible values for the ratio P_c/P_g (Table 3).

The comparison made here shows that the apparent discrepancy between P_c as determined by oxygen accumulation and P_n as measured by ^{14}C fixation does not sit well with the rest of the available evidence.

One argument that could be advanced against the calculations presented here is that, because most of them are strongly biomass-dependent, they could be seriously biassed if the autotrophic (pigment) biomass had been seriously underestimated. Suppose, for example, that there existed a significant autotrophic biomass of organisms whose size was smaller than the pore size of the finest filters in routine field use. This is a possibility that we have given serious consideration. Indeed, it is a hypothesis not easy to exclude.

One reason for entertaining this idea seriously is optical in basis. The submarine irradiance decreases faster with depth than would be expected from the observed pigment biomass. For example, with $\xi = 0.03\ m^{-1}$, $\kappa = 1.6 \times 10^4 m^2 (kg\ Chl\ \underline{a})^{-1}$ and $B = 5 \times 10^{-8} kg\ Chl\ \underline{a}\ m^{-3}$ we would estimate a nominal euphotic depth (1% surface light) of 149m, much greater than that observed in the field (SIO, 1975). The discrepancy might be attributed to absorption by a population of pigmented cells of very small size, sufficiently small to pass through the membrane filters normally used in sampling for phytoplankton biomass. Because of their small size, such cells would have unusually high optical cross sections (Morel and Bricaud, 1981). Their presence could lead to the "new" carbon fixation for which Schulenberger and Reid (1981) claim to see the accumulated oxygen equivalent.

The most persuasive argument we have found for excluding this hypothesis, but which by itself is not at all conclusive, is that

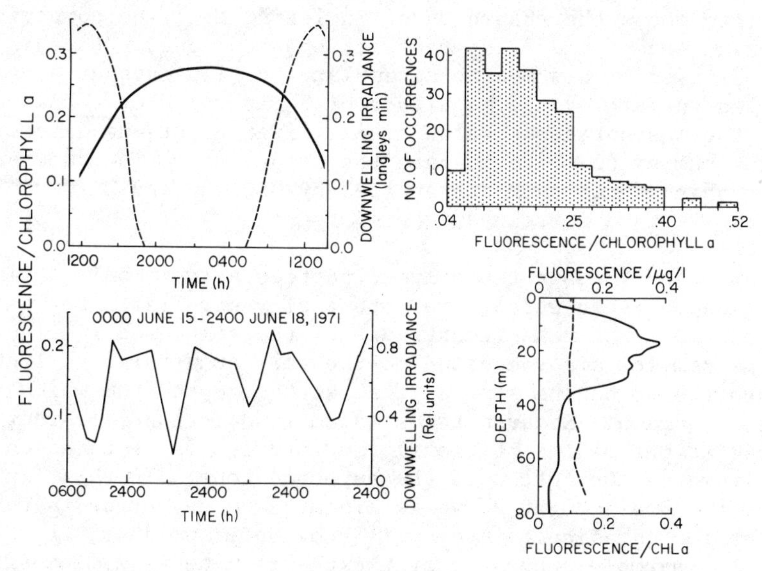

Fig. 11. Fluorescence per chlorophyll for natural samples from various locations in the Pacific (see text for further explanation). Data from Keifer (1973).

in vivo chlorophyll fluorescence in the central gyre (Figure 11) is not atypically high compared to the extracted chlorophyll (Kiefer, 1973) as it would be if an unsampled population of small, pigmented cells were living there.

SAMPLING VARIABILITY AND FUTURE REQUIREMENTS

Our synthesis in the last section of important pelagic fluxes in the central north Pacific has not led to a resolution of the problem posed by Shulenberger and Reid (1981). It is implicit in their paper that, if there is a discrepancy between the oxygen and ^{14}C data, the error must lie with the latter. We have found, however, that the ^{14}C results appear to be consistent with most of other flux data available for the area. It is then required that we consider possible ways in which Shulenberger and Reid's calculations might themselves be in error.

First, we note that by restricting themselves to data sets where the oxygen concentration and ^{14}C fixation were measured from samples taken on the same bottle cast, they have limited themselves to only a small portion of the data that are relevant to the discussion. This limitation is unnecessary and even undesirable. Their oxygen accumulation rates are based on an averaging time of 120 days. The equivalent horizontal length-scale, calculated either from the typical mean flow (10 cm s^{-1}) or the typical horizontal diffusivity ($10^7 cm^2 s^{-1}$), is $\sim 10^3$ km, of the same order, coincidentally, as the dimension of the gyre itself (McGowan and Hayward, 1978). This brings into focus the absurdity of insisting that the comparison cannot be made except under the restrictions mentioned. Because the oxygen observed is assumed to accumulate for 120 days at a rate determined in part by the magnitude of P_g averaged over the same time period, the instantaneous value of ^{14}C fixation at one discrete depth is a weak number on which to base the comparison. The "unusual oceanographic event" such as reported by McGowan and Hayward (1978) over the gyre is also relevant here.

Again the ^{14}C data themselves, particularly if only a single discrete depth is selected, are potentially unreliable for the following reason. The measurements are of the simulated *in situ* type where the samples are incubated on the ship at a relative light level that purports to be the same as that at the depth from which the samples are drawn. Because the stratum in question is rather deep (e.g. 66m at one of the stations) the interest is centered on conditions towards the bottom of the euphotic zone. In this light range the photosynthesis-light curve is linear, and any error in the simulation of the irradiance level will show up in productivity estimates. For example suppose that the photic zone is 75m, corresponding to a vertically averaged attenuation coefficient of $\sim 0.06 m^{-1}$. Then if the extinction coefficient is in error by 10% or if the screens used to modulate the solar input are in error by 10%, the estimate of light available at 65m could be high by 30% or low by 33%. This error would propagate directly to the productivity estimates.

On the other hand, we know that the density structure in the ocean is not fixed in the vertical but can, and does, execute organised vertical oscillations. We can estimate the amplitude of these excursions in the gyre by comparing different density profiles from the same cruise report (SIO, 1975). An amplitude of 50m does not seem too high, judging from these data. Taking this as the extreme range, the 65m sample might have found itself at either the 40% light level or the 0.1% light level, or somewhere in between, at another point in the vertical excursion cycle. The potential error in productivity is considerable.

These considerations argue strongly for data that are better averaged in space and time. The ocean is too variable in the horizontal and too energetic in the vertical for it to be justified to compare the difference in oxygen between an instantaneous concentration measured at a discrete depth and a supposed saturation value 120 days earlier with an instantaneous ^{14}C fixation rate measured once at the same discrete depth. The range of values for the apparent oxygen accumulation rate between stations in the central gyre of the north Pacific by Shulenberger and Reid (1981) is but one manifestation of this variability.

CONCLUDING REMARKS

We have shown that by making simple calculations based on thermodynamic first principles, it is possible to limit the range of values that can be accepted for the primary production of the oligotrophic ocean. Necessarily the approach we have taken reflects our personal philosophy of science and we do not claim that it is the last word on the subject, or even the only way of attacking the problem.

Our calculations show that, by taking the ^{14}C fixation rates to be equivalent to P_n, the net production of small particles (Peterson, 1980), an internally consistent picture can be developed for the pelagic ecosystem of the central gyre of the north Pacific that accomodates most of the published information on the biological fluxes. This picture corresponds to the stereotype expounded by Pomeroy (1974), Williams (1981) and others: most of the metabolic activity is concentrated in very small cells (say 10 μm and less) and the nitrogen demand can be met by recycling _in situ_ with only a modest supplement of "new" nitrogen (cf. Jackson, 1980). The corresponding amount of nitrogen exported as particles would also be modest, and not consistent with some estimates of the equivalent accumulation rate of oxygen in the photic zone. The possibility remains open that this inconsistency will disappear when data that are suitably averaged in space and time become available.

REFERENCES

Atlas, D., and Bannister, T. T., 1981, Dependence of mean spectral extinction coefficient of phytoplankton on depth, water color, and species, Limnol.Oceanogr., 25:157.

Bannister, T. T., 1974, Production equations in terms of chlorophyll concentration, quantum yield and upper limit to production, Limnol.Oceanogr., 19:1.

Beers, J. R., Reid, F. M. H., and Stewart, G. L., 1975, Microplankton of the North Pacific central gyre. Population structure and abundance, June 1973, Int.Revue Ges.Hydrobiol., 60:607.

Beers, J. R., Reid, F. M. H., and Stewart, G. L., 1982, Seasonal abundance of the microplankton population in the North Pacific central gyre, Deep-Sea Res., 29:227.

Blasco, D., Packard, T. T., and Garfield, P. C., 1982, Size dependence of growth rate, respiratory electron transport system activity, and chemical composition in marine diatoms in the laboratory, J.Phycol., 18:58.

Carpenter, E. J., and McCarthy, J. J., 1975, Nitrogen fixation and uptake of combined nitrogenous nutrients by Oscillatoria (Trichodesmium) thieboutii in the western Sargasso Sea, Limnol.Oceanogr., 20:389.

Carpenter, E. J., and Price, IV, C. C., 1977, Nitrogen fixation, distribution, and production of Oscillatoria (Trichodesmium) spp. in the western Sargasso Sea and Caribbean Sea, Limnol. Oceanogr., 22:60.

Conover, R. J., 1978, Transformation of organic matter, in: "Marine Ecology," O. Kinne, ed., John Wiley and Sons, London.

Dugdale, R. C., and Goering, J. J., 1967, Uptake of new and regenerated forms of nitrogen in primary productivity, Limnol. Oceanogr., 12:196.

Eppley, R. W., 1980, Estimating phytoplankton growth rates in the central oligotrophic oceans, in: "Primary Productivity in the Sea," P. G. Falkowski, ed., Plenum Press, New York.

Eppley, R. W., and Peterson, B. J., 1979, Particulate organic matter flux and planktonic new production in the deep ocean, Nature, 282:677.

Eppley, R. W., Renger, E. H., Venrick, E. L., and Mullin, M. M., 1973, A study of plankton dynamics and nutrient cycling in the central gyre of the North Pacific Ocean, Limnol.Oceanogr., 18:534.

Eppley, R. W., and Sharp, J. H., 1975, Photosynthetic measurements in the central North Pacific: The dark loss of carbon in 24 hour incubations, Limnol.Oceanogr., 20:981.

Eppley, R. W., Sharp, J. H., Renger, E. H., Perry, M. J., and Harrison, W. G., 1977, Nitrogen assimilation by phytoplankton and other microorganisms in the surface waters of the central North Pacific Ocean, Mar.Biol., 39:111.

Falkowski, P. G., 1980, Light-shade adaptation in marine phytoplankton, in: "Primary Productivity in the Sea," P. G. Falkowski, ed., Plenum Press, New York.

Fenchel, T., 1974, Intrinsic rate of natural increase: The relationship with body size, Oecologia, 14:317.

Fogg, G. E., 1982, Nitrogen cycling in sea waters, Phil.Trans.R.Soc. Lond., B296:511.

Gallegos, C. L., and Platt, T., 1981, Photosynthesis measurements on natural populations of phytoplankton: numerical analysis, in: "Physiological Bases of Phytoplankton Ecology," T. Platt, ed., Can.Bull.Fish.Aquat.Sci. 210, Ottawa.

Gargett, A. E., 1976, An investigation of the occurrence of oceanic turbulence with respect to fine structure, J.Phys.Oceanogr., 6:139.

Garrett, C., 1979, Mixing in the ocean interior, Dynamics of Atmospheres and Oceans, 3:239.

Gieskes, W. W. C., Kraay, G. W., and Baars, M. A., 1979, Current ^{14}C methods for measuring primary production: gross underestimates in oceanic waters, Neth.J.Sea Res., 13:58.

Gundersen, K. R., Corbin, J. S., Hanson, C. L., Hanson, M. L., Hanson, R. B., Russel, D. J., Stollar, A., and Yamada, O., 1976, Structure and biological dynamics of the oligotrophic ocean photic zone off the Hawaiian Islands, Pacif.Sci., 30:45.

Hall, D. O., and Rao, K. K., 1972, "Photosynthesis," Edward Arnold, London.

Hemmingsen, A. M., 1960, Energy metabolism as related to body size and respiratory surfaces, and its evolution, Reports of the Steno Memorial Hospital, Copenhagen, 9(Part 2): 110p.

Herbland, A., 1978, The soluble fluorescence in the open sea: distribution and ecological significance in the equatorial Atlantic Ocean, J.Exp.Mar.Biol.Ecol., 32:275.

Herbland, A., and LeBouteiller, A., 1981, The size distribution of phytoplankton and particulate matter in the Equatorial Atlantic Ocean: Importance of ultraseston and consequences, J.Plankt.Res., 3:659.

Ivanenkov, V. N., Sapozhnikov, V. V., Chernyakova, A. M., and Gusarova, A. N., 1972, Rate of chemical processes in the photosynthetic layer of the tropical Atlantic, Oceanology, 12:207.

Jackson, G. A., 1980, Phytoplankton growth and zooplankton grazing in oligotrophic oceans, Nature, 284:439.

Jassby, A. D., and Platt, T., 1976, Mathematical formulation of the relationship between photosynthesis and light for phytoplankton, Limnol.Oceanogr., 21, 540.

Jenkins, W. T., 1980, Tritium and ^{3}He in the Sargasso Sea, J.Mar. Res., 38:533.

Johnson, K. M., Burney, C. M., and Sieburth, McN. J., 1981, Enigmatic marine ecosystem metabolism measured by direct diel ΣCO_2, and O_2 flux in conjunction with DOC release and uptake, Mar.Biol., 65:49.

Johnson, P. W., and Sieburth, McN. J., 1979, Chroococcoid cyanobacteria at sea: A ubiquitous and diverse phototrophic biomass, Limnol.Oceanogr., 24:928.

Kiefer, D. A., 1973, Fluorescence properties of natural phytoplankton populations, Mar.Biol., 22:263.

Knauer, G. A., Martin, J. H., and Bruland, K. W., 1979, Fluxes of particulate carbon, nitrogen and phosphorus in the upper water column of the northeast Pacific, Deep-Sea Res., 26A:97.

Mague, T. H., Mague, F. C., and Holm-Hansen, O., 1977, Physiology and chemical composition of nitrogen-fixing phytoplankton in the central North Pacific Ocean, Mar.Biol., 41:213.

Malone, T. C., 1980, Size-fractioned primary productivity of marine phytoplankton, in: "Primary Productivity in the Sea," P. G. Falkowski, ed., Plenum Press, New York.

Martinez, L. A., King, J. M., Silver, M. W., and Alldredge, A. L., 1982, A new system of nitrogen fixation in oligotrophic oceanic waters, (Abstract), EOS., 63:101.

McCarthy, J. J., and Carpenter, E. J., 1979, Oscillatoria (Trichodesmium) thieboutii (Cyanophyta) in the central North Atlantic Ocean, J.Phycol., 15:75.

McGowan, J. A., and Hayward, T. L., 1978, Mixing and oceanic productivity, Deep-Sea Res., 25:771.

McLellan, H. J., 1965, "Elements of Physical Oceanography," Pergamon Press, Oxford.

Morel, A., and Bricaud, A., 1981, Theoretical results concerning light absorption in a discrete medium and applications to specific absorption of phytoplankton, Deep-Sea Res., 28:1375.

Mullin, M. M., Perry, M. J., Renger, E. H., and Evans, P. M., 1975, Nutrient regeneration by oceanic zooplankton: a comparison of methods, Mar.Sci.Comm., 1:1.

Munk, W. H., 1966, Abyssal recipes, Deep-Sea Res., 13:707.

Parker, R. R., 1981, A note on the so-called "soluble fluorescence" of chlorophyll a in natural waters, Deep-Sea Res.,28A:1231.

Peterson, B. J., 1980, Aquatic productivity and the ^{14}C-CO_2 method: A history of the productivity problem, Ann.Rev.Ecol.Syst., 11:359.

Platt, T., 1969, The concept of energy efficiency in primary production, Limnol.Oceanogr., 14:653.

Platt, T., and Denman, K. L., 1977, Organisation in the pelagic ecosystem, Helgolander wiss.Meeresunters, 30:575.

Platt, T., and Denman, K. L., 1978, The structure of pelagic marine ecosystems, Rapp.Proc.-verb.Reun.Cons.perm.int.Explor.Mer, 173:60.

Platt, T., Denman, K. L., and Jassby, A. D., 1977, Modelling the productivity of phytoplankton, in: "The Sea: Ideas and Observations on Progress in the Study of the Seas," E. D. Goldberg, ed., John Wiley, New York.

Platt, T., and Gallegos, C. L., 1980, Modelling primary production, in: "Primary Productivity in the Sea," P. G. Falkowski, ed., Plenum Press, New York.

Platt, T., Gallegos, C. L., and Harrison, W. G., 1980, Photoinhibition of photosynthesis in natural assemblages of marine phytoplankton, J.Mar.Res., 38:687.

Platt, T., and Jassby, A. D., 1976, The relationship between photosynthesis and light for natural assemblages of coastal marine phytoplankton, J.Phycol., 12:421.

Platt, T., and Silvert, W., 1981, Ecology, physiology, allometry and dimensionality, J.Theor.Biol., 93:855.

Platt, T., and Subba Rao, D. V., 1975, Primary production of marine microphytes, in: "Photosynthesis and Productivity in Different Environments," J. P. Cooper, ed., Cambridge University Press, Cambridge.

Pomeroy, L., 1974, The ocean's food web, a changing paradigm, Bioscience, 24:499.

Postma, H., and Rommets, J. W., 1979, Dissolved and particulate organic carbon in the North Equatorial Current of the Atlantic Ocean, Neth.J.Sea.Res., 13:85.

Raven, J. A., and Beardall, J., 1981, Respiration and photorespiration, in: "Physiological Bases of Phytoplankton Ecology," T. Platt, ed., Can.Bull.Fish.Aquat.Sci. 210, Ottawa.

Riley, G. A., 1939, Plankton studies. II. The western north Atlantic, May-June, 1939, J.Mar.Res., 2:145.

Riley, G. A., 1951, Oxygen, phosphate and nitrate in the Atlantic Ocean, Bull.Bingham.Oceanogr.Coll., 13:1.

Riley, G. A., 1970, Particulate organic matter in seawater, Adv.Mar. Biol., 8:1.

Riley, G. A., Stommel, H., and Bumpus, D. F., 1949, Quantitative ecology of the plankton of the Western North Atlantic, Bull. Bingham.Oceanogr.Coll. 12:1.

Ryther, J. H., 1954, The ratio of photosynthesis to respiration in marine plankton algae and its effect upon the measurement of productivity, Deep-Sea Res., 2:134.

Schmitt, R. W., and Evans, D. L., 1978, An estimate of the vertical mixing due to salt fingers based on observations in the North Atlantic Central Water, J.Geophys.Res., 83:1913.

Sheldon, R. W., Prakash, A., and Sutcliffe, W. H., 1972, The size distribution of particles in the ocean, Limnol.Oceanogr., 17:327.

Sheldon, R. W., Sutcliffe, W. H., 1978, Generation times of 3h for Sargasso Sea microplankton determined by ATP analysis, Limnol. Oceanogr., 23:1051.

Shulenberger, E., and Reid, J. L., 1981, The Pacific shallow oxygen maximum, deep chlorophyll maximum, and primary productivity, reconsidered, Deep-Sea Res., 28A:901.

S.I.O., 1975, Data Report, Climax II Expedition SIO Reference 75-6.

Smith, W. O., 1977, The respiration of photosynthetic carbon in euphotic areas of the ocean, J.Mar.Res., 35:557.

Steemann-Nielsen, E., 1954, On organic production in the oceans, J.Cons.Explor.Mer., 19:309.

Suess, E., 1980, Particulate organic carbon flux in the oceans-surface productivity and oxygen utilization, Nature, 288:260.

Talling, J. F., 1957, The phytoplankton population as a compound photosynthetic system, New Phytol., 56:133.

Tijssen, S. B., 1979, Diurnal oxygen rhythm and primary production in the mixed layer of the Atlantic Ocean at 20°N, Neth.J.Sea Res., 13:79.

Venrick, E. L., 1974, The distribution and significance of Richelia intracellulosis Schmidt in the North Pacific central gyre, Limnol.Oceanogr., 19:437.

Waterbury, J. B., Watson, S. W., Guillard, R. R., and Brand, L. E., 1979, Widespread occurrence of a unicellular marine plankton cyanobacterium, Nature, 277:293.

Williams, P. J. LeB., 1981, Incorporation of microheterotrophic processes into the classical paradigm of the planktonic food web, Kieler Meeresforsch, 5:1.

Worthington, L. V., 1976, "On the North Atlantic Circulation," Johns Hopkins University Press, Baltimore.

THE STRUCTURE OF AQUATIC ECOSYSTEMS AND ITS DEPENDENCE ON ENVIRONMENTAL VARIABLES

J. E. Paloheimo, A. P. Zimmerman
W. G. Sprules and M. A. Gates

Dept. of Zoology
University of Toronto
Toronto, Ontario, Canada M5S 1A1

INTRODUCTION

The traditional approach to ecology, to study different species or ecological processes in isolation or in relation to a rather limited set of environmental or biotic factors, has, in part, given way to a synthesis using coupled process models and techniques of systems analysis (e.g. Patten, 1971, 1972; Platt *et al*., 1981). While reductionist in approach, the synthesis has attempted, by use of simulation and large data bases, to arrive at a more holistic description of the ecosystem. The hope was, and perhaps still is, that by coupling major functional, energetically defined components of ecosystems in order to simulate the known or presumed interactions and material fluxes between the parts of the system, we will arrive at a description which allows a more parsimonious characterization of the ecosystem and its dynamic nature. While there is no question that this exercise has increased our understanding of the functional characteristics of ecosystems, the emergent or holistic properties of large-scale ecological systems models have, to our way of thinking, failed to emerge. The failure to find valid generalizations or ecosystem properties has to do, in part, with our lack of understanding of the homeostatic mechanisms involved, with our lack of understanding of the way nature, as opposed to the human mind, integrates subsystems and, in part, with the inadequacy of data on diverse ecosystems as a whole.

Concurrently, with the use of systems analysis and simulation models in ecology, a few, mostly theoretical ecologists, have developed a different, holistic approach. Here we include the characterization of ecosystems by the use of information theory, by the size spectrum of organisms or by degree of material/energy cycling in the sytem (cf. Platt et al., 1981). Two holistic concepts in particular, ecosystem resilience and resistance, originally proposed by Holling (1973), have gained much attention (Webster et al., 1975; O'Neil, 1976). Resistance refers to the ability of the ecosystem to resist displacement, and is normally considered to result from the accumulated structure of the ecosystem, while resilience refers to the ability of the system to return to its former reference state once it is displaced. Again, the data to test the applicability of these holistic concepts has been mostly lacking.

It is generally recognized that the energy flows within aquatic ecosystems and, hence, the whole dynamics of the system, may be more dependent on body size and feeding patterns than on the taxonomic community structure (e.g. Steele and Frost, 1977). Much of the work done on marine ecosystems is centered around what might be called the size adequacy hypothesis and the particle size spectrum (Sheldon et al., 1972; Kerr, 1974; Platt and Denman, 1978). Evidence accumulating in freshwater systems shows the importance of size selective predation (Hall et al., 1976). Sprules and Holtby (1979) have shown that the functional size characterization of planktonic community structures has a stronger statistical relationship with lake morphology and limnological variables than the taxonomically based one.

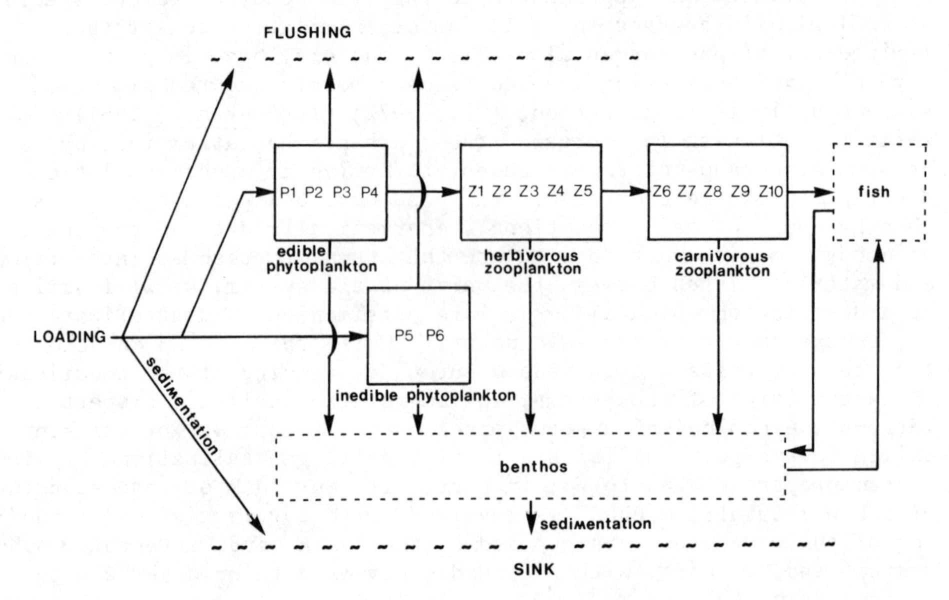

Fig. 1. Schematic presentation of the simplified freshwater ecosystem. Compartments encircled by perforated lines are not included in the present analysis. The symbols P1 - P6 and Z1 - Z10 refer to plankton size classes as described in Table 1.

In this paper we examine freshwater ecosystems on the basis of data on 15-25 lakes sampled over a three year period (see next section), focussing, in particular, on empirical evidence for some of the basic tenets of systems analysis, namely the interconnectivity or coupling of biological compartments and the role of system inputs and the environment on the structure of the ecosystem. The system itself is rather simple, and consists of a number of lumped compartments shown in Figure 1, based on a functional-size categorization of planktonic organisms. The first section examines the statistical redundancies between the system inputs and the environmental variables on the one hand, and the structure of the biotic community on the other. In the second section we study the effect of phosphorus loading on the abundance of phytoplankton in more detail. The extent of the linkages between the biological components is taken up in the third section. Finally, an attempt is made to classify the lakes into functionally distinct categories, and to relate the findings to ecosystem resistance and resilience concepts (Holling, 1973).

THE DATABASE

The realization of the scarcity of data on whole ecosystems prompted a group of us at the University of Toronto to undertake a major (at least in the context of University research) survey of aquatic ecosystems. The Lake Ecosystem Working Group (LEWG), as we call ourselves, includes N. Collins (benthos), H. H. Harvey (fish), R. Knoechel (phytoplankton), G. Sprules (zooplankton), H. A. Regier (fish), A. P. Zimmerman (limnology), B. Shuter (fish), M. A. Gates (protozoa) and J. E. Paloheimo (statistics). At the outset the major hypothesis that we wished to test was that the functional size classification of organisms, i.e. size related feeding ecology, rather than a taxonomic species categorization, was an adequate description of the system for distinguishing major functional lake types. Having collected and recorded the data primarily by functional size classes, the hypothesis is, of course, untestable, although many relevant features of the dynamics of the ecosystem can be answered or examined.

Although we realized that most simulation models as well as most of the theoretical work were dependent on the characterization of fluxes between various biological and material compartments, due to methodological and sampling problems, a decision was made to sample biomass levels only in the described categories. The biomass categories were functional size classes: five spherical phytoplankton size classes, one filamentous class, and five herbivorous and five carnivorous zooplankton size classes. Only fish and some benthos were identified by species. Standard limnological variables were measured, and water chemistry was also sampled, although less frequently. In addition, basic morphological measurements were made or obtained from topographic maps. This list of variables is given in

Table 1. List of Environmental and Biotic Measurements Available for LEWG Lakes 1978-80

Category	Parameter Descriptions
Benthos	not considered
Geology and Meteorology	not considered
Hydrolab (Limnology) 8 cycles May-Sept.	mean epilimnion depth (m) temperature (°C) conductivity (umho/sq cm) all epilimnetic dissolved oxygen (mg/L) oxidation - reduction potential (pE) pH
Morphology	mean lake depth (m) maximum lake depth (m) lake surface ares (km^2) lake volume (cubic km) perimeter (km) percent littoral area drainage basin area (km^2) flushing rate (per year) link order (cf. J.F. Moeller *et al.*, 1979)
Nutrients 2 cycles in 1978 4 cycles in 1979 8 cycles in 1980	14 nutrients were measured only total phosphorus (mg/m^3) is considered here
Species specific fish abundance	not considered

Spherical phytoplankton	includes the following size classes (μm):
	P1 0.00- 2.00
	P2 2.00- 5.00
	P3 5.00-10.00
	P4 10.00-30.00
	P5 30.00+
Filamentous phytoplankton biomass (mg wet wt/m^3) 8 cycles 1978-80	includes the following size classes (μm):
	P6 10.00+
Edible Phytoplankton	P1 + P2 + P3 + P4
Inedible Phytoplankton	P5 + P6
Herbivorous zooplankton biomass (mg wet wt/m^3) 8 cycles 1978-1980	includes the following size classes (mm):
	Z1 0.00- 0.29
	Z2 0.30- 0.49
	Z3 0.50- 0.84
	Z4 0.85- 1.19
	Z5 1.20
Carnivorous zooplankton biomass (mg wet wt/m^3) 8 cycles 1978-1980	includes the following size classes:
	Z6 0.00- 0.49
	Z7 0.50- 0.89
	Z8 0.90- 1.19
	Z9 1.20- 1.49
	Z10 1.50+

Table 1. Benthic data were available only for a subset of the lakes sampled for plankton, and are not included in the present analysis.

The first field crews went out in 1978 and the last in 1981. The data are still being worked on, although most are now available in computer accessible form for 1978-1980. Plankton and limnology were sampled eight times and essentially covered the productive period from May through September. All together there are data on 25 lakes for 1978 and 1979 and 15 for 1980 and 1981. For more detail on the sampling procedures, we refer you to forthcoming publications by individuals in the LEWG group responsible for data collection.

We will deal with planktonic and abiotic variables only for the three year period 1978-1980. All of the analyses of biomass variables are based on the biomass per lake volume, except for phytoplankton, which is given as density in the first 4 meters ("epilimnion"). Data on fish are excluded since the fish data failed to produce any significant results or correlations. Subsequently, we initiated two other independent fish surveys and growth studies which are not yet completed.

STATISTICAL DETERMINANTS OF ECOSYSTEM STRUCTURE

Prerequisites both for both the use of systems analysis and for the holistic approach to ecosystem analysis are that the various parts of the system are interconnected and that the observed links are either stable or change in a predictable manner. While to many the interconnectedness is an obvious fact that need not be questioned, the early ecological research, e.g. in fisheries, did not always assume that. Each commercially important species was studied in isolation with little if any reference to the rest of the ecosystem. The unspecified assumption was that the internal dynamics and exploitation and perhaps a few environmental variables were the major factors governing the productivity, in other words, that each commercially important species occupied a unique niche that could not be usurped or filled by any other species. This view is largely disappearing, although at the level of resolution considered in this and many other papers, there may well be cycling of nutrients within our integrated compartments (Goldman, this volume). This may well have the effect of decreasing the apparent interconnectivity, and, in fact, may lend some credibility to the historical approach.

To elucidate the structure and function of aquatic systems we have examined the nature of the linkages or intercompartmental fluxes and their dependency on the morphological and hydrolab variables (cf. Table 1 for definition of hydrolab and morphological variables) by use of statistical analysis. Not having measured the actual flows, we hypothesize that their magnitude is reflected in the size and assumed turnover rate of recipient compartments.

Statistically it will be difficult to distinguish between active and passive flows, i.e. between the donor and recipient controlled flows that are controlled by both the donor and recipient compartments. Clearly, except for the secondary interactions which can be tracked down, passive flow results in a correlation between the sizes of the donor and recipient compartments. A situation with active flow is more complicated, and can, indeed, result in zero correlation. However, although the statistical analysis will not reveal the nature of linkages, it will, or should, at least establish the degree to which they are present. While the lack of correlation, linear or otherwise, need not imply the lack of interaction, it does imply that there are either counteracting forces or cyclicity in the system, although they are not apparent in our data.

The second major question, beside the extent and nature of linkages, is the role of abiotic factors. May we in fact say that the structure of an ecological system is largely determined by abiotic factors, or do we have to consider the internal dynamics and the species composition as well?

Redundancy analysis (Cooley and Lohnes, 1971; van den Wollenberg, 1977) was used to examine the role of biotic and abiotic factors and the presence of linkages. This kind of analysis can be considered as an extension of multiple correlation analysis for a situation with more than one dependent variable. Given two datasets, X and Y, such as phytoplankton biomasses (Y, six different size classes) and loading and morphology (X, one loading variable plus six morphological variables), the redundancy measures the amount of variability in the Y dataset that the X dataset explains. Loosely speaking, the redundancy measures the overlap between the various datasets. The measure of redundancy in Y given X can also be considered as an average multiple correlation coefficient (R^2) obtained by regressing each of the Y variables on the X variables and averaging the resultant R^2's.

The measures of redundancy that we have calculated are listed in Table 2. Some of the redundancies are rather high. For instance, 73-80% of the variability in six phytoplankton biomass categories can be explained by the hydrolab and morphological variables alone. Perhaps the main conclusion which emerges from the redundancies listed in Table 2 is that in our dataset abiotic variables are more important than the biotic variables, i.e. the linkages between the biomass compartments are relatively weak, and the magnitude of fluxes is more dependent on the physical and chemical environment. In economic jargon: the productivity is more dependent on the infrastructure than on the nature of the production system. Incidentally, if fish are included, all of the redundancies between fish and pelagic variables are more or less zero. Fish abundance seems to be more closely related to the structure of the littoral zone than to the pelagic based productivity of the lake.

Table 2. List of Redundancies Between and Among Morphology, Hydrolab and Biotic Variables

Variables	Redundany (Y/X) in Y given X							
	1978		1979		1980		1978-1980	
	R^2	no. of obs.	R^2	no. of obs.	R^2	no. of obs.	R^2	no. of obs.
HY/MO	.354	22	.365	25	.406	25	.223	72
PH/HY P	.528	20	.590	25	.757	13	.434	58
PH/HY P MO	.803	20	.734	25	*	13	.562	58
PH/P HZ	.574	20	.493	25	.768	13	.364	58
PH/HY P MO HZ	.961	20	.890	25	*	13	.642	58
HZ/HY P	.269	20	.300	25	.442	13	.163	58
HZ/HY P MO	.755	20	.563	25	*	13	.388	58
HZ/PH	.451	20	.496	25	.467	13	.327	58
HZ/PH CZ	.843	20	.757	25	*	13	.537	58
HZ/PH HY P CZ	.969	20	.847	25	*	13	.644	58
CZ/HY P	.292	20	.276	25	.471	13	.222	58
CZ/HY P MO	.623	20	.480	25	*	13	.331	58
CZ/HZ	.479	20	.460	25	.554	13	.397	58
CZ/HY P HZ	.685	20	.613	25	*	13	.193	58

*Values with 2 or less degrees of freedom have been omitted.

MO = morphological variables (mean depth, maximum depth, volume, perimeter, surface area and percent littoral score).

HY = hydrolab variables (mean epilimnion values for the following: temperature, conductivity, pH, dissolved oxygen).

P = mean summer phosphorus.

PH = six phytoplankton size classes (density per epilimnetic volume).

HZ = five herbivorous zooplankton size classes (density per epilimnetic volume).

CZ = five carnivorous zooplankton size classes (density per epilimnetic volume).

While the environmental variables are more significant than the biotic ones, in the sense of explaining higher proportions of between years variability, neither the biotic nor abiotic variables totally explain the year to year changes. For instance, the residuals in phytoplankton given the phosphorus loading and zooplankton still show a significant annual component. The annual component accounts for the lower redundancies in Table 2 when all three years are combined compared with the within year redundancies. The significant year to year component will also be evident from the instability of the ecosystem structure that we will examine in the last Section.

PRIMARY PRODUCTION

All biological systems are open systems depending on external energy and often nutrient input as well. Lakes appear wasteful of essential nutrients that are quickly flushed or sedimented out and, except for the littoral zone and for shallow lakes, become more or less unavailable. Hence, regular input of nutrients is needed to sustain the productivity. Mass balance considerations require that the mean equilibrium phosphorus concentration (P) can be expressed as:

$$(P) = \frac{J/V}{\sigma + \rho}$$

where J is the total annual phosphorus loading, σ the sedimentation rate and ρ the flushing rate (Vollenweider, 1969). Deviations from the above relationship tend to occur because loading estimates are somewhat inaccurate and/or because lakes are not at equilibrium. Phosphorus loading in turn is correlated with ensuing primary production (Sokomoto, 1966; Dillon and Rigler, 1974), or, at least, with mean chlorophyll values. For a set of lakes with a hundredfold variation in the total analytical phosphorus concentrations, the correlation between the phosphorus and chlorophyll on a log-log scale can be as high as 0.95 (Jones and Bachmann, 1976). We expect the correlations to be much lower over lakes with a smaller range of P values.

Despite the high observed correlations, there are a number of uncertainties in projecting or extrapolating from the measured or calculated loading data to the primary production, and not just because the loading data itself may be inaccurate or in a form that is biologically unavailable (e.g. Schaffner and Ogelsby, 1978). The basic mass balance equation given above may be inappropriate in the first place for lakes in the temperate region. During the productive period of the year, the lakes tend to be stratified, and the loading inputs primarily into the epilimnion. In place of a fully mixed homogeneous compartment model assumed by the mass balance equation above, a two or three compartment model should be considered. In fact, better correlations between the loading and chlorophyll are often realized when loading per mixed volume (Ogelsby and Schaffner, 1978), or some other variable such as per surface area (Schindler et al., 1978), is used. In the face of the uncertainties with the loading data and the basic model itself, we have simply used the observed total phosphorus concentrations.

The observed P value in the euphotic zone can be considered as the effective loading into the productive part of the water column; "effective" in the sense that it incorporates corrections for both flushing and sedimentation or transfer of P into hypolimnion, save for P tied up in zooplankton that may undergo diurnal movement and is not usually included in the P determinations.

Squared correlations between phosphorus concentration and phytoplankton range from 0.40-0.47 in a LEWG dataset (Table 3a). Since these values incorporate sampling and measurement errors in both variables, the actual correlations may well be higher. The values only substantiate earlier findings without really adding anything new. Our main interest lies in trying to explain the residuals from the well-known phytoplankton-phosphorus relationship.

Table 3b shows the multiple correlations between the total phytoplankton biomass and the phosphorus + morphological variables. Although the correlations are significantly higher, the improvement is not spectacular nor are they as high as we might have expected on the basis of the redundancy analysis. However, one must bear in mind that the observed mean epilimnetic P integrates the effect of some variables that are essentially morphological in character, e.g. the flushing and sedimentation rate, and, hence, are in this sense already incorporated in measured P and in the correlations shown in Table 3a.

For comparison we have also calculated multiple correlations of the phytoplankton biomass against the mean P and the hydrolab variables (Table 3c). The correlations are somewhat higher. The improvement, however, is somewhat dubious since many of the hydrolab variables (conductivity, DO, ORP) reflect either the abundance or activity of biotic material.

Table 3a. Squared Correlations Between Loading (P) and Edible Phytoplankton (E Phyto = P1 + P2 + P3 + P4, see Table 1), Inedible Phytoplankton (IE Phyto = P5 + P6) and Total Phytoplankton (μg/L)

	1978-1980**		1979-1980**	
E Phyto	0.40	0.47*	0.61	0.58*
IE Phyto	0.35		0.15	
Total Phyto	0.44		0.25	
No. of Lakes	58	38		

*Turnover rate corrected biomass values.
**1978 phosphorus values are based on means of two cycles only; 1979 values on four cycles and 1980 values on eight cycles. From 1978 data set three lakes were excluded because of apparently anomalous phosphorus readings.

Table 3b. Squared Multiple Correlation Coefficients Between Edible, Inedible and Total Phytoplankton Biomass Values with Phosphorus (P) and Morphological variables (MO), 1978-80.

	R^2 biomass/MO*	
E Phyto	0.43	0.50**
IE Phyto	0.54	
Total Phyto	0.59	
No. of Lakes	58	58

*The following morphological variables were included: percent littoral area, mean depth and link order of inflow streams (cf. Moeller *et al*., 1979).
**Turnover corrected biomass.

Table 3c. Squared Multiple Correlations Between Edible, Inedible and Total Phytoplankton Biomass Values with Phosphorus (P) and Hydrolab Variables (HY). 1978-80.

	R^2 biomass P/ HY	
E Phyto	0.05	0.58*
IE Phyto	0.44	
Total Phyto	0.54	
No. of Lakes	61	

*Turnover corrected biomass.

The relationship between primary productivity measured in terms of the biomass, and loading measured by the average P concentration, is dependent on the number of grazers. From the results of the redundancy analysis (Table 2), we observe that the amount of variability in the phytoplankton abundance explained by hydrolab, P, morphology, and herbivorous zooplankton variables is $R^2 = 0.642$ over the three year period, compared with $R^2 = 0.562$ given the values of environmental variables alone. The increase in the R^2 value is statistically significant but not particularly large. This will be dealt with in somewhat more detail in the next section. The statistical evidence that we present will suggest rather loose coupling in the sense that the residuals of the phytoplankton-nutrient (P) relationships are only marginally correlated with the abundance of grazers.

Our lake set spans a ten-fold range of loadings - in contrast with some of the earlier studies that span a hundred-fold difference. If for a moment we equate the primary production with the chlorophyll content of algae, the more extensive datasets (e.g. P. Dillon and F. Rigler, 1974) suggest, to employ economic jargon again, an increasing return to scale, i.e. an increase in nutrient loading brings about a proportionately higher production (chlorophyll). Our data contradict this and suggest diminishing return to scale. When we regress the log phytoplankton mean biomass againast the log phosphorus concentration, the regression coefficient was less than one, that is, in the relationship

$$\text{Phyto biomass} = a * P^{b}$$

the exponent 'b' is less than one. Perhaps as we increase the nutrient input, we are witnessing a "silent" flip of the ecosystem from one state to another. In this case, the "flip" is a change from small, spherical phytoplankton with high turnover rate to a preponderance of large spherical or filamentous type algae with slow turnover rate. The range of loadings observed in our dataset probably is not wide enough for this change in ecosystem status to occur.

BIOLOGICAL LINKAGES

The multilevel, integrated nature of ecosystems is well-accepted both by theoretical and by field-oriented ecologists, although data are largely lacking. While the complete connectivity of an ecosystem where fluxes between compartments depend on the donor or recipient compartment or both, modified by a few environmental parameters, is intuitively and theoretically appealing, its empirical basis is not strong, with the exception of laboratory based studies. Because of the complexity of ecosystems, however, it is difficult to say whether the apparent lack of a strong and empirically well-established relationship between the sizes of the donor and recipient compartments is due to the integrated nature of many of the datasets, due to sampling difficulties, or to the dynamic nature of the interaction not captured by, essentially, linear statistical techniques.

The simple correlations between various functional biomass groupings, namely edible phytoplankton (spherical phytoplankton <30 μm), inedible phytoplankton (spherical >30 μm and filamentous), herbivorous zooplankton and carnivorous zooplankton are presented in Table 4a. For each coupling two values are given, one based on total biomass values, the other on turnover rate corrected biomass values. The turnover rate corrected biomass values are obtained in the first approximation by weighting each size group by its presumed turnover rate - the latter is obtained from a simple allometric relationship where the turnover rate is taken as proportional to the mean body weight to the power -0.25 (Fenchel, 1974). In all cases, when the correlations are significant, the turnover corrected biomass values

Table 4a. Squared Correlation Coefficients Between Biomass Categories

	E Phyto	IE Phyto	Herb. zoo	Carn. zoo
E Phyto	1	0.16	0.07	0.17
IE Phyto	0.21*	1	-0.02 NS	0.01 NS
Herb. zoo	0.14*	-0.02* NS	1	0.32
Carn. zoo	0.24*	0.01* NS	0.35*	1

*Values indicated with an asterisk are based on turnover rate weighted biomasses.

are somewhat more highly correlated. The increases are significant, although not very substantial even though our turnover rate is rather approximate.

In our dataset there are, all together, three linkages: that between nutrient loading and phytoplankton, between phytoplankton and herbivorous zooplankton and between herbivorous zooplankton and carnivorous zooplankton.

Although intuitively it may be obvious that there should be a significant correlation between components that are directly coupled, such as between the predator and prey or between primary producers and nutrient supply, it is not clear whether the correlation should be positive or negative. Equally good arguments could be advanced in favor of either case. In the situation where the predator increases at the expense of the prey, we expect a negative correlation, but in the case where the prey increases and brings about the increase in predator abundance, we would expect a positive correlation. Lotka-Volterra type numeric relationships, to use the somewhat awkward terminology of Solomon (1949), imply that the correlation switches back and forth from positive to negative. A predator-driven system in a temperate region might go through one complete cycle. Seasonally averaged values, the ones considered here, need not necessarily show any correlation. However, the textbook case of cyclicity is rarely observed. In the LEWG dataset, the classical case of cyclicity with the predator lagging behind its prey in abundance is an exception rather than a rule, and when it is observed one year, it is not necessarily observed the next year.

Correlations between all three directly linked biomass categories are significant and positive (Table 4a), although not very high. In other words, by and large, as the resource or prey abundance increases, so does the predator. The control exerted by predators is not complete, it is only partial. If we look at the two way interactions by use of multiple correlations with compartment size

Table 4b. Squared Multiple Correlations Between Biomass Variables and Phosphorus

R^2 E Phyto/P, Herb. Zoo	= 0.53
R^2 Herb. Zoo/E Phyto, Carn. Zoo	= 0.40

regressed against its donor and recipient compartments, shown in Table 4b, the relationships are somewhat improved. Inspection of calculations not shown in the Table indicates that all of the regression coefficients are again positive.

FUNCTIONAL LAKE CLASSIFICATION

The major question, the one that prompted us to undertake this study in the first place is: are there definite lake types that differ from each other significantly in terms of ecosystem structure and dynamics? Indeed it might be argued that the relationships we are trying to establish depend on, or at least are greatly modified by, the lake morphology and the geological basin type, and that the observed correlations depend on the mix of lakes we might have happened to choose. Moreover, if there are definite lake types, the correlations within each lake may well be different and much higher.

Initially we clustered the lakes on the basis of the mean annual biomass densities based on both absolute and relative (percentage) densities using the turnover rate corrected values. A third clustering was made by examining the biomass to nutrient or predator-prey ratios. The "Statistical Analysis System's" (SAS) fast cluster procedure was employed in producing the clusters with the total number of clusters requested being set somewhat arbitrarily at seven. Lakes sampled more than once over the three years were treated by the clustering procedure as different, i.e. the lake identification included both the lake name and year sampled.

All together there were 19 lakes included more than once, and two lakes included only once, giving 50 "generic" lakes to be clustered. Three lakes that showed extreme year to year variations were excluded from the present analysis as well as three lakes in 1978 that showed some anomalous phosphorus concentrations, possibly due to very limited sampling for phosphorus in 1978 (Table 1).

The first two lake clusterings are based on lumped but turnover rate corrected biomass values, namely total edible phytoplankton, total inedible phytoplankton, total herbivorous zooplankton and total carnivorous zooplankton densities. Figures 2 and 3 show the average biomass profiles for each cluster. Visual inspection of

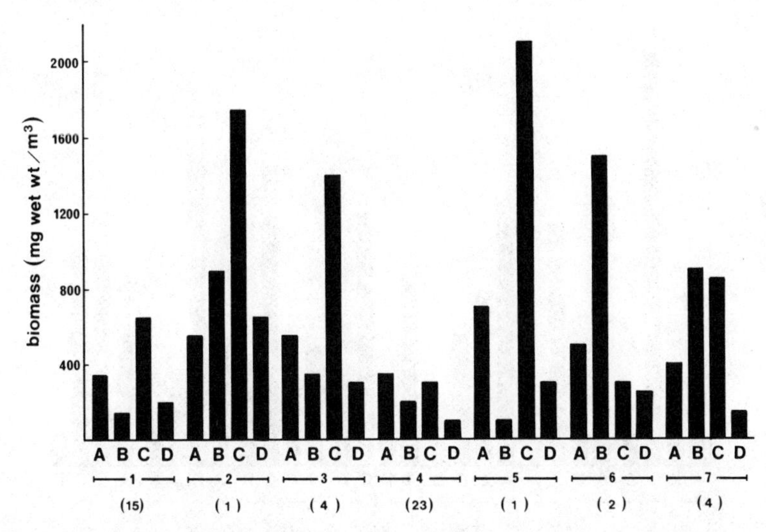

Fig. 2. Lake clustering based on lumped biomass densities. A = turnover rate corrected mean biomass of edible phytoplankton (mg wet wt/m^3); B = inedible phytoplankton mean biomass (mg wet wt/m^3); C = turnover rate corrected mean biomass of herbivorous zooplankton (mg wet wt/m^3); D = turnover rate corrected mean biomass of carnivorous zooplankton (mg wet wt/m^3); X = lake cluster; (#) = number of lakes per cluster.

Figures 2 and 3 show distinctive biomass profiles for each lake "type". Clusters are clearly distinguishable on the basis of either the dominant group (Figures 2 and 3, edible phytoplankton, herbivorous zooplankton, etc.) or on the basis of the overall abundance (Figure 2). Even though the clusters appear quite different in the distribution of the biomass over the four functional groups, out of 19 lakes included more than once, eight (Figure 2) or eleven (Figure 3) change their cluster membership while only 11 or 8 stay in the same cluster during the 2-3 years that they were sampled.

Figure 4 summarizes the clustering based on the ratios of edible phytoplankton to mean phosphorus, herbivorous zooplankton to edible phytoplankton, and carnivorous zooplankton to herbivorous zooplankton. The phosphorus concentration has been normalized so that the overall concentration is one over all lakes for ease of graphical presentation. This normalization was also included in the input data to the clustering procedure. Again we observe very different predator-prey ratios between the different clusters. Yet, 12 out of the 19 lakes included more than once change their cluster membership between years, and only 7 stay within the same cluster.

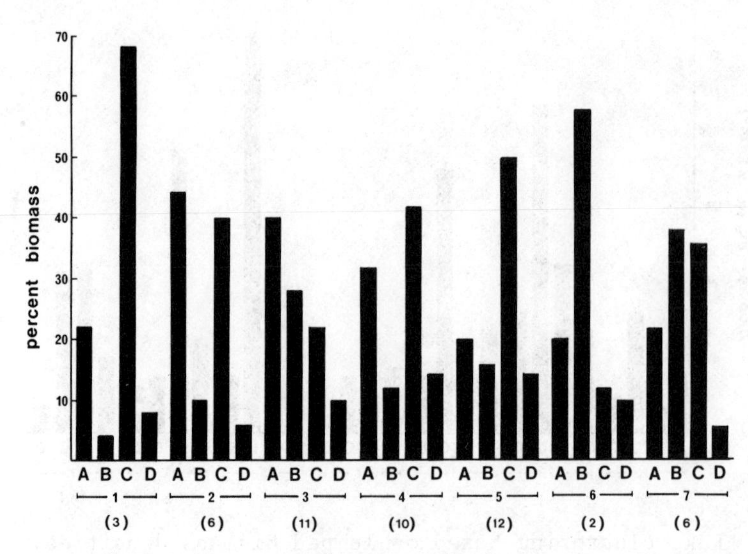

Fig. 3. Lake clustering based on relative (per cent) lumped biomass densities. A = per cent turnover rate corrected mean biomass in edible phytoplankton; B = per cent mean biomass in inedible phytoplankton; C = per cent turnover rate corrected mean biomass in herbivorous zooplankton; D = per cent turnover rate corrected mean biomass in carnivorous zooplankton; X = lake cluster; (#) = number of lakes per cluster.

If for a moment we ignore the never ending excuse, although true in most cases, that we have omitted some very basic variables, what we have here is a lot of instability - the lakes flip from one functional size structure to another as more years of data are included. Although changes in species composition or in size spectrum (Steele and Frost, 1977) have sometimes been predicted for marine species, there does not seem to be any discernible pattern in our data to these flips nor do they appear to be predictable, at least using linear statistical procedures. Discriminant analysis, using cluster membership as the class variable and morphology, annual P and annual mean values for Hydrolab variables as predictors, classifies lakes properly in 36 out of 48 cases for biomass based clusters (Figure 2). This is not too bad; yet only three out of the eleven lakes that changed their class membership were correctly assigned by the discriminant analysis. Similarly, for relative biomass based clusters (Figure 3), only two out of nine lakes that changed their cluster membership at least once in the 2-3 year sampling period were properly assigned by the discriminant analysis even though there were only 12 misclassified out of the total of 50 cases or samples. Perhaps we are in the realm of positive feedback phenomena where small changes in the environment percolate and are magnified by the

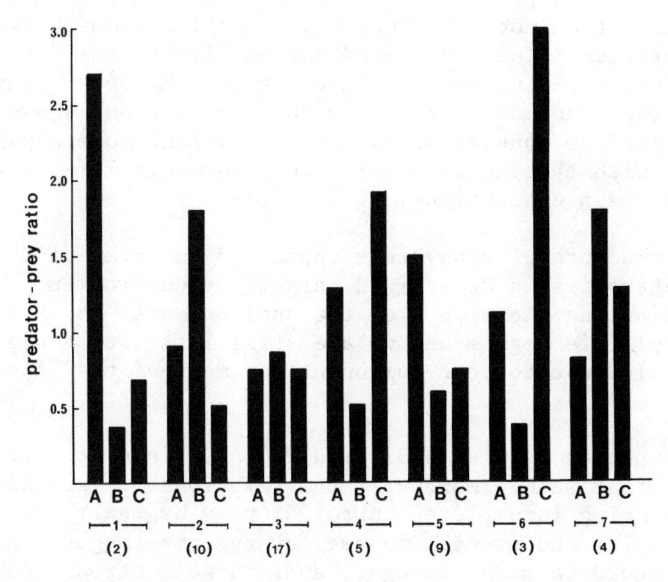

Fig. 4. Lake clustering based on predator to prey ratios. A = ratio of mean edible phytoplankton density to mean summer phosphorus concentration; B = ratio of mean herbivorous zooplankton density to mean edible phytoplankton density; C = ratio of mean carnivorous zooplankton density to mean herbivorous zooplankton density; X = lake cluster; (#) = number of lakes per cluster.

ecosystem. The often claimed ecosystem properties of resilience and resistance do not seem to be valid for freshwater planktonic systems, or at least the zone of stability is relatively small, not even covering natural year to year fluctuations in environmental variables.

DISCUSSION

It is questionable to what extent the conclusions reached in the present study should be taken as firm and generally applicable to all freshwater aquatic systems. For one thing the analysis does not address the sampling errors. Indeed sampling errors have not been examined and, hence, we cannot offer any comment - hopefully this question will be addressed in subsequent publications. Secondly, we have based our conclusions on mean seasonal values and have ignored seasonal patterns. The analysis was restricted to mean seasonal values partly because we wished to focus mainly on the lake to lake and year to year differences, and partly because the seasonal patterns were rather obscure and not particularly consistent between or within lakes. Thirdly, the statistical techniques used presuppose linear rather than nonlinear effects or active flows between the various datasets. The pronounced variability in the data made it somewhat futile to include nonlinear terms. Moreover, the graphs of residuals (not included here) from the correlations shown in Tables 3 and 4 showed no consistent pattern to warrant more sophisticated analysis. With these provisos in mind, we would like to reiterate some of the main conclusions.

The structure of ecosystems expressed in terms of the functional biomass categories is determined largely by environmental variables. After eliminating the effect of the environmental and hydrolab variables by multiple regresion, we are still left with a significant and unexplainable year to year component for many of the lakes in the lake set.

The linkages between biological compartments are significant although rather weak. Moreover, the relationships are always positive, suggesting incomplete control of prey by their predators. Laboratory studies and studies on smaller systems suggest that the linkages should be much stronger, and, indeed, strong coupling is presupposed by any simulation model. We suspect that the apparent weak coupling is at least partly due to both spatial and temporal separation of prey and its predator. While the weak coupling suggests that the lake ecosystems are resistant in the sense that a change in one compartment will not substantially change the rest of the system, this is not borne out by further analyses using clustering techniques.

Lakes tend to "flip" from one functional type or cluster to another, presumably triggered by a change in the environment in a manner that we could not predict. For that reason we propose that planktonic lake ecoystems cannot be considered either resistant or resilient, but perhaps rather adaptive - tracking the changes in nutrient loading and temperature. We suspect that larger forms (e.g. fish) with longer life-spans and with greater nutirent reserves, might show more consistent year to year patterns. However, since most fish species have a planktonic early life stage, their recruitment tends to fluctuate widely and the overall result is an apparent lack of resistance even at the top of the aquatic food chain.

REFERENCES

Cooley, W. W., and Lohnes, P. R., 1971, "Multivariate data analysis," John Wiley and Sons Inc., N.Y.

Dillon, P. J., and Rigler, F. H., 1974, The phosphorus-chlorophyll relationship in lakes, Limnol.Oceanogr., 19:767.

Fenchel, T., 1974, Intrinsic rate of natural increase: The relationship with body size, Oecologia, 14:317.

Hall. D. J., Threlkeld, S. T., Burns, C. W., and Crowley, P. W., 1976, The size efficiency hypothesis and the size structure of zooplankton communities, Ann.Rev.Ecol.Syst., 7:177.
Holling, C. S., 1973, Resilience and stability of ecological systems, Ann.Rev.Ecol.Syst., 4:1.
Jones, J. R., and Bachmann, R. W., 1976, Prediction of phosphorus and chlorophyll levels in lakes, J.Water Polution Control Fed., 48:2176.
Kerr, S. R., 1974, Theory of size distribution in ecological communities, J.Fish.Res.Bd.Can., 31:1859.
Moeller, J. R., Minshall, G. W., Cummins, K. W., Petersen, R. C., Cushing, C. E., Sedell, J. R., Larson, R. A., and Vannotte, R. L., 1979, Transport of dissolved organic carbon in streams of differing physiographic characteristics, Organic Geochem., 1:139.
Oglesby, R. T., and Schaffner, W. R., 1978, Phosphorus loadings to lakes and some of their responses. Part 2. Regression models of summer phytoplankton standing crops, winter total P, and transparency of New York lakes with known phosphorus loadings, Limnol.Oceanogr., 23;135.
O'Neill, R. V., 1976, Ecosystem persistence and heterotrophic regulation, Ecology, 57:1244
Pattern, B. C., ed., 1971, "Systems Analysis and Simulation in Ecology," Vol. I., Academic Press, N.Y.
Pattern, B. C., ed., 1972, "Systems Analysis and Simulation in Ecology," Vol. II., Academic Press, N.Y.
Platt, T., Mann, K. H., and Ulanowicz, R. E., eds., 1981, "Mathematical Models in Biological Oceanography," UNESCO Press, Paris.
Platt, T., and Denman, K., 1978, The structure of pelagic marine ecosystems, Rapp.P.-v.Reun.CIEM, 173:60.
SAS Institute, 1981, SAS 79.5, Changes and enhancements, SAS Technical Report P-115, Cary, North Carolina, U.S.A.
Schaffner, W. R., and Oglesby, R. T., 1978, Phosphorus loadings to lakes and some of their responses. Part 1. A new calculation of phosphorus loading and its application to 13 New York lakes, Limnol.Oceanogr., 23:120.
Schindler, D. W., Fee, E. J., and Ruszczynski, T., 1978, Phosphorus input and its consequences for phytoplankton standing crop and production in the experimental lakes area and in similar lakes, J.Fish.Res.Board Can., 35:190.
Sheldon, R. W., Prakash, A., and Sutcliffe, W. H., 1972, The size distribution of particles in the ocean, Limnol.Oceanogr.,17: 327.
Sokomoto, M., 1966, Primary production of phytoplankton community in some Japanese lakes and its dependence on lake depth, Arch. Hydrobiol., 62:1.
Soloman, M. E., 1949, The natural control of animal populations, J.Animal Ecol., 18:1.
Sprules, W. G., and Holtby, L. B., 1979, Body size and feeding ecology as alternatives to taxonomy for the study of limnetic

zooplankton community structure, J.Fish Res.Board Can., 36:1354.

Steele, J. H., and Frost, B. W., 1977, The structure of plankton communities, Phil.Trans.Roy.Soc.London, B 280:485.

van den Wollenberg, A., 1977, Redundancy analysis an alternative for canonical correlation analysis, Psychometrika, 42:207.

Vollenweider, R. A., 1969, Possibilities and limits of elementary models concerning the budget of substances in lakes, Arch. Hydrobiol., 66:1.

Webster, J. R., Waide, J. B., and Patten, B. C., 1975, Nutrient recycling and the stability of ecosystems, In: "Mineral Cycling in Southeastern Ecosystems," F. G. Howell, J. B. Gentry and M. H. Smith, eds., U.S. Energy Research and Development Admin., Technical Information Center, Office of Public Affairs.

EUTROPHICATION OF A COASTAL MARINE ECOSYSTEM - AN EXPERIMENTAL STUDY USING THE MERL MICROCOSMS

S.W. Nixon, M.E.Q. Pilson, C.A. Oviatt, P. Donaghay, B. Sullivan, S. Seitzinger, D. Rudnick and J. Frithsen

Graduate School of Oceanography, University of Rhode Island, Kingston, RI 02881, U.S.A.

INTRODUCTION

The potential importance of nutrients, especially nitrogen, phosphorus, and silica, in influencing the productivity of the sea has been recognized for over 80 years (Brandt, 1899), though this long awareness has not necessarily produced a thorough understanding. Analytical techniques capable of resolving the concentrations of nutrients usually found in marine waters did not become available until the 1920s and 1930s (for example, Denigés, 1921; Harvey, 1928; Atkins, 1932). Ammonia, the most active form of nitrogen, was not commonly measured until after the publication of Solorzano's direct colormetric technique in 1969. An early enthusiasm for describing, and in some cases quantifying the often dramatic reciprocal seasonal cycles in the concentrations of nutrients and the standing crop of phytoplankton waned as the patterns became better known (for example, Atkins, 1930; Cooper, 1933). It was also perceived early on that rapid nutrient cycles and transformations were probably taking place at rates which could not then be measured. As a result, there was little real advance in the study of nutrients in marine systems for some 25-30 years following the pioneering descriptive work of the 1930s.

While it is true that most, if not all, of the major transformations involving nutrients now thought to occur in marine ecosystems were identified long ago (for example, Johnstone, 1908), our knowledge of the rates and ecological consequences of these processes is recent and incomplete. During the 1960s the use of the ^{14}C technique to measure primary production focused attention on short-term rate processes and thereby encouraged detailed studies of the growth and uptake kinetics of phytoplankton in response to different forms and concentrations of nutrients in laboratory culture. The relatively recent application of ^{32}P (Watt and Hayes, 1963), ^{15}N (MacIsaac and Dugdale, 1969) and ^{29}Si (Goering et al., 1973) to phytoplankton uptake studies has further refined our knowledge of autotrophic nutrient demand. General interest in examining the role of nutrient excretion by pelagic heterotrophs in supplying much of this demand only began with the report of zooplankton excretion measurements by Harris in 1959, and most of the work in this area has only been reported within the past 10-15 years (Martin, 1968; Johannes, 1969; Smith, 1978; Vargo, 1979; etc.). Tracer methods for measuring the more rapid regeneration rates of microzooplankton and pelagic bacteria are only now beginning to be perfected (Harrison, 1978; Caperon et al., 1979; Glibert, 1982).

The role of benthic communities in the element cycling of coastal marine systems was not quantified for any of the nutrients until an annual cycle of nitrogen flux measurements was published six years ago for Narragansett Bay (Nixon et al., 1976). Work on benthic-pelagic exchanges has increased rapidly since then (reviewed by Nixon, 1981; Kemp et al., 1982), including the first direct measurements of denitrification, a major nitrogen cycle process in the sediments, just two years ago (Seitzinger et al., 1980). A first estimate of the loss of nitrogen from this pathway based on an annual cycle of measurements is only now in press (Seitzinger et al.). It follows that even for this element, the most intensively studied of the major nutrients, there is as yet no detailed and well-constrained mass balance or budget for a marine ecosystem (Nixon and Pilson, In press).

It is evident from the dates of the publications cited above that while earlier studies focused on nutrient uptake, much of the recent emphasis has been on nutrient recycling (Harrison, 1980; Nixon, 1981). It is the demonstrated importance of cycling and the as yet unknown inter-actions between nutrient inputs and their cycles that contribute a great deal to our present inability to model or predict the effects of nutrient enrichment in marine systems. There are other areas of ignorance as well, including the relative importance of food supply and predation in regulating different pelagic and benthic trophic levels. But our understanding of the impact of nutrients on primary and secondary production forms a major part of our basic concept of marine ecosystems. It is not surprising that the success of mechanistic numerical simulation models of coastal and open ocean environments has been to a large degree directly proportional to the importance of nutrient inputs and storages in determining their phytoplankton dynamics. Wherever and whenever internal cycles are more important, it becomes increasingly difficult to develop a credible mechanistic model (Kremer and Nixon, 1978).

The reason for this brief historical review and emphasis on present uncertainties in our knowledge is that many of those working in applied environmental management or in other areas of ecology are surprised to find that the sophisticated details from past decades of reductionist laboratory measurements and the documentation of many intensive field surveys have not provided the basis for predicting the behavior of complex marine ecosystems in response to a major perturbation in the input of their basic constituents. This situation has persisted in part because of the great complexity of nature and in part because of a lack of direct experimentation. With very few exceptions, study of the flows of matter and energy in marine ecosystems has involved descriptive (albeit quantitative) field measurements or laboratory experiments using isolated parts (individuals, species, or a limited group of species) of systems.

It is virtually impossible to establish cause and effect in the field because nutrients are almost always associated with inputs of organic matter, fresh water, pollutants of various kinds, or sediments. The input of nutrients is also difficult to determine or control in the field, and the effects of changes in a host of other factors such as temperature and light may confound any response to nutrients. On the other hand, cycling rates are a property of the integrated system and cannot be understood by measurements of the parts alone, however controlled and precise those measurements may be. Almost completely lacking have been experimental studies of whole or reasonably complete ecosystems which fit somewhere "between beakers and bays" (Strickland, 1967). The work on "captured" water masses using large spheres or bags (for example, Antia _et al._, 1963 and many of the papers in Grice and Reeve, 1982) may approach this goal for the plankton assemblages of the surface layers of deep open water systems, but in coastal areas 25-50% of the organic matter is remineralized on the bottom (Nixon, 1981) and the sediments must be included in an experimental system.

Given the long-standing interest in, and importance of, the relationship between nutrients and productivity, it is remarkable that there appear to have been very few systematic attempts to fertilize and study a coastal marine ecosystem. With one exception (Pratt, 1949), all were done with a view toward increasing the yield of fish. The first were really preliminary investigations carried out by Gaarder and Sparck (1931) and others on a Norwegian oyster poll or inlet, and Pratt (1949) confined his attention to the phytoplankton of a small, semi-isolated salt pond. Two others, however, were extensive and serious studies of the effects of repeated nutrient additions on many aspects of the ecology of two Scottish sea lochs (Gross, 1947, 1950 and others). It is unfortunate that the latter efforts are usually remembered for their inconclusive or impractical results in terms of fishery economics rather than as pioneering ventures in experimental ecology. But, in spite of their useful basic scientific results, some of which were quite dramatic, the sea loch experiments suffered from a number of limitations. Principal among these were the lack of a good control, the inability to obtain replicates or to estimate variance, and the limitation of being confined to a single level of treatment.

It has been possible to overcome most of these problems by using microcosms or living models of a coastal marine system rather than a field experiment. In the following sections we report preliminary results from the first 6-12 months of a nutrient addition experiment using large tanks at the Marine Ecosystems Research Laboratory (MERL) at the University of Rhode Island. Although it is simple in concept, we believe it to be a unique experiment, the results of which will be of interest even in the largely descriptive form presented here. The experiment is continuing and further papers will be prepared as the data are more fully analyzed.

THE MERL MICROCOSMS

The MERL facility consists of a laboratory building and 14 large cylindrical tanks maintained outside on the shore of the lower West Passage of Narragansett Bay. Each tank stands 5.5 m high and holds 13.4 tons of water. Sea water is pumped from the bay into the tanks using a diaphram pump (shown to be nondestructive to the plankton) at a rate sufficient to replace the water in the tanks once every 27 days. This approximates the flushing time of the bay and is slow enough that the biological and chemical reactions in each tank determine the conditions which develop. At the same time it also links the microcosms to the larger coastal environment and allows for the input of benthic larvae and other plankton. A sample of the natural benthic community is included in each tank by means of a 40 cm deep tray which covers the bottom and contains sediments and benthos collected from the bay using a large (0.25 m^2) box corer. Great care is taken to preserve the orientation of the sediment as it is placed in the container. A cut-away view showing the configuration is given in Figure 1. Over an annual cycle, the water passes through a temperature range from near 0^oC to about 20^oC, but maintains a relatively constant salinity of 28-32 $^o/_{oo}$. The microcosms have been run in various experimental modes since 1976. They have been maintained over periods of about 2 years, and have been shown to exhibit behavior that is essentially similar to that of Narragansett Bay in terms of species composition and abundance, biological rate processes, and the concentrations of a variety of nutrients, metals, and organic compounds. A number of publications describe the systems in more detail, compare their behavior with that of the bay, discuss variability among the tanks, and give the results of some earlier experiments (Pilson *et al*., 1977, 1979, 1980; Nixon *et al*., 1980; Santschi, 1982; Hunt and Smith, 1982; Elmgren and Frithsen, 1982; Smith *et al*., 1982; Wakeham *et al*., 1982 and others).

THE EUTROPHICATION EXPERIMENT

While a few earlier experiments involving nutrient additions to marine microcosms have been reported (for example, Edmondsen and

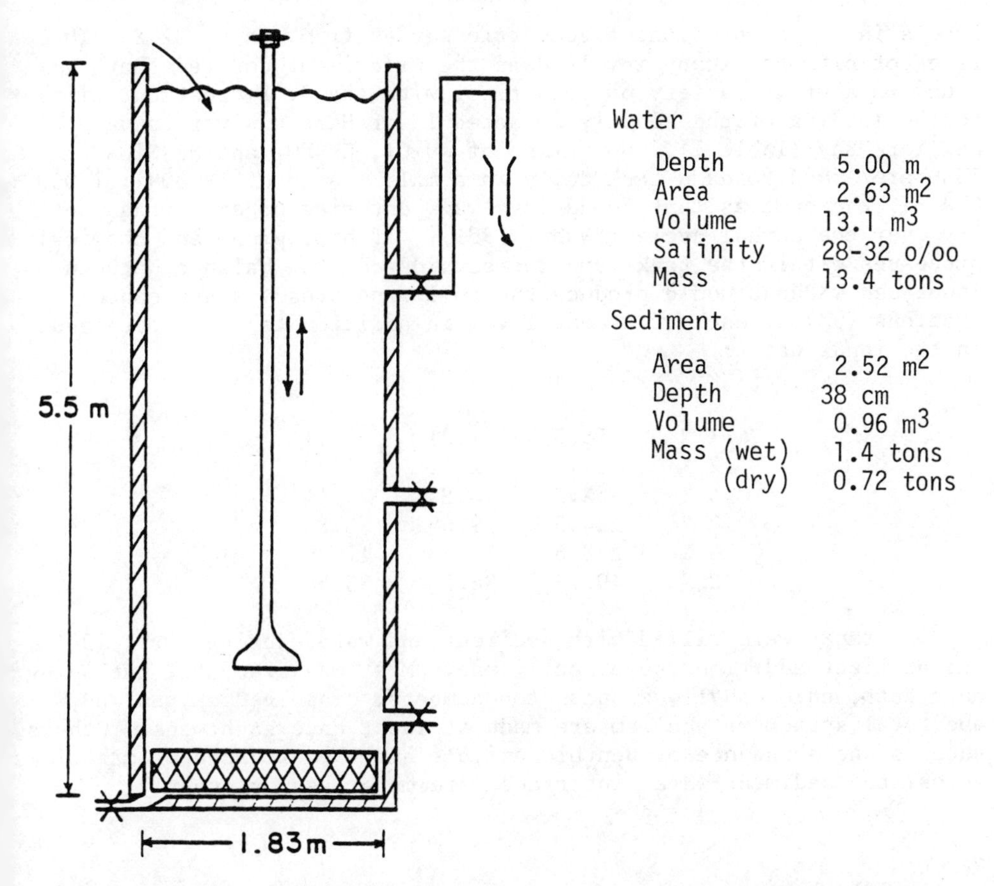

Fig. 1. Cross section of a MERL microcosm tank showing the mixing plunger, sediment container, route of water flow through the tank, and various physical characteristics. From Pilson et al., 1980.

Edmonsen, 1947; Pratt, 1950; Takahashi et al., 1982), their attention was directed exclusively toward plankton dynamics, and sediments were not included. Since no long-term additions involving coupled pelagic-benthic systems had been described, it was decided to carry out a gradient experiment in which replication within treatments would largely be sacrificed in favor of a greater range of treatment levels. Considerable past experience with the microcosms had already provided some description of the variability to be expected in replicate tanks with respect to most of the parameters measured. Of the nine tanks available for this experiment, three were selected to serve as unenriched controls while the others were assigned different treatment levels in six steps along a geometric series from 1 X to 32 X. The range of nitrogen input involved in the experiment covered that found in a great variety of estuaries, with the 32 X treatment similar to the loading of the heavily impacted lower Hudson River estuary of New York Bay (Table 1). Solutions of NH_4Cl, KH_2PO_4 and $Na_2SiO_3 \cdot 7H_2O$ are added to each tank daily in a molar ratio of 12.80N : 1.00P : 0.91Si, the same as that found in sewage entering upper Narragansett Bay over the annual cycle (Nixon, 1981). If biological and chemical processes within the tanks are not considered, the balance between input and washout would produce the following steady-state concentrations (μM) at each treatment level in addition to that contained in the input water:

	N	P	Si
1 X	15.5	1.21	1.11
2 X	31.1	2.42	2.22
4 X	62.2	4.84	4.44
8 X	124.3	9.68	8.88
16 X	248.6	19.36	17.76
32 X	497.3	38.72	35.52

The tanks were filled with sediment and water during April 1981 and nutrient additions begun on 1 June. Most measurements, including nutrients, chlorophyll, zooplankton numbers, dissolved oxygen, pH, and total system metabolism are made at least once each week. Others, such as the abundance of benthic animals and the exchange of materials across the sediment-water interface, are made less frequently.

RESULTS

Dissolved Inorganic Nitrogen

In spite of their significant nutrient inputs, nitrogen concentrations in the 1, 2 and 4 X tanks remained near or below those in the bay and control tanks during June and July (Figure 2). Only with inputs in excess of 2.3 $\mu mol\ L^{-1}\ d^{-1}$ were the uptake processes saturated so that concentrations increased rapidly in the water. However, by the end of summer inorganic nitrogen concentrations had increased dramatically in all of the tanks, and by late fall or early winter the levels were near or somewhat above those calculated for steady-state conditions (Figure 2). The relative abundance of the different forms of nitrogen also differed markedly with treatment level. Ammonia (the form used as input) comprised virtually all of the dissolved inorganic nitrogen pool in all of the tanks until late summer, when nitrite and nitrate began to accumulate (Figure 2). As winter progressed, ammonia accounted for a declining fraction of the nitrogen in the bay and the tanks, but there was a strong treatment effect (Table 2). By the end of January, over 75% of the inorganic nitrogen was in the form of nitrite and nitrate in all but the 8X,

Table 1. Approximate annual input of dissolved inorganic nitrogen (NH_4, NO_2, NO_3) in various estuaries and experimental marine ecosystems.

	N Input	
	m mol m^{-3} y^{-1}	m mol m^{-2} y^{-1}
Experimental Systems		
Loch Craiglin, Scotland (1942)[1]	40	40
Loch Sween, Scotland (1944)[2]	85	340
MERL Microcosms -- 1 X	212	1062
2 X	427	2135
4 X	850	4252
8 X	1701	8505
16 X	3402	17009
32 X	6800	34000
Estuaries [3]		
Chesapeake Bay	80	510
Narragansett Bay, RI	100	950
Potomac River Estuary, MD	140	810
Delaware Bay	140	1300
North San Francisco Bay, CA	290	2010
South San Francisco Bay, CA	310	1600
Raritan Bay, NJ	330	1460
New York Bay	4550	31930

[1]Orr (1947).
[2]Nutman (1950).
[3]Nixon and Pilson (in press).

16 X, and 32 X treatments. Ammonia remained more abundant at the higher rates of input, suggesting that nitrification could not match the higher loading rates. The situation in all of the tanks is complicated by phytoplankton uptake (presumably of ammonia) after early January when the winter-spring bloom began. There was also a marked difference in the ratio of nitrate to nitrite with the more reduced form dominating at higher treatment levels (Figure 3).

In his recent discussion of nitrogen cycle processes, Webb (1981) noted a late summer-early fall nitrite maximum in the York River estuary and in Chespeake Bay that appeared to be increasing during recent years and appearing earlier in the season. His suggestion was that the oxidation of ammonia is favored over that of nitrite at low oxygen levels, thus making the relative abundance of nitrite and nitrate a possible indicator of organic loading associated with pollution. Nutrient additions alone can also influence oxygen con-

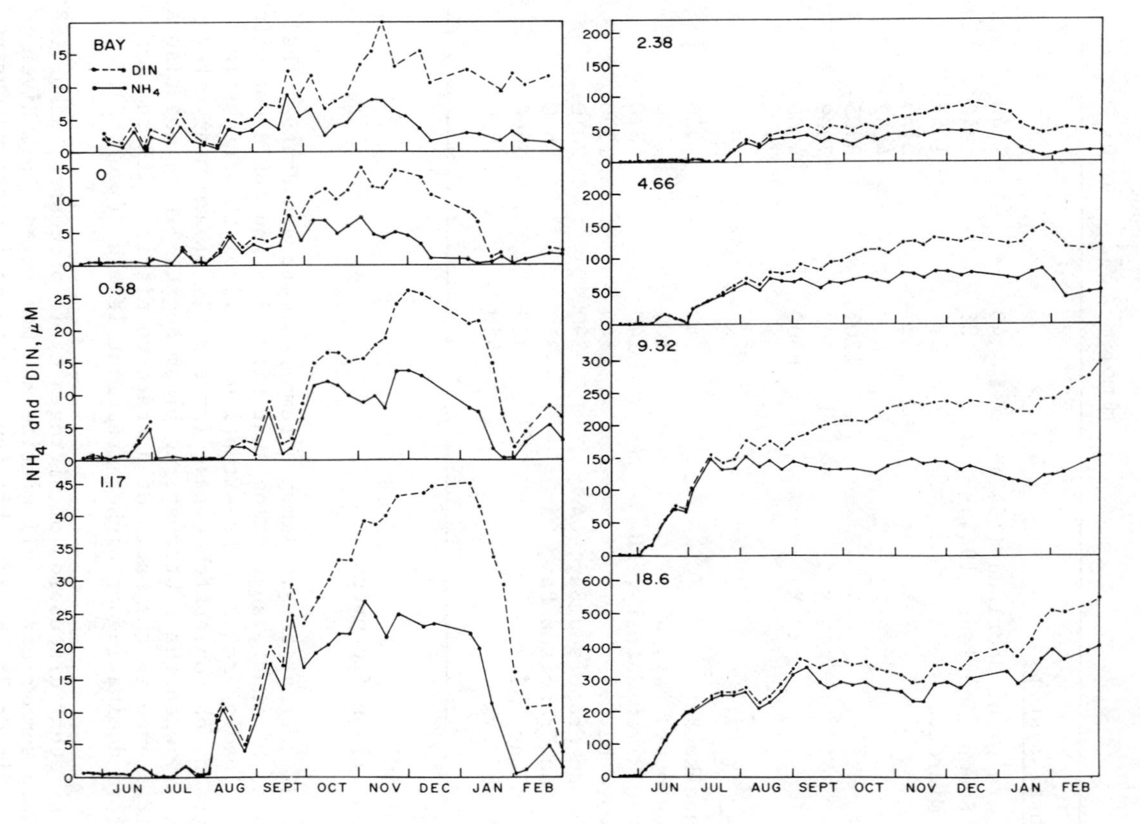

Fig. 2. Concentrations of dissolved inorganic nitrogen (broken line) and ammonia (solid line) in Narragansett Bay and the MERL microcosms during the first 9 months of the eutrophication experiment. The daily ammonia input (μmol L^{-1} d^{-1}) is shown for each treatment. Zero input represents one of the control tanks. Note scale changes.

Table 2. Relative abundance of ammonia and nitrite in Narragansett Bay and the MERL microcosms at various times during the eutrophication experiment

	NH_4 as % of the DIN Pool			
	3 Nov.	8 Dec.	5 Jan.	1 Feb.
Bay	51	25	21	25
Control	40	24	15	43
Control	46	23	6	35
Control	49	24	9	0
1 X	56	50	37	24
2 X	69	53	49	2
4 X	63	55	47	24
8 X	63	58	57	49
16 X	63	58	53	52
32 X	83	82	81	77

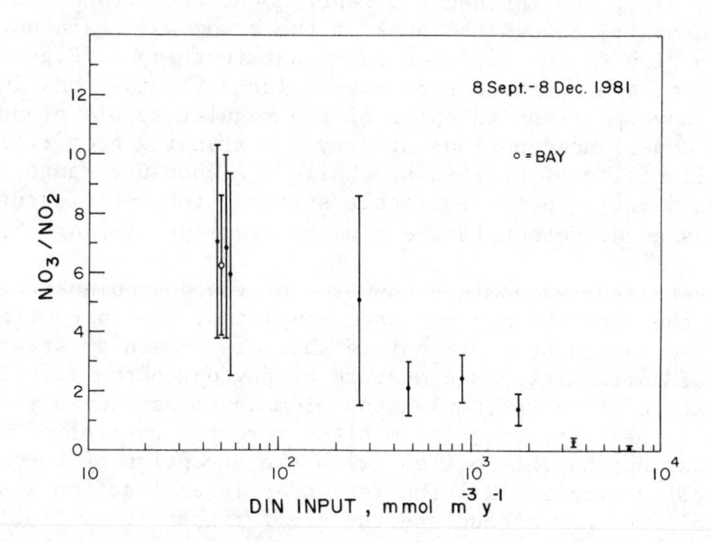

Fig. 3. The mean (±SD) ratio of nitrate to nitrite (molar units) in Narragansett Bay and the MERL microcosms during the first fall season of the eutrophication experiment as a function of the annual ammonia input rate.

centrations, even in well-mixed systems like the MERL microcosms (Figure 4). While oxygen levels were lower in the treated tanks during late summer, it is not clear that oxygen can be related in a simple way to the accumulation of nitrite. Oxygen never became very low (< 4 mg L^{-1}) for long periods in any of the treatments and the concentrations were generally rising during the time nitrite was accumulating. It was also the case, however, that nitrate increased more rapidly as the season progressed, with the result that the nitrate:nitrite ratio in each tank increased through the fall. The suggestion that nitrite may be a useful indicator of anthropogenic impact appears promising, but more work will be required to unravel the story.

Phytoplankton

The mean standing crop of phytoplankton increased with nutrient inputs, though not in strict proportion to the loading (Table 3). During the first six months chlorophyll levels in the 1, 2, 4 and even 16 X treatments remained only moderately higher than those in the bay and controls, while the 8 and 32 X tanks developed large blooms almost immediately (Figure 5). By the time of the characteristic winter-spring bloom in Narragansett Bay (Pratt, 1965; Smayda, 1973), the influence of enrichment was evident in all of the treatments, though the peak of the bloom was again not increased in proportion to the differences in nutrient input (Figure 5, Table 3). All of the microcosms displayed large fluctuations in the biomass of phytoplankton in spite of the regular supply of nutrients and all experienced periods of very low standing crop regardless of the availability of nutrients. While the absolute range in standing crop was greater in the eutrophic systems, the coefficient of variation was no different in the higher treatments (Figure 5, Table 3).

A detailed taxonomic accounting of the phytoplankton population through the year has not yet been completed, but in general there did not appear to be an immediate shift in community structure as a result of enrichment. One measure of phytoplankton structure is the percentage of total phytoplankton biomass (measured as fluorescence) that will pass through a 10 μm nitex screen. Optical counts indicated that during this period (with the exception of the July bloom of a small diatom in 32X) the less than 10 μm fraction was dominated by flagellates and monads and the greater than 10 μm fraction was dominated by diatoms. Based on this measure, during the summer and fall the phytoplankton communities in the bay, controls and lowest treatment were dominated by periodic blooms of diatoms against a background of flagellates (Figure 6). At higher treatment levels, all the major initial blooms were diatom dominated, but by early September diatoms were replaced by flagellates (Figure 6). Diatoms did not again become abundant until the winter-spring bloom, despite adequate silicate levels (Figure 7). The effect of the diatoms can

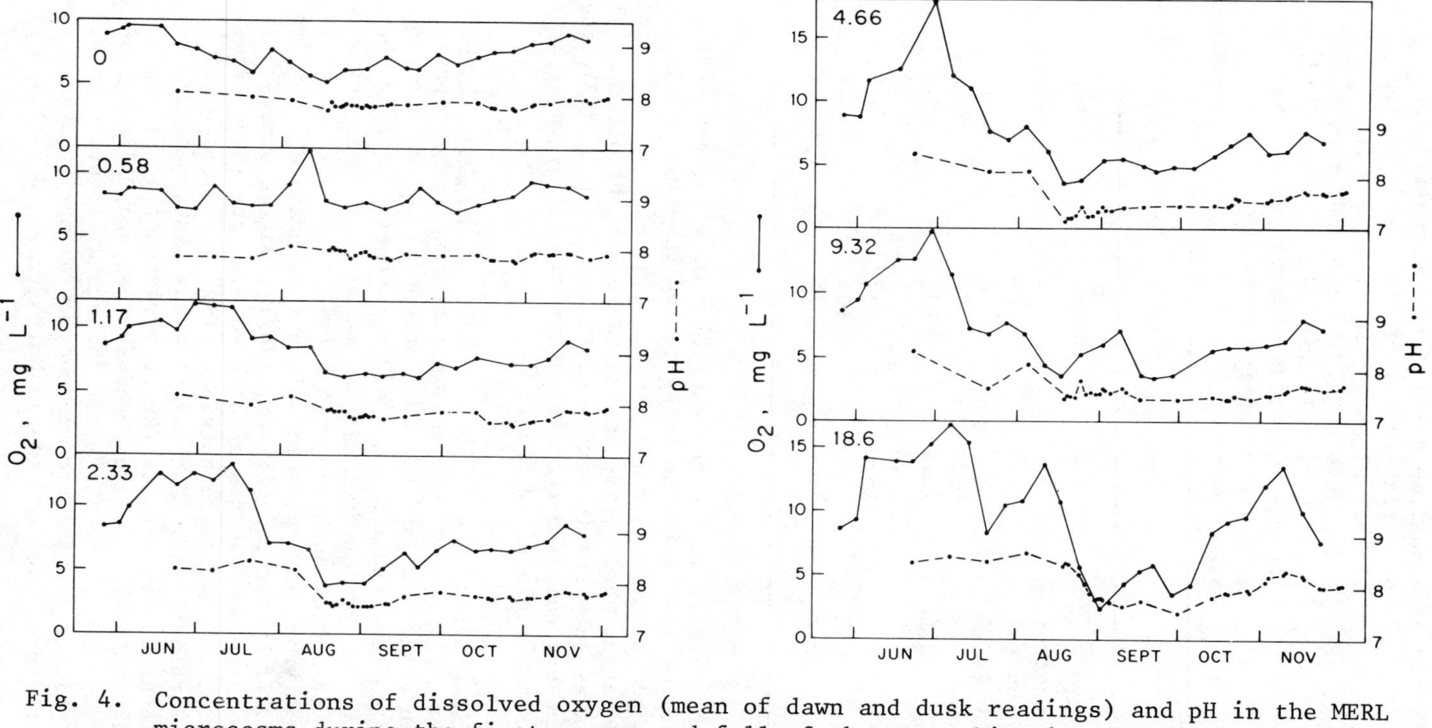

Fig. 4. Concentrations of dissolved oxygen (mean of dawn and dusk readings) and pH in the MERL microcosms during the first summer and fall of the eutrophication experiment. The daily ammonia input (μmol L^{-1} d^{-1}) is shown for each treatment.

Table 3. Annual mean (mg m^{-3}) and coefficient of variation (%) of the standing crop of phytoplankton chlorophyll a in Narragansett Bay and the MERL microcosms during 1 June 1981 - 1 June 1982. The maximum value observed during the 1982 winter-spring bloom (mg m^{-3}) is also shown.

	$\bar{x}$	C.V.	Max.
Bay	2.4	106	15.0
Control	4.1	91	23.3
Control	4.1	79	16.8
Control	4.7	112	24.9
1 X	10.3	107	50.2
2 X	13.9	109	69.7
4 X	24.4	121	91.4
8 X	18.0	143	120.9
16 X	35.6	105	113.9
32 X	61.8	76	118.7

also be seen clearly in the concentration of dissolved silica which was held at low levels in the tanks during the times of high chlorophyll throughout the summer (Figure 7). The silica data further confirm that diatoms were active in the winter-spring bloom at all treatment levels as they are in Narragansett Bay (Pratt, 1959), though the bloom in the bay followed that in the microcosms by about four weeks and is thus not included in Figure 7.

Some may be surprised at the abundance of diatoms throughout the summer, especially in the microcosms receiving large nutrient inputs. It has been suggested (and commonly accepted) on the basis of some laboratory culture studies and field data that shifts in the composition of phytoplankton from diatoms to smaller forms would be a major consequence of anthropogenic nutrient loadings which tend to be rich in nitrogen and poor in silica (for example, Greve and Parsons, 1977; Ryther and Officer, 1981). On the other hand, the results of these experiments might have been predicted from a consideration of the "specific nutrient flux" model discussed by Harrison and Turpin (1982). Missing from any assessment focused solely on phytoplankton, however, is the observation that silica is cycled more rapidly than commonly supposed (Ryther and Officer, 1981). Even though it must return to the water through dissolution rather than excretion, silica fluxes from the sediments during summer have been measured at over 1 mmol m^{-2} h^{-1} in Narragansett Bay (Nixon et al.

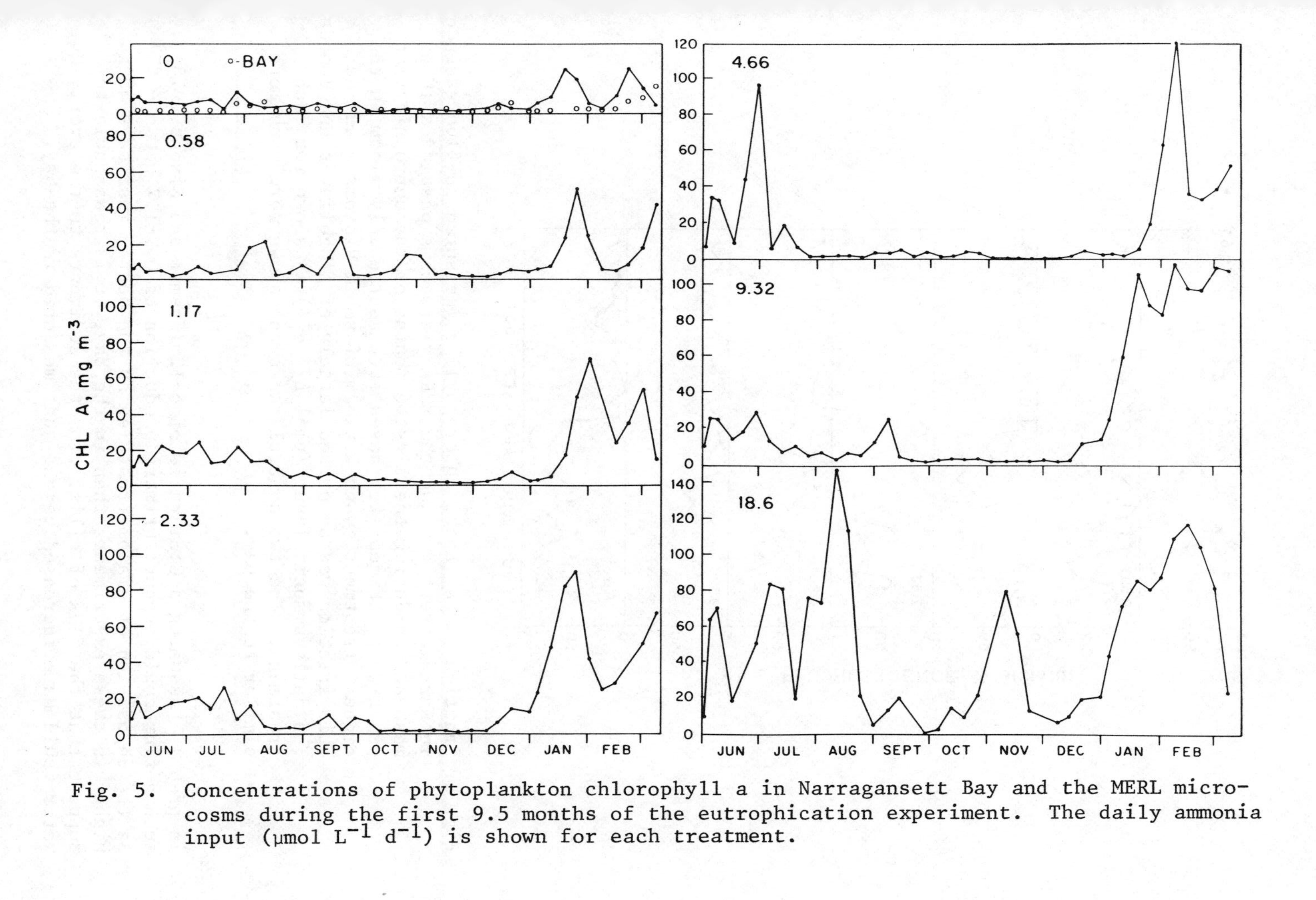

Fig. 5. Concentrations of phytoplankton chlorophyll a in Narragansett Bay and the MERL microcosms during the first 9.5 months of the eutrophication experiment. The daily ammonia input ($\mu mol\ L^{-1}\ d^{-1}$) is shown for each treatment.

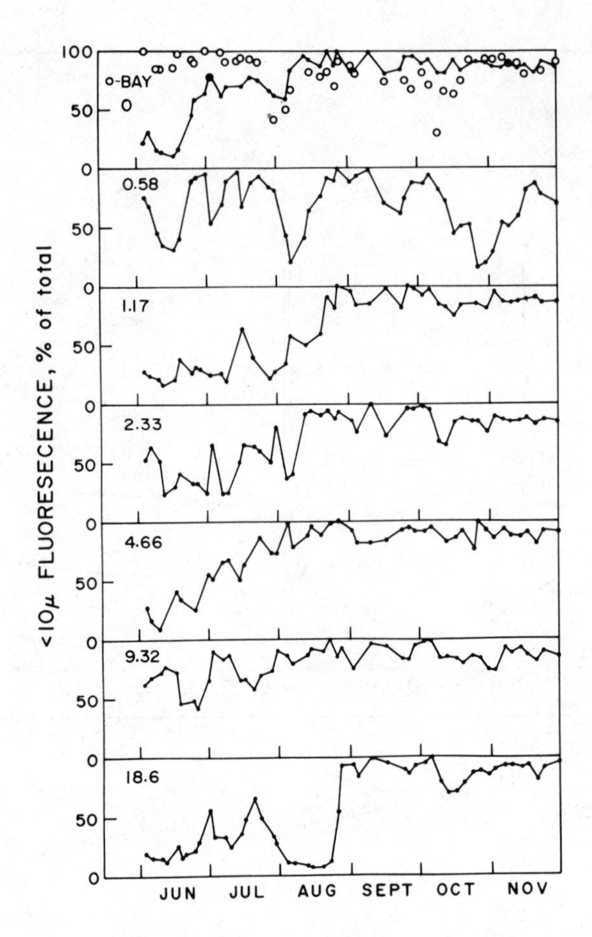

Fig. 6. Fraction of the total chlorophyll a (measured as fluorescence in Narragansett Bay and the MERL microcosms passing a 10 μm mesh screen during the first 6 months of the eutrophication experiment. Except for cases such as the July bloom in the highest treatment, when a bloom of small diatoms occurred, this fraction was comprised largely of flagellates and monads The daily ammonia input (μmol L^{-1} d^{-1}) is shown for each treatment. The two control tanks not shown were much more similar to the bay.

in prep.). Assuming a C:Si ratio of 6-12 (Kremer and Nixon, 1978), such a flux could support primary production of 0.9-1.7 g C m^{-2} d^{-1}. As the deposition of diatom frustules on the bottom must have been higher in these experiments than in the bay, it is reasonable to suppose that the flux of silica from the sediments to the overlying water would have increased beyond that measured in the bay.

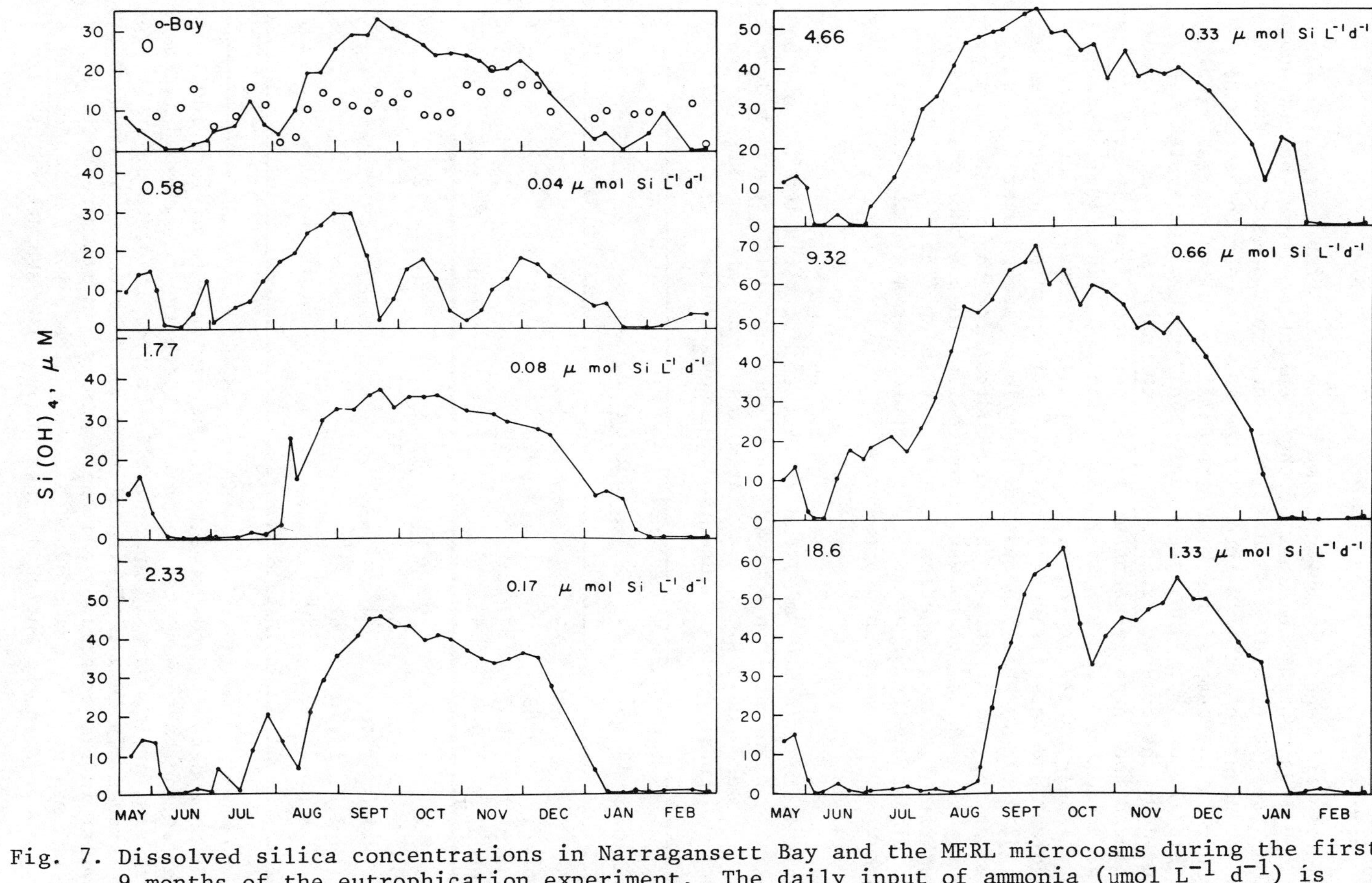

Fig. 7. Dissolved silica concentrations in Narragansett Bay and the MERL microcosms during the first 9 months of the eutrophication experiment. The daily input of ammonia (μmol L^{-1} d^{-1}) is shown for each treatment on the left, daily silica input is on the right.

Zooplankton

The differences in phytoplankton standing crop were not reflected in the mean population sizes of the zooplankton along the eutrophication gradient (Table 4). There were, however, some differences in the abundance of animals through time which appear to be due to a change in the composition of the population at the highest treatment levels (Figure 8). The late summer peak in the 32 X system was caused by the copepod *Pseudodiaptomus coronatus* which almost completely replaced the normally abundant summer copepods (largely *Acartia tonsa*) that dominated the other microcosms (Table 4, Figure 9). Numbers of larval polychaetes in the plankton increased along the nutrient gradient and were frequently the dominant organism in the water of the 32 X tank. While the variability in abundance over time was greater for the *Acartia* species group than for the zooplankton community as a whole, there was no indication that either was more or less variable as a function of nutrient input (Table 4). A more detailed analysis of community structure must await the taxonomic work which is still in progress.

Table 4. Mean (number m^{-3}) and coefficient of variation (%) of the standing crop of total zooplankton, adult *Acartia hudsonica* and adult *A.tonsa* in Narragansett Bay and the MERL microcosms during 1 June 1981 - 1 Nov. 1981.

	$\bar{x}$ (Total)	c.v.	$\bar{x}$ (*A. Hudsonica*)	c.v.	$\bar{x}$ (*A. Tonsa*)	c.v.
Bay	4268	94	241	211	582	173
Control	19289	72	918	163	1923	116
Control	13992	71	412	190	1458	168
Control	9704	64	81	150	594	177
1 X	11993	97	342	130	1767	163
2 X	15381	59	123	158	2052	102
4 X	16462	66	531	234	1004	183
8 X	13200	61	218	187	1305	153
16 X	15176	82	1618	202	1726	126
32 X	22260	130	121	270	141	107

More detailed measurements of the dry weights of individual *Acartia* adults showed that animals in the nutrient enriched microcosms were similar to, or only sporadically heavier than, those in the control tanks (Table 5). The total zooplankton biomass values were also similar in all treatments and the controls. Maximum reproduction rates, measured by egg production, were higher for female *Acartia* of both species in the enriched tanks, though the average rates were fre-

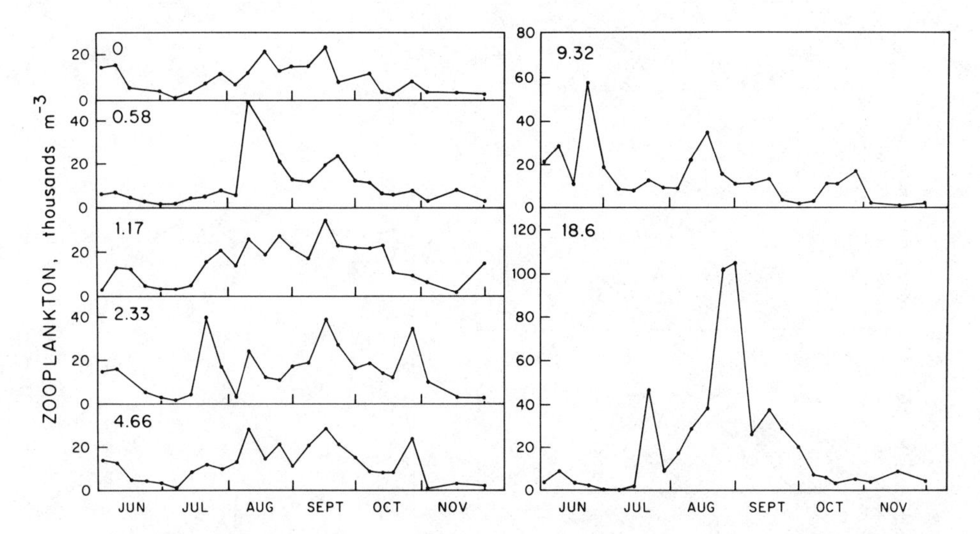

Fig. 8. Numbers of total zooplankton in the MERL microcosms during the first summer and fall of the eutrophication experiment. A single control with the lowest values has been plotted. Data for the other controls are summarized in Table 4. The daily input of ammonia (μmol L^{-1} d^{-1}) is shown for each treatment. Data courtesy P. Lane.

quently found not to differ between treated and control systems (Table 6). These fluctuations in egg production rates were apparently unrelated to changes in the dominance of the phytoplankton by diatoms or flagellates (Sullivan et al., in prep.). The potential for increased zooplankton production over that in controls existed at times in all treatments, but was not associated with significant increases in zooplankton abundances.

The lack of apparent response of the zooplankton to even marked changes in food supply is difficult to explain with the data thus far in hand. While such an outcome presents a considerable challenge to our present understanding of plankton food chain dynamics, it was also observed earlier in the first Scottish sea-loch fertilization experiments (Marshall, 1947). Fertilization of the second sea loch, Kyle Scotnish, however, reportedly produced a great increase in zooplankton (Gauld, 1950). Regardless of the causal factors, the descriptive evidence from the MERL experiment is unambiguous in showing that nutrient enrichment need not necessarily lead to a predominance of small phytoplankton and that copepod population size is not as closely tied to the phytoplankton biomass as expected.

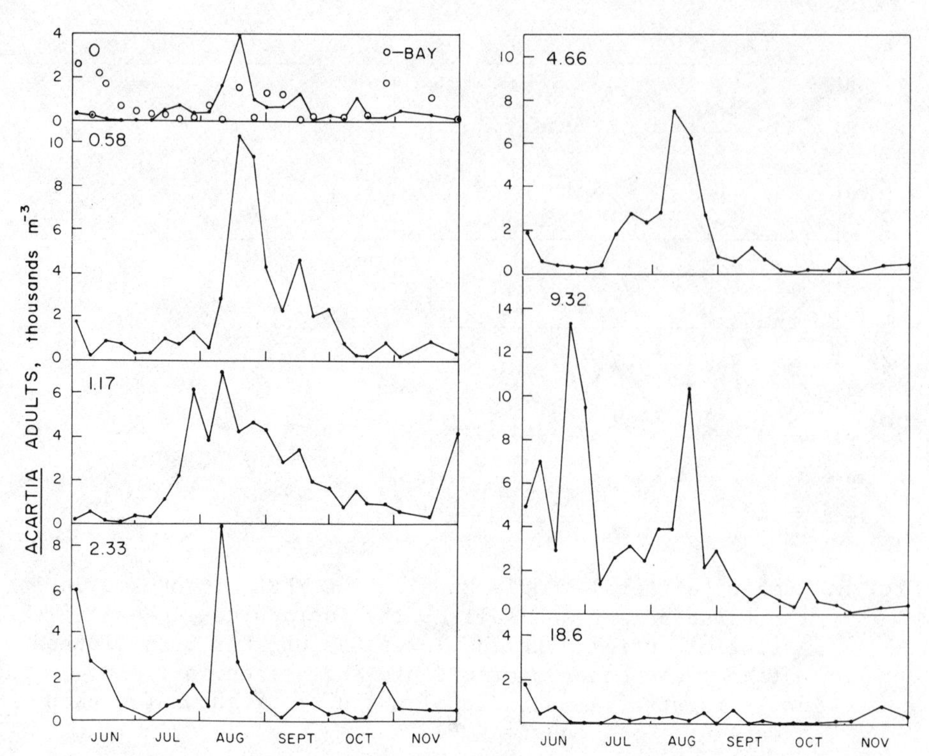

Fig. 9. Numbers of Acartia hudsonica (formerly clausi) and A. tonsa in Narragansett Bay and the MERL microcosms during the first summer and fall of the eutrophication experiment. A single control with the lowest values has been plotted. Data for the other controls are summarized in Table 4. The daily input of ammonia (μmol L^{-1} d^{-1}) is shown for each treatment. Data courtesy P. Lane.

Benthos

The most dramatic response to nutrient enrichment, at least during the first 3 months, occurred in the benthos. While the numbers and taxonomic composition of all of the microcosms were very similar at the start of the experiment, by September an almost linear response (r^2 = 0.97) to treatment was evident in the total numbers of macroinfauna (Figure 10). With one exception, this response was confined to the polychaetes, especially Mediomastus ambiseta, Polydora ligni, and Streblospio benedicti. Together their numbers increased some 15-fold in response to the 32-fold increase in nutrient input. The only significant increase in bivalves took place in the 8 X treatment, where the opportunistic filter feeder Mulinia lateralis became abundant. The normally dominant bivalve Nucula annulata showed no statistically significant response, though it was the only bivalve

Table 5. The results of t-tests using equal variances for differences in the dry weight of female <u>Acartia</u> in the MERL microcosms at different treatment levels compared with control tanks during the first summer of nutrient input. Level of significance (≤ 0.10) is shown. Other comparisons were not significantly different (NS) or not carried out (-).

	6/14/81	6/17/81	7/1/81	7/14/81
Acartia hudsonica				
2 X	NS	NS	NS	NS
16 X	NS	+ 0.05	--	NS
32 X	--	NS	--	--

	7/21/81	8/18/81	3/31/81	9/17/81	9/29/81
Acartia tonsa					
2 X	NS	+ 0.05	+ 0.05	NS	+ 0.01
8 X	NS	NS	NS	+ 0.10	--
16 X	+ 0.01	+ 0.10	NS	+ 0.10	+ 0.05
32 X	NS	--	--	--	--

Table 6. The results of t-tests using unequal variances for differences in the egg production of female <u>Acartia</u> in the MERL microcosms at different treatment levels compared with control tanks during the first summer of nutrient input. Levels of significances (≤ 0.10) are shown. Other comparisons were not significantly different (NS) or not carried out (-). Positive or negative differences are noted (+,-).

	6/4/81	6/17/81	7/1/81	7/14/81
Acartia hudsonica				
2 X	+0.01	NS	--	NS
16 X	NS	+ 0.07	- 0.07	NS
32 X	+0.05	NS	--	--

	7/21/81	8/18/81	8/31/81	9/17/81	9/29/81
Acartia tonsa					
2 X	+ 0.001	+ 0.05	NS	--	+ 0.01
8 X	+ 0.025	NS	+ 0.06	NS	--
16 X	+ 0.02	+ 0.002	+ 0.08	+ 0.06	+ 0.001
32 X	+ 0.003	--	--	+ 0.08	--

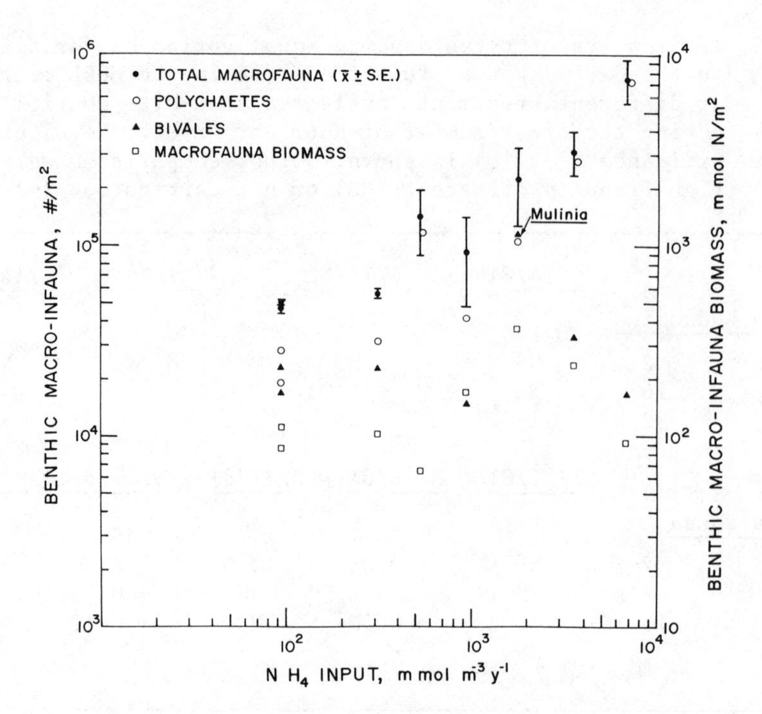

Fig. 10. Abundance of benthic macroinfauna in the MERL microcosms in September as a function of the annual ammonia input rate. Control tanks (only two have been plotted) were assumed equal to the bay in terms of input.

species which persisted in the highest treatment level. An increase in the biomass of polychaetes was also evident, though as a group they weighed so little compared to bivalves that no increase in total infauna biomass was found (Figure 10).

It seems clear from later samples that the benthic response was not complete by September, but the analyses required for a full description are still underway. A longer-term view may be afforded by some results of the sea-loch experiments, where marked increases in the benthos were also found following fertilization of Loch Craiglin (Raymont, 1947) and the much larger and more marine Kyle Scotnish at Loch Sween (Raymont, 1950). The average benthic population recorded from eight stations in Kyle Scotnish over four years (three of them during fertilization less intense than that of the MERL experiment, Table 1, and one post fertilization) is shown in Figure 11. The numbers themselves cannot strictly be compared with the MERL data because of differences in sampling technique (for example, 1 mm mesh screens instead of the 0.3 mm screen used at MERL), but the secular trend through the seasonal cycles is impressive. There is also evidence that the benthic enhancement persisted for some time after direct fertilization of the water column had ceased. As a final point, it should be noted that while polychaetes and some bivalves

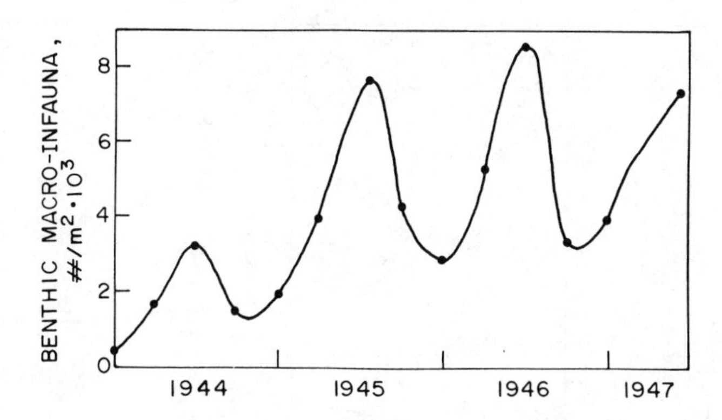

Fig. 11. Mean abundance of benthic macroinfauna from 8 regular stations in Kyle Scotnish, Loch Sween, Scotland. The loch was fertilized repeatedly in 1944, 1945, and 1946 (see Table 1). From Raymont (1950).

increased in the loch, the most dramatic response in that case was by the amphipods.

Productivity and Respiration

The long residence time of a defined water mass in each tank makes it possible to measure the total system metabolism of each microcosm, including the water column and benthos, by following diel changes in dissolved oxygen. Changes in oxygen are corrected for diffusion across the air-water interface using an empirical exchange coefficient (Nixon et al., 1980). In this way, the entire system is treated as a giant "light bottle" during day and a "dark bottle" at night. The contribution of the benthos to observed oxygen changes is measured directly by separate determinations using a large chamber which covers the entire tray of sediment.

Maximum rates of apparent production during the day increased by a factor of 2-3 in the more enriched tanks, and they were accompanied by higher rates of oxygen uptake at night (Figure 12). Production was much more variable than dark respiration in all treatments, though this was most evident in the 8X and 32 X systems. When integrated over the first 6.5 months of the experiment, these data show remarkably little metabolic response relative to the very large increases in nutrient loading. Total biological energy flow through the systems appeared to increase by a factor of 3 or less in the face of a 32-fold increase in loading (Figure 13). However, the production data from at least the 8 X, 16 X and 32 X tanks are underestimates because we were unable to measure the production of oxygen which was removed from those systems in bubbles. It is impossible

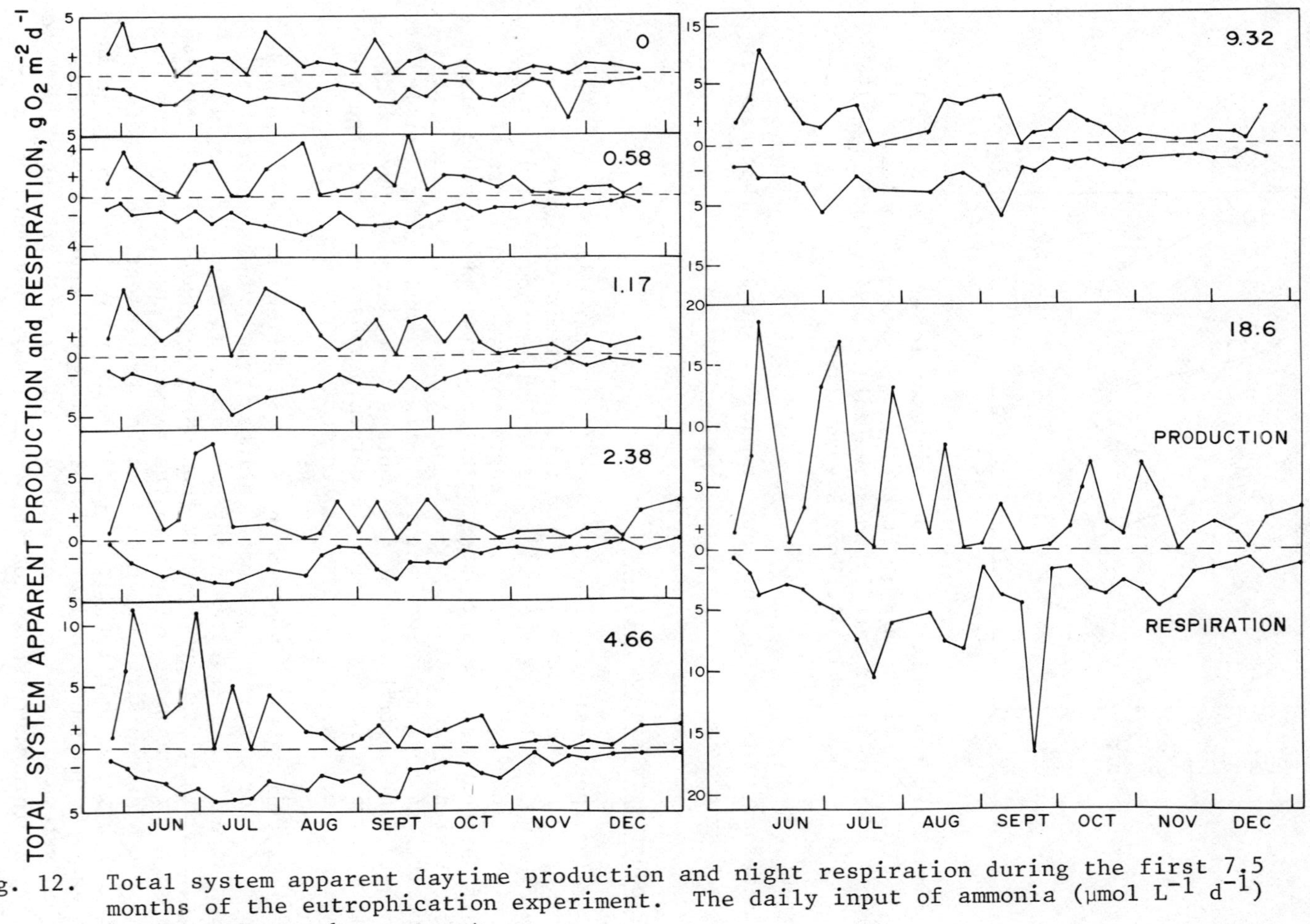

Fig. 12. Total system apparent daytime production and night respiration during the first 7.5 months of the eutrophication experiment. The daily input of ammonia (µmol L^{-1} d^{-1}) is shown for each treatment.

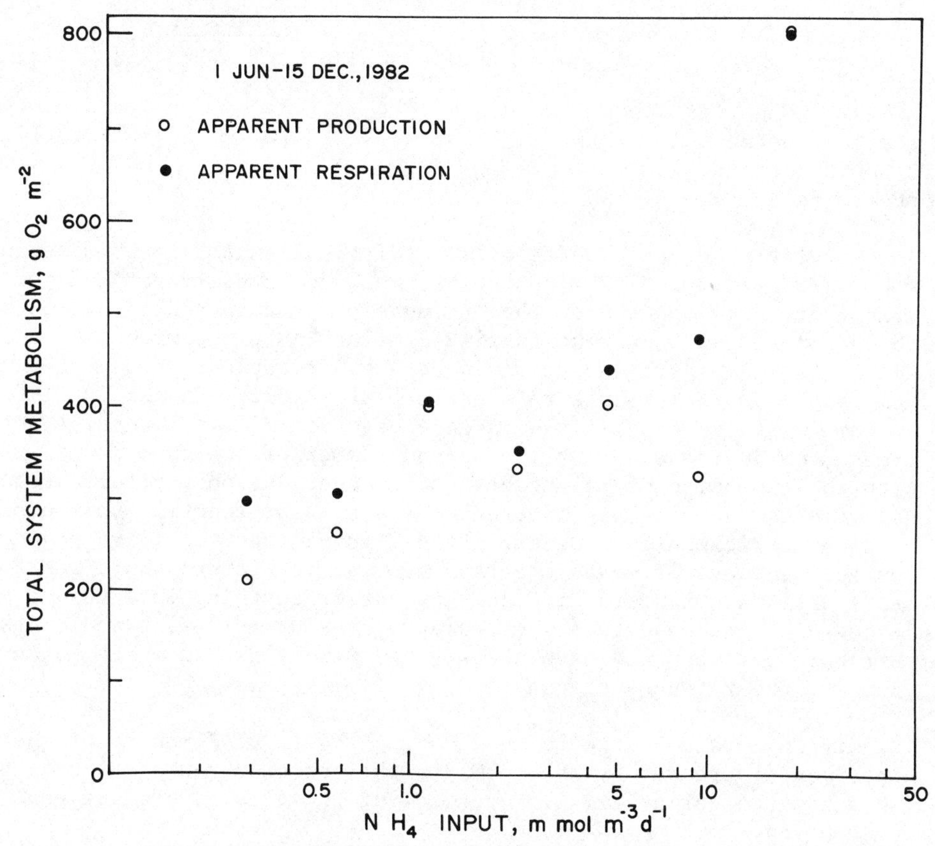

Fig. 13. Total system apparent daytime production and night respiration in the MERL microcosms, integrated over the first 6.5 months of the eutrophication experiment, as a function of the daily ammonia input.

to quantify the loss at this point, though bubbling was confined to some daylight periods during summer, and the close coupling of production and respiration in all of the other treatment and control tanks suggests the actual production may not have been very much larger than measured. The common slight excess of night respiration over daytime production is probably an error caused by the diffusion correction scheme and the oxygen sampling schedule. Nevertheless, the results suggest that in well-mixed systems which seldom experience low oxygen conditions (Figure 4), eutrophication may not lead to a rapid accumulation of organic matter. It is not clear how long this apparent coupling of production and decomposition might persist. Organic carbon analyses of the upper 4 cm of sediments provided by J. Gearing show a large enrichment of the top 1 cm in the 32 X tank in samples collected on 21 April 1982, following the winter-spring bloom (Figure 5) :

	Treatment					
	1 X	2 X	4 X	8 X	16 X	32 X
Excess organic carbon with respect to the mean of the 3 control tanks, g m^{-2} in the surface 1 cm.	13	2	10	28	17	112

Interpretation of these data will be difficult until later measurements are available. The control tanks were actually sampled on 16 June, after the input from the spring bloom had an additional 56 days to decompose, and experiments with similar benthic communities and sediments from Narragansett Bay have shown a rapid increase in decomposition as a response to organic matter additions (Kelly, 1982). It remains to be seen if the large excess at 32 X will persist, increase, or decrease during the second summer of the experiment. But, with the exception of the highest tank, even the post-bloom carbon data did not show a large increase as a regular function of increasing nutrient enrichment. However, it is evident that the large and quite variable amounts of organic carbon present as background in the sediments will make it very difficult to detect anything less than a very large accumulation with any certainty. The organic rich sediments of eutrophic environments have usually had many years in which to accumulate their excess carbon.

Oxygen uptake by the benthos during the first 7 months of the experiment increased by about 60% in the enriched tanks (Figure 14) and accounted for a variable fraction of the total system oxygen consumption:

	Benthic O_2 Uptake, % of Total System Night Respiration
Control	32
1 X	45
2 X	33
4 X	34
8 X	37
16 X	35
32 X	18

These preliminary values are an overestimate, since we have assigned a constant 50% of the 24 h benthic oxygen uptake to the night. While they should not, therefore, be accepted in an absolute sense, they are nevertheless useful in comparing the different treatments.

Denitrification

Since earlier direct measurements of the production of nitrogen gas by Narragansett Bay sediments had shown that denitrification was

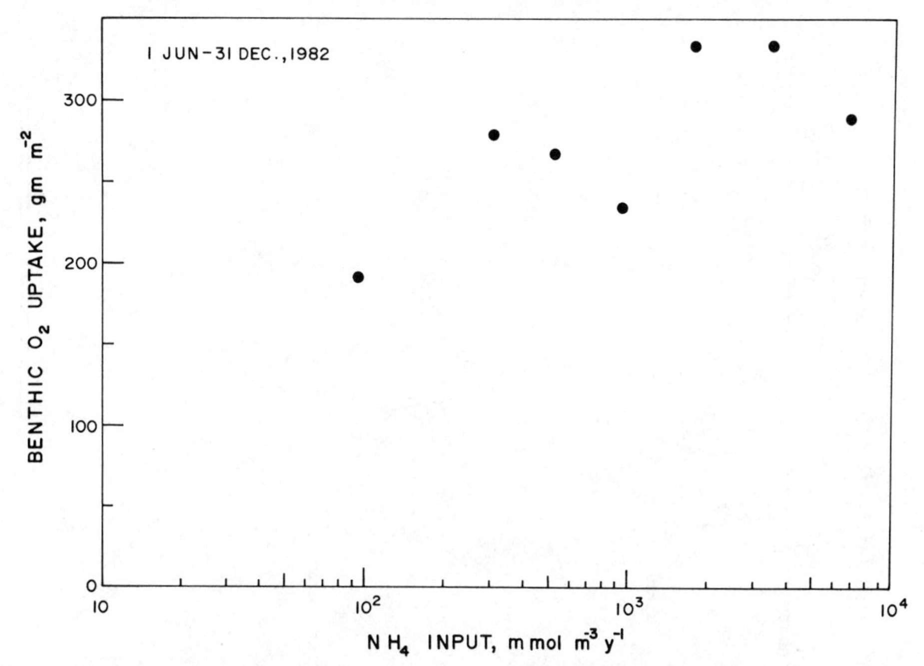

Fig. 14. Oxygen consumption by the benthos in the MERL microcosms integrated over the first 7 months of the eutrophication experiment as a function of the annual ammonia input.

an important part of the bay nitrogen budget (Seitzinger et al., 1980; Seitzinger, 1982), it seemed desirable to see if there was an increase in nitrogen removal with higher inputs in the microcosms. Measurements made at the end of the first summer, after three months of enrichment, showed a clear enhancement of denitrification at the higher treatment levels (Figure 15). However, the increase in nitrogen removal, as with all of the other responses described in this report, was considerably less than the increase in nitrogen input. At the lower treatment levels, denitrification was capable of removing over 100% of the nitrogen addition, while in the 32 X system only some 20% could be eliminated.

CONCLUSION

As part of his introduction to the collection of papers reporting the results of the second series of Scottish sea-loch experiments, Fabius Gross (1950) made an observation that is worth quoting at length:

> If, in retrospect, much of what has been established as response of the marine communities to an increase of basic nutrients appears as just what one would have anticipated

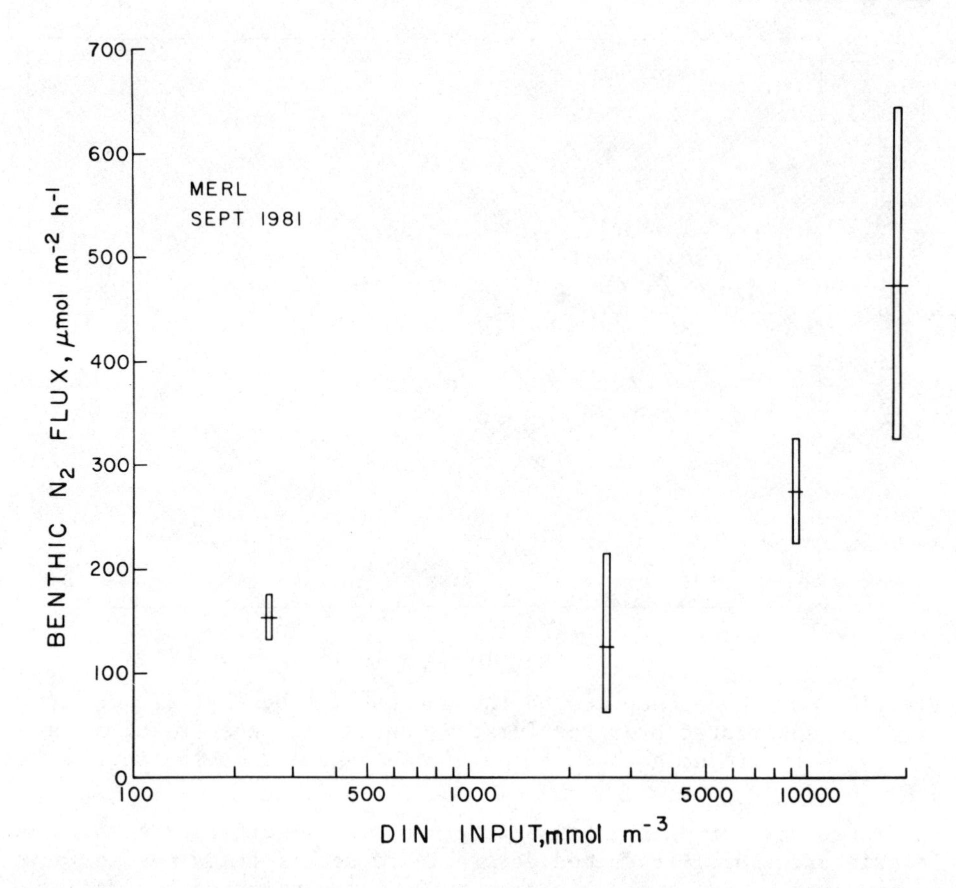

Fig. 15. Denitrification in the sediments of some of the MERL microcosms measured in September, after the first summer of the eutrophication experiment. Rates are plotted as a function of the cumulative ammonia input since the start of the experiment. The mean range of each set of measurements is shown.

on general grounds . . ., it may be mentioned that when the scheme was first submitted in 1941 to several leading authorities in marine biology, some of their comments expressed serious doubts not so much as to its economy but as to first principles. Would not most of the fertilizer be adsorbed to the mud and thus rendered ineffective ? If phytoplankton production does increase, would not the growth of forms be encouraged which were of no use at all for larval food or the growth of bottom fauna? Would not bottom animals competing for food with fish increase out of all proportion to the increase in the amount of useful food organisms? . . . .

It is, of course, precisely to answer such questions that the experiment must be undertaken. Our concern in the MERL studies has not necessarily been directed toward the increased production of fish, and we have made some effort in the introduction to this paper to emphasize the relatively modest level of our current understanding of marine ecosystems. Many of the same questions asked in Scotland 40 years ago were raised again at the start of this experiment. We played the game among ourselves and with numerous colleagues of trying to predict in as much detail as possible the outcome of the nutrient additions. Elaborate causal webs were often spun, but confidence in any particular scenario was always low. Such caution was well advised. Even the preliminary and largely descriptive results given here contain a number of surprises. For example, the detailed numerical model of plankton dynamics developed by Steele and Frost (1977) as well as the analysis of laboratory culture and field data presented by Greve and Parsons (1977) would have predicted very different results from those actually found.

The behavior of equations can be fascinating and instructive, as can the study of laboratory cultures and the description of natural communities. But our capability for experimental perturbations and manipulations of natural ecosystems or suitably complex living models of natural systems must be developed and exploited if marine ecology is to move forward as a vigorous and relevant science.

Acknowledgements

The ecosystem level experiments at MERL are carried out by many people. Those involved in the work reported here included G. Almquist, V. Beronsky, B. Buckley, C. Doremus, J. Gearing, J. Grassle, J. F. Grassle, M. Hutchins, A. Keller, S. Kelly, K. Kopchynski, P. Lane, P. Ritacco, P. Sampou, V. Weiss. Financial support came largely from the U.S. Environmental Protection Agency (Grants CR-810265 and CR-807795), with additional contributions from the Office of Marine Pollution Assessment (Grants NA-80-RA-D-00046; NA-81-RA-D-00002) and the Office of Sea Grant, both a part of NOAA.

REFERENCES

Atkins, W.R.G., 1930, Seasonal variations in the phosphate and silicate content of sea-water in relation to the phytoplankton crop. Pt. V. J. Mar. Biol. Assoc., N.S., 16: 821.

Atkins, W.R.G., 1932, Nitrate in sea-water and its estimation by means of diphenylbenzidine, J. Mar. Biol. Assoc., U.K., N.S. 9, XVIII: 167.

Antia, N.J., McAllister, C.C., Parsons, J.R., Stevens, K., and Strickland, J.D.H., 1963, Further measurements of primary production using a large volume plastic sphere, Limnol. Oceanogr., 8: 166.

Brandt, K., 1899, Uber den Stoffwechsel im Meer. Wissenschaftliche Meeresuntersuchunger, N.F. Abt. Kiel, Bd. IV.

Caperon, J., Schell, D., Hirota, J., and Laws, L.E., 1979, Ammonium excretion rates in Kaneohe Bay, Hawaii, measured by a ^{15}N isotope dilution technique, Mar. Biol., 54: 33.

Cooper, L.H.N., 1933, Chemical constituents of biological importance in the English Channel, November, 1930, to January 1932, Part I. Phosphate, silicate, nitrate, nitrite, ammonia, J. Mar. Biol. Assoc., U.K., N.S. 18: 677.

Deniges, G., 1921, Determination quantitative de plus faibles quantites de phosphates dans les produites biologiques par la method ceruleomolybique, Compt. Rend. Biol. Paris, 84: 875.

Edmondson, W.T., and Edmondson, Y.H., 1947, Measurements of production in fertilized salt water, J. Mar. Res., 6: 228.

Elmgren, R., and Frithsen, J.B., 1982, The use of experimental ecosystems for evaluating the environmental impact of pollutants: A comparison of an oil spill in the Baltic Sea and two long-term, low-level oil addition experiments in mesocosms, In: "Marine Mesocosms," Grice, G.D. and Reeve, M.R. eds., Springer-erlag, NY.

Gaarder, T., and Sparck, R., 1931, Biochemical and biological investigations of the variations in the productivity of the West Norwegian oyster polls, Rapp. et Proc. - verbaux des Reunions, LXXV : 47.

Gauld, D.T., 1950, A fish cultivation experiment in an arm of a sea-loch. III. The Plankton of Kyle Scotnish, Proc. Roy. Soc. Edinburgh, LXIV, Sect. B: 36.

Glibert, P.M., 1982, Regional studies of daily, seasonal, and size fraction variability in ammonium remineralization, Mar. Biol. 70: 209.

Goering, J.J., Nelson, D.M., and Carter, J.A., 1973, Silicic acid uptake by natural populations of marine phytoplankton, Deep-Sea Res., 20: 777.

Greve, W. and Parsons, T.R., 1977, Photosynthesis and fish production: Hypothetical effects of climatic change and pollution, Helogolander wiss. Meeresunters., 30: 666.

Grice, G.D., and Reeve, M.R., eds., 1982, "Marine Mesocosms," Springer Verlag, NY.

Gross, F., 1947, An experiment in marine fish cultivation: I. Introduction. V. Fish growth in a fertilized sea loch, Proc. Roy. Soc. Edinburgh, B LXIII: 1-2, 56.

Gross, F., 1950, A fish cultivation experiment in an arm of a sea loch. I. Introduction, Proc. Roy. Soc. Edinburgh, LXIV, B: 1.

Harris, E., 1959, The nitrogen cycle in Long Island Sound, Bull. Bingham Oceanogr. Coll., 17: 31.

Harrison, W.G., 1978, Experimental measurements of nitrogen remineralization in coastal waters, Limnol. Oceanogr., 23: 684.

Harrison, W.G., 1980, Nutrient regeneration and primary production in the sea, In: "Primary Productivity in the Sea," Falkowski, P.G., ed., Plenum Press, NY.

Harrison, P.J., and Turpin, D.H., 1982, The manipulation of physical, chemical, and biological factors to select species from natural phytoplankton communities, In: "Marine Mesocosms," Grice, G.D., and Reeve, M.R. eds. Springer-Verlag, NY.

Harvey, H.W., 1928, Concerning methods for estimating phosphates and nitrates in solution in sea water. Section II. Estimation of nitrates and nitrites, Conseil. Int. l'Exp. de la Mer. Rapp et Proc. -verbaux, 53: 96.

Hunt, C.D., and Smith D.L., 1982, Controlled marine ecosystems --A tool for studying stable trace metal cycles: long-term response and variability, In: "Marine Mesocosms," Grice G.D., and Reeve, M.R., eds., Springer-Verlag, NY.

Johannes, R.E., 1969, Nutrient regeneration in lakes and oceans, In: "Advances in the Microbiology of the Sea, Vo. 1," Droop, M.R., and Wood, E.J.F., eds. Academic Press, NY.

Johnstone, J., 1908, "Conditions of Life in the Sea," Cambridge University Press. Reprinted by Arno Press, 1977, NY.

Kelly, J.R., 1982, Benthic-pelagic coupling in Narragansett Bay, Ph.D. Thesis, University of Rhode Island, Kingston, RI.

Kemp, W.M., Wetzel, R.L., Boynton, W.R., D'Elia, C.F., and Stevenson, J.C., 1982, Nitrogen cycling and estuarine interfaces: some current concepts and research directions, In: "Estuarine Comparisons," Kennedy, V.S., ed. Academic Press, NY.

Kremer, J.N., and Nixon, S.W., 1978, "A Coastal Marine Ecosystem, Simulation and Analysis," Ecological Studies 24, Springer-Verlag, NY.

MacIsaac, J.J., and Dugdale, R.C., 1969, The kinetics of nitrate and ammonia uptake by natural populations of marine phytoplankton, Deep-Sea Res., 16: 45.

Marshall, S.M., 1947, An experiment in marine fish cultivation: III. The plankton of a fertilized loch, Proc. Roy. Soc. Edinburgh, LXIII, Part I, Sec. b: 21.

Martin, J.H., 1968, Phytoplankton-zooplankton relationships in Narragansett Bay. III. Seasonal changes in zooplankton excretion rates in relation to phytoplankton abundance, Limnol. Oceanogr., 13: 63.

Nixon, S.W., 1981, Remineralization and nutrient cycling in coastal marine ecosystems, In: "Estuaries and Nutrients,", Neilson, B.J., and Cronin, L.E., eds., Humana Press, Clifton, NJ.

Nixon, S.W., and Pilson, M.E.Q., In press, Nitrogen in estuarine and coastal marine ecosystems, In: "Nitrogen in the Marine Environment,", Carpenter, E.J. and Capone, D.G., eds. Academic Press, 1983.

Nixon, S.W., Oviatt, C.A., and Hale, S.S., 1976, Nitrogen regeneration and the metabolism of coastal marine bottom communities, In: "The Role of Terrestrial and Aquatic Organism in Decomposition Processes," Anderson, J.M., and Macfadyen, A., eds. Blackwell Scientific Publications, London.

Nixon, S.W., Alonso, D., Pilson M.E.Q., and Buckley, B.A., 1980, Turbulent Mixing in aquatic microcosms, In: "Microcosms in Ecological Research," Geisy, J.P., ed., DOE Symposium Series, Augusta, GA, Nov. 8-10, 1978, Conf., 7811-1, NTIS.

Nixon, S.W., Kelley, J.R., Furnas, B.N., Oviatt, C.A., and Hale, S.S., 1980, Phosphorus regeneration and the metabolism of coastal marine bottom communities, In: "Marine Benthic Dynamics,", Tenore, K.R., and Coull, B.C., eds. University of South Carolina Press, Columbia, SC.

Nutman, S.R., 1950, A fish cultivation experiment in the arm of a sea-loch. II. Observations on some hydrographic factors in Kyle Scotnish, Proc. Roy. Soc. Edinburgh, LXIV, Sect. B: 5.

Orr, A.P., 1947, An experiment in marine fish cultivation: II, Some physical and chemical conditions in a fertilized sea-loch (Loch Craiglin, Argyll), Proc. Roy. Soc. Edinburgh, B. LXIII:3.

Pilson, M.E.Q., and Nixon, S.W., 1980, Marine microcosms in ecological research, In: "Microcosms in Ecological Research," Giesy, J.P., ed. DOE Symposium Series, Augusta, GA, Nov. 8-10, 1978, Conf. 781101 NTIS.

Pilson M.E.Q., Oviatt, C.A., and Nixon, S.W., 1980, Annual nutrient cycles in a marine microcosm, In: "Microcosms in Ecological Research," Giesy, J.P., ed. DOE Symposium Series, Augusta, GA, Nov. 8-10, 1978, Conf., 781101, NTIS.

Pilson, M.E.Q., Oviatt, C.A., Vargo, G.A., and Vargo, S.L., 1979, Replicability of MERL microcosms: Initial observations. In: "Advances of Marine Environmental Research," Jacoff, F.S., ed., Report EPA-600/9-79-035, Environmental Protection Agency, Narragansett, RI.

Pilson, M.E.Q., Vargo, G.A., Gearing, P.J., and Gearing, J.N., 1977, Investigation of effects and fates of pollutants, In: Proceedings Second Nat'l. Conf. Interagency Energy/Environment R & D Program June 6-7, 1977, Washington, DC, EPA-600/9-77-012, Environmental Protection Agency, Narragansett, RI.

Pratt, D.M., 1949, Experiments in the fertilization of a salt-water pond, J. Mar. Res., 8: 36.

Pratt, D.M., 1950, Experimental study of the phosphorus cycle in fertilized salt water, J. Mar. Res. 9: 29.

Pratt, D.M., 1959, The phytoplankton of Narragansett Bay, Limnol. Oceanogr., 4: 425.

Pratt, D.M., 1965, The winter-spring diatom flowering in Narragansett Bay, Limnol. Oceanogr., 40: 173.

Raymont, J.E.Q., 1947, An experiment in marine fish cultivation: IV. The bottom fauna and the food of flatfishes in a fertilized sea-loch (Loch Craiglin), Proc. Roy. Soc. Edinburgh, LXIII, Part I: 34.

Raymont, J.E.G., 1950, A fish cultivation experiment in an arm of a sea-loch. IV. The bottom fauna of Kyle Scotnish, Proc. Roy. oc. Edinburgh, LXIV, Sect. B: 65.

Ryther, J.H., and Officer, C.B., 1981, Impact of nutrient enrichment on water uses. In: "Estuaries and Nutrients," Neilson, B.J., and Cronin, L.E. eds., Humana Press, Clifton, NJ.

Santschi, P.H., 1982, Application of enclosures to the study of ocean chemistry, In: "Marine Mesocosms,", Grice, G.D., and Reeve M.R., eds., Springer-Verlag, NY.

Seitzinger, S.P., 1982, The importance of denitrification and nitrous oxide production in the nitrogen dynamics and ecology of Narragansett Bay, Rhode Island, Ph.D. Thesis, University of Rhode Island, Kingston, RI.

Seitzinger, S., Nixon, S.W. and Pilson, M.E.Q. in press, Denitrification and nitrous oxide production in a coastal marine ecosystem. Limnol. Oceanogr.

Seitzinger, S., Nixon, S., Pilson, M.E.Q., and Burke, S., 1980, Denitrification and N_2O production in near-shore marine sediments, Geoch. et Cosmochem. Acta., 44: 1853.

Smayda, T.J., 1973, The growth of Skeletonema costatum during a winter-spring bloom in Narragansett Bay, Rhode Island, Norw. J. Bot., 20: 219.

Smith, S.L., 1978, The role of zooplankton in the nitrogen dynamics of a shallow estuary, Est. and Coastal Mar. Sci., 7: 555.

Smith, W., Gibson, V.R., and Grassle, J.F., 1982, Replication in controlled marine systems: presenting the evidence, In: "Marine Mesocosms," Grice, G.D. and Reeve, M.R., eds. Springer-Verlag, NY.

Solorzano, L., 1969, Determination of ammonia in natural waters by the phenolhypochlorite method, Limnol. Oceanogr., 14: 799.

Steele, J.H., and Frost, B.W., 1977, The structure of plankton communities, Philos. Trans. Roy. Soc. Lon. B., 280: 485.

Strickland, J.D.H., 1967, Between beakers and bays, New Sci., 33: 276.

Takahashi, M., Koike, I., Iseki, K., Bienfang, P.K., and Hattori, A., 1982, Phytoplankton species' response to nutrient changes in experimental closures and coastal waters, In: "Marine Mesocosms," Grice, G.D. and Reeve, M.R., eds. Springer-Verlag, NY.

Vargo, G.A., 1979, The contribution of ammonia excreted by zooplankton to phytoplankton production in Narragansett Bay, J. Plankton Res., 1: 78.

Wakeham, S.G., Davis, A.C., and Goodwin, J.T., 1982, Biogeochemistry of volatile organic compounds in marine experimental ecosystems and the estuarine environment -- Initial results, In: "Marine Mesocosms, Grice, G.D. and Reeve, M.R., eds. Springer-Verlag, NY.

Watt, W.D., and Hayes, F.R., 1963, Tracer study of the phosphorus cycle in seawater, Limnol. Oceanogr., 8: 276.

Webb, K.L., 1981, Conceptual models and processes of nutrient cycling in estuaries, In: "Estuaries and Nutrients,", Nielson, B.J. and Cronin, L.E., eds., Humana Press, Clifton, NJ.

OCEANIC NUTRIENT CYCLES

Joel C. Goldman

Woods Hole Oceanographic Institution
Woods Hole, Massachusetts 02543, USA

INTRODUCTION

The perception that the euphotic zone of nutrient-impoverished oceanic waters is a steady state system has been based, to a large extent, on the choice of temporal and spatial scales upon which the pertinent biological and chemical measurements typically are made. For example, subsamples obtained from well-mixed, large volume samples (liters to tens of liters) commonly are used at sea both for a variety of analytical measurements and for long-term (tens of hours) bottle incubations to determine rates of primary production and nutrient turnover. The results of such measurements consistently demonstrate that nutrient concentrations in oligotrophic surface waters are below detection limits (Carpenter and McCarthy, 1975; McCarthy and Carpenter, 1979), and that both standing stocks of primary and secondary producers and rates of biological activity are uniformly low (Eppley *et al.*, 1973).

One of the major limitations in using discrete, large volume samples to represent water column biological activity is that the chosen temporal and spatial scales are based more on convenience than they are on the best representation of the real world of organisms (Wangersky, 1978). Lost in such an approach is both the patchiness in biomass that is known to exist on scales extending beyond meters and hours (Kelley, 1976; Platt and Denman, 1980) and the patchiness that may be important on scales of microns to millimeters and seconds to minutes (Allen, 1977; Harris, 1980). In addition, it is tenuous to presume that the rate processes measured during a small bottle incubation are accurate simulations of biological activity in the water column (Goldman *et al.*, 1981a).

The notion that the microenvironment surrounding a phytoplankter is an important determinant of nutrient availability was advanced several years ago by my colleague, J. J. McCarthy and myself (McCarthy and Goldman, 1979). We showed that, depending on the nutritional status of a cell, rapid uptake of nutrients far in excess of the quantity required for balanced growth occurred over short (5 min) intervals. We then hypothesized that rapid uptake was a nutritional strategy of phytoplankton residing in nutrient-poor waters to exploit microzones of higher nutrient concentrations that conceivably could result from grazer excretions and microbial aggregations; such microzones of nutrients, although possibly of crucial importance to the survival of individual cells, would never be detected with standard sampling procedures. An additional hypothesis we developed was that growth rates of marine phytoplankton in oligotrophic waters were close to the physiological maxima (Goldman et al., 1979).

In waters seemingly void of nutrients and consistently low in biomass, and in which nutrient regeneration is thought to be the major mode of nutrient supply (Eppley and Peterson, 1979; Harrison, 1980), rapid phytoplankton growth would have to be balanced by rapid herbivore grazing and, concomitantly, nutrient remineralization would also be rapid. Quantitatively, this relationship would be expressed as:

$$\mu \approx G \approx R \rightarrow \hat{\mu} \qquad \text{Equ. 1}$$

in which μ and $\hat{\mu}$ are, respectively, the specific and maximum specific growth rates of phytoplankton, G is the specific grazing rate of herbivores and R is the specific nutrient regeneration rate, all with units of reciprocal time (t^{-1}).

Equation (1) represents the steady state system often implied in descriptions of oligotrophic waters. Yet, from the perspective of an individual cell, steady state on the important biological scales may likely be the exception rather than the rule. Ignoring for the moment the practical limitations in the choice of spatial and temporal scales available to the experimentalist, the major questions then are, "do the summed events experienced by individual cells on the microscale equal the integrated events we typically measure at larger scales" or "are microscale responses of phytoplankton to their environment non-integratable?" In this paper I will address these questions and offer some nonconventional ideas on how nutrient cycling in oceanic waters may occur.

CONTEMPORARY CONCEPTS OF NUTRIENT DYNAMICS

The concept that nutrient regeneration by herbivores and bacteria is the major source of nutrients for phytoplankton growth in oceanic waters has been well-established for a long time, dating back to the classical studies of Cooper (1933) and Redfield et al. (1937). Current estimates are that about 80 to 90% of nutrients used by phytoplankton in the euphotic zone of oceanic waters is derived from regenerative sources(Eppley and Peterson, 1979; Harrison, 1980). For the most part, these estimates have been based either on comparisons of independent measurements of nutrient uptake by phytoplankton and nutrient excretion by zooplankton (Harrison, 1980), or by comparisons of NO_3^- and NH_4^+ uptake by phytoplankton in relation to total primary production (Eppley and Peterson, 1979). Whereas these types of measurements typically demonstrate tight coupling between uptake and excretion of nutrients, they invariably provide no clue as to how fast these processes are proceeding relative to the physiological maxima possible.

The Chemical Composition of Phytoplankton

One indication of how fast these processes may be occurring can be found in the circumstantial argument we recently advanced (Goldman et al., 1979). We hypothesized that, whereas the chemical composition of particulate matter in oceanic surface waters often is in the Redfield proportions of $C_{106}N_{16}P_1$, the cellular chemical composition of phytoplankton under laboratory growth conditions varies tremendously and approaches the Redfield ratio only when nitrogen or phosphorus begins to become nonlimiting and μ is close to $\hat{\mu}$. I (Goldman, 1980) extended this hypothesis further by showing that changes in the cellular chemical composition of phytoplankton occur as a function of relative growth rate (that is, as a function of the ratio $\mu:\hat{\mu}$) and not of the absolute values of μ (Figure 1A).

Thus cellular chemical composition data can be used in a diagnostic fashion to estimate the physiological state of a cell population, but they provide no information on the absolute magnitude of $\hat{\mu}$ which, not only is an intrinsic characteristic of a particular species, but also is influenced strongly by temperature and light.

Considerable confusion exists over this latter point. For example, Laws and Bannister (1980) claimed that the relationship between the cellular N:C ratio and μ of the diatom *Thalassiosira fluviatilis* grown in continuous culture was influenced as much by light limitation as it was by nitrogen or phosphorus limitation; they found that these two limiting conditions led to opposite effects on the cellular N:C ratio as μ varied with convergence towards the Redfield proportion when $\hat{\mu}$ was approached (Figure 2A). This conclusion was primarily based on the demonstration of large slopes in the curves of N:C versus $\mu:\hat{\mu}$ when either light or nutrients were limiting. However, when the data of Laws and Bannister are summarized using the more conventional C:N ratio as a function of $\mu:\hat{\mu}$ (Figure 2B), it is apparent that under light limitation the Redfield proportions are

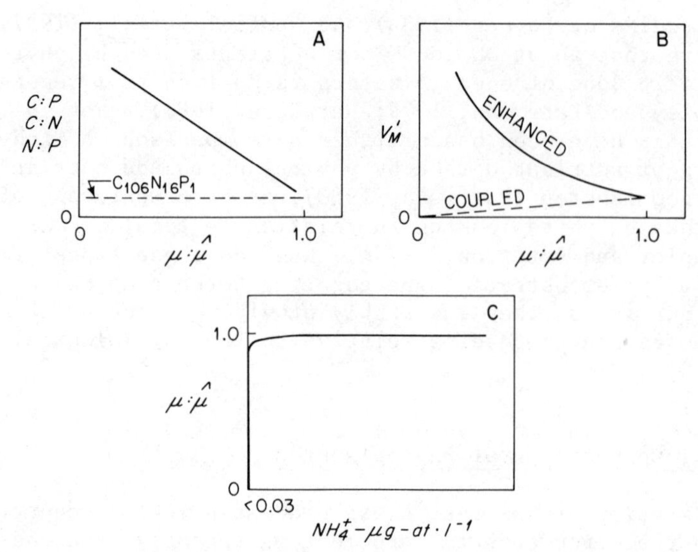

Fig. 1. Relationships between the physiological state ($\mu:\hat{\mu}$) of marine phytoplankton grown in laboratory continuous cultures under steady state conditions and different biochemical and kinetic parameters: A. Phytoplankton cellular chemical composition expressed as the ratios of C:N:P; B. Short-term nutrient uptake rate, V_M', in units of time^{-1}. Dashed line represents uptake rate coupled to growth rate and solid line indicates enhanced uptake rate exceeding growth rate; C. Effect of residual limiting nutrient. Limit of detection of limiting nutrients such as NH_4^+ is 0.03 μg atoms· liter^{-1}.

maintained at all growth rates, and that only nutrient limitation leads to the expected large changes in the cellular C:N ratio. Clearly, light availability controls the magnitude of $\hat{\mu}$, but it does not influence the relationship between cellular chemical composition and relative growth rate.

Another limitation of the above approach is that phytoplankton biomass in natural marine waters is only some fraction of total particulate material (Menzel, 1974). Thus the presence of a large detrital component in particulate matter limits, to some degree, the use of chemical composition data in estimating the relative growth rate of a natural population. However, as I stressed earlier (Goldman 1980), the Redfield proportion of $C_{106}N_{16}P_1$, represents the approximate lower limit in the C:N ratio (~6:1 to 7:1) that is possible in

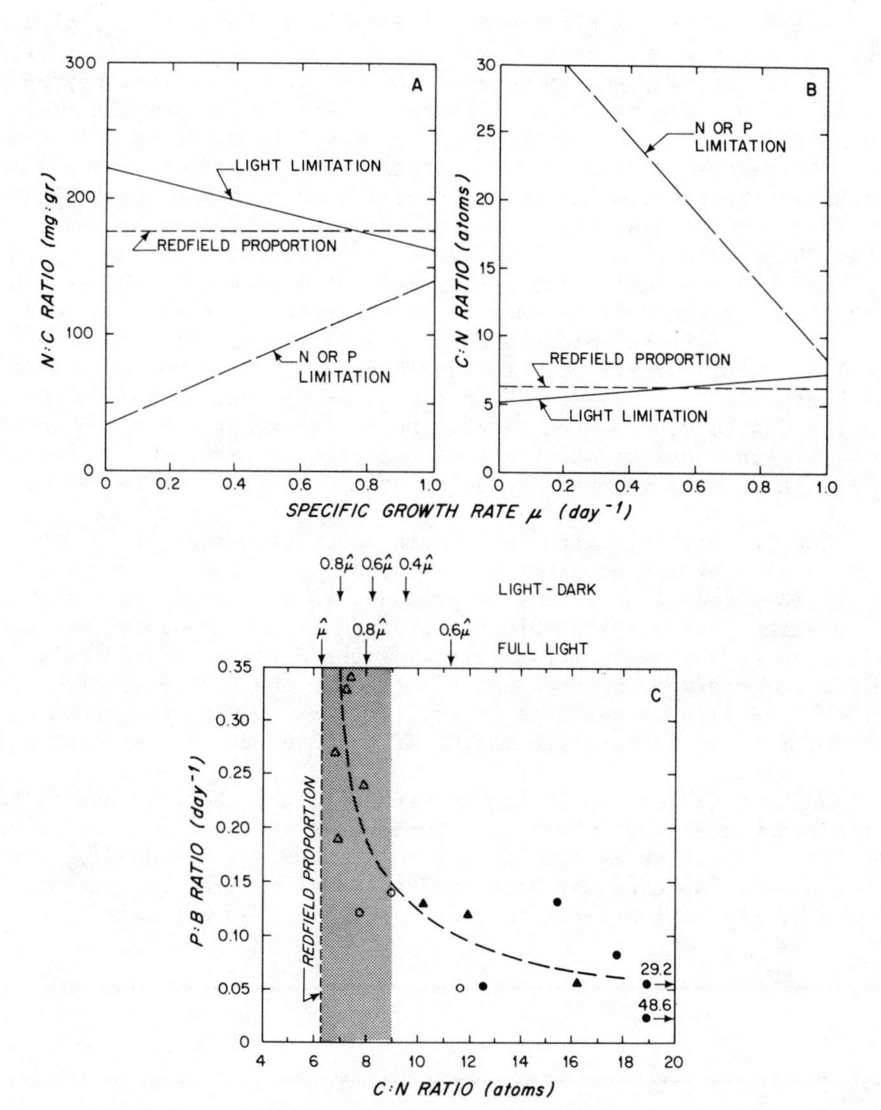

Fig. 2. Recasting of cellular and particulate chemical composition data from recent studies: A. Summary of data of Laws and Bannister (1980) for relationships between cellular N:C ratio and steady state growth rate of marine diatom Thalassiosira fluviatilis grown in continuous ulture under nutrient (N or P) or light limitation. The Redfield proportion of $C_{106}N_{16}P_1$ is represented by a dashed horizontal line; B. Same as in Figure 2A but with C:N ratio used in place of the N:C ratio; C. Recasting of the original N:C data of Sharp et al. (1980) to show the influence of the (continued)

phytoplankton from a biochemical standpoint. For example, although the C:N ratio in the common amino acids varies from 1.5:1 (arginine) to 9:1 (tyrosine), the amino acid composition of healthy phytoplankton is such that, typically, the C:N ratio of cellular protein (which constitutes a maximum of 50 to 60% of total biomass) is less than 4:1 (Gallager and Mann, 1981). Bacteria, in contrast, contain much higher proportions of protein and nucleotides than do phytoplankton so that the cellular C:N ratios of these microbes can be lower than 4:1 (Churchward et al., 1982; Esener et al., 1982). However, the bulk of the chemical composition data for marine particulate matter has been generated from analyses of filtered material collected on glass-fiber filters which have nominal pore sizes only down to ~0.8 to 1.0 μm. By comparison, Azam and Hodson (1977) have shown that 90% of bacterial activity occurs in the ≤1 μm size class and 70% occurs in the 0.4 to 0.6 μm size class. Hence, there probably has not been a significant weighting effect of bacterial biomass on the chemical composition data reported to date for marine particulate matter.

One is thus left with the simple conclusion that if the chemical composition of marine particulate matter is in the approximate Redfield proportions it must represent, to a large degree, the sum of healthy phytoplankton biomass growing at a reasonably fast rate relative to the maximum possible plus the detritus of recently living plankton that were metabolically active at the time of death. No other compelling hypothesis is available as to what the Redfield proportions in particulate matter of surface waters represent.

Jackson (1980), in trying to reconcile the above idea with his conclusion that phytoplankton growth rates in the oligotrophic central North Pacific must be low on an absolute basis (~0.1 day^{-1}), suggested that oceanic species may have evolved lower maximum growth rates than those of the typical weed species that have been maintained in culture

---

Fig. 2. (cont.)
C:N ratio of particulate organic material on the ratio of productivity ($grC \cdot l^{-1} \cdot day^{-1}$) to organic carbon biomass ($grC \cdot l^{-1}$) from four cruises to the Central North Pacific Ocean: △-Tasaday II cruise, o-Tasaday 1, ▲-Climax 8, ● -Southtow 13. Vertical dashed line represents C:N ratio equivalent to Redfield proportion (6.6), and shaded area represents regions of expected C:N ratio equivalent to ≥0.75 $\hat{\mu}$ for fully lighted N-limited continuous cultures and ≥0.50 $\hat{\mu}$ for similar cultures on 12 h light - 12 h dark cycles (Goldman, 1980). Broken line curve was drawn by inspection to indicate trend only.

collections for years and are commonly used by experimentalists. Jackson's argument is appealing, but it is not consistent with the recent results of Brand and Guillard (1981), who determined that no major differences exist between the growth rate potential of numerous recent isolates from the oligotrophic Sargasso Sea and that of the laboratory week counterparts. Whether these isolates have the same growth rate potential in natural waters as they do in the laboratory remains an unanswered question.

Sharp *et al.*(1980), in criticizing the chemical composition hypothesis, maintained that use of chemical composition data was far too simplistic to be of any real value in trying to gauge the growth rates of natural populations. They cited the conclusions of Ketchum *et al.* (1958) as a supportive claim of their position that oligotrophic waters had to be nutrient deficient and that there were far too few analyses of particulate chemical composition data for natural populations to draw generalizations regarding the nutritional status of marine waters.

Whereas in 1958 it was true that sufficient chemical composition data were lacking, today (1982) such data for numerous water bodies are available, and one is still struck by the remarkable consistency of those data in demonstrating proportions of carbon, nitrogen and phosphorus in particular matter from surface waters that approach the Redfield ratio. Many of those data I summarized earlier Goldman (1980). More recent chemical composition data of Bishop *et al.*(1978, 1980), Gordon *et al.*(1979), Morris *et al.*(1981) and Herbland and LeBouteiller (1981), to list a few, provide further evidence for the overall consistency of this chemical ratio.

The simplicity of the chemical composition hypothesis clearly does not detract from its applicability. Sharp *et al.* (1980), in fact, purported to demonstrate that phytoplankton growth rates in the central North Pacific were considerably lower than maximal by showing that the ratio of productivity to phytoplankton biomass (P:B), although a function of the composite data of N:C ratio for particulate material collected during four cruises, never reached the Redfield proportion. Yet, when recasting their P:B data as a function of the more conventional C:N ratio (by atoms), and including in this plot a comparative scale of the expected magnitude of $\mu:\hat{\mu}$ as a function of the cellular C:N ratio taken from summary plots for laboratory continuous cultures grown under both continuous light and 12 hr light-12 hr dark cycles(Goldman, 1980), a different interpretation is possible. As seen in Figure 2C, all of the C:N data from one cruise (Tasaday 11) represent phytoplankton growth rates that are greater than 60 to 80% of $\hat{\mu}$ and two of the three datum points from the Tasaday 1 cruise indicate that $\hat{\mu}:\mu:<0.5$ was possible. Interestingly, the Tasaday 11 data, suggesting high $\mu:\hat{\mu}$, and the Southtow 13 data, indicating very low $\mu:\hat{\mu}$, were obtained one year apart at the same station, but during similar months (February to

March). Thus one could easily conclude from this exercise that if the chemical composition hypothesis has validity, high relative growth rates in a severely oligotrophic water are indeed possible.

Rapid Nutrient Uptake

The ability of phytoplankton to take up nutrients at rates far in excess of the concentrations required for balanced growth over short (5 min) durations appeared, based on our results with the marine diatom *Thalassiosira pseudonana* (3H), (McCarthy and Goldman, 1979), to be restricted to cell populations growing at relatively low growth rates. In those experiments maximum NH_4^+ uptake rates (V_M') were found to be far in excess of the rates required for balanced growth ($V_M'=\mu$) at low relative growth rates, but this enhanced uptake capacity decreased as $\mu:\hat{\mu}$ increased until V_M' was equal to $\hat{\mu}$ when $\mu:\hat{\mu}$ approached unity (Figure 1B).

At the same time we (Goldman and McCarthy, 1978) showed that *T. pseudonana* (3H) had such a strong affinity for NH_4^+ that it was impossible to detect residual NH_4^+ in steady state continuous cultures at all growth rates below 0.87 $\hat{\mu}$. Thus, as seen in Figure 1C, measurement of undetectable NH_4^+ (<0.03 μg atoms $\cdot l^{-1}$) in natural waters would not provide information as to whether NH_4^+ was limiting growth or not.

The extention of these ideas to the hypotheses that phytoplankton residing in oligotrophic waters see and exploit undetectable micropatches of nutrients that arise from grazer excretions and other small scale nutrient sources while still maintaining high growth rates led to criticisms by Jackson (1980) and Williams and Muir (1981). They argued that diffusion away from point sources would be so rapid that enhanced uptake would be precluded. In particular, they took issue with the possibility that an individual phytoplankter could, over a 5 min period, exploit concentrations of NH_4^+ higher than the ambient in the microzone surrounding a copepod (the example of McCarthy and Goldman,(1979)).

In our original study (McCarthy and Goldman, 1979) we arbitrarily chose a 5 min incubation period to represent a "short" uptake response and we used the diatom *T. pseudonana* (3H) only out of convenience. More recently, however, my colleague, P. M. Glibert and I have shown that both natural phytoplankton populations (Glibert and Goldman, 1981) and a variety of species cultured in the laboratory (Goldman and Glibert, 1982), when pulsed with saturating NH_4^+ under nitrogen limiting conditions, take up this NH_4^+ over time in a non-linear fashion, that is, highest uptake rates are found over the shortest incubation periods. Consequently, depending on the species and over incubation periods much shorter than 5 min, we were able to measure NH_4^+ uptake rates in great excess of those possible with *T. pseudonana*

(3H). For example, the specific NH_4^+ uptake rate of the diatom Phaeodactylum tricornutum, measured over the first 15 sec of incubation (the shortest incubation technically feasible in our experiments with ^{15}N isotopes), was 0.095 min^{-1}. When compared to the alga's steady state growth rate of 0.00014 min^{-1}, this enhanced uptake rate is 684 times that required for balanced growth over a 15 sec period. Accordingly, under the above imposed conditions this organism would need to be exposed to saturating NH_4^+ for only 0.14% of a growth period to meet its nutritional requirements for nitrogen. Three other test species, Chaetoceros simplex, Dunaliella tertiolecta and Thalassiosira weissflogii, likewise displayed enhanced uptake rates for very short intervals; these rates, although not as great as those found in P. tricornutum, were still in great excess of that required for balanced growth (Table 1).

The trends of the time-course data for NH_4^+ uptake we presented earlier (Figure 6 in Goldman and Glibert (1982)) indicate that considerably greater uptake rates would have been possible over incubation periods less than 15 sec, were such measurements possible. The upper bound to very short-term uptake would be controlled by molecular diffusion of NH_4^+ from the aqueous environment to the cell surface. Koch (1971) eloquently demonstrated that when diffusion was the only limiting factor the maximum specific uptake rate by a spherical cell could be determined for any nutrient from the simple expression:

$$V_{D_o} = 180\, D_o R^{-2} \qquad \text{Equ 2}$$

in which V_{D_o} is the diffusion-controlled specific uptake rate (min^{-1}), D_o is the molecular diffusion coefficient ($cm^2\ sec^{-1}$), and R is the cell radius (cm). As seen in Figure 3, in dilute aqueous environments for which the diffusion coefficient is approximately 10^{-5} cm^2 sec^{-1} diffusion limits on the rate of nutrient uptake are inversely related to the diameter squared of a spherical cell. Over the range of phytoplankton cell sizes found in marine water (1 to tens of microns) V_{D_o} decreases by a couple of orders of magnitude; but even for large cells, the magnitude of V_{D_o} is tremendously higher than the highest values of V_M' that have been measured for the common nutrients such as nitrogen and phosphorus. For example, the highest rate of V_M' for NH_4^+ we have measured, 0.095 min^{-1} by the diatom P. tricornutum, is lower by a factor of at least 10^3 than V_{D_o} (Table 1). Even more dramatically, the tremendous phosphorus uptake rate (0.83 min^{-1}) reported for the giant oceanic dinoflagellate Pyrocystis noctiluca (Rivkin and Swift, 1982) (which has a diameter up to 450 μm (Swift et al. 1976)) is still less than V_{D_o} by a factor of four. Hence, as we pointed out earlier (Goldman and Glibert, 1982), the rapid uptake potential of marine phytoplankton we envision may be more a response to frequent, but very short-lived pulses of higher nutrient concentrations arising from a variety of discrete point

Table 1. Summary of short-term NH_4^+ uptake results from studies of marine phytoplankton grown in NH_4^+-limited continuous cultures together with maximum possible NH_4^+ uptake rates controlled by diffusion.

Species	Approximate Dimensions μm	Temperature °C	μ day^{-1}	$\mu:\hat{\mu}$	Incubation Duration sec	V_M min^{-1}	$V_M:\mu$	V_{Do} min^{-1}	$V_{Do}:V_M$
Phaeodactylum tricornutum	2 x 30	16	0.21	0.17	15	0.095[a]	684	8.0×10^{2}[c]	8.4×10^{3}
Chaetoceros simplex	5 x 6	16	0.21	0.17	15	0.014[a]	96	2.0×10^{4}[c]	1.4×10^{6}
Dunaliella tertiolecta	6 x 11	16	0.20	0.22	15	0.002[a]	14	6.0×10^{3}[c]	3.0×10^{6}
Thalassiosira weissflogii	10 x 14	16	0.58	0.60	60	0.004[a]	10	3.7×10^{3}[c]	1.9×10^{6}
		25	0.26	0.09	60	0.025[b]	138	3.7×10^{3}[c]	1.5×10^{5}
		25	0.80	0.27	60	0.018[b]	33	3.7×10^{3}[c]	2.1×10^{5}

[a] Data presented in Goldman and Glibert (1982)
[b] Unpublished data
[c] Determined using Equation (2). For each species the maximum dimension was used as the equivalent diameter. Such an approach is conservative in underestimating V_{D_o}.

sources, than it is to a 5 min exposure to nutrients in the wake of a swimming copepod. Under such conditions oligotrophic phytoplankton species may overcome diffusion-controlled barriers to the persistence of nutrient micropatches with relative ease. Physiological transport and/or cellular adsorptive processes (Goldman and Glibert, 1982; Wheeler *et al.*, 1982), rather than diffusion at the cell surface would set the upper limit to enhanced uptake potential.

In our earlier study (McCarthy and Goldman, 1979) it appeared that rapid uptake was limited to cells growing at a low relative growth rate (Figure 1B). Thus, the argument that phytoplankton could, on the one hand, grow rapidly (based on the chemical composition hypothesis described in Figure 1A), and still, on the other hand, maintain rapid uptake capabilities (based on results with *T. pseudonana*) obviously was inconsistent. This apparent dichotomy was addressed in our recent study on rapid NH_4^+ uptake, of which the data in Table 1 is included (Goldman and Glibert, 1982). As it turns out, *T. pseudonana*, when compared on a relative basis with the species listed in Table 1, has rather poor NH_4^+ uptake capabilities at all relative growth rates. In fact, *P. tricornutum*, *T. weissflogii*, and *C. simplex*, when first cultured at growth rates 80 to 95% of $\hat{\mu}$, are capable of sustaining NH_4^+ uptake rates over 5 min periods that are 4 to 8 times greater than required for balanced growth. Moreover, enhanced NH_4^+ uptake capability is a function not only of cell physiological state ($\mu:\hat{\mu}$), but also of the incubation period. Totally different perspectives thus emerge as to how uptake and growth are related depending on the temporal scale chosen (Figure 4). For example, although we do not as yet have any data demonstrating very rapid uptake at high $\mu:\hat{\mu}$ over very short (sec) incubation periods, we know from existing 5 min exposure data that for *P. tricornutum*, at least, $V_M':\hat{\mu}$ is about 2.5 times as great when $\mu:\hat{\mu}$ is 0.2 then it is when $\mu:\hat{\mu}$ is 0.85 (Goldman and Glibert, 1982). If this enhancement ratio holds as well for a 15 sec incubation period, then *P. tricornutum* when growing at $0.85\hat{\mu}$ should still be capable of taking up NH_4^+ at a rate almost 75 times the rate necessary for balanced growth. Clearly, over periods of seconds to minutes relative uptake rates, although decreasing with increasing $\mu:\hat{\mu}$, are high enough at growth rates approaching the maximum to allow cells to both grow fast and still exploit pulses of nutrients. Only for longer exposure periods extending to hours would coupling between uptake and growth begin to be realized.

STEADY STATE AND NUTRIENT UPTAKE

The steady state continuous culture has the unique feature that growth rate is not dependent on the flux of limiting nutrient dispensed to the culture vessel, but rather is equal to the dilution rate, or percentage displacement of culture volume per unit time (Herbert *et al.*, 1956). Thus it is possible to achieve a range of

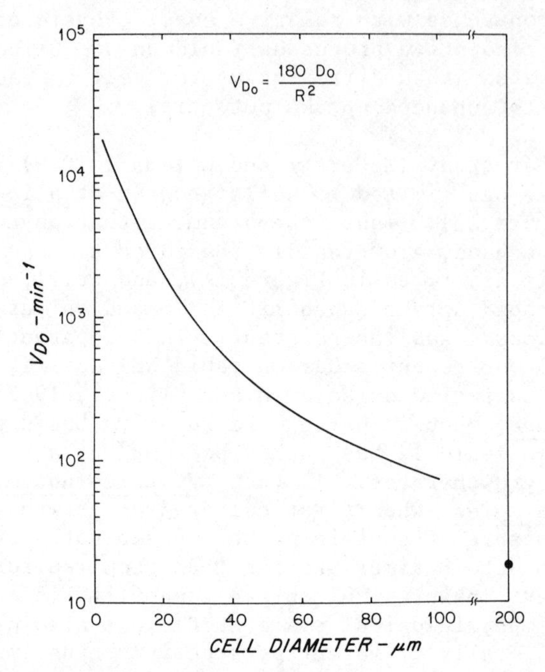

Fig. 3. Maximum possible nutrient uptake rate, V_{Do}, possible for a spherical phytoplankton cell as a function of cell diameter when diffusion of nutrient towards the cell surface is the only limiting factor. Curve is drawn according to Equation (2), developed by Koch (1971).

steady state biomass concentrations at any growth rate simply by setting the flow rate on the medium supply pump and varying the concentrations of limiting nutrient in the medium.

Eppley et al. (1973) were the first to draw the analogy between the steady state continuous culture and a segment of the open ocean, in their case the central gyre of the North Pacific Ocean. In each system both biomass and ambient nutrients remain invariant over time. Nutrient input and biomass output are regulated by mechanical pumping in the continuous culture and, analogously, in a given portion of the open ocean these processes are controlled by the combination of grazing and nutrient regeneration (Figure 5). The analogy is appealing because it invokes in quantitative terms the notion that for surface waters, in which there is little seasonal variation in the depth of the mixed layer so that external inputs of nutrients are minimal, it is possible to have rapid growth rates of autotrophs

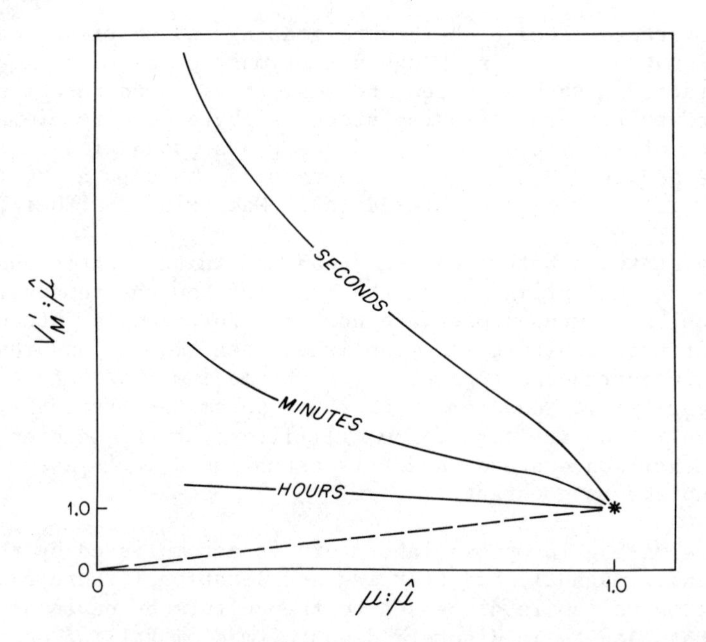

Fig. 4. Conceptual scheme of how time of exposure influences the relationship between transient specific nutrient uptake rate and steady state growth rate of a phytoplankton culture. Dashed line represents coupled nutrient uptake and growth rates, culminating in $V'_M:\hat{\mu} = \hat{\mu}:\mu = 1$ when the maximum growth rate $\hat{\mu}$ is attained.

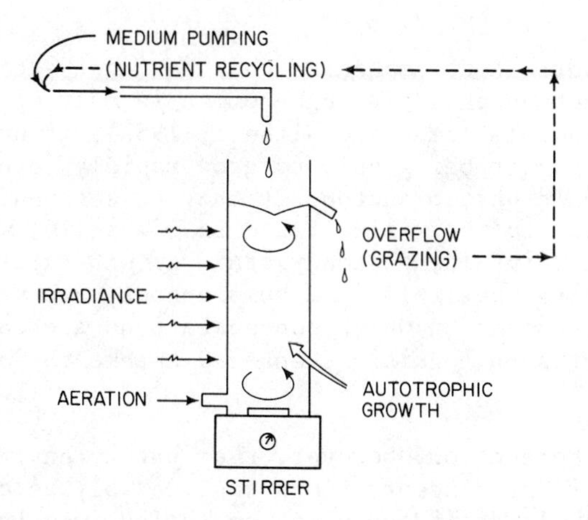

Fig. 5. Schematic diagram of how a continuous culture can represent a segment of the open ocean.

and equally rapid rates of herbivore grazing and nutrient regeneration (Equation 1), even though net primary production might be low. In fact, in such a system, because there is no relationship between growth rate and standing stock of phytoplankton biomass, the growth rate of individual cells and the gross production rate of the entire population need not be connected. This is a critical point that is not well appreciated (c.f. Maestrini and Bonin, 1981).

Steady state in both the open ocean and in the continuous culture is defined primarily on the basis of how the temporal and spatial scales of measurement are chosen. For example, steady state in the continuous culture is established when biomass and residual nutrient concentrations together with the medium flow rate are invarient over time. Measurement of these parameters typically involves sampling of at least a few milliliters at frequencies of hours to days. Once steady state is established it is assumed that nutrient uptake is constant.

Culture mixing in my own laboratory is accomplished by the combination of magnetic bar stirring and aeration (Figure 5). At the same time medium is dispensed to the culture by peristaltic pumping that results in discrete drops of medium falling on the culture surface at intervals of about 5 to 50 sec, depending on the desired growth rate. The time required for complete mixing of an individual drop of medium, based on dye studies (Figure 6), is about 25 sec. Hence, those cells nearest the culture surface at the time a drop of medium strikes the culture will see and exploit with enhanced uptake a microzone in which the nutrient concentration near the surface is much greater than the overall ambient concentration (the latter which most likely is undetectable). From this perspective, an individual phytoplankter clearly is never at steady state.

It is important to emphasize that having the capacity for enhanced nutrient uptake, although common in many of the phytoplankton species examined (Goldman and Glibert, 1982), is not a precondition for a phytoplankter being able to grow rapidly in nutrient poor waters. Oceanic phytoplankton, if they possess nutrient uptake systems similar to that illustrated in Figure 1C, would have no difficulty in maintaining steady state growth rates that approach the maximum rates possible in a homogeneous, but yet undetectable nutrient environment in which concentrations approached saturating levels. Under such conditions enhaced uptake obviously would be unnecessary.

It is my contention, however, that patchy environments that encompass nutrient concentrations considerably below to greater than saturating levels do indeed exist on scales important to phytoplankton. Consequently, enhanced uptake capability among successful

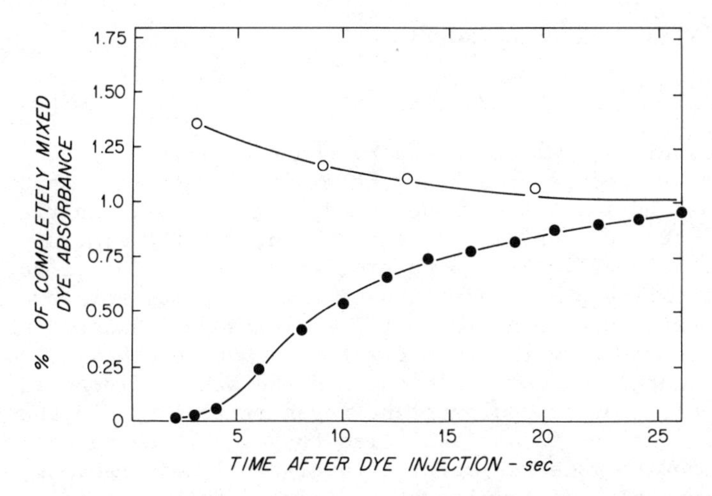

Fig. 6. Mixing time in 0.5 liter continuous culture of the type described in Goldman et al. (1981b). Data represent percentage of completely mixed dye absorbance from samples taken from the surface (o) and bottom (•) of the culture vessel after a drop of Rhodomine B dye was added at the surface.

oligotrophic species may simply be an evolutionary adaptation to life in an otherwise nutrient-deprived environment. Although all of our studies dealing with enhanced NH_4^+ uptake have been performed with the common laboratory weed species, there is some evidence that rapid NH_4^+ uptake potential, at least in the cosmopolitan species T. pseudonana is significantly more well developed in an oligotrophic clone (13-1) than it is in a coastal clone (3H) (McCarthy and Goldman, 1979; Goldman and Glibert, 1982). One would suspect that true oligotrophic species would have evolved the necessary biochemical machinery, including rapid nutrient uptake capability, to cope with frequent nutrient impoverishment (Koch, 1979; Poindexter, 1981). In this regard, Turpin et al. (1981) made the important point that the ability to respond to a pulse of nutrients might be of crucial importance to a phytoplankton cell in overcoming very low-level threshold concentrations of nutrients, below which nutrient uptake is precluded. A similar argument could be made that enhanced uptake potential is an important way for cells to overcome diffusion limitations on uptake at very low nutrient concentrations (Gavis, 1976). Lehman and Scavia (1982 a,b) more recently demonstrated, both experimentally and theoretically, that micropatches of nutrients arising from zooplankton excretions can be exploited successfully by those phytoplankton cells in closest proximity to individual zooplankters; competitive stresses amongst phytoplankton species thus would be far different in patchy versus homogeneous nutrient environments.

NUTRIENT CYCLING IN OLIGOTROPHIC WATERS

Importance of Microplankton Processes

The oceanic system described by Equation 1 is akin to a wheel spinning at a speed approaching the maximum possible. Abiotic factors such as light and temperature, together with intrinsic characteristics of resident phytoplankton, set the limits on the speed of this wheel. From a physiological standpoint alone, it is clear that marine phytoplankton have the potential to grow rapidly under the perceived conditions of low biomass and undetectable nutrients common to oligotrophic waters. Whether they do so or not depends primarily on how efficient are the other components of the wheel, namely grazing and nutrient regeneration. Given the technological limitations in our abilities to perform the necessary experiments and to make the critical analytical measurements, there is, unfortunately, no simple way to address this problem.

During recent years, however, some revolutionary insights have been made into the functioning of oceanic food chains. Among these new ideas is the notion that microheterotrophic processes involving protozoa and bacteria, rather than direct macrozooplankton grazing, regulate the flow of energy and nutrients beyond the autotroph step (Pomeroy, 1974; Sieburth et al., 1978; Williams, 1981). Pertinent to this argument is the growing realization that very small (<10μm) organisms, both autotrophic and heterotrophic, make up a very significant fraction of the plankton in oceanic waters (Pomeroy and Johannes, 1966; 1968; Berman, 1975; Sieburth et al., 1978; Throndsen, 1979; Herbland and LeBouteiller, 1981; Beers et al., 1982; Booth et al., 1982). Difficulties in preserving and counting these populations, which often are dominated by very small (<5 μm) and fragile flagellated types, have been, and still are, a major problem in assessing their true significance (Sorokin, 1981; Parker and Tranter, 1981; Beers et al., 1982; Booth et al., 1982). The recent discovery that marine cyanobacteria (1 to 2 μm in size) are ubiquitous in the world's oceans, sometimes in significant numbers, adds still another dimension to the importance of the very small size class of microorganisms in pelagic food chains (Waterbury et al., 1979; Johnson and Sieburth, 1979).

Nutrient Regeneration by Microplankton

Concomitant with the recent findings that very small autotrophs and heterotrophs are present in oceanic waters in numbers far greater than previously thought, is the discovery, through size fractionation studies with the ^{15}N isotope dilution technique, that a very large percentage of the ammonium regenerated by heterotrophs occurs in this same small size class (Harrison, 1978; Caperon et al., 1979; Glibert, 1982). Moreover as shown by Glibert (1982), rates of NH_4^+

uptake and regeneration by the less than 10 μm size class in oceanic as well as in coastal waters are in virtual balance, at least on a diurnal basis. This important finding provides the first quantitative evidence for the very tight coupling in oceanic waters between autotrophic and heterotrophic organisms depicted by Equation 1.

There is, however, some uncertainty in extending the results of short-term (hrs) incubation studies such as those of Glibert (1982) to a generalization that nutrient uptake and regeneration rates are always in balance. Intuitively, it is difficult to envision such a highly efficient system in which an inorganic nutrient like NH_4^+ is first synthesized into the cellular protein of a phytoplankter, the whole photoplankton cell is then ingested by a predator of the same relative size as the prey, and finally the ingested protein is released back to the aqueous environment in the original form, NH_4^+, all at comparable rates. Obviously, such a process must operate at less than 100% efficiency due to a variety of possible losses, (e.g. incorporation into predator biomass, excretion of dissolved organic nitrogen and urea, incomplete digestion of autotroph biomass, and predation by macrozooplankton).

Virtually nothing, however, is known of the interactions between very small autotrophs and heterotrophs, particularly the relative importance of bacteria and protozoa in recycling nutrients. To date, what meager data that are available have been obtained primarily from studies dealing with the decomposition of detritus by model microbial systems consisting of bacteria and ciliates (Johannes, 1965; Barsdate *et al*., 1974; Fenchel and Harrison, 1976). Although some controversy exists as to which group of these organisms plays the key role in nutrient regeneration (Johannes, 1965; Barsdate *et al*., 1974), the most contemporary ideas are that the bulk of nutrient regeneration is performed by bacteria and that the ciliates, being bacterivorous, primarily serve to control the population of bacteria (Fenchel and Harrison, 1976). Sherr *et al*. (1982), in a recent study on freshwater decomposition of dead dinoflagellate cells, found that bacterial breakdown of carbohydrate material in the presence of microflagellates was enhanced relative to the action of bacteria alone.

Unfortunately, the above results are not directly applicable in addressing the question of how nutrient uptake and regeneration rates by organisms $\leq$ 10 μm can be equal. Such a system is not detrital so that direct grazing on actively growing autotrophs through phagotrophic predation by heterotrophs smaller than ciliates must be occurring. Although phagotrophic predation on bacteria by microflagellates smaller than 10 μm is well documented (Haas and Webb, 1979; Sieburth, 1979; Fenchel, 1982 a,b),only some qualitative evidence is available to indicate that these small predators also are capable of grazing on equally small autotrophs (Raymont and Adams, 1958; Goldman and Stanley, 1974; Sieburth, 1979; Kimor, 1981; Haas, unpubl.).

Recently, Fenchel (1982b) showed that clearance rates of filter-feeding microflagellates decreased as the square of particle food diameter, and that, in general, such organisms when feeding on bacteria had specific clearance rates 10 to 50 times greater than comparable rates in ciliates. These findings led Fenchel (1982b) to conclude that microflagellates are the important bacterial consumers in oceanic food chains. Using similar reasoning as did Fenchel, one could argue that these voracious predators select their food primarily on the basis of particle size and are thus relatively nondiscriminent in feeding on both bacteria and small autotrophs. This idea is consistent with the spinning wheel hypothesis because, as shown by Fenchel (1982b), maximum grazing and growth rates of marine microflagellates are considerably greater than growth rates of phytoplankton (e.g. $\hat{\mu}$ of 6 day^{-1} for the heterotrophic microflagellate Actinomonas mirabilis at 20°C (Fenchel, 1982b) compared to a typical $\hat{\mu}$ value of <2 day^{-1} for the fastest growing phytoplankton under comparable conditions (Eppley, 1972)). Thus, the rate limiting step in Equation 1 would not be G (the grazing rate). Equally consistent with the above ideas are the observations of Pomeroy and Johannes (1966: 1968) that flagellates <10 μm were responsible for the bulk of plankton respiration in a variety of oceanic waters.

Thus even though a strong circumstantial argument can be made for a tightly knit microplankton food chain involving autotrophs, bacteria, and microflagellates, nothing is known of the possible ways by which nutrients are processed in such a system. At present we can only guess that microflagellate grazing of both autotrophs and bacteria, together with bacterial hydrolysis of macromolecular compounds leads to extraordinarily efficient nutrient cycling, consistent with the results of Glibert (1982) and others. Small losses of nutrients could readily be balanced by inputs of external nutrients (Eppley and Peterson, 1979), so as not to disturb the appearance of steady state.

Aggregation and Microenvironments

If we accept the spinning wheel hypothesis as a reasonable description of oceanic food chain dynamics, then we must consider the possible environmental conditions under which such a rapid and continuous sequence of events could occur. At one end of the spectrum is a generally prevailing view of oceanic life that microbes are randomly dispersed and truly planktonic. In such a system autotrophs would be forced to depend on undectably low nutrient concentrations that were homogenously distributed in time and space through molecular diffusion (Jackson, 1980); similarly, grazers (microflagellates) would rely on chance encounters with their prey (bacteria and autotrophs). Intuitively, such a bleak, desert-like environment would not seem capable of supporting rapidly growing and nutritionally sufficient microbial populations, given the sparse numbers of microorganisms generally believed to be present in the euphotic zone of nutrient impoverished waters.

Another view of oceanic microbial life, first advanced by Pomeroy and Johannes (1968) and Riley (1970), is that microscopic-size, flocculated masses of organic material, upon which small autotrophs, protozoa and bacteria colonize, are ever present in marine waters. These amorphous aggregates, first described by Riley (1963), range in size from a few microns to several millimeters, although their median range is from 25 to 50 μm (Riley, 1970). Pomeroy and Johannes (1968) attributed most of the plankton respiration in oceanic waters to be occurring within these aggregates by colonized bacteria and microflagellates.

The formation of these microscopic organic aggregates, although a subject of some controversy (summarized in Riley, 1970), originally was shown by Baylor *et al.*, (1962) and Sutcliffe *et al.* (1963) to occur by adsorption of dissolved organic matter onto seawater bubbles followed by bubble dissolution. Johnson (1976) and Johnson and Cooke (1980) more recently confirmed this mode of organic aggregate formation. Research on other types of known microscopic aggregates, ranging from the remains of decomposing phytoplankton to inorganic particles coated with organics, has been summarized by Wangersky (1977).

More recently, with the advent of SCUBA techniques, great interest has been focused on macroscopic organic aggregates, commonly referred to as "marine snow", that are present throughout the water column (Silver *et al.*, 1978; Trent *et al.*, 1978; Silver and Alldredge, 1981). These aggregates, formed from the mucus structures and remains of a variety of gelatinous zooplankton, vary in size from ∿0.5 mm to several meters and have the common characteristics of being highly amorphous and extremely fragile (Silver *et al.*, 1978; Trent *et al.*, 1978). They are sites of colonization by a variety of microbes in both surface waters (Silver *et al.*1978; Caron *et al.*, in press) and in bathypelagic regions (Silver and Alldredge, 1981). Concentrations of autotrophs, protozoa and nutrients within the aggregates are up to orders of magnitude greater than in the surrounding water (Silver *et al.*, 1978; Trent *et al.*, 1978; Shanks and Trent, 1979; Caron *et al.*, 1982), suggesting that they are self-contained microenvironments of very intense microbial activity. Recent evidence for enhanced primary productivity within marine snow particles relative to the surrounding water column supports this latter conclusion (Alldredge and Cox, 1982).

Other types of larger-scale biological aggregations are commonly found in pelagic waters, ranging from floating mats of the diatom Rhizosolenia (Carpenter et al., 1977; Alldredge and Silver, 1982) and the blue green alga Trichodesmium (Carpenter and Price, 1977), to surface micro-layers (Harvey, 1966; Norkrans, 1980). The presence of these widely-varied aggregations leads to the speculation that the formation of patchy microenvironments within nutrient impoverished waters is not entirely by chance; rather, it may reflect evolutionary adaptations by successful organisms for survival in such a harsh environment. By living in close proximity to one another, autotrophs and heterotrophs can concentrate and store nutrients from the surrounding water and provide each other with the necessary ingredients for life.

Microbial Adhesion

Microbial adhesion to surfaces is a well-studied, if not satisfactorily understood phenomenon (Harris and Mitchell, 1973; Marshall, 1976; Rutter, 1980; Ellwood et al., 1982). Overt manifestations of this general microbial characteristic occur either through direct attachment to solid surfaces or flocculation of individual organisms into larger aggregates, or both, in diverse aquatic environments, e.g. the human mouth, on ship bottoms and submerged structures, at sediment-water interfaces, and in biological wastewater treatment systems. Most is known of bacterial aggregation and adhesion, but probably many of the principles of colloidal chemistry and physics, adsorption phenomena, and polymer chemistry governing bacterial-surface interactions (Ellwood et al., 1982) are common to autotroph and protozoan adhesion as well.

There is strong evidence that among bacteria, at least, attachment to surfaces is more pronounced in cultures maintained under nutrient limitation than when nutrients are abundant, (Jannasch, 1979; Dawson et al., 1981; Kjelleberg et al., 1982; Ellwood et al., 1982). The rationale for this phenomenon is that attachment to a surface provides a way for a bacterium to have access to a relatively nutrient-enriched microenvironment and thus avoid starvation in the aqueous void (Kjelleberg et al.,1982). Much controversy exists as to whether marine bacteria in nature are free-living or attached (Wiebe and Pomeroy, 1972; Jannasch and Pritchard, 1972; Seki, 1972; Paerl, 1975; Azam and Hodson, 1977). Some of these differences might be attributed to the possibility that attached bacteria, although frequently found to be in far lower concentrations than free-living forms, are considerably larger and more active biologically (Jannasch and Pritchard, 1972; Wangersky, 1977; Linley and Field, 1982). Another possible cause of this discrepancy might be due to problems in experimentally differentiating between "free-living" and "particle-bound" organisms. For example, the extraordinarily fragile nature of marine snow aggregations (Silver et al.,

1978) probably precludes accurate estimates of firmly attached versus loosely bound or particle-associated organisms. In this regard, Azam (in press;this volume) recently has suggested that bacteria, while not bound to particles, cluster around sustained sources of dissolved organic matter such as phytoplankton cells or aggregates.

Although little is known about the adhesive properties of phytoplankton and protozoa, there is ample evidence that heterotrophic microflagellates in marine waters are predominently attached to surfaces, presumably to allow for more efficient filtering of water in order to enhance feeding (Fenchel 1982a).

Phytoplankton also adhere to surfaces under a variety of conditions. I observed a good illustration of this phenomenon in an earlier, unpublished study of mine (Figure 7). A culture of the diatom T. pseudonana (3H), first grown on batch, was switched to a continuous mode at a dilution rate slightly greater than the alga's maximum growth rate which was 1.7 day^{-1} under the imposed temperature and light intensity. Initially there was a decrease in cell number over time identical to the physical washout rate (Figure 7A). Between days 4 and 9 a basal level of biomass persisted in the culture, as seen by the more sensitive measurement of relative fluorescence (Figure 7B). Concurrent with the first visual evidence of algal growth on the culture sidewalls at day 9 was a dramatic increase in phytoplankton biomass which continued through day 12 even though the dilution rate still was $>\hat{\mu}$. At day 12 the entire culture volume was switched to an identical, but clean vessel. Rapid biomass washout followed as before until at day 18 wall growth together with an increase in culture biomass again was observed. Clearly, the effect of increasing wall growth during washout was to establish a reservoir of phytoplankton cells growing at $\hat{\mu}$ that were continuously released into the culture. Similar wall growth effects have been documented for bacterial continuous cultures (Larsen and Dimmick, 1964; Topiwala and Hamer, 1971). Whether or not such stickiness is an attempt to avoid physical washout from the culture is highly speculative. The results do serve, however, to demonstrate that phytoplankton, as well as other microbes, can maintain high growth rates while attached to surfaces.

Microenvironments and the Spinning Wheel

One simple way to reconcile the seemingly incompatible concept that phytoplankton may be growing fast in oligotrophic waters with observations that indicate these waters are incapable of supporting such rapid growth is to suggest that small amorphous aggregates of organic matter in the water column represent self-contained microhabitants in which all the processes described by Equation 1 are carried out. Each aggregate then is a floating "oasis" in the

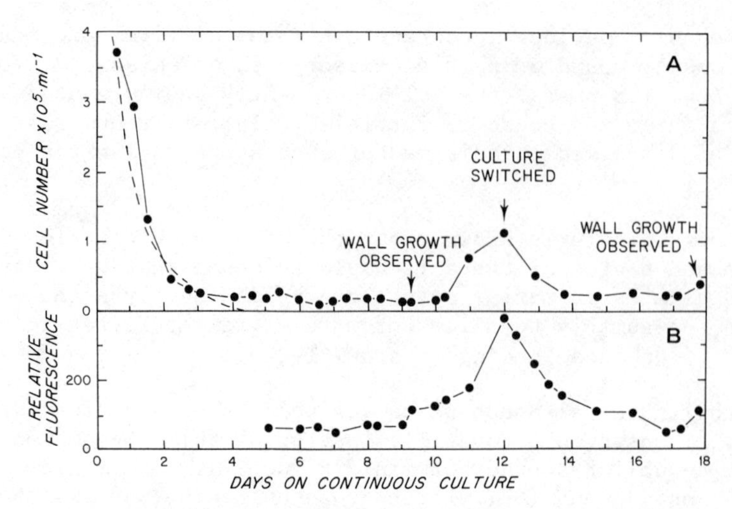

Fig. 7. Demonstration of wall growth effects in continuous culture of Thalassiosira pseudonana (3H) during washout (dilution rate $>\hat{\mu} = 1.7\ day^{-1}$). A - cell number. Dashed line represents expected washout from physical dilution; B - relative fluorescence.

desert, serving as a life-support system for its resident population of autotrophs and heterotrophs. The ability of microbes to adhere or stay within close proximity to these aggregates is an adaptive mechanism that allows them to cope most advantageously in an otherwise nutrient-poor environment.

The preceeding idea is not entirely novel - others deserve credit for its inception (Pomeroy and Johannes, 1968; Riley, 1970; Wangersky, 1977; 1978). My contribution is to suggest that within small aggregates of the type described by Riley (1970), the wheel may be spinning at a very rapid rate. Conceptually, I suggest that within these microscopic aggregates (which, as described by Riley (1970), form primarily by way of physical-chemical processes involving dissolved organics and surface turbulence) colonization by bacteria, autotrophs and microflagellates quickly develops. The dominant size class of these organisms would be ≤ 5 μm. For aggregates on the order of 25 to 50 μm in length the available surface area and internal space would probably limit the size of the resident microbial population.

The wheel, representing the turning over of biomass and nutrients (Figure 8), probably would start slowly as inorganic nutrients were absorbed and concentrated in a newly formed aggregate from the surrounding aqueous environment and also generated through bacterial oxidation of organics within the aggregate. The rapidly spinning wheel within a mature microhabitat obviously would be driven by photosynthetic incorporation of both radiant energy (PAR) and nutrient

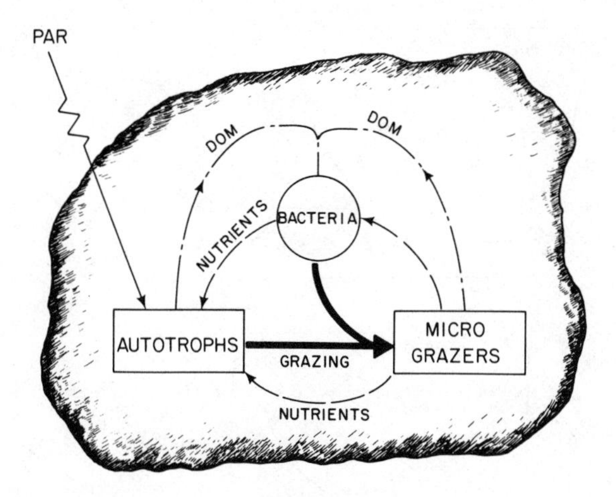

Fig. 8. Conceptual scheme of a microbial food chain within a discrete organic aggregate. All aggregate organisms would be $\leq$ 5 µm in size.

by autotrophs growing close to $\hat{\mu}$. Microflagellates would graze on both autotrophs and bacteria at rates (G) comparable to $\hat{\mu}$ and nutrients would be regenerated in some fashion by the combined activities of the bacteria and microflagellates, also at rates (R) approximating $\hat{\mu}$. Excreted organics from both the autotrophs and the microflagellates would fuel bacteria growth.

Although the above description suggests that the wheel is spinning at a constant rate (steady state), the system probably is far from steady state on temporal scales important to individual organisms. A possible scenario in which the autotrophs and microflagellate grazers interact is shown in Figure 9. Oscillations in biomass of both autotroph and grazer (Figure 9A) and dissolved nutrients (Figure 9B) most likely would occur in a classical predator-prey pattern. Of particular interest in this type of system is the possibility that, whereas autotroph biomass would be regulated by microflagellate grazing, a rapid crash in microflagellate population would follow through some form of cell lysis after the autotroph population was reduced to a threshold feeding level. Such a mode of grazer control is purely speculative, but it is consistent with the few observations that these species simply disappear when stressed (Haas, unpubl.; Parker and Tranter, 1981). Following the crash of microflagellates, there would be a lag period (lag a in Figure 9) before the nutrient pool within the aggregate reached a peak level. Nutrient uptake by the remaining autotrophs

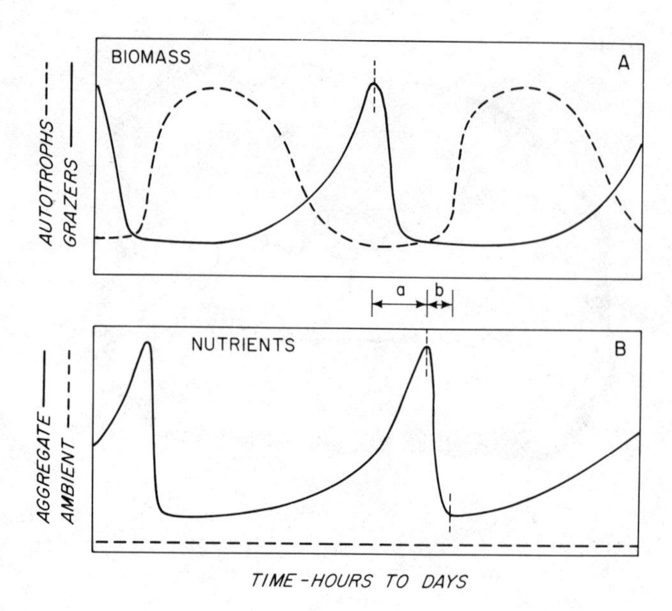

Fig. 9. Conceptual scheme of the cycling of biomass (A) and nutrients (B) that might occur within a discrete aggregate food chain shown in Figure 8.

then would be very rapid and uncoupled from growth (lag b in Figure 9). Growth of the autotrophs would follow and the cycle would be repeated.

Adhesion of participating organisms to surfaces complicates tremendously the expected predator-prey relationships in such a system (Curds, 1971; van den Ende, 1973). For example, based on his laboratory predator-prey studies, van den Ende (1973) suggested that adhesion of bacteria to the sidewalls of continuous cultures was a means of avoiding protozoan predation. For this and other reasons I've made no attempt to quantify the amplitudes and frequencies of these cycles, other than to suggest that hrs to days may be the appropriate temporal scale (Figure 9). My perception that the crash of the microflagellate would occur with dramatic speed (on the order of hours) is based on the unpublished results of Dr. L. Haas who, while carrying out research in my laboratory, observed the complete disappearance of a population of the microflagellate *Paraphysomonas vestita* within one day after it had decimated an outdoor mass culture of the diatom *Phaeodactylum tricornutum* by phagotrophic predation.

Regardless of how the autotrophs and heterotrophs interact within these aggregates, the important point, as stated by Wangersky (1977), is that nutrient regeneration in the open ocean is most efficient where surfaces exist. The 25 to 50 μm size aggregate can thus be thought of as a basic biological unit, providing the necessary surfaces for successful microbial life in an oligotrophic environment. Spontaneous formation of larger aggregates through physical-chemical processes would lead to even greater surface areas for microbial colonization. The abundance and distribution of these aggregates in oligotrophic waters is not well documented (Riley, 1970; Sheldon et al., 1972). Gordon (1970), using histological staining techniques, was able to distinguish amorphous and very fragile aggregates, which exhibited strong carbohydrate staining, from scale-like particles (flakes) which when stained appeared to be highly proteinaceous and not fragile. Unfortunately, most of our knowledge of these microscopic aggregates is based on filtration methods for collection. Because of their extraordinarily fragile nature, the integrity of most of these aggregates is probably destroyed with current methods of collection and observation. The actual surfaces upon which organisms attach are probably no more than a matrix of high molecular weight organic compounds, perhaps colloidal in nature (Cauwet, 1978), which have a viscosity much greater than seawater and thus act like glue to concentrate organic exudates, nutrients and microbes together into a very complete and efficient, but also fragile microbial food chain.

CONCLUSIONS

The attractiveness of the aggregate-spinning wheel concept is that it reconciles some of the conflicting contemporary arguments on biological activity in oligotrophic waters, as based on current methodologies involving large volume samples. For example:

1. It allows for low measured biomass and nutrient concentrations (aggregate dilution).
2. It allows for low measured primary production (aggregate dilution).
3. It allows for high expected growth rates from chemical composition data (aggregate concentration).
4. It provides an important role for bacteria in converting algal exudate back into particle organic material.
5. It eliminates diffusion restrictions on rapid nutrient uptake by autotrophs and provides an important role for this process.
6. It is consistent with contemporary data showing that <10μm size class of organisms are important both in numbers and biological activity.
7. It is consistent with recent data demonstrating coupled nutrient uptake and regeneration rates by <10 μm size class of organisms.

The major restriction of the hypothesis is that it currently is non-testable. At present we do not have the necessary methodologies for quantifying the concentrations and distributions of these aggregates and their resident microbial populations, and for measuring the important biological rate reactions within individual aggregates. Newly developed holographic techniques (Carder, 1979; Carder *et al*., 1982) may solve the first part of the problem, but performing rate measurements of primary production, respiration, and nutrient regeneration of microscopic-size particles poses seemingly insurmountable obstacles.

The aggregate-spinning wheel hypothesis, if not currently testable, is at least consistent with contemporary ideas on oceanic food chain dynamics which are rapidly replacing the classical phytoplankton-zooplankton-fish concept (Figure 10A). An appealing new approach, described in detail by Silver and Alldredge (1981) and summarized in Figure 10B, is that aggregations of small mibrobes, rather than individual organisms, not only are a major food source for macrozooplankton in oceanic waters, but also play an important role in the transport of organic matter out of the euphotic zone. The microaggregates I envision probably are close to being neutrally buoyant, and as such, would not be important sources of particulate matter that might be transported out of the euphotic zone through sinking. In this respect they would be distinctly different from the larger "marine snow" particles. Scavenging by larger particles could be an effective way to concentrate the smaller aggregates, however. The main point, thus, is that microaggregates might have a much longer residence time in the euphotic zone than their larger "marine snow" counterparts, perhaps on the order of days to weeks. Given my perceptions of "short" duration microbial cycles (Figure 9) and "long" residence periods of these small aggregates, it should be possible for active microbial activity to develop and persist during the life of each aggregate.

In conclusion, I return to the problem of whether or not microscale events experienced by individual cells are adequately integrated by the larger-scale measurements we routinely perform. For the continuous culture system I described previously, in which individual cells possibly see tremendous variations in nutrient concentrations during the few seconds between the falling of drops of medium into the culture, such events are, for all practical purposes. adequately integrated on the larger temporal and spatial scales commonly used to define steady state conditions. We know this because, by our definition of steady state (i.e. no changes in culture biomass over periods of hours to days), the average specific growth rate of the population is equal to the dilution rate (Spicer, 1955). The procedure works only because the system is vigorously mixed and all cells are homogeneously dispersed within the culture confines. In addition, a typical sample volume of tens of ml represents a relatively large percentage of the total volume of a 0.5 to 2 liter continuous culture, and as such, provides a good description of that system because all micro events experienced by individual cells are averaged out. If, on the other hand, some of

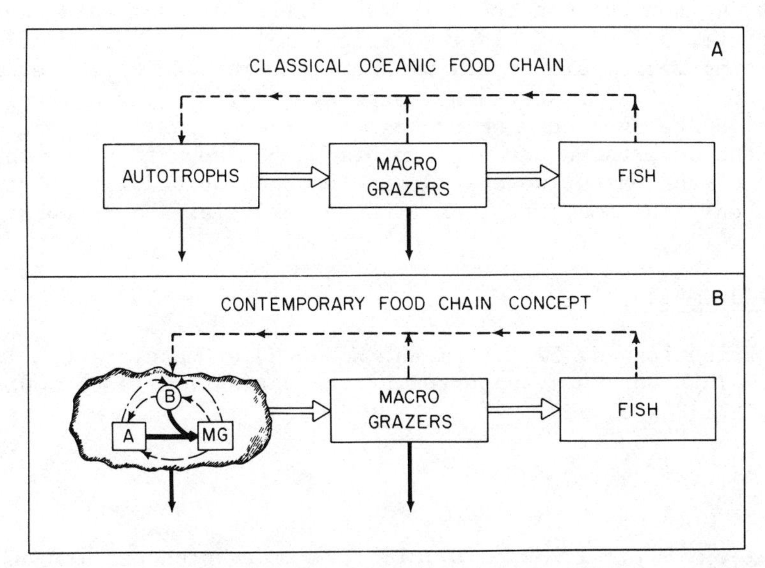

Fig. 10. Comparison of the classical oceanic food chain (A), consisting of phytoplankton-macrograzers-fish, and the more contemporary concept that aggregates comprising complete microbial food chains, as shown in Figure 8, are the major food source for macrograzers (B). Size of solid vertical arrows indicate the relative importance of particulate transport out of euphotic zone. Dashed lines represent nutrient regeneration.

the cells adhere to the glass walls and continue to divide then it is impossible to attain steady state and a 10 ml sample would no longer represent the entire culture population very satisfactorily, particularly if adhered cells had different physiological characteristics than dispersed cells (Ellwood _et al._, 1982).

Analogously, in the open ocean we are faced with a similar problem of adequately characterizing the spatial and temporal distribution of particulate matter. But unlike the well-mixed continuous culture system of finite volume that can readily be sampled with large volumes relative to the total volume of the system, the samples we take in the open ocean represent only an infinitesimally small fraction of the larger system we wish to characterize. For example, even on spatial scales of <5 m tremendous patchiness of particulate matter is revealed with conventional Niskin bottle sampling (Wangersky, 1978). When we consider all the levels of known small-scale patchiness that exist in the open ocean from the organic aggregates of Riley (1970) to the marine snow particles of

Silver et al. (1978), together with our suspicions that microbial activity within these patches is very different from rates measured on discrete, well-mixed Niskin bottle samples, the impossibility of integrating microscale events with current methodologies becomes readily evident. As stated by Wangersky (1978), "our methods of collection are such that we average out the organisms in the oases and in the deserts in between; before we can hope to study the ecology of the oases, we must devise methods which will locate and collect only the small regions rich in particles and organisms."

Acknowledgements

Contribution No. 5206 from the Woods Hole Oceanographic Institution. This work was supported by the National Science Foundation Grant OCE81-24445.

REFERENCES

Alldredge, A.L., and Cox, J.L., 1982, Primary productivity and chemical composition of marine snow in surface waters of the Southern California Bight, J. Mar. Res., 40: 517.

Alldredge, A.L., and Silver, M.W., 1982, Abundance and production rates of floating diatom mats (Rhizosolenia castracanei and R. imbricata var. shrubsolei) in the Eastern Pacific Ocean, Mar. Biol. 66: 83.

Allen, T.F.H., 1977, Scale in microscopic algal ecology: a neglected dimension, Phycologia, 16:253.

Azam, F., In press, Measurement of growth of bacteria in the sea and the regulation of growth by environmental conditions, in:"Heterotrophic Activity in the Sea," J. Hobbie and P. J. LeB. Williams, eds., Plenum Press, New York.

Azam, F., and Hodson, R.E., 1977, Size distribution and activity of marine microheterotrophs, Limnol. Oceanogr., 22: 492.

Barsdate, B.J., Prentki, R.T., and Fenchel, T., 1974, Phosphorus cycle of model ecosystems: significance for decomposer food chains and effect of bacterial grazers, Oikos, 25: 239.

Baylor, E.R., Sutcliffe, W.H., and Hirschfield, D.S., 1962, Adsorption of phosphate onto bubbles, Deep Sea Res., 9:120.

Beers, J.R., Reid, F.M.H., and Stewart, G.L., 1982, Seasonal abundance of the microplankton population in the North Pacific central gyre, Deep Sea. Res., 29:227.

Berman, T., 1975, Size fractionation of natural aquatic populations associated with autotrophic and heterotrophic carbon uptake, Mar. Biol., 33:215.

Bishop, J.K.B., Ketten, D.R. and Edmond, J.M., 1978, The chemistry biology and vertical flux of particulate matter from the upper 400 m of the Cape Basin in the southeast Atlantic ocean, Deep Sea Res.,25:1121.

Bishop, J.K.B., Collier, R.W., Ketten, D.R., and Edmond, J.M., 1980, The chemistriy, biology and vertical flux of particulate matter from the upper 1500 m of the Panama Basin, Deep Sea Res., 27:615.

Booth, B.C., Lewin, J. and Norris, R.E., 1982, Nanoplankton species predominant in the subartic Pacific in May and June 1978, Deep Sea Res., 29:185.

Brand, L.E., and Guillard, R.R.L., 1981, The effects of continuous light and light intensity on the reproduction rates of twenty-two species of marine phytoplankton, J. exp. mar. Biol. Ecol., 50:119.

Caperon, J., Schell, D., Hirota, T., and Laws, E., 1979, Ammonium excretion rates in Kaneohe Bay, Hawaii, measured by a ^{15}N isotope dilution technique, Mar. Biol., 54: 33.

Carder, K.L., 1979, Holographic microvelocimeter for use in studying ocean particle dynamics Opt. Eng., 18: 524.

Carder, K.L., Steward, R.G., and Betzer, P.R., 1982, In situ holographic measurements of the sizes and settling rates of oceanic particulates, J. Geophys. Res., 87:5681.

Caron, D.A., Davis, P.G., Madin, L.P., and Dieburth, J. McN., 1982 Heterotrophic bacteria and bacterivorous protozoa in oceanic macroaggregates, Science.

Carpenter, E.J., and McCarthy, J.J., 1975, Nitrogen fixation and uptake of combined nitrogenous nutrients by Oscillatoria (Trichodesmium) thiebautii in the western Sargasso Sea, Limnol. Oceanogr., 20: 389.

Carpenter, E.J., and Price, C.C., 1977, Nitrogen fixation, distribution, and production of Oscillatoria (Trichodesmium) spp. in the western Sargasso and Caribbean Seas, Limnol. Oceanogr., 22: 389.

Carpenter, E.J., Harbison, G.R., Madin, L.P., Swanberg, N.R., Biggs, D.C., Hulbert, E.M., McAlister, V.L., and McCarthy, J.J., 1977, Rhizosolenia mats, Limnol. Oceaogr., 22: 739.

Cauwet, G., 1978, Organic chemistry of sea water particulates. Concepts and developments, Oceanologica Acta, 1: 99.

Churchward, G., Bremer, H. and Young, R., 1982, Macromolecular composition of bacteria, J. Theor. Biol., 94: 651.

Cooper, L.H.N., 1933, Chemical constituents of biological importance in the English Channel. Pt. I. Phosphate, silicate, nitrate, nitrite, ammonia, J. Mar. Biol. Assoc. U.K., 23: 171.

Curds, C.R., 1971, A computer-simulation study of predator-prey relationships in a single-stage continuous-culture system, Water Res., 5: 793.

Dawson, M.P., Humphrey, B.A., and Marshallm K.C., 1981, Adhesion: A tactic in the survival strategy of a marine vibrio during starvation, Current Microbiol., 6: 195-199.

Ellwood, D.C., Keevil, C.W., Marsh, P.D.. Brown, C.M., and Wardell, J.N., 1982, Surface-associated growth, Phil. Trans. R. Soc. Lond., B297: 517.

Eppley, R.W., 1972, Temperature and phytoplankton growth in the sea, Fish. Bull., 70:1063.

Eppley, R.W., and Peterson, B.J., 1979, Particulate organic matter flux and planktonic new production in the deep ocean, Nature, 282: 677.

Eppley, R.W., Renger, E.H., Venrick, E.L., and Mullin, M.M., 1973, A study of plankton dynamics and nutrient cycling in the central gyre of the North Pacific Ocean, Limnol. Oceanogr., 18: 534.

Esener, A.A., Roels, J.A., and Kossen, N.W.F., 1982, Dependence of the elemental composition of K. pneumoniae on the steady-state specific growth rate, Biotechnol. Bioeng., 24: 1445.

Fenchel, T., 1982a, Ecology of Heterotrophic microflagellates. I. Some important forms and their functional morphology, Mar. Ecol. Prog. Ser. 8:211.

Fenchel, T., 1982b, Ecology of heterotrophic microflagellates. II. Bioenergetics and growth, Mar. Ecol. Prog. Ser., 8: 225.

Fenchel, T., and Harrison P., 1976, The significance of bacterial grazing and mineral cycling for the decomposition of particulate detritus, in: "The Role of Terrestrial and Aquatic Organisms in Decomposition Processes," J.M. Anderson and A. Macfadyen, eds., Blackwell Scientific Publications, Oxford.

Gallager, S.M., and Mann, R., 1981, The effect of varying carbon/nitrogen ratio in the phytoplankter Thalassiosira pseudonana (3H) on its food value to the bivalve Tapes Japonica, Aquaculture, 26: 95.

Gavis, J., 1976, Munk and Riley revisited: nutrient diffusion transport and rates of phytoplankton growth, J. Mar. Res., 34: 161.

Glibert, P.M., 1982, Regional studies of daily, seasonal, and size fraction variability in ammonium remineralization, Mar. Biol., 70: 209.

Glibert, P.M. and Goldman, J.C., 1981, Rapid ammonium uptake by marine phytoplankton, Mar. Biol. Lett., 2: 25.

Goldman, J.C., 1980, Physical processes, nutrient availability, and the concept of relative growth rate in marine phytoplankton ecology, in: "Primary Productivity in the Sea," P.G. Falkowski, ed. Plenum Press, New York.

Goldman, J.C., and Glibert, P.M., 1982, Comparative rapid ammonium uptake by four species of marine phytoplankton, Limnol. Oceanogr., 27: 814.

Goldman, J.C., and Stanley, H.I., 1974, Relative growth of different species of marine algae in wastewater-seawater mixtures, Mar. Biol., 28: 17.

Goldman J.C., and McCarthy, J.J., 1978, Steady state growth and ammonium uptake of a fast-growing marine diatom, Limnol. Oceanogr., 23: 695.

Goldman, J.C., McCarthy, J.J., and Peavey, D.G., 1979, Growth rate influence on the chemical composition of phytoplankton in oceanic waters, Nature, 279:210.

Goldman, J.C., Taylor, C.D., and Glibert, P.G., 1981a, Nonlinear time-course uptake of carbon and ammonium by marine phytoplankton, Mar. Ecol. Prog. Ser., 6: 137.

Goldman, J.C., Dennett, M.R., and Riley, C.B., 1981b, Inorganic carbon sources and biomass regulation in intensive microalgal cultures, Biotechnol. Bioeng., 23: 995.
Gordon, D.R. Jr., 1970, A microscopic study of organic particles in the North Atlantic Ocean, Deep Sea Res., 17:175.
Gordon, D.C. Jr., Wangersky, P.J., and Sheldon, R.W., 1979, Detailed observations on the distribution and composition of particulate organic material at two stations in the Sargasso Sea, Deep Sea Res., 26: 1083.
Haas, L.W., and Webb, K.L.,1979, Nutritional mode of several non-pigmented microflagellates from the York River Estuary, Virginia, J. exp. mar. Biol. Ecol., 39: 125.
Harris, G.P., 1980, Temporal and spatial scales in phytoplankton ecology. Mechanisms, methods, models, and management, Can. J. Fish. Aquat. Sci., 37: 877.
Harris, R.H., and Mitchell, R., 1973, The role of polymers in microbial aggregations, Ann. Rev. Microbiol., 27: 27.
Harrison, W.G., 1978, Experimental measurements of nitrogen remineralization in coastal waters, Limnol. Oceanogr., 23: 684.
Harrison, W.G., 1980, Nutrient regeneration and primary production in the sea, in: "Primary Productivity in the Sea," P.G. Falkowski, ed., Plenum Press, New York.
Harvey, G.W., 1966, Microlayer collection from the sea surface: a new method and initial results, Limnol. Oceanogr., 11: 608.
Herbert, D., Elsworth, R., and Telling, R.C., 1956, The continuous culture of bacteria; a theoretical and experimental study, J. Gen. Microbiol., 14: 601.
Herbland, A., and LeBouteiller, A., 1981, The size distribution of phytoplankton and particulate organic matter in the Equatorial Atlantic Ocean: Importance of ultraseston and consequences, J. Plankton Res., 3: 659.
Jackson, G.A., 1980, Phytoplankton growth and zooplankton grazing in oligotrophic waters, Nature, 284: 439.
Jannasch, H.W., 1979, Microbial ecology of aquatic low nutrient habitats, in: "Strategies of Microbial Life in Extreme Environments", M. Shilo, ed., Dahlem Konferenzen, Berlin.
Jannasch, H.W., and Pritchard, P.H., 1972, The role of inert particulate matter in the activity of aquatic microorganisms, Mem Ist Ital. Idrobiol. Suppl., 29: 289.
Johannes, R.E., 1965, The influence of marine protozoa on nutrient regeneration, Limnol. Oceanogr., 10: 434.
Johnson, B.D., 1976, Nonliving organic particle formation from bubble dissolution, Limnol. Oceanogr., 21: 444.
Johnson, B.D., and Cooke, R.C., 1980, Organic particle and aggregate formation resulting from the dissolution of bubbles in seawater, Limnol. Oceanogr., 25: 653.
Johnson, P.W., and Sieburth, J. McN., 1979, Chroococcoid cyanobacteria in the sea: A ubiquitous and diverse phototrophic biomass, Limnol. Oceanogr., 24: 928.

Kelley, J.C., 1976, Sampling the sea, in: "The Ecology of the Seas, D.H. Cushing and J.J. Walsh, eds., W.B. Saunders Co, Philadelphi

Ketchum, B.H., Ryther, J.H., Yentsch, C.S., and Corwin, N., 1958, Productivity in relation to nutrients, Cons. Internat. Explor. Mer. Rapp. and Proces. Verb., 144: 132.

Kjelleberg, S., Humphrey, B.A., and Marshall, K.C., 1982, Effect of interfaces on small, starved marine bacteria, Appl. Environ. Microbiol., 43: 1166.

Kimor, B. 1981, The role of phagotrophic dinoflagellates in marine ecosystems, Kieler Meeresforsch. Sonderh., 5: 164.

Koch, A.L., 1971, The adaptive responses, of Escherichia coli to a feast or famine existence, Adv. Microbiol. Ecol., 6: 147.

Koch, A.L., 1979, Microbial growth in low concentrations of nutrient in: "Strategies of Microbial Life in Extreme Environments," M. Shilo, ed., Dahlem Konferenzen, Berlin.

Larsen, D.H., and Dimmick, R.L.,1964, Attachment and growth of bacteria on surfaces of continuous-culture vessels, J. Bacteriol 88: 1380.

Laws, E.A., and Bannister, T.T., 1980, Nutrient- and light-limited growth of Thalassiosira fluviatilis in continuous culture, with implications for phytoplankton growth in the ocean, Limnol. Oceanogr., 25: 457.

Lehman, J.T., and Scavia, D., 1982a, Microscale patchiness of nutrie in plankton communities, Science, 216: 729.

Lehman, J.T., and Scavia, D., 1982b, Microscale nutrient patches pro- duced by zooplankton, Proc. Natl. Acad. Sci. USA, 79: 5001.

Linley, E.A.S., and Field, J.G., 1982, The nature and ecological significance of bacterial aggregation in a nearshore upwelling ecosystem, Estuar. Coast. Shelf Sci., 14:1

Maestrini, S.Y., and Bonin, D.J., 1981, Competition among phyto- plankton based on inorganic macronutrients, in:"Physiological Bases of Phytoplankton Ecology," T. Platt, ed., Bulletin 210, Canadian Bulletin of Fisheries and Aquatic Sciences, Ottawa.

Marshall, K.C., 1976, "Interfaces in Microbial Ecology," Harvard University Press, Cambridge, Mass.

McCarthy, J.J.,and Carpenter, E.J., 1979, Oscillatoria (Trichodesmiu thiebautii (cyanophyta) in the central North Atlantic Ocean, J. Phycol., 15: 75.

McCarthy, J.J., and Goldman, J.C., 1979, Nitrogenous nutrition of marine phytoplankton in nutrient-depleted waters, Science, 203: 670.

Manzel, D.W., 1974, Primary productivity, dissolved and particulate organic matter, and the sites of oxidation of organic matter, in: "The Sea, Vol. 5, Marine Chemistry," E.D.Goldberg, ed., John Wiley and Sons, New York.

Morris, I., Smith, A.E., and Glover, H.E., 1981, Products of photo- synthesis in phytoplankton off the Orinoco River and in the Caribbean Sea, Limnol. Oceanogr., 26: 1034.

Norkrans, B., 1980, Surface microlayers in aquatic environments, Adv. Microbial Ecol., 4: 51.

Paerl, H.W., 1975, Microbial attachment to particles in marine and freshwater ecosystems, Microb. Ecol., 2: 73.
Parker, R.R., and Tranter, D.J., 1981, Estimation of algal standing stock and growth parameters using in vivo fluorescence, Aust. J. Mar. Freshwater Res., 32:629.
Platt, T., and Denman,, K., 1980, Patchiness in phytoplankton distribution, in:"The Physiological Ecology of Phytoplankton," I Morris, ed., Blackwell Scientific Publications, Oxford.
Poindexter, J.S., 1981, Oligotrophy. Fast and famine existence, Adv. Microbial Ecol., 5: 63.
Pomeroy, L.R., 1974, The ocean's food web, a changing paradigm, BioScience, 24: 499.
Pomeroy, L.R. and Johannes, R.E., 1966, Total plankton respiration, Deep Sea Res., 13: 971.
Pomeroy, L.R. and Johannes, R.E., 1968, Occurance and respiration of ultraplankton in the upper 500 meters of the ocean, Deep Sea Res., 15: 381.
Raymont, J.E.G., and Adams, M.N.E., 1958, Studies on the mass culture of Phaeodactylum, Limnol. Oceanogr., 3: 119.
Redfield, A.C., Smith H.P., and Ketchum, B., 1937, The cycle of organic phosphorus in the Gulf of Maine, Biol. Bull., 73: 421.
Rifkin, R.B., and Swift, E., 1982, Phosphate uptake by the oceanic dinoflagellate Pyrocystis noctiluca, J. Phycol., 18: 113.
Riley, G.A., 1963, Organic aggregates in seawater and the dynamics of their formation and utilization, Limnol. Oceanogr. 8: 372.
Riley, G.A., 1970, Particulate organic matter in sea water, Adv. mar. Biol., 8: 1.
Rutter, P.R., 1980, The physical chemistry of the adhesion of bacteria and other cells, in: "Cell Adhesion and Motility," A.S.G. Curtis and J.D. Pitts, eds, Cambridge University Press, London.
Seki, H., 1972, The role of microorganisms in the marine food chain with reference to organic aggregate, Mem. Ist. Ital. Idrobiol. Suppl., 29: 245.
Shanks, A.L., and Trent, J.D., 1979, Marine snow: Microscale nutrient patchiness, Limnol. Oceanogr., 24:850
Sharp, J.H., Perry, M.J., Renger, E.H., and Eppley, R.W., 1980, Phytoplankton rate processes in the oligotrophic waters of the central North Pacific Ocean, J. Plankton Res., 2: 335.
Sheldon, R.W., Prakash, A., and Sutcliffe, W.H. Jr., 1972, The size distribution of particles in the ocean, Limnol. Oceanogr., 17: 327.
Sherr, B.F., Sherr, E.B., and Berman, T., 1982, Decomposition of organic detritus: A selective role for microflagellate protozoa, Limnol. Oceanogr., 27: 765.
Sieburth, J. McN., 1979, "Sea Microbes," Oxford University Press, New York.
Sieburth, J.McN. Smetacek, V., and Lenz, J., 1978, Pelagic ecosystem structure: Heterotrophic compartments of the plankton and their relationship to plankton size fractions, Limnol. Oceanogr., 23: 1256.

Silver, M.W., and Alldredge, A.L., 1981, Bathypelagic marine snow: deep-sea algal and detrital community, J. Mar. Res., 39: 501.
Silver, M.W., Shanks, A.L., and Trent, J.D., 1978, Marine snow: Microplankton habitat and source of small-scale patchiness in pelagic populations, Science, 201: 371.
Sorokin, Y.I., 1981, Microheterotrophic organisms in marine ecosystems, in: "Analysis of Marine Ecosystems," A.R. Longhurst, ed., Academic Press, London.
Spicer, C.C., 1955, The theory of bacterial constant growth apparatus, Biometrics, 11: 225.
Sutcliffe, W.H., Baylor, E.R., and Menzel, D.W., 1963, Sea surface chemistry and Langmuir circulation, Deep Sea Res., 10: 233.
Swift, E., Stuart, M., and Meunier, V., 1976, The in situ growth rates of some deep-living oceanic dinoflagellates: Pyrocystis fusiformis and Pyrocystis noctiluca, Limnol. Oceanogr., 21: 418.
Throndsen, J., 1979, The significance of ultraplankton in marine primary production, Acta Bot. Fennica, 110: 53.
Topiwala, H.H., and Hamer, G., 1971, Effect of wall growth in steady-state continuous cultures, Biotechnol. Bioeng., 13: 919.
Trent, J.D., Shanks, A.L. and Silver, M.W., 1978, In situ and laboratory measurements on macroscopic aggregates in Monterey Bay, California, Limnol. Oceanogr., 23: 626.
Turpin, D.H., Parslow, J.S., and Harrison, P.J., 1981, On limiting nutrient patchiness and phytoplankton growth: A conceptual approach, J. Plankton Res., 3: 421.
van den Ende, P., 1973, Predator-prey interactions in continuous culture, Science, 181: 562.
Wangersky, P.J., 1977, The role of particulate matter in the productivity of surface waters, Helgolander wiss. Meeresunters, 30: 546.
Wangersky, P.J., 1978, The distribution of particulate organic matter in the oceans: ecological implications, Int. Revue ges. Hydrobiol., 63: 567.
Waterbury, J.B., Watson, S.W., Guillard, R.R.L., and Brand, L.E., 1979, Widespread occurance of a unicellular, marine, planktonic, cyanobacterium, Nature, 277: 293.
Wheeler, P.A., Glibert,P.G., and McCarthy, J.J., 1982, Ammonium uptake and incorporation by Chespeake Bay phytoplankton: Short-term uptake kinetics, Limnol. Oceanogr., 27: 1113.
Wiebe, W.J., and Pomeroy, L.R., 1972, Microorganisms and their association with aggregates and detritus in the sea: a microscopic study, Mem. Ist. Ital. Idrobiol. Suppl.,29: 325.
Williams, P.J., LeB., 1981, Incorporation of microheterotrophic processes into the classical paradigm of the planktonic food web, Kieler Meeresforsch., Sondeh., 5: 1.
Williams, P.J. LeB. and Muir, L.R., 1981, Diffusion as a constraint on the biological importance of microzones in the sea, in: "Ecohydrodynamics", J.C.J. Nihoul, ed. Elsevier Scientific Publishing Co., Amsterdam.

STATE-OF-THE-ART IN THE MEASUREMENT OF PRIMARY PRODUCTION

Winfried W. Gieskes and Gijsbert W. Kraay

Netherlands Institute for Sea Research
1790AB Texel
The Netherlands

INTRODUCTION

In this paper on the state-of-the-art in the methodology of measuring primary production we will focus on primary production by phytoplankton, not only because we do not want to present a comprehensive review of methods for measuring production in the aquatic environment, but also because on a global scale the production by phytoplankton is by far the most important in lakes, seas and oceans (de Vooys, 1979). In shallow areas, of course, benthic production by unicellular and macro-algae is predominant, but such areas comprise only a small part of the world's marine regions. Primary production in the deep sea by chemosynthetic bacteria is another example of a productivity that is of no more than local importance. At the end of this chapter readers may be left with the conclusion that one can hardly speak of an "art" when the methods for the measurement of primary aquatic productivity are discussed. This at least is our conclusion after having been in this field of research for nearly 10 years. However, the concluding remarks will be more optimistic, namely that even methods that have long been in common use may not be so bad after all, provided they are done in the proper way. We will illustrate this chapter with examples based on research at the Netherlands Institute for Sea Research, not because we think that we master the art, but because we have the impression that we are in the forefront of at least trying.

In the past, thousands and thousands of measurements of primary production have been done, yet very few of all these measurements seem to be reliable, judging from the recent debate on their value (Peterson, 1980). Some participants in the debate even go so far that they regard most measurements of the last 30 years as untrustworthy and not even of any relative value. It is in fact not easy to see how old data can be compared with new ones when important aspects of the methodology have changed so much over the years.

The reason why we were invited to write this chapter on primary production methodology is probably in part that we are members of a group of Dutch scientists that not long ago challenged quite directly the conventional technique of measuring primary production while at the same time offering attractive alternatives. Our results were rather shocking since they indicated that with the common ^{14}C methodology the productivity of phytoplankton could be underestimated by up to 10 to 20 times. We found these dramatic underestimates when measuring in open oligotrophic tropical waters with different, independent methods: we compared the rate of incorporation of ^{14}C in organic matter during incubations in small and in large bottles (Gieskes et al., 1979) with in situ changes in oxygen (Tijssen, 1979) and in particulate organic carbon (Postma and Rommets, 1979). The assumption underlying the comparison was that it is possible to use the diurnal variation in POC and in O_2 as an estimate of the primary production responsible for this variation.

Differences in results when different methods are used are of course nothing new in any field of science. In the case of primary production methodology there was right from the start in 1952, when Steemann Nielsen introduced the ^{14}C method, a debate not only about what this method measures, but in the first place about the enormous discrepancy between results obtained with this method (which is now used all over the world and is the most accepted one) and with older methods, e.g. those in which oxygen changes as a measure of production are used. And yet it all seems so very simple with the ^{14}C method. A water sample is divided into one darkened and one clear-glass bottle; then, labeled dissolved inorganic carbon, ^{14}C-CO_2, is added. The bottles are then incubated for some time to give the enclosed algae the chance to photosynthesize and take up CO_2. The uptake of labeled CO_2 is a measure of production. In the calculations, discrimination against the heavier ^{14}C, which is taken up more slowly the ^{12}C, is accounted for. Dark bottle uptake is a kind of blank value; dark levels should be less than 10% of light bottle values (Qasim et al., 1973); when this percentage is higher the reliability of the light bottle values is accordingly lower. The activity of the ^{14}C in the algae is counted nowadays with a liquid scintillation counter.

Steeman Nielsen used this ^{14}C-CO_2 method on the Galathea expedition back in the early fifties. His estimate of total annual primary production of oceans and seas was around 15 x 10^9 tons of carbon, which was nearly 10 times less than estimates of the late thirties by workers such as Gordon Riley (126 x 10^9 tons), who based their estimate on the oxygen method, including the oxygen light-and-dark bottle method. Work by Russians in the early seventie

(Ivanenkov et al., 1972) again indicated that the ^{14}C method results in far lower values of primary production than methods in which oxygen changes in the water are used as a measure of photosynthetic rate. Also in the early seventies other, more indirect, arguments emerged in the literature that supported the view that primary production was far greater than ^{14}C data made believe. Sieburth et al. (1977) summarized the argument by stating that the demand of marine microbial heterotrophs (bacteria, protozoa) is often far higher than the supply of decomposable organic matter if this supply figure is based on primary production values obtained with the common ^{14}C method. In order to check the value of our ^{14}C productivity data collected in the North Sea we have during the last few years also made estimates of consumption of organic matter by monitoring oxygen changes in the water and converting these to carbon equivalents. In North Sea areas not influenced by allochthonous input of organic materials from rivers and other sewers, both in the central and the southern part, we often find a neat balance between primary production and consumption of organic matter, so in the North Sea at least our primary production figures, based on the normal ^{14}C method of Steeman Nielsen (during periods of Phaeocystis abundance in the version of Schindler et al., 1972), do not seem so bad.

However, there are a number of studies, some 30 years old, some new, that suggest that in many areas the ^{14}C method under-estimates the true production by up to 10 times (Sheldon and Sutcliffe, 1978; Shulenberger and Reid, 1981). This is so much that it makes the debate on whether the ^{14}C method measures gross or net production, or what to do with dark bottle values, etc. seem futile. Even under-estimates due to excretory losses or photorespiration and other respiratory processes can hardly reconcile such a difference although some workers claim that respiration by the phytoplankton itself can be very high indeed and thus may be responsible for large differences between gross and net (Joiris et al., 1982). But the single most important reason why primary production may in the past often have been under-estimated in large parts of the oceans is probably the confinement in glass bottles of too small samples for too long periods of time. Because of this, changes in the enclosed ecosystem can occur (Venrick et al., 1977; Gieskes et al., 1979) - possibly simply due to leakage of poisonous substances from the glass or from addition of metals with the ampoules (Steeman Nielsen and Wium-Andersen, 1970; Carpenter and Lively, 1980; Fitzwater et al., 1982); but even if this is not the case it must of necessity be near impossible to determine the primary production of phytoplankton because of the unknown pathways of radiocarbon flux and the flow dynamics of the primary products of photosynthesis in a microcosm enclosed in a bottle for any given length of time. What actually happens in a bottle is depicted in Figure 1. This Figure is not presented in the "energy circuit language" of Odum but it is abundantly clear that the flow pattern of labeled carbon can be very complicated in a sample within an incubation container. Modelling

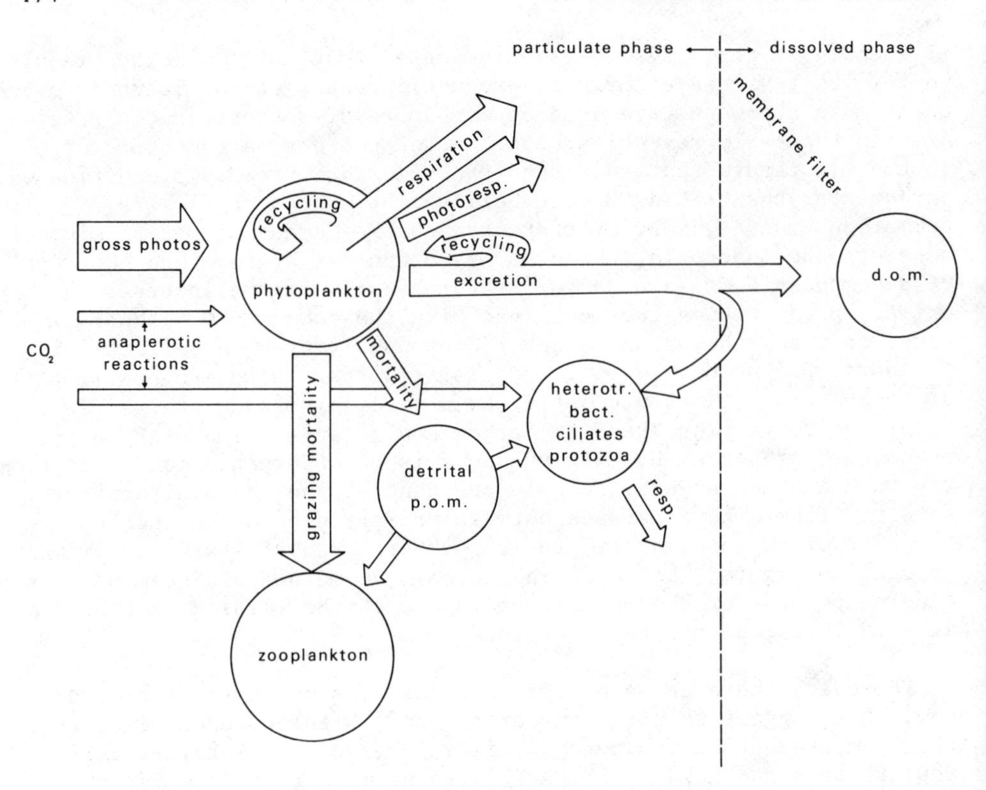

Fig. 1. Major pools and fluxes of particulate and dissolved matter in a seawater sample incubated in a bottle. Scheme from Anon., 1981 (drawn by Dr. J. -P. Mommaerts).

of this scheme with computers is a modern method to determine the importance of different compartments of the system. It is therewith possible to evaluate for example the effect of copepod grazing on the assessment of primary production for different incubation periods assuming different specific algal growth rates (G. Jackson, 1983. Larsson and Hagström (1979) have dealt with one of the major pathways in the scheme represented in Figure 1: the assimilation and respiration by bacteria of dissolved organic matter excreted by the algae. One of the problems of researchers in this field is that in the ocean the size of algae is often similar to that of bacteria (Johnson and Sieburth, 1982), so how is uptake of ^{14}C by autotrophs and heterotrophs to be distinguished (cf. Larsson and Hagström, 1982)

However, this may be, it is easy to understand that it has often been argued that when incubations last too long relative to the rate of all processes depicted in Figure 1 the ^{14}C may have spread through out the organic pool of living algal cells, detritus, bacteria, protozoa, dissolved organic matter, etc., before the end of the incubation, giving no indication of real primary production. Even this might be useful information, at least on the assumption that the activity of the microcosm in the bottle is the same as before enclosure. It should be kept in mind that in the various compartments the rate of incorporation of label may not be linear during the incubation (cf. Goldman et al., 1981), which makes the interpretation of an analysis of label found in the organic pool extremely difficult.

SOME RESULTS OF RECENT MEASUREMENTS

We have argued above that enclosure in bottles may be the most important factor influencing the results of measurements of primary production. In order to avoid "enclosure effects" as much as possible we incubated (in situ) during our last cruise to the central Atlantic Ocean* in large bottles of 5 liters that were cleaned very rigorously (24 hours shaking with 6N HCl, 0.1 N HCl left in the bottles for 2 months) in order to avoid heavy metal contamination and leakage of other substances from the bottle walls. The ^{14}C stock solution was also ultra-clean. It was made with only the highest-grade pure chemicals, and added to the samples in a diluted form: 1 ml to 5000 ml of seawater. Compare this with the procedure of 2 ml of ampoule contents added to 100 ml. of sample described by Strickland and Parsons (1968). Samples were not brought into contact with rubber or metals. Long incubations of a full light day were compared with series of short incubations, and with series of incubations of increasingly longer duration. Labelling of the suspended matter's various size fractions was followed during the time-course experiments. In Figures 2 and 3 and Table 1 results are shown. A series of short incubations gave lower results than those with longer incubations (Figure 3). This discrepancy can possibly be explained in terms of loss of ^{14}C to the dissolved organic pool and subsequent recovery in particulate form after some time lag, namely through uptake by bacteria (cf Larsson and Hagström, 1979; 1982). The constancy of the distribution of ^{14}C in the various size fractions (Table 1) indicates that ^{14}C was not transfered rapidly to larger "grazers" during the time of incubation. Notice that the rate by which the suspended matter was labeled at the end of the incubation was the same as during the beginning (see Figure 3, a), so apparently there was no more nutrient limitation in the bottle towards the end of the incubation than at the beginning. Not even the amount of light limited production in most of the euphotic zone: the relative constancy of production from 15 to 70 m depth (see

---

*NECTAR III, North Atlantic Current Trans Atlantic Research, March, 1982.

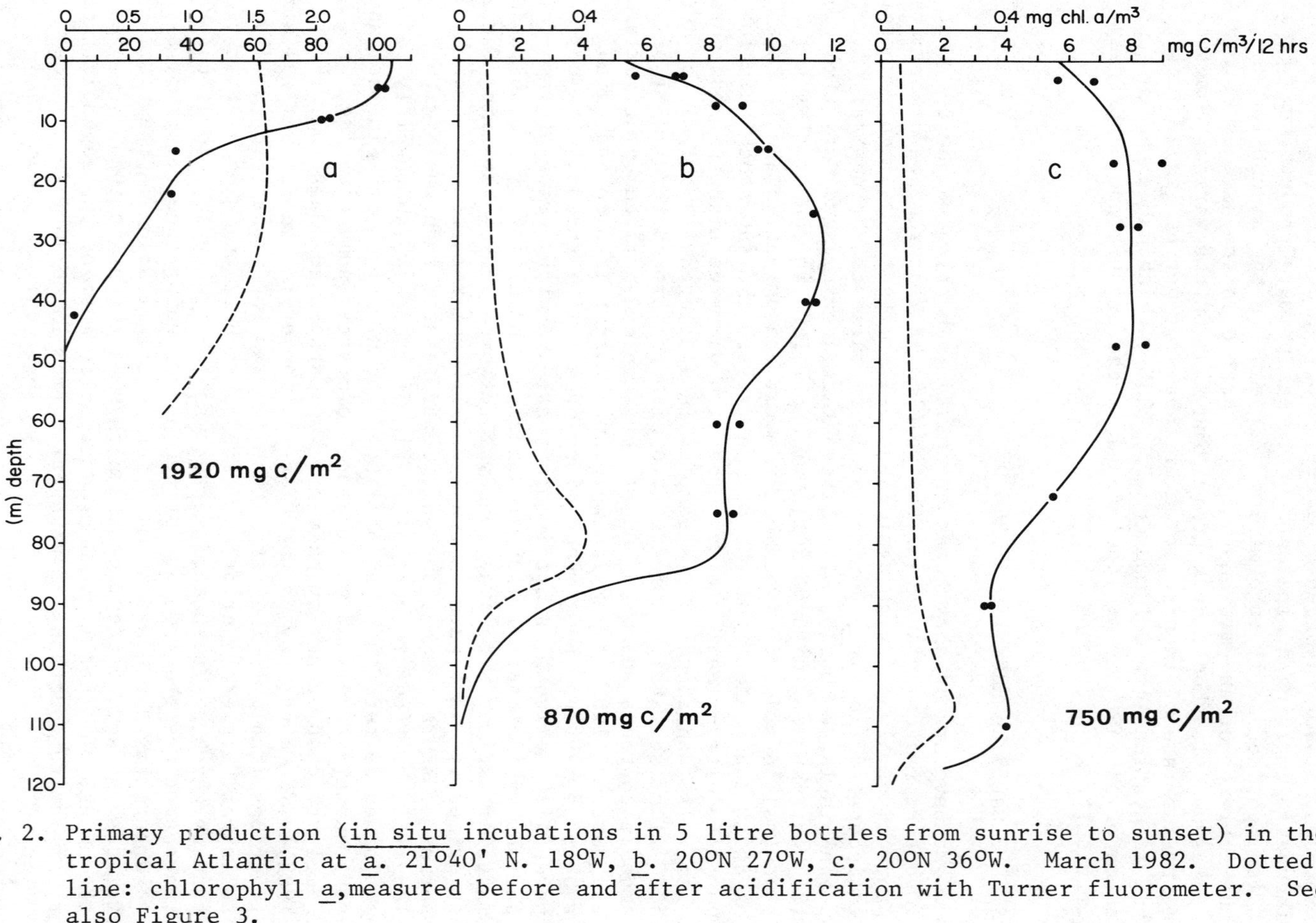

Fig. 2. Primary production (*in situ* incubations in 5 litre bottles from sunrise to sunset) in the tropical Atlantic at *a*. 21°40' N. 18°W, *b*. 20°N 27°W, *c*. 20°N 36°W. March 1982. Dotted line: chlorophyll *a*, measured before and after acidification with Turner fluorometer. See also Figure 3.

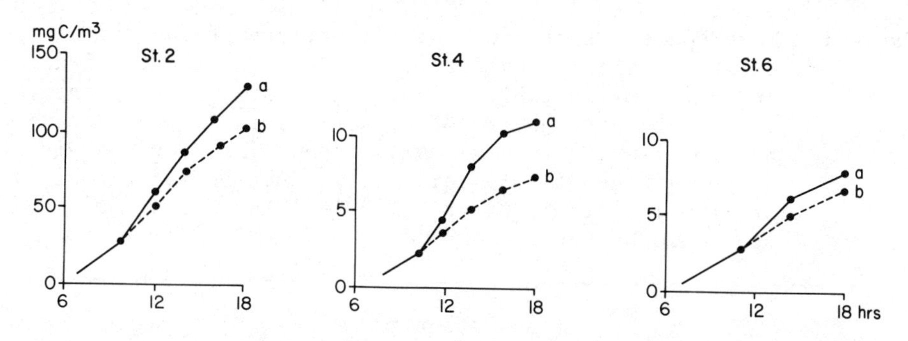

Fig. 3. Time course experiments at stations mentioned in Figure 2. Samples from 5 m depth inclubated in 5 litre bottles on deck (simulated in situ incubation), samples receiving 30% of surface irradiance. Drawn line (a): incubations lasting increasingly longer (b): cumulative from series of short incubations. Only light bottle values are given; dark bottle values have not been substracted to allow comparison between a and b. Dark values were no more than a few percent of light bottle values.

Figure 2, b and c) indicates that the density of the algal crop, which apparently produced at a maximum rate, set limits to production in the open oligotrophic Atlantic. Notice that primary production per unit chlorophyll was unexpectedly high, higher than in most eutrophic locations. The idea that primary production in oligotrophic water is nutrient-limited must clearly be reconsidered.

To summarize: with the "clean" but otherwise common ^{14}C method that we used we were able to measure a very high production in the oligotrophic ocean. Such figures are in agreement with recent ideas about the dynamics in the plankton system of these areas (McCarthy and Goldman, 1979). We measured this high production just because we avoided "enclosure effects". Apparently, the tiny algae of the open ocean (most are between 0.5 and 1.0 μm, see Table 1) are very delicate, and sensitive to handling. In the past workers have certainly more often than not overlooked this important characteristic of open ocean phytoplankton - and as a consequence primary production has been grossly underestimated in vast areas of the world's oceans.

We have been discussing incorporation of label in organic matter. Even if we know that this organic matter is phytoplankton biomass, and not incorporation in bacteria or grazers, even so we must be careful in the interpretation of results of the measurements. The group of Ian Morris has frequently drawn attention to

Table 1. Time course experiments of ^{14}C primary production of different size fractions in samples from the tropical Atlantic, March 1982. Simulated in situ incubations in 5 litre bottles at 30% of surface irradiance, incubations lasting increasingly longer. Productions given as percentage of total primary production (cf Figure 3). Station locations: St. 2, 21°40'N 18°W; St. 4, 20°N 27°W; St. 6, 20°N 36°W.

St. 2	incubation period (local time)				
	08-10	08-12	08-14	08-16	08-18
0.4-1μ	11%	12%	9%	9%	8%
1-3μ	17%	13%	18%	16%	14%
3-8μ	18%	13%	17%	16%	15%
8-50μ	21%	21%	17%	19%	19%
>50μ	39%	41%	40%	40%	44%

St. 4	incubation period (local time)				
	08-10	08-12	08-14	08-16	08-18
0.4-1μ	60%	60%	57%	50%	60%
1-3μ	16%	19%	19%	17%	18%
3-8μ	12%	11%	12%	10%	10%
>8μ	11%	10%	11%	22%	11%

St. 6	incubation period (local time)		
	07-11	07-14.30	07-18.30
0.4-1μ	40%	42%	42%
1-3μ	17%	16%	12%
3-8μ	20%	22%	15%
>8μ	23%	19%	31%

the fact that the rate of carbon uptake in carbohydrates, lipids and proteins can be very different in the course of a lengthy incubation (Morris and Skea, 1978; Morris et al, 1981). It is rather common that algae accumulate carbohydrates in the daytime that they use up for cell maintenance during the night. The algal carbohydrate pool thus looses ^{14}C during the night, the protein pool is labeled continuously from the beginning. The group Mur, at the University of Amsterdam, has reported a rapid increase of freshwater algal biomass in the hours of daylight by carbohydrate synthesis, a biomass that is low at the end of the night and the beginning of the day because of cellular sugar burning in the night (Loogman, 1982). My colleagues Postma and Rommets are accumulating evidence that this happens in the open tropical Atlantic. They relate production with in situ POC changes throughout diurnal cycles, measured at the same time as we do ^{14}C measurements. In situations where we found hardly any changes in algal cell numbers or phytoplankton pigment concentrations they found very considerable diel POC fluctuations (Postma and Spitzer, 1982). Net productions of algal biomass over 24 hours may therewith be small compared to the accumulation of organic matter during the hours of daylight. Of course, the rate of flow of labeled carbon into the end products of photosynthesis will not be the same in all systems, and in a tropical system these end products will be different from those in an antarctic system (Smith and Morris, 1980).

It is clear that if one wants to assess the amount of "new production" (Eppley and Peterson, 1979), incubations should not last too short, but then, with long incubations, it becomes increasingly difficult to decide how much of the ^{14}C has in the course of an incubation come in the heterotroph or detrital pool, or has recycled otherwise, causing underestimates of the true level of gross and even net primary production. Obviously, net production can only be measured with the ^{14}C method when the incubation period is longer than the turnover time of the phytoplankton population in the sample. A short incubation could overestimate net production because in respiration carbon may be used that has not become labeled yet. Comparison of one long incubation with a series of short incubations and a series of increasingly longer incubations (see Figure 3) may elucidate what exactly is being measured: gross, net, or "new" production.

ALTERNATIVES TO THE ^{14}C METHOD

We have until now paid much attention to the problem of deciding what length of incubation to choose. Of course this is only one aspect of the whole primary production methodology. One may ask, if there are so many problems with this one aspect alone, why not use quite other methods? This is exactly what we, at the Netherlands Institute for Sea Research, have been doing during the last few years. We use different methods side-by-side in order to check the reliability of each of them, and to learn what we are measuring with each. We have mentioned already the studies of the diurnal POC and oxygen changes in seawater and the relation of these with ^{14}C data. Doing all these comparisons of different, independent methods we have come to the conclusion that even the common ^{14}C method can sometimes give trustworthy results. In Figure 4 an example is presented of a comparison of the ^{14}C with the oxygen light-and-dark bottle technique. Oxygen was measured with a method that is nearly as sensitive as the ^{14}C method. It is a version of the Winkler titration described by Bryan et al. (1976). We use this technique in the version described

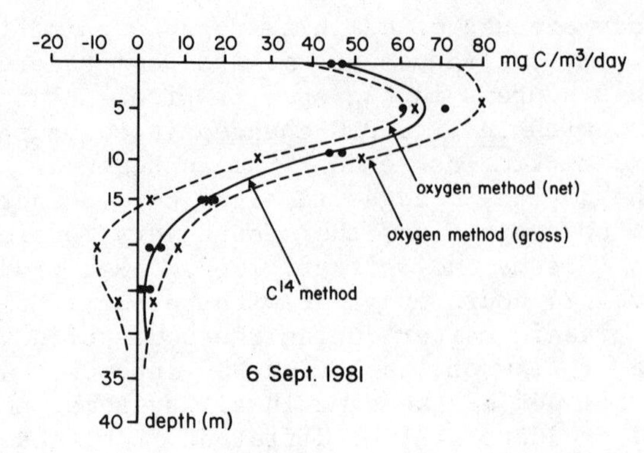

Fig. 4. Primary production profiles, central North Sea, September 1981. Incubation in situ in 100 ml bottles from sunrise to sunset. ^{14}C results plotted in between oxygen method results, i.e. assuming a photosynthetic quotient of 1.2 (1 µg C equivalent with 0.2 µgAt O). Net production with oxygen method: light bottle value minus initial concentration Gross production: light bottle value minus dark bottle value at end of incubation. Notice negative net production (as measured with oxygen technique) below 15 m - information not obtainable with the ^{14}C method.

by Tijssen (1981). The endpoint determination of the titration is done colorimetrically. At sea a precision of 0.25 µgAt/litre is obtained. The standard error is only 0.03%.

The difference between net and gross primary production was apparently rather small (see Figure 4). The ^{14}C profile has been drawn in between the net and gross production profile obtained with the oxygen method. Oxygen data can be converted to carbon production figures, and ^{14}C and O_2 profiles can accordingly be matched, assuming a value of the so-called photosynthetic quotient. In the case of Figure 4 we have chosen a value of 1.2. If we had chosen 1.8 the ^{14}C profile would have been much the same as the gross production profile. The photosynthetic quotient (P.Q.) is the rate of oxygen production divided by the rate of CO_2 consumption. It is in theory 1.0 (molar;on the basis of weight 1.37) since the general equation of photosynthesis is

$$6CO_2 + 6H_2O \rightleftarrows C_6H_{12}O_6 + 6O_2$$

This equation is of course far too simple. In reality the P.Q. varie between 0.3 and 2.5, values depending on rate of photo-respiration, availability of nutrients (Williams et al., 1979), oxygen concentration (Burris, 1981), light conditions (Falkowski and Owens, 1978). When the end product of photosynthesis is lipid rather than carbohydrates the P.Q. is > 1 (lipids are more reduced compounds); when more oxidized compounds, such as organic acids, are the end-products the P.Q. is < 1.

Assuming our choice of P.Q. (see Figure 4) was correct we can conclude that results obtained with the oxygen light-and-dark method were in fair agreement with those obtained with the ^{14}C method. Moreover, since we measured only particulate production we may also conclude that dissolved organic matter production was quite low (cf Figure 5). Notice that, as mentioned above, labeled DOM may be recovered by bacteria and protozoa soon after excretion. This recovery is often very efficient, so that one can measure excreted DOC in particulate matter after a very short time. The problem remains, of course, what the ^{14}C data indicate: are they closer to net or to gross primary production (Savidge, 1978)?

The time course experiment shown in Figure 5 also indicates that the ^{14}C method used, which is basically the original one of Steeman Nielsen, agreed with the oxygen light-and-dark method. During the same cruise colleagues measured the diurnal variation in POC and in oxygen in the water, which was marked with drogues for the time of observation. When these figures become available they will help in interpreting the ^{14}C data. This back-up of incubation data with in situ chemical data is now a routine in the ocean-going programmes at our Institute: we believe that only by dealing with the measurement of primary production with a variety of methods at the same time we can learn more about the meaning of the word "production" and interpret it in terms of gross, net, and "new" production. Microbiologists and zooplanktologists in our team gather the knowledge on heterotrophic activity necessary for this information (cf Figure 1); geologists are hired to study sinking flux rates with sediment traps; and optical oceanographers gather data on spectral irradiance and in situ light absorption (Spitzer and Wernand, 1979, 1981) for calculations of light utilization efficiencies of the phytoplankton that is sampled.

It must be realized that comparisons of in situ changes of chemical constituents in the water column with primary production of algae in bottles will not necessarily result in agreement between such basically different methods. During the spring bloom in the northern North Sea Weichart (1980) measured a long-term, 10-day primary production of 20 g $C^{-2}m$, i.e. a mean of 2 g C per day. He based this figure on changes of CO_2 in the water column. We found, at the same station in the same period, a primary production of 2 g $C^{-2}m$ day^{-1} based on ^{14}C measurements. This agreement is only apparent not real: the daily amplitude of the CO_2 changes in the water during

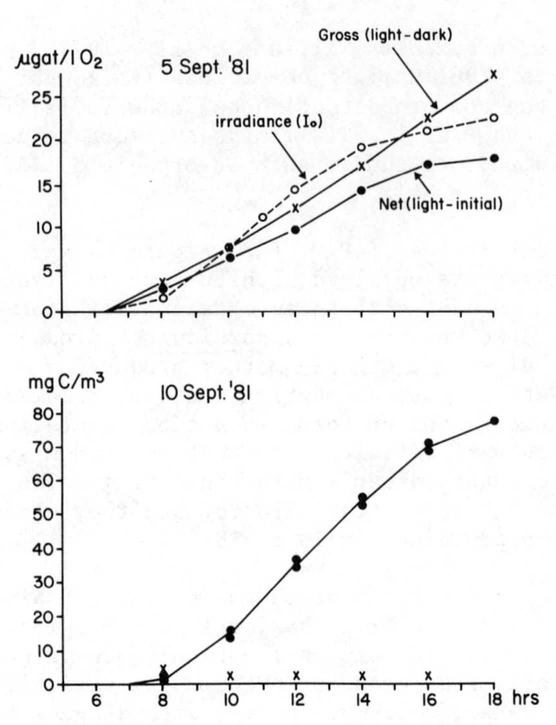

Fig. 5. Time course experiments, central North Sea, September 1981 (cf Figure 4). Incubations lasting increasingly longer: 2 hrs. 4 hrs. 6 hrs. etc. In situ incubations in 100 ml bottles at depth of 30% surface irradiance (I_o). Upper figure: oxygen light-and-dark bottle technique. Evolution of I_o during the day is also indicated (total irradiance on 5th Sept. 1245 joule/cm^2). Lower figure: ^{14}C method crosses represent production of dissolved organic carbon (filtrate ^{14}C activity). Dark bottle values increased from 0.38 to 1.15 mg C/m^3.

the period of observation (see Figure 8 in Weichart's paper) suggests that net primary production alone was at least 2 times higher than our ^{14}C data made believe. One should bear in mind that the translation of CO_2 changes into primary production figures may be a disputable enterprise (cf. Johnson et al., 1981).

Another example: our colleague Tijssen measured diel oxygen changes in the water column during the Phaeocystis spring bloom in the southern North Sea (Figure 6) and concluded that net primary production must have been 2.4 g $C^{-2}m$ day^{-1} (gross production up to 3.2 g $C^{-2}m$ day^{-1}) to account for the oxygen fluctuation in the water

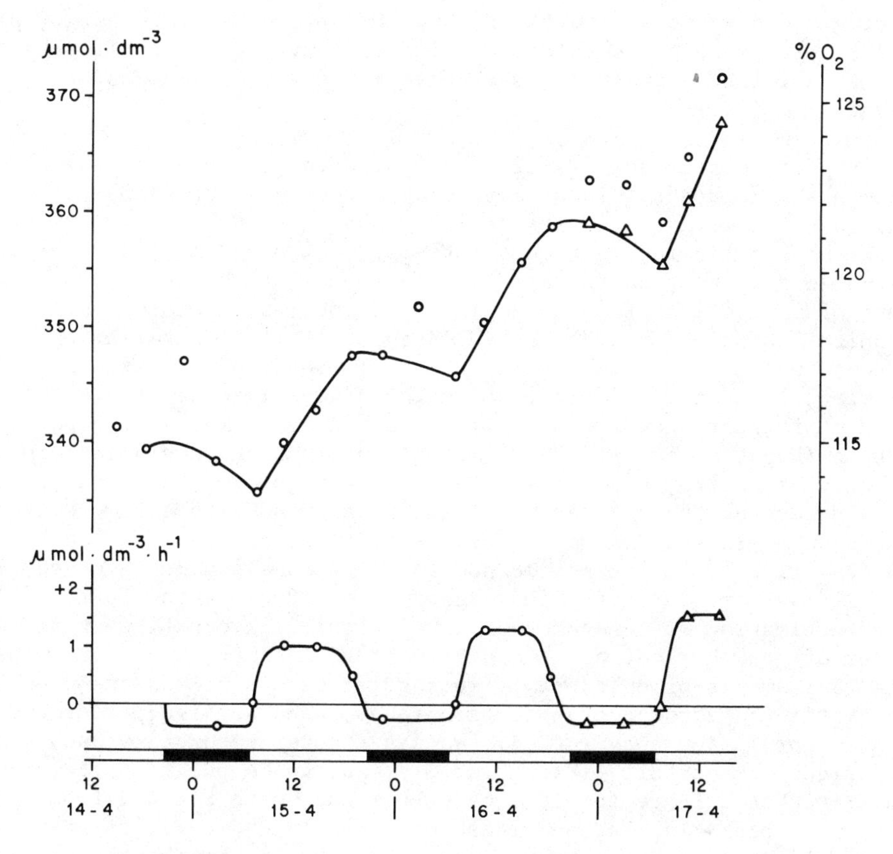

Fig. 6. In situ measurements of oxygen concentration in the North Sea (central Southern Bight) during a Phaeocystis bloom, April 1980. The rate of change in the concentration of oxygen is also indicated (lower part of Figure). From Tijssen and Eijgenraam (1982).

(Tijssen and Eijgenraam, 1982). Interestingly, we measured with the ^{14}C method, albeit in different years, a production close to Tijssen's values: between 3 and 4 g C m^2 are typical values during the Phaeocystis-dominated spring bloom in the southern North Sea when measurements are made with the ^{14}C method (in the version of Schindler et al., 1972, which measures production of particulate plus dissolved organic matter). A major source of error in translating diurnal oxygen changes in the water into primary production figures is probably that processes by which oxygen is consumed are not the same during the day and during the night. Such processes are, e.g., photo-respiration, and photo-chemically-induced oxygen losses (Gieskes and Kraay, 1982), both processes occuring during the hours of daylight.

Use of methods for measuring primary production other than the common ^{14}C method should of course be encouraged for comparative purposes. However, all alternative methods appear to have their own inherent shortcomings. Some of these are of a purely practical nature e.g., the technique of track-autoradiography for the measurement of production of individual cells requires a considerable sophistication of methodology (Wildschut, 1981). Measuring the production of individual cells by manual isolation (Rivkin and Seliger, 1981) is only practical with large-celled species. The method of counting dividing cells or dividing nuclei in a population (Hagström _et al_., 1979) is only attractive when algae can be distinguished from non-algae. The technique of recording changes in fluorescence before and after addition of DCMU to a sample cannot be used in regions where natural fluorescence is high (Carlson and Shapiro, 1981). A new method in which labeling of chlorophyll is used as a measure of primary production (Redalje and Laws, 1981), i.e. estimation of the turnover of an algal cell constituent as an index of growth rate, has the disadvantage that it is hard to measure the real amount of chlorophyll in natural waters unless one uses sophisticated chromatographic techniques, such as high-performance liquid chromatrography (HPLC). We have in our laboratory used this technique successfully (Kraay and Gieskes, to be published): we have incubated large samples, both from the North Sea and from the tropical open Atlantic, and measured not only ^{14}C activity in organic matter fixed during incubation, but also the ^{14}C in the various pigments separated by HPLC. Herewith accurate values of the carbon-to-chlorophyll _a_ ratio can be obtained of the algae that are photosynthetically active (see Redalje and Laws, 1981), assuming of course that there is no differential loss of chlorophyll during incubation. We found C: chlorophyll _a_ ratios of 99.6 and 104 for North Sea populations of late spring. The ratio was 107 in the oligotrophic tropical Atlantic in March 1982. In the latter area the ratio was lower (67) at the depth of the deep chlorophyll maximum layer (between 75 and 110 m).

It is clear that every single method has its own drawbacks. We have not until now even mentioned the basic problem of how to choose stations and at these stations, incubation depths when algae are distributed in a patchy way - which is more the rule than the exception. An example of _vertical_ inhomogeneity of algal crops is presented in Figure 7. One method of obtaining insight in the _horizontal_ distribution of patches, an insight that is imperative

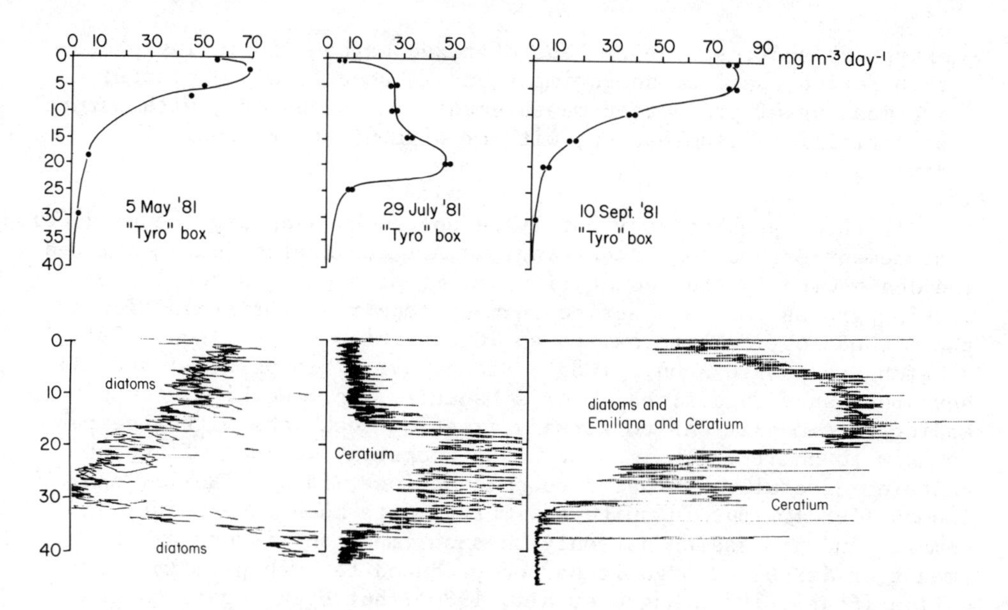

Fig. 7. Primary production (^{14}C method) and chlorophyll in the central North Sea. Vertical distribution of chlorophyll indicated by Variosens profiles. Bottom at 45 m. Most important species are also indicated.

when planning a station network for measurements in an area, is the use of Remote Sensing images. A lot of work remains to be done in the domain of development of good algorithms for relating these images to sea truth observations.

CONCLUDING REMARKS

In this article we have not mentioned yet any of the relevant problems discussed during the last 10 years regarding:

- diurnal variations in photosynthetic potential and diel variability in physiological state of populations;
- light inhibition, the effect of turbulence and light adaptations;
- the difference between measurements in bottles fixed at one depth (which is common practice) and measurements in bottles moving up and down in the water column for simulation of vertical mixing;
- what to do with dark fixation values and how to do "time-zero determinations";
- ways of filtering samples;

- excretory and respiratory loss of fixed carbon, including photo-respiration, said to be so important in so-called C-3 plants;
- the meaning of production measurements in incubators, with natural or artificial illumination, with or without other simulation devices.

All these problems may partly be solved by adapting the production measurement methodology accordingly. All the topics just mentioned are dealt with in the recent literature, or receive attention of working groups that are active in many countries for evaluation of the methodology. There exists an ICES working group (Anon. 1981), a Dutch group (see Anon., 1983), Scandinavian groups etc.; and all have published Guidelines, User's Manuals, etc. Working groups keep adding recommendations to already existing cookbooks with receipes on "how to do it". The most recent "cookbook" we know of, which contains carefully evaluated recommendations, is by O'Reilly and Thomas (1982). But in spite of improvements much uncertainty remains and discussion not only on such important issues as the amount of dissolved organic carbon produced through primary production (Sharp, 1977; Mague *et al.*, 1980), but even regarding the simplest parts of the ^{14}C method, such as how exactly to filter and handle the filters after incubation, or indeed whether or not to filter to all (Schindler *et al.*, 1972; Legendre, pers. comm.).

Now that we approach the end of this chapter some readers may be disappointed that we have not given some kind of review. However, others have done that recently (Peterson, 1980). We merely intended to indicate that there is still much uncertainty and much to be discussed and learnt. In fact is is near impossible to write an essay on the state of the art of measuring primary production since it clearly has yet to *become* an art. The task of writing on the methodology is particularly difficult because every single step in the procedure of measuring primary production with *any* method has been or is being hotly debated in the literature these days, so the pile of recent papers of high quality on this topic is gigantic.

Let us conclude that the time of simply doing standard measurements is past since workers have realized that there exists no "standard" procedure yet. Decisions on the type of method by which primary production should be measured must depend on knowledge of the structure and function of the ecosystem that is sampled. Only when the task of measuring primary production is undertaken by a team of scientists from a variety of disciplines, ranging from physical oceanography through microbiology to plant physiology, real progress will be made in the field of primary production research.

REFERENCES

Anon., 1981. Report of the working group on the methodology of primary production. ICES, C.M. 1981/L.

Anon., 1983. Hydrobiol. Bull., In press.

Bryan, J.R., Riley, J.P., and Williams, P.J. LeB., 1976, A Winkler procedure for making precise measurements of oxygen concentrations for productivity and related studies, J. exp. mar. Bio. Ecol., 21: 191.

Burris, J.E., 1981, Effects of oxygen and inorganic carbon concentrations on the photosynthetic quotients of marine algae, Mar. Biol. 65: 215.

Carlson, R.E., Shapiro, J.S., 1981, Dissolved humic substances: a major source of error in fluorometric analyses involving lake waters, Limnol. Oceanogr., 26: 785.

Carpenter, E.J. and Lively, J.S., 1980, Review of estimates of algal growth using ^{14}C tracer technique, Brookhaven Symp. Biol., 31: 161.

Eppley, R.W. and Peterson, B.J., 1979, Particulate organic matter flux and planktonic new production in the deep ocean, Nature, 282: 677.

Falkowski, P.G., and Owens, T.G., 1978, Effects of light intensity on photosynthesis and dark respiration in six species of marine phytoplankton, Mar. Biol., 45: 289.

Fitzwater, S.E., Knauer, G.A. and Martin, J.H., 1982, Metal contamination and its effect on primary production measurements, Limnol. Oceanogr., 27: 544.

Gieskes, W.W.C., Kraay, G.W. and Baars, M.A., 1979, Current ^{14}C methods for measuring primary production: gross underestimates in oceanic waters, Neth. J. Sea Res., 13: 58.

Gieskes, W.W.C. and Kraay G.W., 1982, Effect of enclosure in large plastic bags on diurnal change in oxygen concentration in tropical ocean water, Mar. Biol., 70:99

Goldman, J.C., Taylor, C.D., and Glibert, P.M., 1981, Nonlinear time, course uptake of carbon and ammonium by marine phytoplankton, Mar. Ecol. Progr. Ser., 6: 137.

Hagström, A., Larsson, U., Horstedt, P. and Normark, S., 1979, Frequency of dividing cells, a new approach to the determination of bacterial growth in aquatic environments, Appl. Environm., Microbiol. 37: 805.

Ivanenkov, V.N., Sapozhnikov, V.V., Chernyakova, A.M. and Gusarova, A.N., 1972, Rate of chemical processes in the photosynthetic layer of the tropical Atlantic, Oceanology, 12: 207.

Jackson, G.A., 1983, Zooplankton grazing effects on C^{14}-based production measurements : a theoretical study , J. Plank. Res., 5:83.

Johnson, K.M., Burney, C.N., and Sieburth, J. McN., 1981, Enigmatic marine ecosystem metabolism measured by direct diel ΣCO_2 and O_2 flux in conjunction with DOC release and uptake, Mar. Biol. 65: 49.

Johnson, P.W., and Sieburth, J. McN., 1982, In situ morphology and occurance of eucaryotic phototrophs of bacterial size in the picoplankton of estuarine and oceanic waters, J. Phycol., In press.

Joiris, C., Billen, G., Lancelot, C., Daro, M.H., Mommaerts, J.P., Hecq, J.H., Bertels, A., Bossicart, M., and Nijs, J., 1982, A budget of carbon cycling in the Belgian coastal zone: relative roles of zooplankton, bacterioplankton and benthos in the utilization of primary production. Neth. J. Sea Res.: 16:260

Larsson, U., and Hagström, A., 1979, Phytoplankton exudate release as an energy source for the growth of pelagic bacteria, Mar. Biol., 52:199.

Larsson, U. and Hagström, A., 1982, Fractionated phytoplankton, primary production, exudate release and bacterial production in a Baltic eutrophic gradient, Mar. Biol. 67: 57.

Loogman, J.G., 1982, Ph.D. thesis, University of Amstersdam, the Netherlands.

Mague, T.H., Friberg, E., Hughes, D.H. and Morris, J., 1980, Extracellular release of carbon by marine phytoplankton: a physiological approach, Limnol. Oceanogro., 25:262.

McCarthy, J.J., and Goldman, J.C., 1979, Nitrogenous nutrition of marine phytoplankton in nutrient depleted waters, Science, N.Y., 103: 670.

Morris, I., and Skea, W., 1978, Products of photosynthesis in natural populations of marine phytoplankton from the Gulf of Maine, Mar. Biol. 47: 303.

Morris, I., Smith, A.E., and Glover, H.E., 1981, Products of photosynthesis in phytoplankton off the Orinoco River and in the Caribbean Sea, Limnol. Oceanogr. 26: 1034.

O'Reilly, J.E., and Thomas J.P., 1982, A manual for the measurement of total daily primary productivity using ^{14}C simulated in situ sunlight incubation, Biomass (SCAR-SCOR-IABO-ACMRR) Handbook No. 10.

Peterson, B.J., 1980, Aquatic primary productivity and the $^{14}C-CO_2$ method: a history of the productivity problem, Ann. Rev. Ecol. Syst. 11:359.

Postma, H., and Rommets, J.W., 1979, Dissolved and particulate organic carbon in the North Equatorial Current of the Atlantic Ocean, Neth. J. Sea Res. 13: 85.

Postma, H., and Spitzer, D., 1982, Possible daily variation in the color of the blue ocean, Appl. Optics, 21: 12.

Qasim, S.Z., Bhattathiri, P.M.A., and Devassy, V.P., 1973, Some problems related to the measurements of primary production using radio-carbon technique, Int. Rev. ges. Hydrobiol. 57:535.

Redalje, D.G., and Laws, E.A., 1981, A new method for estimating phytoplankton growth rates and carbon biomass, Mar. Biol., 62:73.

Riley, G.A., 1939, Plankton studies II. The western North Atlantic, May-June, 1939, J. mar. Res. 2: 145.

Rivkin, R.B., and Seliger, H.H., 1981, Liquid scintillation counting for ^{14}C uptake of single algal cells isolated from natural samples, Limnol. Oceanogr. 26: 780.

Savidge, G., 1978, Variations in the progress of ^{14}C uptake as a source of error in estimates of primary production, Mar. Biol. 49: 295.

Schindler, D.W., Schmidt, R.V., and Reid, R.A., 1972, Acidification and bubbling as an alternative to filtration in determining phytoplankton production by the ^{14}C method, J. Fish. Res. Bd Can. 29: 1627.

Sharp, J.H., 1977, Excretion of organic matter by marine phytoplankton: do healthy cells do it? Limnol. Oceanogr. 22: 381.

Sheldon, R.W. and Sutcliffe, W.H., 1978, Generation times of 3 h for Sargasso Sea microplankton determined by ATP analysis, Limnol. Oceanogr. 23: 1051.

Shulenberger, E. and Reid, J.L., 1981, The pacific shallow oxygen maximum, deep chlorophyll maximum, and primary productivity, reconsidered, Deep-Sea Res. 28A: 901.

Sieburth, J. McN, Johnson, K.M., Burney, C.M., and Lavoie, D.M., 1977, Estimation of in situ rates of heterotrophy using diurnal changes in dissolved organic matter and growth rates of picoplankton in diffusion culture, Helgoländer wiss. Meeresunters. 30:565

Smith, A.E., and Morris, J., 1980, Synthesis of lipid during photosynthesis by phytoplankton of the southern ocean, Science (N.Y.) 207: 197.

Spitzer, D., and Wenand, M., 1979, Photon scalar irradiance meter, Appl. Optics 18:1698.

Spitzer, D. and Wenand, M., 1981, In situ measurements of absorption spectra in the sea, Deep-Sea Res. 28A: 165.

Steemann Nielsen, E., and Wium-Andersen, S., 1970, Copper ions as poison in the sea and in fresh water, Mar. Biol. 6: 93.

Steeman Nielsen, E., 1952, The use of radio-active carbon (C^{14}) for measuring organic production in the sea, J. Cons. perm. int. Explor. Mer 18: 117.

Strickland, J.D.H., and Parsons, T., 1968, A practical handbook of seawater analysis, Bull. Fish. Res. Bd Can., 167: 1.

Tijssen, S.B., 1979, Diurnal oxygen rhythm and primary production in the mixed layer of the Atlantic Ocean at 20°N, Neth. J. Sea Res. 13:79.

Tijssen, S.B., 1981, Anmerkungen zur photometrische Winkler-Sauerstofftitration und ihre Anwendung zur Schatzung der Primärproduktion im Meer, Int. hydromikrobiol. Symp., Smolenice (C.S.S.R.):343.

Tijssen, S.B. and Egenraam, A., 1982, Primary and community production in the Southern Bight of the North Sea deduced from oxygen concentration variations in the spring of 1980, Neth. J. Sea Res., 16:247.

Venrick, E.L., Beers, J.R., and Heinbokel, J.F., 1977, Possible consequences of containing microplankton for physiological rate measurements, J. exp. mar. Biol. Ecol. 26: 55.

Vooys, C.G.N. de, 1979 Primary production in aquatic environment, In: "The Global Carbon Cycle", B. Bolin, E.T. Degens, S. Kempe, and P. Ketner, Wiley, New York.

Weichart, G., 1980, Chemical changes and primary production in the Fladen Ground area (North Sea) during the first phase of a spring phytoplankton bloom, "Meteor" Forsch. -Ergebn. A, 22: 79.

Wildschut, M.A., 1981, Onderzoek naar de betrouwbaarheid van de ^{14}C-methode met behulp van track-autoradiografie, Int. Versl. NIOZ Texel, The Netherlands, 1981-6: 1.

Williams, P.J. LeB., Raine, R.C.T., and Bryan, J.R., 1979, Agreement between the ^{14}C and oxygen methods of measuring phytoplankton production: reassessment of the photosynthetic quotient, Oceanol. Acta 2: 411.

FUNCTIONAL TYPES OF MARINE PLANKTONIC PRIMARY PRODUCERS AND THEIR RELATIVE SIGNIFICANCE IN THE FOOD WEB

Malte Elbrächter

Biologische Anstalt Helgoland, Litoralstation List
Hafenstr. 3, D- 2282 List/Sylt; F R G

INTRODUCTION

The process of primary production is restricted to cells with special organelles containing chlorophyll. In the marine pelagic ecosystem the great bulk of organisms showing the ability of primary production are unicellular plants usually called phytoplankton to which this review is restricted. Other types of primary producers, e.g. floating beds of the large brown alga Sargassum or chemosynthetic organisms which might be locally important for the flow of energy and matter in the pelagic ecosystem are not dealt with.

There are more than 10,000 phytoplankton species belonging to about 13 classes of algae, according to the adopted system. For practical reasons the 'blue-greens' are regarded here as algae despite the fact that they are now commonly regarded as belonging to the bacteria. Each of the species represents a 'functional type' of its own. In addition different stages in the life cycle as well as physiologically different stages during ontogenetic development make it nearly hopeless to ally an organism to a particular 'functional type'.

Man in his search for order has classified this natural chaos of differently behaving units in several more or less meaningful categories since the early days of marine planktology, e.g. Schütt (1892). The attempt to produce reasonable models of the marine ecosystem in our days has renewed the desire to classify the tremendous variety of primary producers into functional types. Some recent reviews and books on various aspects (Falkowski, 1980; Morris, 1980; Platt, 1981; Sournia, 1982) demonstrate the interest in this topic.

The present paper cannot cover all the aspects dealt with extensively in the literature but will discuss only the most commonly accepted 'functional types' of primary producers as these are and will be used by modellers. Only very few 'laws' applied in ecology have been tested experimentally and the chance is high, therefore, that correlations found in nature do not mean direct causality but only temporal coincidence, if we consider that many factors vary simultaneously in the sea. Progress in culture technique for primary producers during the last two decades will permit such experiments although techniques for cultivating the bulk of truely oceanic phytoplankton species and those with highly complicated structures still have to be developed.

I have considerable problems in accepting some of the generalizations leading to the classification of 'functional types' hence I will pinpoint some problems and exceptions being aware, however, that this subjective view might not facilitate but complicate the job of ecologists and modellers. But perhaps some exceptions are more widespread than the rule?

THE ORGANISMS

The concept of functional types is based on the assumption that species are adapted to their environment. In consequence, this means that each separate species represents a separate functional type. Therefore this chapter first has to deal with the organisms. No taxonomical review will be given, as several textbooks on algae exist which give a good overview, e.g. Bold and Wynne (1978), Ettl (1980) and others. A good synopsis of the algal classes present in the phytoplankton is provided by Sournia (1981). Nevertheless some aspects of the organisms are dealt with which are thought to be important for classifying the functional types.

1) Organisation types:

Phytoplankton species may be unicellular and solitary or several cells may be aggregated into larger entities. First single cells are dealt with, then the different types of cell aggregations.

Unicellular Organisation

1) The flagellate organisation is widespread and includes many different types. The number of flagella can vary from 1 (some Chrysophyta) to 2 (most Dinophyta, Chrysophyta, Cryptophyta, Euglenophyta) or more (some Chrysophyta, Prasinophyta) up to 8 (some Chlorophyta). Besides their number, their length and behaviour as well as their location in respect to cell morphology and the location of the flagellar base inside the cell exhibit a large diversity. The flagella of Chlorophyta and most Prymnesiophyta are apparently all smooth. In most other groups they bear either one

or two rows of hairs and filaments and their distribution is also highly variable from group to group (Bouck, 1971, 1972) and thus within the algal groups there is a large variety of functional types (Taylor, 1976). In addition, mineralized scales are present on the flagella in Prasinophyceae and some species of Chrysophyta and Dinophyta. The Prymnesiophyceae have in addition to the flagella a flagellum-like structure of variable length - the haptonema - with ultrastructure very different from that of the flagella. Up to now we can describe the morphology and behaviour of these different structures but we are far from understanding their functional meaning and ecological importance. An important and successful attempt to understand the function of these different and differentiated structures has been made by Fenchel (1982 a, b) for nonphotosynthetic flagellates, where they are partly used for feeding. Some aspects of flagellar activity will be discussed below (see paragraph on movement).

Flagellates are ubiquitous in their geographic distribution and present in all habitats. Size varies to a large extent, many of them are tiny, less than 10 μm long, but there are a few large ones with a length of more than 200 μm (mostly dinoflagellates). A large fraction of the small flagellates will be damaged by routine fixing procedures for quantitative phytoplankton counting and it can be only guessed therefore that flagellates are an important fraction in the marine food web. But the quota they contribute to the flow of energy and matter in the food web is up to now unknown. There is increasing evidence that small organisms including flagellates provide the greatest part of primary production (Throndsen, 1979).

2) The coccoid organisation includes unicellular, nonflagellated organisms with a more or less rigid cell wall. All diatoms belong to this group as do species of all other groups. Size and shape is variable, size range is from less than 1 μm to more than 1 cm. The shape may be spherical to obovate or much more complicated. Many coccoid species, in particular diatoms and dinophytes, may have large vacuoles. It has been speculated that vacuoles may serve for regulation of buoyancy (see below). The cell covering of coccoid species differs according to the systematic position of the species. Many species produce mucilage whereas this is less known for flagellates. Also more coccoid species form colonies, compared to flagellates.

Coccoid species are ubiquitous in the plankton. They represent at different localities and different seasons the bulk of primary producers, e.g. the spring phytoplankton blooms of neritic temperate waters are mostly formed by diatoms. There is increasing evidence that a fraction of small (coccoid?) organisms contribute an important part to primary production in the planktonic food web.

3) The rhizopodial organisation includes unicellular organisms with no rigid cell wall but with the ability to form pseudopodia like amoeba. Phytoplankton species of this organisation type are mainly

known from fresh water but a few chrysophytes and dinophytes are also known from the marine environment. They may easily confused with amoeba having ingested food bodies if preserved at all. The phytoplankton of the neuston and epineuston is more or less unexplored in the marine environment, that habitat from which several rhizopodial species have been recorded in the fresh water. Nothing is known on the distribution and importance of these organisms in the marine plankton but remember that oceanographers usually take their 'surface' samples at 0.5 to 1 m water depth and that phytoplankton ecologists will not count 'amoeba' as phytoplankton.

Multicellular Organisation

1) The trichoid organisation is rare in marine plankton except for the blue greens like *Trichodesmium* which might form blooms in tropical waters. In brackish waters like the Baltic Sea, other trichoid blue greens of the genera *Aphanomizemonon* and *Nodularia* may be bloom forming. As some blue greens are able to fix nitrogen these species may be very important in those regions. Very little is known about species feeding on blue greens and the food value of these organisms. Perhaps a large part of primary production including the nitrogen compounds of trichoid blue greens may often not be used directly in the food web by zooplankton but rather after decomposition.

Pseudotrichal organisation is common in chain forming diatoms (see below).

2) Siphonaceous organisation is of no importance in the marine phytoplankton.

3) Colonial organisation is very important in marine phytoplankton. Two to ten thousands of cells may be aggregated into colonies. The size of the colonies range from a few μm to several centimeters. The shape may be quite diverse. Many coccoid species form colonies whereas flagellates are mostly solitary. Several types of cell aggregations exist representing a more or less intensive connection between the cells. In addition the geometric arrangement of the cells in a colony may form structures of higher order, e.g. the coiled colonies of the diatom *Eucampia zoodiacus* Ehrenb.. The connection of several colonies to an aggregation of colonies, e.g. in *Chaetoceros socialis* Lauder or *Thalassiosira partheneia* Schrader, also represents an organisation of higher order. The connection between single cells may be more or less intimate so that colonies either may disintegrate very easily into single cells, or cells may be connected so intimately that they can hardly be segregated. The question arises, at least for the latter, whether these aggregates of cells represent a higher entity, that is that a colony is more than the sum of single cells. In several *Chaetoceros* species

(diatoms) special terminal setae are formed inside a chain at the presumptive place where the colonies divide (Husted, 1930). The chain length of many species, e.g. Thalassiosira rotula Meunier is not random but may be described by the sequence 2^{n-1} (n = 1,2,3...). Light may influence the chain length (Schöne, 1972). In Th. partheneia stress factors such as nutrient depletion or high light intensities can favour disintegration of the colonies (Elbrächter and Boje, 1978).

Pseudotrichoid organisation is found in many colonies of diatoms, e.g. Rhizosolenia alata Brightw. and these colonies may aggregate to large algal mats (Aldredge and Silver, 1982). These large aggregates may prevent feeding by many filter feeders. Their use in the food web has to be investigated more thoroughly. Dinoflagellate species like Ceratium lunula (Karst.) Jörg., C. hexacanthum Gourret and others can live as single cells or in colonies of up to more than 20 cells. In these chains all the cells have a short apical horn except the most anterior cell. This cell has a long apical horn like the solitary cells. A direct cytoplasmatic connection may be present between cells as in other chain forming dinoflagellates such as Pyrodinium bahamense Plate (Steidinger et al. 1980). The chains swim as do single cells which means that the flagella movement has to be coordinated. All these and many other observátions suggest that some colonies are units of higher order also in the physiological sense. If so, how does such a colony react on mechanical destruction when parts of it are consumed by zooplankton? This and many other questions now can be investigated experimentally as many colony forming species can be cultivated. The physiology of the single cells may be changed by the process leading to colony formation. The haptophyte Phaeocystis pouchetii (Harriot) Lagerh. occurs as single celled flagellate. Colonies are built up by thousands of single non-motile cells lying in a peripheral mucus layer; normally no cells occur centrally in the spherical or elongated colonies which may be several centimeters long. In nature as well as in culture both the single cells and the nonflagellated colonies are present at the same time, apparently in changing quantities. This organism will be ideal for experiments on the induction of colony formation. Phaeocystis is a further example demonstrating that organisms can change their organisation type. There are other examples such as the chain forming Ceratium species already mentioned.

Colony forming species are ubiquitous. About 70% of plankton diatoms may be colony forming species, hence being very important in the food chain. Colony forming flagellates are less common although some toxin producing dinoflagellates are colonial; thus colony forming flagellates may be locally very important.

Reproduction

In many phytoplankton species reproduction is differentiated into a vegetative and a sexual cycle. In addition, resting spore formation may complicate the life cycle. As phenomena related to the life cycle may influence considerably population dynamics, this aspect is reviewed briefly. Vegetative reproduction may be by binary fission, being obligate for diatoms and common in many other groups, e.g. Chrysophyta, Cryptophyta, Dinophyta or Euglenophyta. Flagellates may divide either in the motile stage or in special 'dividing cysts' after the flagella are shed. The cell covering of the mother cell may be shared by the daughter cells (diatoms, several dinoflagellates) or both offspring may produce a new cell wall (several dinoflagellates). There is no explanation for this difference. Even species of closely related genera like the dinoflagellates *Peridinium* Ehrenb. and *Protoperidinium* Bergh behave differently. The amount of energy involved in the formation of these thecae may be variable due to the chemical composition of the cell covering, it's thickness and to the uptake kinetics of the cell wall material. It might be that in some groups the organisms deposit a cell covering to get rid of metabolites (see below). Nevertheless, this does not explain why dinophytes behave so differently in production of their theca, made up of organic carbon.

Vegetative reproduction may be by multiple fission, e.g. in *Chalmydomonas* or *Chlorella*. This mode of reproduction is also described for some dinoflagellates, e.g. *Goniodoma* and *Prorocentrum* (Silva, 1965) or *Pyrophacus*, but sexuality could not be excluded. Multiple fission will result in a higher number of offsprings per division. If division takes place in phase or synchronously, cell number increases to a multiple of the original value in a short time period. The question, whether in principle species with a multiple fission have a higher mean doubling time of the population than equal sized species reproducing by binary fission has not been tackled to my knowledge.

During vegetative reproduction, many diatoms reduce their size. This process cannot continue indefinitively, therefore at a critical size a cyclic enlargement has to take place. This is normally by auxospore formation, related to sexuality (Drebes, 1977), but vegetative cell enlargement can take place also. The auxospore formation is favoured by several environmental conditions, e.g. temperature, light and others. Hence sudden changes in the environment and subsequent auxospore formation can induce a dramatic change of the size spectrum of the phytoplankton in a short time. If auxospore formation is suppressed, cells become too small to form auxospores and cells will die after several further divisions. These processes may influence species distribution and succession.

In many diatoms, there is an obligate alteration of vegetative and sexual reproduction. In other groups, sexual reproduction is rarely encountered or entirely unknown (blue greens). Nevertheless, gamete formation may influence species abundance to a large extent. In Ceratium, for instance, male gametes are formed by a 'depauperating' division (Stosch, 1964) resulting in very small gametes with a short survival time. In Kiel Bight, about 80% of Ceratium tripos (Muller) Ehrenb. cells forming a dense bloom, started to form male gametes on one day. In 5 days the population decreased to only 5% by this process. The same result has been encountered in culture for several Ceratium species (Elbrächter, unpubl.).

Zygote formation may result in resting cyst formation, at least in several dinophytes and chrysophytes. These resting cysts may have a very high sinking rate compared to vegetative cells. Thus a population may disappear very quickly from the plankton if zygote formation takes place simultaneously. On the other hand, germination of dormant resting cysts induced by environmental changes may result in the sudden appearance of high populations of vegetative cells. Blooms of species, in particular of toxic dinoflagellates, may be initiated by this process. The asexual formation or germination of resting spores of other groups, e.g. in diatoms, can have the same effect for population dynamics of phytoplankton. It may be noted that the formation of spores is restricted to distinct species. Even in the same genus species may or may not form resting cysts and spores.

In conclusion, life cycle phenomena can influence species abundance, species succession and size class distribution of the phytoplankton to a large extent. Processes like resting cyst formation with increased sinking rates can influence the vertical transport of organic matter as can germination of cells from resting cysts or spores. Resting cysts and spores are commonly resistant to extreme environmental conditions. They will not be digested even if ingested by zooplankton.

Life cycle stages are rarely dealt with during routine phytoplankton investigations. The relative importance of these phenomena to the flow of energy and matter in the pelagic ecosystem is therefore more or less unknown.

Cell Covering

The cytoplasm is enclosed by a continuous membrane, the plasmalemma. This membrane represents the sole cell covering in only a few organisms such as Dunaliella and some gamentes and zoospores. Most phytoplankton cells have additional layers of various materials forming a more or less rigid cell covering. These layers may be inside or outside the plasmalemma, resulting in an endo- or exoskeleton.

As most or all parts of the cell covering are formed inside vesicles originally located in the cytoplasm, this separation is somewhat arbitrary and partly depends on the time of observation. A good example is provided by diatoms (Schmid and Schultz, 1979; Schmid *et al.*, 1982), in which silica deposition starts in the silicalemma, inside the plasmalemma. In the final stage, the plasmalemma is inside the frustule which is composed of several elements (for terminology see Anonymous, 1975). The valve of diatoms is formed by a thin layer of amorphous silica on which additional silica is deposited in a regular pattern, characteristic of the species. True perforations of the silica layer are present only in special structures, e.g. the strutted and labiated processes in *Thalassiosira eccentrica* (Ehrenb.) Cleve. The situation in the girdle region is unclear. Silification patterns may be quite different, some species show no silification at all whereas other species are heavily silicified.

In dinophytes, the cell covering or amphiesma is composed of several layers. The outermost membrane represents the plasmalemma beneath which is a layer of vesicles which are surrounded by membranes. In thecate species, wall material of compounds similar to cellulose is deposited inside these vesicles. The size of the vescles is variable from species to species as are the size of the thecal plates. Between the vesicles trichocyst pores or small channels are located. In the flagellar pore region the cytoplasm apparently is covered only by the plasmalemma (Dürr, 1979). Plates, if present may be of quite different number and size and they may be smooth or heavily ornamented. Solid spines or horns may be present.

The calcareous scales (coccoliths) of the coccolithoporids are preformed inside the plasmalemma in special vesicles and later transported to the cell surface. One cell may construct different coccoliths, so that some species may have 3 or even more different types of coccoliths. Many coccoliths together build up the complicated and species specific cell wall. The function of these structures is unknown.

In other groups, calcareous or silicified scales may be present. Cells of some species may be enclosed in mucus layers. In several species the single cells are loosely connected by mucoid compounds which are destroyed very easily, particularly during fixation. Thus the large diatom *Coscinodiscus granii* Gough can be surrounded by such mucus and several cells may be embedded in mucus. *C. nobilis* Grunow can produce so much mucus that fishing trawls are clogged (Boalch and Harbour, 1977).

The significance of the cell covering and the mucus is not clearly understood. Heavy cell walls will give a good protection against mechanical forces. At the same time they enhance the specific weight causing higher sinking rates. Mucus layers result in an enlargement of the volume. It may be suggested that the specific weight of the mucus is more or less identical to that of the surrounding medium and, therefore, mucus will reduce the specific weight of the organism. Large mucus layers may prevent grazing of zooplankton and, in addition, parasites may not penetrate through a mucus layer. Mucus layers may absorb nutrients and act as a nutrient reservoir if cells are living in nutrient poor waters.

Solid layers of silica like those of the diatom valve might prevent uptake of nutrients as diatom valves are very effective semi-permeable systems, as pointed out by Stosch (1956). He demonstrated a permiability for water, NH_3 and at least for one of the gaseous components of the air, but they were impermeable for methanol and other organic molecules and for salts. Additional membranes beneath the plasmalemma like those of the vesicles in dinophytes may slow down nutrient uptake and augment energy needed as two more membranes have to be passed. The cell covering may also reduce light availability. Thus a rigid cell covering may not be an advantage. Nevertheless, species with thick cell coverings of different chemical composition and naked cells live in all environments. Even in sandy beaches where sand grains are permanently moved by waves, species with all sorts of cell covering types included 'naked' ones flourish. As Stosch (1980) speculates for the endochiastic areolae of diatoms, the cell wall 'might be a means for the cell to get rid of formative energy' or of metabolites.

Size

Size of phytoplankton species varies to a large extent as already mentioned. Single cells may be from less than 0.5 µm to more than 1 cm, or, expressed as organic carbon per cell, from less than 1 pg C to more than 200,000 pg C. The size of colonies may range from 1 µm to several centimetres. The influence of size on various aspects of life provoked several generalizations leading also to a classification of functional types of primary producers. Recently, Malone (1981) gave a comprehensive review on algal size. Here only few aspects are mentioned which might be important for modelling.

One aspect is size and sinking rate, although shape cannot be separated in discussing this topic (see below). Sinking velocity of spherical cells increases with increasing cell size (Smayda, 1970; Bienfang, 1980), but, proportionally to their size, larger cells sink slower (Smayda, 1970). Cell morphology may explain this for centric diatoms. Most centric diatoms have a large central vacuole whereas the cytoplasm only covers the cell wall as a thin layer. Thus the percentage of cytoplasm to total volume may decrease with increasing cell size. If the specific weight of cytoplasm is higher than that of the vacuole fluid, sinking velocity will be relatively reduced for larger cells.

Other groups (such as most photosynthetic dinoflagellates, chrysophytes etc.) have no large vacuole. Thus the sinking rate of equal sized and shaped species may be different from that of diatoms. Sink ing rate may be also influenced by physiological processes, e.g. production of reserve material like oil or lipids. Cell wall may be deposited in different quantities according to the physiological stat of the cells thus giving further possible changes in specific weight and sinking rate. In consequence, no strict correlation between cell size and sinking rate exists. An extensive review of sinking and floating of phytoplankton is given by Walsby and Reynolds (1980).

Growth is generally accepted to be correlated with size (Banse, 1976; Sournia, 1981), smaller cells growing faster than larger ones. Although this may be true for many diatoms, this generalization does not apply for all of them and in dinoflagellates such a correlation has to be denied. Thecate dinoflagellates of medium size as Prorocentrum micas Ehrenb., Ceratium lineatum (Ehrenb.) Cleve as well as other small Ceratium species like C. furca (Ehrenb.) Clap and Lach. C. fusus (Ehrenb.,) Duj. or C. teres Kofoid all have the same Mean Doubling Time (MDT) of two days under optimal conditions. The large C. tripos (O.F. Müller) Nitzsch also has a MDT of two days but has 25 times as much carbon content per cell as C. lineatum and the other smaller species mentioned above. This implies that the production rate of organic carbon per unit organic carbon is constant in various dinoflagellates species and not dependent on size as is apparently the case for many diatoms. Diatoms seem to grow faster e.g. the large Coscinodiscus granii has the same carbon content per cell as Ceratium tripos and a 25 times higher carbon content per cell than the smaller dinoflagellates mentioned above but has a MDT of only 1 day. Even the large Conscinodiscus concinnus W. Smith with a carbon content 50 times higher than that of Ceratium lineatum has a MDT of 32 hours whereas that of C. lineatum is 48 hours (Table 1).

Diatoms undergo a cyclic size change during growth, being largest after auxospore formation and smallest just before auxospore formatic Some diatoms show an influence of cell size on MDT if large and small populations of the same clone are investigated under constant conditions. Others do not even if they are closely related species of the same genus, see Table 2, after Baars (1981). The question arises whether those species showing no influence of cell size on MDT have different carbon contents per cell if regarded small and large cells. Does there exist a correlation between carbon content per cell and MDT even if there is no correlation between size and MDT? This has to be taken into consideration if using conversion factors for size to carbon content.

Theoretically, size should be correlated to nutrient uptake rate at least in spherical cells, as nutrients are transported via the surface of the plasmalemma into the cell. Therefore cells with a

Table 1. Correlation between sizes, Mean Doubling Time (MDT) and production of organic carbon for various phytoplankton species; 1) organism; 2) cell diameter in μm; 3) organic carbon content per cell in pg; 4) MDT in hours at optimal culture conditions; 5) number of days after which 10^8pg org. C dm^{-1} is obtained if starting with 1 cell dm^{-3}; 6) number of cell divisions to obtain this; 7) number of days after which 10^8pg org. Cdm-3 is obtained if starting with 1 cell m^{-3}; 8) number of cell divisions to obtain this. Whereas in diatoms obviously size influences growth and production, in dinoflagellates growth and production is apparently independent of size.

1	2	3	4	5	6	7	8
Flagellate	3	1	4	4.5	27	6	37
<u>Thalassiosira rotula</u>	45	1000	10	6.5	17	11	26
<u>Thalassiosira eccentrica</u>	90	5000	16	10	15	16	24
<u>Coscinodiscus granii</u>	150	25000	24	12	12	22	22
<u>Coscinodiscus concinnus</u>	220	50000	32	14	11	28	21
<u>Prorocentrum micans</u>	50 x 30	1500	48	32	16	52	26
<u>Ceratium lineatum</u>		1000	48	33	17	52	26
<u>Ceratium tripos</u>		25000	48	24	12	44	22

high ratio of surface to volume can have high uptake rates as paths to the enzymatic reaction centres are shorter than in larger cells. In nutrient saturated environments size should be less important for growth than in nutrient limited environments. Malone (1980) discusses this topic in detail, in particular contradictionary experimental results. Recently, Goldman and Glibert (1982) found no

Table 2. Thalassiosira nordenskioeldii and Th. rotula; Influence of cell diameter (size) on Mean Doubling Time (MTD) at various temperatures, light intensities and light-dark-cycles (LDC). In Th. nordenskioldii MDT is size dependent. in Th. rotula MDT is size independent. Data after Baars (1981).

Thalassiosira nordenskioeldii

Temp (°C)	Light (lux)	LDX (h)	Size (μm)	MDT (h)
-1.5	1700	14-10	13.5	67.4 (65.3 - 69.7)
			22.7	76.8 (73.7 - 80.2)
			36.2	135.4 (130.0 - 141.3)
0.0	900	14-10	20.2	82.5 (75.5 - 90.9
			35.7	98.5 (95.1 - 102.0)
0.0	1700	10-14	20.2	35.6 (33.7 - 37.7)
			35.7	55.6 (53.5 - 57.9)
0.0	1700	14-10	18.2	32.3 (31.3 - 33.3)
			30.4	46.1 (45.0 - 47.3)
			41.8	81.0 (78.6 - 83.5)
6.0	4000	6-18	16.1	25.7 (24.8 - 26.6)
			29.4	38.4 (37.0 - 40.0)
6.0	1700	14-10	17.2	22.1 (21.1 - 23.2)
			21.0	24.0 (23.0 - 25.0)
			31.6	26.6 (25.8 - 27.4)
			43.5	31.4 (30.5 - 32.4)

Thalassiosira rotula

Temp (°C)	Light (lux)	LDX (h)	Size (μm)	MDT (h)
6.0	1700	6-18	22.7	62.0 (60.8 - 63.2)
			36.5	61.5 (60.1 - 63.0)
6.0	1700	10-14	23.0	41.5 (40.7 - 42.2)
			37.9	41.2 (39.8 - 42.6)
6.0	1700	14-10	24.5	28.8 (28.0 - 29.7)
			38.4	28.8 (27.8 - 29.9)
6.0	1700	18-6	20.3	27.7 (26.9 - 28.6)
			46.6	28.2 (27.6 - 28.9)
18.0	4000	14-10	21.6	12.7 (12.3 - 13.1)
			36.2	11.8 (11.0 - 12.9)

correlation between cell size and rapid nutrient uptake rates in short time experiments in four plankton species. Extensive additional experiments have to be carried out before generalizations should be made.

The size of phytoplankton cells is very important in respect to grazing (Frost, 1980). There is a positive correlation between particle diameter and probability of contact with a filter feeder. On the other hand, most grazers can feed only on particles of a discrete size range. For optimal transfer of energy and material in the planktonic food web, the size of the primary producers has to match that which grazers can use. Drastic changes of size classes, e.g. by auxospore formation of diatoms or gamete fromation may influence the transfer of primary producers to consumers drastically. The size range of primary producers regulates the feeding efficiency of consumers whereas consumers regulate the size class distribution of the primary producers.

Shape

Shape is one of the most commonly used features for taxonomic species separation. This illustrates how variable the shape of phytoplankton cells can be. Grouping into functional types has been made as early as 1892 by Schütt, later by Gran (1912) and others. Besides small spheres the following main types were discerned:
a) bladder type: large cells with a thin peripheral cytoplasm layer and a large central vacuole (e.g. *Ethmodiscus* Castr., *Pyrocystis noctiluca* Murray *ex* Haeckel);
b) ribbon type: flattened, thin walled cells which may be joined together by their valves to form colonies (e.g. *Fragilaria* Lyngby)
c) hair type: narrow, elongated cells (e.g. *Rhizosolenia alata* Brigh) or colonies (e.g. *Nitzschia seriata* Cleve);
d) branched type; cells with long horns or spines (e.g. *Chaetoceros* Ehrenb., *Bacteriastrum* Shadb.) or colonies in which individual cells are arranged so that they project out at various angles (e.g. *Thalassionema nitzschioides* Grunow).
One may also add the mucous type, where cells are embedded in a mucus layer (e.g. *Coscinodiscus* species; Boalch and Harbour, 1977).

These types are represented either by single cells or by colonies and in nearly all size classes. Organisms of all shape categories just mentioned are present in all environmental habitats and in most cases present in the same sample. Despite this fact, many speculations and 'explanations' of the importance of these different morphological types have been made. One of the most common ones is that of sinking prevention. The development of large vacuoles is said to reduce the specific weight if the vacuolar sap has a lower specific weight than the cytoplasm (Kahn and Swift, 1978, see also above). In this context it is noteworthy that, except in Cyanophyta, no gas vacuoles have been developed in marine phytoplankton. Gas vacuoles regulate buoyancy very effectively and are present in various marine organisms including seaweeds and protozoa. For a detailed discussion of this topic see Walsby and Reynolds (1980).

Each deviation from spherical shape decreases the ratio of surface to volume. Cell appendages like horns, spines, setae, or bristles are assumed to prevent sinking or assist floating (Walsby and Reynolds, 1980). Physically, several effects may have to be considered which are partly contradictary, although hydrodynamics of cells with intermediate Reynold numbers are hardly predictable, up to now. These different effects may be listed:

1) appendages (usually siliceous, calcareous, cellulosic) enhance specific weight, hence increasing the sinking rate;
2) appendages increase drag resistance, decreasing the sinking rate;
3) cell orientation during movement may be modified, influencing drag resistance, resulting either in increase or decrease of sinking rate;
4) passive entrainement by surrounding water is enhanced.

Bienfang *et al*. (1982) demonstrated that sinking rates of various shaped phytoplankton alga were influenced seriously by nutrient availability,limitation of different nutrients resulting in different sinking rates. Hence physiological regulation and adaptation of specific weight is much more important and effective than shape. It is noteworthy that shape varies to a much larger extent in marine than in freshwater phytoplankton species. On the other hand, both specific weight and viscosity are much lower in fresh water than in marine waters. This demonstrates the ability of phytoplankton to regulate the specific weight physiologically rather than by shape.

A further hypothesis is that enhanced surface area due to shape variation (including horns, bristles and other appendages) will enhance nutrient uptake. This view may be correct for those appendages covered by the plasmalemma through which nutrients are taken up, or if nutrients can diffuse through the cell wall material coming into contact to the plasmalemma. The setae of many diatoms, e.g. *Bacteriastrum*, *Corethron* Castr. or that of many *Chaetoceros* species are apparently not connected with the plasmalemma as this lies inside the frustule (Schmid *et al*. 1982). The situation in the many dinoflagellates with elaborate wings (e.g. *Ornithocerus* Stein or *Histioneis* Stein; Figure 1) is not certain. Nutrients may adsorb to the cell wall appendages although the transport mechanisms into the cell remain unclear. Mucilage production may also result in enhanced nutrient adsorbtion but also in adsorbtion of toxic substances like heavy metals - a possibility of self-conditioning the environment.

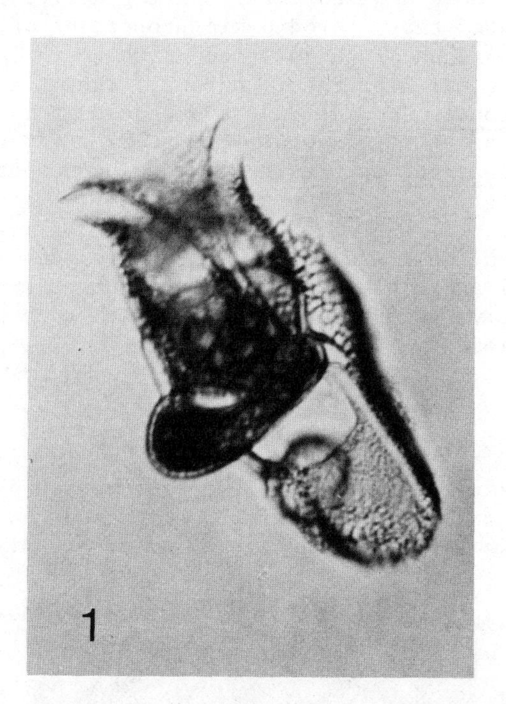

Fig. 1. Histioneis mitchellana Murray and Whitting (Dinophyta) from formalin preserved samples; note the elaborate wings and lists.

The effect that cell shape may have on the passive drifting of cells in currents, turbulence and other water movements is partly the same as discussed for sinking (see above). Cell shape may enhance passive drifting in turbulent waters and this may facilitate water exchange around the cells, an advantage for nutrient uptake, particularly in nutrient poor waters. Nutrients may be depleted very quickly around the cells, as Goldman and Glibert (1982) reported cases of very rapid nutrient uptake. The behaviour of variously shaped cells in the water current generated by filter feeders is more or less unknown. Only recently, Paffenhöfer et al. (1982) have demonstrated that by microcinemato-graphic methods this question can be tackled. Have cells of any peculiar shape a better chance of not being caught by filter feeders or, in contrast, do long setae of Chaetoceros species or horns of Ceratium species enhance the probability of being caught?

The shape of some tropical dinoflagellates like Ornithocercus, Citharistes Stein or Histioneis (Figure 1) are really fantastic. Although these organisms are motile, they have elaborate wings and lists which should be a disadvantage during active swimming. Again it has to be stated that hydrodynamic properties of organisms in this size range are hardly understood up to now. The observations (Taylor, pers. comm.; Elbrächter, unpubl,) that the hypotheca of at least some Histioneis species can be embedded in mucus in the living stage, surrounding all appendages (Figure 2) are noteworthy. In contrast, Ornithocercus species never have been observed to possess such a mucus layer.

In conclusion, cell and colony shape have been correlated to various physical and biological phenomena but at the moment I cannot see an unequivocal experimental proof for such relations between shape and function.

Movement

Passive sinking and floating have been reviewed by Walsby and Reynolds (1980) and have been briefly discussed above with size and shape. Here some aspects of active swimming are discussed.

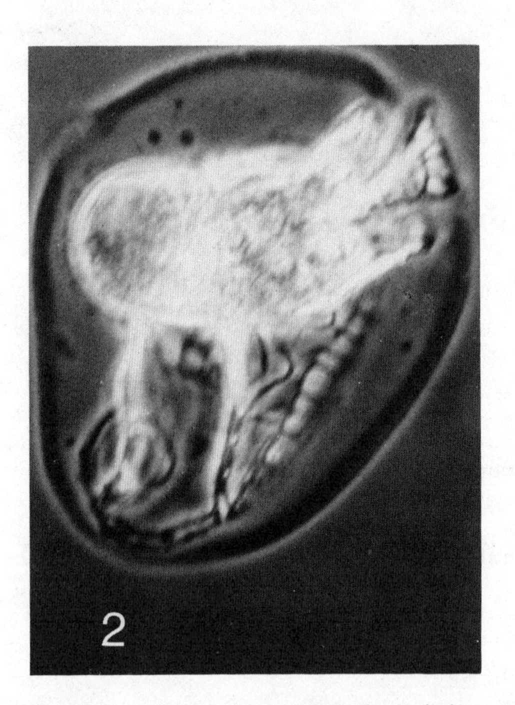

Fig. 2. Histioneis dolon Murray and Whiting (Dinophyta), living; note the mucus layer, enclosing the elaborate wings and lists.

Flagellates are widespread in all environments and in most algal groups. There is evidence that flagellates swim permanently, interrupted only by short intervals. To my knowledge, resting is not reported for flagellates, but several dinoflagellates can form temporary cysts in which they remain for longer periods (Kofoid and Swezy, 1921). Flagellar motion and swimming costs energy and this loss of energy has to be compensated. Throndsen (1973) reported a swimming speed for 12 nanoflagellates between 75 to 370 μm sec^{-1}, corresponding to about 1 m h^{-1}. Vertical migration in the field and in experiments seem to be in the same order of magnitude (Hasle 1954; Eppley et al., 1968; Staker and Bruno, 1980). Hence swimming speed is higher than average sinking rate. Apparently, swimming is not (only ?) used to compensate sinking.

During flagellar movement the exchange of water surrounding the cell is enhanced. Hence, nutrient uptake, particularly in nutrient poor waters, may be augmented. Aspects of water flow have recently been discussed by Fenchel (1982) for zooflagellates and are dealt with in this volume by Fenchel. Leblond and Taylor (1976) have discussed water flow for dinoflagellates with large lists and wings, e.g. Ornithocercus. The junction of cingulum and sulcus is the spot on the cell surface in contact with the greatest volume of external medium in a given time. In this region the plasmalemma is in direct contact with the cytoplasm, not separated by 2 additional membranes including thecal plates. Thus nutrient uptake may be facilitated.

The phenomenon of vertical migration is well documented for a number of flagellates, in particular for dinoflagellates, e.g. Staker and Bruno (1980). It is unknown whether this migration represents geotaxis or phototaxis. A cell organelle for perception of gravitation is unknown although the sacpusule of dinoflagellates may act as such an organelle if the fluid of the sacpusule has a specific weight different from that of the cytoplasm. Most of the species showing vertical migration have no eyespot although normal chromatophores may also allow phototactic behaviour. Phototactic induced migration is most probable as this phenomenon has been divided - according to the time in the light-dark-cycle - into dawn ascending and dusk ascending migration. In addition, endogenous rhythms are involved (Sournia, 1974; Forward, 1976; Kamykowski, 1981). Kamkowski (1981) provided a computer model showing that a vertical migration phase provides little selective advantage for shallow populations and for populations in low turbidity water columns. A strong selective advantage was computed for deep populations or for those in high turbidity waters. Horstmann (1981) claimed that Peridinium cf quinquecorne Abe shows a tidally controlled vertical migration in a shallow lagoon and claimed that this behaviour should prevent the population being washed out into the open sea.

A disadvantage of vertical migration for the whole population will be that cells concentrate in a defined water layer. This makes filter feeding of herbivores much more efficient in comparison to randomly distributed cells in the water column. Nutrient uptake in discrete water layers is another aspect of vertical phytoplankton migration. Hence vertical migration complicates measuring, predicting and modelling processes related to primary production.

Nutrition

a) Nutritional types:

Photosynthesis is one mode of nutrition but many photosynthetic organisms have the capability of additional modes of nutrition and, therefore a short review on nutritional types and their relative significance in the pelagic ecosystem is provided. A detailed review of nutritional types for dinophytes is provided by Morey-Gaines and Elbrächter (in press). Chemosynthesis is not referred to in this review, being important in the pelagic ecosystem only in the Black Sea and near hot vents in the deep sea, so that we are dealing with only autotrophic organisms using energy derived from sunlight, and water as the electron donor. Conversely heterotrophy is the term used in this paper to describe the use of organic compounds, taken up from the environment, involving a benefit for survival of the species during a complete life cycle. As several phytoplankton species are able to share photosynthetic and heterotrophic nutrition, it is important to distinguish the degree to which a species relies upon heterotrophy over autotrophy, and the form in which the nutrients are obtained. These two factors largely describes the diversity of nutrition in phytoplankton species.

It is important to stress that heterotrophy is growth or increased survival of the organism by using organic substances. Uptake of organic substances is not sufficient to establish true heterotrophy (Richardson and Goff, 1982). There are numerous examples of efficient uptake of organic substances by algae without any evidence of increased growth or survival (Hellebust and Lewin, 1977; Baden and Mende, 1978).

Species are called autotrophic if nutrition and all metabolic steps do not require uptake of any external organic compound but are supported only by photoassimilation. True autotrophy is very difficult to establish and only very few phytoplankton species are now regarded as true autotrophs. Most photosynthetic species require vitamins or other organic substances, and with more sophisticated culture techniques, more and more phytoplankton species are proving to be auxotrophs (see below). Thus only few marine diatoms are thought to be true autotrophs (Hellebust and Lewin, 1977). Only two truely autotrophic marine dinoflagellates are known (Morey-Gaines and Elbrächter, in press). Although little or nothing is known on nutritional requirements of the other flagellates, in particular of the so called nanoplankton and picoplankton, it can be guessed that the great bulk of marine phytoplankton is auxotrophic.

Species are called auxotrophic if nutrition and cell metabolism is supported mainly by photoassimilation of inorganic compounds but some specific external organic compounds are required in small amounts (e.g. vitamins for catalytic processes). Most of the phytoplankton species known to have chromatophores seem to be auxotrophic. Thus this group of organisms is the most important group of primary producers. They rely on the presence of the substances they need and therefore, the presence of vitamins can influence species abundance, species succession, and primary production in the sea. For further discussions see section on "Nutrient uptake' (below) and Swift (1980).

Species with chromatophores are called mixotrophic if their nutrition is alternatively or simultaneously by photoassimilation and by uptake of organic compounds. Mixotrophy is difficult to establish as it has to be proved that the organic substances taken up are used as energy source. Mixotrophy can be presumed in the case a photosynthetic organism which feeds phagotrophically, but even in these cases auxotrophy cannot be excluded and it is possible that the prey is used only as a "vitamin pill". Mixotrophic uptake of dissolved organic substances has to be proved by experiments. Whereas several dinophytes are mixotrophic (Morey-Gaines and Elbrächter, in press) mixotrophy in other marine phytoplankton groups is less well investigated except for diatoms (Hellebust and Lewin, 1977). As it is very difficult to establish mixotrophy, the role, mixotrophs play in the marine pelagic ecosystem is more or less unknown.

For completeness, the last large group of nutritional types is briefly mentioned, that is organotrophy in which the nutrition of the organisms is exclusively by uptake of external organic compounds, obligately for organisms without chromatophores. In various groups of phytoplankton algae there are species lacking chromatophores. Whereas there are only very few such diatoms, about half of the dinophytes are considered to be organotrophic and some chrysophytes, cryptophytes and euglenophytes have no chromatophores, although they are generally classified as phytoplankton. Planktologists counting phytoplankton with the inverted microscope should take this fact into consideration as well as modellers. It is common practice to regard all dinoflagellates as phytoplankton in the sense of primary producers although apparently about 50% of the species have no chromatophores.

The method of food uptake provides another way of classifying nutritional types. Uptake of dissolved organic substances is called osmotrophy although this uptake is in most cases by active uptake, not by the process of osmosis as may be suggested by the term. Although this term is misleading, it is in common use and no better one has been suggested to my knowledge. How much osmotrophic nutrition supplements photosynthesis in phytoplankton species with chromatophores is unknown but it may be suggested that in most pelagic ecosystems concentrations of dissolved organic carbon are so low that unicellular algae cannot compete with bacteria. Thus only in special conditions like during a bloom in the break down phase or in rock pools osmotic mixotrophy may be of importance.

Uptake of particulate matter is called phagotrophy. Species with chromatophores which ingest other organisms in addition to photosynthesising are well documented in dinoflagellates. Even Ceratium species, commonly regarded as primary producers, are said to feed on particles - for a discussion see Morey-Gaines and Elbrächter (in press). This way of nutrition may be more widespread than hitherto known. The part phagotrophic mixotrophs contribute to the primary production is unknown.

It should be just mentioned that some parasitic dinoflagellates have chromatophores and apparently have to rely on photosynthesis during part of their life cycle, as recently demonstrated for Dissodinium pseudolunula Swift ex Elbr. and Dreb. (Drebes, pers. comm.).

Last but not least one group of primary producers has to be mentioned, i.e. those organisms which have photosynthetic endosymbionts. The ciliate Mesodinium rubrum, together with its endosymbiotic cryptomonad (Hibberd, 1977) is a photosynthetic entity which can form large blooms, apparently with a high primary production rate (Packard et al., 1978). Other such associations are those of foraminifera and dinoflagellates or radiolarians and their endosymbionts. The part these photosynthetic active symbionts contribute to total primary production has not been calculated to my knowledge. The direct influence on the flow of energy and matter through the food web may be low but indirectly they may support growth or survival of their host organisms. As some of these are voracious feeders (Bé et al., 1977, for foraminifera) the indirect contribution to that flow may be higher than expected.

b) Nutrient uptake pattern and competition:

During photosynthesis, inorganic compounds are incorporated into the cell. Availability of these so called nutrients like nitrogen, phosphorus and silicon or trace metals like iron, cobalt etc. regulate growth. The strategies of uptake of these nutrients may be different in different species. As most phytoplankton species are auxotrophs, that is they need vitamins, the availability of these compounds may influence primary production. Again several species excrete organic substances which might influence the growth of other species. These factors together largely influence phytoplankton abundance, primary production and species succession. Some aspects of these complex interactions should be briefly mentioned.

1) Inorganic nutrient uptake:

Nutrient uptake in general has been reviewed by Raven (1980) and will be dealt with in the present volume by Goldman. Uptake of nutrients like nitrogen, phosphorus, silicon and others has been described by Michaelis-Menton equation and it has been proved, that different species may have different demands, affinities, and uptake speeds for these nutrients (e.g. McCarthy, 1980 for nitrogen; Nalewajko and Lean, 1980, for phosphorus; Paasche, 1980 for silicon; Huntsman and Sunda, 1980, for trace metals). MacIsaac and Dugdale (1972) investigated the uptake of nitrate and ammonium as a fraction of light which can be adequately described by the Michaelis-Menton equation. They defined a constant K_{Lt} which is analogous to a half-saturation constant and determined that this value usually occurs at the subsurface depth to which 1 to 14% of surface light is penetrating. Bates (1976) suggested that different K_{Lt} values for Skeletonoma *costatum* (Grev.) Cleve and an unidentified coccoid chlorophyte may explain the alternating seasonal abundance of these two species. Falkowski (1977) described a theoretical model which describes phytoplankton growth based on kinetics of nitrogen and light. Different affinities to phosphorus and nitrogen have been demonstrated for *Skeletonema costatum* and *Thalassionema nitzschioides* Grunow by Fedorow and Kustenko (1972). Changes in the ratio of nutrients in mixed cultures caused subsequent changes in biomass of the species. Together with luxury consumption (e.g. Droop, 1974, 1975), uptake kinetics of nutrients may influence drastically growth patterns and biomass production, at least in multi-species-cultures, but perhaps also in nature. Multi-species batch cultures (Elbrächter, 1977) showed quite different cell production patterns depending on the species combined. Three major types can be distinguished:

1) No effect on either of the species co-cultivated. The biomass production expressed as cell number, was the same for each species compared to that in monospecific cultures;
2) No effect on one of the species but on the other:
   a) biomass less than the sum of those in monospecific cultures,
   b) biomass more than the sum of those in monospecific cultures;
3) All species effected:
   a) biomass less than the sum of those in monospecific cultures,
   b) biomass more than the sum of those in monospecific cultures;

As nutrient addition to stationary phase cultures caused further growth, these effects are ascribed to different nutrient uptake patterns. Different uptake types and rates for the nutrients by the different phytoplankton species may be responsible for species composition and succession in the natural environment, at least in regions or seasons in which nutrients are limiting for primary production. In future, more sophisticated methods have to establish species specific nutrient uptake rates in the natural environment. Only then their importance for primary production, species composition and succession and other phenomena can be established.

Lack of nutrients caused variation of sinking rates (Bienfang et al., 1982). In three of four species investigated, depletion of N and P caused lower sinking rates but Si depletion caused enhanced sinking rates. This mechanism may show how complicated the interactions of nutrient depletion may be for the plankton community.

2) Vitamins and pheromones:

Most photosynthetic phytoplankton species are auxotrophic, relying on external sources of vitamins. Recently, Swift (1980) provided an extensive review on this topic. Here only some aspects related to species interactions and mutual influence on production capacity are mentioned. Growth of phytoplankton organisms may be influenced drastically by the availability of vitamins. Different species may have the need of different vitamins (Swift, 1980). Phytoplankton species which do not require an external source of one special vitamin may synthesize and excrete that particular vitamin during growth as do some bacteria. Phytoplankton species excreting one vitamin may have a need for an other one. Carlucci and Bowes (1970) demonstrated that in bialgal cultures the coccolithophorid *Emiliania huxleyi* (Lohm.) Kampt. produced vitamin B_{12} which was used by the diatom *Thalassiosira pseudonana*, and *Skeletonema costatum* could produce biotin for *Amphidinium carterae*. Pintner and Altmyer (1973,1979) demonstrated that phytoplankton can excrete organic substances which for example bind the vitamin B_{12}. As cells do not take up bound vitamin B_{12}, species excreting such binding factors may play an important role in species composition and succession as well as in primary production. These processes have been little investigated up to now and the same applies to the influence of excreted antibiotics and pheromones (for a review see Aubert et al., 1980).

ENVIRONMENTAL FACTORS

Each organism is adapted in one way or the other to the environment it inhabits. Numerous publications exist dealing with various aspects of the interactions of the environment and the organisms (for reviews see e.g. Falkowski, 1980; Morris, 1980; Lange et al., 1981; Platt, 1981). Here only some reactions to environmental factors are dealt with briefly, which might show that generalization should be made only with caution. First some commonly adopted biotopes in which planktonic phytoplankton species are abundant are dealt with, then some physical factors.

Habitats

Many species are restricted to special habitats whereas others seem to be ubiquitous. Thus environment suggests possibilities of classifying the phytoplankton into functional types. The reason and mechanism for such separation of the species to habitats is not known but special habitats could even be classified by their species communities. Here only some different habitats are mentioned and biogeographical aspects are not dealt with. A critical discussion of some of the terms used here is provided by Smayda (1958).

Oceanic species and genera are those which only occur in the true oceanic waters, not influenced in any way by the shore. Oceanic species may be transported to near shore waters by currents but they do not thrive in such neritic waters. Apparently there are only few diatom genera like *Ethmodiscus* Castrac., which can be classified as oceanic whereas dinoflagellate genera like *Histioneis* (Figures 1,2), *Ornithocercus*, *Amphisolenia* Stein and others seem to be exclusively oceanic.

Neritic species and genera are those living in near-shore waters. Most of the well known species are neritic and more or less all species which are cultivated are neritic or even rock pool species. These species may form dense blooms and are abundant in shallow waters thus forming the main nutritional source for most of the animals important for fisheries. Also commercial shellfish fishery and most aquaculture systems rely directly or indirectly on neritic primary producers. Toxic red tides are formed by neritic species (Taylor and Seliger, 1979).

In rock-pools and tidal-pools another group of unicellular algae live and from this environment most of the algae were isolated on which the majority of culture work has been done until recently. Species of the genera *Monochrysis*, *Prymnesium*, *Dunaliella*, *Chlorella*, *Chlamydomonas*, *Nannochloris*, *Phaeodactylum* (Droop, 1954) are examples as is *Amphidinium carterae* (Hulburt, 1957). These species contribute nothing or little to the flow of energy and matter in the pelagic ecosystem, being absent from that habitat . Nevertheless most physiological data produced or used in order to understand the oceanic pelagic ecosystem are taken from experiments with these species.The question, how far such data should be used as representative for modelling the pelagic oceanic ecosystem should be discussed seriously.

Estuaries are inhabited by special phytoplankton communities with species often forming large blooms as they are tolerant to low salinity and/or high nutrient concentrations. Being washed out permanently into the open sea, this estuarine plankton contribute an important part in shelf ecosystems.

Neustonic are called thosed species living in the upper millimeters of the water or even attached to the surface of the water. Those species sitting on the surface of the water are called epineustonic. Norkrans (1980) provided a review of this environment. The question, whether true neustonic unicellular algae exist in the marine environment has hardly been tackled, although in freshwater, species of all major phytoplankton groups are specialized to this strange habitat. Wandschneider (1979) investigated this habitat but did not look at the living samples. Fragile, unarmored flagellates and rhizopodial species may have escaped observation. Hardy and Valett (1981) found several species predominant in the neuston compared to subsurface layers. In particular, in regions where "slicks" occurred neuston plankton was abundant. Neustonic copepods are more or less exclusively carnivores thus it is unlikely that there is a high primary production in the neuston although colourless flagellates and ciliates may act as consumers (Sieburth et al., 1976) The indirect influence the neustonic primary producers may have on the primary production in subsurface layers by changing the physical and chemical properites (e.g. light penetration, gas exchange) of the water-air interface is hardly understood up to now.

Shade-plankton species are those which are adapted to very low light intensities, living exclusively near the lower boundary of the euphotic zone. Apparently, the part these organisms contribute to total primary production is not very high, although special investigations obviously are lacking. Sournia (1982) has prepared a critical review on this topic.

Light

Light is the energy source for photosynthetic primary producers and is highly variable in the environment (Jerlov 1968; Steeman-Nielsen, 1975; Lüning, 1981). Photosynthesis and growth in different species varies depending on light intensity and light quality. Here only a few aspects will be mentioned. In general, increase in light intensity causes an increase of both photosynthesis and growth until light saturation is reached. A further increase above light saturation levels does not result in further increase of photosynthesis or growth and there is evidence that light saturation for growth is different from that for photosynthesis. A drastic increase above light saturation level results in light-damage. Different species may have quite different levels for light saturation and light-damage. In addition, populations of a given species may be able to adapt to light and shade (Perry et al., 1981) but species may have different strategies (Falkowski and Owens, 1980). Phytoplankton cells are exposed to short time fluctuations of light quantity and quality by physical changes of the environment (sunshine - clouds; orbital movement due to turbulences; internal waves). The effect of fluctuating light on photosynthesis has recently attracted attention, e.g. by Gallegos *et al*. (1980) and Savidge (1980) but to my knowledge no study has been published on the influence of fluctuating light intensities on phytoplankton growth rate.

Hundreds of photoperiodic responses are known for higher plants (Vince-Prue, 1975) and recently Lüning (1981) documented some examples for sea-weeds. For phytoplankton species apparently no unequivocal experimental proof exists. There is a hint of photoperiodic responses in *Ceratium hexacanthum* Gourret; this species does not grow in culture if the light phase lasts more than 12 hours (Elbrächter, unpubl.). It has to be proved by special experiments whether photoperiodic phenomena are widespread also in phytoplankton species as such phenomena are important for calculating and modelling primary productivity from biomass data.

Temperature

Besides light, temperature is the second main factor influencing photosynthesis and growth. Photosynthesis as a photochemical reaction should not be dependent on temperature but as enzymatic processes are involved in photoassimilation, also this process is temperature dependent. Growth is highly influenced by temperature. Eppley (1972) provided a good review on temperature and phytoplankton growth and Li (1980) reviewed temperature adaptation on a cellular basis. Baars (1981) provided many data on the mean doubling time of 50 marine diatoms at various temperatures. In general, each species has its temperature optimum for growth. At lower temperatures growth slows down and most species tolerate for weeks and months temperatures far below the optimum. Increase of temperature above the optimal range very soon results in temperature stress and death. Few phytoplankton species tolerate 35°C. Few algal species are able to adapt to various temperatures and have nearly equal photosynthetic and growth rates over a broad range of temperatures (JΦrgensen, 1968; Steeman-Nielsen, 1975 for *Skeletonema costatum*). Diatoms in northern seas seem not to be very well adapted to temperatures below 0°C even if thriving in nature at these low temperatures (Baars, 1981). In contrast, Antarctic diatoms are well adapted to temperatures below 0°C, growing fast at these temperatures and showing no temperature dependence of MDT from -15°C to at least 2°C (Elbrächter, unpubl.). Genetic variability of temperature responses in different clones of a species as demonstrated by Brand *et al*. (1981) has to be taken into consideration for modelling primary production, if proved to be widespread.

Turbulence

Apparently turbulence can have an important influence on primary production rates although this aspect has been neglected by most investigators. Recently, Pollingher and Zemel (1981) demonstrated that cell division is influenced by strong turbulences in *Peridinium cinctum* fo. *westii* (Lemm.) Lef. The short occurence of turbulence during a parcticular period in the cell division cycle of the species can suppress cell division of the whole population. As cells divide only in phase during a particular time of the light-dark-cycle, cell division can occur only 24h later. Such gating phenomena can influence very much cell production and population dynamics. Although *P. cinctum* is a fresh water dinoflagellate the same applies to marine species as White (1976) demonstrated for cultures of *Gonyaulax excavata* (Braarud) Balech. Such gating phenomena and other endogenous circadian rhythms, reviewed by Sournia (1974), may influence primary production very seriously as well as population dynamics but have attracted little interest up to now. Some results on the influence of wind induced waves on phytoplankton are provided by Schöne (1970).

REFERENCES

Aldredge, A.L., and Silver, M.W., 1982, Abundance and production rates of floating diatom mats (*Rhizosolenia castracanei* and *R. imbricata* var. *shrubsolei*) in the Eastern Pacific Ocean, *Mar. Biol.*, 66: 83.

Anonymous, 1975, Proposals for a standarization in diatom terminology and diagnosis, *Nova Hedwigia*, *Beih.*, 53: 323.

Aubert, M., Gauthier, M. and Bernhard, P., 1980, Les systemes d' information des microorganismes marins, *Revue Intern. L'Océan. Médic.*, 21: 1.

Baars, J.W.M., 1981, Autecological investigations on marine diatoms. 2. Generation times of 50 species, *Hydrobiol. Bull.*, 15: 137.

Baden, D.G., and Mende, T.J., 1978, Glucose transport and metabolism in *Gymnodinium breve*, *Phytochemistry*, 17: 1553.

Banse, K., 1976, Rates of growth, respiration and photosynthesis of unicellular algae as related to cell size - A review, *J. Phycol.* 12: 135.

Bates, S.S., 1976, Effects of light and amonium on nitrate uptake by two species of estuarine phytoplankton, *Limnol. Oceanogr.*, 21: 212.

Bé, A.W.H., Hemleben, C., Anderson, O.R., Spindler, M., Hacundra, J., and Tuntivate, S., 1977, Laboratory and field observations of living planktonic foraminifera, *Micropaleont.*, 23:155.

Bienfang, P.K., 1980, Phytoplankton sinking rates in oligotrophic waters off Hawaii, USA, *Mar. Biol.*, 61: 69.

Bienfang, P.K., Harrison, P.J., and Quarmby, L.M., 1982, Sinking rate responses to depletion of nitrate, phosphate and silicate in four marine diatoms, *Mar. Biol.*, 67: 295.

Boalch, G.T., and Harbour, D.S., 1977, Unusual diatom off the coast of South-West England and its effect on fishing, *Nature* 269: 687.

Bold, H.C., and Wynne, M.J., 1978, "Introduction to the algae," Prentice Hall, Englewood Cliffs, N.J.

Bouck, G.B., 1971, The structure, origin, isolation and composition of the tubular mastigonemes of the Ochromonas flagellum, J. Cell. Biol., 50: 362.

Bouck, G.B., 1972, Architecture and assembly of mastigonemes, Adv. Cell Molec. Biol., 2: 237.

Brand, L.E., Murphy, L.S., Guillard, R.R.L., and Lee, H. -t., 1981, Genetic Variability and differentiation in the temperature niche component of the diatom Thallassiosira pseudonana, Mar. Biol., 62: 103.

Carlucci, A.F., and Bowes, P.N., 1970, Vitamin production and utilization by phytoplankton in mixed culture, J. Phycol., 6: 394.

Drebes, G., 1977, Sexuality, in: "The biology of diatoms," D. Werner, ed., Blackwell, Oxford.

Droop, M.R., 1954, A note on the isolation of small marine algae and flagellates for pure cultures, J. Mar. Biol. Ass. U.K., 33: 511.

Droop, M.R., 1974, The nutrient status of algae in continuous culture, J. Mar. Biol. Ass. U.K., 54: 825.

Droop, M.R., 1975, The nutrient status of algal cells in batch culture, J. Mar. Biol. Ass. U.K., 55: 541.

Dürr, G., 1979, Elektronenmikroskopische Untersuchungen am Panzer von Dinoflagellaten II Peridinium cinctum, Arch. Protistenk., 122: 88.

Elbrächter, M., 1977, On population dynamics in multi-species cultures of diatoms and dinoflagellates, Helgoländer wiss. Meeresunters., 30: 192.

Elbrächter, M., and Boje, R., 1978, On the ecological significance of Thalassiosira partheneia in the Northwest African upwelling area, in:"Upwelling ecosystems," R. Boje and M. Tomczak, eds., Springer, Berlin, Heidelberg, New York.

Eppley, R.W., 1972, Temperature and phytoplankton growth in the sea, Fishery Bull. 70: 1063.

Eppley, R.W., Holm-Hansen, O., and Strickland, J.D.H., 1968, Some observations on the vertical migration of dinoflagellates, J. Phycol. 4: 333.

Ettl, H., 1980, "Grundriß der allgemeinen Algologie," G. Fischer, Stuttgart.

Falkowski, P.G., 1977, A theoretical description of nitrate uptake kinetics in marine phytoplankton based on bisubstrate kinetics, J. theor. Biol., 64: 375.

Falkowski, P.G., and Owens, T.G., 1980, Light-shade adaptation. Two strategies in marine phytoplankton, Pl. Physiol., 66: 592.

Fedorov, V.D., and Kustenko, N.G., 1972, Competition between marine planktonic diatoms in monoculture and mixed culture, Oceanology, 12: 91.

Fenchel, T., 1982, a, Ecology of heterotrophic microflagellates. I. Some important forms and their functional morphology, Mar. Ecol. Prog. Ser. 8: 211.

Fenchel, T., 1982 b, Ecology of heterotrophic microflagellates. II Bioenergetics and growth, Mar. Ecol. Prog. Ser. 8: 225.

Forward, Jrn., R.B., 1976, Light and diurnal vertical migration: Photobehaviour and photophysiology of plankton, in: "Photochemical and Photobiological Reviews," K.C. Smith, ed., Plenum Press, New York.

Frost, B.W., 1980, Grazing, in: "The physiological ecology of phytoplankton," I. Morris, ed., Blackwell, Oxford.

Gallegos, C.L., Hornberger, G.M., and Kelly, M.G., 1980, Photosynthesis-light relationships of a mixed culture of phytoplankton in fluctuating light, Limnol. Oceanogr., 25: 1082.

Goldman J.C., and Glibert P.M., 1982, Comparative rapid ammonium uptake by four marine phytoplankton species, Limnol. Oceanogr., 27: 814.

Gran. H.H., 1912, Pelagic plant life, in: "The depth of the ocean," J. Murray and H. Hjort, ed., MacMillan, London.

Hardy, J.T., and Valett, M., 1981, Natural and microcosm phytoneuston communities of Sequin Bay, Washington, Estuarine, Coastal and Shelf Science, 12: 3.

Hasle, G.R., 1954, More on phototactic diurnal migration in marine dinoflagellates, Nytt Magasin for Botanikk, 2: 139.

Hellebust, J.A., and Lewin, J., 1977, Heterotrophic nutrition, in: "The biology of diatoms," D. Werner, ed., Blackwell, Oxford.

Hibberd, D.J., 1977, Observations on the ultrastructure of the cryptomonad endosymbiont of the red-water ciliate Mesodinium rubrum, J. Mar. Biol., Ass. U.K., 57: 45.

Horstmann, U., 1981, Observatıons on the peculiar diurnal vertical migration of a red tide Dinophyceae in tropical shallow waters, J. Phycol., 16: 481.

Hulburt, E.M., 1957, The taxonomy of unarmored Dinophyceae of shallow embayments on Cape Cod, Massachusetts, Biol. Bull. 112: 196.

Huntsman, S.A., and Sunda, W.G., 1980, The role of trace metals in regulating phytoplankton growth, in:" The physiological ecology of phytoplankton," I. Morris, ed., Blackwell, Oxford.

Hustedt, F., 1930, Die Kieselalgen Deutschlands, Osterreichs und Scweiz mit Berücksichtigung der übrigen Länder Europas sowie der angrenzenden Meeresgebiete, in "Dr. L. Rabenhorst's Kryptogamenflorra," Bd. 7, L. Rabenhorst, ed., Akad. Verl. -Ges., Leipzig.

Jerlov, N.G., 1968, "Optical Oceanography," Elsevier Oceanography Series, 5, Elsevier, Amsterdam.

Jϕrgensen, E.G., 1968, The adaptation of plankton algae. II. Aspects of the temperature adaptation of Skeletonema costatum, Physiol. Plant., 21: 423.

Kahn, N., and Swift, E., 1978, Positive buoyancy through ionic control in the nonmotile marine dinoflagellate Pyrocystis noctiluca Murray ex Schütt, Limnol. Oceanogr., 23: 649.

Kamykowski, D., 1981, Dinoflagellate growth rate in water columns of varying turbidity as a function of migration phase with daylight, J. Plankton Res. 3: 357.

Kofoid, C.A., and Swezy, O., 1921, "The free-living unarmored Dinoflagellata," Mem. Univ. Calif., 5: 1.

Lange, O.L., Nobel, P.S., Osmond, C.B., and Ziegler, H., Eds., 1981, "Physiological Plant Ecology I. Responses to the physical environment," Springer, Berlin, Heidelberg, New York.

Leblond, P.H., and Taylor, F.J.R., 1976, The propulsive mechanisms of the dinoflagellate transverse flagellum reconsidered, Biosystems, 8: 33.

Li, W.K.W., 1980, Temperature adaption in phytoplankton: Cellular and photosynthetic characteristics, in: "Primary productivity in the sea," P.G. Falkowski, ed., Plenum, New York, London.

Lüning, K., 1981, Light, in: "The Biology of sea weeds," C.S. Lobban and M.J. Wynne, ed., Blackwell, Oxford.

MacIsaac, J.J., and Dugdale, R.C., 1972, Interactions of light and inorganic nitrogen in controlling nitrogen uptake in the sea, Deep Sea Res., 19: 209.

Malone, T.C., 1980, Algal size, in:"The physiological ecology of phytoplankton," I. Morris, ed., Blackwell, Oxford.

McCarthy, J.J., 1980, Nitrogen, in: "The physiological ecology of phytoplankton," I. Morris, ed., Blackwell, Oxford.

Morey-Gaines, G., and Elbrächter, M., in press, in: "The biology of dinoflagellates," F.J.R. Taylor, ed., Blackwell, Oxford.

Morris, I. Ed., 1980, "The physiological ecology of phytoplankton," Blackwell, Oxford.

Nalewajko, C., and Lean, D.R.S., 1980, Phosphorus, in: "The physiological ecology of phytoplankton," I. Morris, ed., Blackwell, Oxford.

Norkrans, B., 1980, Surface microlayers in aquatic environments, in: "Advances in microbial ecology," Vol. 4, M. Alexander, ed., Plenum, New York, London.

Paasche, E., 1980, Silicon, in: "The physiological ecolgy of phytoplankton," I. Morris, ed., Blackwell, Oxford.

Packard, T.T., Blasco, D. and Barber, R.T., 1978, Mesodinium rubrum in the Baja California Upwelling system, in: "Upwelling systems," R. Boje and M. Tomczak, eds., Springer, Berlin, Heidelberg, New York.

Paffenhöfer, G. -A., Strickler, J.R., and Alcaraz, M., 1982, Suspension feeding by herbivorous calanoid copepods: a cinematographic study, Mar. Biol., 67: 193.

Perry, M.J., Talbot, M.C., and Alberte, R.S., 1981, Photoadaption in marine phytoplankton, responses to the photosynthetic unit, Mar. Biol., 62: 91.

Pintner, I.J., and Altmyer, V.L., 1973, Production of Vitamin B_{12} binder by marine phytoplankton, J. Phycol. (suppl.) 9: 13.

Pintner, I.J., and Altmyer, V.L., 1979, Vitamin B_{12} binder and other algal inhibitors, J. Phycol., 15: 391.

Platt, T. Ed., 1981, "Physiological bases of phytoplankton ecology," Can. Bull. Fish. Aquat. Sci., 210: 1.

Pollingher, U., and Zemel, E., 1981, In situ and experimental evidence of the influence of turbulence on cell division processes of Peridinium cinctum forma westii (Lemm.) Lefèvre, Br. phycol. J., 16: 281.
Raven, J.A., 1980, Nutrient transport in microalgae, in: "Advances in microbial physiology," A.H. Rose and J.G.Morris, eds., Academic Press, London.
Richardson, K., and Fogg, G.E., 1982, The role of dissolved organic material in the nutrition and survival of marine dinoflagellates, Phycologia 21: 17.
Savidge, G., 1980, Photosynthesis of marine phytoplankton in fluctuating light regimes, Mar. Biol. Letters, 1: 295.
Schmid, A.M., and Schulz, D., 1979, Wall morphogenesis in diatoms : deposition of silica by cytoplasmic vesicles, Protoplasma, 100: 267.
Schmid, A.M., Borowitzka, M.A., and Volcani, B.E., 1982, Morphogenesis and Biochemistry of diatom cell walls, in: "Cell biology monographs," Vol. 8, O. Kiermayer, ed., Springer, Wien, New York.
Schöne, H., 1970, Untersuchungen zur ökologischen Bedeutung des Seegangs für das Plankton mit besonderer Berücksichtigung mariner Kieselalgen, Int. Revue ges. Hydrobiol. 55: 595.
Schöne, H.K., 1972, Experimentelle Untersuchungen zur Okologie der marinen Kieselalge Thalassiosira rotula. I. Temperatur und Licht, Mar. Biol., 13: 284.
Schütt, F., 1982, Das Pflanzenleben der Hochsee, Ergebnisse der Plankton-Expedition der Humboldt-Stiftung, 1 (A): 243.
Sieburth, J. McN., Willis, P.J., Johnson, K.M., Burney, C.M., Lavoie, D.M., Hinga, K.R., Caron, D.A., French, F.W., Johnson, P.W., and Davis, P.G., 1976, Dissolved organic matter and heterotrophic Microneuston in the surface microlayers of the North Atlantic, Science 194: 1415.
Silva, E.S., 1965, Note on some cytophysiological aspects in Prorocentrum micans Ehrenb. and Goniodoma pseudogoniaulax Biech., Notas e Est. do I.B.M., 30: 3.
Smayda, T.J., 1958, Biogeographical studies of marine phytoplankton. Oikos, 9:158.
Smayda, T.J., 1970, The suspension and sinking of phytoplankton in the sea, Oceanography and Marine Biology 8: 353.
Sournia, A., 1974, Circadian periodicities in natural populations of marine phytoplankton, Adv. Mar. Biol., 12: 325.
Sournia, A., 1981, Morphological bases of competition and succession, Can. Bull. Fish. Aquat. Sci., 210: 339.
Sournia, A., 1982, Form and function in marine phytoplankton, Biol. Rev., 57: 347.
Sournia, A., 1982, Is there a shade flora in the marine plankton? J. Plankt. Res., 4: 391
Staker, R.D., and Bruno, S.F., 1980, Diurnal vertical migration in marine phytoplankton, Bot. Marina, 23: 167.
Steeman Nielsen, E., 1975, "Marine photosynthesis," Elsevier Oceanography Series, 13, Elsevier, Amsterdam.

Steidinger, K.A., Tester, L.S., and Taylor, F.J.R., 1980, A redescription of Pyrodinium bahamense var. compressa (Bohm) stat. nov. from Pacific red tides, Phycologia, 19: 329.

Stosch, H.A., von, 1956, Abgeschlossene Hohlraumsysteme mit semipermeablen Wänden als Strukturelemente von Diatomeenschalen, Ber. Deutsche Bot. Ges. 69: 99.

Stosch, H.A., von, 1964, Zum Problem der sexuellen Fortpflanzung in der Peridineengattung Ceratium, Helgol. wiss. Meeresunters. 10: 140.

Stosch, H.A. von, 1980, The "endochiastic areola", a complex new type of siliceous structures in a diatom, Bacillaria, 3: 21.

Swift, D.G., 1980, Vitamins and phytoplankton growth, in: "The physiological ecology of phytoplankton," I. Morris, ed., Blackwell, Oxford.

Taylor, D.L., and Seliger, H.H., Eds., 1979, "Toxic dinoflagellate blooms," Elsevier, New York, Amsterdam, Oxford.

Taylor, F.J.R., 1976, Flagellate phylogeny : a study in conflicts, J. Protozool. 23: 28.

Throndsen, J., 1973, Motility in some marine nanoplankton flagellates, Norw. J. Zool., 21: 193.

Throndsen, J., 1979, The significance of ultraplankton in marine primary production, Acta Bot. Fennica, 110: 53.

Vince-Prue, D., 1975, "Photoperiodism in plants," McGraw-Hill, London.

Walsby, A.F., and Reynolds, C.S., 1980, Sinking and floating, in: "The physiological ecology of phytoplankton," I. Morris, ed., Blackwell, Oxford.

Wandschneider, K., 1979, Vertical distribution of phytoplankton during investigations of a natural surface film, Mar. Biol. 52: 105.

White, A.W., 1976, Growth inhibition caused by turbulence in the toxic marine dinoflagellate Gonyaulax excavata, J. Fish. Res. Board Can., 33: 2598.

MEASURING THE METABOLISM OF THE BENTHIC ECOSYSTEM

Mario M. Pamatmat

Tiburon Center for Environmental Studies
San Francisco State University
P.O. Box 855, Tiburon, California 94920

INTRODUCTION

Determining and modeling the flow of energy and matter in the food web is perceived to be a key to our understanding of the productivity and other properties of the ecosystem (Platt *et al.*, 1981). The measurement of metabolic rates has been a major part of our undertaking to determine energy flow. The oxygen uptake and carbon dioxide production of individual organisms as well as representative samples of communities have been our primary measures of metabolic rates. In the last decade, the activity of the electron transport system (Packard, 1971) has gained popularity and has even been preferred by some (Båmstedt, 1980).

Our current knowledge and thinking about community metabolism in all parts of the ecosystem including the benthos derive mostly from measurements with the three mentioned methods. The status of knowledge regarding benthic community metabolism has been reviewed (Pamatmat, 1977), including its relation to other benthic processes (Zeitzschel, 1980, 1981; Nixon *et al.*, 1980). More recent benthic respiration measurements have given Kemp and Boynton (1981) and Boynton *et al.* (1981) some insight into the dynamics of community metabolism.

We now realize, however, that respiration measurements do not represent benthic community metabolism (McCave, 1976: p 303). Energy flow models assume that respiration measurements in the laboratory or in situ represent natural rates of total energy transformations by living organisms. My task is to discuss the problems with our measurements and some developments in ecological energetics.

METABOLISM, RESPIRATION, ENZYME ACTIVITY: EXPECTATION AND REALITY

In effect, metabolism, the sum-total of all chemical and physical changes taking place in living matter, has commonly been measured in terms of a change in a substance involved in intermediary metabolism or in respiration. Oxygen is converted into water in the last step in the chain of respiratory electron transport and the rate of oxygen uptake is taken as a measure of overall metabolic rate. Carbon dioxide is produced in different steps of intermediary metabolism and its rate of production is assumed to represent total metabolism. Chemical reactions in the course of metabolism are catalyzed by a wide variety of enzymes and the electron transporting activity of respiratory enzymes is used as a measure of metabolic rate. These three measures are undoubtedly related to overall metabolic rate. The uncertainty lies in the conversion of the measured parameters into their equivalent energy units. There are undetermined variables that affect the quantity of energy released by total metabolic activity per unit amount of oxygen consumed, carbon dioxide produced, or unit activity of a respiratory enzyme. Therefore, the same amount of oxygen consumed by different organisms, or by the same individual at different times, may not represent the same energy transformation. The same may be said of carbon dioxide production and enzyme activity. The possible variability has been ignored and tacitly assumed to be within acceptable limits in comparison with other unknown sources of error in energy budget estimations.

The energy content, and energy change during transformation, of a substance are a function of the amounts and identity of the substance and its reaction products. We do not determine the kinds and amounts of all the reactants and products in metabolism at the time of each respiration experiment. Rather, we rely on the universal features of metabolism (Florkin, 1960) and assume that 1) all organisms metabolize organic matter that varies in energy content only within narrow, or acceptable, limits, and 2) the aerobic oxidation of organic matter yields the same amount of energy per unit amount of oxygen and produces the same kinds of metabolic end-products. The mass of evidence that we base these assumptions on come from experimental results with warm-blooded animals (Kleiber, 1962). There are similar evidences for poikilotherms (Peakin, 1973; Pamatmat, 1978) but there are also some large deviations, shown primarily by benthic organisms that undergo anaerobiosis (Grainger, 1968; Pamatmat, 1978, in press; Hammen, 1980).

In the benthic ecosystem, we are dealing with a complex community of 1) obligate aerobes, 2) aerobes that may not be fully aerobic all the time, even when they are taking up oxygen, 3) facultative anaerobes, 4) fermenters that utilize different substrates and produce different end-products, 5) chemolithotrophs that use oxygen to oxidize reduced sulfur compounds, 6) bacteria that oxidize, reduce, precipitate, or dissolve iron compounds, 7) obligate anaerobes that use sulfate, nitrate, or carbon dioxide as terminal electron acceptors, and 8) many other kinds of microbial metabolic types. There may be an underlying unity in the biochemistry and energetics of all these organisms, but there is no single measurable chemical parameter common to all of them that will give us an unbiased measure of their combined metabolic rate. For this reason none of the traditional measurements of metabolism commonly applied to aerobic organisms is valid for measuring soft-bottom community metabolism, or wherever there is a mixture of aerobes and anaerobes.

Anaerobes are ubiquitous (von Brand, 1946), even in oxic environments (Jorgensen, 1977a). Aerobic organisms may harbor anaerobes in their gut. While it may be acceptable to measure the aerobe's metabolism in terms of oxygen uptake, this technique would leave out the metabolism of gut microflora and fauna, which are part of the ecosystem. Strictly speaking then, none of the methods that measure fluxes or turnover of matter is accurate for measuring community metabolism anywhere in the ecosystem. The magnitude of the error, e.g. is oxygen uptake values, which remains to be determined, will certainly vary with different conditions and parts of the ecosystem; it may be expected to range from zero in the absence of anaerobes to greatest in organically rich, soft sediments, to be smaller in deep-sea than in near-shore sediments, and in exposed high-energy beaches than in tideflats.

METABOLISM AND HEAT PRODUCTION

All chemical reactions and transformations of matter during metabolism by all living organisms are accompanied by energy change. Some chemical energy is conserved, for shorter or longer periods, but there is always a net loss of energy manifested as heat loss whether organic matter is completely mineralized or not. The heat flow from living organisms regardless of metabolic type is a direct result of metabolism. The total heat flow from a mixed community of metabolic types is the result of their total metabolism. It seems. therefore, that the measurement of heat flow, or direct calorimetry, is a suitable technique for measuring system metabolism, not only of soft-bottom communities, but other parts of the ecosystem as well. Direct calorimetry has the unique feature of giving results directly in the desired energy units. The comparison between respiration and heat flow measurements for different ecological conditions and parts of ecosystems would seem to be an important undertaking in ecological energetics.

SEDIMENT HEAT PRODUCTION AND BENTHIC ENERGY FLOW

Direct calorimetry measures heat effects of energy and matter transformations. When applied to sediments, direct calorimetry does not distinguish metabolic heat production from heat effects of extracellular chemical reactions. Since all chemical and physical transformations are accompanied by energy changes, we should presume that all geochemical processes, such as redox reactions, complex formation, co-precipitation, solution, diffusion, ionization, neutralization, etc. that have been identified and could be taking place in sediments are contributing to the total heat flux that we measure. It is also likely that at least some of these reactions are coupled with metabolic activity, e.g. the rate of sulfide oxidation is related to the rate of sulfate reduction and sulfide formation, making it impossible to partition metabolic from non-metabolic heat by killing the organisms. Thus, if sulfate reducers are killed a dynamic equilibruim could be disturbed and the subsequent rate of sulfide oxidation could be lower than the undisturbed rate. On the other hand, there is the added problem that metabolic poisons react with the sediment to produce tremendous quantities of heat (Pamatmat, 1982a).

If we cannot distinguish metabolic heat production from total sediment heat production, then what good is the latter. In ecological energetics we are concerned with the fate of all radiant energy that is absorbed by primary producers, part of which is stored as chemical energy that moves through the food web. This chemical energy is dissipated in other ways besides metabolic processes. The total heat production by sediments represents all those combined processes, including metabolism, by which chemical energy of organic matter is dissipated into heat. It was suggested, therefore (Pamatma 1982b), that:

Total sediment heat production = benthic energy flow

The rationale for this equation, in essence, is that measurable heat production, whether metabolic or not, is ultimately traceable to the activity of living organisms and to the energy of organic matter that was originally fixed by photosynthesis. Some sediments have no measurable heat production. Anoxic sediments as well as sediments maintained under normal surface aerobic condition, show a decrease in heat production to vanishingly low rates with time, implying that in the absence of measurable metabolic heat there is no measurable extracellular heat production either.

It is desirable, of course, to know the relative magnitude of heat contribution by the separate geochemical and biochemical processes at the time when they are proceeding simultaneously. Non-metabolic heat production, just like metabolic heat, is an energy loss to the benthic ecosystem, and together they represent the energy cost of maintaining the benthic community and the ecosystem.

The current status is that we do not have a method for measuring benthic community metabolism. Direct calorimetry of sediments measures benthic energy flow, including community metabolism and non-metabolic heat production. It is important in ecological energetics to separate these two components and to develop a better understanding of both.

INSTRUMENT AND TECHNIQUES

Case studies are presented here to illustrate the kind of information and understanding of ecological energetics that can be gained from heat flow measurements. The results have been obtained with a double-twin heat-flow calorimeter. The instrument, techniques, and associated problems of measurements have been described (Pamatmat, 1978, 1979, 1980, in press).

The relation between heat production by the sample, the heat flow through the sensors, and thermopile voltage signal by the instrument is illustrated in Figure 1. At any instant, the thermopile voltage is proportional to actual heat flow through the thermopile sensors. A step change in heat production rate is followed by an exponential change in heat flow as reflected by thermopile voltage, the lag representing the heat capacity of the sample and its container. The response lag makes the instrument unsuitable for fast chemical kinetics studies. Actually, the response time is often faster than indicated in Figure 1. The response time becomes shorter as the heat source increases in size relative to the thermal mass of the whole system.

Thermograms viewed with caution can give us some idea of instantaneous changes in energy conversions taking place in the system. In spite of the response lag in the instrument average rates of heat production and total amounts of heat liberated over time periods can be accurately determined.

The measurement of heat flow requires containment of the sample or organism in a closed system. Direct calorimetry of open systems is possible, and there are continuous-flow calorimeters in the market, but it is more complex and requires differential determinations of the energy content of inflow and outflow. Consequently, closed-system calorimetry is preferable whenever valid. There are ways to avoid continuous-flow calorimetry.

The most common reason for having a continuous-flow system is to maintain oxygen tension at a constant level. To achieve practically constant oxygen tension in a closed system for extended periods, sufficiently large reservoirs of air can be included. Reliance has been placed on the organisms' swimming, crawling, and ventilating activities to oxygenate the water faster than diffusion alone would supply. The volume of water is minimized as much as possible. A large volume of water in combination with a rapid rate of oxygen uptake causes the oxygen tension of the water in equilibrium

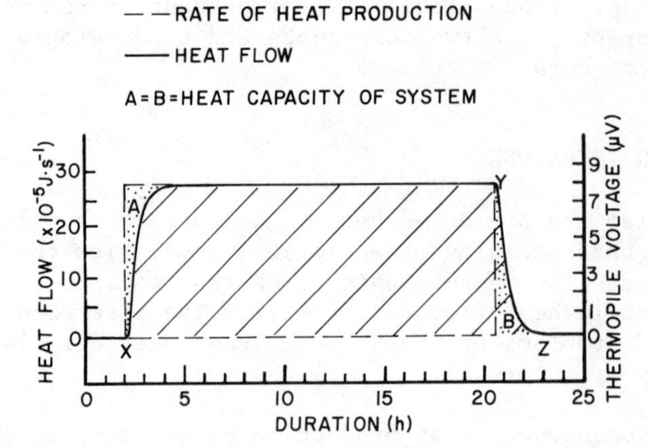

Fig. 1. Relationship between heat production rate, heat flow, and thermopile voltage output. A step change in rate of heat production by switching electric current on or off through a resistance heater contained inside a metabolic chamber is followed by an exponential increase or decrease in heat flow through the thermopile sensors, producing a corresponding voltage signal. The lag is proportional to the thermal mass and conductivity of the sample. The total amount of heat flow through the sensors (area XYZ) is equal to the heat produced (area bounded by dashed lines) but instantaneous rates of heat production are underestimated or overestimated when the thermopile voltage is rising or falling, respectively.

with the air to remain relatively steady but at a low level. To raise oxygen tension of the water the air can be mixed with pure oxygen. A pO_2 electrode immersed in the water (the Radiometer electrode does not require stirring) can monitor oxygen tension during calorimetry and the oxygen content of the overlying air can be raised or lowered as needed.

The accumulation of carbon dioxide and other metabolites could be a problem, though evidently less so in sediments where substances are recycled by different metabolic types. The seriousness of this problem can be determined by flushing the water out periodically, without removing the metabolic chamber from the calorimeter. The rate of heat production often remains the same after flushing, indicating this is not normally a problem.

Turbulence is believed to be an important factor in some situations (Pamatmat, 1971; Boynton *et al.*, 1981). The effect of turbulence on animal metabolism and benthic community energy flow can now be studied by direct calorimetry. The instrument can be placed in gimbals (or aboard a rolling ship) which could be rocked to any desired frequency to simulate natural turbulence.

EXPERIMENTAL MEASUREMENTS OF INFAUNAL METABOLISM

In situ metabolism of benthic infauna proceeds under some or all of the following conditions: 1) in the presence of natural sediment with all its physical and chemical properties, 2) in the presence of other biota that are continuously affecting sediment properties, 3) with fluctuating food supply, 4) in the presence of competitors and predators, 5) over a continuously changing range of temperatures, 6) with periodic exposure to air and submergence in water, 7) with periodic and aperiodic occurrence of currents and turbulence, and 8) with seasonal changes in salinity. The effects of some of these factors plus endogenous rhythms in metabolism have been widely studied, but it is not always clear that they have been incorporated into energy flow models. The first four conditions have been largely overlooked, or deliberately avoided because of the complications that they create in metabolism experiments. Unfortunately, they appear to be vital for the accuracy of energy flow models, as the following results will show.

Metabolism of *Neanthes virens*: Effect of Sediment Deprivation

Neanthes virens shows a steady average metabolic rate in sediment (Figure 2A) in comparison with large fluctuations while in water only (Figure 2B). The thermograms give the average metabolic rates in each medium and also show the nature of the differences in metabolic activity. The irregular large fluctuations in heat production by the worm in water signify irregular variations in its physical activity. It is not possible to observe the animal while inside the calorimeter but outside the calorimeter in water only, it is seen to alternately swim, settle, crawl, and rest on the bottom. It may be inferred that the animal was swimming above the bottom during periods of highest metabolic rates. When it became tired it settled to the bottom. The lowest metabolic rates represent periods of rest.

The fluctuations in metabolic activity in sediment, of much smaller amplitude that those in water, are probably related to the worm's ventilation movements. Outside the calorimeter, the worm could be seen with its entire body sometimes next to the bottom of the dish, and sometimes the anterior end was just below the sediment surface. Sometimes it emerged from its burrow but only once was it seen to come out and crawl around on the sediment surface. The thermogram showed no evidence of a circadium rhythm in anaerobiosis

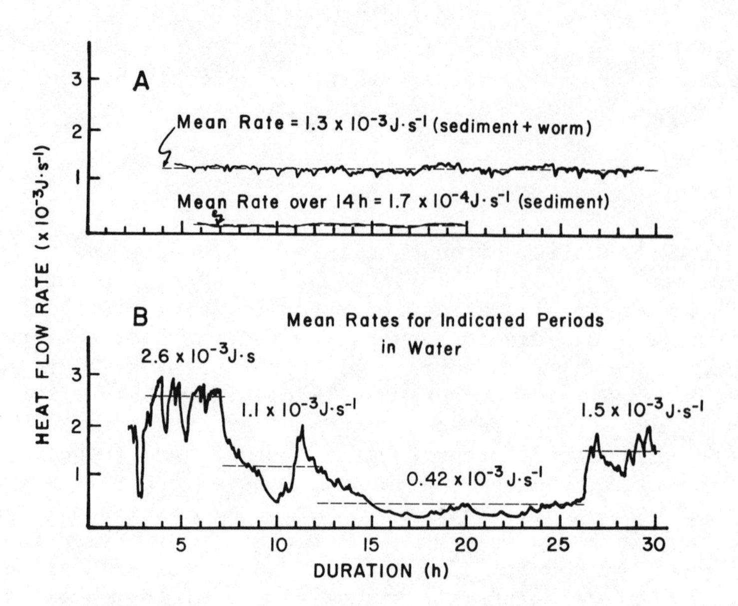

Fig. 2. A) Differential thermogram produced by duplicate sediment samples, at 20°C, one in the experimental and the other in the compensation chamber, indicating that the former had an average of 1.7×10^{-4} J.s^{-1} higher rate of heat production than the latter. The addition of one Neanthes virens (3.27g live weight) to the sediment in the experimental chamber raised its rate of heat production to an average of 1.3×10^{-3} J.s^{-1}. The difference between initial and final rates, 1.1×10^{-3} J.s^{-1}, could be an overestimate of the worm's metabolic rate because of the possible stimulation of the sediment's heat production as a result of the worm's aeration of its burrow.
B) Thermogram of the same individual after its transfer to water only, versus water only in the compensation chamber. Metabolic rate is highly variable and unpredictable. Average rates over several hours duration could overestimate metabolic rate relative to rate in sediment.

that Scott (1976) had observed. Another individual Neanthes virens, however, placed in a different sediment (Figure 3) appeared to undergo periods of anaerobiosis, but nothing like a circadian, or endogenous tidal rhythm such as can be readily seen in the thermograms of Uca Pugnax (Pamatmat, 1978 and unpublished).

Measuring the metabolic rate of a burrowing polychaete in sediment is complicated not only by the presence of other organisms but also by the probability that the worm's burrowing and ventilating activities affect the sediment's rate of oxygen uptake,

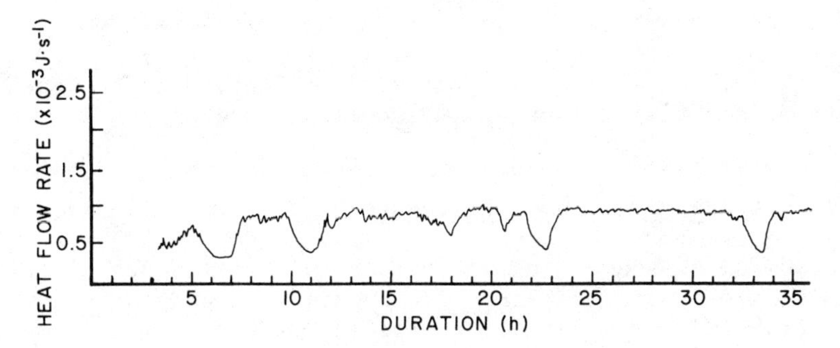

Fig. 3. Thermogram of Neanthes virens (3.38g live weight, 20°C) from an industrially polluted site, in its native sediment. The 2-h periods of reduced metabolic activity indicate cessations in burrow ventilation from time to time, with consequent anaerobiosis.

respiratory as well as chemical, with the result that,

(Heat production rate by sediment with worm)

\- (Heat production rate by sediment without worm)

\> (Heat production rate by worm)

If the presence of the worm stimulates the sediment's heat production rate then the difference is an overestimate of the worm's true metabolic rate in sediment. The mean rate of 1.3×10^{-3} J.s^{-1} for Neanthes and sediment, corrected by subtracting 1.7×10^{-4} J.s.$^{-1}$ for the sediment, gives and estimate of 1.13×10^{-3} J.s^{-1} for Neanthes alone. Even though this may be an overestimate it is clear that the polychaete's metabolic rate in water can greatly exceed this value for extended periods. An interesting question is, which is closer to the true rate, 1.13×10^{-3} J.s^{-1} or 0.42×10^{-3} J.s^{-1}, the latter representing the average metabolic rate during periods of rest in water (Figure 2B). It is also interesting that the average rate in sediment is the same as the average rate for the entire 27h in water. Does this mean that if the measurement is carried out long enough to encompass periods of unusual physical activity and periods of compensatory rest we obtain a good estimate of natural metabolic activity ? Do the results then also indicate that the polychaete's effect on sediment metabolism is quite small ? These are all relevant questions pertaining to our attempt to understand not only benthic community metabolism but the metabolic activity of each member of the community.

We do not know the answers to these questions yet, but it should be clear from Figure 2 that it helps to know not only the average values of metabolic rates but also some indication of the nature of the metabolic activity that is measured and averaged.

Heat Flow versus Oxygen Uptake Measurements

The common closed-chamber respirometry technique should give satisfactory measures of metabolism by fully aerobic organisms with steady metabolic rates, but not by species with widely fluctuating metabolic rates. Many replicate measurements and calculation of mean and standard deviation do not adequately reveal the nature of variability in what has been averaged. From the nature of the thermograms of *Neanthes virens* there is no question that its metabolic activity is very different when in water than when in sediment. On the other hand, it is clear from Figure 2B that the average rates of oxygen uptake over several hours, as it is often measured, could be shown to be the same as, lower, or higher than the average rate in sediment. We would rely on the statistical significance of these tests and it would lead us astray in our thinking. It is not that metabolic rate in water is the same as, lower, or higher than in sediment, but that because of the unnatural condition without sediment, the animal's metabolic activity becomes unpredictable, if not abnormal.

In general, when dealing with species that show highly variable rates of oxygen uptake, much understanding can be gained from direct calorimetry. We need to know not only the average metabolic rates of different species but the full range of their metabolic rates. This range would seem to be an indication of a metabolic potential of some survival value to the species. Also, by studying thermogram patterns under a variety of experimental conditions, we may eventually be able to identify those times when the species is exhibiting the average in situ rate.

The question may be raised as to how good are the values of respiration rate in the literature for energy flow models. The estimation of total macrofaunal metabolism from biomass and respiration measurements have sometimes exceeded measured total sediment respiration (Nichols, 1975), leading us to suspect that something is wrong somewhere.

Other species of infauna show much increased metabolic rates when deprived of sediment (Vernberg *et al*., 1978; Pamatmat, 1982a). The presence of sediment while measuring the metabolic rate of infauna is a necessary, but is not a sufficient, condition to assure accurate estimation of in situ metabolic rates, as shown in the next results.

Metabolism of *Capitella capitata*: Effect of Starvation

The experimental system consisted of 100 worms in 100 cc of clean beach sand that had passed a 0.3-mm mesh sieve and flooded with 1-μ-filtered estuarine water (25% salinity). The system at 20°C showed a heat production rate of 2.1×10^{-4} $J.s^{-1}$ initially, decreasing exponentially to 1.2×10^{-4} $J.s^{-1}$ over a period of 20 h (Figure 4A). The worms weighted 125 mg. so their averate rate decreased from 1.68×10^{-6} to 0.96×10^{-6} $J.s^{-1}$ mg^{-1} over this period.

Many independent measurements of *Capitella* respiration by Leon Camen and Kenneth Tenore (unpublished) in terms of oxygen uptake, carbon dioxide production, and weight loss show a wide range of values. Their equivalent heat production rates range from 1.4×10^{-6} to 12.6×10^{-6} $J.s^{-1}$ mg^{-1} (assuming RQ=1, 1 ml O_2=20.1 J, 8.4% of wet weight = g carbon). The results of direct calorimetry tend to be lower than those of the other techniques. The difference may be explained mainly by the duration of the experiments. Camen's and Tenore's measurements lasted for shorter durations and my own measurement clearly showed decreasing metabolic rate with time.

The metabolic rates of non-feeding *Capitella* are much lower than those of feeding worms (Figure 4B). A batch of 20 worms + 12 mg of Pablum (Gerbers Mixed Cereal), a baby food of known chemical composition used in faunal feeding experiments (Tenore, 1981), was placed in a dish with sand. One dish was prepared each day for 3 successive days, giving three replicate runs. Each dish was kept in flowing water the first day and given another 12 mg of Pablum the next day just before placing it inside the calorimeter. The heat production rates were 3.5×10^{-4}, 3.8×10^{-4}, and 4.2×10^{-4} $J.s^{-1}$ for the three dishes containing 37.0, 44.1, and 49.2 mg of *Capitella*, respectively. These heat production rates, while showing fluctuations, are steady on the average in comparison with that of non-feeding polychaetes.

If it is assumed that the difference in heat production rate between any two dishes is due mainly to the difference in worm biomass, then the difference of 0.3×10^{-4} $J.s^{-1}$ can be divided by the corresponding difference in worm biomass of 7.1 mg, giving 4.2×10^{-6} $J.s^{-1}$. Likewise, the corresponding difference between two other dishes give 7.8×10^{-6} and 5.7×10^{-6} $J.s.^{-1}$ mg^{-1}. These three estimates give an average of 5.9×10^{-6} $J.s^{-1}$ mg^{-1} for feeding worms versus a range of 0.96×10^{-6} to 1.68×10^{-6} $J.s^{-1}$ mg^{-1} for non-feeding worms. The difference between feeding and non-feeding worms is readily explained by the work of feeding plus the calorigenic effect of food (Kleiber, 1965). The value of 5.9×10^{-6} $J.s^{-1}$ mg^{-1} lies well within the range of 1.4×10^{-6} to 12.6×10^{-6} $J.s^{-1}$ mg^{-1} obtained by other methods for non-feeding worms; this may be viewed as an acceptable argument in favor of respiration measure-

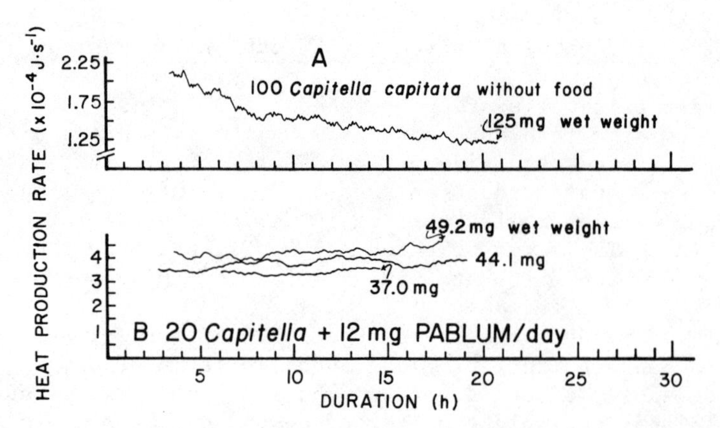

Fig. 4. Thermograms at 20°C of Capitella capitata with and without food. The exponential decline in heat production rate of worms without food is the result of starvation. The wet weights of the worms at the end of the experiment are indicated.

ments. On the other hand, it is also an indication of the effects of other factors that could lead to chance agreement.

All species should tend to have lower metabolic rates with increased duration of starvation. This tendency may be masked by the effects of other factors on metabolic rates, such as sediment deprivation, for example. It also seems that the higher the metabolic rate, relative to the species's energy reserve, the faster will be the rate of metabolic decline. Consequently, the problem could be more severe with meiofauna and small macrofauna than with the large species, and with physically active as compared to sedentary species.

METABOLISM OF MICROBES : INTERACTIONS WITH MEIOFAUNA

Benthic community metabolism is the sum of the metabolic activity of the component species. Interest in the relative contribution of different species to community metabolism has led to a number of attempts to estimate each species population's metabolism (Pamatmat, 1968; Smith, 1973; Nichols, 1975). Besides the problems with sediment deprivation and starvation, our job of partitioning community metabolism is made difficult by interactions between species. If one species has a significant effect on another species' metabolic activity, then we may presume that separating them would alter both their metabolic rates. Then the sum of individually measured metabolic activity will no longer add up to equal whole community metabolism. The magnitude of this interaction effect is revealed by experiments with artificial systems.

Effect of Food Addition on Microbial Growth and Metabolism

Dry beach sand flooded with filtered water showed a low level of heat production (Figure 5) which increased until the 16th h to 1.2 x 10^{-4} $J.s^{-1}$ before decreasing to undetectable level. The initial heat production was probably mostly physical-chemical but the ensuing 7-8 h period of increase was probably due to bacterial growth on available organic matter on the sand particles as well as dissolved organics in the estuarine water. As the bacterial substrate was used up the population declined, as reflected by a corresponding decrease in metabolic heat production rate.

The addition of 12 mg of Pablum to flooded sand causes the heat production rate to increase faster and reach a higher peak (7.3 x 10^{-4} $J.s^{-1}$) sooner (Figure 5). The first 4 h of the thermogram is not usable because of the temporary disturbance in the instrument's thermal equilibrium. The dashed lines are reasonable forward extrapolations to indicate the course of heat energy conversion during the unobservable period. The difference between the first and second thermograms, the area shown by hash marks (=29.9 J) is an estimate of the total potential chemical energy of Pablum that was converted into heat during this period. Since Pablum has a heat of combustion of 4.0 cal mg^{-1} (=16.7 J mg^{-1}; Tenore, 1981), in the 22.5-h period shown the microbial community oxidized nearly 15% of the chemical energy in 12mg.

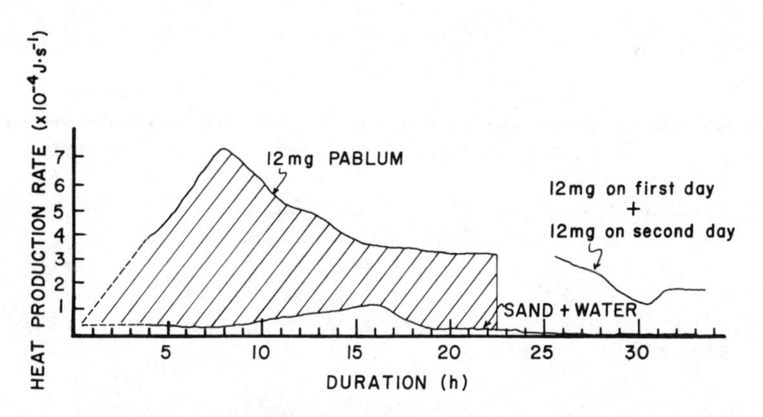

Fig. 5. Thermograms at 20°C of flooded beach sand, flooded sand with 12 mg Pablum, and flooded sand with 12 mg Pablum on day 1 plus 12 mg on day 2 added just before placing it inside the calorimeter. The hashed area, equivalent to 29.9 J, estimates the chemical energy of Pablum that was metabolized in 22.5 h (=15% of heat of combustion of 12 mg Pablum).

In the third part of this experiment, 12 mg Pablum was added to the flooded sand. The dish was kept for 24 h in an open aquarium with flowing filtered water at 20°C. On the second day another 12 mg of Pablum was added to the dish before placing it inside the calorimeter. This sample did not show the increase resulting from an initial addition of Pablum. Instead it showed what appears to be the second day continuation of the previous experiment's thermogram, and the second batch of 12 mg Pablum resulted in only a relatively minor increase to 1.9×10^{-4} J.s.$^{-1}$.

The thermogram of a batch culture of microorganisms has been shown to be correlated with the growth and decline of the population (Forrest et al, 1961; Pamatmat et al., 1981). The experiment indicates the growth of microbes on Pablum during the first day but an inability to grow on the same amount of Pablum during the second day. This growth trend is compared next with a similar system to which nematodes or polychaetes had been added.

Effects of Nematodes on Microbial Growth

Sand with 7000 nematodes (*Diplolaimella chitwoodi*) without food shows barely detectable heat production (Figure 6). The nematode culture fluid plus 12 mg of Pablum shows slowly decreasing heat production rate from 3×10^{-4} J.s.$^{-1}$ leveling off to 2×10^{-4} J.s^{-1} by the 24th h. The presence of 7000 nematodes with 12 mg Pablum shows an increase to 7.1×10^{-4} J.s^{-1} by the 12th h decreasing thereafter. For comparison the thermogram of sand plus Pablum (Figure 5) is shown again in Figure 6. There is a 4-h difference in time lag, and the thermogram of the sample with nematodes is lower than that of the sample without nematodes.

From the difference in total heat production curves nematodes appear to affect the rate of bacterial population growth in two ways: they prolong the lag phase in bacterial population growth and control the population size to a lower level as well. However, there appears to be something in the old nematode culture fluid, some metabolites perhaps, that is inimical to microbes. The nematodes had been washed but there may have been some substances carried over from the culture to the sand to affect bacterial growth. Thus, the possibility exists that the lag and depression described before were not purely the effects of nematodes.

The free-living nematode (*Diplolaimella chitwoodi*),at the same density that we have used in our experiment, tripled the rate of mineralization (measured in terms of CO_2 production) of *Gracilaria detritus* (Findlay and Tenore, 1982). The increased CO_2 production was much greater than nematode respiration alone, indicating a synergistic effect by the nematodes on the microbes. Likewise,

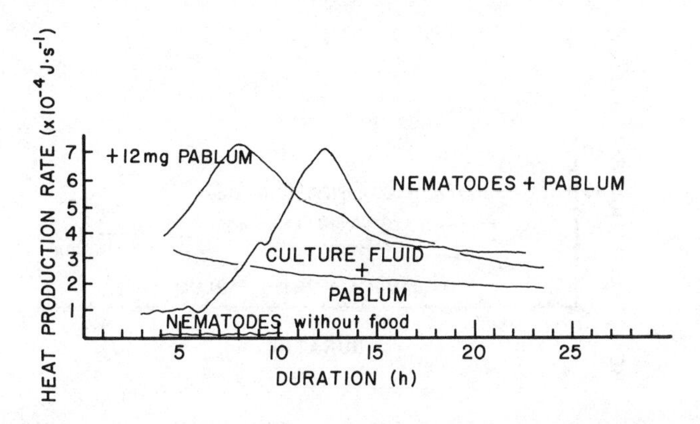

Fig. 6. Thermograms at 20°C of flooded sand with 7000 washed nematodes, flooded sand with old nematode culture fluid plus 12 mg Pablum, and flooded sand with 7000 washed nematodes plus 12 mg Pablum. The thermogram of flooded sand with 12 mg Pablum shown in Figure 5 is also shown for comparison. The effect of nematodes is to prolong the lag phase of microbial growth and prevent microbes from attaining their maximum population size.

nematodes increased oxygen consumption and decomposition of sludge (Abrams and Mitchell, 1980) and increased carbon utilization and nitrogen and phosphorus mineralization in soils (Anderson et al., 1981). My data show a definite effect of nematodes on microbial growth and metabolism, but it is opposite to that found by others. It is possible that the interaction effects between meiofauna and microbes depends on the amount and kind of food used in the experiment.

Effects of Capitella capitata on Microbial Growth

In the first experiment (Figure 7) 20 plychaetes (=2700 m^{-2}) and 29 mg of Pablum (=3.9 g m^{-2} d^{-1}) were added to a dish, which was kept submerged in a tray with flowing filtered estuarine water for ca. 24 h. The following day 29 mg of Pablum was again added to the dish before transferring the dish inside the calorimeter. The thermogram on the second day shows an average heat production rate of 10.3 x 10^{-4} J.s^{-1} decreasing to 9.2 x 10^{-4} J.s^{-1}. With 100 Capitella + 29 mg Pablum/day the heat production rate the second day was initially lower (5 x 10^{-4} J.s^{-1}) than with 20 Capitella, but increased slowly. Total heat production rate with 29 mg Pablum but different amounts of worms was obviously very different in the two cases during the

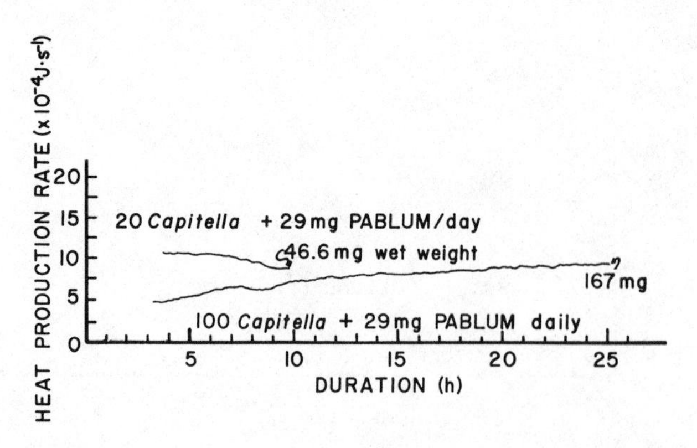

Fig. 7. Thermograms at 20°C of Capitella capitata and microbes in sand showing the effect of worm biomass on microbial population and metabolic activity. Pablum was added to the dish on day 1 and the same amount again on day 2 just before placing inside the calorimeter. The trends shown by the thermograms indicate an initially higher but decling microbial population with 20 Capitella and an initially lower but increasing microbial population with 100 Capitella.

first day but converged during the second day to about 9 x 10^{-4} J.s^{-1} The same processes (Capitella growth and metabolism, microbial growth and metabolism) occurred in both experiments but followed different time courses of development. The group of 20 Capitella (food ration equivalent to 0.62 mg Pablum mg^{-1} worm) gained weight while the batch of 100 (o.17 mg Pablum mg^{-1} worm) lost weight. Thus the decrease in system energy flow with 20 worms indicates slowly decaying microbial population. On the other hand, the increase in system energy flow with 100 evidently starving Capitella means that the worms initially kept microbial populations down but slowly the microbes increased in number.

Applying the value of 5.9 x 10^{-6} J.s^{-1} mg^{-1} of feeding worms (previous experiment) to the worms in Figure 7, we can estimate that the 20 Capitella (46.6 mg) accounted for a rate of 2.8 x 10^{-4} J.s^{-1}; thus, 10.3 x 10^{-4} J.s^{-1} minus 2.8 x 10^{-4} J.s^{-1}, or 7.5 x 10^{-4} J.s^{-1}, should represent average microbial metabolism alone during that same period. The same calculations with the 100 Capitella (167 mg) gives a value of 9.8 x 10^{-4} J.s^{-1}. This rate is higher than the measured rate of 100 Capitella plus microbes (8.8 x 10^{-4} J.s^{-1}). The weight loss by these worms indicate that their average metabolic rate was probably lower than 5.9 x 10^{-6} J.s^{-1} mg^{-1} for well-fed worms but higher than 1.7 x 10^{-6} J.s-1 mg^{-1}, the highest average measurable rate shown by the unfed worms. The 20 Capitella with 29 mg Pablum showed an average weight increase of 23% during the two-day experimen while 20 Capitella with 12 mg Pablum did not grow significantly.

Microbial metabolism was highest at the highest ration level, and lowest but slowly increasing at the lowest ration level.

If the value of 5.9×10^{-6} $J.s^{-1}$ mg^{-1} of feeding worms is applied to the worms in Figure 4B (weighing 37.0, 44.1 and 49.2 mg) their total heat production rate will be 2.2×10^{-4}, 2.6×10^{-4}, and 2.9×10^{-4} $J.s^{-1}$, respectively. Subtracting these values from the corresponding measured rates for microbes plus worms, we obtain 1.3×10^{-4}, 1.2×10^{-4}, and 1.3×10^{-4} $J.s^{-1}$ for average microbial metabolic rate in each of the three dishes. These rates are lower than microbial (without polychaetes) heat production on the second day (= 1.9×10^{-4} $J.s^{-1}$, Figure 5).

The known effects of polychaetes on community metabolism are variable. _Nereis succinea_, a large, active, errant polychaete, caused a doubling of the rate of detritus oxidation due to bioturbation (Briggs _et al_., 1979). _Abarenicola pacifica_, a more sedentary and permanent burrow dweller, also enhanced detrital decomposition, possibly through a 'gardening' effect on microorganisms (Hylleberg, 1975). An enchytraeid (_Cognettia sphagnetorum_) increased microbial activity and organic matter decomposition (Standen, 1978). On the other hand, _Capitella capitata_, a tiny worm that does not physically disturb the sediment as it moves about, did not increase the rate of detritus oxidation (Tenore, 1977) in contrast to a much larger species, _Nephtys incisa_, which did (Tenore _et al_., 1977). Our findings with _Capitella_ show that it affects microbial metabolism, and, therefore, the rate at which the microbial community oxidizes organic matter. The quantitative effect is variable, depending at least on the number of worms and the amount of available food. The greater the amount of added food for about the same amount of worms the higher the microbial heat production rate (by a correspondingly higher population size). If food is limited, microbial metabolism (and population size) is actually depressed, but the combined metabolic heat production by worms and microbes, being relatively steady, is initially higher, then lower and finally higher than that of the microbes alone.

One important implication of the foregoing experiments is that microbial metabolism must be measured in the presence of the organisms that affect their growth. This speaks in favor of the radiotracer techniques with electron acceptors (Jorgensen, 1977b; Howarth and Teal, 1979) but all other methods that require isolation and long-term incubation will, all other things being equal, tend to give overestimates because of the absence of competition and grazing, but especially with added organic substrates.

HEAT PRODUCTION BY NATURAL SEDIMENT

The heat production by natural sediment may appear quite steady (Figure 8), decrease, or increase with time. An apparently steady

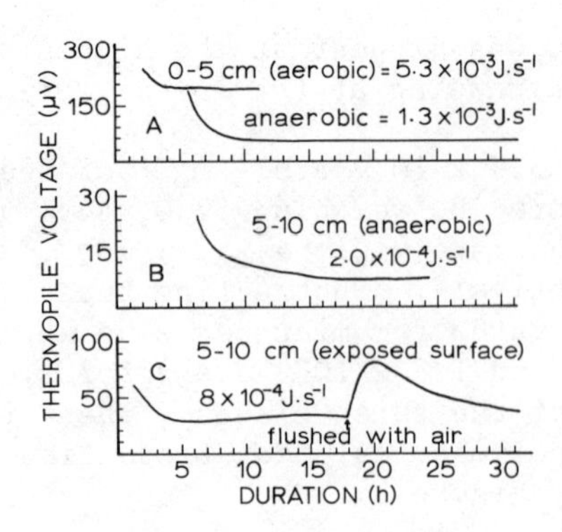

Fig. 8. Thermograms at 20°C of sediment cores from an intertidal mudflat. A) Surface of 0-5 cm layer exposed to air and after covering the surface with a Plexiglas plate. B) 5-10 cm anaerobic layer. C) The same 5-10 cm layer after exposing the surface to air and following injection of more air inside the metabolic chamber.

heat production rate merely reflects a very slow rate of decline. The decrease in heat production rate is to be expected as internal energy of the closed system naturally decreases with time by the amount of heat energy liberated. Why some sediments show much faster decline in energy flow than others upon transfer into a closed system remains to be studied. The accuracy of calorimetric measurements is more uncertain with the latter than the former kind. Sediments show increasing heat production rate following some kind of perturbation (physical disturbance, addition of organic matter, or increase in oxygen tension).

Effect of Anoxia on Surface Layer

When the sediment core in Figure 8A was taken out of the calorimeter, opened, and covered with a plexiglas plate which rested right the sediment surface, excluding any air or water, heat production ra dropped to a steady rate of 1.3×10^{-3} $J.s^{-1}$. The 75% drop in heat flow is an estimate of heat production equivalent of aerobic respiration plus chemical oxidation. The heat flow of a similar core whose surface was treated with formaldehyde dropped by about 55%. Assuming that formaldehyde killed mainly aerobes we can say that hea of oxidation is 75% minus 55% or 20% of total heat flow. However, we do not know how much of the heat flow after poisoning is the result of reaction between formaldehyde and reduced substances, whic is substantial when formaldehyde is injected directly into reduced sediment. Obviously, downward diffusion of formaldehyde is quite slow, indicating that it will also take a long time for poison simpl applied to the surface to kill the anaerobes. Furthermore, we cannot be certain that poison kills all aerobes at once. Many species are burrowing facultative anaerobes, whose natural reaction would be to withdraw deeper into the mud where they may live for days in the anaerobic state.

Anaerobic Heat Production by Sublayers

The 5-10 cm anaerobic layer shows a heat production rate that is about 4% of that of the aerobic 0-5 cm layer and 15% of the 0-5 cm layer under anaerobic condition (Figure 8B). It took 6-7 h longer for the 5-10 cm alyer to show a steady rate than the anaerobic 0-5 cm layer. These samples contain a large thermal mass (weighing ca. 500 g but of uncertain specific heat), and it appears that the larger the thermal mass and the lower the rate of heat production the longer it takes to reach a state of thermal equilibrium. Note the difference in length of time for the 0-5 cm layer to show a stable reading when it was aerobic and anaerobic. It is possible that the slow decline from the 10th to the 16th h in Figure 8B is real but this needs to be ascertained.

When the 5-10 cm layer was taken out of the calorimeter and its plate cover removed, thus exposing the surface to air inside the canister, it showed a heat production rate of ca. 8×10^{-4} $J.s^{-1}$, or four times the anaerobic rate. When air was injected into the canister through a fine-bore plastic tubing without opening the calorimeter there was a rapid increase in heat production rate to a peak of 2×10^{-3} $J.s^{-1}$ decaying exponentially afterwards. The disturbance effect on the instrument's thermal equilibrium was negligible. Evidently, by the time the instrument regained thermal equilibrium after the sample was returned in the former experiment oxygen tension had become limiting and the additional oxygen injected through the tubing caused a higher rate of heat production. The rapid increase over a 2-h period must have been the result of subsurface chemical oxidation at a higher oxygen pressure. The growth of sulfide oxidizing bacteria would be enhanced also under this condition but their growth is probably too slow to produce the change indicated by the thermogram.

By experiments such as these we can reach some understanding and develop testable hypotheses about the major causes of variation in heat production by sediment. For the finer details of second and third order sources of heat we will have to perform chemical analyses as well, putting this work more in the field of thermodynamic geochemistry than benthic ecology.

HOLISTIC VERSUS REDUCTIONIST MEASUREMENTS

At least in so far as the benthic ecosystem is concerned, the results of direct calorimetry give us reason to be much more cautious than we have been about assuming that our sampling, handling and experimental manipulations of organisms have negligible effects on their metabolic rate. We cannot assume that the usual temperature and other corrections are all that is needed to convert laboratory results to in situ rates.

Present ecosystem models deal with interactions between populations in terms of predation, grazing, or competition for food. There may be other forms of interactions. In the benthic system, where organisms of the same as well as different species are in relatively intimate physical contact with each other, the metabolic consequences of such intimacy could be significant. We have not begun to investigate the energetics of commensalism, for example, and other forms of symbiosis. The separation of a species from the community could have a substantial effect on its metabolic rate as well as that of the community. The magnitude of this effect is an indication of interspecific interactions that could be overlooked by reductionist approach to ecosystems. On the other hand, continuous heat flow measurements of intact communities will include all kinds of interaction effects.

It should be clear that direct calorimetry does not simply give us a measure of average metabolic rate to be used in simulation models for determining the step average changes in state variables. Present heat flow measurements already indicate that a system respons to food input will depend on species composition, thus giving us hope that calorimetry might contribute to our understanding of the quantitative relation between community structure and function. Heat flow is the total manifestation of all energy changes taking place in the system as a consequence of food supply, growth, predation, reproduction, competition, etc. The reductionist approach would be to sort out the energetics of each of these processes and put them all together into a simulation model of system energy flow. The validation of the reductionist model with holistic measurements could point to the existence of unsuspected interaction or disturbance effects, including artifacts in some measurements.

Up to now energy has been used in ecological models primarily for the purpose of expressing state variables in the common unit of the calorie. Little or nothing is gained by using constant calorific coefficients to convert measured material fluxes into energy units. Both material and energy fluxes must be measured simultaneously if we hope to really understand the quantitative relation between material flow and energy flux in ecosystems.

THE NEED FOR A DIFFERENT APPROACH TO ECOSYSTEM STUDY

The most common approach to ecological study involves the assumption, whether this is recognized or not, that the ecosystem is at steady state, not in the strict sense, but that it is fluctuating quite slowly around some average condition that persists year after year (Slobodkin, 1961). Diel and seasonal fluctuations are mere perturbations about the mean. Many ecosystem models describe the annual mean condition for state variables. Our sampling schemes (e.g. once a week, often much less frequently; a transect or grid of stations; selected depths) assume, in effect, that the sample (and whatever rates of processes we measure for that sample) is representative of a larger space and time span.

Strictly speaking, the ecosystem or any part of it, rather than being at steady state, is continuously changing. The scientist's perception of the significance of this fact depends on what part of the ecosystem and which group of organisms he is studying. Thus a microbiologist (Jannasch, 1969) views microbial populations as being in transient states, although, surprisingly, some microbiologists evidently pay no heed to this observation and take their samples no differently from macrofauna workers.

The question is, would a change in attitude and approach to the study of ecosystems give us better understanding and perception of how the ecosystem works than we now have? The results that I see by direct calorimetry make me a firm believer in the need to change our approach. Without denying the value of knowing the average states of the ecosystem and its parts, we must begin to look at how the system or subsystems respond on a whole range of time scales to continually occurring physical, chemical, and biological perturbations. It is the very nature of average values to smooth out differences, whereas the variations from those average values may be the key to understanding how natural systems operate. Thus, two different ecosystems might have the same annual energy flow on the average but very different short-term perturbations and mode of response to such events. The long-term average values may be interesting but they are not enough. The other properties of the system that concern us, e.g. the productivity of fish populations, may be related not only to long-term average states but to the nature of short-term fluctuations in one or more processes.

A change in approach means not only much more intensive sampling but in many cases continuous monitoring of the system. It may mean tracking migrating populations. It certainly means much more cooperation between scientists and coordination of disciplines than the lip service that we all hear on this matter. While continuous field monitoring is the ideal, unfortunately, the present state of the art of calorimetry precludes field measurements. This means continuous measurements of some other parameters besides heat flow, coupled with laboratory simulations and a running correlation between field measurements and laboratory experiments.

When I finally see a simulation model, including all the major processes and compartments, that can accurately reproduce the short-term changes in energy flow following one kind of perturbation or another, then I will have to admit that the modeler has finally understood the dynamics of the ecosystem and described it well.

ACKNOWLEDGEMENTS

Contribution No. 4 from the Tiburon Center for Environmental Studies. My research has been supported by the National Science Foundation (Grant No. OCE-8003197) and the National Oceanic and Atmospheric Administration, Office of Marine Pollution Assessment (Grant No. NA81RAD00028). I thank Dr. Curtiss O. Davis for his constructive criticisms of the manuscript.

REFERENCES

Abrams, B.I., and Mitchell, M.J., 1980, Role of nematode-bacterial interactions in heterotrophic systems with emphasis on sewage sluge decomposition, Oikos, 35:404.

Anderson, R.V., Coleman, D.C., Cole, C.V., and Elliott, E.T., 1981, Effect of the nematodes Acrobeloides sp. and Mesodiplogaster lheritieri on substrate utilization and nitrogen and phosphorus mineralization in soil, Ecology, 62:549.

Bamstedt, U., 1980, ETS activity as an estimator of respiratory rate of zooplankton populations. The significance of variations in environmental factors. J. exp. mar. Biol. Ecol., 42:267.

Boynton, W.R., Kemp., W.M., Osborne, C.G., Kaumeyer, K.R., and Jenkins, M.C., 1981, Influence of water circulation rate on in situ measurements of benthic community respiration, Mar. Biol. 65:185.

Briggs, K., Tenore, K.R., and Hanson, R., 1979, The role of microfauna in detrital utilization by the polychaete, Nereis succinea (Frey and Leuckhart), J. exp. mar. Biol. Ecol., 36:225.

Findlay, S., and Tenore, K.R., 1982, Effect of free-living marine nematode (Diplolaimella chitwoodi) on detrital carbon mineralization, Mar. Ecol. Progr. Ser., 8:161.

Florkin, M., 1960, "Unity and diversity in biochemistry. An introduction to chemical biology," Pergamon, London.

Forrest, W.W., Walker, D.J., and Hopgood, M.F., 1961, Enthalpy changes associated with the lactic fermentation of glucose, J. Bacteriol., 82:685.

Grainger, J.N.R., 1968, The relation between heat production, oxygen consumption and temperature in some poikilotherms in: "Quantitative Biology of Metabolism," A. Locker, ed., Springer-Verlag, New York.

Hammen, C.S., 1980, Total energy metabolism of marine bivalve mollusks in anaerobic and aerobic states, Comp. Biochem. Physiol. 67A:617.

Howarth, R.W., and Teal, J.M., 1979, Sulfate reduction in a New England salt marsh, Limnol. Oceanogr., 24:999.
Hylleberg, J., 1975, Selective feeding by Abarenicola pacifica with notes on Abarenicola vagabunda and a concept of gardening in lugworms, Ophelia, 14:113.
Jannasch, H.W., 1969, Current concepts in aquatic microbiology, Verh. Internat. Verein. Limnol., 17:25.
Jørgensen, B.B., 1977a, Bacterial sulfate reduction within reduced microniches of oxidized marine sediments, Mar. Biol., 41:7.
Jørgensen, B.B., 1977b, The sulfur cycle of a coastal marine sediment (Limfjorden, Denmark), Limnol. Oceanogr., 22:814.
Kemp, W.M., and Boynton, W.R., 1981, External and internal factors regulating metabolic rates of an estuarine benthic community, Oecologia (Berl), 51:19.
Kleiber, M., 1962, "The fire of life. An introduction to animal energetics," John Wiley, New York.
Kleiber, M., 1965, Respiratory exchange and metabolic rate in: "Handbook of physiology. A critical, comprehensive presentation of physiological knowledge and concepts, vol. II, sec. 3, Respiration," W.O. Fenn and H. Rahn, eds., Am. Physiol. Soc., Washington, D.C.
McCave, I.N., ed., 1976,"The benthic boundary layer," Plenum Press, New York.
Nichols, F.H., 1975, Dynamics and energetics of three deposit-feeding benthic invertebrate populations in Puget Sound, Washington, Ecol. Monogr., 45:57.
Nixon, S.W., Kelly, J.R., Furnas, B.N., Oviatt, C.A., and Hale, S.S., 1980, Phosphorus regeneration and the metabolism of coastal marine bottom communities in: "Marine Benthic Dynamics," K.R. Tenore, and B.C. Coull, eds., Univ. S. Carolina Press, Columbia.
Packard, T.T., 1971, The measurement of respiratory electron-transport activity in marine phytoplankton, J. mar. Res., 29:235.
Pamatmat, M.M., 1968, Ecology and metabolism of a benthic community on an intertidal sandflat, Int. Rev. ges. Hydrobiol., 53:211.
Pamatmat, M.M., 1971, Oxygen consumption by the seabed. VI. Seasonal cycle of chemical oxidation and respiration in Puget Sound, Int. Rev. ges. Hydrobiol., 56:769.
Pamatmat, M.M., 1977, Benthic community metabolism: a review and assessment of present status and outlook in: "Ecology of Marine Benthos," B.C. Coull, ed., Univ. S. Carolina Press, Columbia.
Pamatmat, M.M., 1978, Oxygen uptake and heat production in a metabolic conformer (Littorina irrorata) and a metabolic regulator (Uca pugnax), Mar. Biol., 48:317.
Pamatmat, M.M., 1979, Anaerobic heat production of bivalves (Polymesoda caroliniana and Modiolus demissus) in relation to temperature, body size, and duration of anoxia,Mar. Biol., 53:223.
Pamatmat, M.M., 1980, Facultative anaerobiosis of benthos in: "Marine Benthic Dynamics," K.R. Tenore, and B.C. Coull, eds., Univ. S. Carolina Press, Columbia.
Pamatmat, M.M., 1982a, Direct calorimetry of benthic metabolism

in: "The Dynamic Environment of the Ocean Floor," K.A. Fanning and F.T. Manheim, eds., Lexington Books, D.C. Heath, Lexington, Mass.

Pamatmat, M.M., 1982b, Heat production by sediment: ecological significance, Science, 215:395.

Pamatmat, M.M., in press, Simultaneous direct and indirect calorimetry in: "Polarographic Oxygen Sensors: Aquatic and Physiologica Applications," E. Gnaiger and H. Forstner, eds., Springer-Verlag, Berlin.

Pamatmat, M.M., Graf, G., Bengtsson, W., and Novak, C.S., 1981, Heat production, ATP concentration and electron transport activity of marine sediments, Mar. Ecol. -Prog. Ser., 4:135.

Peakin, G.J., 1973, The measurement of the costs of maintenance in terrestrial poikilotherms: a comparison between respirometry and calorimetry, Experientia, 29:801.

Platt, T., Mann, K.H., and Ulanowicz, R.E., eds., 1981, "Mathematical Models in Biological Oceanography," The UNESCO Press, Paris.

Scott, D.M., 1976, Circadian rhythm of anaerobiosis in a polychaete annelid, Nature, 262:811.

Slobodkin, L.B., 1961, "Growth and regulation of animal populations," Holt, Rinehart and Winston, New York.

Smith, K.L., Jr., 1973, Respiration of a sublittoral community, Ecology, 54:1065

Standen, V., 1978, The influence of soil fauna on decomposition by microorganisms in blanket bog litter, J. Anim. Ecol., 47:25.

Tenore, K.R., 1977, Utilization of aged detritus derived from different sources by the polychaete, Capitella capitata, Mar. Biol., 44:51.

Tenore, K.R., 1981, Organic nitrogen and caloric content of detritus I. Utilization by the deposit-feeding polychaete, Capitella capitata, Est. Coast. Shelf Sci., 12:39.

Tenore, K.R., Tietjen, J., and Lee, J., 1977, Effect of meiofauna on incorporation of aged eelgrass detritus by the polychaete, Nephtys incisa, J. Fish. Res. Board Can., 34:563.

Vernberg, W.B., Coull, B.C., and Jorgensen, D.D., 1977, Reliability of laboratory metabolic measurements of meiofauna, J. Fish. Res. Board Can., 34:164.

von Brand, T., 1946, "Anaerobiosis in invertebrates," Biodynamica, Normandy, Missouri.

Zeitzschel, B., 1980, Sediment-water interactions in nutrient dynamics in: "Marine Benthic Dynamics," K.R. Tenore and B.C. Coull, eds., Univ. S. Carolina Press, Columbia.

Zeitzschel, B., 1981, Field experiments on benthic ecosystems in: "Analysis of Marine Ecosystems," A.R. Longhurst, ed., Academic Press, London.

THE MEASUREMENT OF THE ENTHALPY OF METABOLISM IN MARINE ORGANISMS

Pierre Lasserre*

University of Bordeaux I, Institute of Marine Biology
33120 Arcachon, France.*New address: Station D'Oceanologie
Et De Biologie Marine University of Paris VI & C.N.R.S.
(LP4601) 29211 Roscoff, France

INTRODUCTION

In marine ecosystems the high variety of species-specific patterns of populations living under aerobic and low oxygen conditions, is a very complex indicator of firstly, the biochemical analyses of the level and time evolution of metabolism and, secondly, the energetic activity of populations within a given ecosystem.

A deeper understanding of how marine ecosystems function has been enhanced in recent years, by reductionist approaches involving techniques of measuring rates of feeding and element recycling, assimilation rates, excretion, growth and respiration (review by Mann and Smith 1981).

From the viewpoint of energy metabolism, living organisms consume energy from food, converting or degrading this energy, and then passing it on to the environment. This energy metabolism requires consideration of (1) the amount of energy intake, (2) the rate of energy consumption or energy output, and (3) the general utilization of energy within and between organisms.

Part of the energy obtained from food is thermodynamically destined for conversion to heat since most end products have higher entropies than that of the initial substrates; but as pointed later, some end products can be highly ordered and possess low entropy. Of course, the former always dominate the latter. In other words, the heat energy content and free energy content of food are not necessarily identical, the latter being generally somewhat less.

In the thermodynamic relation:

$$\Delta G = \Delta H - T\,\Delta S, \qquad \text{Equ. 1}$$

where ΔG is the free energy change (or usable enrgy), ΔH is the enthalpy change, ΔS is the entropy change, and T is the absolute temperature, it can be shown that ΔG is the maximum amount of useful energy which can be extracted from a biological process of known temperature and pressure, while ΔH represents the total heat plus useful energy change, again at constant temperature and pressure. $T\Delta S$ is the change in unavailable energy.

There is no direct way of measuring free energy change. Furthermore, the experimental determination of the entropy of material of unknown composition requires access to low temperature equipment (Wiegert, 1968) and, therefore, has no direct ecological application. In contrast, changes in enthalpy can be evaluated with calorimetric approaches.

In the application of the thermodynamic laws to organism energeti and ecological problems, a somewhat free interpretation of the classi cal thermodynamics and thermodynamics of irreversible processes can explain differences of opinion. Schrödinger (1945) was one of the first to use the entropy concept for characterising order and organisation in biological processes. Odum and Pinkerton (1955), Patten (1959) were among the first to discuss the application of thermodynamic theory to ecology. Margalef (1968, 1980) felt it of value to discuss ecological energetics in terms of entropy and information. Margalef, Odum and many others have given examples of ecological succession, the latter appears as the tendancy to achieve the maintenance of a maximum biomass for a minimum change in metabolic energy Slobodkin (1962) and other authors pointed out that the concept of entropy and free energy "don't seem to apply to ecological systems in any direct way". This negative point of view is comprehensible, since community evolution can be masked by periods of instability thresholds, undergoing transitional states of increasing entropy production (see discussion).

An ecosystem can be considered as a system driven far from equilibrium by continuous energy dissipation. According to Glansdorff and Prigogine (1971), the processes involved can be regarded as a product of a generalized flux J and a thermodynamic force X. Platt (1981) has rightly pointed out that the definition of biological thermodynamic forces are still obscure and, in the present state of knowledge, according to Ulanowicz (1981, page 67) "it may be worth exploring how far one may go in describing community behaviour without having to invoke forces". Therefore, if we talk about thermodynamic quantities, metabolism and growth, all essential figures in ecological energetics will be transformed into equivalent amounts of energy flux measurements will include the determination of heat production by direct and indirect calorimetry.

THE MEASUREMENT OF THE ENTHALPY : DIRECT AND INDIRECT APPROACHES

The direct measurement of energy output has the advantage of being based only on conservation of energy and not upon any assumption about the physiology of energy metabolism (energy pathways, oxygen utilization, which food is being oxidized to provide the energy etc.). Thus, its accuracy is limited only by technical skill in construction of the calorimeter and the heat measuring instruments.

The application of direct calorimetric measurements of the loss of energy in the form of heat resulting from biochemical oxidations and various motor activities is not recent. Calorimetric measurements of heat production in all kind of living organisms were reported as early as 1780 (Lavoisier and Laplace). During the following 100 years there was not much progress in the field. At the end of the 19th century and the beginning of this century, combustion calorimetry was developed and applied to animal materials, food products and biological waste products; applications of the bomb calorimeters to marine organisms and ecological problems are reviewed by Crisp (1971). In the 50's, the Frenchman Prat using Calvet calorimeters based on the heat conduction principle, made a large number of measurements on living organisms, which have been of importance for later development of modern direct calorimetry (Calvet and Prat, 1963).

In the field of ecological energetics, direct calorimetry, although appealing for many reasons, has hitherto been regarded as impracticable, notably because of the low rate of heat output and the other poor resolution and drift of the available apparatus.

It was considered more convenient to use indirect calorimetric measurements. Reliable and sensitive respirometers, both closed and flow systems, have been designed, or modified for ecological energetics on marine organims. The laboratory and *in situ* methods and equipment range from the very simple to the sophisticated, the choice of a given method being guided by the desired objective.

Among the most reliable methods we can cite : Winkler and micro-Winkler titration, Scholander gasometric technique and polarographic methods using Clark electrodes (review by Crisp, 1971). For very small organisms (ciliates, plankton, meiofauna), the Cartesian diver microrespirometer and its various modifications remains a reference technique (review by Lasserre, 1976).

Another good reason for selecting to privilege the oxygen uptake approach is that respiration is considered to be directly proportional to heat production. In the steady state, the release of energy from food is associated with the consumption of oxygen and the production of carbon dioxide. The relation between energy release and gas volume is considered as stoichiometric for any particular reaction. This conversion of gas exchange to energy is hampered, however, by serious limitations.

First, if part of the energy was released through anaerobic processes, the oxygen uptake would give too low a value of energy loss. We know that anaerobiosis is common among marine microorganisms and that intertidal and estuarine ciliates and metazoans, including invertebrates and fish, can display facultative anaerobiosis (Vernberg and Vernberg, 1981).

Secondly, it is assumed that all the energy released when oxygen is consumed is converted into heat or work. One underestimates the possible re-utilisation of the energy released by oxidation of metabolites to form high energy chemical bonds, the products so formed being stored in the organism. In this condition, the energy actually released as heat at the time will be less than the quantity calculated from the product of oxygen uptake and the normal oxy-calorific equivalent (Crisp,1971).

Over the last ten years, the technical development of calorimetry has reached a stage that even very slow processes connected with a low rate of heat output, such as occur in biological systems, may be detected. Several different calorimetric principles and a wide variety of particular designs have been used in modern biological calorimetry applied to molecular, cellular, multicellular, and complex systems studies (general review by Spink and Wadsö, 1976, Lamprecht and Zotin, 1978; Beezer, 1980).

The most reliable instruments for biological applications are based on the heat conduction principle ("Calvet" type, review by Wadsö, 1974). In such "microcalorimeters", the heat evolved is allowed to pass through a thermopile wall before it is absorbed by a heat sink. For steady state processes and for slow processes, the thermopile voltage (V) is directly proportional to the heat effect evolved dQ/dt with:

$$V = \text{const.} \times dQ/dt \qquad \text{Equ. 2.}$$

The calorimeter thus acts as a watt-meter and the resulting voltage - time curve, under these conditions, is directly related to the kinetics of the processes. Modern instruments have sensitivities of approximately 0.05 - 0.06 microvolt per microwatt. With some technical adaptations, especially for a straight and constant zero line, one may resolve 1 microvolt or even less. A resolution of 1 microvolt represents a heat flow of 10 microwatts. Limit of detectability with static (ampoule) calorimeters is 0.50 microwatt, with water flow calorimeters, 1.0 microwatt (200 microjoules), with stability of ±0.[illegible] microwatt over 8 hours. For example, with a flow microcalorimeter only 10^5 bacteria are necessary to obtain the routine lower detectable limit of 1 microwatt.

Static ampoule and circulation (flow) microcalorimeters have their own advantages (Figure 1). Both models utilized by the author of this review have proved quite satisfactory (LKB, Sweden, models 2107, with LKB 10200 Perpex pumps). With these systems, calorimetric experiments can be maintained for several days, under controlled conditions of temperature, oxygen tension, pH etc. A recent calorimeter (LKB model 2277) with up to four parallel channels allows us to run comparative studies simultaneously along with a control experiment. Other models built in the laboratory by the experimenter may be adequate. A prototype of microcalorimeter for *in situ* measurements at sea is presently being developed in our laboratory, at Arcachon, in cooperation with a Manufacturer.

The output of a calorimetric experiment is a plot of power (energy evolution per unit time) expressed in watts as a function of time. We recommended using the term power-time-curve (PTC) in place of the common but rather vague "thermogram" (Beezer, 1980). By integration of a PTC one obtains the total heat evolved $\int Q$ which is nearly equal to the total change in enthalpy ΔH. The heat quantity measured by microcalorimetry is the total enthalpy change of the growth process (ΔH). Energy consuming reactions, in the biosynthesis of new cells (ΔHb) will subtract from the energy yielding reactions of catabolism (ΔHc).

For heterotrophic organisms, ΔHb would be expected to be a small fraction of ΔHc (at most 3 per cent of ΔHc). In such cases, the quantity measured by microcalorimetry is mainly the enthalpy change of catabolism. This does not necessarily hold true for autotrophic organisms (Dessers et al., 1970). Finally, it is often useful to derive values for thermal value produced per living cell for a defined mass (dry weight, ATP content etc.) of cells.

ORGANISM'S ENERGY EXCHANGE

In the living organisms, all dissipative processes are determined by aerobic and anaerobic processes of metabolism, since in the final analysis, the organism's energy exchange is determined by these two processes. We have:

$$\dot{q}O_2 + \dot{q}an = \dot{q}M \qquad \text{Equ. 3.}$$

measured by indirect calorimetry and where :

$\dot{q}O_2$ is the specific respiration intensity (dO_2/dt per defined biomass)
$\dot{q}an$ is the specific anaerobic process intensity (e.g. glycolysis)
$\dot{q}M$ is the specific rate of metabolism

On the other hand, the intensity of heat production of an organism $\dot{q}$ is measured by direct calorimetry. Generally, for most aerobic

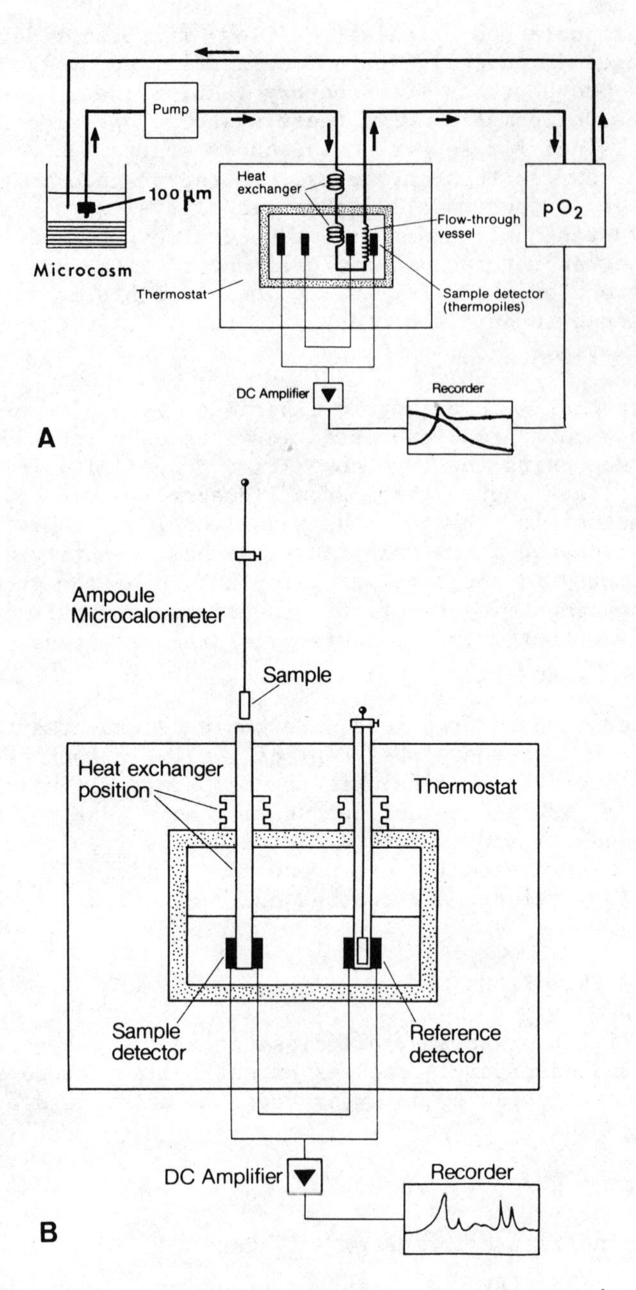

Fig. 1. A. Diagrammatic representation of a circulation (flow) microcalorimeter and its connection to a marine microcosm (From Lasserre, 1980, modified), B. Diagrammatic view of ampoule microcalorimeter (modified from LKB technical information report).

organisms the specific rate of metabolism (obtained by indirect calorimetry) can be, to a certain approximation, equated to the intensity of heat production ($\dot{q}_M = \dot{q}$). In very small organisms and in developing oocytes and larvae, the level of heat production is so elevated, due to a very high ratio between the surface and volume that, even in a stationary state, $\dot{q}_M$ is greater than $\dot{q}$.

According to Zotin (1972), the "bound dissipation function":

$$\psi_u = \dot{q}_M - \dot{q} \text{ (with } \dot{q}_M > \dot{q}\text{)}, \qquad \text{Equ 4.}$$

will describe the part of energy fixed inside the living system (to form presumably high energy chemical bonds).

The data collected by Zotin et al. (1978) show that the largest values of the bound dissipation function (with $\dot{q}_{O_2} > \dot{q}$) are found in aerobic bacteria. Moreover, in our recent experiments dealing with small aerobic marine metazoans (meiobenthic copepods), we have found respiration rate ($\dot{q}_{O2}$) higher than heat production ($\dot{q}$) (unpublished results). Similar observations have been made during the development and in the early stages of growth of plants, microorganisms and metzoans both invertebrates and vertebrates (Zotin, 1972; Zotin et al., 1978). Although such data ought to be supplemented, the values are closely related to organism size. In the examples given by Zotin et al. (1978) the ψ_u function is close to zero ($\dot{q}_M$ almost equal to $\dot{q}$) in animals weighing more than 20-30 g wet weight. It is noteworthy that, in small organisms, the heat production cannot be calculated from the respiratory data in all cases. This result must be kept in mind when calculating the energy balance of such organisms.

These observations have thermodynamic implication. In accordance with the Second Law, any irreversible process is accompanied by the production of dissipation heat. In large animals and in adults near the equilibrium, the irreversible processes proceed slowly and the dissipation heat may leave the system completely ($\dot{q}_M = \dot{q}$). Conversely, the rate of the dissipation heat production inside small organisms or during the process of development and growth, is high, and not all the dissipation heat will leave the system, a part of it may be utilised for some irreversible processes.

ASPECTS OF MICROCALORIMETRY APPLIED TO ECOLOGICAL ENERGETICS

Calorimetric methods are completely nonspecific, this holistic property is very advantageous for the detection of unknown phenomena. The connection between a recorded heat effect curve and the underlying metabolic processes is not usually well understood, especially for the transient periods. Nevertheless, the technique can be employed successfully on the empirical level, provided that the measurements precision is good enough and that no artefact influences the results. Obviously, reductionist approaches dealing with measuring rates of growth, assimilation, excretion, population dynamics, physiological adaptations, are necessary adjuncts to explain the phenomenon shown by direct calorimetry.

In complex living processes, such as the succession patterns of populations colonising waters and sediments, a detailed knowledge of the underlying metabolic activities is not easily obtained. The continuous microcalorimetric recording of heat production, over period of several hours to several days or weeks, detects integrated fluctuations of the vital functions and, furthermore, the combined techniques of static (ampoule vessel) and flow system can be used to demonstrate the emergence of biological patterns : population successions, metabolic regulation and adaptation following a controlled perturbation etc., no matter how complex the process occuring. Here microcalorimetry can be used to obtain an overall picture of the catabolic properties, since under carefully controlled experimental conditions, a reproducible "fingerprint" is obtained.

Recent experiments dealing with microcalorimetric measurements (1) on marine bacteria (Wagensberg *et al*., in Barcelona, Spain) and (2) on complex populations of marine microorganisms colonising the water-sediment interface submitted to controlled perturbations (Lasserre *et al*., in Arcachon, France) led to some results of ecological and thermodynamic significance which will be described, in more detail, in the following paragraphs.

EVOLUTION OF MARINE BACTERIA STRAIN

Microorganisms differ greatly in their metabolism, their nutritional requirements and their adaptation to the environment. It has been observed experimentally that heat dissipation in microbial cultures is systematically reproducible and characteristic of a particular strain. This makes it possible to differentiate between organisms by simply comparing their heat evolution. For exampl seventeen species of *Enterobacteria* have been characterised by Boling *et al.* (1973) by direct ampoule calorimetry, providing a rapid (8-14 h) and specific characterization of bacteria in a "fingerprint".

Wagensberg *et al*. (1978) have studied the heat production of marine bacteria from strains of *Vibrio* and *Flavobacterium* isolated, at depths of 200 and 20 meters respectively, in an upwelling region of north-west Africa. He used a static closed ampoule calorimeter. Inoculum of 10^7 individual cells were introduced into 6 ml ampoule containing an appropriate nutritional medium. The technique reveals itself as a powerful means of direct detection of biological behaviour of bacterial strains.

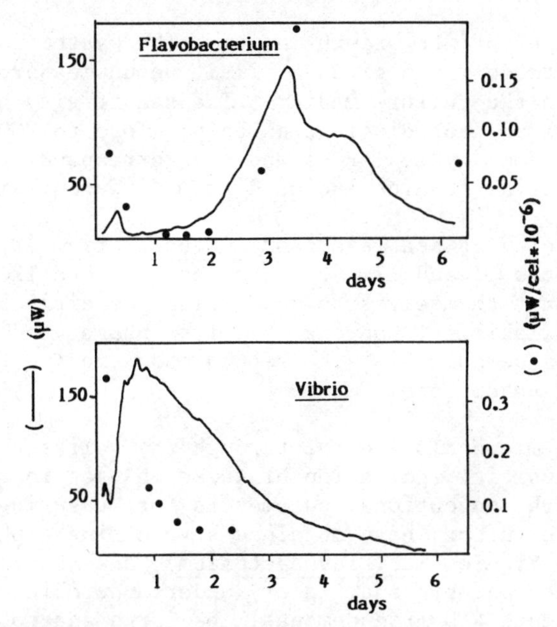

Fig. 2. Biphasic power-time curves (solid line) and specific heat dissipation (dotted line) of Flavobacterium and Vibrio strains sampled at 20 m and 200 m depths respectively, in an upwelling region of north-west Africa. (From Wagensberg et al., 1978).

Figure 2 shows typical PTCs for Vibrio and Flavobacterium strains. The first heat dissipation peak is connected with aerobic conditions, the second and large heat dissipation peaks are associated with anaerobic conditions. The succession of two heat dissipation phases has been described already for many non marine bacteria (Monk and Wadsö, 1975).

Two remarks can be made:
1) According to the specific heat dissipation (dQ/dt per weight), the Flavobacterium strain shows a maximum efficiency during the aerobic phase (the first two days). During the anaerobic phase, the process is inverted. For such a bacterial population it is energetically expensive to live in anaerobic conditions.
2) The Vibrio strain was isolated from a low oxygen water mass (200 m depth). The specific heat dissipation gradually decreases, until it reaches a maximum of efficiency, during anaerobic phase.

In an other work (Castell et al., 1981) the heat dissipation producted by strains of Enterobacter, Flavobacterium and Pseudomonas were studied by means of an ampoule microcalorimeter of the heat conduction type. The 10^6/ml bacterial samples were enclosed in a 6 ml ampoule with appropriate growth medium. The system is equipped with an oxygen electrode which gives a simultaneous measurement of oxygen availability in the culture medium. The sensitivity here is 7 microwatts, the maximum heat dissipation being close to 300 microwatts. Figure 3 gives four typical PTCs and the corresponding oxygen tension curves of bacterial strains isolated from different sites of the Biscay Gulf:

- Enterobacter at sea level and Flavobacterium at 20 m depth,
- Pseudomonas II and III at depths of 1000 and 1200 m respectively

The strains adapt themselves to new ambient conditions, according to the oxygen availability. During the first phase, the bacterial population develops aerobically, the second phase in the PTC corresponds with anaerobic conditions.

Therefore, microcalorimetry, through its holistic nature, provides a rapid and sensitive indication of these changes in growth rate, not easily seen with conventional parameters for measuring growth. In some cases, the authors have obtained atypic single phase PTC (Figures 4a and b). In Figure 4 a (Flavobacterium), one observes simultaneousl two phases under aerobic and, later, under anaerobic conditions. Conversely, in Figure 4 b (Pseudomonas) the large anaerobic phase is preceeded by a short adaptation period.

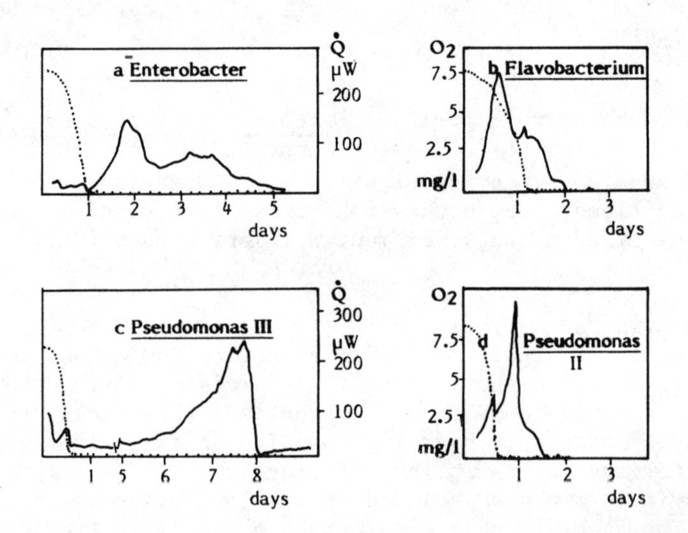

Fig. 3. Heat dissipation produced by isolated bacterial populations of Enterobacter, Flavobacterium, Pseudomonas II and III, found respectively at sea level, 20 m, 1000 m, and 1200 m depths. Power-time curve (in solid line), oxygen tensions in the calorimetric ampoule (in dotted line; From Castell et al., 1981).

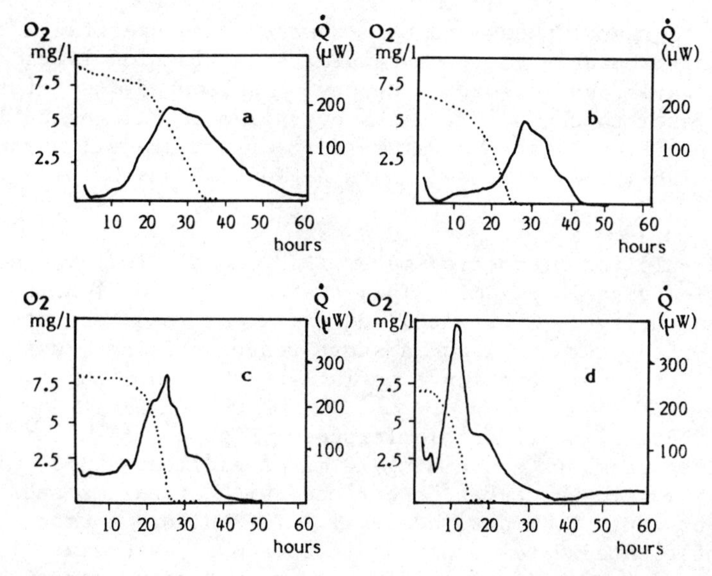

Fig. 4. Atypical monophasic power-time curves obtained from strain of *Flavobacterium* (a), after 6 months in culture, and *Pseudomonas II* (b). In (c) and (d) heat production generated by mixed populations of *Flavobacterium* and *Pseudomonas II*. Power-time curve in solid line, oxygen tension in dotted line (From Castell *et al*., 1981).

Moreover, the case of Figure 4 a shows that a response change in the strain of *Flavobacterium* occurs with time : the experiments made on the same strain and given on Figures 3 b and 4 a are separated by a period of 6 months. This strain shows an interesting pattern of temporal organisation. Its "fingerprint" was derived from a typical biphasic aerobic *versus* anaerobic metabolism toward a mixed behavioural pattern.

Heat production generated by mixed populations of *Flavobacterium* and *Pseudomonas* II have been studied (Figures 4 c,d). Inoculum were respectively 10^{-5} (Figure 4 c) and 10^{-3} (Figure 4 d). The PTC are different from isolated bacterial populations. The responses, however, can be compared to an atypical response as shown in Figures 4 a and 4 c (in these cases one observes that aerobic and anaerobic phases are simultaneous).

METABOLIC EVOLUTION AT THE SEAWATER-SEDIMENT INTERFACE

Flow microcalorimetry is well suited as a method for quantifying energetic changes occuring in experimental ecosystems ("microcosms"), in order to reveal phenomena not shown by more specific methods. This is particularly true for complex situation involving the presence of water and sediment. We have developed a flow microcalorimetric system to characterise the overall metabolic heat flux occuring specifically at the water-sediment interface (Lasserre, 1980; Tournié, 1981; Lassierre and Tournié, Tournié and Lasserre, in preparation).

A circulation microcalorimeter (LKB, model 2107) was equipped with a flow vessel of 1 ml. This system proved well suited to its high sensitivity (1 microwatt) and excellent long term stability (5 microwatts over 24 h) and a short response delay (less than 2 minutes), the equilibration time is less than 15 minutes.

The routine experimental microcosm (Figure 1 A) consisted of 35 ml glass chambers containing 10 ml of sediment and 25 ml of overlying sea water, which can be enriched with organisms and/or nutrients or drugs to a desired level. The culture chamber containing the reconstituted water-sediment interface is maintained in controlled environmental conditions, at a constant temperature (19°C) with a daylight cycle of 6 hours. All the calorimetric measurements are made in darkness to eliminate heat production associated with photosynthesis. The sea water is pumped continuously from the microcosm at a distance of 5 mm above the sediment surface, at 23 ml h^{-1} (LKB Perpex pump). A 100 μm filter prevents passage of course detritus and metazoans, while allowing free circulation of microflora and microfauna. This circulating interface is taken through the microcalorimeter and from there to an oxygen electrode (Radiometer) and is returned to the microcosm. Other types of microcosm have been tested : for example, cores containing undisturbed sediment and water. Basically, we submit the microcosms to experimental perturbations by introducing known quantities of organic matter or nutrients or creating a bioturbation in the sediment by introducing meiobenthos or small deposit feeders (*Hydrobia*).

Heat production in the circulating interface has been recorded after introduction of peptone (Difco), a nutritive mixture of proteins, at a concentration of 4 mg per ml sea water. These experimenta eutrophications resulted in a significant increase of heat production of between 20 and 180 microwatts ml^{-1} of circulating interface (Figures 5 & 6). Under carefully standardized test conditions reproducible "fingerprints" are obtained after the introduction of peptone solution into the water of the microcosm. The standardized test conditions used allowed comparative studies on the kinetics of the effects of experimentally controlled perturbations applied to microcosms. It is worthwhile noting the similarities between these power-time curves and curves obtained from mixed marine bacteria populations (Castell *et al.*, 1981, and Figure 4).

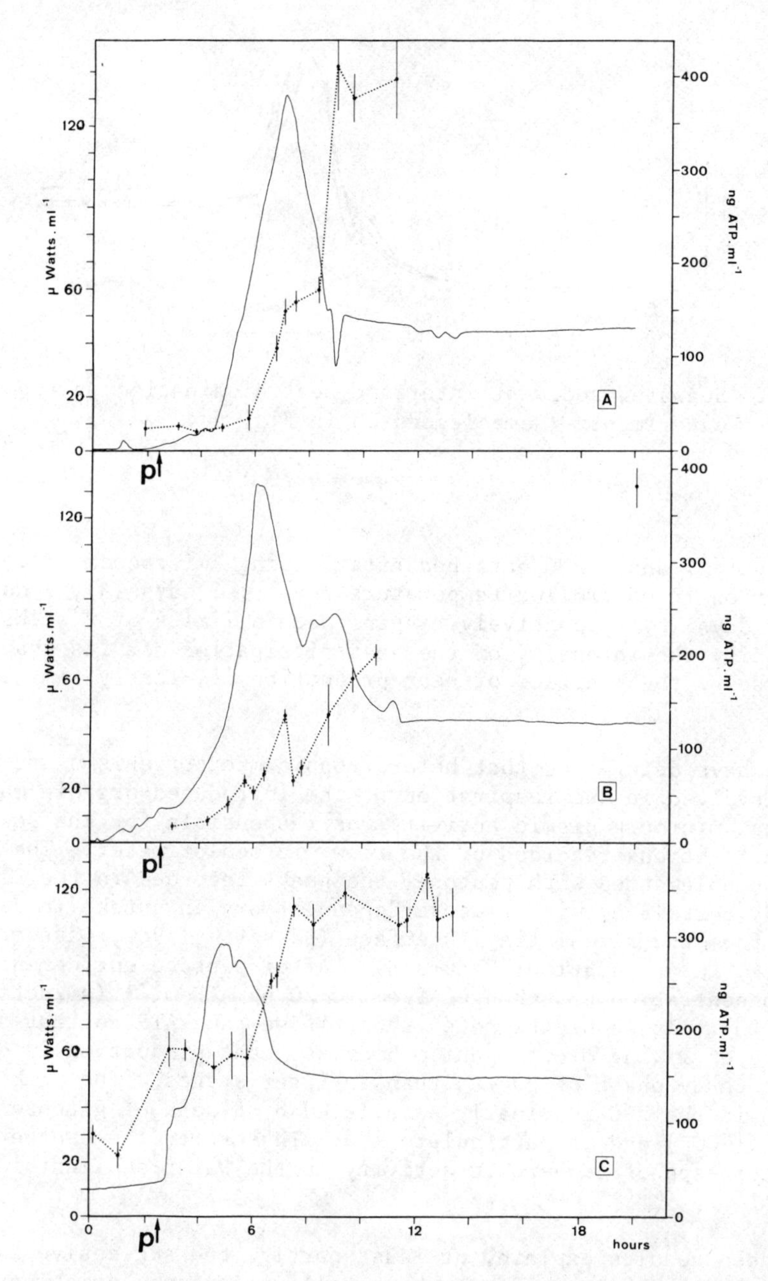

Fig. 5. Three typical power-time curves (solid line) and ATP content evolution (dotted line) generated at the water-sediment interface of marine microcosms, after experimental eutrophication (p, peptone enrichment); A- intertidal beach, Arcachon lagoon; B- mud flat, Gironde estuary; C-fish pond, Arcachon lagoon (July, 1979). (From Lasserre, 1980).

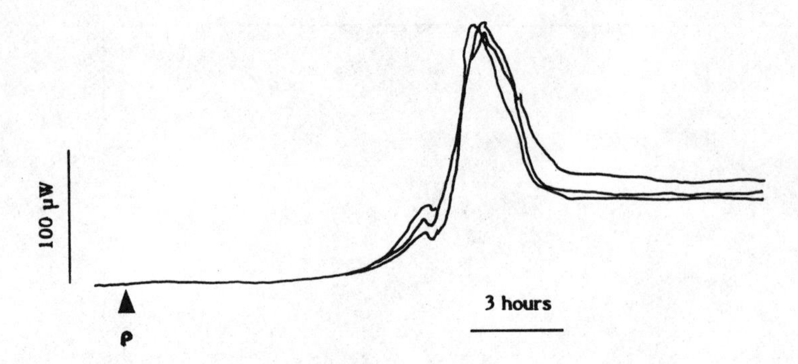

Fig. 6. Seawater-sediment interface heat dissipation in replicate experiments (same legend as in Fig. 5).

Figure 7 shows PTC obtained after "aging" microcosms (i.e. pre-incubation in controlled temperature room) : 2 days (A), 7 days (B) and 16 days (C) respectively. Aging within limits of 2 weeks does not modify the intensity of the heat dissipation nor its shape, conversely, the kinetics of heat production is clearly time-correlated.

We have determined that heterotrophic microorganisms are indeed involved in the respiration of the introduced organic material and that microorganismic activity was responsible for the power-time deflections recorded by the flow microcalorimeter. The microcosms supplemented with peptone, the peaks recorded in the PTC are clearly correlated with glucose-dependent oxygen uptake in aliquots taken from the circulating interface (Lasserre, 1980, Lasserre and Tournié, in preparation; Figure 8). After peptone enrichment, the ATP concentration has changed from 10-20 ng ATP ml^{-1} (a factor of 40 to 20). It should be noted that, if we take ATP content as a measure of living microorganism biomass, heat production is greater in the early phase of growth than in later stages. The peak recorded in the PTC is clearly paralleled by a peak of glucose respired ($^{14}CO_2$) and of particulate ^{14}C. Therefore, the exothermic peak corresponds to aerobic activity in the water-sediment interface.

These results explain, at least partly, the successive evolution in the PTC. The heat output which follows peptone enrichment is a metabolic process occuring at the water-sediment interface. The metabolic activity is not totally parallel with ATP and biomass increase, nor with oxygen depletion. These results are quite reproducible. Figure 9 shows the effect of formalin addition on the final steady state plateau of a PTC, after peptone enrichment a complete inhibition of heat dissipation occurs within 10 minutes.

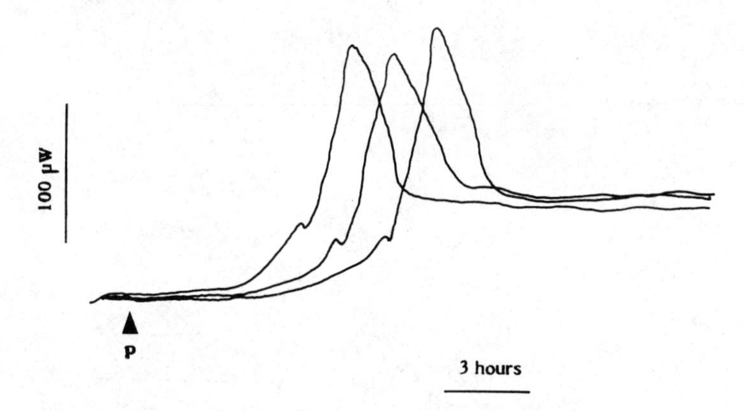

Fig. 7 Effect of "aging" on microcosms preincubated 2 days (A), 7 days (B) and 16 days (C) respectively (From Lasserre and Tournié, in preparation).

Using the same experimental approach, microcosms were prepared every month from cores of sediment and water collected during three consecutive years : we have observed seasonal trends in oxygen tension and heat production in experimentally eutrophied microcosms (Tournié and Lasserre, in preparation):

Considering oxygen tension evolution, it is apparent that oxidative capabilities of the sea-water interface can be divided into 3 characteristic periods:

- one period of 4 winter months (January-April) where oxygen depletion is never total. The pO_2 plateau is reached 12 hours after peptone treatment,

- one period of 4 summer months (July-October) where oxygen depletion is total, 8 hours after peptone treatment,

- two transitional periods : one in May-June and one in November-December.

The power-time curves are also correlated with seasonal trends. The reactivity is higher in summer than in winter, but the final steady state plateau is more elevated in winter. Generally, spring PTC are bimodal, showing a transition step between typical winter and summer curves.

The three characteristic evolutionary types of pO_2 and heat production curves are summarized in Figure 10 (winter type in a, spring type in b, summer type in c).

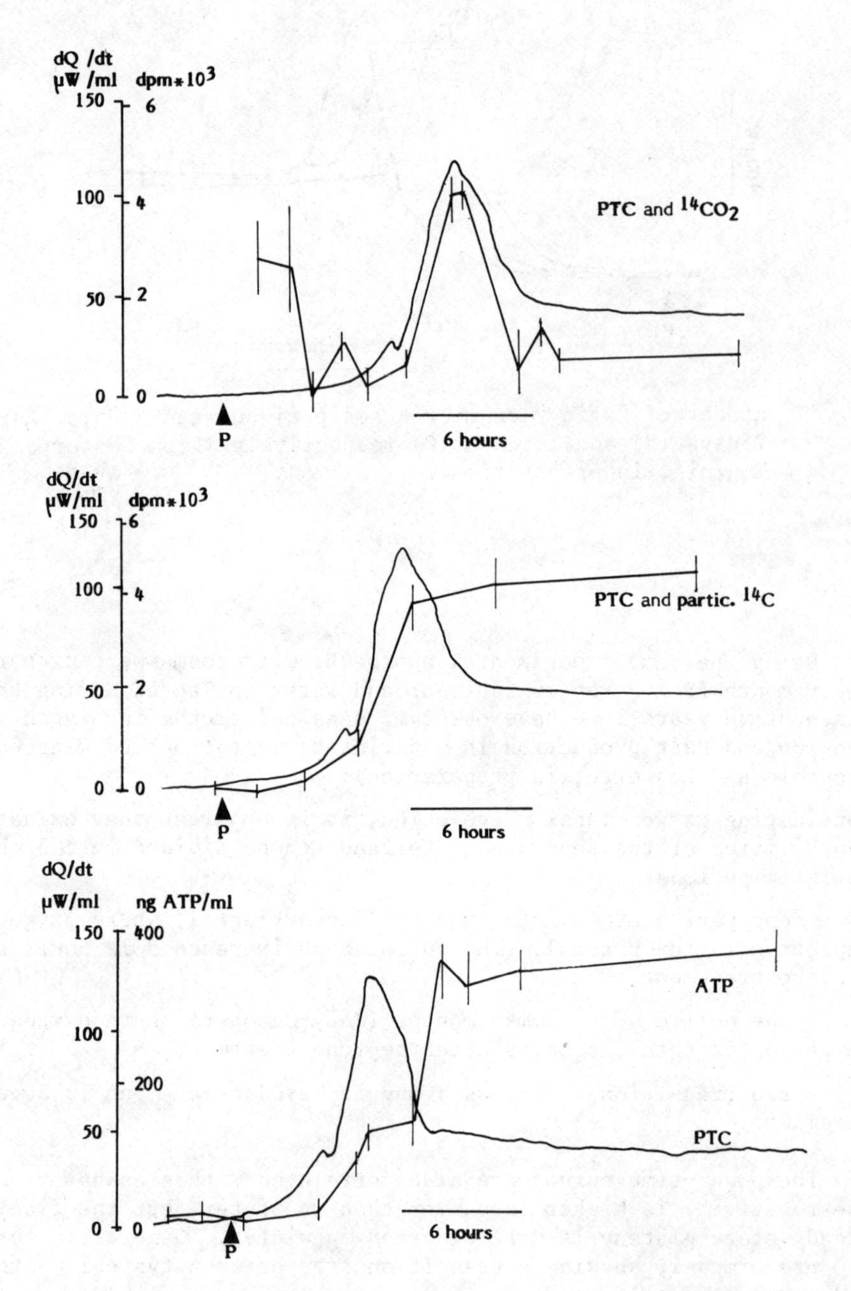

Fig. 8 Flow microcalorimetric experiment on seawater-sediment interface and ATP, $^{14}CO_2$ and particulate ^{14}C, after peptone enrichment (P, 4mg ℓ^{-1}) (Frome Lasserre and Tournié, in preparation).

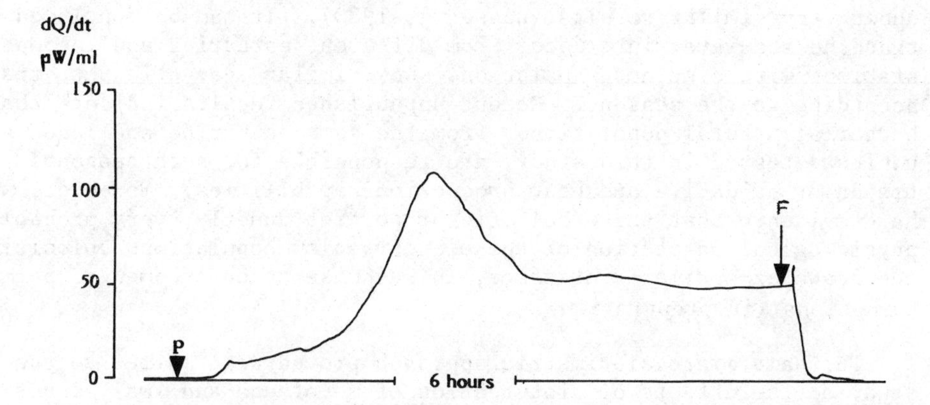

Fig. 9. Effect of formalin addition on the final steady state plateau of a power-time curve (From Lasserre and Tournié, in preparation).

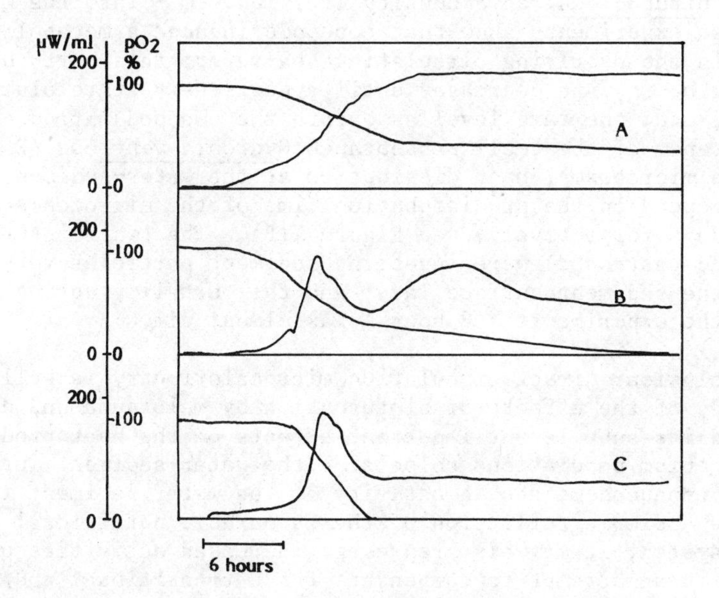

Fig. 10. Evolutionary types of heat production and pO_2 at the sea-water-sediment interface, after peptone enrichment : A, winter type ; B, spring type ; C, summer type (From Tournié and Lasserre, in preparation).

These experiments refer to one routine field station (an intertidal beach in the lagoon of Arcachon). Experiments done with samples of sediments and water from contrasting field stations have shown very similar results (Lasserre, 1980). It can be concluded that the sea-water interfaces from different estuarine and lagoonal stations with distinct populations show similar energetic patterns, according to the seasons. Recent unpublished results indicate that bacterial natural populations, from the same estuarine and lagoonal habitats tested in this study, are responsible for such seasonal trends of oxidative and heat production capabilities. Moreover, we have evidence that these holistic properties underly, very probably, physiological adaptation of the microorganism populations colonizing the seawater-sediment interface, in response to environmental perturbations (in preparation).

The same microcalorimetric approach proved well suited to the study of the effects of bioturbation of meiofauna and small deposit-feeders.

Figure 11 shows the heat dissipation produced at the water-sediment interface by the introduction of the meio-epibenthic copepod, *Eurytemora hirundoides*, at a density of 2 ind. cm^{-2} into the microcosm. These experiments show that copepods induced a metabolic perturbation in the overlying circulating sea-water; this perturbaction lasts a few hours, and decreases until a new steady state plateau is reached, i.e. the same level as before the copepod introduction. In the presence of the small prosobranch *Hydrobia ventrosa* (2 ind. cm^{-2} in the microcosm), heat dissipation at the water-sediment interface will depend on the pre-incubation time of the microcosms (2, 7, 13 and 16 days respectively, see Figure 11B). The two "bioturbators" (copepod and gastropod) were ingesting sediment particles very actively at the sediment surface layer and the mortality during the course of the experiments (48 hours) was almost nil.

In conclusion, direct circulation microcalorimetry is well suited to the study of the effects of bioturbation by meiofauna and deposit-feeders and its subtile but important effects on the heat production of microorganism populations colonizing the water-sediment interface. The enhanced metabolic activity at the water-sediment interface may be a simple reflection of the favourable nutritional status of the ecosystem, it may also represent increased activities of microbial forms in an attempt to compensate for a transitional and/or adverse impact of some perturbations or stress. It is noteworthy that given time and constant nutrient conditions, successions will produce stable heat production with stable biomass and metabolism per unit area (or volume) of a sea-water interface (Lasserre, 1980).

DISCUSSION AND CONCLUSION

Ideally, when calculating the energy balance of organisms, the heat production should be calculated using both direct and indirect calorimetric approaches. Indirect calorimetry will measure changes

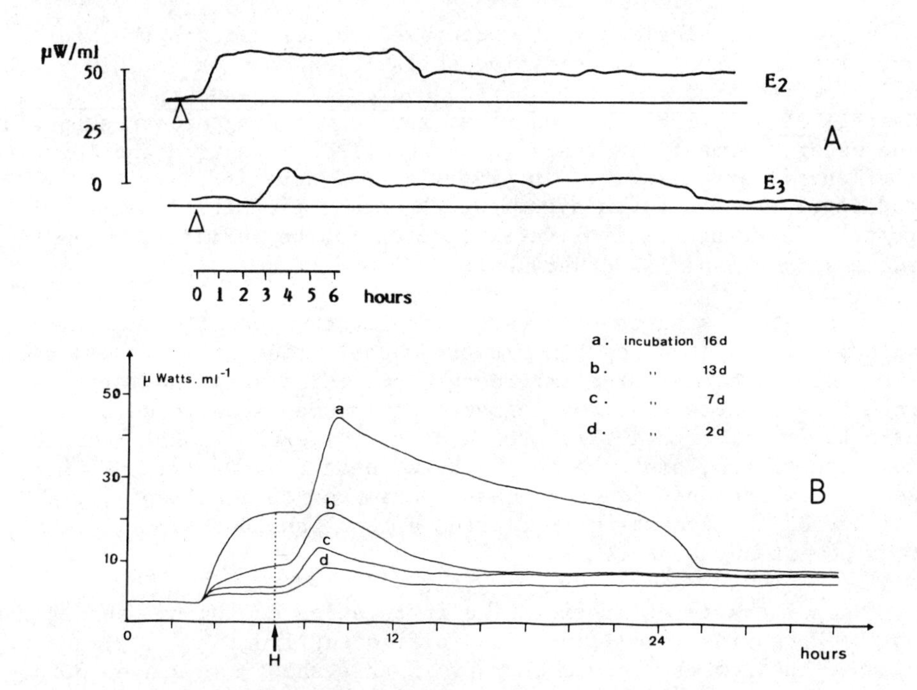

Fig. 11 Microbial heat dissipation, at the seawater-sediment interface, generated by the introduction of estuarine epibenthic feeders : the copepod Eurytemora hirundoides (A) and the gastropod Hydrobia ventrosa (B) in different microcosms. In (B), the microcosms have been "aged" 2 days, 7 days, 13 days and 16 days respectively. H: Hydrobia (From Lasserre, 1980, modified).

in gas volume, substrate concentration, oxygen tension etc. and will evaluate their contribution to energy giving an amount of free enthalpy liberated at the same time. Direct calorimetry will provide an unbiaised and holistic measurement of the energy flow, both aerobic and anaerobic.

Application of direct calorimetry to the field of ecological energetics is quite recent. The energetics of a freshwater invertebrate (oligochaetes) and of developing fish eggs have been studied in relation to anoxibiosis (Gnaiger, 1979 & 1980). Heat production of marine invertebrates, including molluscs and crustaceans, have been studied during aerobic-anoxic transitions by Pamatmat (1978). Heat production in marine sediments maintained in closed calorimetric vessels has been reported also by Pamatmat (1975, 1980). A review of these latter approaches is given in this book.

Flow microcalorimetry has been used to demonstrate the microbial activity in normal and acidified soils (Ljungholm et al., 1979 a,b). Calorimetric experiments on marine bacteria (Wagensberg et al., 1978; Castell et al., 1981) and on mixed marine microorganisms colonising the water-sediment interface (Lasserre, 1980; Lasserre and Tournié, the Tournié and Lasserre, in preparation), have indicated that, at and level of population dynamics, the power-time curves (the "thermograms") describe dissipative structures in the form of temporal succession or spatial orderings.

The latter studies have established the value of direct microcalorimetry in studying homeostatic capabilities at a global level. The temporal behaviour of microorganisms colonising the water-sediment interface of estuarine and lagoonal environments was studied to compare their relative ability to recover from experimental perturbations (eutrophication, bioturbation). Here, heat flow variations with respect to seasons, is an adequate parameter of resilience, i.e. the ability of the ecosystem to return, after transitory oscillations, to a new steady state.

One of the chief difficulties in relating ecological succession to the Prigogine (1967) and Glansdorff and Prigogine (1971) thermodynamic theorem of minimum entropy is that the first stages of a succession in an "immature" ecosystem does not develop, apparently, in a linear way and displays periods of increase in entropy (heat production peaks in the power-time curves). A characteristic feature of the microcalorimetric PTC, in Lasserre's experiments (see above), is the increase in the specific heat production during the growth stage followed by a decrease, indicating subsequent functional adaptational changes in populations, and marked by a final stationary state.

While the increase in dissipation during the early stage of development has often been interpretated as the inadequacy of irreversible thermodynamics to account for biological phenomena, it has been considered, recently, by Lurié and Wagensberg (1979), on a theoretical basis, that no incompatibility exists between the observed increase in heat dissipation and Prigogine's theorem of minimum entropy production, based on linear irreversible thermodynamics. These authors exhibit a linear model for which the time evolution of the different terms in the entropy balance is parallel to that of the evolution of living systems.

REFERENCES

Beezer, A.E., ed., 1980 "Biochemical calorimetry," Academic Press, London.

Boling, E.A., Blanchard, G.C., and Russel, W.J., 1973, Bacterial identification by microcalorimetry, Nature (London), 241: 472.

Calvet, E. and Prat, H., 1963, "Recent progress in microcalorimetry," Pergamon Press, Oxford.

Castell, C., Wagensberg, J., Tejero, A. and Vallespinos, F., 1981, Identificacion de las fases metabolicas en termogramsas de cultivos bacterianos, Inv. Pesq., 45: 291.

Crisp, D., 1971, Energy flow measurements, In: "Methods for the study of marine benthos," eds. N.A. Holme and A.D. McIntyre, Blackwell Scientific, Oxford.

Dessers, A., Chiang, C., Landelot, H., 1970, Calorimetric determination of free efficiency in Nitrobacter winogradskyi, J. gen. Microbiol., 64: 71.

Glansdorff, P. and Prigogine, I., 1971, "Thermodynamic theory of structure, stability and fluctuation," Wiley, New York.

Gnaiger, E., 1979, Direct calorimetry in ecological energetics. Long term monitoring of aquatic animals, Experientia, Suppl. 27: 155.

Gnaiger, E., 1980, Energetics of invertebrate anoxibiosis: direct calorimetry in aquatic oligochaetes, FEBS Letters, 112: 239.

Lamprecht, I., and Zotin A.I., ed., 1978, "Thermodynamics of biological processes," Walter de Gruyter, Berlin and New York.

Lasserre, P., 1976, Metabolic activities of benthic microfauna and meiofaunca, In: "The Benthic Boundary Layer," ed. I.N. McCave, Plenum Press, New York.

Lasserre, P., 1980, Energetic role of meiofauna and epifaunal deposit-feeders in increasing level of microbial activity in estuarine ecosystems, at the water-sediment interface, In: "Biogéochimie de la matière organique à l'interface eau-sédiment marin," ed. R. Daumas, Actes Colloq Int. CNRS, Paris, 293: 309.

Lasserre, P. and Tournié, T., Use of microcalorimetry for the characterization of marine metabolic activity, at the water-sediment interface, Submitted to Oikos.

Lavoisier, A., and Laplace, P., 1780, Memoir on heat, Reprinted in "Great experiments in biology," ed. M.L. Gabriel and S. Fogel, (1955), Printice-Hall, New Jersey.

Ljungholm, K., Norén, B., Sköld, R., and Wadsö, I., 1979, a, Use of microcalorimetry for the characterization of microbial activity in soils, Oikos, 33: 15.

Ljungholm, K., Norén, B., and Wadsö, I., 1979 b, Microcalorimetric observations of microbial activity in normal and acidified soils, Oikos, 33: 24.

Lurié, D., and Wagensberg, J., 1979, Non-equilibrium thermodynamics and biological growth and development, J. Theor. Biol., 78: 241.

Mann, K.H., and Smith D.F., 1981, Physiological rates and ecological fluxes, In: "Mathematical models in biological oceanography," T. Platt, K.H. Mann and R.E. Ulanowicz, The Unesco Press, Paris.

Mann, K.H., 1969, Dynamics of aquatic ecosystems, Ad. Ecol. Res., 6: 1.

Margalef, R., 1968m "Perspectives in ecological theory," Univ. Chicago Press, Chicago.

Margalef, R., 1980, "La Biosfera, entre la termodinamica y el juego," Ediciones Omega, Barcelona.

Monk, P., and Wadsö, I., 1975, The use of microcalorimetry for bacterial classification, J. Appl. Bact. 38: 71.
Odum, H.T., and Pinkerton, R.C., 1955, Time's speed regulator: the optimum efficiency for maximum power output in physical and biological systems, Am. Sci., 43: 331.
Pamatmat, M.M., 1975, In situ metabolism of benthic communities, Cah. Biol. Mar., 16: 613.
Pamatmat, M.M., 1978, Oxygen uptake and heat production in a metabolic conformer (Littorina irrorata) and a metabolic regulator (Uca pugnax), Mar. Biol. 48: 317.
Pamatmat, M.M., 1980, The annual mineralization of organic matter in sediments. Present knowledge, persistent technical problems and uncertainties in our measurements, In: "Biogéochimie de la matière organique à l'interface eau-sédiment marin," R. Daumas, ed., Actes Colloq. Int. CNRS, Paris, 393: 309.
Patten, B., 1959m An introduction to the cybernetics of the ecosystem trophicdynamic aspect, Ecology, 40: 221.
Platt, T., 1981, Ecological application of irreversible thermodynamics, In: "Mathematical models in biological oceanography," T. Platt, K.H. Mann and R.E. Ulanowicz, eds., The Unesco Press, Paris.
Prigogine, I., 1967, "Introduction of thermodynamics of irreversible processes," 3rd ed., Wiley, New York.
Schrödinger, E., 1944, "What is life?," Cambridge University Press, New York.
Slobodkin, L.B., 1962, Energy in animal ecology, In: "Advances in ecological research," vol. 1, J.N. Cragg, ed., Academic Press, London and New York.
Spink, C., and Wadsö, I., 1976, Calorimetry as an analytical tool in biochemistry and biology, In: "Methods of biochemical analyses" vol. 23, D. Glick, ed., Wiley, New York.
Tournié, T., 1981, Contribution à l'étude microcalorimétrique de l'activité biologique d'interfaces eau-sédiment en milieu marin, Thèse 3ème cycle, Université de Bordeaux I, France, (unpublished thesis).
Tournié, T. and Lasserre, P., in preparation, Microcalorimetric characterization of seasonal metabolic trends in marine microcosms, Submitted to Oikos.
Ulanowicz, R.E., 1981, Information theory applied to acosystem structure. In: "Mathematical models in biological oceanography," T. Platt, K.H. Mann and R.E. Ulanowicz, eds., The Unesco Press, Paris.
Vernberg, F.J., and Vernberg, W.B., eds., 1981, "Functional adaptations of marine organisms," Academic Press, New York.
Wadsö, I., 1974, A microcalorimeter for biological analyses, Science Tools, 21: 18.
Wagensberg, J., Castell, C., Torra, V., Rodellar, J., and Vallespinos F., 1978, Estudio microcalorimétrico del metabolismo de bacterias marinas: deteccion de procesos ritmicos, Inv. Pesq., 42:172.

Wiegert, R.G., 1968, Thermodynamic considerations in animal nutrition, Amer. Zool., 8: 71.
Zotin, A.I., 1972, "Thermodynamic aspects of developmental biology," Karger, Basel.
Zotin, A.I., Konoplev, V.A., and Grudnitzky, V.A., 1978, The questions of non-linearity for using criterion of ordeliness, In: "Thermodynamics of biological processes," I. Lamprecht and A.I. Zotin, Water de Gruyter, Berlin and New York.

BACTERIAL PRODUCTION IN THE MARINE FOOD CHAIN: THE EMPEROR'S NEW SUIT OF CLOTHES ?

Peter J. leB. Williams

Bigelow Laboratory for Ocean Sciences
McKown Point, West Boothbay Harbor
Maine 04575, U.S.A.

INTRODUCTION

Writers on biological oceanography and modellers of the marine food chain have been hesitant and uncertain of the importance of bacterial processes, indeed of microbial processes in general. There are good reasons for this. Marine bacteria are most probably among the smallest free-living organisms in the biosphere. Until comparatively recently the determination of their numbers and biomass has been difficult. The measurement of bacterial activity is even more problematic, the techniques specialized and their interpretation difficult if not obscure. Thus, the acceptance by non-microbiologists that bacteria may play a significant role in the marine food chain has been, and to some extent still is, an act of faith. Understandably and probably quite rightly biological oceanographers in the past have been cautious in incorporating a significant microbial component into conceptual or other models of the pelagic food chain.

There would however appear to be at the present a mood in marine ecology to re-evaluate the role of bacteria in the food chain. The aim of the present review is to look as objectively as possible at the existing evidence and to attempt to come to some generalizations regarding the quantitative aspects of bacterial biomass and its production in the pelagic food chain; then to consider them in an ecosystem context. In assembling the factual information I have drawn very heavily upon three recent reviews (Sorokin, 1978; Joint and Morris, 1982; van Es and Meyer Reil, 1982).

Although I will restrict my considerations to the productive part of the pelagic water column, this is not to imply that bacterial processes in other parts of the marine ecosystem (i.e. the deeper parts of the ocean and the sediments) are of lesser importance, however I find it is difficult at the present to be at all specific of the quantitative role of bacteria in these environments.

BACTERIAL ABUNDANCE AND BIOMASS

Early attempts to determine the abundance of bacteria in the sea used culture techniques and they appear to have given numbers low by two to three orders of magnitude. Direct staining techniques, which originally used erythrosin but subsequently fluorescence stains such as acridine orange, have been adopted as virtual standards. The measurements of cell dimensions for the eventual calculation of biovolume and biomass, may be done either on the fluorescent image or by parallel measurements with a scanning electron microscope. Despite the now wide use of these techniques and their essential simplicity, there are still remarkably few published studies of the distribution of bacterial abundance and biomass in ocean waters. Table 1 is a compilation of data abstracted principally from the reviews of van Es and Meyer-Reil (1982) and Sorokin (1978) with additions from other sources. Only data from truly marine environments have been included. Those from estuaries and enclosed seas have been omitted because of the possibility that the food chain in these areas may be structured quite differently. Bacterial numbers themselves are unsatisfactory for many comparative purposes; biomass is generally more useful. The conversion of abundance data to biovolume and subsequently to biomass relies upon accurate estimation of the linear dimensions of the organism. It is a very uncertain calculation to make: the biomass may often be related to the cube of the linear dimension of the organism and so the final figure is very sensitive to comparatively small errors in determining the size of cells which are close to the limit of resolution of the optical miscroscope. Fuhrman (1981) has discussed the problem of this type of measurement. The uncertainties of calculating bacterial biomass or volume is no doubt the main reason why the majority of authors have simply reported abundance data. The available data on bacterial biomass is so scarce and so pertinent to the present argument that I have felt justified in attempting to convert the abundance data to biomass, whilst at the same time recognizing limitations of the calculation. Although many marine bacteria are as small as 0.2 μm, Fuhrman's calculations (Fuhrman, 1981) imply that as far as volume is concerned, in the samples he examined, the median equivalent spherical diameter was slightly over 0.6 μm. His calculations were for the California Bight; I have adopted a figure of 0.5 μm which is intended to take note of the observations that there is generally a decrease in bacterial size seaward (Ferguson and Rublee, 1976; Hoppe, 1976; Sieburth, 1983). Assuming the carbon biomass of the organism to be 10% of the wet biomass, this gives an individual cell biomass of

6.5 fg, similar to values estimated or adopted by other workers for offshore and oceanic communities (Hoppe, 1976; Watson et al., 1977; Fuhrman and Azam, 1980, 1982).

There is a wide spread of data, as may be expected from an emergent research area. However, it would appear that in coastal waters bacterial biomasses in the range 5-10 μgC ℓ^{-1} are commonly encountered, falling to 2-5 μgC ℓ^{-1} in non-oligotrophic ocean water and 1-3 μgC ℓ^{-1} in oligotrophic oceans. All these data refer to the productive part of the water column.

The rapid division rates characteristic of small organisms, in conjunction with the substantial bacterial biomasses indicated in Table 1, taken at face value imply high bacterial productivity. However, as well as exhibiting the highest metabolic activity of all free-living organisms, bacteria paradoxically can also exhibit the lowest. This inevitably leads to very real uncertainties when attempting to anticipate metabolic rates from bacterial biomass measurements.

Microbiologists are well aware of this problem and have used a variety of biochemical and radiochemical techniques designed to determine the "active" or viable fraction of the population. It is recognized that there will be no sharp distinction between the "active" and "non-active" states; thus, although the precise percentage of "active" bacteria probably has little exact meaning, the observations do caution against the automatic assumption that all bacteria in a particular sample are equally active. In their review, van Es and Meyer-Reil assembled the published data of the active proportion of the bacterial population, which characteristically fell in the range of 10-50% of the total bacterial community.

BACTERIAL PRODUCTION AND ACTIVITY

There are two general approaches to the problem of determining biomass production: direct measurement of some aspect of metabolic activity or calculations from biomass observations using measurements or best guesses of growth rate. Again the subject is in an active phase of development and one can do no more at this stage than offer a snap-shot of the present status.

Bacterial Activity Measurement

The measurement of bacterial activity and the results of these measurements have been reviewed by Sorokin (1978); Williams (1981b); Joint and Morris (1982) and van Es and Meyer-Reil (1982). A variety of techniques have been used to determine microbial activity, some of them (e.g. electron transport activity measurement, oxygen con-

Table 1. Bacterial abundance and biomass (abstracted with additions from van Es and Meyer-Reil, 1982; Sorokin, 1978)

Area	Sampling Depth (m)	No. of Observations	No. of Cells (x $10^9 \ell^{-1}$)	Reported or Calculated* Bacterial Biomass (μgC ℓ^{-1})	Reference
Shelf & Coastal Waters					
U.S. East Coast	0 - 20	3	0.55 - 0.68	5.2 - 6.0	Ferguson & Rublee (1976)
U.S. East Coast & Shelf	0 - 50	96	$\bar{x}$ = 1.85	12*	Ferguson & Palumbo (1979)
U.S. East Coast & Shelf	5 & 50	2	1.8 - 4.95	0.57 & 1.6	Johnson & Sieburth (1979)
S. Californian Bight	n.d.	98	0.6 - 21	$\bar{x}$ = 4.8	Fuhrman *et al.* (1980)
S. Californian Bight	0.50	c.30	0.5 - 15	3 - 10*	Fuhrman & Azam (1982)
Sea of Japan	euphotic zone	-	1 - 5	20 - 200**	Sorokin (1978)
Antarctic	surface	13	0.07 - 1.0	7.0 - 9.1	Furhman & Azam (1980)
English Channel	0 - 24	15	0.39 - 1.65	3.2 - 35	Holligan *et al.* (in press)
Georges Bank	0 - 50	9	0.75 - 1.25	5 - 8*	Rublee (pers. communication)
Offshore & Oceanic Water					
[illegible]	[illegible] - 65	4	0.02 - 0.03	0.9 - 2.6	Liebezeit et al.

N. Central Pacific	1 & 75	2	0.14 & 0.5*	0.9 - 3.3*	Carlucci & Williams (1978)
N. Central Pacific	0-100	3	2 - 10	13 - 65*	Williams _et al_. (1980)
S.E. Atlantic	surface	n.d.	1 . 6	10*	Hobbie _et al_., (1977)
Peru Upwelling	euphotic zone	-	1 - 5	50 - 200*	Sorokin (1978)
Pacific Ocean, equatorial divergence	euphotic zone	-	0.3 - 1.0	10 - 40**	Sorokin (1978)
Oligotrophic gyre	euphotic zone	-	0.05 - 0.15	1 - 3**	Sorokin (1978)

* assuming 1 cell = 6.5 fgC;

** assuming C-biomass is 10% wet biomass;

n.d. - no data available.

Table 2. Estimates of bacterial or presumed bacterial biomass production (abstracted with addition from van Es & Meyer-Reil, 1982)

Area	Method	Depth (m)	Production Rate ($\mu gC\ \ell^{-1} day^{-1}$)	Reference
English Channel, summer	^{14}C-amino acid assimilation	0 - 70	0.4 - 1.8*	Andrews & Williams (1971)
California, coastal	^{14}C-amino acid assimilation	25	0.2 - 1.6	Williams *et al.* (1976)
Florida Straits	^{14}C-amino acid assimilation	0 - 50	0.26	Williams & Yentsch (1976)
Sea of Japan, coastal	dark $^{14}CO_2$ fixation	-	40 - 70	Vyshkvartsev (1980)
Sea of Japan, coastal	increase in biovolume	-	18 - 160	Vyshkvartsev (1980)
California, coastal	increase in biovolume	n.d.	10 - 34	Fuhrman & Azam (1980)
Californian, coastal	DNA synthesis rate	n.d.	0.7 - 53	Fuhrman & Azam (1980)
Antarctic, coastal	DNA synthesis rate	n.d.	0 - 2.9	Fuhrman & Azam (1980)
U.S. East Coat	frequency of dividing cells	surface	19 - 178	Newell & Christian (1981)
Caribbean	RNA synthesis rate	450	1.9 - 5.5	Karl (1979)
N. Atlantic	ATP increase in diffusion culture	50	122	Sieburth *et al.* (1977)

* calculated from amino acid respiration; n.d. - no data available

sumption) measure mineralization rather than production and are only relevant in the present context in as much as they provide an estimate of the scale of bacterial processes generally. Williams (1983), in a recent review of plankton respiration, concluded from the evidence available that bacteria accounted for a major component of overall plankton respiration.

Table 2 gives a summary of measurements of microbial activity which relate to the production of bacterial biomass. The data in the main refer to the productive period of the year and part of the water column. Again only data from truly marine systems have been included. No single method is expected to give an unequivocal answer and there are good reasons to expect a spread of values: for example the methods using calculations from the uptake of ^{14}C-organic substrates and those determining DNA production are expected to give underestimates. Collectively, the data do provide some concensus: rates commonly fall in the range of 1-50 $\mu gC\ \ell^{-1}$ day.

Biomass Derived Estimates of Production

In principle, given data for the biomass of bacteria and a measurement, estimate or best guess of the generation time, a production rate may be calculated. In their review, van Es and Meyer-Reil summarized the published data on the growth rate determinations made on natural communities of marine bacteria. Table 3 is abstracted from their review. Again the data are scarce and the scatter large, but the body of data suggest overall generation times in the region of one to one half of a day for bacteria in the upper part of the water column. This, in most cases, represents a mean for the whole community: thus if only 20% or so of the community is active and dividing, then their generation times would be in the region of 2-4 hours.

Table 4 is an attempt to put together the biomass and generation time data in order to derive estimates of bacterial biomass production for three planktonic biomes: temperate coastal water, oligotrophic and non-oligotrophic ocean waters. The data again considers the active part of the water column during the productive period of the year. Included in this table are essentially independent estimates of bacterial production rates for these water types taken from Sorokin's review. There is reasonable agreement between the two sets of estimates for the oceanic water types but my estimate of bacterial production in coastal water (10 $\mu C\ \ell^{-1}\ day^{-1}$) is at the low end of Sorokin's range. Other considerations, such as the projected and observed bacterial respiration rates for coastal areas, would suggest that my estimate is at the low end of the range and Sorokin's estimate would seem to be more realistic. The reason for the discrepancy may be the value for generation times adopted in

Table 3. Bacteria generation times - samples from euphotic zone (abstracted from van Es 1982).

Area	Method	Generation Time (hr)	Temp (^{o}C)	Reference
Shelf Coastal Waters				
N. Atlantic	wash out from continuous pure culture	52 - 170	24	Jannasch (1967)
N. Atlantic	frequency of dividing cells	8 - 37	8 - 37	Newell & Christian (1981)
Sea of Japan	rate of increase of numbers	$\bar{x}$ = 24	-	Vyshkvartsev (1980)
N. Pacific	rate of increase of an isolate	3	20	Carlucci & Shimp (1974)
		25	5	
Californian Bight	rate of thymidine incorporation	8 - 48	n.d.	Fuhrman & Azam (1982)
Oceanic Waters				
N. Pacific	rate of increase of an isolate	22.6	14	Carlucci & Williams (1978)
		18.8	11	
Caribbean	RNA synthesis	2.5 - 6.5	10 - 25	Karl (1979)

Table 4. Calculated bacterial production rates for various water types (attempted median values abstracted from Tables 1 & 3)

Area	Bacterial Biomass ($\mu gC\ \ell^{-1}$)	Generation Time (hr)	Bacterial Production Rate ($\mu gC\ \ell^{-1} day^{-1}$)	Bacterial Respiration Rate ($\mu gO_2\ \ell^{-1}\ day^{-1}$)
Temperate coastal water	10	24	10*	-
Non-oligotrophic oceans	5	12	10	-
Oligotrophic oceans	2	12	4	-
Data abstract from Sorokin (1978)**				
Temperate coastal waters	20 - 200	-	10 - 70	200 - 500
Peru upwelling	50 - 200	-	20 - 50	140 - 500
Equatorial divergence, Pacific	10 - 40	-	10 - 30	60 - 200
Oligotrophic ocean	1 - 3	-	2 - 4	10 - 30

* see comment in text; ** calculated assuming C-biomass equals 10% of wet biomass

Table 5. Estimates of community biomass distribution (all values μgC ℓ^{-1} except where indicated).

	CEPEX (inlet, British Columbia)	English Channel E5 (shelf break)	English Channel M (shelf, mix)	English Channel F (front)
Autotrophs				
Diatoms	231	c.0	95	0
Dinoflagellates	22	c.0	c.0	1434
Others	33	11	7	21
Total autotrophs	286	11	102	1455
Heterotrophs				
Bacteria	26	4.3	8	12
Zooflagellates	33	2.4	6.6	18
Ciliates	8.6	n.d.	n.d.	n.d.
Micrometazoa (larvae)	21 (161***)	3.5	3.5	5.3
Mesometazoa (herb.)	16	19.5	14.2	9.2
Mesometazoa (carn.)	n.d.	n.d.	n.d.	n.d.
Total heterotrophs	104 (266***)	29.7	32.3	44.5
Time Span of Observations	44 days	station	station	station
Source(s)	Williams (1982)	Holligan et al. (in press)	Holligan et al. (in press)	Holligan et al. (in press)

* units cal m^{-3}; ** units cal m^{-2}; *** data in parenthesis included brief bloom of gastropod veleger larvae.

Californian Coastal Water	North Pacific Gyre	Tropical Oligo-trophic*	Tropical Meso-trophic**	Sea of Japan (coastal)
10	0.43			
20	0.34	} 15	} 1000	} 50
10	2.93			
40	3.70	15	1000	50
96	3	10.8	1000	35
4	3.5	n.d.	n.d.	50
0.9	0.37	0.05	80	20
4.0	0.25	2.14	320	7.5
20	4	9.6	730	50
n.d.	n.d.	4.1	630	n.d.
34.5	11.1	26.7	2760	162.5
station	various	several stations	several stations	-
Strickland (1970) Fuhrman *et al.* (1980)	Beers *et al.* (1975) Beers (pers. commn) Carlucci & Williams (1978) Williams *et al.* (1980) Shulenberger & Reid (1981)	Vinogradov *et al.* (1973)	Vinogradov *et al.* (1972	Sorokin (1978)

Table 4 for coastal water which does seem to be long in relation to that adopted for ocean waters. By comparison, one might expect a value nearer 6 hours for coastal water. One simply awaits more data on bacterial growth rates in marine systems. In passing, one may note that the estimate for the oligotrophic environment is calculated to fall within the upper closure condition of biological production for the N. Central Pacific calculated by Platt et al. in an accompanying article in this volume.

BACTERIA AND BACTERIAL PROCESSES IN AN ECOSYSTEM CONTEXT

It is perhaps profitable now to consider bacteria in relation to other planktonic organisms and the wider implications of bacterial processes. Four matters will be discussed:

1. bacterial biomass and activity in comparison with other members of the plankton.
2. supply of material for bacterial processes.
3. the fate of bacterial activity and production.
4. bacteria, bacterial activity and models of plankton processes.

Bacterial Biomass and Activity in Comparison with Other Planktonic Groups.

Over the past decade improvements in the methods for the determination of biomass of the members of the plankton, especially the smaller forms, has made it possible to produce reasonable profiles of biomass distribution: certainly from adult copepods downwards to bacteria. The collecting of the data and its processing are time-consuming and this no doubt accounts for the small number of published profiles.

In Table 5 I have calculated or collated biomass profiles of a broad spectrum of planktonic groups for a number of locations. Some compilations represent means over a period of a month or so, whereas others are data obtained from discrete stations. The Soviet workers have a precedent in this type of activity and two profiles from Vinogradov's work (Vinogradov et al., 1972; Vinogradov et al., 1973) and one abstracted from Sorokin's review are included. In two cases (the N. Pacific Gyre and the Californian coast) the profiles in the table have been compiled from separate studies. This is unfortunate but unavoidable. Finally one should note the difficulty in obtaining reliable data on the biomass of the zooflagellates. In some cases (e.g. Beers et al., 1975) no distinction was made in the publication between the photosynthetic and non-photosynthetic forms. In this and other cases it has been necessary to rework the original data and to make some arbitrary assumptions in order to produce a figure for the biomass of the heterotrophic flagellates. The general conclusion from the data is that the biomass of the single celled heterotrophic organisms (bacteria, non-photosynthetic flagellates, and ciliates) is comparable to that of the metazoa. Unless the major proportion of microbial community is in a dormant state one must inevitably infer that their contribution to the overall metabolic activity will be substantial.

At present, there is no general solution to the problem of making direct measurements of the distribution of metabolic activity of the various components of the food chain. Williams (1981a) combined size fractionation with respiration measurements in an attempt to examine the distribution of metabolic activity within the planktonic community. The conclusion was that 50%, or on occasions more, of plankton respiration was associated with organisms of bacterial to microflagellate size. A limited study using electron transport activity as a measure of respiration gave a similar size distribution (Packard and Williams, 1981).

Although one can gain insight into the very general pattern of the distribution of metabolism from this type of approach, the methods presently available to physically or biochemically separate the various components of the plankton are crude and cannot give the type of resolution one is really seeking. We can, however, achieve an adequate resolution in determining the abundance of planktonic forms. Thus, at the present one needs to determine how much information on distribution of metabolism can be wrung from numerical abundance observations. It is, in principle, possible to obtain a calculated distribution of activity from biomass profiles using allometric equations which relate specific metabolic activity to individual biomass. Williams (1982) undertook such calculation with the CEPEX data set and obtained a profile of respiratory activity distribution which was a reasonable match to field observations of respiration rate. The results of the calculations are given in Table 6. Included also in Table 6 are data, abstracted from Sorokin's review and Vinogradov's papers, of their estimates of production distribution within planktonic communities. Platt et al (this volume) have made a very careful set of calculations of the size distribution of plankton respiration for the N. Central Pacific Gyre Community. All things considered this is a fair measure of concordance between the estimates.

A range of exponents have been proposed for the allometric equations. (See Banse, 1982, for a very thoughtful discussion of the allometric equations). Commonly, the exponents lie between 0.7 and 0.75 (see Platt et al., in this volume for the form of the equation) and the simple inference is that metabolism is more closely related to surface area than either length or volume. Thus, if one is principally concerned with an assessment of the distribution of activity within the planktonic community, rather than absolute measurements of activity, then there may be some argument for short

Table 6. Estimations of production and respiration within planktonic communities; upper section gives absolute rates; lower rates are expressed as a percentage of total heterotrophic activity.

	CEPEX*	Peru upwelling**	Sea of Japan**	Tropical Pacific	
				Mesotrophic***	Oligotrophic****
Phytoplankton	135	87,000	4,750	1,000	15
Bacteria	138	14,000	3,600	1,000	16.2
Protozoa	4.8	} 2,300	} 1,800	} 80	} 0.2
Micrometazoa (larvae)	20				
Mesometazoa (herb.)	4.9	1,900	300	550	4.3
Mesometazoa (carn.)	n.d.	600	460	750	0.5
Bacteria	82	74	58	42	76
Protozoa	3	12	29	3	1
Micrometazoa (larvae)	12				
Mesometazoa (herb.)	3	10	5	23	20
Mesometazoa (carn.)	n.d	3	7.5	32	2.5

n.d. - no data available; * from Williams, 1982, respiration rates as $\mu gO_2\ \ell^{-1}\ day^{-1}$; ** from Sorokin, 1978, production rates reported as Kcal $m^{-2}\ day^{-1}$ but presumed to be gcal $m^{-2}\ day^{-1}$; *** from Vinogradov *et al.*, 1972, production rates as gcal $m^{-2}\ day^{-1}$ over upper 200m; **** from Vinogradov *et al.*, 1973, production rates as gcal $m^{-3}\ day^{-1}$ over upper 150 m.

circuiting the exercise and to simply calculate the distribution surface area. As far as I am aware Azam was the first to suggest this approach. The principle uncertainty is that it infers that the calculated surface is active. At first sight this would seem to be more problematic with the metazoa, however, when it comes to the interpretation of the data, it turns out that the main problem lies with the bacteria. In the case of the metazoa, the surface area I have calculated is that of the sphere circumscribing the animal. Detailed calculations based on the morphology of *Calanus finmarchicus* shows this to overestimate surface area. Regardless of the details of what fraction of the surface of a zooplankter is "active", their contribution to the overall planktonic surface area is small and a maximum estimate is quite adequate for present purposes and will not materially alter the conclusions. In the case of the bacteria, the point has been made already that a percentage of the population will be in a non-active state and a calculation of total bacterial surface area will over-estimate their "active" surface. It was noted earlier that the published data indicate that the active fraction of the population appeared to lay between 10 and 50% of the total; thus, for the sake of the present calculations I have adopted a median figure of 20%

The results of the calculations are given in Table 7. The most interesting outcome of the calculation is that the variability in the profiles for different ecosystems is reduced as compared with the biomass data; furthermore, some of the "untidy" aspects of the biomass calculations are reduced or eliminated. For example, in the CEPEX data (see Table 5) a brief bloom of gastropod velegers virtually trebled the mean beterotroph biomass calculated for a two month period; one intuitively suspects that this is way out of proportion to their metabolic impact. Their overall effect on the total heterotrophic surface area is insignificant.

One may extend Azam's suggestion and put up the proposition that if surface area does give us a useful insight into the distribution of activity then in a balanced food chain (i.e. when production matches consumption) the surface area of the producers may be expected to be comparable to that of the consumers. With the exception of the frontal station F, which may not meet the criterion of a balanced situation, the calculated autotrophic and heterotrophic surface areas do appear to be comparable. However, one must caution that the assumption has been made that only 20% of the bacteria are active, and since they dominate the heterotroph surface area, this assumption is of course critical. Whether the calculation of surface area of planktonic populations serves any useful function beyond being a simple illustration of the potential impact of the microbial population remains to be seen.

Table 7. Estimates of community surface area distribution (all values cm^2 ℓ^{-1}; sources of data as Table 5.

	CEPEX (inlet, British Columbia)	English Channel E5 (shelf break)	English Channel M (shelf mixed)	English Channel F (front)	Californian coastal water	North Pacific Gyre
Autotrophs						
Diatoms	6.5	c.0	3.4	c.0	0.29	0.025
Dinoflagellates	0.37	c.0	c.0	30	0.39	0.27
Others	1.93	0.9	0.56	1.7	1.18	0.38
	8.8	0.9	4	32	1.85	0.70
Heterotrophs						
Bacteria	4.92 (24.6)*	1.3 (6.5)*	2.4 (12)*	6(30.5)*	1.64(8.2)*	0.56(2.8)*
Zooflagellates	1.43	0.19	0.044	1.8	0.47	0.3
Ciliates	0.26	n.d	n.d.	n.d.	0.075	0.034
Micrometazoa (larvae)	0.27(0.127)**	0.0092	0.0043	0.01	0.1	0.04
Mesometazoa (herb.)	0.04	0.15	0.13	0.23	0.27	0.03
	6.9	1.65	2.54	8.0	2.2	0.93

*value in parenthesis is total calculated surface area, assuming 0.6 μm diameter cell; value outside is 20% of calculated surface area, i.e. the presumed active area, see Text. ** value in parenthesis is the calculated surface of the gastropod veleger larvae; c.f. Table 5

Supply of Organic Material for Bacterial Activity

In offshore and in many coastal environments, phytoplankton production will be essentially the ultimate source of organic material for bacterial processes. The routes from the algae to the bacteria are known in principal, the essential question is whether they are of sufficient scale to provide for the relative high rates of bacterial metabolism claimed by the marine microbiologists. There is no doubt but that this is a complex issue. Williams (1981b) made a preliminary attempt to examine the problem, and using what he argued to be conservative estimates for the supply to the bacteria, concluded that it was possible to produce a scheme of organic flow that was consistent with present day estimates of bacterial activity. The two important conclusions from the exercise were that events prior to the ingestion of algae by the zooplankton (i.e. exudation of organic material by the algal, cell spoilage) were more important routes than the various animal organic excretions. Secondly, the bacteria may pass on twice as much organic material to the next trophic level as the herbivorous zooplankton. The scheme of the flow of carbon produced by Williams is given in Figure 1. It is important to stress that it summarizes the results of an exercise and should not be regarded as a stereotype of organic flow to marine bacteria. In the present article there would seem to be little to be gained by repeating or attempting to extend the exercise, but it would seem more profitable to determine to what extent it is possible to come to a conclusion over the fate of bacterial production within the plankton.

Fate of Bacterial Activity and Production

Two issues will be considered at this stage, first the growth yield or conversion efficiency of marine bacteria, secondly, the fate of bacterial biomass production.

The question of growth yield of bacteria is critical. Traditionally, bacteria were regarded to grow very inefficiently and were accordingly ascribed the role of mineralizers. Up until the late 1960's this view was endorsed by the mainstream microbiological literature. Payne noted, in his review of bacterial growth yields (Payne, 1970), that the earlier observations on bacterial conversion efficiencies came from studies with non-proliferating cultures. The more recent work, reviewed by Payne (1970), Calow (1977) and Payne and Wiebe (1978) has generally given much higher yields. Characteristically, conversion efficiencies in the region of 50-80% are obtained with pure and mixed cultures of bacteria growing on simple and complex, but not refractory substrates. The introduction of radiochemical techniques to the study of marine microbial ecology enabled the conversion efficiency of single organic compounds to be studied with relative ease. It was argued by Williams (1973)

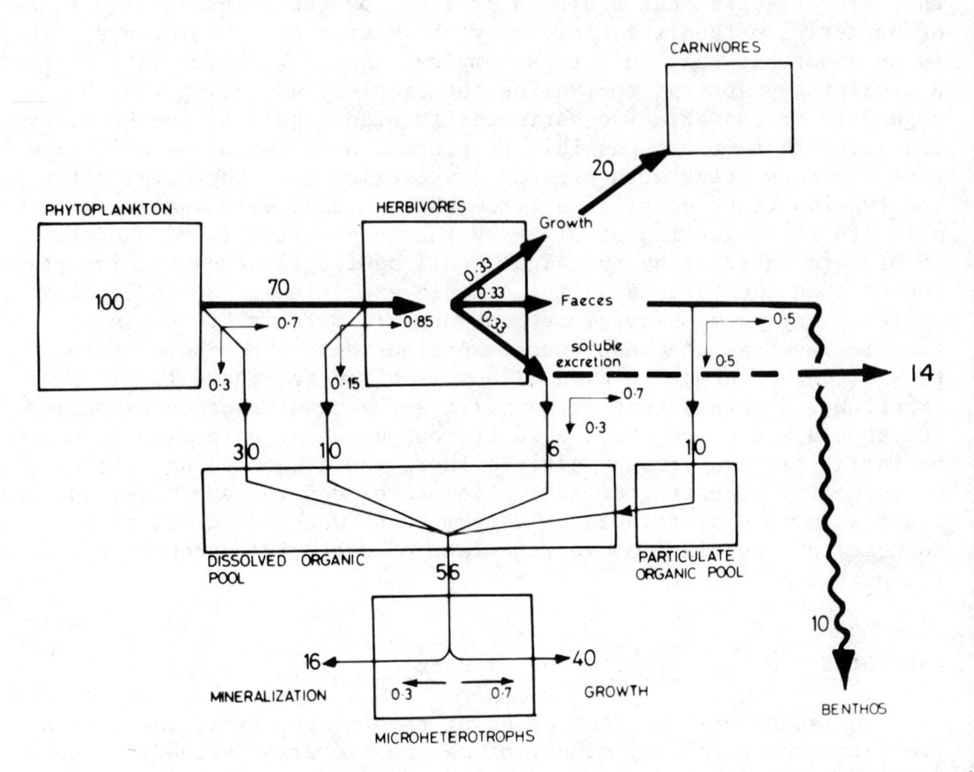

Fig. 1. Calculation example of the flow of organic material through the planktonic system. Details are given in Williams (1981 The large numbers refer to the calculated percentage of phytoplankton production passing along a particular route. The smaller numbers, normally associated with finest lines, indicate the fractional flow along alternative pathways.

that this may give insight into the efficiency of bacterial growth in the sea. If this is the case, the data then suggested that the notion of marine bacteria as mineralizers needed reconsidering (see discussion in Williams, 1981b). The need for caution in interpreting these conversion efficiencies as growth yields is recognized. With the exception of a set of measurements by Billen et al. (1980) surprisingly similar conversion efficiencies have been obtained by different groups of workers for a variety of environments. The conversion efficiencies (see Table 8) are generally high (50-80%) suggesting very efficient growth as compared with metazoa. Williams (1981b) noted that this was consistent with the observations of Dewey (1976), that although the specific metabolic activity and respiration rate increased with decrease in size, respiration as a proportion of metabolism fell with decrease in size, i.e. growth yield would increase with decrease in size. Iturriaga and Zsolnay (1981) followed respiration and incorporation of simple organic substrates over a period of three days. With the single exception of glycollate, the initial conversion efficiency was high (i.e. in excess of 60%), but this fell after a period of time. At this stage, however, the external ^{14}C-labelled substrate was nearing exhaustion and the falling conversion efficiency could have arisen from a dilution of the external isotope pool and a turnover and respiration of the previously assimilated isotope.

Newell's group (see Stuart et al., 1981; Newell et al., 1981) using classical animal ecology techniques have made a series of studies of conversion efficiency of natural organic detritus (kelp and phytoplankton debris) by bacteria and have obtained much lower values; characteristically in the region of 10%. The difference between these two sets of observation is crucial when attempting to come to a conclusion over the role of the bacteria in the sea and the consequence of their activity. There is evidence that the biochemical nature of the organic substrate and the experimental conditions can give rise to variations in the observed conversion efficiency. In detailed studies of conversion efficiency of individual amino acids, three groups (Crawford et al., 1974; Williams et al., 1976 and Keller et al., 1982) noted high conversion efficiencies (i.e. c. 90%) with amino acids such as leucine, <u>iso</u>-leucine, valine and phenylalanine and lower (i.e. c. 40-60%) with glutamate, aspartate, alanine, serine and glycine. In a general way this, as the difference in conversion efficiency between glucose and an amino acid mixture observed by Williams (1970) and Williams and Yentsch (1976), can be accounted for by their anticipated metabolism. The effect of experimental conditions can be seen in the study of Parsons et al. (1980) who used both a tracer and a mass balance technique to determine the conversion efficiency of glucose. The mass balance procedure gave conversion efficiencies about one-half of the radioisotope procedure (see Table 8). They also noted a significant reduction in conversion efficiency when the ambient substrate concentration was high. To some extent this might explain the differences

Table 8. Summary of reported organic substrate conversion efficiencies by marine populations (abstracted with additions from Joint and Morris, 1982).

Area	Substrate(s)	Range of Observed Conversion Efficiencies (%)	Reference
English Channel	glucose & amino acid mixture	66 - 76	Williams (1970)
Mediterranean	glucose & amino acid mixture	54 - 86	Williams (1970)
N.E. Atlantic	glucose & amino acid mixture	51 - 92	Williams (1970)
Bahamas	glucose & amino acid mixture	75 - 80	Williams & Yentsch (1976)
Californian coast	individual amino acids	70 - 98	Williams *et al*. (1976)
North Sea	malate, glucose, acetate, lactate	62 - 71	Gocke (1976)
Baltic Sea	phytoplankton exudates	69 - 83	Iturriaga & Hoppe (1977)
North Sea	glucose, acetate, lactate, amino acids	15 - 60	Billen *et al*. (1980)
Canadian coast	glucose (tracer experiment)	54 - 90	Parsons *et al* (1980)
(CEPEX)	glucose (mass balance)	34 - 45	
New England coast	individual amino acids	31 - >90	Keller *et al* (1982)

between the tracer techniques and the approach used by Newell. This puts one in something of an apparent dilemma: it is not immediately obvious what are the most appropriate set of conditions and substrates to adopt. Indeed, it is without question naive to imagine that a single conversion efficiency can be used to cover all eventualities. One may perhaps suppose that the conversion efficiencies derived from incubations of ^{14}C-labelled substrates and mixtures may be appropriate to a situation in which there is rapid turnover of organic material within the planktonic community. If this is the case, and the data on conversion efficiency of algal organic exudates obtained by Iturriaga and Hoppe (1977) lend support, then the adoption of a high conversion efficiency for the type of calculation made by Williams (1981b) does seem justified. The conversion efficiencies obtained by Newell and his co-workers seem appropriate to a situation in which an undergrazed algal population has entered a moribund state, due, for example, to nutrient exhaustion.

Leaving this issue for the moment. The overall routes of material flow from marine bacteria were considered in outline by Williams (1981b) who made no attempt to come to a conclusion regarding their relative importance. I would like to open up this discussion here.

One needs to consider first the question of the extent to which planktonic metazoa use marine bacteria in their diet. There is good reason to believe that some adult planktonic metazoa may derive a substantial part of their diet from bacteria (see e.g. King et al., 1980). Whether grazing by adult or larval metazoa represents a significant sink for bacterial production is far from clear. It may turn out that the spatial organization of the bacterial community has some considerable bearing upon this. It would appear from experimental studies and theoretical considerations (Jorgensen, 1983; Fenchel, 1982b) that certainly most adult metazoa would encounter fluid mechanical problems when attempting to graze upon randomly distributed free-living bacteria. Non-randomly distributed, and more especially particle-associated populations, would be amenable to metazoan grazing. Although the present evidence does appear to suggest that marine bacteria are predominantly free-living, arguments for non-random distributions recently proposed by Azam and Ammerman and Goldman in this volume need serious consideration. Until these issues are resolved, we are left to presume that the metazoa do not extensively graze down bacterial production and thus by default the major part of bacterial production must be expected to go to the protozoa. It should be noted in passing that the observations made by Parsons et al. (1980), that glucose assimilation presumably by bacteria was transferred to the larger size fractions (> 100 μm) without any sign of incorporation into intermediate size fractions (12-35 μm), do argue for direct assimilation of bacteria by metazoa.

The data summarized in Table 5 suggest that protozoan biomass appears to range from one tenth to equal to that of the bacteria.

Sieburth (1983) when generalizing on his classifications of the microplankton suggests that the picoheterotrophs (i.e. the bacteria) are approximately comparable in biovolume to the nanoheterotrophs (protozoa). The zooflagellates are presently the most difficult marine microbes to enumerate and it may be that the low protozoan biomasses seen in some of the profiles in Table 5 result from an inability to observe the zooflagellate biomass. Where careful attention has been paid to zooflagellates, the data do suggest their biomass to be comparable to that of the bacterial.

It is perhaps profitable, at this stage, to consider more generally the fate of bacterial production. If we presume that it predominantly passes onto the flagellates, this then raises the question of the fate of flagellate production itself. If we were to suppose that this is to the ciliates and the information we have on the ciliate relative to flagellate biomass structure would seem to be consistent with this (see Table 6). The ciliates can also prey directly upon certain size fractions of the primary producers (see Sieburth, 1983; Banse, 1982) and so may act as a focal point in the trophic relationships at the microbial level. The ciliates, one may suppose, are of sufficient size to be grazed upon by the herbivores. If we assume a conversion efficiency of 40% for the protozoa (Fenchel, 1982a; Heinbokel, 1978) and 50% for the bacteria then some 8% of the organic material used by the bacteria and 16% of bacterial production would reach the metazoa. If the conversion efficiency at the bacterial level were 10% then the former figure would be reduced from 8 to 1.6%; the latter figure would be unchanged. Thus, the microbial community (i.e. bacteria, zooflagellates and ciliates) would be mainly responsible for remineralization and although bacterial production may be very substantial, unless there is direct utilization of bacteria by the metazoa, the organic material incorporated by the bacteria is lost through mineralization in the protozoan food chain. This notion is far from new: it was illustrated experimentally by Johannes (1965). If this is the case, then the present uncertainty over bacterial conversion efficiencies, effectively only has bearing on whether the mineralization occurs at the bacterial or protozoan level. In passing it should be noted that observations on phosphate remineralization made by Barsdate et al. (1974) were opposite to those of Johannes: they concluded that the bulk of the mineralization occurred at the bacterial rather than the protozoan level.

If we accept the type of microbial biomass profile discussed above, then the conversion efficiency at the bacterial level will have a marked bearing on our anticipated production rates within the microbial community. If the bacterial conversion efficiencies are high then the bulk of the organic material used by them will pass on to the protozoa. If the biomasses of these two communities are comparable, then it would mean that the protozoan biomass will be turning over quite rapidly, i.e. similar to that of the bacterial. If, on the other hand, the bacterial conversion efficiencies are low and the major part of bacterial activity results in remineralization, bacterial production being low, then the relatively large microbial (i.e. bacterial plus protozoan) biomasses presently reported must be turning over very slowly. Banse (1982) came to the conclusion that ciliate growth rates in the sea would be low, held down by the scarity of their food.

Bacteria, Bacterial Processes and Models of the Pelagic Food Chain

To turn finally to the question: should a microbial component be incorporated into food chain models and, if so how does one go about it?

If the preceding inference, that only a small part of bacterial production passes onto the metazoa is accepted, then in some respects the questions are not too difficult to answer. The answer to the first naturally depends mainly upon the aims and structure of the model. It may be seen that any comprehensive model of the food chain which considers material (either organic or inorganic) flow will certainly need a microbial component. However, unless there is extensive utilization of bacteria production by metazoans in the pelagic food chain, it would appear appropriate and convenient to regard the microbial community as a mineralizing unit. This being the case, the principal aspect of marine microbial metabolism which needs to be considered is mineralization (respiration, regeneration). Thus, food chain models which have nutrient recycling built into them would need to pay very serious attention to microbial processes. Again, if metazoan utilization of bacterial production is of little significance, then models of metazoan production (e.g. zooplankton to fish) in which primary production, where needed, is set as a boundary condition, would need to pay little attention to microbial processes. It is, of course, important to note that these generalizations only apply to models of the pelagic food chain and from a distance it would appear that direct utilization of bacteria by metazoa in the benthic food chain may be of greater significance.

Having argued the need to incorporate microbial processes into general models of the pelagic food chain one has to own up to the present lack of any mechanistic (rational) model of bacterial production. Empirical models of bacterial oxidation of organic material in estuaries have been evolved for sanitary engineering purposes but these seem to offer the worst of all alternatives. During the CEPEX Foodweb 1 experiment, a serious attempt was made to produce a mechanistic model of bacterial production, but realistically little progress was made. The main problem was the state of development of the understanding of the factors controlling microbial activity in the sea. The subject is developing very rapidly and one can hope for some considerable inroads to be made over the next 3 to 5 years. In the meantime, all one can offer the modeller is our estimates of the rates of microbial activity. Although this is recognized to be less acceptable and less flexible than a mechanistic model, it is important that the models take note of these processes. If overlooked, there is a temptation for the modeller to try to make the model run with zooplankton as the only heterotrophs. The danger here is obvious.

Finally, it would seem timely to ask the question: why does the marine food chain apparently sustain a large microbial community, which seems to be consuming comparatively large quantities of organic material but contributing very little towards food chain production? One line of argument may be suggested. It may be supposed that in an oceanic environment the supply of phosphorus (and also to a large extent nitrogen) to the pelagic food chain is limited by the dynamics of the ocean (see Redfield, 1958; Eppley and Peterson, 1979). Thus, in the long run the potential planktonic biomass in any single pelagic environment will be controlled by the extent to which the mineralization of particulate material is effected in the upper mixed zone. One should note that this argument is not in contradiction with the gist of the paper by Eppley and Peterson. The main loss of particles from the food chain will be due to episodic events on various scales, ranging from the formation of feces to the collapse and sinking of phytoplankton blooms. The residence time of these particles in the mixed layer, in most cases, will be perhaps no more than a few days, and unless there is very rapid colonization by bacteria or grazing by zooplankton, the nutrients bound up with the particles will be lost from the euphotic zone. Regardless of whether the particles are consumed and remineralized by micro-organisms or metazoa, the residence time in the warm mixed layer is sufficiently short that, unless the heterotrophic community is in an active state of metabolism it will not be able to effect much decomposition within the time available. Thus, in a very general way one could envisage a need to maintain an active microbial community for such eventualities. One may argue that the penalty for doing this is not too great: it will require diverting some energy and nutrients from the grazing chain to the micro-organisms. If the microbial community is active, then the nutrients will not be bound up for any length of time but will be rapidly released and so are not lost. The need to conserve energy in the long run may not be as important priority for the pelagic food chain as the need to conserve nutrients. This type of argument of course presupposes that community evolution in the oceans is ultimately more powerful than species evolution.

CONCLUSIONS

The first conclusion must be that despite the availability of some of the methods of studying micro-organisms (e.g. Acridine Orange direct staining technique) for over half a decade, the amount of data available is small. This inevitably sets a limit to the extent and confidence with which one can generalize. The evidence would seem to be that the bacterial biomass is often the major component of the microbial heterotroph biomass and this in its turn appears to be comparable to, and at times greater than, the metazoan biomass. Various considerations lead to the conclusion that a substantial proportion of plankton production eventually reaches the bacteria. The exact fate of this material whether it is mineralized by bacteria or whether it is used for production of bacterial biomass, is not resolved because of the present uncertainty over the conversion efficiencies of bacteria. However, if it turns out that herbivorous zooplankton do not extensively graze upon bacteria, then it may be argued that the microbial community will substantially mineralize the material it receives and whether achieved at the bacterial or protozoan level is probably of little consequence as far as many models of the food chain are concerned. The implication is that bacterial production probably gives rise to little supplementation of the planktonic metazoan food chain and that its omission from models of copepod production would not be dire. This would clearly not be the case for any model in which nutrient regeneration played an important role: there are signs that the metazoa may play a minor role in the cycling of nutrients in comparison with the microbial community in the oceanic environment (see review by Harrison, 1980).

If it turns out, however, that there is extensive metazoan utilization of microbial production then the above generalizations will need revising. These generalizations only apply to the planktonic environment, the greater number densities of bacteria in sediments make the direct utilization of bacteria by metazoa more probable and accordingly their impact on the metazoan food chain more significant.

Acknowledgements

I should like to record my appreciation to Drs. Rublee, Ferguson, Palumbo and Buckley for permission to include their unpublished data into Table 1, and Drs. Meyer-Reil, van Es, Joint and Morris for providing me with draft copies of their reviews. I wish to thank Prof. Karl Banse and Dr. Hugh Ducklow for their critical reading of the manuscript.

REFERENCES

Andrews, P., and Williams, P.J. leB., 1971, Heterotrophic utilization of organic compounds in the sea. III. Measurement of the oxidation rates and concentrations of glucose and amino acids in seawater, J. Mar. Biol. Ass. U.K., 51:111.

Azam, F. and Ammerman, J.W., 1982, Cycling of organic material by bacterioplankton in marine ecosystems: microenvironmental considerations, This volume.

Banse, K., Cell volumes, maximal growth rates of unicellular algae and ciliates, and the role of ciliates in the marine pelagial. Limnol. Oceanogr., 27: 1059.

Barsdate, R.J., Prentki, R.T., and Fenchel, T., 1974, Phosphorus cycle in model ecosystems: significance for decomposer food chain and effect of bacterial grazers, Oikos, 25: 239.

Beers, J.R., Reid, F.M.H., and Stewart, G.L., 1975, Microplankton of the North Central Pacific Gyre. Population structure and abundance, June 1973, Int. Revue Ges. Hydrobiol., 60: 607.

Billen, G., Joiris, C., Winant, J., and Gillain, G., 1980, Concentration and metabolism of small organic molecules in estuarine, coastal and open sea environments of the Southern North Sea, Est. Coastal Mar. Sci., 11: 279.

Calow, P., 1977, Conversion efficiencies in heterotrophic organisms, Biol. Rev., 52: 385.

Carlucci, A.F., and Shimp, S.L., 1974, Isolation and growth of a marine bacterium in low concentrations of substrate, in; "Effect of the ocean environment on microbial activities," R.R. Colwell and R.Y. Morita, eds., Univ. Park Press, Baltimore.

Carlucci, A.F., and Williams, P.M., 1978, Simulated in situ growth rates of pelagic marine bacteria, Naturwiss., 65: 541.

Crawford, C.C., Hobbie, J.E., and Webb, K.L., 1974, The utilization of dissolved free amino acids by estuarine micro-organisms, Ecology, 55:551.

Dewey, J.M., 1976, Rates of feeding, respiration and growth of a rotifer Branchionus plicatilis in the laboratory, Ph.D. University of Washington.

Eppley, R.W., and Peterson, B.J., 1979, Particulate organic matter flux and planktonic new production in the deep ocean. Nature, 282:677.

Es, F.B. van, and Meyer-Reil, L.A., 1982, Biomass and metabolic activity of heterotrophic marine bacteria, To appear in Adv. in Microbial Ecology.

Fenchel, T. 1982a Ecology of heterotrophic microflagellates. II. Bioenergetics and growth, Mar. Ecol. Prog. Ser., 8: 225.

Fenchel, T., 1982b, Suspended marine bacteria as a food source, This volume.

Ferguson, R.L., and Palumbo, A.V., 1979, Distribution of suspended bacteria in neritic waters south of Long Island during stratified conditions, Limnol. Oceanogr., 24: 697.

Ferguson, R.L., and Rublee, P., 1976, Contribution of bacteria to standing crop of coastal plankton, Limnol. Oceanogr., 22:141.

Fuhrman, J.A., 1981, Influence of method on the apparent size distribution of bacterioplankton cells: epifluorescence microscopy compared to scanning electron microscopy, Mar. Ecol. Prog. Serv., 5: 103.

Fuhrman, J.A., Ammerman, J.W.A., and Azam, F., 1980, Bacterioplankton in the coastal euphotic zone; distribution activity and possible relationships with phytoplankton, Mar. Biol., 60: 201.

Fuhrman, J.A., and Azam, F., 1980, Bacterioplankton secondary production estimates for coastal waters of British Columbia, Antarctica and California, Appl. Environ. Microbiol., 39: 1085.

Fuhrman, J.A. and Azam, F., 1982, Thymidine incorporation as a measure of heterotrophic bacterioplankton production in marine surface waters: evaluation and field results, Mar. Biol., 66:109.

Gocke, K., 1976, Respiration von gelosten organischen Verbindungen durch naturliche Mikroorganismen-Populationen. Ein Vergliech swischen verschiedenen Biotopen, Mar. Biol., 35: 375.

Goldman, J., 1982, The marine nutrient cycle, This volume.

Harrison, W.G., 1980, Nutrient regeneration and primary production in the sea, in: "Primary Productivity in the Sea," P. Falkowski ed. Plenum Press, New York.

Heinbokel, J.F., 1978, Studies on the functional role of tintinnids in the Southern California Bight 1. Grazing and growth rates in laboratory cultures, Mar. Biol., 47: 177.

Hobbie, J.E., Daley, R.J., and Jasper, S., 1977, Use of Nuclepore filters for counting bacteria by fluorescence microscopy, App. Environ. Microbiol., 33: 1225.

Holligan, P.M., Harris, R.P., Harbour, D.S. Head, R.N., Tranter, P.R.G., Weekley, C.M., Newell, R.C., Linley, E.A.S., and Lucas, M.I. The partitioning or organic carbon in mixed, frontal and stratified waters of the English Channel, in press.

Hoppe, H. -G., 1976, Determination and properties of actively metabolizing heterotrophic bacteria in the sea, investigated by means of microautoradiography, Mar. Biol., 36: 291.

Iturriaga, R., and Hoppe, H.-G., 1977, Observations of heterotrophic activity on photoassimilated organic matter, Mar. Biol., 40: 100

Iturriaga, R., and Zsolnay, A., 1981, Transformations of some dissolved organic compounds by a natural heterotrophic population, Mar. Biol., 62: 125.

Jannasch, H.W., 1967, Growth of marine bacteria at limiting concentrations of organic carbon in seawater, Limnol. Oceanogr., 12: 264.

Joint, I.R., and Morris, R.J., 1982, The role of bacteria in the turnover of organic matter in the sea, Oceanog. Mar. Biol. Ann. Rev., 20: 65.

Johannes, R.E., 1965, Influence of marine protozoa on nutrient regeneration, Limnol. Oceanogr., 10: 434.

Johnson, P.W., and Sieburth, J. McN., 1979, Chroccoid cyanobacteria in the sea: a ubiquitous and diverse phototrophic biomass, Limnol. Oceanogr., 24: 938.

Jorgensen, C.B., 1983, Effect of grazing: metazoan suspension feeders, o appear in "Heterotrophy in the Sea", J.E. Hobbie and P.J.leB. Williams, eds., Plenum Press, New York.

Karl, D.M., 1979, Measurement of microbial activity and growth in the ocean by rates of stable ribonucleic acid synthesis, App. Environ. Microbiol., 38: 850.

Keller, M.D., Mague, T.M., Badenhausen, M., and Glover, H.E., 1982, Seasonal variations in the production and consumption of amino acids by coastal microplankton, Est. Coast. Mar. Sci., 15:301.

King, K.R., Hollibaugh, J.T., and Azam, F., 1980, Predator-prey interactions between the larvacean Oikopleura dioica and bacterioplankton in enclosed water columns, Mar. Biol., 56:49.

Liebezeit, G., Bolter, M., Brown, I.F., and Dawson, R., 1980, Dissolved free amino acids and carbohydrates at pyncocline boundaries in the Sargasso Sea and related microbial activity, Oceanol. Acta, 3:357.

Newell, R.C., Lucas, M.I., and Linley, E.A.S., 1981. Rate of degeneration and efficiency of conversion of phytoplankton debris by marine micro-organisms, Mar. Ecol. Prog. Series, 6: 123.

Newell, S.V., and Christian, R.R., 1981, Frequency of dividing cells as an estimator of bacterial productivity, App. Environ. Microbiol. 42:23.

Packard, T.T., and Williams, P.J.leB. 1981, Respiration and respirator electron transport activity in sea surface seawater from the northeast Atlantic, Oceanol. Acta, 4:351.

Parsons, T.R., Albright, L.J., Whitney, F., Wong, C.S., and Williams, P.J.leB., 1980, The effect of glucose on the productivity of seawater: an experimental approach using controlled equatic ecosystems. Mar. Environ. Res., 4:229.

Payne, W.T., 1970, Energy yields and growth of heterotrophs, Ann. Rev. Microbiol., 24:17.

Payne, W.T., and Wiebe, W.J., 1978, Growth yield and efficiency in chemosynthetic micro-organisms, Ann. Rev. Microbiol., 32:155.

Platt, T., Lewis, M., and Geider, R., 1983, Thermodynamics of the pelagic ecosystem: elementary closure conditions for biological production in the open ocean, this volume.

Redfield, A.C., 1958, The biological control of the chemical factors in the environment, Amer. Sci., 46:205.

Rublee, P., Ferguson, R.L., Palumbo, A.V., and Buckley, E., personal communication.

Shulenberger, E., and Reid, J.L., 1981, The Pacific shallow oxygen maximum, deep chlorophyll maximum, and primary production, reconsidered, Deep-Sea Res., 28: 901.

Sieburth, J. McN., 1983, Grazing of bacteria by protozooplankton in pelagic marine waters, in: "Heterotrophy in the Sea," J.E. Hobbie and P.J. leB. Williams, eds., Plenum, New York.

Sieburth, J. McN., Johnson, K.M., Burney, C.M., and Lavoie, D.M., 1977, Estimated in situ rates of heterotrophy using diurnal change in dissolved organic matter and growth rates of picoplankton in diffusion cultures, Helgolander wiss Meeresunters. 30:565.

Sorokin, Y.I., 1978, Decomposition of organic matter and nutrient regeneration, in: "Marine Ecology," Vol. IV, O. Kinne ed., Wiley Interscience, Chichester.

Strickland, J.D.H., 1970, The ecology of the plankton off La Jolla, California, in the period April through September, 1967, Bull. Scripps Inst. Ocean., Volume 17.

Stuart, V., Lucas, M.I., and Newell, R.C., 1981, Heterotrophic utilisation of particulate matter from the kelp Laminaria pallida, Mar. Ecol. Prog. Ser., 4: 337.

Vinogradov, M.Y., Krapivin, V.F., Menshutkin, V.V., Fleyshman, B.S., and Shushkina, E.A., 1973, Mathematical model of the functions of the pelagic ecosystem in tropical regions (from 50th voyage of the R/V Vityaz), Oceanology, 13: 704.

Vinogradov, M.E., Menshutkin, V.V., and Shushkina, E.A., 1972, On a mathematic simulation of a pelagic ecosystem in tropical waters of the ocean, Mar. Biol., 16:261.

Vyshvartsev, D.I., 1980, Bacterioplankton in shallow inlets of Poyeta Bay. Microbiology, 48:603.

Watson, S.W., Novitsky, T.M., Quinby, H.L., and Valois, F.W., 1977, Determination of bacterial number and biomass in marine environments, App. Environ. Microbiol., 33:940.

Williams, P.J. leB., 1970, Heterotrophic utilization of dissolved organic compounds in the sea. I. Size distribution of population and relationship between respiration and incorporation of growth substances, J. Mar. Biol. Ass. U.K., 50: 859.

Williams, P.J. leB. 1973, On the question of growth yields of natural heterotrophic populations, in: "Modern Methods in the Study of Microbial Ecology," T. Rosswall, ed., Bull. Ecol. Res. Comm. (Stockholm) 197. Swedish Natural Science Research Council.

Williams, P.J. leB., 1981a, Microbial contribution to overall marine plankton metabolism: direct measurements of respiration, Oceanol. Acta, 4: 359.

Williams, P.J.leB., 1981b, Incorporation of microheterotrophic processes into the classical paradigm of the planktonic food web, 15th European Symposium on Marine Biology, Kiel F.G.R. Kieler Meereforsch, 5: 1.

Williams, P.J. leB., 1982, Microbial contribution to overall plankton community respiration - studies in CEE's in: "Marine Mesocosms: Biological and Chemical Research in Experimental Ecosystems," G.D. Grice and H.R. Reeve, eds., Springer-Verlay, Berlin.

Williams, P.J. leB., 1983, A review of measurements of respiration rates of marine plankton communities, in: "Heterotrophy in the Sea," J.E. Hobbie, and P.J. leB. Williams, eds., Plenum Press, New York.

Williams, P.J. leB., Berman, T. and Holm-Hansen, O., 1976, Amino acid uptake and respiration by marine heterotrophs, Mar. Biol. 35: 41.

Williams, P.J. leB. and Yentsch, C.S., 1976, An examination of phytosynthetic production, excretion of photosynthetic products, and heterotrophic utilization of dissolved organic compounds with reference to results from a coastal subtropical sea, Mar. Biol., 35: 31.

Williams, P.M., Carlucci, A.F., and Olson, R., 1980, A deep profile of some biologically important properties in the central North Pacific gyre, Oceanol. Acta, 3: 471.

SUSPENDED MARINE BACTERIA AS A FOOD SOURCE

Tom Fenchel

Department of Ecology and Genetics
University of Aarhus
DK-8000 Aarhus C, Denmark

INTRODUCTION

During the last decade, evidence has accumulated which undermine the classical picture of planktonic food chains. The general belief was that phytoplankton is consumed by herbivorous zooplankters with an efficiency approaching 100% (Steele, 1976). It has more recently been found that the largest fraction of heterotrophic metabolic activity can be attributed to bacteria rather than to the herbivorous zooplankton. Studies based on a variety of methods suggest that bacterial biomass has turnover times ranging from <0.5 to a few days and that this represents a production which amounts to 10-30% of the primary production. The reduced carbon sustaining this productivity derives mainly from exudates of phytoplankton cells, but leachates from dead cells and from herbivores as well as detrital material contribute, so that as much as 20-40% of the primary production turns up as dissolved organic matter to be utilized by bacteria (Azam and Hodson, 1977; Hagström _et al._, 1979; Larsson and Hagström 1979; Fuhrman _et al._, 1980; Rheinheimer, 1981; Williams, 1981; Stuart _et al._, 1982; Wolter, 1982). To this heterotrophic production of bacteria, a photosynthetic production of unicellular cyanobacteria, now known to be an ubiquitous component of plankton, must be added (Sieburth, 1979).

While bacterial numbers of seawater are known to fluctuate somewhat on a yearly basis as well as with characteristic time scales of weeks, days or hours, they remain relatively constant when compared to the productivity attributed to them. In oceanic and off-shore waters the range is $2 \times 10^5 - 10^6$ cells ml^{-1}; in more productive waters closer to land, bacteria occur within the range $1 - 3 \times 10^6$ ml^{-1} (Ferguson and Rublee, 1976; Meyer-Reil _et al._, 1979; Fenchel,

1982d). Consequently bacteria must be removed at a rate which on the average equals their productivity. Although lysis of cells and subsequent bacterial recycling may play a role, grazing is usually assumed to be the most important sink for planktonic bacteria. This process in general and the identification of the grazers in particular has, however, been a rather open question in the literature.

Although bacterial biomass usually represents 5 - 20% of that of the phytoplankton, corresponding to 5 - 10 μg C l^{-1} for the least, to 50 - 100 μg C l^{-1} for the most productive waters (Ferguson and Rublee, 1976; Fuhrman et al., 1980), the dispersion of this resource into very small (0.3 - 1μm) particles constrains the efficiency by which grazers can concentrate them. In the following, we will first consider the quantification of suspension feeding. Thereafter, very simple hydrodynamical considerations will be given to show that the size ratio of food particles to grazers in general limits efficiency and that the efficiency of filter feeding is limited by small particle size. These considerations suggest that in plankton the bacterial grazers will be found among the smallest phagotrophs. Experimental evidence on animals feeding on suspended bacteria will be reviewed. It can be concluded that only heterotrophic nanoplankton flagellates can be held responsible for controlling bacterial numbers in seawater and that these organisms constitute a necessary link in the food chain between bacteria and zooplankton organisms such as ciliates, rotifers and small crustaceans.

THE QUANTIFICATION OF PARTICLE CAPTURE

Many studies on food interrelationships in nature are based either on purely qualitative observations or they represent direct or indirect measurements of particle uptake, but the data cannot readily be interpreted or used for comparative purposes. The concept of clearance, that is the volume of water which an organism can clear for particles per unit time has been widely used by students of filter feeding animals (Jørgensen, 1966; 1975). This measure of uptake rate divided by food concentration can readily be applied to all types of feeding on suspended particles (or the uptake of organic solutes) and permits meaningful comparisons. The uptake of particles $U(x)$ as a function of particle concentration, x, will always show some sort of satiation effect, that is, clearance, $F(x) = U(x)/x$ will decrease with increasing particle concentration. This is because at high concentrations of food particles the ingestion or phagocytosis rate limits the consumption. For many metazoan suspension feeders, the exact form of the functional response is not known. In some cases, experimental data seem to fit a variety of possible models equally well (Mullin et al., 1975). The subject may be complicated by behavioral idiosyncracies of different species and many empirical data are difficult to interpret due to experimental details which have not been sufficiently analyzed. In suspension feeding ciliates and flagellates the functional response very closely fits a hyperbolic function of the Michaelis-Menton form. This can be rationalized in physiological terms as the result of two limiting processes. At low particle concentration the uptake rate is determined by the structure and function of the organelles which concentrate the particles from the water (rate of water propulsion, area of some filter, etc.). This rate of processing of water, or the maximum clearance rate may depend on particle size or other properties, but is assumed to be invariant with particle concentration. At higher particle concentration the rate at which particles can be phagocytized becomes limiting and eventually sets an upper limit to consumption rate. In terms of Michaelis-Menten kinetics, the uptake of particles is given by $U(x) = U_m x/(x+K)$. The "half-saturation constant", K, is an ad hoc parameter which equals U_m/F_m and U_m is the reciprocal of the time taken to ingest one food particle. For $x \to 0$, it is seen that $U \simeq xF_m$, where the maximum clearance is a true measure of the ability of an organism to compete for food particles at low concentrations and will be considered as such in the sequel. In order to compare organisms of very different sizes, we will, rather than measuring clearance as absolute volume of water cleared per individual per unit time (dimension L^3T^{-1}) consider a "specific clearance", that is the number of body volumes cleared (dimension: T^{-1}). It is recognized, that the metabolic demand of organisms is not proportional to weight, but rather to weight$^{0.75}$ (so that among two organisms, one being say 1000 times larger than the other and both feeding on the same resource, then they would perform equally well if the larger species has a specific clearance some 0.56 times smaller than that of the smaller species). Nevertheless, specific clearance remains the best basis for comparison.

BASIC MECHANISMS FOR CONCENTRATING PARTICLES

Suspension feeding organisms possess adaptations which serve to concentrate particles. It will be useful first to consider the mechanisms which are at all possible within the constraints of physical laws and physiological limitations. First it should be mentioned, that the dimensions and velocities in question yield Reynolds numbers <<1, so that all mechanisms based on inertial forces can be ruled out. The majority of organisms which can be considered as potential grazers of bacteria owe their motility to flagella or cilia. The velocities of water currents produced by these organelles are rather invariant (within the range 50-200 $\mu m\ sec^{-1}$ for individual flagella and up to around 1000 $\mu m\ sec^{-1}$ for ciliary fields or membranelles) as discussed by Sleigh and Blake (1977). Furthermore, the hydrostatic pressure which cilia can support is limited and usually well below 100 dyn cm^{-2} (1 mm H_2O), see Foster-Smith (1976) and Fenchel (1980 c).

In the following, we will consider simplified models of particle capture. In spite of their unrealistic details, they give a general feeling for the possible mechanisms and in particular for the significance of the relative dimensions of grazers and food particles. The mechanisms available can be grouped into three categories: sieving, direct interception and diffusion; in the terminology of fisheries they have their analogues in trawling, spearing and fish traps, respectively. The two first mentioned mechanisms depend on the motility of the grazer, the latter on the motility of the food particles.

In organisms which depend on sieving (and assuming a 100% particle retention) clearance is the product of filter area and water volocity, v, that is proportional to R^2v, where R is cell radius. Therefore specific clearance $\propto R^{-1}v$, and decreases inversley with body size. In more realistic models, one cannot apply the head on, far field water velocity, but must take the velocity gradient around the cell into account. This is given by the solution to Stoke's equations with the boundary condition that the tangential component is zero at the surface (Figure 1, A,B). Furthermore, the velocity field differs according to whether the cell is freely swimming or whether it is attached, the latter giving somewhat more favorable conditions (Lighthill, 1976). Taking such details into consideration and using direct measurements of the far field water velocity and morphological data, Fenchel (1980, c,d, 1982a) found reasonably good agreement with measurements of clearance based on particle uptake from suspensions with known concentrations for some filter feeding ciliates and microflagellates.

It would seem that the efficiency of this type of feeding is independent of particle size, but this is not so because the pressure drop over the filter sets a lower limit to porosity. At low Reynolds numbers, the hydrostatic pressure of a filter is proportional to the water velocity. In filters consisting of a parallel array of cylinders, the pressure drop also relates to the ratio between the diameter of the cylinders and the distance between them, and when this ratio approaches unity, pressure drop increases very strongly (Chen, 1955; Tamada and Fujikawa, 1957). In filter feeding ciliates the filter consists of a parallel array of eukaryote cilia (diameter: 0.25 μm); the minimum distance between cilia found is around 0.3 μm and this is associated with very low water velocities (around 50 μm sec^{-1} or less). In ciliates with coarser filters (>2-5 μm), water velocities and thus values of clearance some 10-20 times higher are found. In all cases, the pressure drop over the filter, as calculated by the model of Tamada and Fujikawa (1957), works out to be around 30 dyne cm^{-2} (Fenchel, 1980c,d; Fenchel and Small, 1980). In filter feeding microflagellates (choanoflagellates, helioflagellids) the filter is built of pseudopodia with diameters smaller than that of cilia and this allows for a lower porosity. Also in these forms, water velocity through the filter is correlated with porosity (Fenchel

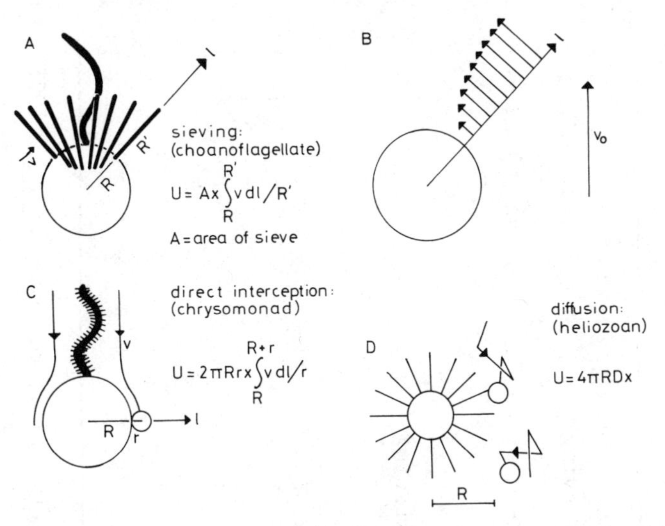

Fig. 1. The rate of uptake of particles, U, from a suspension with concentration, x, for three different mechanisms of concentrating suspended particles (A,C and D). In B, corresponding to A, the perpendicular velocity gradient along a pseudopodium is shown for the case where the flagellate swims "downwards" with the superficial velocity, $-V_o$. If the flagellate was attached, the water flow through the filter would, assuming similar flagellar beat parameters as when swimming, be more favorable.

1982a,b). With respect to metazoan filter feeders, the effect will be discussed in the following section.

By "direct interception" is meant that food particles carried along the flow lines are directly intercepted by the surface of the grazer which releases some mechanism leading to the ingestion or phagocytosis of the particle. (This type of feeding is mostly referred to as "raptorial", but this term somehow infers a sophistication beyond the basis of this mechanism for concentrating food particles, which is random collision between suspended particles of which at least one type is motile.) In this type of feeding a critical flow line exists within which particles will be intercepted (Figure 1 C). In a simplistic model for a spherical grazer (radius: R) feeding on other particles (radius: r), the transectional area of the water flow delimited by the critical flow line would be $2\pi Rr$ ($R >> r$) so the specific clearance would become $3/2R^{-2}r$ v. This does not account for the velocity gradient around the grazer. For a spherical collector moving through the water, specific clearance be-

Table 1. Maximum specific clearance rate as function of food particle size

Organism	Approximate volume, ml	Clearance, h^{-1}	Mechanism	Minimum particle size retained, μm
Microflagellates[a]				
Monosiga	2×10^{-11}	9.8×10^{4}	sieving	∿0.2
Actinomonas	7.5×10^{-11}	1.1×10^{6}	-	1-2
Paraphysomonas	1.9×10^{-10}	9.1×10^{4}	interception	no sharp limit
Pseudobodo	9×10^{-11}	1.1×10^{5}	-	-
Ciliates[b]				
Cyclidium	5×10^{-10}	6×10^{3}	sieving	∿0.3
various species		$5 \times 10^{4} - 10^{5}$	-	>2
Metazoa				
Mycale[c]		7.7×10^{2}	-	∿0.2
Daphnia[d]	4×10^{-4}	2.5×10^{3}	-	0.2 - 0.5 ?
Oikopleura[e]	$10^{-6} - 10^{-4}$	2.2×10^{4}	-	0.2 - 0.5 ?
Geukensia[f]	1 - 10	10^{2}	?	>1 ?
typical bivalve[g]	0.5 - 10	$1\text{-}2 \times 10^{3}$	?	<2
copepods[h]	10^{-4}	$1\text{-}2 \times 10^{4}$	?	>4-5

References: a, Fenchel (1982 b); b, Fenchel (1980 b); c, Reiswig (1974); d, Peterson *et al*. (1978); e, King *et al*. (1980); f, Wright *et al*. (1982); g, Møhlenberg and Riisgård (1978) ; h, Jørgensen (1975).

comes proportional to $R^{-3}r^2$ rather than to $R^{-2}r$, a much harder dependence on size ratios (Spielman, 1977). If the spherical grazer is attached and produces a velocity field with its flagellum, this is somewhat relaxed (Lighthill, 1976), but the two expressions probably delimit the possibilities.

Fenchel (1982 a) applied such models to nearly spherical flagellates and compared the results with clearance values obtained by measuring the rate of bacterial uptake in cell suspensions with known concentrations. The hydrodynamical considerations yield comparable, but too low values. In addition to the fact that the geometry is somewhat more complex, there are other mechanisms which may explain the discrepancy. These include van der Waals forces, particle rotation which may tend to make the particles in a velocity gradient migrate towards a surface, and finally Brownian movement of the particles. All three may lead to interception of particles from somewhat outside the critical flowline as defined above (Spielman, 1977; Jørgensen, 1981). With the exception of Brownian movement (see below), these effects cannot be evaluated quantitatively. Nevertheless, it is obvious that this type of feeding is highly dependent on size ratios: with respect to bacterivory the mechanism can only be expected among the very smallest phagotrophic organisms.

The last possibility to consider is "feeding by diffusion". By this is meant that the food particles perform Brownian movements and due to this may be intercepted by a motionless grazer. It is formally identical to the mechanism by which a bacterial cell takes up organic molecules from a solution. It can be shown (Koch, 1971; Roberts, 1981), that at low particle concentration (diffusion-limited case in which particle concentration at the surface of the grazer is zero), the uptake of a spherical cell is given by the expression $U = 4\pi RDx$, where D is the diffusion coefficient and x the bulk concentration; maximum specific clearance is then $3\ R^{-2}D$, that is, it decreases with the second power of the radius (Figure 1 D). Consider a suspension of non-motile, spherical bacteria with a diameter of 1 μm. Their Brownian movement is then described by the Einstein relation, $D = kT/(6\pi\mu R)$, where k is Boltzmann's constant, T is absolute temperature and μ is water viscosity. At 20°C and $\mu = 0.01$ Poise, we get $D = 5 \times 10^{-9}\ cm^2 sec^{-1}$. A spherical, motionless protozoan with a diameter of 10 μm would then have a specific clearance of about $8 \times 10^2 h^{-1}$ with respect to this bacterial suspension, 2-3 orders of magnitude lower than the values expected from and found for any of the other mechanisms described. Brownian movement is therefore not expected to play a role in this example.

This picture, however, changes if motile bacteria are considered. Assume a bacterium which performs three-dimensional random walk with a velocity, v, and an average path length, l. This gives $D = lv/6$; with the realistic values, $l = 30$ μm and $v = 30\ \mu m\ sec^{-1}$, then $D=1.5 \times 10^{-6}\ cm^2\ sec^{-1}$. Specific clearance for the above mentioned protozoan would then be $2.7 \times 10^5\ h^{-1}$, a realistic value on which to base life on naturally occurring concentrations of bacteria. The protozoan could even improve this if it gave up its spherical shape and grew long pseudopodia. This, of course, requires that most planktonic bacteria are motile, but empirically this does not seem to be the case (Sieburth, 1979).

The main conclusion of this section is that we would expect only the smallest phagotrophs to use diffusion or direct interception as a means to concentrate bacteria since efficiency decreases with the square of their length dimensions. For organisms with a radius in the range 2-5 μm, direct interception and sieving work about equally well. With a reasonable choice of parameter values, specific clearances in the range of 10^5 - 10^6 h^{-1} are predicted. The efficiency of sieving is less affected by increase in body size and must be the only mechanisms available for larger bacterivores. We may, however, expect that filter feeders which retain particles of bacterial size have relatively lower values of clearance and may only perform well in environments with very high concentrations of bacteria.

BACTIVORY IN REAL ORGANISMS

A variety of suspension feeding organisms have been described as bacterivorous; however, only few cases have been sufficiently studied to assess the quantitative significance. Among the protozoans heterotrophic microflagellates, which are within the size range of 3 - 9 μm have been suggested to play a large role as bacterial grazers in seawater (Sieburth et al., 1978; Sieburth, 1979; Haas and Webb, 1979; King et al., 1980). A number of types, notably choanoflagellate and colorless chrysomonads are of ubiquitous occurrence in seawater and a few reports indicate densities of the order of 10^3 ml^{-1} (for references, see Sieburth, 1979 and Fenchel, 1982d). Fenchel (1982 a-d carried out investigations on various aspects of the ecology and physiology of such forms. They are all bacterivorous and catch their food particles either by direct interception or by sieving through a pseudopodial filter. Bioenergetic studies in pure cultures revealed that their net and gross growth efficiencies are about 60% and 40% respectively (on C basis) and these efficiencies are invariant with growth rate. Minimum doubling time at 20°C is within the range of 3-4 h (for six studied species), but balanced growth could be maintained at doubling times as long as 20-24 h: such generation times are found for bacterial concentrations around 10° ml^{-1}. With a food bacterium measuring about 1.3 x 0.7 μm, most forms have a maximum specific clearance of around 10^5 h^{-1}, only the helioflagellid, Actinomonas, with a relatively coarse filter (porosity: 1-2 μm) has a clearance of around 10^6 h^{-1}. Choanoflagellates, the most important filter feeding forms, have a porosity around 0.2 μm (Table 1).

The qualitative and quantitative composition of the fauna of nanoplankton flagellates was studied in the waters of Limfjorden,

Denmark during the summer. Choanoflagellates alone account for around 50% of the individuals, and colorless chrysomonads are second in importance followed by bicoecids; other types of flagellates are much rarer. Altogether, from around 200 to 3 x 10^3 cells ml^{-1} were found. Based on laboratory data on clearance, it could be calculated that between 12 and 67% (average over a month: 20%) of the water column is filtered by the flagellates per day. Based on the bacterial concentrations (ranging from about 10^6 to 3 x 10^6 ml^{-1} during the period) typical generation times for the flagellates would be of the order of 24 h. These studies suggest that microflagellates are the dominant grazers of bacteria in the water column and that their activity can account for the turnover of suspended bacteria. This is also supported by Sorokin's (1977) observations on the protozoan succession following a bacterial bloom in the Japan Sea. The fact that microflagellates are the dominant grazers of bacteria in laboratory microcosms based on seawater and decomposing organic material also indicates the significance of these organisms (Fenchel and Harrison, 1976; Fenchel and Jφrgensen, 1977; Linley _et al_., 1981; Newell _et al_., 1981; Robertson _et al_., 1982).

Among rhizopods, small amoebae have been shown to occur regularly in coastal as well as oceanic waters (Davies _et al_., 1978 Fenchel, 1982 d); they are undoubtedly bacterivorous and are likely to depend on microbial cells attached to suspended detrital material or water films. Their quantitative role is unknown. This also applies to ubiquitous small heliozoans, e.g., _Ciliophrys_. It is an example of a "diffusion feeder" which may partly or wholly depend on bacteria; it is also possible, however, that it feeds on motile microprotozoans.

The ciliates comprise a very diversified group of organisms with respect to ecology in general and to feeding biology in particular, and bacterivorous suspension feeders are only represented by a restricted number of orders. Among the primitive ciliate orders, the free living representatives predominantly feed on large particles (microalgae, other protozoans) although a few, quantitatively unimportant groups independently have evolved adaptations for suspension feeding on bacteria. The polyhymenophoran cilicates (spirotrichs) are suspension feeders, but the structure of their filter, arrays of parallel ciliary membranelles, does not seem to permit the retention of particles smaller than about 2 μm. Bacterivorous suspension feeding ciliates are mainly found among the oligohymenophorans (hyenostomes _sensu lato_ and peritrichs). These forms have values of clearance which are low relative to those of ciliates feeding on larger particles (Fenchel, 1980 a-d; see also Table 1). In the benthos and in special environments rich in bacteria, bacterivorous ciliates are numerous (Fenchel, 1968), but their role in marine plankton is disputable. Certain forms, in particular _Uronema_ and relatives, are easy to isolate by enriching seawater with bacteria (e.g. Hamilton and Preslan, 1969). However, these ciliates are "opportunists" depending on local patches rich in bacteria such as decomposing carrion and after the feast these ciliates encyst (Fenchel, 1968). The quantitatively dominating ciliates in the plankton belong to the polyhymenophoran (tintinnid or non-tintinnid) oligotrichs which select particles much larger than bacteria and mainly depend on phytoplankton cells (Spittler, 1973; Rassoulzadegan, 1977; Heinbokel, 1978, Fenchel, 1982 d). It is conceivable, however, that in very productive waters such as estuaries, ciliates may play a quantitative role as consumers of bacteria. Such a role may possibly also be attributed to the epifaunal peritrich colonies associated with copepods and other zooplankton, but no evaluation of this exists.

Among metazoans, many benthic forms, that is detritus feeders and representatives of the meiofaua, seem at least in part to depend on bacterial diet (for a review, see Fenchel and Jϕrgensen, 1977). With respect to suspension feeding on bacteria, convincing evidence is scarce. Many studies have failed to consider whether measured clearance values suggest any bioenergetic significance. Laboratory studies have often involved the use of bacterial stains grown on nutrient broth or similar media, and such cells are significantly larger than those belonging to the predominant size fraction of natural waters (Ferguson and Rublee, 1976; Sieburth, 1979). This is especially important, since the exact identification of the filter and the understanding of function is incomplete in some important groups of suspension feeders (see Jϕrgensen, 1981).

The majority of metazoan filter feeders do not seem to retain particles smaller than 1-2 μm; this applies to most crustaceans, and to bryozoans, bivalves, ascidians, rotifers and to various larval forms (Jϕrgensen, 1966, 1975; Pourriot, 1977; Mϕhlenberg and Riisgard, 1978; Randlϕv and Riisgard, 1979). Sponges are an exception to this, since the collar cells, which are structurally and functionally quite identical to choanoflagellates, have a porosity of 0.2 - 0.3 μm. In accordance with this, clearance of sponges is quite low due to the low velocity of the water passing the collar (Reiswig, 1974, 1975; see also Table 1).

Wright *et al*., (1982) recently found that the estuarine bivalve *Geukensia* retains bacteria. However, the low specific clearance calculated from their data (around 10^2 h^{-1}) which is about one order of magnitude lower than that otherwise reported from bivalves, renders this result disputable. If we assume 10^{-4} g dw bacteria l^{-1} (the authors reported 1.5×10^6 bacteria ml^{-1} is the estuarine water), then the bivalves would only gain around 1 o/oo of their body weight per day from bacteria, indeed a modest contribution to their energetic requirements. Wright *et al*. (1982) found no bacterial uptake in the common mussel, *Mytilus*, whereas some older studies (Zobell and Landon, 1937; Zobell and Feltham, 1937-38) found that mussels can grow on a bacterial diet. Peterson *et al*. (1978) convincingly demonstrated the filtration of a natural assemblage of bacteria by (freshwater) daphnia. The calculated clearance value is about half of that found when the daphnia were filtering yeast cells, but the results indicate that at very high bacterial concentrations, bacterivory would be of some significance for these planktonic animals.

Thus far, the most convincing example of a metazoan suspension feeder utilizing bacteria is the appendicularian *Oikopleura* (King *et al.*, 1980). In this case the measured values of clearance suggest that these animals may obtain a significant contribution to their energetic requirements from natural concentrations of bacteria in seawater. The structure of the catching net, as studied by Flood (1978) may explain not only the fine porosity of the filter, but also the relatively high water velocity required to accord with the clearance rate, since the filaments making up the filter have an unusually small diameter.

The role of planktonic bacteria for ciliate and metazoan suspension feeders may be somewhat larger than the above discussion suggests if a substantial part of the bacteria occur in conglomerates or are associated with suspended detrital material. Some authors (Seki, 1972; Sorokin, 1978; Prieur, 1981) believe that this is of significance. The extent to which bacteria are found associated with larger particles in seawater is not quite clear. In laboratory microcosms the formation of such bacterial aggregates has been demonstrated (Robertson *et al.*, 1982) and while some studies (Hobbie *et al.*, 1972; Azam and Hodson, 1977) found that in the open sea by far the most bacteria are freely suspended, it might be that bacterial aggregates have a higher turnover. In shallow waters with higher concentrations of suspended detrital material, attached bacteria are very likely to be significant for suspension feeders (e.g. Fenchel *et al.*, 1975).

DISCUSSION AND CONCLUSIONS

The present considerations suggest in a general way that in pelagic food chains there are limits to size ratios between prey and predator species. This in turn lends support to models such as that of Platt and Denman (1977) for explaining the characteristic smooth size spectrum of pelagic communities. With special respect to bactivory, we would expect only the very smallest phagotrophic cells to be important as grazers. This picture is mainly supported by empirical evidence and the only convincing exception, so far, of a bacterivorous "baleen whale" is in appendicularians. The explanation for this generalization is to be found in simple physical and physiological principles which thus determine the structure of pelagic communities. Such simple principles do not apply in benthic systems where microbial concentrations on surfaces and in conjunction with dead particulate organic material occurs.

REFERENCES

Azam, F., and Hodson, R.E., 1977, Size distribution and activity of marine microheterotrophs, Limnol. Oceanogr., 22: 492.

Chen, Y.C., 1955, Filtration of aerosols by fibrous media, Chem. Rev., 55: 595.

Davis, P.G., Caron, D.A., and Sieburth, J.McN., 1978, Oceanic amoebae from North Atlantic: culture, distribution and taxonomy, Trans. Am. Micros. Soc., 96: 73.

Fenchel, T., 1968, The ecology of marine microbenthos. II. The food of marine benthic ciliates, Ophelia, 5: 73.

Fenchel, T., 1980 a, Suspension feeding in ciliated protozoa: functional response and particle size selection, Microb. Ecol.,6: 1.

Fenchel, T., 1980 b, Suspension feeding in ciliated protozoa: feeding rates and their ecological significance, Microb. Ecol., 6: 13.

Fenchel, T., 1980 c, Relation between particle size selection and clearance in suspension feeding ciliates, Limnol. Oceanogr. 25: 733.

Fenchel, T., 1980 d, Suspension feeding in ciliated protozoa: structure and function of feeding organelles, Arch. Protistenk., 123: 239.

Fenchel, T., 1982a Ecology of heterotrophic microflagellates. I. Some important forms and their functional morphology. Mar. Ecol. Prog. Ser., 8: 211.

Fenchel, T., 1982b, Ecology of heterotrophic microflagellates, II. Bioenergetics and growth, Mar. Ecol. Prog. Ser., 8:215.

Fenchel, T., 1982c, Ecology of heterotrophic microflagellates, III, Adaptations to heterogenous environments, Mar. Ecol. Prog. Ser., 9: 25.

Fenchel, T., 1982 d, Ecology of heterotrophic microflagellates. IV. Quantitative occurrence and importance as consumers of bacteria. Mar. Ecol. Prog. Ser., 9: 35.

Fenchel, T., and Harrison, P., 1976, The significance of bacterial grazing and mineral cycling for the decomposition of particulate detritus, in: "The Role of Terrestrial and Aquatic Organisms in Decomposition Processes," Anderson, J.M. and Macfadyen, A., eds. Blackwell, Oxford.

Fenchel, T., and Jϕrgensen, B.B., 1977, Detritus food chains of aquatic ecosystems: the role of bacteria, Adv. Microb. Ecol. 1: 1.

Fenchel, T., Kofoed, L.H., and Lappalainen, A., 1975, Particle size-selection of two deposit feeders: the amphipod Corophium volutator and the prosobranch Hydrobia ulvae, Mar. Biol., 30: 119.

Fenchel, T., and Small, E.B., 1980, Structure and function of the oral cavity and its organelles in the hymenostome ciliate Glaucoma, Trans. Am. Micros. Soc., 99: 52.

Ferguson, R.L., and Rublee, P., 1976, Contribution of bacteria to the standing crop of coastal plankton, Limnol. Ocreanogr., 21:141.

Flood, P.R., 1978, Filter characteristics of appendicularian food catching nets, Experientia, 34: 173.

Foster-Smith, R.L. 1976, Pressures generated by the pumping mechanisms of some ciliary filter feeders, J. Exp. Mar. Biol. Ecol., 25: 199.
Fuhrman, J.A., Ammerman, J.W.A., and Azam, F., 1980, Bacterioplankton in the coastal euphotic zone; distribution, activity and possible relationships with phytoplankton, Mar. Biol., 60: 201.
Haas, L.W., and Webb, K.L., 1979, Nutritional mode of several non-pigmented microflagellates from the York River Estuary, Virginia, J. Exp. Mar. Biol. Ecol., 39: 125.
Hagström, A., Larsson, U., Hörstedt, P., and Normark, S., 1979, Frequency of dividing cells, a new approach to the determination of bacterial growth rates in aquatic environments, Appl. Environ. Microbiol., 37: 805.
Hamilton, R.D., and Preslan, J.E., 1969, Cultural characteristics of a pelagic marine hymenostome ciliate, Uronema sp., J. Exp. Mar. Biol. Ecol., 4: 90.
Heinbokel, J.F., 1978, Studies on the functional role of tintinnids in the Southern California Bight. I. Grazing and growth rates in laboratory cultures, Mar. Biol., 47: 177.
Hobbie, J.E., Holm-Hansen, O., Packard, T.T., Pomeroy, L.R., Sheldon, R.W., Thomas, J.P., and Wiebe, W.J., 1972, A study of the distribution and activity of microorganisms in ocean water, Limnol. Oceanogr., 17: 544.
Jφrgensen, C.B., 1966, "Biology of Suspension Feeding," Pergamon, Oxford.
Jφrgensen, C.B., 1975, Comparative physiology of suspension feeding, Ann. Rev. Physiol., 37: 57.
Jφrgensen, C.B., 1981, A hydromechanical principle for particle retention in Mytilus edulis and other ciliary suspension feeders, Mar. Biol. 61: 277.
King, K.R., Hollibaugh, J.T., and Azam, F., 1980, Predator-prey interactions between the larvacean Oikopleura dioica and bacterioplankton in enclosed water columns, Mar. Biol., 56: 49.
Koch, A.L., 1971, The adaptive responses of Escherichia coli to a feast and famine existence, Adv. Microb. Physiol., 6: 147.
Larsson, U., and Hägstrom, A., 1979, Phytoplankton exudate release as an energy source for the growth of pelagic bacteria, Mar. Biol. 52: 199.
Lighthill, J., 1976, Flagellar hydrodynamics, SIAM Rev., 18:161.
Linley, E.A.S., Newell, R.C., and Bosma, S.A., 1981, Heterotrophic utilisation of mucilage released during fragmentation of kelp (Eckonia maxima and Laminaria pallida). I. Development of microbial communities associated with the degradation of kelp mucilage, Mar. Ecol. Prog. Ser., 4: 31.
Meyer-Reil, L. -A., Bölter, M., Liebezeit, G., and Schramm, W., 1979, Short-term variations in microbiological and chemical parameters, Mar. Ecol. Prog. Ser., 1:1.
Mφhlenberg, F., and Riisgård, H.U., 1978, Efficiency of particle retention in 13 species of suspension feeding bivalves, Ophelia, 17: 239.

Mullins M.M., Stewart, E.F., and Fuglister, F.J., 1975, Ingestion by planktonic grazers as a function of concentration of food, Limnol. Oceanogr., 20: 259.
Newell, R.C., Lucas, M.I., and Linley, E.A.S., 1981, Rate of degradation and efficiency of conversion of phytoplankton debris by marine micro-organisms, Mar. Ecol. Prog. Ser., 6: 123.
Peterson, B.J., Hobbie, J.E., and Haney, J.F., 1978, Daphnia grazing on natural bacteria, Limnol. Oceanogr., 23: 1039.
Platt, T., and Denman, K., 1977, Organisation in the pelagic ecosystem, Helgoländer wiss. Meeresunters., 30: 575.
Pourriot, R., 1977, Food and feeding habits of rotifera, Arch. Hydrobiol. Beitr., 8: 243.
Prieur, D., 1981, Experimental studies of trophic relationships between marine bacteria and bivalve molluscs, Kieler Meeresforsch. Sonderh., 5: 376.
Randløv, A., and Riisgård, H.U., 1979, Efficiency of particle retention and filtration rate in four species of ascidians, Mar. Ecol. Prog. Ser., 1: 55.
Rassoulzadegan, F., 1977, Evolution anuelle des ciliés pelagiques en Méditerranée nord-occidentale: ciliés oligotriches "non-tintinnides" (Oligotrichina), Ann. Inst. Océanogr. Paris, 53: 125.
Rassoulzadegan, F., and Etienne, M., 1981, Grazing rate of the tintinnid Stenosomella ventricosa (Clap & Lachm.) Jörg. on the spectrum of naturally occurring particulate matter from a Mediterranean neritic area, Limnol. Oceanogr., 26: 258.
Reiswig, H.M., 1974, Water transport, respiration and energetics of three tropical marine sponges, J. Exp. Mar. Biol. Ecol., 14: 231.
Reiswig, H.M., 1975, The aquiferous systems of three marine demospongiae, J. Morph., 145: 493.
Rheinheimer, G., 1981, Investigations on the role of bacteria in the food web of the Western Baltic, Kieler Meeresforsch. Sonderh., 5: 284.
Roberts, A.M., 1981, Hydrodynamics of protozoan swimming, in: "Biochemistry and Physiology of Protozoa," IV, 2nd ed., M. Levandowsky and S.H. Hutner eds, Academic,New York.
Robertson, M.L., Mills, A.L., and Zieman, J.C., 1982, Microbial synthesis of detritus-like particles from dissolved organic carbon released by tropical seagrasses, Mar. Ecol. Prog. Ser., 7: 279.
Seki, H., 1972, The role of microorganisms in the marine food chain with reference to organic aggregations, Mem. Ist. Ital. Idrobiol. 29 Suppl.: 245.
Sieburth, J., 1979, "Sea Microbes," Oxford Univ. Press, New York.
Sieburth, J., Smetacek, V., and Lenz, J., 1978, Pelagic ecosystem structure: heterotrophic compartments of the plankton and their relationship to plankton size fractions, Limnol. Oceanogr., 23: 1256.
Sleigh, M.A., and Blake, J.R., 1977, Methods of ciliary propulsion and their size limitation, in: "Scale Effects in Animal Locomotion," T.J. Pedley ed., Academic, London.

Sorokin, Yu. I., 1977, The heterotrophic phase of plankton succession in the Japan Sea, Mar. Biol., 41: 107.

Sorokin, Yu. I., 1978, Decomposition of organic matter and nutrient regeneration, in: "Marine Ecology, 4," O. Kinne ed., Wiley, Chichester.

Spielman, L.A., 1977, Particle capture from low-speed laminar flows, Ann. Rev. Fluid. Mech., 9: 297.

Spittler, P., 1973, Feeding experiments with tintinnids, Oikos, Suppl., 15: 128.

Steele, J.H., 1976, "The Structure of Marine Ecosystems," Harvard Univ. Press, Cambridge, Mass.

Stuart, V., Newell, R.C., and Lucas, M.I., 1982, Conversion of kelp debris and fecal material from the mussel Aulacomya ater by marine microorganisms, Mar. Ecol. Prog. Ser., 7: 47.

Tamada, K., and Fujikawa, H., 1957, The steady two-dimensional flow of viscous fluid at low Reynolds numbers passing through an infinite row of equal parallel circular cylinders, Quart. Mech. Appl. Math., 10: 425.

Williams, P.J.leB., 1981, Incorporation of microheterotrophic processes into the classical paradigm of the planktonic food web, Kieler Meeresforsch., Sonderh., 5: 1.

Wolter, K., 1982, Bacterial incorporation of organic substances released by natural phytoplankton populations, Mar. Ecol. Prog. Ser., 7: 287.

Wright, R.T., Coffin, R.B., Ersing, C.P., and Pearson, D., 1982, Field and laboratory measurements of bivalve filtration of natural marine bacterioplankton, Limnol. Oceanogr., 27: 91.

ZoBell, C.E., and Feltham, C.B., 1937-38, Bacteria as food for certain marine invertebrates, J. Mar. Res., 1: 312.

ZoBell, C.E., and Landon, W.A., 1937, The bacterial nutrition of the California mussel, Proc. Soc. Exp. Biol. Med. N.Y., 36: 607.

THE BIOLOGICAL ROLE OF DETRITUS IN THE MARINE ENVIRONMENT

R. C. Newell

Institute for Marine Environmental Research
Prospect Place
Plymouth, PL1 3DH

INTRODUCTION

The heterotrophic fate of detrital material in the marine environment has attracted widespread interest since the pioneer work of Teal (1962) and others established that primary production by wetland vegetation in coastal saltmarshes of Georgia, U.S.A., greatly exceeded direct consumption by grazing herbivores. Such material was thus available for decomposition or export as a potential trophic resource for consumer organisms in the shallow waters bordering such wetland ecosystems (see Darnell, 1967a,b; Odum, 1971; Keefe, 1972; Day _et al_., 1973; Gosselink and Kirby, 1974; Woodwell _et al_., 1977; for reviews, Newell, 1979; Nixon, 1980). Much the same situation exists on many rocky shores which are dominated by kelp beds. In contrast to the situation in the open sea, where much of the primary production is thought to be consumed directly by herbivores (Steele, 1974), the discrepancy between the primary production and the material which is directly removed by grazers near to kelp beds suggests that as in wetland ecosystems, large quantities of photo-assimilated material may become available to consumer organisms after fragmentation and partial decomposition (see Mann, 1972, 1973; Field _et al_., 1977; Newell _et al_., 1982). This situation, in which energy flow is dominantly through the decomposer pathway rather than through a typical plant-herbivore-carnivore food chain is widespread in other ecosystems including tropical rain forests and perhaps even in open savannah environments.

It is therefore of considerable importance to our understanding of energy flow through heterotrophic communities to study the extent of energy dissipation or 'remineralisation' of materials through the microheterotrophic decomposer food web. Clearly, the rate and extent of mineral recycling by microbial activity is of central importance in supporting further primary production whilst the amount of energy 'immobilised' or incorporated into microbial biomass controls the potential significance of micro-organisms as a trophic resource for higher consumer organisms.

Rather surprisingly, there has until recently been little information on the ecological energetics of such decomposer communities partly because the chemistry of the structural material available for decomposition is difficult to investigate and partly because little is known of the trophodynamics of the smallest members of the microheterotrophic succession, or of the extent to which the digestive physiology of larger consumer organisms can be adapted to exploit fragmented detrital material as a direct carbon resource. It is the purpose of this review to present the results of recent work both on the physiology of detritus utilisation by consumer organisms, and on the energetics of marine microbial decomposer organisms, which suggests that the microheterotrophic community may have a significant role both in the mineralisation processes necessary to support further primary production and as a trophic resource for those consumer organisms which are capable of exploiting bacterio-organic complexes in the water column.

ECOLOLOGICAL ENERGETICS OF DETRITUS DECOMPOSITION

One approach to the investigation of incorporation and mineralisation of materials by marine microheterotrophs is to study the kinetics of uptake of simple ^{14}C-labelled substrates by natural assemblages of marine bacteria (see Wright and Hobbie, 1965; Hobbie, 1967; Hobbie and Crawford, 1969; Gordon _et al_., 1973; Crawford _et al_., 1974; Azam and Hodson, 1981; Wolter, 1982; for reviews, Fenchel and Jørgensen, 1976; Joint and Morris, 1982). Exposure to labelled substrates allows an estimate of total incorporation but depends to a large extent on measurement of uptake of label prior to release of $^{14}CO_2$ through respiration. This value for initial uptake will obviously be higher than the net incorporation into microbial biomass following longer periods of incubation when $^{14}CO_2$ losses have occurred.

In practice it is often more satisfactory to calculate energy flow through the microheterotrophs from longer-term ^{14}C uptake from which $^{14}CO_2$ losses have already occurred, without measuring initial ^{14}C incorporation. Williams (1970, 1973) has pointed out that the net incorporation following longer-term exposure to ^{14}C-labelled substrates plus losses through respiration equals the initial or total uptake. Provided that dissolved organic losses are small, therefore, the 'apparent growth yield' (α), or amount of organic carbon incorporated into growth per unit of carbon consumed is:

$$\text{Apparent growth yield } (\alpha) = \frac{\text{Net } ^{14}\text{C incorporated}}{\text{Net } ^{14}\text{C incorporated + respired}} \times 100 \quad (1)$$

Growth yields and conversion efficiencies of a variety of heterotrophs have been reviewed by Calow (1977) and Payne and Wiebe (1978). They have pointed out that the mean conversion efficiency for growing cultures of bacteria is approximately 60% although earlier observations on cells in steady state cultures gave variable conversion efficiencies which are generally less than 50%. Rather variable values have also been reported by Billen _et al._ (1980) for heterogeneous populations of marine bacteria in the southern North Sea. One of the problems with the extrapolation of results from experiments based on simple ^{14}C-labelled substrates is that such substances commonly comprise only 10-25% of the total organic matter available for bacterial decomposition and may even represent as little as 0.1% (Wiebe and Smith, 1977; Larsson and Hagström, 1979; see also Andrews and Williams, 1971; Yurkovsky, 1971; Ogura, 1972, 1975; Allen, 1973; Ogura and Gotoh, 1974; for review, see Joint and Morris, 1982). Bacterial growth yields based on the utilisation of the complex mixture of 'labile' dissolved and more refractory components which comprise natural detrital material are therefore very difficult to determine by the ^{14}C method, unless the mixed detrital substrate could be uniformly labelled with ^{14}C.

An alternative approach, and one which we have used in a series of studies on the microbial communities associated with the degradation of detritus from kelp (Newell _et al._, 1980; Lucas _et al._, 1981; Linley and Newell, 1981; Linley _et al._, 1981; Newell and Lucas, 1981; Stuart _et al._, 1981; Stuart, Newell and Lucas, 1982; Koop _et al._, 1982a, b), phytoplankton (Newell _et al._, 1981) and saltmarsh grasses (Newell _et al._, 1983), is to measure the increase in biomass of microheterotrophs which colonise detrital material incubated in seawater freshly collected from the locality where the decomposition products are liberated. The numbers of micro-organisms can be estimated by acridine orange direct counts (AODC; see Hobbie _et al._, 1977; Daley, 1979; Linley _et al._, 1981) whilst the size of the cells, and hence mean cell volume can be estimated either from the AODC material (Fuhrman, 1981) or from scanning electron microscopy (Dempsey, 1981; Linley _et al._, 1981). The dry biomass and carbon equivalent of the wet biomass can then be calculated from the cell numbers and volumes using values for the specific gravity cited by Calkins and Summers (1941) and Luria (1960) and the coefficients 0.2 and 0.1 for the dry biomass and carbon equivalent of the wet biomass respectively (Luria, 1960; Troitsky and Sorokin, 1967; Sorokin and Kadota, 1972; see also Ferguson and Murdoch, 1978).

Direct observations on the biovolume of the succession of decomposer micro-organisms which colonise detrital material has the advantage that estimates of bacterial production are based on the natural mixture of dissolved and particulate organic matter which comprises organic detritus, rather than on selected labile components of the dissolved fraction alone. Such estimates can also be made over a period of days rather than the short-term incubations required by the ^{14}C methods, and allow some estimates of energy transfer through the second trophic step of the decomposer food chain from bacteria to microflagellates.

Microbial Communities Associated with Detritus

Although there are now many studies on microbial communities associated with the decomposition of plant detritus in terrestrial habitats, freshwater and in brackish water, there is less information for marine ecosystems. Decomposition of allochthonous material in soils and freshwaters is initially accomplished by a fungal succession (Kaushik and Hynes, 1971; Park, 1972; Suberkropp and Klug, 1974, 1976; Anderson, 1975; Lousier, 1982). In marine and brackish waters, however, primary microbial colonisation by bacteria rather than by fungi has commonly been reported. Robb *et al.* (1979) showed that bacteria rather than fungi were responsible for the initial stages of decomposition of the brackish water sago pond weed (*Potamogeton pectinatus*) and similar results have been obtained on decomposing kelp (*Ecklonia maxima*) on the strandline (Koop *et al.*, 1982a, b).

Experiments on the microbial colonisation of plant detritus from a variety of sources including seagrasses (Fenchel, 1970; Robertson *et al.*, 1982; Newell *et al.*, 1983), macrophyte algae(Linley *et al.*, 1981; Linley and Newell, 1981; Stuart *et al.*, 1981; Stuart, Newell and Lucas, 1982) and phytoplankton (Newell *et al.*, 1981) incubated in seawater also results in an initial colonisation by bacteria which are subsequently replaced by a complex community of Protozoa. A typical microbial succession which colonises powdered leaf debris from the saltmarsh grass Juncus roemerianus incubated at 10°C in local estuarine water is shown in Figure 1.

It will be noticed that there is an initial phase of up to 7 days in the case of the *Juncus* debris when bacteria were the dominant members of the microbial community. Such bacteria are initially free-living forms which utilise the dissolved organic leachates from the detrital source and are commonly replaced by a dense community of bactivorous flagellates, such as *Monas*, *Oikomonas*, *Bodo* and *Rhynchomonas*. Following the decline in free-living bacteria, a complex community of residual attached bacteria utilising the particulate debris, as well as Protozoa including flagellates, ciliates and choanoflagellates becomes established on

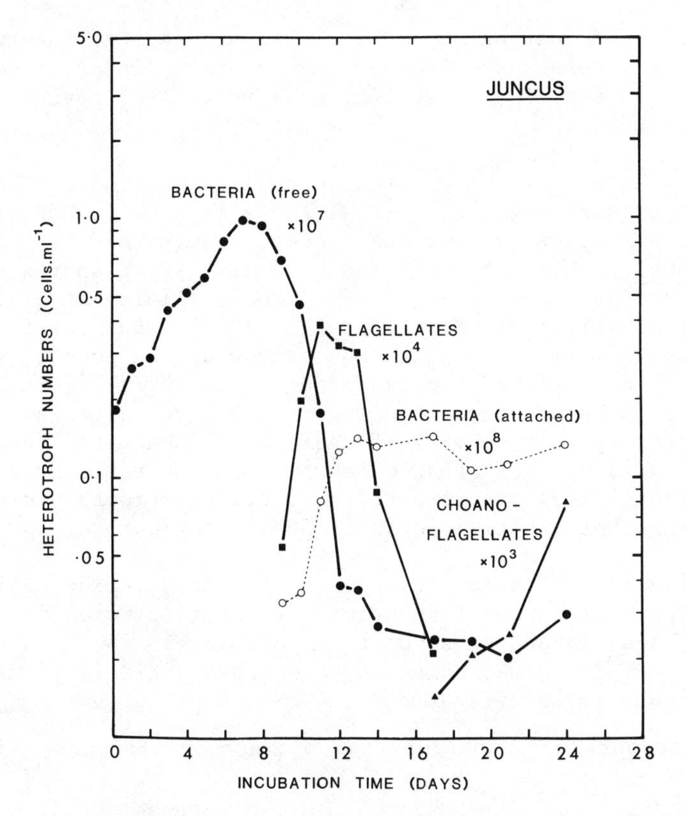

Fig. 1. The numbers of microheterotrophs (cells ml^{-1}) colonising 14.25 mg $C.l^{-1}$ powdered leaf litter from the seagrass *Juncus roemerianus* incubated at 10°C in local estuarine water from Beaufort, North Carolina, USA (data from Newell *et al.*, 1983).

plant detritus from a variety of sources (Linley and Newell, 1981; Newell *et al.*, 1981; Newell *et al.*, 1983).

The First Trophic Step: Bacterial Production and Growth Yields Based on Detritus

The important point from the microbial succession shown in Figure 1 is that there is a brief phase, sometimes lasting for up to 7 days for *Spartina* and *Juncus* debris (Newell *et al.*, 1983.), but generally lasting less than 3 days with algal debris (Linley *et al.*, 1981; Newell *et al.*, 1981) depending on the proportion of dissolved to particulate organic matter, when the bacteria enter the logistic phase of growth. Under these conditions, provided that mortality is negligible, the increase in bacterial biomass per unit time gives the bacterial production based on a natural detrital substrate and measured for several days, rather than hours as in experiments involving incubation of ^{14}C-labelled substrates (for review, see Joint and Morris, 1982).

It should be pointed out that occasional microflagellates must be present even during this initial logistic growth phase of the bacteria since flagellates become common enough to count after 9 days in the *Juncus* incubation media and generally after approximately 3 days in the presence of algal cell debris (Linley *et al.*, 1981; Linley and Newell, 1981; Newell *et al.*, 1981). But careful examination of the samples from which the bacteria were counted has revealed them in densities not exceeding 1.2×10^3 ml^{-1} (with ratios of phototrophs to heterotrophs ranging from 3:1 to 28:1), and despite the high grazing rate of micro-flagellates on bacteria (see next section), it is unlikely that consumption of bacteria by such occasional flagellates significantly affects the growth yield of bacteria during their initial logistic phase of growth.

In practice it is more useful to base energy flow estimates on the proportion of carbon from primary production which is incorporated into bacteria and subsequent trophic levels in the decomposer food chain. This value, which is the growth yield in terms of carbon or carbon conversion efficiency:

$$[\text{carbon incorporated into bacterial biomass. carbon used}^{-1}] \times 100$$

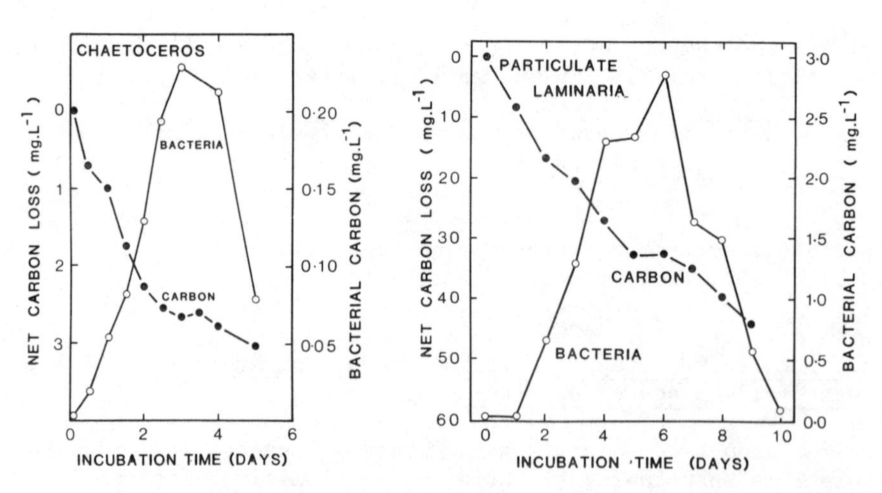

Fig. 2. The carbon utilisation and corresponding increase in the carbon equivalent of bacterial biomass in incubation media containing 111.0 mg $C.1^{-1}$ dried particulate matter from the kelp *Laminaria pallida* and 13.02 mg $C.1^{-1}$ cell debris from the diatom *Chaetoceros tricornutum* (data from Stuart *et al.*, 1982a and Newell *et al.*, 1981).

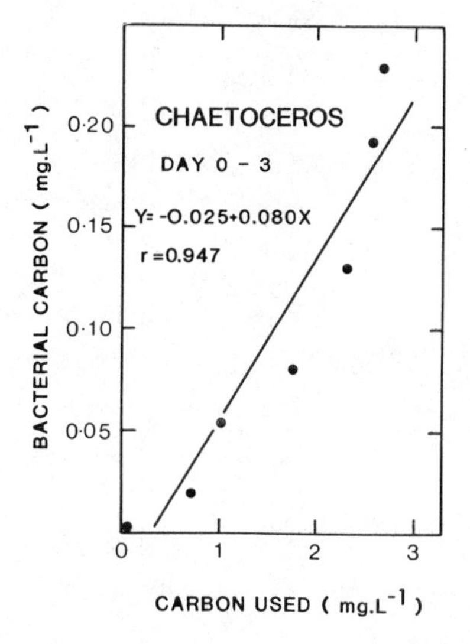

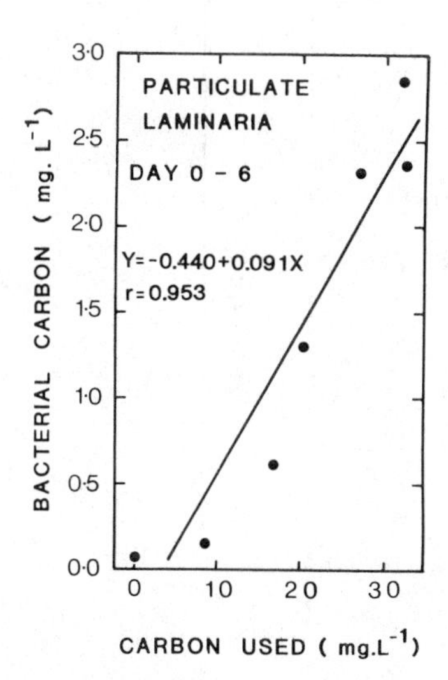

Fig. 3. Bacterial carbon incorporation plotted as a function of carbon utilisation from incubation media containing 111.0 mg $C.1^{-1}$ dried particulate matter from the kelp *Laminaria pallida* and 13.02 mg $C.1^{-1}$ cell debris from the diatom *Chaetoceros tricornutum*. Plotted from Figure 2.

can be obtained from the synchronous measurement of the utilisation of carbon from media containing known additions of detrital material and the simultaneous increase in bacterial biomass and its carbon equivalent. The utilisation of carbon and the corresponding increase in the carbon equivalent of bacterial biomass in media containing kelp exudates and phytoplankton debris incubated in sea-water at 10°C is shown in Figure 2. From this it is apparent that if bacterial carbon incorporation is plotted as function of carbon utilisation as in Figure 3, the slope of the regression gives the growth yield in terms of carbon for bacteria on that detrital source.

Values for both the bacterial carbon production (mg $C.d^{-1}$) and growth yields in experimental media at 10°C in the presence of plant detritus from a variety of sources are summarised in Table 1. Bacterial carbon production is, however, largely dependent on the concentration and ease of degradation of the detrital material available for colonisation. The mean bacterial production at 10°C based on a variety of phytoplankton sources is 0.084 mg $C.1^{-1}$ by estuarine bacteria utilising powdered *Spartina* and *Juncus* leaf litter. When kelp debris is allowed to decompose directly on the strandline, however, a vast production of bacteria, based mainly on the exudation of mannitol occurs (Koop *et al.*, 1982a). High production values based on the incubation of soluble components of kelp mucilage have also been recorded by Lucas *et al.* (1981).

Table 1. The bacterial production (mg $C.d^{-1}$) and growth yield in terms of carbon $\frac{\text{mg C in bacteria}}{\text{mg used from substrate}} \times 100$ from a variety of detrital sources incubated in non-enriched local water at 10°C. Values in parentheses are for systems in which the nitrogen was high (for details see text).

Detrital Source	Concentration mg $C.l^{-1}$	Production mg $C.l^{-1}.d^{-1}$	Growth Yields	Reference
(a) Diatoms				
Thalassiosira angstii	12.12	0.127	13.48	Newell, Lucas and Linley, 1981
Chaetoceros tricornutum	13.02	0.079	8.04	"
Skeletonema costatum	11.86	0.079	8.25	"
(b) Dinoflagellates				
Scrippsiella trochoidea	10.76	0.061	11.99	"
Isochrysis galbana	12.0	0.072	7.60	"
		$\bar{X}$ 0.084	$\bar{X}$ 9.90	
(c) Macroalgae				
Laminaria pallida			9.17	
Mucilage (winter)	9.0	2.5	6.55	Lucas, Newell and Velimirov, 1981
Mucilage (summer)	9.0	-	14.70	"
Particulate	110	0.550	9.11	Stuart, Newell and Lucas, 1982a
Ecklonia maxima	11.0	2.1	6.90	
Mucilage (winter			6.20	Lucas, Newell and Velimirov, 1981
Fronds (on shore)		(34.4)	(28.0)	Koop, Newell and Lucas, 1982a
		$\bar{X}$ 1.717	$\bar{X}$ 8.80	

(d) Seagrasses				
Spartina alterniflora	13.03	0.100	13.20	Newell, Linley and Lucas, 1983
Juncus roemerianus	14.25	0.033	6.12	"
		$\bar{X}$ 0.067	$\bar{X}$ 9.70	
(e) Faeces				
Aulacomya ater	62	(0.425)	(23.7)	Stuart, Newell and Lucas, 1982

In contrast, bacterial carbon yields are relatively uniform despite wide differences in the concentration and ease of degradation of the detrital source. It can be seen that bacterial carbon yields give an average of 9.9% for experimental incubations based on phytoplankton debris, 8.8% for those based on kelp debris and 9.7% for those based on Juncus and Spartina debris. These data are in agreement with those of Robertoson et al. (1982) who estimated the bacterial biomass supported by experimental degradation of powdered leaves of the seagrass Thalassia testudinum and Syringodium filiforme. They found that the carbon equivalent of bacterial biomass represented 4% of the detrital carbon used from Thalassia detritus and 16.9% of that from Syringodium detritus.

These values for natural assemblages of heterotrophic marine bacteria utilising detritus are much lower than those reported for pure strains of bacteria in nutrient enriched media (see Payne, 1970). It is of interest, however, that in a microcosm on the strandline where ammonia values were high (Koop et al., 1982a, b) and also in incubations of mucus-rich fecal material from the mussel Aulacomya ater (Stuart, Newell and Lucas, 1982) the carbon conversion efficiency reached values of 28.0% and 23.7% respectively. High values for carbon conversion have also been reported by Haines and Hanson (1979) for bacterial yields based on saltmarsh plant debris incubated in nitrogen-rich media. This suggests that in the open sea where nitrogen is limiting, the bacteria may oxidise comparatively large amounts of detrital carbon to fix the carbon and nitrogen required for biosynthesis and growth. Where the nitrogen content of the detritus is high, however, or where ammonia regeneration from the sediments supplements the nitrogen available in the detrital source, less carbon is required to incorporate the nitrogen required for biosynthesis and the carbon conversion efficiency, or growth yield, reaches a higher value of approximately 30%.

Conversion Through the Second Trophic Step in the Decomposer Food Chain: Consumption and Growth Yields of Protozoa

One of the characteristic features of the microbial succession which colonises detrital material is that the free-living bacteria decline sharply with the establishment of heterotrophic microflagellates. Although the significance of microflagellates and other Protozoa as potential consumers of marine bacteria has been widely recognised (see Pomeroy and Johannes, 1968; Lighthart, 1969; Chretiennot, 1974; Sorokin, 1977; Haas and Webb, 1979; Sieburth, 1979; Hollibaugh et al., 1980), there has until recently been surprisingly little information on the consumption requirements and energy flow through these organisms which form the second step in the microbial decomposer food chain. Burkill (1978) studied the daily consumption of bacteria by the ciliate Uronema marinum and estimated that a large ciliate of 1000 μm^3 consumes approximately 6 x the body weight per day. Very high consumption rates of up to 500-600 bacteri

Table 2. The uptake of bacteria (mean volume 0.6 μm^3) and cell yield of a variety of heterotrophic micro-flagellates. Also shown is the volume of the flagellates during their growth phase and their daily consumption of bacteria ([volume of of bacteria consumed.volume of flagellates^{-1}].24h^{-1}) as well as the percentage yield of flagellates from their bacterial prey ([volume of flagellates produced. volume of bacteria consumed^{-1}] x 100). From this it can be calculated that the flagellates consume an average 16.98 ± 3.14 times their body volume of bacterial prey per day and that 27.26 ± 4.26% of the bacterial prey volume emerges into the next trophic level as flagellate biovolume (data recalculated from Fenchel, 1982d).

Species	Cell volume during growth (μm^3)	Maximum uptake of bacteria h^{-1}	$\frac{\text{Vol. bact. consumed}}{\text{Vol. flag.}}$ $24h^{-1}$	Numerical Yield $\left(\frac{\text{No. flag.}}{\text{No. bact. consumed}}\right)$	Volumetric Yield $\left(\frac{\text{Vol. flag.}}{\text{Vol. bact. consumed}} \times 100\right)$
Monosiga	20	27	19.44	6.2×10^{-3}	20.67
Pleuromonas	50	54	15.55	3.0×10^{-3}	25.00
Actinomonas	75	107	20.54	2.3×10^{-3}	28.75
Pseudobodo	90	84	13.44	1.8×10^{-3}	27.00
Paraphysomonas	190	254	19.25	9.1×10^{-3}	28.82
Ochromonas	200	190	13.68	1.0×10^{-3}	33.33
			$\bar{X}$ 16.98 ± 3.14		$\bar{X}$ 27.26 ± 4.26

per hour have been reported for Tetrahymena during the log phase of growth (Fenchel, 1975). Linley and Newell (1981) estimated that as much as 35 x the body weight per day of bacteria could be consumed by the small micro-flagellates of 12 μm^3 body volume which colonise incubation media containing kelp exudates. Although this value is almost certainly too high due to a combination of factors including difficulties in estimating consumption rates during a reduction in bacterial prey density at high predator density (see Fenchel, 1982a, b, c), the results do indicate that large quantities of bacteria, ranging from 10-30 x the body weight may be consumed daily by marine microheterotrophs and it is clear that the grazing Protozoa are likely to have a major impact on bacterial biomass in natural waters.

More recently, Fenchel (1982d) has made a detailed study of the feeding rates and bioenergetics of six genera of marine microflagellates of differing cell sizes. His data represent the first quantitative estimates of grazing rates and cell yields through the microflagellates. Some of his data are abstracted into Table 2 which also shows the volume of bacteria consumed. 24 h^{-1} relative to volume of flagellate predators and the yields both in terms of cell numbers and volume of flagellates from bacterial numbers and volumes. It can be seen that despite a wide range of flagellate body volume from 20-200 μm^3, the daily consumption rate of bacterial prey was approximately 17 x that of the flagellate predator volume. Of this 20-33% (mean 27.3 ± 4.3 s.d) was incorporated into flagellate biovolume. This increase in ecological efficiency at higher trophic levels accords well with what is known for other ecosystems and in the grazing Protozoa to some extent offsets the reduced energy available for conversion at higher trophic levels (see McNeill and Lawton, 1970; Humphreys, 1979).

The energetics of decomposition of natural plant debris thus suggests that under the nitrogen-limited conditions which commonly occur in the sea, 10-15% of the detrital carbon is likely to become incorporated into bacterial biomass and of this, 27% is incorporated into microflagellates. Carbon oxidation through the decomposer system based on detritus is thus as much as 85% by bacteria and a further 11% by micro-flagellates, giving a total carbon oxidation of 96% through the first two steps of the microheterotrophic decomposer food chain.

DETRITUS AS A CARBON RESOURCE FOR NEARSHORE BENTHIC COMMUNITIES

The 'energy cascade' through two steps in the detrital food chain from plant debris to bacteria and from bacteria to microflagellates allows some basic estimates of the role of decomposers in the ecological energetics of nearshore communities. A good deal

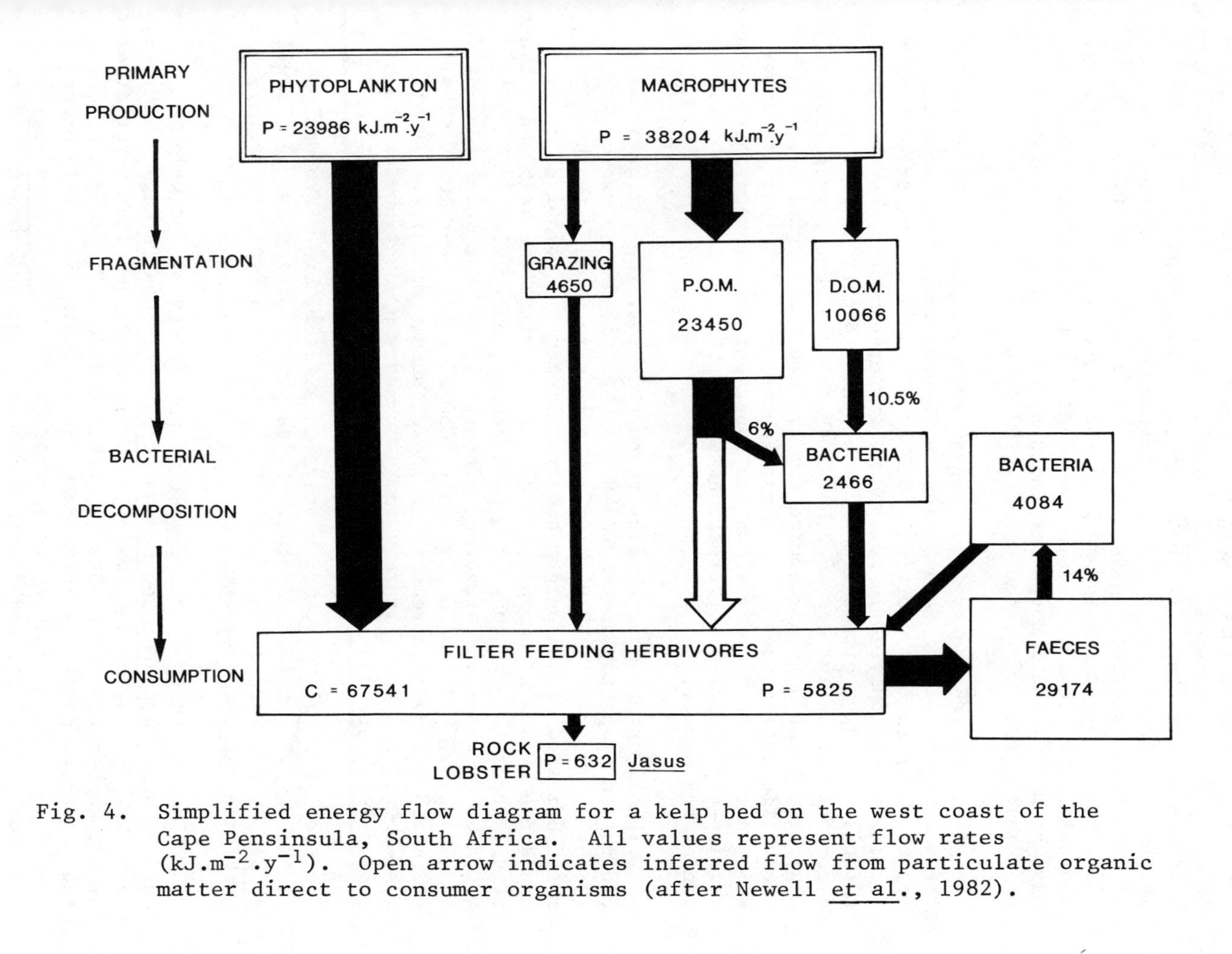

Fig. 4. Simplified energy flow diagram for a kelp bed on the west coast of the Cape Pensinsula, South Africa. All values represent flow rates ($kJ.m^{-2}.y^{-1}$). Open arrow indicates inferred flow from particulate organic matter direct to consumer organisms (after Newell *et al.*, 1982).

is now known, for example, of both the primary production and consumer requirements in the kelp bed communities of the Cape Peninsula, South Africa (see Newell et al., 1982; Newell and Field, 1982). This system may be rather different from detritus communites of saltmarshes and estuaries in that the fronds of kelp fragment from the tip and this material is available in a freshly-fragmented form to the consumer organisms for much of the year, whereas in saltmarshes and estuaries the particulate component may be more refractory and have been recycled through the feces several times by deposit feeding organisms (for review, see Newell, 1979).

A simplified energy flow diagram for an idealised kelp bed off the west coast of the Cape Peninsula, South Africa, is shown in Figure 4. A small proportion (4650 kJ m^{-1} y^{-1}) of the primary production by macrophytes is removed by grazers and the rest of the material can flow either via fragmentation to microheterotrophic decomposer organisms or, together with the resident phytoplankton, could represent a food resource for the consumer organisms. Conversion efficiencies used for the particulate and dissolved fractions of kelp, and for feces, are shown and represent the average carbon conversion efficiencies for experimental incubations carried out in winter and summer.

The first and most obvious feature is that the sum of primary production by phytoplankton (23986 kJ m^{-2} y^{-1}) and by macrophytes (38204 kJ m^{-2} y^{-1}) = 62190 kJ m^{-2} y^{-1} is very similar to the estimated combined carbon consumption requirements of 67541 kJ m^{-2} y^{-1} of the dense consumer community which characterises the kelp bed. Even if all the kelp were converted to bacteria, only 2466 kJ m^{-2} y^{-1} would be obtained and, together with the yield from fecal degradation of 4084 kJ m^{-2} y^{-1} would give only 6550 kJ m^{-2} y^{-1} or 10.5% of all estimated energy requirements of the consumer community. It will also be noticed from Figure 4 that secondary production for the consumers which are dominated by the mussel *Aulocomya ater*, is 5825 kJ m^{-2} y^{-1} or 9.3% of primary production and that the rock lobster *Jasus lalandii*, which preys on *A. ater*, has a production of 632 kJ m^{-2} y^{-1} or 10.8% of that of its prey (for review, see Newell et al., 1982) much as we would anticipate for a typical plant-herbivore-carnivore system.

Direct studies on the digestive physiology of the kelp bed mussels *Aulocomya ater* and *Choromytilus meridionalis* suggest that these bivalves are indeed able to utilise kelp detrital material as a direct carbon resource. Stuart, Field and Newell (1982) have shown that although bacterial cultures based on isolates of kelp bacteria can be absorbed with an efficiency of 67-70% by *Aulocomya ater* (see also Sorokin, 1972), the detritus itself is also absorbed with an efficiency of approximately 50%. Again, Seiderer et al. (1982) has shown that the crystalline style of the mussels *Choromytilus meridionalis* and *Perna perna* possess an array of digestive enzymes including α-amylase, cellulase, laminarinase and alginate lyase which are capable of liberating sufficient carbon from the ingested detrital food to meet the carbon requirements of the mussels.

Table 3 shows the components of the carbon budget for *Choromytilus* and *Perna* together with the carbon released by the enzyme activity of whole styles. From this, the style turnover time required to meet the carbon requirements of the mussels was estimated to be approximately 25h for *C. meridionalis* and 136h for *P. perna*, both of which accord well with the turnover time estimated from loss of ^{14}C from labelled styles (Seiderer *et al.*, 1982).

The implication from both experimental studies on the energetics of microbial conversion through the decomposer food chain and from the ecological energetics of the community as a whole is, therefore, that the significance of bacteria as a carbon resource for consumer organisms is small compared with that available from freshly-fragmented substrate. The filter-feeders which characterise kelp beds are thus analogous to herbivores in the sense that energy flow is likely to be dominantly from plants even though it is in a fragmented form, and the consumer bivalves which dominate the community evidently possess the necessary digestive enzymes to quantitatively meet their carbon requirements by digestion of the detrital material.

Several studies of the standing stock of bacteria in sediments have also shown that the carbon requirements of the consumer organisms exceed that available in bacteria. Cammen *et al*. (1978) and Cammen (1980a, b), for example, have pointed out that bacterial carbon is generally only 1-3% of total sediment organic carbon (Dale, 1974; Sorokin, 1978; Kepkay and Novitsky, 1980) and this is calculated to be capable of meeting only 25% of the estimated carbon requirements of the polychaete *Nereis succinea* (see also Baker and Bradnam, 1976; Tunnicliffe and Risk, 1977; Wetzel, 1977; Jensen and Siegismund, 1980). Production to biomass ratios for bacteria may, however, be high and Hargrave (1970a, b, 1971) has estimated that ingestion of less than 10% of the daily microflora production is sufficient to meet the energy requirements of the amphipod *Hyalella azteca*.

Bacterial production may also be of considerable importance in supplementing the immobilised nitrogen available to consumer organisms, especially since the protein concentration in detritus is often low (see also Newell, 1965, 1979) and the conversion of nitrogen from detritus into bacteria appears to be achieved with an efficiency of as much as 94% by the heterogeneous population of marine bacteria which colonise kelp debris (Koop *et al.*, 1982b). Newell and Field (1983) have, for example, estimated that the microbial population associated with the decomposition of kelp debris could contribute as much as 69% of the nitrogen requirements of the consumer community.

Table 3. Components of the carbon budget of the mussels *Choromytilus meridionalis* and *Perna perna* and the corresponding carbon liberated by the enzyme activity of whole styles at 18°C. From this, the style turnover time required to meet the carbon requirements of the mussels can be calculated and agrees well with independent estimates of style turnover (data from Seiderer *et al*., 1982).

Shell Length mm	Respiration (R) (mg $C.h^{-1}$ at 18°C)	Absorbed Ration (A) (mg $C.h^{-1}$ at 18°C)	Carbon liberated by style at 18°C (mg $C.h^{-1}$)	Style turnover time (h) required to meet carbon requirements (A)
Choromytilus meridionalis				
20	0.0348	0.0470	0.780	16.59
40	0.1179	0.1593	4.768	29.93
60	0.2089	0.2823	8.756	31.02
80	0.3589	0.4850	12.744	26.28
100	0.5357	0.7239	16.731	23.11
				$\bar{X}$ 25.39 ± 5.82
Perna perna				
20	0.0120	0.0162	4.011	247.59
40	0.0501	0.0677	9.215	136.12
60	0.1156	0.1562	14.410	92.25
80	0.2089	0.2823	19.610	69.47
				$\bar{X}$ 136.35 ± 79.14

THE SIGNIFICANCE OF MICROHETEROTROPHS IN ENERGY FLOW THROUGH OFFSHORE PELAGIC SYSTEMS

The energy flow model which I have described related primarily to a freshly-fragmented kelp detrital system. Unfortunately there is much less information on detritus formation and consumer requirements in other ecosystems despite the intensive studies which have been made on Saltmarsh-estuarine ecosystems (for review, see Nixon, 1980) and planktonic food webs (see Steele, 1974; Joint and Morris, 1982). The apparent uniformity of values which are obtained for the bacterial carbon yield based on plant detritus (see Table 1) and the similar uniformity of conversion by micro-flagellates of a wide variety of sizes (see Fenchel, 1982d; also Table 2), however, suggest that some common principles may be usefully applied to the energetics of microbial conversion of plant detritus in other systems.

If we take, for example, a planktonic system which is dominated by herbivores, we could assume that as much as 80% of primary production is consumed by grazing herbivores, perhaps 10% of the non-grazed particulate matter sediments below the thermocline and a further 10% is decomposed in the surface waters. Now the absorption efficiency of the herbivores is likely to be approximately 50% whilst that of the carnivores is likely to be higher at approximately 70%. We can thus estimate the fecal production and calculate the conversion of both plant debris and fecal material through the microbial decomposer food chain, and can also calculate the oxidation of carbon through such a herbivore-dominated system.

A carbon flow diagram for such a system is shown in Figure 5. The important feature is that although energy flow is through a typical plant-herbivore-carnivore system, microbial oxidation of the carbon from primary production is approximately 49%. It will also be noted that despite the fact that energy flow is initially into herbivores, production by the bacteria exceeds that for the herbivores if we use a value of 23% for the cell yield (see above) and is comparable with that of the herbivores even if we use the low values of 10-15% for the carbon conversion efficiency of bacteria on plant detritus (see Table 1). Again because the micro-flagellate yield is as much as 27% of the bacterial prey (see Table 2), it follows that even in a herbivore dominated system production by micro-flagellates is likely to be of comparable significance or even exceed that of first-order carnivores.

These features of a herbivore dominated system such as occurs in open oceanic waters are of considerable interest in view of the observations of Sorokin (1971a, b, 1973; see also Williams, 1981) that the biomass and production of bacteria in the tropical Pacific are very high. Although these results have been criticised (see Banse, 1974), other workers have reported that bacterial production may be much greater than is generally assumed. Sieburth *et al.*

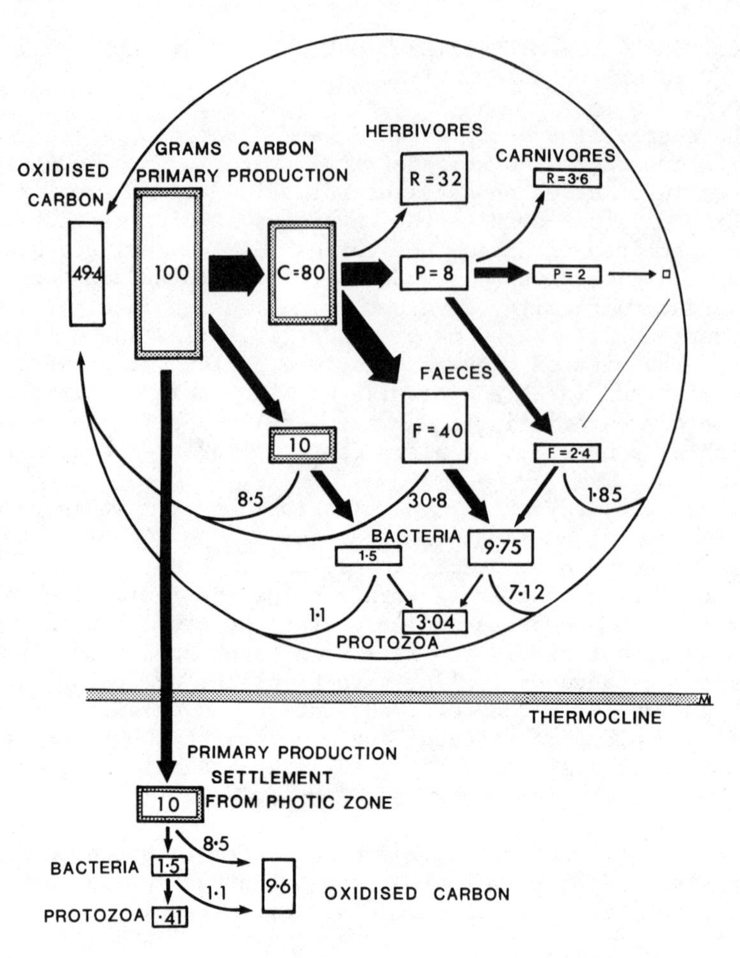

Fig. 5. Schematic flow diagram showing the heterotrophic conversion of carbon from primary production in a herbivore dominated pelagic system. Of the 100g carbon from primary production , 80g is assumed to be consumed directly by herbivores, 10g of the non-grazed particulate matter sinks below the thermocline and 10g is decomposed in the surface layers. Absorption efficiency of the herbivores is assumed to be 50%, and that of the carnivores 70%. Bacterial carbon conversion of plant detrital material is 23% (see Table 1) whilst the efficiency of transfer from bacteria to flagellates is 27%. It will be noted that the feces have been assumed to decompose in the surface waters, but it is recognised that a variable proportion of the fecal material may sediment below the thermocline. C in the figure = consumption; R = respiratory losses; P = production; and F = feces.

(1976) suggested that 30-40% of the plankton biomass in the North Atlantic is composed of bacteria. A recent survey of the carbon budget for the water column of the western approaches to the English Channel has also shown that the standing stocks of bacteria and microflagellates are often high compared with those of the larger consumers and may actually exceed those of copepods (Holligan *et al.*, in press).

Because the metabolism of heterotrophs is related to surface area (see Hemmingsen, 1960), it follows that carbon flow through the microheterotrophic community is likely to be much greater than that through a similar biomass of larger heterotrophs. It is of interest, therefore, to calculate whether carbon flow through the first two steps of the microheterotrophic decomposer food chain could, in fact, be met from *in situ* primary production by the phytoplankton. From Figure 1 it can be seen that the maximum flagellate numbers in detritus-enriched cultures were 0.4 x 10^3 cells.ml^{-1}. Their mean volume was 10 μm^3 (see also Linley *et al.*, 1981; Linley and Newell, 1981) and the carbon equivalent of their wet biomass is estimated to be 0.44 µg C.1^{-1}. From Table 2 we see that their consumption requirements are approximately 17 x their body volume of bacteria per day, or approximately 7.48 µg C.1^{-1}.d^{-1}. Since the bacterial carbon conversion efficiency (Table 1) is approximately 10% from the detrital substrate, a daily detrital carbon production of 74.8 µg C1^{-1}.d^{-1} would be required to support the population of flagellates in the culture. In an upwelling region outside a kelp bed on the west coast of the Cape Peninsula, South Africa, phytoplankton primary production ranged from 240-3600 µg C.1^{-1}.d^{-1} during January, with a mean annual value of 225 g C.1^{-1}.d^{-1} (Carter, 1982). It is thus clear that in upwelling regions, primary production is well in excess of the likely requirements of the microheterotrophic consumer community. Even in regions where primary production is lower, and where the biomass of heterotrophic microflagellates is correspondingly lower, it seems likely that carbon fixation by primary production is in excess of the carbon requirements of the microheterotrophic community, even though as much as 96% of the carbon is oxidised in the first two steps of the decomposer food chain.

The estimates for microheterotrophic production based on the carbon flow summarised in Figure 5 thus suggest that the high values for bacterial production which have been recorded in the water column could be sustained even from the relatively low yields obtained from direct estimates of the growth of bacteria on algal debris. The role of bacteria as a potential food resource for higher trophic levels in pelagic systems is thus likely to be at least as important as that of herbivores even in a herbivore-dominated system, especially if they are associated with aggregated particulate material and can thus be retained by the filtration structures of larger consumer organisms (see also Linley and Field, 1982; Robertson *et al.*, 1982).

In more eutrophic waters the episodic blooms of phytoplankton may exceed consumer requirements and under these conditions a greater proportion of primary production flows directly into the decomposer system (see for example Bodungen et al., 1981). In this case bacterial production greatly exceeds herbivore production because the bacteria also receive feces as a substrate in addition to the decomposing phytoplankton cells. An interesting feature of this contra ing situation is, however, that carbon oxidation is also approximately 50% because in both cases carbon flow is predominantly through the decomposers whether via the feces or directly following a phytoplankton bloom situation.

It thus seems likely that in pelagic systems approximately 50% of the carbon from primary production is incorporated into the heterotrophic community as a whole despite wide differences in the proportion of larger grazing consumer organisms in the water column. The main factor which will control whether the products of remineralisation can be used to promote further primary production is whether significant decomposition can be achieved before the detritus sinks below the thermocline. Certainly in the case of the 30-40% which comprises the 'dissolved' fraction of algal cells (Newell et al., 1980, 1981) carbon oxidation and associated mineralisation of plant nutrients is likely to be achieved within days or even less (see Larsson and Hagström, 1979; Lucas et al., 1981; Newell and Lucas, 1981; Joint and Morris, 1982).

To a large extent, therefore, differences in primary production in near-shore waters and in oligotrophic oceanic systems may be related to differences in the site of microbial regeneration rather than in the rate or efficiency of conversion by microbial decomposer organisms. In shallow coastal waters, embayments and lagoons, much of the material regenerated from the microbial decomposer succession is potentially available to support further primary production, whilst in deeper waters only the more rapidly decomposed labile fraction is likely to be remineralised in the surface waters. In both systems, however, the ecological energetics of microbial decomposition suggests that the yield of bacteria and microflagellates in the water column is comparable to, or may even exceed that of the herbivores and first-order carnivores respectively. Apart from their ability to oxidise at least 50% of the carbon through two steps in the decomposer food chain, the microheterotrophic community thus represents a significant food resource for those consumer organisms capable of exploiting small particles or bacterio-organic complexes in the water column.

Acknowledgements

This work was supported by a Senior Research Fellowship of the Royal Society. I wish to thank my colleagues Dr. M. I. Lucas and

Ms E. A. S. Linley for their helpful discussions and permission to cite some of our unpublished data. I am also grateful to Professor T. Fenchel for kindly making some of his unpublished data available to me and for his permission to cite these in Table 2, and to Professor John Gray and Dr. Peter Williams for their helpful comments of the mauscript.

REFERENCES

Allen, H. L. , 1973, Dissolved organic carbon: patterns of utilisation and turnover in two small lakes, Int.Revue Ges.Hydrobiol., 58:617.

Anderson, J. M., 1975, Succession, diversity and trophic relationships of some soil animals in decomposing leaf litter, J.Anim. Ecol., 44:475.

Andrews, P., and Williams, P. J. le B., 1971, Heterotrophic utilization of dissolved organic compounds in the sea. III. Measurements of the oxidation rate and concentration of glucose and amino acids in seawater, J.Mar.Biol.Ass.U.K, 51:111.

Azam, F., and Hodson, R. E., 1981, Multiphasic kinetics for D-glucose uptake by assemblages of natural marine bacteria, Mar.Ecol. Prog.Ser., 6:213.

Banse, K., 1974, On the role of bacterioplankton in the tropical ocean, Mar.Biol., 24:1.

Baker, J. H., and Bradnam, L. A., 1976, The role of bacteria in the nutrition of aquatic detritivores, Oecologia (Berl.), 24:95.

Billen, G., Joiris, C., Wijnant, J., and Gillian, G., 1980, Concentration and microbial utilization of small organic molecules in the Scheldt Estuary the Belgian coastal zone of the North Sea and the English Channel, Estuar.Coast.Mar.Sci., 11:279.

Bodungen, B. V., Bröckel, K. V., Smetacek, V., Zeitzschel, B., 1981, Growth and sedimentation of the phytoplankton spring bloom in the Bornholm Sea (Baltic Sea), Kieler Meeresforsch., 5:49.

Burkill, P. H., 1978, Quantitative aspects of the ecology of marine planktonic ciliated protozoans with special reference to Uronema marinum Dujardin, Ph.D Thesis, Univ. of Southampton.

Calkins, G. N., and Summers, F. M., (eds.), 1941, in: "Protozoa in Biological Research," Columbia University Press, New York.

Calow, P., 1977, Conversion efficiencies in heterotrophic organisms, Biol.Rev., 52:385.

Cammen, L. M., 1980a, Ingestion rate: an empirical model for aquatic deposit feeders and detritovores, Oecologia (Berl.), 44:303.

Cammen, L. M., 1980b, The significance of microbial carbon in the nutrition of the deposit feeding polychaete Nereis succinea, Mar.Biol., 61:9.

Cammen, L. M., Rublee, P., and Hobbie, J. E., 1978, The significance of microbial carbon in the nutrition of the polychaete Nereis succinea and other aquatic deposit feeders, Sea Grant Pub. UNC-SG-78-12, North Carolina State University, Raleigh.

Carter, R. A., 1982, Phytoplankton biomass and production in a Southern Benguela kelp bed system, Mar.Ecol.Prog.Ser., 8:9.

Chretiennot, M. J., 1974, Nanoplancton de flaques supralittorales de lat region de Marseille, Protistologica, 10:477.

Crawford, C. C., Hobbie, J. E., and Webb, K. L., 1974, The utilization of dissolved free amino acids by estuarine microorganisms, Ecology, 55:551.

Dale, N. C., 1974, Bacteria in intertidal sediments, factors related to their distribution, Limnol.Oceanogr., 19:509.

Daley, R. J., 1979, Direct epifluorescence enumeration of native aquatic bacteria: uses, limitations and comparative accuracy, in: "Native Aquatic Bacteria: enumeration, activity and ecology," J. W. Costerton and R. R. Colwell, eds., Amer. Soc. for Testing and Materials.

Darnell, R. M., 1967 , The organic detritus problem, in: "Estuaries," G. H. Lauff, ed., Publs. Am. Ass. Advant Sci.

Darnell, R. M., 1967b, Organic detritus in relation to the estuarine ecosystem, in: "Estuaries," G. H. Lauff, ed., Publs. Am. Ass. Advant Sci.

Day, J. W., Smith, W. G., Wagner, P. R. and Stowe, W. C., 1973, Community structure and carbon budget of a salt marsh and shallow bay estuarine system in Louisiana, Center for Wetland Resources, Louisiana State Univ. Publ. LSU-SG-72-04.

Dempsey, M. J., 1981, Marine bacterial fouling: a scanning electron microscope study, Mar.Biol., 61:305.

Fenchel, T., 1970, Studies on the decomposition of organic detritus derived from the turtle grass Thalassia testudinum, Limnol. Oceanogr., 15:14.

Fenchel, T., 1975, The quantitative importance of the benthic microfauna of an arctic tundra pond, Hydrobiologia, 46:445.

Fenchel, T., 1982a, Ecology of heterotrophic microflagellates. I. Some important forms and their functional morphology, Mar. Ecol.Prog.Ser., 8:211.

Fenchel, T., 1982b, Ecology of heterotrophic microflagellates. II. Bioenergetics and growth, Mar.Ecol.Prog.Ser., 8:225.

Fenchel, T., 1982c, Ecology of heterotrophic microflagellates. III. Adaptations to heterogeneous environments, Mar.Ecol.Prog.Ser., 9:25.

Fenchel, T., 1982d, Ecology of heterotrophic microflagellates. IV. Quantitative occurrence and importance as consumers of bacteria, Mar.Ecol.Prog.Ser., 9:35.

Fenchel, T. M., and Jorgensen, B. B., 1976, Detritus food chains of aquatic ecosystems and the role of bacteria, Adv.Microb. Ecol., 1:1.

Ferguson, R. L., and Murdoch, M. B., 1978, Microbiology of the Newport River Estuary. Atlantic Fisheries Center Ann. Rept. to Atomic Energy Commission, NOAA National Marine Fisheries Service, Beaufort, N.N. 63-83.

Field, J. G., Jarman, N. G., Dieckmann, G. S., Griffiths, C. L., Velimirov, B., and Zoutendyk, P., 1977, Sun, waves, seaweed and lobsters: the dynamics of a west coast kelp bed, S.Afr.J. Sci., 73:7.

Fuhrman, J. A., 1981, Influence of method on the apparent size distribution of bacterioplankton cells: Epifluorescence microscopy compared to scanning electron microscopy, Mar.Ecol. Prog.Ser., 5:103.

Gordon, G. C., Robinson, G. G. C., Hondzel, L. L., and Gillespie, D. C., 1973, A relationship between heterotrophic utilization of organic acids and bacterial populations in west Blue Lake, Manitoba, Limnol.Oceanogr., 18:264.

Gosselink, J. G., and Kirby, C. J., 1974, Decomposition of salt marsh grass, Spartina alterniflora Loisel, Limnol.Oceanogr., 19:825.

Haas, L. W., and Webb, K. L., 1979, Nutritional mode of several non-pigmented microflagellates from the York River Estuary, Virginia, J.Exp.Mar.Biol.Ecol., 39:125.

Haines, E. B., and Hanson, R. B., 1979, Experimental degradation of detritus made from the salt marsh plants Spartina alterniflora L., and Juncus roemerianus Scheele, J.Exp.Mar.Biol.Ecol., 40:27.

Hargrave, B. T., 1970a, The utilization of benthic microflora by Hyalella azteca (Amphipoda), J.Anim.Ecol., 39:427.

Hargrave, B. T., 1970b, The effect of a deposit-feeding amphipod on the metabolism of benthic microflora, Limnol.Oceanogr., 51:21.

Hargrave, B. T., 1971, An energy budget for a deposit feeding amphipod, Limnol.Oceanogr., 16:19.

Hemmingsen, A. M., 1960, Energy metabolism as related to body size and respiratory surfaces and its evolution, Rep.Steno.Mem. Hosp.Copenhagen, 9:7.

Hobbie, J. E., 1967, Glucose and acetate in freshwater: concentrations and turnover rates, in: "Chemical environment in aquatic habitat," H. L. Gotterman and R. S. Clymo, eds., North Holland Publishing Co., Amsterdam.

Hobbie, J. E., and Crawford, C. C., 1969, Respiration corrections for bacterial uptake of dissolved organic compounds in natural waters, Limnol.Oceanogr., 14:528.

Hobbie, J. E., Daley, R. T., and Jasper, S., 1977, Use of nuclepore filters for counting bacteria by fluorescence microscopy, Appl.Environ.Microbiol., 33:1225.

Hollibaugh, J. T., Fuhrman, J. A., and Azam, F., 1980, Radioactively labeling of natural assemblages of bacterioplankton for use in trophic studies, Limnol.Oceanogr., 25:172.

Holligan, P. M., Harris, R. P., Head, R. N., Linley, E. A. S., Lucas, M. I., Newell, R. C., Tranter, P. R. G., and Weekley, C. M., The partitioning of organic carbon in mixed, frontal and stratified waters of the English Channel, Mar. Ecol.Prog.Ser., (in press)

Humphreys, W. F., 1979, Production and respiration in animal populations, J.Anim.Ecol., 48:427.

Jensen, K. T., and Siegismund, M. R., 1980, The importance of diatoms and bacteria in the diet of Hydrobia species, Ophelia 19 (Suppl.):193.

Joint, I. R., and Morris, R. J., 1982, The role of bacteria in the turnover of organic matter in the sea, Oceanogr.Mar.Biol.Ann. Rev., 20:65.

Kaushik, N. K., and Hynes, H. B. N., 1971, The fate of dead leaves that fall into streams, Arch.Hydrobiol., 68:465.

Keefe, C. W., 1972, Marsh production: A summary of the literature, Contrib.Mar.Sci., 16:163.

Kepkay, P. E., and Novitsky, J. A. 1980, Microbial control of organic carbon in marine sediments: coupled chemoautotrophy and heterotrophy, Mar.Biol., 55:261.

Koop, K., Newell, R. C., and Lucas, M. I., 1982a, Biodegradation and carbon flow based on kelp (Ecklonia maxima) debris in a sandy beach microcosm, Mar.Ecol.Prog.Ser., 7:315.

Koop, K., Newell, R. C. and Lucas, M. I., 1982b, Microbial regeneration of nutrients from the decomposition of macrophyte debris on the shore, Mar.Ecol.Prog.Ser., 9:91.

Larsson, V., and Hagström, A., 1979, Phytoplankton exudate release as an energy source for the growth of pelagic bacteria, Mar. Biol., 52:199.

Lighthart, B., 1969, Planktonic and benthic Bactivorous protozoa at eleven stations in Puget Sound and adjacent Pacific Ocean, J.Fish.Res.Bd.Can., 26:299.

Linley, E. A. S., and Field, J. G., 1982, The nature and ecological significance of bacterial aggregation in a nearshore upwelling ecosystem, Est.Coastal Shelf Sci., 14:1.

Linley, E. A. S., and Newell, R. C., 1981, Microheterotrophic communities associated with the degradation of kelp debris, Kieler Meeresforsch., 5:345.

Linley, E. A. S., Newell, R. C., and Bosma, S. A., 1981, Heterotrophic utilisation of mucilage released during fragmentation of kelp (Ecklonia maxima and Laminaria pallida). 1. Development of microbial communities associated with the degradation of kelp mucilage, Mar.Ecol.Prog.Ser., 4:31.

Lousier, J. D., 1982, Colonisation of decomposing deciduous leaf litter by Testacea (Protozoa, Phizopoda): Species succession abundance, and biomass, Oecologia (Berl.), 52:381.

Lucas, M. I., Newell, R. C., and Velimirov, B., 1981, Heterotrophic utilisation of mucilage released during fragmentation of kelp (Ecklonia maxima and Laminaria pallida). 2. Differential utilisation of dissolved organic components from kelp mucilage, Mar.Ecol.Prog.Ser., 4:43.

Luria, S. E., 1960, The bacterial protoplasm: composition and organisation, in: "The bacteria," Vol. 1, I. C. Gunsalus and R. Y. Stanier, eds., Academic Press, New York.

Mann, K. H., 1972, Macrophyte production and detitus food chains in coastal waters, Mem.1st.Ital.Idrobiol., 29 (Suppl.):353.

Mann, K. H., 1973, Seaweeds: their productivity and strategy for growth, Science, N.Y. , 182:975.

McNeill, S., and Lawton, J. M., 1970, Annual production and respiration in animal populations, Nature, Lond., 225:472.

Newell, R. C., 1965, The role of detritus in the nutrition of two marine deposit-feeders, the prosobranch Hydrobia ulvae and the bivalve Macoma balthica, Proc.Zool.Soc.Lond., 144:25.

Newell, R. C., 1979, "Biology of intertidal animals," Marine Ecological Surveys, Faversham, Kent.

Newell, R. C., and Field, J. G., 1983, The contribution of bacteria and detritus to carbon and nitrogen flow in the benthic community, Mar.Biol.Letters, 4:23.

Newell, R. C., Field, J. G., and Griffiths, C. L., 1982, Energy balance and significance of micro-organisms in a kelp bed sommunity, Mar.Ecol.Prog.Ser., 8:103.

Newell, R. C., Linley, E. A. S., and Lucas, M. I., 1983, Microbial production and carbon conversion based on saltmarsh plant debris, Est.Coast Shelf Sci., (in press).

Newell, R. C., and Lucas, M. I., 1981, The quantitative significance of dissolved and particulate organic matter released during fragmentation of kelp in coastal waters, Kieler Meeresforsch., 5:356.

Newell, R. C., Lucas, M. I., and Linley, E. A. S., 1981, Rate of degradation and efficiency of conversion of phytoplankton debris by marine micro-organisms, Mar.Ecol.Prog.Ser., 6:123.

Newell, R. C., Lucas, M. I., Velimirov, B., and Seiderer, L. J., 1980, Quantitative significance of dissolved organic losses following fragmentation of kelp (Ecklonia maxima and Laminaria pallida), Mar.Ecol.Prog.Ser., 2:45.

Nixon, S. W., 1980, Between coastal marshes and coastal waters - a review of twenty years of speculation and research on the role of salt marches in estuarine productivity and water chemistry, in: "Estuarine and Wetland Processes," P. Hamilton and K. B. MacDonald, eds., Plenum Publ. Corp., N.Y.

Odum, E. P., 1971, "Fundamentals of Ecology," 3rd edn., W. B. Saunders, San Francisco.

Ogura, N., 1972, Rate and extent of decomposition of dissolved organic matter in surface seawater, Mar.Biol., 13:89.

Ogura, N., 1975, Further studies on decomposition of dissolved organic matter in coastal seawater, Mar.Biol., 31:101.

Ogura, N., and Gotoh, T., 1974, Decomposition of dissolved carbohydrates derived from diatoms of Lake Yuno-Ko, Int.Revue.Ges. Hydrobiol., 59:39.

Park, D., 1972, Methods of detecting fungi in organic detritus in water, Trans.Br.Mycol.Soc., 58:281.

Payne, W. J., 1970, Energy yields and growth of heterotrophs, Ann. Rev.Microbiol., 24:17.

Payne, W. J., and Wiebe, W. J., 1978, Growth yield and efficiency in chemosynthetic microorganisms, Ann.Rev.Microbiol., 32:155.

Pomeroy, L. R. and Johannes, R. E., 1968, Occurrence and respiration of ultraplankton in the upper 500 metres of the ocean, Deep-Sea Res., 51:381.

Robb, F. T., Davies, B. R., Cross, R., Kenyon, C., and Howard-Williams, C., 1979, Cellulolytic bacteria as primary colon-

izers of Potamogeton pectinatus L. (Sago pond weed) from a brackish wouth-temper-te coastal lake, Microb.Ecol., 5:167.
Robertson, M. L., Mills, A. L., and Zieman, J. C., 1982, Microbial synthesis of detritus-like particulates from dissolved organi carbon released by tropical seagrass, Mar.Ecol.Prog.Ser., 7:279.
Seiderer, L. J., Newell, R. C., and Cook, P. A., 1982, Quantitative significance of style enzymes from two marine mussels (Choromytilus meridionalis Krauss and Perna perna Lamark) in relation to diet, Mar.Biol. Letters, 3:257.
Sieburth, J. McN., 1979, "Sea Microbes, " Oxford Univ. Press, N.Y.
Sieburth, J. McN., Willis, P. J., Johnson, K. M., Burney, C. M., Lavoie, D. M., Hinge, K. R., Caron, D. A., French, F. W., Johnson, P. W., and Davies, G., 1976, Dissolved organic matte and heterotrophic microneuston in the surface microlayers of the North Atlantic, Science, N.Y., 194:1415.
Sorokin, Y. I., 1971a, On the role of bacteria in the productivity o tropical oceanic waters, Int.Revue,Ges.Hydrobiol., 56:1.
Sorokin, Y. I., 1971b, Bacterial populations as components of oceanic ecosystems, Mar.Biol., 11:101.
Sorokin, Y. I., 1972, Bacteria as food for coral reef fauna, Oceanology, 12:169.
Sorokin, Y. I., 1973, Data on biological productivity of the western tropical Pacific Ocean, Mar.Biol., 20:177.
Sorokin, Y. I., 1977, The heterotrophic phase of plankton succession in the Japan Sea, Mar.Biol., 41:107.
Sorokin, Y. I., 1978, Microbial production in the coral-reef community, Arch.Hydrobiol., 83:281.
Sorokin, Y. I., and Kadota, H. eds., 1972, "Techniques for the assessment of microbial production and decomposition in fresh water," I.B.P. Handbook (23), Blackwell, Oxford.
Steele, J. H., 1974, "The Structure of Marine Ecosystems," Harvard University Press, Cambridge, Mass.
Stuart, V., Field, J. G., and Newell, R. C., 1982, Evidence for the absorption of kelp detritus by the ribbed mussel, Aulocomya ater (Molina), using a new ^{51}Cr-labelled microsphere technique, Mar.Ecol,Prog.Ser., 9:263.
Stuart, V., Lucas, M. I., and Newell, R. C., 1981, Heterotrophic utilisation of particulate matter from the kelp Laminaria pallida, Mar.Ecol.Prog.Ser., 4:337.
Stuart, V., Newell, R. C., and Lucas, M. I., 1982, Conversion of kelp debris and faecal material from the mussel Aulocomya ater by marine microorganisms, Mar.Ecol.Prog.Ser., 7:47.
Suberkropp, K., and Klug, M. J., 1974, Decomposition of deciduous leaf litter in a woodland stream. 1. A scanning electron microscope study, Microb.Ecol., 1:96.
Suberkropp, K., and Klug, M. J., 1976, Fungi and bacteria associated with leaves during processing in a woodland stream, Ecology, 57:707.
Teal, J. M., 1962, Energy flow in the salt marsh ecosystem of Georgia, Ecology, 43:614.

Troitsky, A. S., and Sorokin, Y. I., 1967, On the methods of calculation of the bacterial biomass in water bodies, Trans.Inst. Biol. Inland Waters Acad.Sci. USSR, 19:85.
Tunnicliffe, V., and Risk, M. J., 1977, Relationships between the bivalve Macoma balthica and bacteria in intertidal sediments: Minas Basin, Bay of Fundy, J.Mar.Res., 35:499.
Wetzel, R. L., 1977, An experimental radio-tracer study of detrital carbon utilisation in a Georgia salt marsh, Ph.D Thesis, University of Georgia.
Wiebe, W. J., and Smith, D. F., 1977, Direct measurement of dissolved organic carbon release by phytoplankton and incorporation by microheterotrophs, Mar.Biol., 42:213.
Williams, P. L. Le.B., 1970, Heterotrophic utilization of dissolved organic compounds in the sea. I. Size distribution of population and relationship between respiration and incorporation of growth substrates, J.Mar.Biol.Ass.U.K., 50:859.
Williams, P. J. LeB., 1973, On the question of growth yields of natural heterotrophic populations, Bull.Ecol.Res.Comm. (Stockholm), 17:400.
Williams, P. J. LeB., 1981, Incorporation of microheterotrophic processes into the classical paradigm of the planktonic food web, Kieler Meeresforsch., 5:1.
Wolter, K., 1982, Bacterial incorporation of organic substances released by natural phytoplankton populations, Mar.Ecol.Prog. Ser., 7:287.
Woodwell, G. M., Whitney, D. E., Hall, C. A. S., and Houghton, R. A., 1977, The Flax Pond ecosystem study: exchanges of carbon in water between a saltmarsh and Long Island Sound, Limnol. Oceanogr., 22:833.
Wright, R. T., and Hobbie, J. E., 1965, The uptake of organic solutes in lake water, Limnol.Oceanogr., 10:22.
Yurkovsky, A. K., 1971, Results of fraction investigation of the organic substances in the Baltic Sea, Proc. Joint Oceanog. Assem. (Tokyo, 1970), 1971:466.

CYCLING OF ORGANIC MATTER BY BACTERIOPLANKTON IN PELAGIC MARINE ECOSYSTEMS: MICROENVIRONMENTAL CONSIDERATIONS

Farooq Azam and James W. Ammerman

Institute of Marine Resources
Scripps Institution of Oceanography
University of California, San Diego
La Jolla, California 92093

INTRODUCTION

Recently developed methods for measuring production rates of heterotrophic bacteria have shown that the bacterioplankton is a major route for the flux of material and energy in marine ecosystems (Hagström et al., 1979; Fuhrman and Azam, 1980, 1982; Williams, 1981). Even conservative estimates (Fuhrman and Azam, 1980, 1982) show that the measured bacterial productivity corresponds to 10-50% of the primary productivity.

The discovery that bacteria utilize a large fraction of primary productivity is in apparent contradiction with the conventional wisdom that dissolved organic matter (DOM) is too dilute to support significant bacterial growth (Stevenson, 1978). Bacteria, however, can only utilize DOM; particulate organic matter (POM), as well as dissolved polymers, must first be hydrolyzed to small molecules, generally monomers. Fuhrman and Azam (1982) used microautoradiography and size-fractionation to show that free-living bacteria (rather than those attached to POM) are responsible for most of the bacterial secondary productivity.

Concentrations of most DOM components in seawater range from 10^{-12}M to 10^{-8}M (Mopper and Lindroth, 1982; Ammerman and Azam, 1981; Mopper et al., 1980. DOM is, however, produced at a discrete loci in seawater. High DOM concentrations are expected near a phytoplankter that is exuding organic matter or is autolyzing. Loss of intracellular pools of the prey damaged during predation (Lampert, 1978; Copping and Lorenzen, 1980) may also create microzones rich in DOM. Organic detritus colonized by bacteria is also expected to be a source of DOM, and high concentrations may exist in close proximity to such particles. These considerations raise questions regarding the nature of coupling between DOM production and its utilization by bacteria.

This paper will discuss whether new DOM is utilized at high concentrations by bacteria within the production microzone or whether much of it diffuses into the (exceedingly dilute) bulk-phase before being utilized by bacteria. We propose a conceptual model of bacterial cycling of organic matter which takes into account the possibility of microenvironmental nutrient gradients and the ability of bacteria to respond to them to maximize nutrient uptake.

SOURCES AND MODE OF DOM PRODUCTION

Figure 1 shows the main sources and mechanisms which introduce DOM into seawater. Most DOM comes from phytoplankton [via exudation, autolysis, and sloppy-feeding (cellular pool spillage during zooplankton feeding)], zooplankton (via excretion), and organic particles (via depolymerization). These processes channel about one-half of the primary productivity into the DOM pool (Williams, 1981). Bacteria can directly utilize some DOM components such as sugars and amino acids ("utilizable DOM"; UDOM). Some other components are utilizable after chemical modification (e.g. dissolved proteins and polysaccharides are hydrolyzed before uptake). The above distinction is important in considering the spatial and temporal coupling between

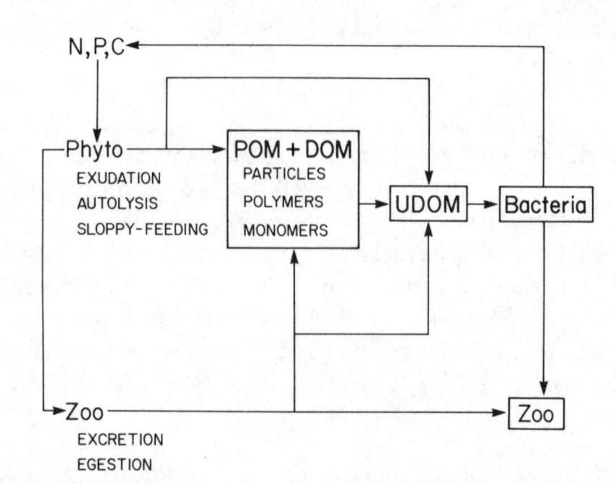

Fig. 1. Sources and pathways of utilizable dissolved organic matter (UDOM) in seawater. DOM = dissolved organic matter; POM = particulate organic matter; sloppy-feeding = spillage of algal cell contents during feeding by zooplankton; phyto = phytoplankton; zoo = zooplankton; N, P, C = inorganic nitrogen, phosphorus, and carbon, respectively.

DOM production and utilization. The additional step from DOM to UDOM may profoundly affect the nature and tightness of coupling.

STRUCTURED NUTRIENT FIELDS IN BACTERIAL MICROENVIRONMENTS

Various sources of DOM, suspended in the dilute bulk-phase, will create enriched microzones of variable persistence. Episodic events such as sloppy-feeding will produce short-lived concentration gradients ("episodic gradients"). These gradients dissipate by molecular diffusion in seconds to minutes (Jackson, 1980). Metabolically driven mechanisms (exudation, autolysis, detritus hydrolysis) in contrast produce "sustained gradients". This is illustrated in Figure 2, which also depicts DOM production events involving sustained oscillatory patterns. A source of sustained UDOM production will thus create a "structured nutrient field", i.e. a region in which there are sharp gradients of bacterial nutrients. We suggest that the "structure" of the nutrient fields will vary in response to biological uptake of nutrients, and physical forces such as molecular diffusion and microturbulence.

If bacteria were randomly distributed in seawater, what fraction of bacteria would be situated within the structured nutrient fields? This will depend on the abundance of sustained DOM sources and their (average) spatial extent. The total volume of the enriched microzone can be calculated as follows: In oceanic surface waters algal abundance is on the order of 10^3 cells per ml (with perhaps a comparable abundance of colonized POM). Consider only algae for the moment, and

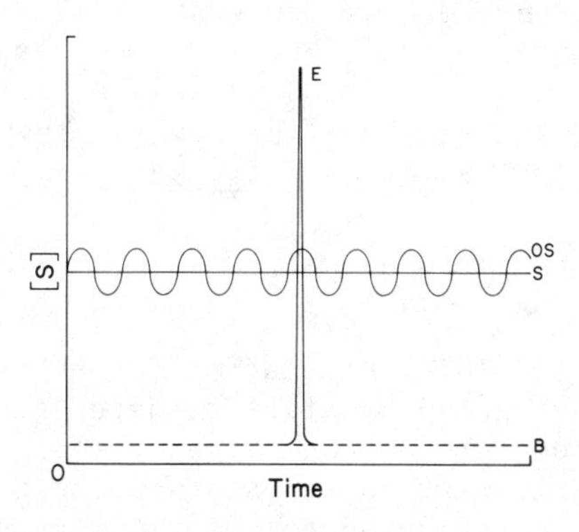

Fig. 2. Time dependence of the concentration [S] of the utilizable dissolved organic matter (UDOM) in the microenvironment of a bacterium, resulting from different modes of UDOM production. B = constant bulk-phase concentration; E = episodic production event; S = sustained production event; OS = oscillating sustained production event.

(a) assume the average cell radius is 10 μm; and (b) define the extent of the enriched zone as the distance where the exuded substrate concentration drops to 10% of that at the cell surface. With these assumptions the enriched zone will be a sphere of 100 μm radius (from the cell surface the nutrient concentration drops as 1/r, where r equals the radius of the sphere). Each microzone will occupy about $4/3(\Pi)\ (100\ \mu m)^3$ or roughly $4 \times 10^6\ \mu m^3$. Thus the microzones around 10^3 algae in each ml will occupy only 0.004 ml or 0.4% of the total volume. Even if we double the microzone volume to account for other sources of sustained gradients, still <1% of bacteria would be in an enriched microzone. Thus, if bacteria were randomly distributed an average bacterium would spend <1% of its time in an enriched microzone.

NON-RANDOM DISTRIBUTION OF BACTERIA

Might bacteria sense the presence of enriched microzones and swim to a nutritionally optimum position within the microzone? To do so, the bacteria must (1) be motile; (2) be chemotactic towards some component of newly produced DOM; and (3) be close enough to a DOM source to sense the chemoattractant concentration gradient and swim rapidly to the enriched microzone.

Motility

Most marine bacterial isolates are motile (ZoBell, 1946; Rheinheimer, 1971). Since isolates may not reflect natural assemblages, we examined concentrates of natural bacterial assemblages for motility (Azam and Ammerman, unpublished). We concentrated the <1 μm fraction of plankton on a 0.2 μm Nuclepore filter, strained live bacteria with acridine orange, and observed the samples by epifluorescence microscopy. A significant fraction of the bacteria were motile with swimming speeds estimated at 20-40 μm sec^{-1}. The estimated swimming speeds are similar to those of *Escherichia coli*.

Chemotaxis

Bell and Mitchell (1972) demonstrated that isolates of marine bacteria were chemoattracted by algal exudates (and proposed the existence of a "phycosphere" around algal cells). Chet and Mitchell (1976) found that enrichment cultures of marine bacteria grown on albumin or casein were attracted to the nutrient on which they were grown. Wellman and Paerl (1981) showed that a bacterium isolated from tar bells in the Atlantic was chemoattracted towards Kuwait crude oil. Young and Mitchell (1973) demonstrated negative chemotaxis by marine bacteria to heavy metals and hydrocarbons. Sieburth (1968) suggested that algae may excrete substances which inhibit bacterial attachment to algae. Although these studies deal only with isolates, they do support the general notion that marine bacteria, like the enterics, are capable of chemotaxis.

Proximity to the UDOM Source

Table 1 shows that bacteria and algae are the most abundant organisms in seawater. Each microliter (mm^3) of coastal surface seawater, for example, may contain 1 algal cell and 10^3 bacteria (these values are approximations reflecting most abundances reported in the literature). Table 1 also points out that the total (microbial) biomass occupies a mere 10^{-6} of the volume in which it occurs. Despite the apparent sparsity of biota (on a volume basis), the numerical abundance of phytoplankton and bacteria (due to their small size) means that average distances between algae and bacteria are quite small. At 10^3 algal cells ml^{-1}, each algal cell could be visualized as being in the center of a cube of side 1 mm (Figure 3) along with 10^3 bacteria. Thus all bacteria are within about 500 μm of an algal cell. Given an equal number of other DOM sources, all bacteria will be within a few hundred μm of a source of sustained DOM production. The purpose of this rough calculation is to point out that bacteria do not have to swim very far before they encounter an enriched microzone. If a bacterium swims at 20 μm sec^{-1} and is 500 μm from the nearest UDOM source it would traverse this distance in 25 seconds of swimming in a straight line. Since bacteria follow gradients by a biased random walk it will take longer than 25 seconds but perhaps no more than a few minutes. Most bacteria are closer than 500 μm to the source and hence need less time.

Table 1. Carbon, abundance, and volume of the biota in surface seawater. Estimates are also given for particulate organic matter (POM) and dissolved organic matter (DOM). These estimates are not precise and show only the general order-of-magnitude reported in the literature.

	Amount ml^{-1} seawater		
Group	Carbon (ng)	Individuals	Volume (ml)
Algae	100	10^3	10^{-6}
Bacteria	10	10^6	10^{-7}
Microzooplankton	1	10^2	10^{-8}
Zooplankton	10	10^{-1}	10^{-7}
Larger Animals	<1	$<<10^{-1}$	$<<10^{-7}$
POM	10-100	10^3	10^{-7}-10^{-6}
DOM	1000	-	-

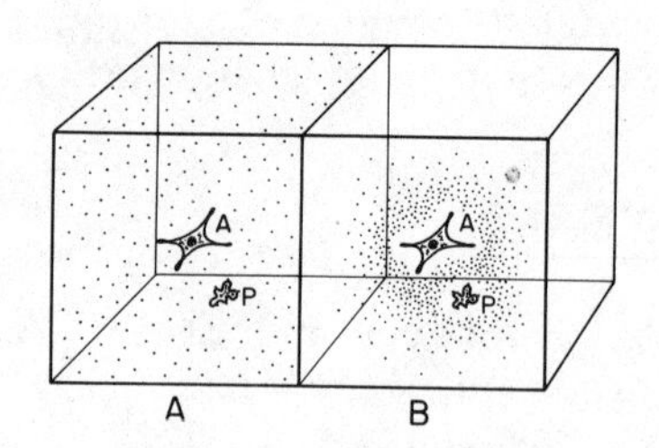

Fig. 3. Distribution of the bacteria in a 1 mm cube of seawater. (A) Random distribution; (B) Clustering of bacteria around an algal cell and a colonized particle. A = algal cell; P = colonized particle.

CLUSTER HYPOTHESIS

The above observations lead us to suggest the following framework for nutritional interactions within the bacterium's microenvironment. We hypothesize that the bacterial microenvironment is "structured" with respect to nutrient concentration, and that bacteria respond to this structure via chemotaxis and motility to optimize their position in the nutrient field. The result is a non-uniform distribution of bacteria and the formation of "bacterial clusters" in the vicinity (5-100 μm) of the sustained sources of DOM (depicted in Figure 3B).

FACTORS AFFECTING THE PERSISTENCE OF CLUSTERS

Two factors may tend to disrupt bacterial clusters, shear and movement of the DOM source. At the short distances (5-100 μm) between bacteria and the DOM source, shear should be insignificant except during the rare periods of strong turbulence (Lehman and Scavia, 1982). Bacteria can probably swim to maintain their positon. However, the movement of the source itself at speeds greater than the swimming speeds of bacteria is bound to disrupt the association. Thus fast swimming algae such as dinoflagellates would not develop stable bacterial clusters. Slow movement of the source, such as sinking and buoyant rising (in diatoms) will distort the clusters and induce "cluster dynamics" with bacteria tending to return to their optimal positions.

EXPERIMENTAL EVIDENCE

Kinetics of Nutrient Uptake

It is generally believed that marine bacteria have evolved very low K_m transport systems for efficient acquisition of nutrients from the nanomolar concentrations found in the (bulk-phase) seawater

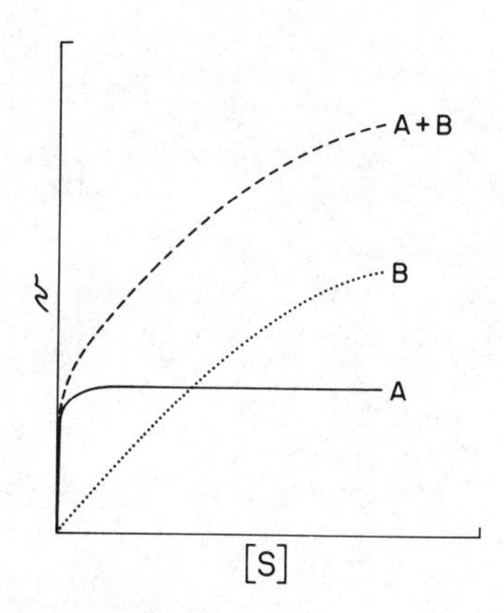

Fig. 4. Dependence of uptake rate (v) on substrate concentration [S]. A = low K_m, low V_{max} uptake system; B = high K_m, high V_{max} uptake system; A + B = sum of the two uptake systems.

(Wright and Burnison, 1979). However, the low K_m, low V_{max} systems will saturate at low substrate concentrations. If bacteria were adapted to utilizing DOM from high microenvironmental concentrations then one would predict that high K_m and V_{max} transport systems will also be found in marine bacteria. A combination of low and high K_m systems should provide metabolic flexibility in a heterogeneous and fluctuating environment (Figure 4). Conversely the kinetic properties of nutrient uptake systems could serve as a "bioassay" for

Table 2. Kinetic parameters for glucose assimilation by a natural microbial assemblage in a seawater sample from the end of Scripps Pier. Data computed from the kinetic plots in Figure 5. [A] = range of added glucose; K_t = uptake constant; S_n = ambient glucose concentration; V_{max} = maximum uptake velocity; T_t = turnover time for the ambient glucose pool. Reprinted from Azam and Hodson, 1981, p.219, by courtesy of Inter-Research.

[A] (molar)	$(K_t + S_n)$ (molar)	V_{max} (nmol $l^{-1}h^{-1}$)	T_t (h)
2.5 x 10^{-9} - 10^{-8}	2.3×10^{-9}	0.64	4.5
10^{-8} - 10^{-7}	3.2×10^{-8}	1.86	15
10^{-7} - 10^{-6}	1.5×10^{-7}	3.60	47
10^{-6} - 10^{-5}	2.4×10^{-6}	11.5	205
10^{-5} - 10^{-4}	5.8×10^{-5}	66	940
10^{-4} - 10^{-3}	5.9×10^{-4}	300	2210

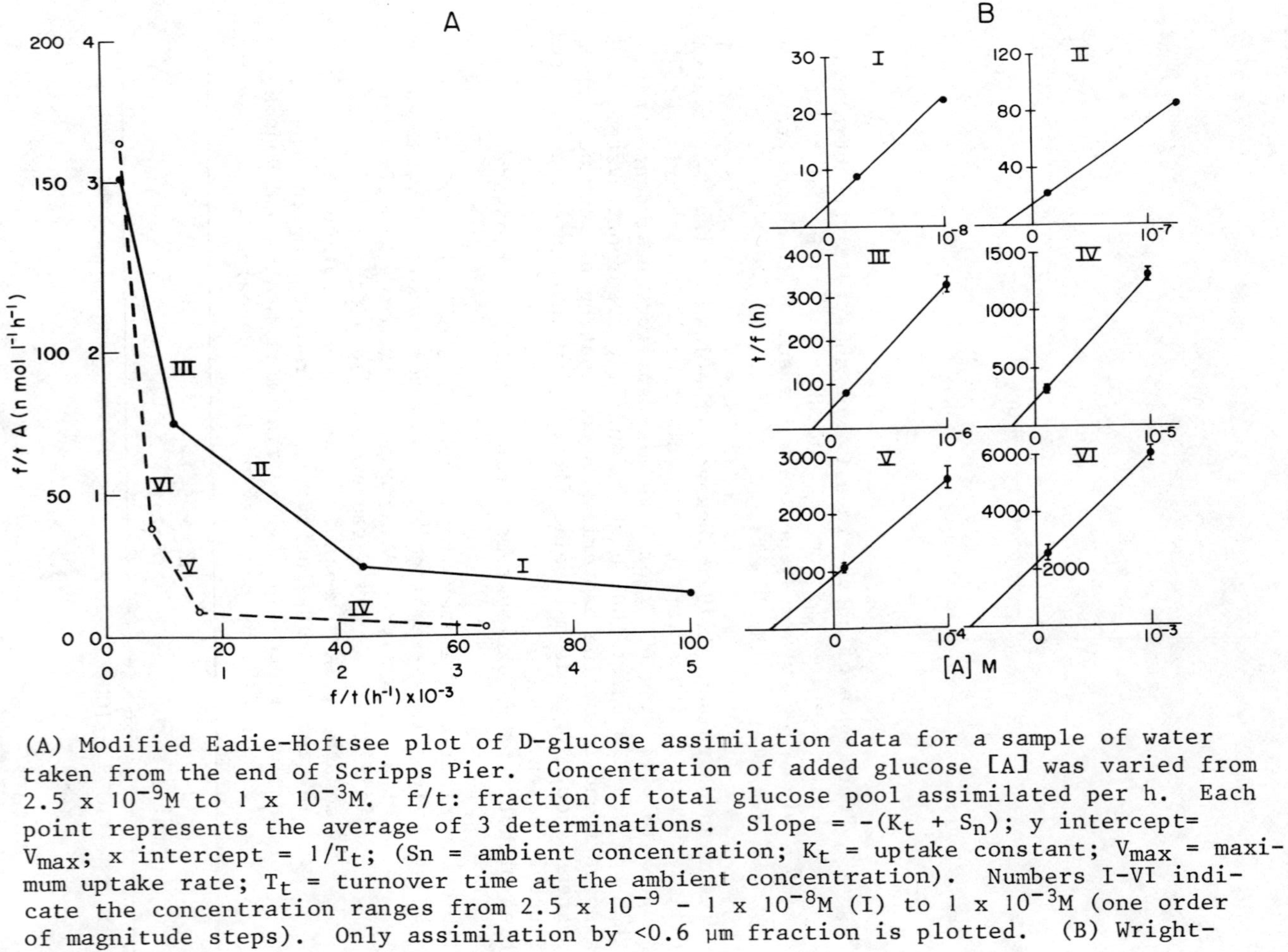

Fig. 5. (A) Modified Eadie-Hoftsee plot of D-glucose assimilation data for a sample of water taken from the end of Scripps Pier. Concentration of added glucose [A] was varied from 2.5×10^{-9}M to 1×10^{-3}M. f/t: fraction of total glucose pool assimilated per h. Each point represents the average of 3 determinations. Slope = $-(K_t + S_n)$; y intercept= V_{max}; x intercept = $1/T_t$; (Sn = ambient concentration; K_t = uptake constant; V_{max} = maximum uptake rate; T_t = turnover time at the ambient concentration). Numbers I-VI indicate the concentration ranges from $2.5 \times 10^{-9} - 1 \times 10^{-8}$M (I) to 1×10^{-3}M (one order of magnitude steps). Only assimilation by <0.6 μm fraction is plotted. (B) Wright-

nutrient concentration regimes encountered by bacteria in their microenvironments (Azam and Hodson, 1981). These authors found clear evidence of multiphasic kinetics for D-glucose uptake in assemblages of marine bacterioplankton (Figure 5). Surprisingly broad ranges of K_m (10^{-9} 10^{-4}M) and V_{max} (0.6-300 nmol $l^{-1}h^{-1}$) were obtained (Table 2). High K_m uptake components were in fact saturable (and not artifacts of superposition of uptake by low K_m systems and simple diffusion). Thus there are bacteria in seawater which have K_m values many orders of magnitude higher than the bulk-phase substrate concentrations. The existence of very high K_m transport systems is presumptive evidence that at least some bacteria experience high substrate concentrations in their microenvironments (Azam and Hodson, 1981).

The existence of multiphasic uptake in assemblages of marine bacteria may mean that some bacteria have high-affinity uptake systems while some others have low-affinity systems. This may lead to niche separation within the structured nutrient microenvironments. Individual species may also have multiphasic or multiple systems, giving the species metabolic flexibiltiy in a nutritionally fluctuating microenvironment.

Hodson et al. (1979) found that a marine isolate LNB-155 transports D-glucose by a low-capacity high-affinity (K_m = 7 x 10^{-9}M) and a high-capacity low-affinity system (K_m = 1.1 x 10^{-7}). Nissen et al. (unpublished) working with the same isolate (but measuring uptake kinetics over a broader range than Hodson et al.) found four kinetic phases; the highest phase yielded a millimolar K_m. Nissen et al. argue, based on the abruptness of breaks in kinetic plots, that the four phases are in fact due to a single transport system (multiphasic rather than multiple). This transport system may undergo (allosteric?) transitions at defined substrate concentrations and change its affinity and capacity. They suggest that a multiphasic transport system would provide metabolic flexibility at a low cost in terms of membrane proteins, since the same transport system would be used with a variable K_m and V_{max}.

It is not feasible to directly measure the substrate uptake rates at different concentrations experienced by a bacterium within its microenvironment. We can, however, simulate a variety of concentration regimes by exposing subsamples of an entire bacterial assem-

←

Fig. 5. (cont.)
Hobbie plots. Data points for each order of magnitude change in [A] are plotted on a separate graph, for clarity; representing all points on one graph compresses all but the highest concentration points, thus making it difficult to discern non-linearity. Linearity between each two points is assumed. Numbers I-VI refer to concentration ranges as in (A). Reprinted from Azam and Hodson, 1981, p.217, by courtesy of Inter-Research.

Table 3. Turnover time (T_t) and rate of D-glucose uptake (v, v'), at different added concentrations [A], for a natural assemblage of bacteria sampled from the pier at Scripps Institution of Oceanography, La Jolla, California. We calculated v and v' from [A] and T_t values given in Figure 5B.

[A] (mol l^{-1})	T_t (h)	v (g C $cell^{-1}$ h^{-1})	v' (% cell C h^{-1})[a]	Doubling time (h)
2.5×10^{-9}[b]	9.5	1.9×10^{-17}	0.2	528
1×10^{-8}	22	3.3×10^{-17}	0.3	307
1×10^{-7}	85	8.0×10^{-17}	0.9	118
1×10^{-6}	320	2.3×10^{-16}	2.3	43
1×10^{-5}	1300	5.0×10^{-16}	5.4	19
1×10^{-4}	2700	2.7×10^{-15}	26.7	4
1×10^{-3}	6000	1.2×10^{-14}	120.8	1

[a]Assuming 1×10^{-14}g C $cell^{-1}$ (Fuhrman and Azam, 1980).
[b]Actual value will be slightly higher since S_n was not measured.

blage to varied nutrient concentrations. D-glucose assimilation data from one such experiment (Figure 5B) is used here to calculate (Table 3) the assimilation rate per cell h^{-1} (v) and % cell carbon assimilated h^{-1} (v'). Assimilation rate per bacterium increased about 600-fold, from 1.9×10^{-17} g C at 2.5 nM D-glucose to 1.2×10^{-14} g C at 1 mM D-glucose. At near-ambient D-glucose concentration the cell doubling time (on glucose alone) was 22 days. However, if the bacterium were in a 1 μM D-glucose microzone the doubling time would be 1.8 days, and only 1 h in a 1 mM D-glucose concentration microzone. This calculation applies to the "average bacterium"; individual bacteria could vary greatly in their uptake rates.

Bacterial Uptake of Algal Exudates

Clustering of bacteria around algae might be inferred from bacterial uptake of ^{14}C-labeled algal exudate ("exuded organic carbon", EOC). Recent reports show that bacteria take up EOC very rapidly, and, inexplicably, much more rapidly than added radiolabeled substrates. Lancelot (1979), found that 15 - 39% of the EOC was assimilated per hour at stations in the North Sea. Since some EOC must have been respired, the actual rates of uptake were even higher. Other investigators (Derenbach and Williams, 1974; Larsson and Hagström, 1979; Wiebe and Smith, 1977) found similarly high rates of EOC tunrover. Table 4 compares the EOC assimilation rates with uptake rates of added substrates for the same samples (Lancelot, 1979;

Table 4. Turnover times of algal exudates and added organic substrates at various locations.

Location	Turnover Rate (% h^{-1})	
	Exudates	Organic Substrates
Hansweert	27[a]	0-11[b]
Ostende	39[a]	0.1-3[b]
Calais	16[a]	0[b]
N. Atlantic	15-34[a]	ND
Southampton		3[c]
English Channel	15-30[d]	1.7-4.6[c]
N. Atlantic	(range 1-30)	0.2-4.8[c]
Baltic	27[e]	
South Pacific	10-12[f]	

[a]Lancelot, 1979; [b]Billen *et al.*, 1980; [c]Williams, 1970; [d]Derenbach and Williams, 1974; [e]Larsson and Hagström, 1979; [f]Wiebe and Smith, 1977. ND - not determined.

Billen et al., 1980). The differences are dramatic. The Ostende and Calais stations showed an order of magnitude faster assimilation of EOC than of the added substrates.

This discrepancy may be explainable by the cluster hypothesis. If bacteria utilized EOC from the high concentrations existing in close proximity to the exuding alga, the rate of uptake per bacterium should be higher than in the bulk-phase (Table 3). However, the rapid disappearance of EOC cannot be explained on this basis. If we assume the enriched microzone to be within 100 μm of the algal cells then, as shown above, only 0.4% of the bacterial assemblage will be within the enriched microzone. The uptake rate per bacterium will increase by an order of magnitude if a bacterium in nanomolar glucose is moved into a zone of micromolar glucose (Table 3). Since only 0.4% of the bacteria are present in the microzone this will enhance the rate by only 4%. To explain the rapid rate of EOC uptake, therefore, a significant fraction of the bacterial assemblage must be clustered within the enriched microzone.

CHEMICAL CUES IN CLUSTER FORMATION

Algal exudation is believed to be one of the major sources of UDOM for bacterioplankton (Hagström *et al.*, 1979; Williams, 1981;

Wiebe and Smith, 1977). Bacteria are attracted by one or more components of the exudates (Bell and Mitchell, 1972) and repelled by some algal products (Sieburth, 1968). It has been suggested that a metabolic feedback between heterotrophic bacteria and algae might exist (Smith and Higgins, 1978). However, the mechanisms regulating algal excretion, and the identity of any chemical cues exchanged between the algae and bacteria remain unknown. It is an interesting (and common) observation that healthy algal cells are generally free of attached bacteria. In our experiments (Azam and Ammerman, unpublished) we mixed healthy and senescent cultures of *Thalassiosira weissflogii* and found that bacteria became attached to senescent cells but not to the healthy cells. Microscopic observations suggested a zone of a few microns around healthy algae which the bacteria appeared to avoid.

While the identity of metabolic effectors, both positive and negative, in algal exudates is largely unknown, it is interesting that a ubiquitous metabolic regulator, 3'5'-cyclic AMP (cAMP), was recently found in exudates of freshwater algae (Francko and Wetzel, 1980, 1981). Ammerman and Azam (1981, 1982) consistently found dissolved cAMP (1-30 pM) in a large number of seawater samples. They found diel variations of cAMP similar to reported variations of amino acids and polysaccharides (Mopper and Lindroth, 1982; Burney *et al*., 1981). Significantly, marine bacterial assemblages transport cAMP by a highly specific, active transport system with a picomolar K_m. This system can double the bacterial intracellular cAMP pool in minutes to hours (Ammerman and Azam, 1982).

It is tempting to speculate that, in algal-bacterial interactions, algae excrete cAMP and bacteria take it up as a metabolic cue. The near absolute specificity of the bacterial cAMP uptake system, its energy requirement, and its ability to substantialy enhance the intracellular cAMP pool within minutes, suggests a regulatory role for cAMP. Since cAMP occurs in pM concentrations (100 times lower than AMP which is not transported by the cAMP transport system), it is unlikely that the cAMP transport system in marine bacteria evolved to take up this compound as a carbon or energy source. Cyclic AMP is intimately involved in regulation of the synthesis of catabolic enzymes. If cAMP is exuded along with bacterial nutrients, it may act as a unifying signal. We hasten to point out that there is no evidence at present to support this speculation. It should be interesting, however, to see if marine bacteria are chemotactic towards cAMP. Although no bacteria are known to be chemoattracted by cAMP, myxobacteria are attracted by a closely related compound, cGMP (Botsford, 1981).

IMPLICATIONS OF THE CLUSTER HYPOTHESIS

Bacterial clustering around algae is hypothesized here to be a coevolved commensal relationship in which the algae provide the heterotrophic bacteria with the organic matter for material and energy. Bacterial mineralization of DOM within the microenvironment maintains high concentrations of plant nutrients within the microenvironment, thus allowing rapid growth of the alga. The bacterial-algal association is viewed functionally as a coordinated metabolic unit exhibiting rapid photoautotrophic and heterotrophic metabolism.

The persistence of high concentrations of DOM and of mineralized plant nutrients within the clusters has implications for the models of nutrient-limited growth of both algae and bacteria. Bulk-phase nutrient concentrations may not be a valid basis for considering the relationships between the nutrient regime and plant growth. Goldman et al. (1979) argued that rapid growth of phytoplankton in oligotrophic oceans was inconsistent with the low concentrations of the limiting nutrient, ammonia. They suggested that algae might encounter high ammonia concentrations in microscale patches. Zooplankton excretion of ammonia and phosphate is believed to produce high concentration plumes (Goldman et al., 1979; Lehman and Scavia, 1982), but these plumes may diffuse too rapidly to account for nutrient-sufficient growth of phytoplankton (Jackson, 1980).

The cluster hypothesis suggests a more plausible mechanism for maintaining high concentrations of plant nutrients in the microenvironment of the alga. An attractive feature of the above explanation, suggested by our hypothesis, is that the clusters (as well as the nutrient patches) are created and sustained in part by the metabolic activity of the alga itself. This means that the extent of clustering could be coupled to the nutrient status of the algal cell. It is noteworthy in this context that algal exudation (carbon exuded as % of carbon fixed may be inversely related to the inorganic nitrogen concentration (Joiris et al., 1982).

The high abundance of bacteria within the clusters could be important in density-dependent predation on bacteria. Bacteriovores such as phagotrophic flagellates and ciliates might obtain a greater fraction of their rations within the clusters than in the bulk-phase. For example, Fenchel (1980) concluded that (given the bulk-phase bacterial abundance) ciliates could not play a role as grazers of bacteria in the open ocean. Predaceous bacteria of the genus Bdellovibrio occur consistently in seawater although the bulk-phase bacteria prey density may be too low for their proliferation (Varon and Shilo, 1980). If, however, only 10% of the bacteria in a planktonic assemblage were within 100 μm of the algal cells, then the bacterial abundance within the clusters would be 2.5×10^7 ml^{-1} (cf. 1×10^6 ml^{-1} in the bulk-phase); and the above organisms might become effective bacterial predators.

The cluster hypothesis provides a general framework for considering the bacteria-organic matter interaction in the ocean. In studies of bacterial utilization of organic matter a distinction is usually made between DOM and POM. Although convenient, this distinction is simplistic. Since bacteria do not directly utilize POM (or even dissolved polymers) growth on particles simply refers to a site for bacterial growth but does not define the UDOM regime on and in the vicinity of the particle. Colonized organic particles may become the sites of sustained depolymerization. The resulting DOM may be utilized by attached bacteria, or by free-living bacteria in clusters, or in the bulk-phase. Therefore, a demarcation in space between POM and DOM is unrealistic.

Acknowledgements

We thank G. Jackson for help in calculation of diffusion gradients. We thank M. Ogle for providing assistance in manuscript preparation. This research was supported by an NSF grant (OCE79-26458) and a Department of Energy contract (DE-AT03-82-ER60031).

REFERENCES

Ammerman, J. W., and Azam, F., 1981, Dissolved cyclic adenosine monophosphate (cAMP) in the sea and uptake of cAMP by marine bacteria, Mar.Ecol.Prog.Ser., 5:85.

Ammerman, J. W., and Azam, F., 1982, Uptake of cyclic AMP by natural populations of marine bacteria, Appl.Environ.Microbiol., 43: 869.

Azam, F., and Ammerman, J. W., 1982, Growth of free-living marine bacteria around sources of dissolved organic matter, EOS, 63:54.

Azam, F., and Hodson, R. E., 1977, Size distribution and activity of marine microheterotrophs, Limnol.Oceanogr., 22:492.

Azam, F., and Hodson, R. E., 1981, Multiphasic kinetics for D-glucose uptake by assemblages of natural marine bacteria, Mar.Ecol. Prog.Ser., 6:213.

Bell, W., and Mitchell, R., 1972, Chemotactic and growth responses of marine bacteria to algal extracellular products, Biol.Bull., 143:265.

Billen, G., Joiris, C., Winant, J., and Gillian, G., 1980, Concentration and metabolism of small organic molecules in estuarine, coastal and open sea environments of the Southern North Sea, Est.Coast.Mar.Sci., 11:279.

Botsford, J. L., 1981, Cyclic nucleotides in procaryotes, Microbiol. Rev., 45:620.

Burney, C. M., Johnson, K. M., and Sieburth, J. McN., 1981, Diel flux of dissolved carbohydrate in a salt marsh and a simulated estuarine ecosystem, Mar.Biol., 63:175.

Chet, I., and Mitchell, R., 1976, An enrichment technique for isolation of marine chemotactic bacteria, Microbial Ecol., 3:75.

Copping, A. E., and Lorenzen, C. J., 1980, Carbon budget of a marine phytoplankton-herbivore system with carbon-14 as a tracer, Limnol.Oceanogr., 25:873.

Derenbach, J. B., and Williams, P. J. leB., 1974, Autotrophic and bacterial production: Fractionation of plankton populations by differential filtration of samples from the English Channel, Mar.Biol., 25:263.

Fenchel, T., 1980, Suspension feeding in ciliated protozoa: Feeding rates and their ecological significance, Mircrobial Ecol., 6:13.

Francko, D. A., and Wetzel, R. G., 1980, Cyclic adenosine -3':5'-monophosphate: Production and extracellular release from green and blue-green algae, Physiol.Plant, 49:65.

Francko, D. A., and Wetzel, R. G., 1981, Dynamics of cellular and extracellular cAMP in Anabaena Flos-Aquae (cyanophyta): Intrinsic culture variability and correlation with metabolic variables, J.Phycol., 17:129.

Fuhrman, J. A., and Azam, F., 1980, Bacterioplankton secondary production estimates for coastal waters of British Columbia, Antarctica, and California, Appl.Environ.Microbiol., 39:1085.

Fuhrman, J. A., and Azam, F., 1982, Thymidine incorporation as a measure of heterotrophic bacterioplankton production in marine surface waters: Evaluation and field results, Mar.Biol., 66: 109.

Goldman, J. C., McCarthy, J. J., and Peavey, G. D., 1979, Growth rate influence on the chemical composition of phytoplankton in oceanic waters, Nature, 279:210.

Hagström, A., Larsson, U., Hörstedt, P., and Normark, S., 1979, Frequency of dividing cells, a new approach to the determination of bacterial growth rates in aquatic environments, Appl.Environ.Microbiol., 37:805.

Hodson, R. E., Carlucci, A. F., and Azam, F., 1979, Glucose transport in a low nutrient marine bacterium, Abstr.Annu.Meet.Am.Soc. Microbiol., N 59:189.

Jackson, G. A., 1980, Phytoplankton growth and zooplankton grazing in oligotrophic oceans, Nature, 284:439.

Joiris, C., Billen, G., Lancelot, C., Daro, M. H., Mommaerts, J. P., Bertels, A., Bossicarta, M., Nijs, J., and Hecq, J. H., 1982, A budget of carbon cycling in the Belgian coastal zone: Relative roles of zooplankton, bacterioplankton and benthos in the utilization of primary production, Neth.J.Sea Res., 16: 260.

Lampert, W., 1978, Release of dissolved organic carbon by grazing zooplankton, Limnol.Oceanogr., 23:831.

Lancelot, C., 1979, Gross excretion rates of natural marine phytoplankton and heterotrophic uptake of excreted products in the Southern North Sea, as determined by short-term kinetics, Mar.Ecol.Prog.Ser., 1:179.

Larsson, U., and Hagström, A., 1979, Phytoplankton exudate release as an energy source for the growth of pelagic bacteria, Mar. Biol., 52:199.

Lehman, J. T., and Scavia, D., 1982, Microscale patchiness of nutrients in plankton communities, Science, 216:729.

Mopper, K., Dawson, R., Liebezeit, G., and Ittekkot, V., 1980, The monosaccharide spectra of natural waters, Mar.Chem., 10:55.

Mopper, K., and Lindroth, P., 1982, Diel and depth variations in dissolved free amino acids and ammonium in the Baltic Sea determined by shipboard HPLC analysis, Limnol.Oceanogr., 27:336.

Rheinheimer, G., 1971, "Aquatic Microbiology," J. Wiley and Sons, London.

Sieburth, J. McN., 1968, Observations on planktonic bacteria in Narragansett Bay, Rhode Island: a resume, Misaki.Mar.Biol. Inst., Kyoto Univ., 12:49.

Smith, D. F., and Higgins, H. W., 1978, An interspecies regulatory control of dissolved organic carbon production by phytoplankton and incorporation by microheterotrophs, in: "Microbial Ecology," M. W. Loutit and J. A. R. Miles, eds., Spring-Verlag, Berlin.

Stevenson, L. H., 1978, A case for bacterial dormancy in aquatic systems, Microbial Ecol., 4:127.

Varon, M., and Shilo, M., 1980, Ecology of aquatic bdellovibrios, in: "Adv. Aquatic Microbiol.," M. R. Droop and H. W. Jannasch, eds., Academic Press, London.

Wellman, A. M., and Paerl, H. W., 1981, Rapid chemotaxis assay using radioactively labeled bacterial cells, Appl.Environ.Microbiol., 42:216.

Wiebe, W. J., and Smith, D. F., 1977, Direct measurement of dissolved organic carbon release by phytoplankton and incorporation by microheterotrophs, Mar.Biol., 42:213.

Williams, P. J. leB., 1981, Incorporation of microheterotrophic processes into the classical paradigm of the planktonic food web, 15th European Symposium on Marine Biology, Kiel F.R.G., Kieler Meeresforsch., Sonderh., 5:1.

Wright, R.T., and Burnison, B. K., 1979, Heterotrophic activity measured with radiolabeled organic substrates, in: "Native Aquatic Bacteria: Enumeration, Activity, and Ecology," J. W. Costerton and R. R. Colwell, eds., Am. Soc. for Testing and Materials, STP 695.

Young, L. Y., and Mitchell, R., 1973, Negative chemotaxis of marine bacteria to toxic chemicals, Appl.Microbiol., 25:972.

ZoBell, C. E., 1946, "Marine Microbiology. A Monograph on Hydrobacteriology," Chronica Botanica Co., Waltham, Mass.

AN OVERVIEW OF SECONDARY PRODUCTION IN PELAGIC ECOSYSTEMS

R. Williams

Natural Environment Research Council
Institute for Marine Environmental Research
Prospect Place, The Hoe, Plymouth
United Kingdom

INTRODUCTION

For the purpose of this general view of secondary production in the pelagic environment I shall consider that secondary production is almost synonymous with zooplankton production with fish referred to as tertiary producers. Although I shall deal principally with zooplankton, to accurately quantify the level and rate of secondary production in a pelagic ecosystem requires a complete understanding of the functioning of all trophic levels and the interaction of the processes within the ecosystem. I repeat the sentiment expressed by Mann (1969) that the average ecosystem is so complex that ecologists have tended to concentrate their attention on the processes involving single species or isolated food chains. Over a decade later this is still true but positive steps have been taken towards gaining insights to the structure and functioning of ecosystems using the holistic approach of modelling biological systems (Nihoul, 1975; Platt *et al.*, 1981). The cycle of production of material in the sea is usually described in simple terms starting with the incorporation of solar energy into autotrophic production in the euphotic zone. The amount of primary production is restrained by the availability of nutrients and grazing activities of the herbivores/omnivores. The secondary production is then passed onto the tertiary consumers, such as pelagic fish, cephalopods and mammals with material loss (dissolved and particulate organic matter) at all levels to the benthic communities and the decomposers or remineralisers. Energy is the 'currency' of ecological processes but the rates at which processes occur are controlled in natural systems by nutrient or element availability. It has emerged in the last decade that patterns of energy flow and element cycling in ecological systems cannot be interpreted independently without the chance of introducing erroneous conclusions about ecosystem functioning (Reichle et al., 1975).

In this overview, I shall consider the earlier work on production in aquatic ecosystems reviewed by Mullin (1969) and especially the valuable contribution by Mann (1969), then consider the research carried out during the decade (1964 to 1974) of the International Biological Programme and finally assess the developments in research in the post-IBP era. The aim of the IBP was to research 'the biological basis of productivity and human welfare' and to study organic production on land, in fresh-water and in the sea (Clapham et al., 1976). The most active years of the programme, which provided great stimulus to research on productivity, were 1968 to 1973 and culminated in a series of International meetings and publications in the early to mid-seventies.

An important outcome of the IBP was the formulation of methodologies and procedures for measuring productivity (Ricker, 1968; Winberg, 1971; Edmondson and Winberg, 1971). The problems attached to measurements of productivity were seen by many research workers as responsible for inhibiting the progress of our understanding of the structure and functioning of marine ecosystems. I shall deal with these problems in detail in this overview. They range from the limitations of the trophodynamic concept (Lindeman, 1942) of energy flow in biological systems (Rigler, 1975; Reichle et al., 1975) to the general awareness of more complex food networks, than simple food chains, in the marine ecosystems. Associated with these food networks are the feeding patterns of the individual species which, in many instances, change throughout the development and life histories of the species, and make it very difficult to allocate species to particular 'trophic' levels. There is paucity of information in the literature on the flux of energy and material through systems and there are immense difficulties in obtaining accurate estimate of population densities by sampling a three-dimensional, biologically heterogeneous environment. From the mid-seventies marine research has been concentrated on certain of these problems although Mann, thirteen years after his review on productivity in aquatic systems, states (in Platt et al., 1981) that the lack of data on the flux of energy and material through biological systems is still the most important stumbling block to our understanding of the functioning of ecosystems.

MEASUREMENT OF PRODUCTION

Biological productivity was first studied in the culture of pond-fish and was defined as the average annual production of fish taken from a given body of water in weight units. This definition was developed through the work of Tansley (1935), Winberg (1936), Juday (1940), Lindeman (1942) and Ivlev (1945) and led to research in the flow of energy and organic matter in the food chain or, as Lindeman (1942) stated, the transfer of energy from one part of the ecosystem to another. Ivlev (1945) refined this concept and defined production as the sum of all organic matter added to the stock or product (or other defined organic unit) in a unit of time regardless of whether or not it remained alive (i.e. part of the stock) at the end of that time. Four types of study were identified by Ivlev to quantify the flow of energy from solar radiation to a given end product. These were, a quantitative estimate of primary production, the identification of the paths of energy transformation that lead to a chosen end product, the determination of the ecotrophic coefficients (Ivlev, 1945) for each step in the food pyramid from primary organic matter to the end product and finally, the determination of the energy coefficient of growth, i.e. the fraction of consumed food that is converted into body substances for each trophic step. The production of any organism is equal to the quantity of food it ingests multiplied by the growth coefficient. In his discussion Ivlev (1945) made a statement which is as true now as it was 37 years ago: "considering the tremendous difficulties inherent in problems involving the quantitative analysis of production processes, difficulties associated with obtaining numerical values of the statistics describing any of the four necessary elements, it must be recognised that a phenomenon as complex by nature as the biological production of a body of water cannot be resolve by simple methods".

Carbon Balance Approach

The trophodynamic model of ecosystems formulated in the forties was further refined in attempts to evaluate the pathways of energy in various systems (Odum, 1957; Teal, 1957, 1962; Slobodkin, 1962; Macfadyen, 1964 and Mann, 1964). In his review of production in aquatic systems, Mann (1969) suggested that the development of the understanding of the dynamics of ecosystems has consisted of a progressive clarification of the processes of production and those of food capture, ingestion, assimilation and metabolism.

The basic relationship of production of heterotrophic organisms, which was adopted by the International Biological Programme (Ricker, 1968), was

$$P_S = C - (F + U_S + R_S) \quad , \qquad (1)$$

where P_S = production by the specified heterotroph (growth + reproduction), C = food ingested, F = feces egested, U_S = excretion, R_S = respiration by the specified heterotroph. The equation may be evaluated in terms of energy (Kilocalories or Joules) or in terms of carbon and nitrogen.

Mann (1969) defined three different categories of secondary production which must be considered: production to man (often called yield), production removed by predators and production removed by decomposers as well as the increase in weight i.e. growth and reproduction. Dagg (1976), modified the basic relationship given in equation (1) to take account of materials excreted as end products of metabolic pathways and for the loss of 'useful' material by leakage. He concluded that his expanded equation was equally useful for energy or material budgets in determining growth.

$$P_G = C - (PR + M + F + L + T) \quad , \tag{2}$$

where P_G = growth, C = ingestion, PR = reproduction, M = metabolism, F = defecation, L = leakage, T = moults. Using this equation Dagg (1976) constructed a complete budget for the amphipod *Calliopius laeviusculus* in terms of carbon and nitrogen and at three temperatures for all life stages of this zooplankton species. Although budgets have been constructed for other crustaceans (Lasker, 1966; Corner *et al*., 1967; Harris, 1973) this was the first study in which all parameters were measured, with no parameter being calculated by difference. If all parameters are measured accurately the budget should balance; an unbalance budget would indicate that at least one parameter was inaccurately measured.

It can be argued (Tranter, 1976) that food which passes undigested through the gut is a direct flux from algae to detritus and that 'assimilation' is a better basis for production estimates than ingestion, such that

$$P = A - R_S \quad , \tag{3}$$

where A = food assimilated (= $C - F - U_S$) and A/C = Assimilation Efficiency. In routine estimates of production it is easier to determine coefficients for growth efficiency which can be applied either to ingestion (K_1 = gross growth efficiency) or to assimilation (K_2 = net growth efficiency).

$$P = K_1 C \text{ and } P = K_2 A \quad . \tag{4}$$

The budget for the production of a zooplankton herbivore is shown in Figure 1. This method of determining secondary production has been referred to as the 'carbon-balance' or 'carbon-budget' approach by Mullin (1969) and Tranter (1976) and has been used for pelagic species by Curl (1962), Lasker (1966), Smayda (1966) and Petipa (1967) to mention a few studies, and in freshwater systems by Paloheimo *et al*. (1982).

A method for calculating secondary production using a weight dependent production model is given by Conover and Huntley (1980). The authors concluded that it was possible to accurately measure all the parameters in the carbon-budget equation and to predict the effect of grazing by a zooplankton community on a given particulate environment, and to determine the efficiency of energy transfer between trophic levels by comparing growth rates of zooplankton and phytoplankton, respectively. Secondary production can be calculated from:

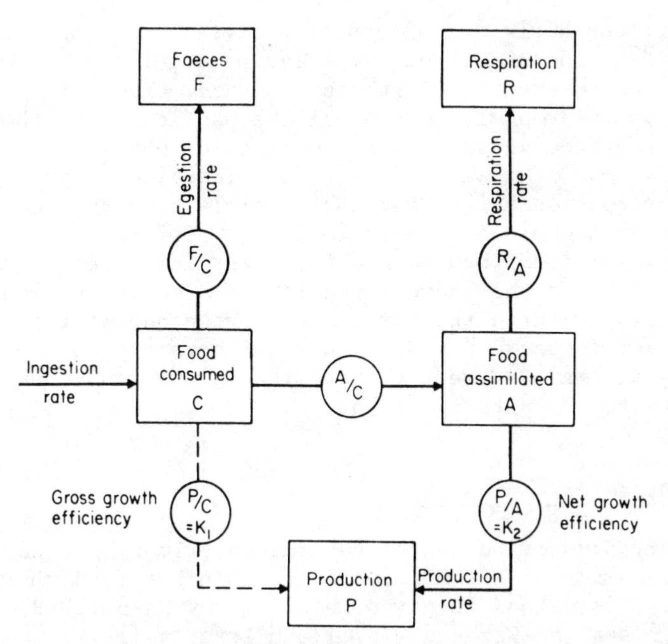

Fig. 1. Compartmentalized budget for the production of a plankton herbivore. The arrows are the fluxes; the boxes are the 'stocks'; and the circles are the efficiencies (from Edmondson and Winberg, 1971 modified by Tranter, 1976).

$$P = \sum_{i=1}^{j} N_i \left[a\left(\sum_{s=1}^{h} f \log(V_s/V_o) e^{0.17T} \overline{W}_i^{\ 0.82} C_s\right) - k\overline{W}_i^{\ m} \right] \quad (5)$$

where N_i = the number of animals in weight class i, j = the number of weight classes of animals in the community; $\overline{W}_i$ = the geometric mean weight of an animal in the i'th weight class (mg dry); a = the assimilated efficiency (assumed to be 0.7); f = the feeding constant (4.85); V_s = the geometric mean volume of the particle in size class s; V_o = the volume of the smallest filterable particle (approx. 25 μm^3); h = the number of size classes of particles which are fed upon; k = the respiration constant ($k = 0.349\ e^{0.056T}$, where T is temperature, °C); m = the respiration exponent ($m = 0.92\ e^{-0.016T}$, where T is temperature, °C); and C_s = the carbon concentration of particles in size category s ($\mu g\ C\ ml^{-1}$).

Conover and Huntley (1980) state that it is theoretically possible to obtain an estimate of secondary production from a single plankton tow, one carbon sample, a water sample for evaluation of particle size and concentration, and a temperature reading. From tank experiments carried out at Scripps Institute, Huntley and Brooks (pers. comm.) compared estimates of production of the copepod

Calanus pacificus derived from cohort analyses and from the weight dependent production model and very similar results were obtained. There are a few reservations with this method. Food, for instance, is measured as chlorophyll content of the particles and the carbon content derived from an assumed ratio of chlorophyll to carbon. No assessment is made of what is food to the individual species or populations under consideration from the particle spectrum measured by Coulter Counter techniques and assimilation efficiency is assumed to be constant. This is a new method for determining secondary production using the carbon budget technique and merits serious consideration. To evaluate the parameters of the budget requires accurate measurements to be taken in the laboratory and even if these measurements are made correctly there remains the question of their applicability to the natural situation.

Population Dynamics Approach

Secondary production can be determined from a knowledge of the population dynamics of the herbivore and values for birth rate and death rate. This method is given in detail by Mann (1969), Winberg (1971), Edmondson and Winberg (1971) and Tranter (1976).

The population dynamics approach relies on the estimation of growth and mortality of individuals of all ages in the population and the measurement of weights of individuals at various stages in their development. This technique is applicable to the dynamics of a single population but little information is gained from this approach on the dynamics of the feeding process (Mullin, 1969). The population dynamics approach with estimates of growth and mortality, were used to determine production in the copepods such as *Diaptomus salinus* (Yablonskaya, 1962), *Acartia clausi* and *Centropages kroyeri* (Greze and Baldina, 1964), *Acartia tonsa* (Heinle, 1966) and *Calanus plumchrus* (Parsons *et al.*, 1969). These are some of the studies carried out in the sixties and more comprehensive accounts are given in the reviews of Mullin (1969), Mann (1969) and Winberg (1971). There are difficulties in estimating production using this method because of the complex life histories of the planktonic invertebrates, often involving metamorphosis and long periods of almost continuous reproduction.

A technique developed by Boysen Jensen (1919) for measuring secondary production of marine benthic animals laid the foundations of this method. Production of an age class in time interval t_1 to t_2 is given by

$$P = (N_1 - N_2)\ \frac{(\overline{W}_1 + \overline{W}_1)}{2}\ , \tag{6}$$

where N_1 and N_2 are the numbers per unit area at times t_1 and t_2 and $\overline{W}_1$ and $\overline{W}_2$ are the mean weights of the class at times t_1 and t_2

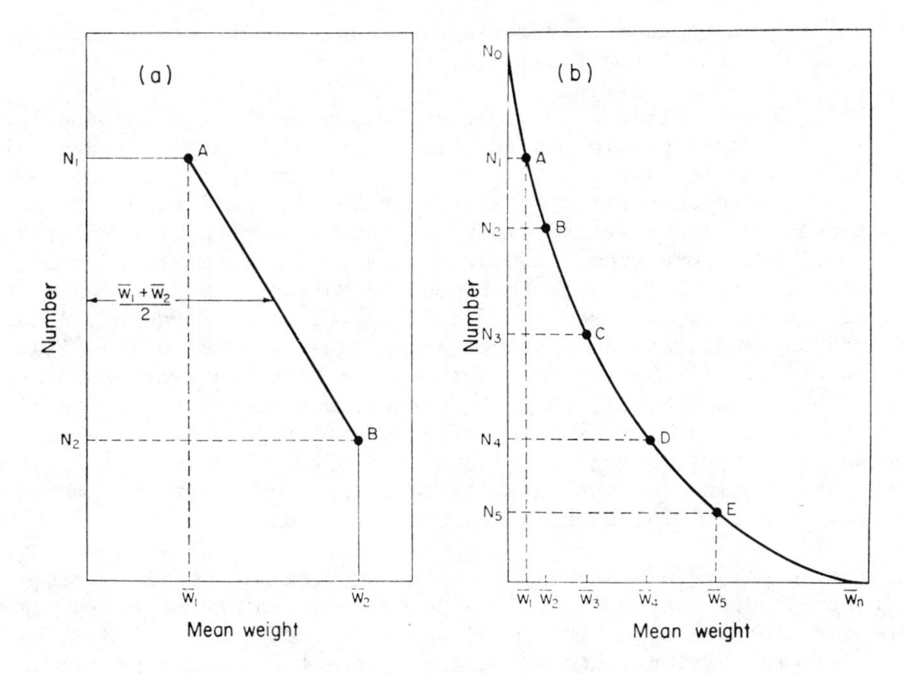

Fig. 2. a) Numbers plotted against mean weight on two occasions. Production is given by $N_1 - N_2.\frac{1}{2}(\overline{W}_1 + \overline{W}_2)$. b) Hypothetical curve relating number to mean weight on five occasions (from Mann, 1969).

(Figure 2a). The production of the population throughout the life history of the age class would be given by the area under the curve $N_o\overline{W}_n$ (Figure 2b) which is

$$\int_{N_o}^{N_n} \overline{W}dN \quad . \tag{7}$$

A similar approach was used by Allen (1951) to measure the production of an age class of fish stock over a short period of time. Production being the product of the average numbers present N and the average weight increment $W_1 - W_2$. The total production for the year class throughout the life being

$$\int_{W_o}^{W_n} NdW \quad . \tag{8}$$

The Allen Curve can be drawn by counting the numbers lost from a population in a specified interval of time and multiplying this by the average weight of those lost, alternatively, by counting the numbers surviving at a particular time and multiplying this by the average weight increment in a specified interval of time. The first estimate is the production removed by predation; the other estimate is the production of the survivors.

There are problems with these methods of estimating production. The most important problem is the assumption of uniform exponential population dynamics where rates of growth and mortality of the cohorts are in the same ratio at all times during the interval t_1 to t_0; it need not necessarily remain constant throughout the interval. Failure of this condition can result in an under-estimate of production (Ricker, 1968, 1978; Le Blond and Parsons, 1978). The difficulty in analysing time series of development stage frequency data, with varying aggregation of stages and frequency of sampling, led Parslow et al., (1979) to apply a technique known as systems identification (Sage and Milsa, 1971) to copepod population data. This technique involves specification of a dynamic population model with unknown parameters; using least squares analysis the model is fitted to the data series and the parameters and production are estimated (Parslow et al., 1979; Sonntag and Parslow, 1981).

The technique was used by Sonntag and Parslow (1981) to estimate secondary production in the following way. Assuming the estimates of stage duration and mortality rates are known, estimates of M_i, the integrated recruitment into age class i, i = 1 ...N, are produced when an age-structure model is fitted. Given the mean weights in each age class, total production can be estimated by

$$P = \sum_{i=1}^{N=1} (M_i - M_{i+1}) . W_i + M_N W_N \quad , \qquad (9)$$

where P = estimate of total production of cohort, M_i = estimate of total recruitment into age class i, i = 1 ... N, N = mature age class (adults), W_i = mean weight of age class i, i = 1 ... N.

Sonntag and Parslow (1981) gave an example where two cohort population models were described and tested against data generated by a simulation model with known production and from this the authors concluded that the technique of systems identification offered a reliable method for estimating herbivore production. Equally good estimates of production were obtained from Winberg's method (1971) when good estimates of stage duration were available and all stages were identified. Classical techniques for estimating secondary production (Edmondson and Winberg, 1971) can be used with time series data alone if stage duration and good numerical abundance data are available. Sonntag and Parslow (1981) point out that this is rarely the case in natural systems and the estimate of production deteriorates if the age class residence time is long compared with changes in recruitment; this is probably the case for most multivoltine species. The systems identification approach has provided reliable estimates of production from the population data measured by Sonntag and Parslow (1981) from experiments carried out in the CEPEX (Controlled Ecosystems Pollution Experiment) enclosures (Grice et al., 1980). The method has been used also by Harris et al., (1982) to estimate production of the grazing herbivores (mainly Paracalanus and Psuedocalanus) in the CEPEX Food Web I experiments; in this experiment the observed data were fitted to model predictions using the pattern of population growth given by Edmondson (1974).

There are problems in estimating the production rate of a plankton population where the population size and age structure appears to be similar on two successive sampling dates (Mann, 1969). The population is either static with a low level of production, or it may be in a state of flux with mortality balanced by recruitment and a high level of production. A method to distinguish between these two conditions was devised by Edmondson (1960) and involved the determination of birth and death rates in a population. The following is taken from Mann (1969). The method of Edmondson involves counting the number of females and eggs and determining the eggs: female ratio. To obtain an estimate of the absolute birth rate the duration of development of the eggs is required. These are determined in the laboratory and the birth rate is calculated from

$$B = E/D \quad , \qquad (10)$$

where B = rate of eggs laid per female per day, E = number of eggs per female in the samples, D = duration of the embryonic stage in days.

The corresponding instantaneous birth rate b' is calculated from B by the formula

$$b' = \ln (B + 1) \quad . \qquad (11)$$

If continuous reproduction is assumed and that the characteristics of a population remain constant for the period between observations, population growth is given by

$$N_t = N_o e^{r't} \quad , \qquad (12)$$

where N_o = the initial population size, N_t = the population size at time t, r' = the instantaneous coefficient of population increase. r' can be calculated from population counts on two successive occasions.

$$r' = \frac{\ln N_t - \ln N_o}{t} \quad , \qquad (13)$$

and from this the mortality rate d' can be calculated as

$$d' = b' - r' \quad , \qquad (14)$$

and the product of martality rate and biomass data gives the production. Further derivations of this formula are given by Caswell (1972), Paloheimo (1974), Tranter (1976), Keen and Nassar (1981), and Taylor and Slatkin (1981) and are discussed by Edmondson (1974, 1977 and 1979).

The ratio of production to biomass (P/B) is an important characteristic of the population of a given species under known conditions. This coefficient is constant for a population with a constant age structure and biomass, but changes in a fluctuating population; it is only a mean value true for a defined period of time. A graphical

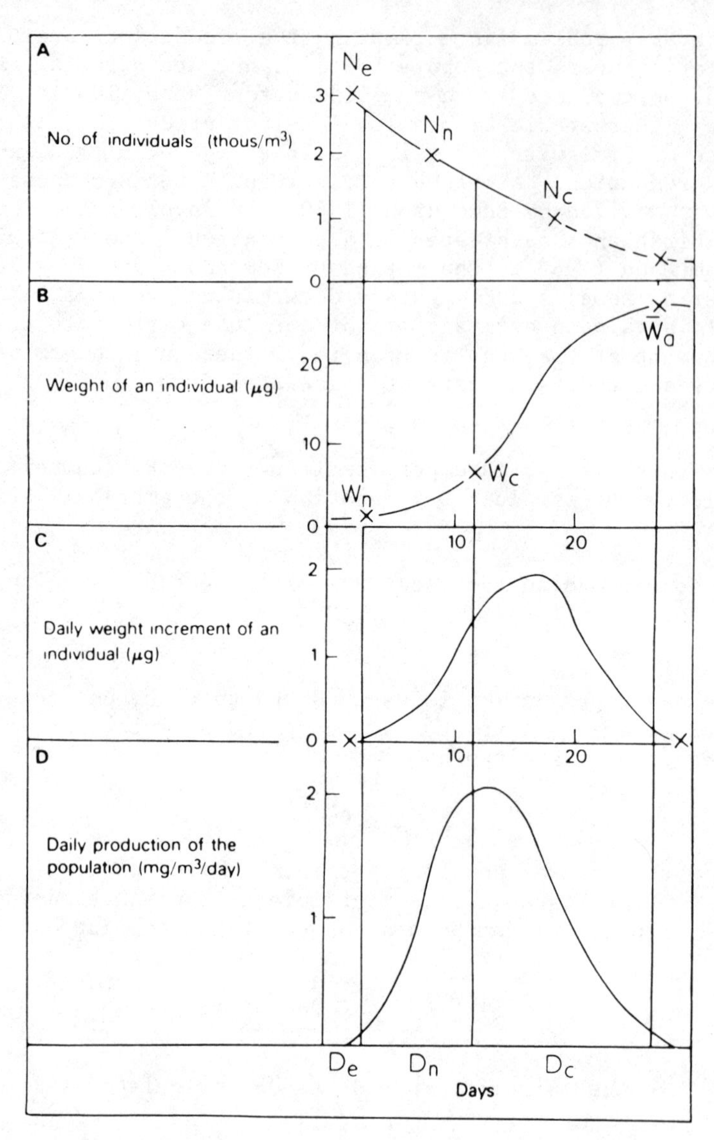

Fig. 3. Graphical method of estimating the production rate of a steady state population N = number of individuals, e = egg; n = nauplius; c = copepodite; a = adult; D = development time; W = initial weight; $\overline{W}$ = mean weight. (From Edmondson and Winberg, 1971).

method of estimating the production rate of a steady-state copepod population is given in Figure 3. The production of the population is given by

$$P = \frac{N\Delta\overline{W}}{t} \qquad (15)$$

N is the number of individuals in the population and $\Delta\overline{W}$ is their mean individual growth over time interval t. Because of the different ages of the individuals in the population and their differing growth rates the growth increments need to be determined for a number of instances in the life of the species. The population of a copepod population would be given as

$$P = \frac{N_e \ \Delta W_e}{D_e} + \frac{N_n \ \Delta W_n}{D_n} + \frac{N_c \ \Delta W_c}{D_c} \qquad (16)$$

where e = eggs, n = nauplii, c = copepodite stage and D_e, D_n, D_c are the development times of the stages.

There are of course methods refined for use in freshwater for univoltine and mutivoltine species. One such method, originally proposed by Hynes (1961), uses size-frequency for estimating primary production of entire benthic faunas, permitting the combination of data on all taxa present. An advantage of this method is that single cohorts within the data need not be identified to calculate production. The method was modified by Hynes and Coleman(1968), Hamilton (1969) and again by Benke (1979) and Menzie (1980) although retaining the advantage of being independent of the need to identify single cohorts. Production is estimated by determining the number of individuals removed between successive size categories and their resulting biomass; production is taken as the sum of these losses. Annual production would be obtained by estimating the total number of individuals that developed into each size category during the year and calculating the losses between size categories and the biomass they represent. The method has been used primarily for single species, but it has been also used with benthic faunas where taxa are combined regardless of life cycle or trophic level (Fisher and Likens, 1973). Annual production is given as

$$P = \sum_{j=1}^{i} (N_j - N_j + 1) \quad (W_j \ W_j + 1)^{\frac{1}{2}} \qquad (17)$$

P is annual production and N_j is the number of individuals that developed into a particular size category (j) during the year. When j = i, $N_{j+1} = 0$ i.e. all the animals have died or have been removed from the system. W_j is mean weight of an individual in the jth size category and $(W_j \ W_{j+1})^{\frac{1}{2}}$ is the geometric mean weight between two size classes. The refinements of this equation are given by Menzie (1980) and Krueger and Martin (1980).

Production/Biomass Approach

Secondary production can be determined from biomass data, when growth and mortality patterns of age composition are unknown, using P/B coefficients. Winberg (1971) expressed the view that production may eventually be determined, not from labour-intensive field and experimental studies, but from biomass estimates and established P/B coefficients. He cautioned that the method for obtaining P/B coefficients must be clearly stated, i.e. what period of the year or to what part of the life cycle they apply and to what conditions of temperature the corresponding values of production and biomass are related. The B/P ratio (production /biomass) is determined from the rate of production of a given species population under known conditions. Production is given by multiplying the P/B coefficient by biomass for the corresponding period of time.

The majority of estimates of production and P/B ratios have been made in northern temperate waters. The P/B ratios and daily production for certain zooplankton species, together with production and production efficiencies of some plankton communities, have been collated by Conover (1979, Tables 7 and 8). The P/B ratio is not related to the size of the producer (Conover, 1979; Banse and Mosher, 1980), but there is a trend in the tables showing that daily P/B values greater than 0.2 are associated with species from shallow coastal environments.

There are difficulties in measuring production of zooplankton from the tropics, with methods such as cohort analysis, because of difficulties in identifying and isolating cohorts from amongst the diverse species complexes found in tropical seas. The variations of zooplankton respiration and excretion rates correlate fairly well with the size of the zooplankton particles which led Le Borgne (1978) to propose a method of measuring the total zooplankton production based on metabolic loss and net growth efficiency (K_2) (Le Borgne, 1978, 1981a, b). The net growth efficiency is derived from C:N:P ratios by the method of Ketchum (1962) and further details of the C:N:P: ratios method, for the measurement of K_2, are given by Corner and Davies (1971) and Le Borgne (1978).

Using this technique Le Borgne (1982) has measured the production of two size fractions (50-200 µm, 200-5000 µm) of the zooplankton collected off the coast of West Africa (eastern tropical Atlantic Ocean). The microzooplankton P:B ratios were 2 to 4 times greater than those for the mesozooplankton and indicated very high turnover times; 6.9 to 1.6d for mesozooplankton and 2.9 to 0.4d for microzooplankton. The P:B values derived by Le Borgne (1982) for mesozooplankton agree with those for tropical regions given in the literature, in spite of the different methods, except those of Binet (1979), which Le Borgne considered to be underestimated, and of Malovitskaya (1971) on two large copepod genera (see Le Borgne,

1982, Table 7). There is less data on microzooplankton P:B values but Le Borgne's (1982) results are in agreement with values from Shushkina and Kisliakov (1975) for nauplii and copepodites. The P:B values of the mixed populations of the tropical regions are greater than those of the temperate or arctic regions (see Conover, 1979 and reviews by Bougis, 1974 and Greze, 1978) and raises the question of the value of the P:B ratio for the assessment of production from biomass, especially in the tropics due to their large variability.

The concept of specific production rate (per unit biomass) in relation to the annual production/mean biomass in the pelagic environment has been used more for the comparison of performance of different species under various environmental conditions than to define production occurring in ecosystems. P/B coefficients have not fulfilled Winberg's expectations of their usefulness but they are quoted extensively in the literature. Recently Banse and Mosher (1980) listed the P/B coefficients of 48 temperate invertebrates attempting to find biological generalisations in the data between P/B and mass as suggested by Dickie (1972). They concluded that there was a single power function governing the mass dependence of the P/B ratios of temperate invertebrates which implies an underlying ecological cause as to why species of so different phyla with different life histories adhere so closely to a line after adjustment of adult mass. Banse and Mosher (1980) believe that when this concept is unravelled it will contribute to our understanding of the size dependence of life processes in the sea.

PROBLEMS INHERENT IN THE METHODS

Tropho-Dynamics Concept

There are a number of problems involved in the study of energy flow in ecosystems which include the variety of measurements and methods used to determine the flux of material such as C, N, O_2 or CO_2. Invariably, energy flux in aquatic systems is quoted in kilocalories, but it is rarely measured directly. Usually measurements of dry weight or carbon are converted to energy units by calorific equivalents taken from the literature. As Rigler (1975) pointed out, there is really nothing wrong with the concept of energy flow provided that accurate conversion factors are available and material flow can be measured accurately; he quoted the ^{14}C method as an example. To date, this long-standing method for the determination of primary production is still being validated (Gieskes *et al.*, 1973, 1979; Williams *et al.*, 1979; Peterson, 1980).

The most accurate method for determining the energy content of organisms is by direct calorimetry. It is surprising, therefore, that since the development of a micro-bomb calorimeter capable of handling individual planktonic organisms (Phillipson, 1964), there have been few comprehensive studies on marine zooplankton. Soviet workers still rely on a wet oxidation technique and coefficients to calculate calorific value from the chemical composition of the organism.

Feeding Relationships

A simple food chain is one in which an animal species (herbivore) feeds on a plant species and is itself preyed upon by another animal species (carnivore). Each of these steps in the food chain is regarded as a trophic level and species having the same feeding relationships and habitats are referred to as being on the same ecological level. If energy is transferred up a food chain through each of the trophic levels, at a fixed level of efficiency, then it can be seen from Figure 4 that the end result is not the same for a simple food web as it is for a food chain. The more horizontal links there are in the food web, the more energy is dissipated within a trophic level; this occurs when an animal feeds on more than one trophic level. This has a direct bearing on the application of the trophodynamic concept to pelagic ecosystems. If the feeding webs are extremely complex then a simple compartment model of trophic level energy transfer is very difficult to apply to these systems.

The complexity of a food web can be seen in the feeding relations of the herring (*Clupea harengus*) given in the classic illustration of Hardy (1924) (Figure 5). The feeding relations of herring represented in this figure are, as Wyatt (1976) points out, misleading. The figure summarises data collated for ten months in two consecutive years and from fish collected over a wide area of the

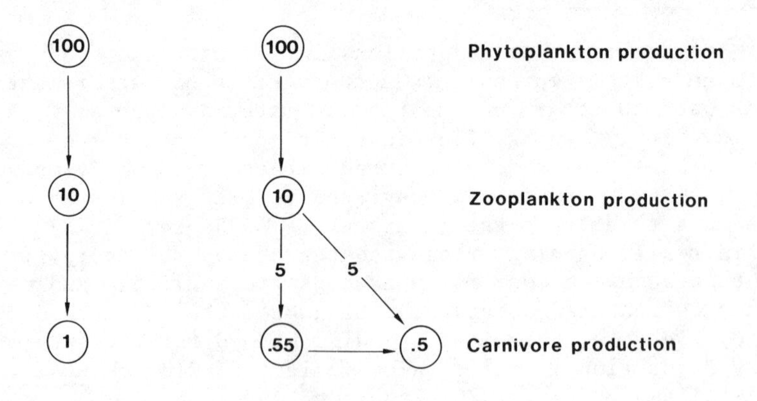

Fig. 4. A simple example of the decreased energy output caused by branching in a food chain, assuming 10% efficiency in transfer through each production level.

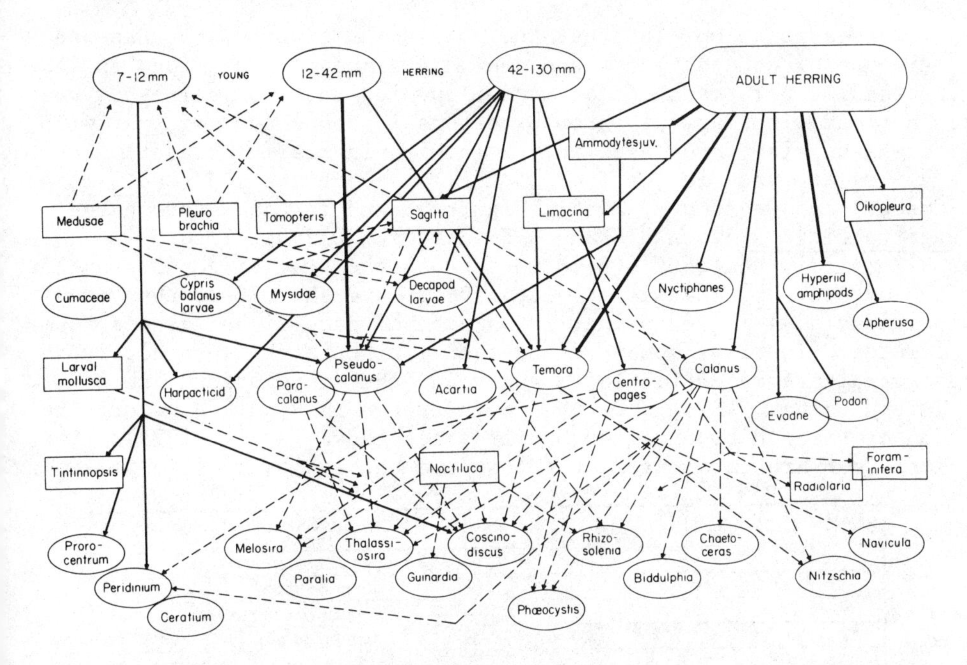

Fig. 5. Feeding relations of the North Sea herring, Clupea harengus, during different stages of its life history. (From Hardy, 1924).

North Sea. The choice of food organisms available to any one size of herring is less than the diagram implies, but it does illustrate the change in diet as the herring matures.

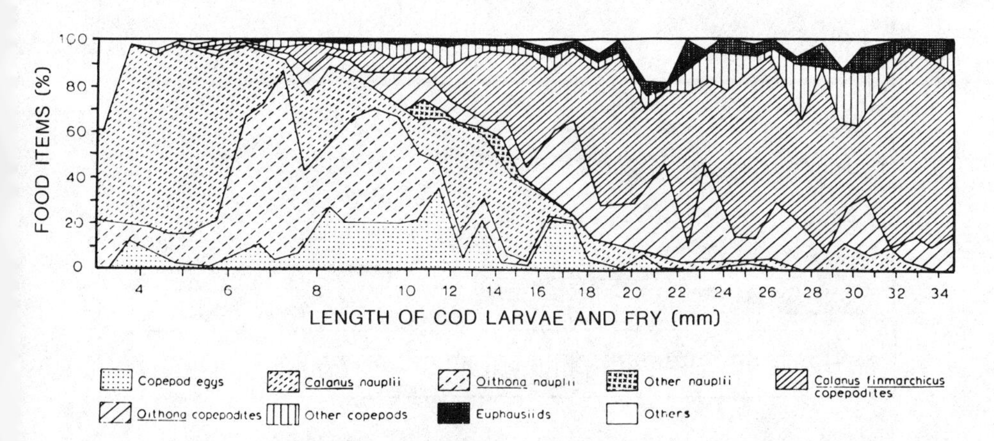

Fig. 6. Changes in the composition of the food of larvae and pelagic fry of cod, Gadus morhua with increasing length. (From Sysoeva and Degtereva, 1965 redrawn by Parsons et al., 1977).

There is an opinion that the complexity of food webs has been over-emphasised and that there are more simple, linear food chains in the feeding relationships of marine species than is realised. What is obvious with many plankton predators is the change of prey with increasing size; prey size increases with increasing predator length. Examples are seen in the food of the larvae of the cod (Gadus morhua) (Sysoeva and Degtereva, 1965), Figure 6, and in the chaetognath Sagitta elegans given by Rakusa-Suszczewski (1969), Figure 7. The smaller individuals of 4 to 8 mm feed on the smaller copepod species such as Oithona. Harris et al., (1982) have shown that the younger individuals <4 mm fed primarily on copepod nauplii but also consumed ciliates. In cases such as this the distinction between herbivore and carnivore is difficult to apply and the designations are of limited value. This is further illustrated in Figure 8, where the food of sixteen tropical euphausiids are shown (Roger, 1973). The first four are carnivores, the last two herbivores and the remaining

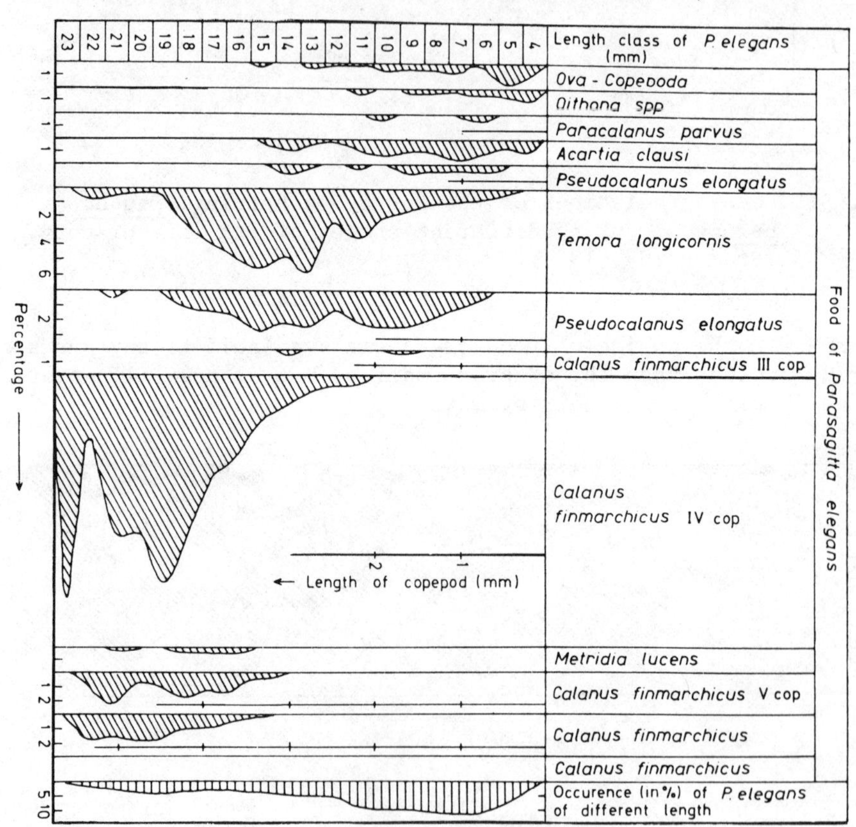

Fig. 7. Percentage composition of food of Sagitta elegans in relation to predator size. The food organisms listed at the top of the diagram are in order of increasing size (from Rakusa-Suszczewski, 1969).

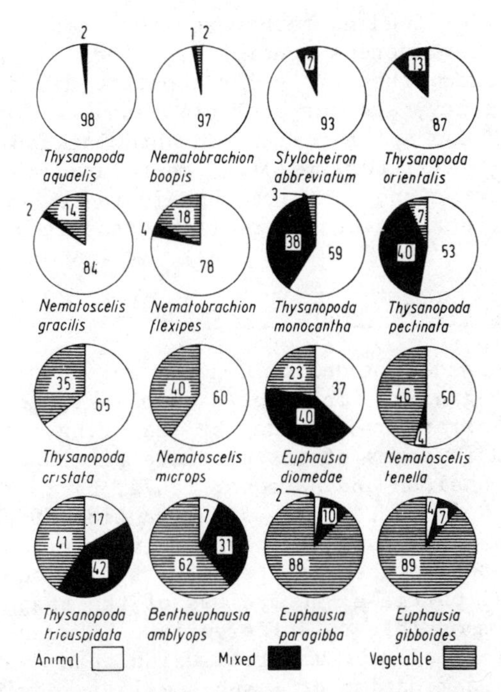

Fig. 8. Relative quantities of phytoplankton and zooplankton eaten by sixteen species of tropical euphausiids. (From Roger, 1973).

species are omnivores. Euphausiids also possess the ability to change their diet from carnivore to herbivore (Parsons and Le Brasseur, 1970), therefore the feeding regimes given by Roger (1973) for the tropical euphausiids are possibly not fixed.

The food web with many trophic levels is probably a more accurate representation of the marine ecosystem than is the simple food chains, although there are many examples of highly efficient single predator-prey relationships in the pelagic environment. Such a specialised example is seen in the feeding of the pteropod mollusc Clione limacina on the mollusc Limacina retroversa, which is primarily herbivorous (Conover and Lalli, 1972, 1974). Wyatt (1976) contends that all food chains are simple and short lived and as animals grow and their diet changes so does the size of their food. It is the summation of these simple chains over long periods of time that gives rise to the apparent complexity of the food web. This principal may be applicable to zooplankton carnivores, but it cannot be applied to the filter-feeding herbivorous copepods. If, through prey size selectivity, herbivorous species selected larger cells from phytoplankton species, then the cells about to divide would suffer preferential mortality, and our present technique for estimating grazing by zooplankton and mortality on phtoplankton would require drastic revision (McAllister, 1970). Age-related changes in the diet of certain zooplankton species, from herbivore to omnivore and then to carnivore, limits the usefulness of the trophic level concept.

Measurement of Natural Food Concentrations

The weight dependent model of Conover and Huntley (1980) derives an estimate of the grazing impact of a zooplankton community on natural food concentrations obtained by measuring the change in concentration of food in terms of its particle size spectrum. The Coulter counter® (Sheldon and Parsons, 1967a, b) is the instrument on which this approach is based. The size fraction of particulate material is analysed by measuring particle volume, and size in linear dimensions is related to the diameter of a sphere equivalent in volume to the original particle; regardless of the shape of the original particle. It is impossible to differentiate between non-living detritus and growing cellular material using this instrument, therefore, microscopic examination of a sub-sample is advisable to complement the technique. The Coulter counter has been used extensively in oceanographic research for determining seston size particle distributions (Sheldon et al., 1972; Haffner and Evans, 1974) and zooplankton grazing (Parsons et al., 1967; Parsons, 1969; Parsons and Le Brasseur, 1970; Poulet, 1973, 1974, 1976; Poulet and Chanut, 1975; Richman et al., 1977; Gamble, 1978). There are limitations to the technique because the volume of particles are measured regardless of particle shape. Particles are retained by the filtering mechanisms of the copepod on the basis of their maximum dimensions. There are other problems in accurately sizing particles (Karuhn et al., 1975; Kachel, 1976), measuring long thin algae or chains (Vanderploeg, 1981) and in interpreting the results from Coulter counter studies (Harbison and McAlister, 1980) but these do not detract from the usefulness of the instrument. Problems also occur in measuring copepod grazing rates, such as, post-capture rejection of particles (Donaghay and Small, 1979), food quality, previous feeding history, presence of non-food particles and alteration of the copepods filtering movements, which have to be considered as possible mechanisms for modification of the filtering pattern (Donaghay, 1980).

Other instuments have been developed and used for the continuous counting of particles in the zooplankton size range (Cooke et al., 1970; Boyd and Johnson, 1969; Boyd, 1973; Mackas, 1976; Pugh, 1978). The data has been used to investigate the small-scale horizontal and vertical distributions of zooplankton (Denman and Mackas, 1978; Pugh, 1978).

Sampling

To determine the productivity of any plankton community is an immensely difficult task; Mullin (1969) considered it to be an intractable problem which we may never have the complete expertise to solve. In Mullin's view the single most important problem in research on production in the late sixties was how to take and evaluate samples from populations of zooplankton on successive occasions. Sequential sampling of a pelagic population is extremely difficult because of the limitations of plankton sampling equipment, water movement, dispersion and plankton patchiness. These problems must be overcome before the study of the population concept can be properly researched because the calculations of growth and mortality rates are generally based on comparisons between successive samples thought to come from the same population of organisms (Mullin, 1969). The accurate sampling of plankton is considered one of the major problems in marine science (Cushing, 1970; Tranter, 1976; Steele, 1978; Conover, 1979; and van der Spoel and Pierrot-Bults, 1979). Steele (1978) concluded that it was impossible to obtain a true time-series of events in the open sea, although approximate time sequences may be possible in special circumstances such as Gulf Stream Rings (Wiebe et al., 1976). An outcome of these difficulties in sampling plankton was the effort committed over the last decade to the study of plankton patchiness which has provided a basis for the understanding of spatial variability and has shown the importance of biological and physical factors in controlling ecosystem structure.

Plankton Patchiness

The variability in zooplankton abundance caused by plankton patchiness is one of the most significant factors contributing to the complexity of interpreting zooplankton distributions and time series data (Haury et al., 1978). A conceptual framework for time-space scales of variability is shown in Table 1 and Figure 9 (from Haury et al., 1978); these illustrate the forms of spatial patterns which can occur and the physical and biological mechanisms which probably create them. The problems of space-time interactions of phytoplankton, zooplankton and fish, where their respective horizontal distributions, life cycles and physical size are considered, are shown in Figure 10 (Steele, 1978). Using the general relationship between equivalent diameters and duration of life, shown by Sheldon et al., (1972), Steele compared the distribution scales of phytoplankton, zooplankton and fish with the types of sampling programmes carried out at sea. To understand the dynamics of these populations, sampling and theoretical studies should be concerned with interactions along the diagonal in Figure 10. Steele (1978) recognised that the former was logistically impossible in that sampling of plankton communities is a compromise only accurate within poorly defined limits. To explain the variability of chlorophyll data and zooplank-

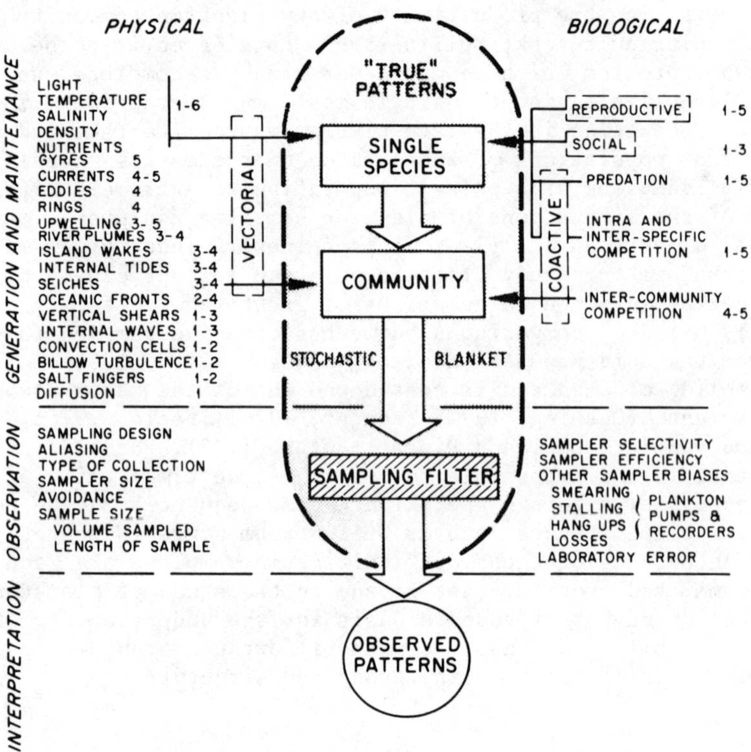

Fig. 9. Factors in the generation, maintenance and observation of plankton patterns. (From Haury et al., 1978).

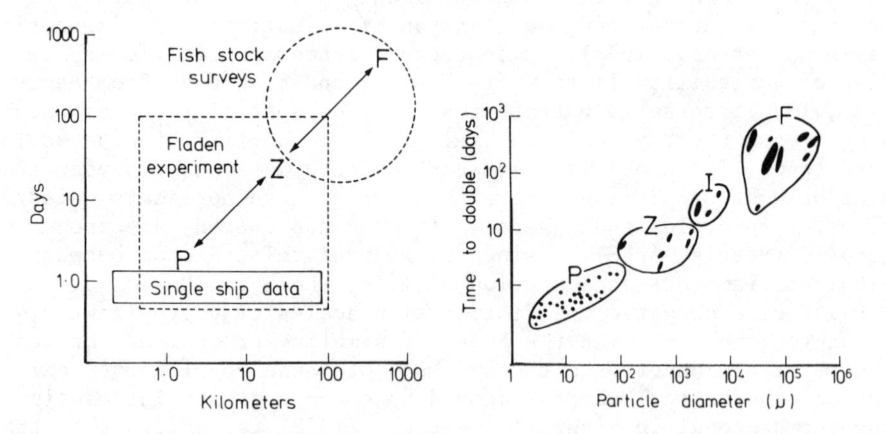

Fig. 10. a) A simplified representation of the time and space scales associated with, phytoplankton (P), herbivorous zooplankton (Z) and pelagic fish (F) with an indication of the various type of sampling programmes.

Table 1. Categories of Scales of Pattern.

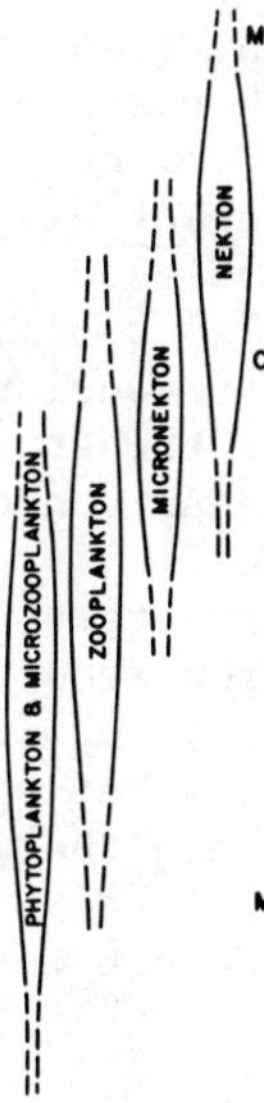

NAME	SPACE SCALE	DOMINANT PATTERN*	HOW BEST LOOKED AT	WHAT WE LEARN
MEGA	10^4 km	Vectorial	Communities Biomass Species	Biogeography Evolutionary history
MACRO	10^3 km	Vectorial Reproductive	Communities Biomass Species	Biogeography Speciation "Best" places to live Ecotones Inter-community competition "Hot spots" within ecosystems
MESO	10^2 km	Vectorial Reproductive	Biomass Species	Faunal boundaries Invasions Nekton ambit Genetic selection Relationship to environmental parameters
COARSE	10 km 1 km	Vectorial Reproductive Coactive Social	Species	Intra-community competition Upwelling responses Micronekton ambit Rclationship to environmental parameters
FINE	100 m 10 m	Vectorial Reproductive Coactive Social	Species	Coexistence, niche partitioning Inter-and intra-species competition Predation Food densities required Zooplankton ambit Relationship to environmental parameters
MICRO	1 m 10 cm 1 cm	Vectorial Social	Species Individual	Inter-and intra-species competition Niche partitioning Relationship to environmental parameters

* Stochastic acts on all scales

ton abundance in temporal sequences of data in terms of the effects of physical conditions, nutrient limitation or grazing is extremely difficult. It must also be realised that the variability of the physical environment extends over the same range of scales. The alternative approach is to characterise and explain the nature of the variability in terms of its structure at different spatial scales using theroretical studies (Denman and Platt, 1976; Fasham, 1978; Horwood, 1978; Steele and Henderson, 1979). The significance of horizontal and vertical variability of zooplankton distributions is reviewed by Longhurst (1981), who concluded that there may be no recurrent pattern in horizontal patchiness, but that there is a strong and consistent vertical pattern in the zooplankton distributions caused by diel, reproductive, ontogenetic and seasonal migrations.

←

Fig. 10. (see opposite)
(From Steele, 1978). b) Relationship between doubling time and particle size from data for phytoplankton (P), herbivores (Z), invertebrate carnivores and omnivores (I) and fish (F). (From Sheldon et al., 1972, redrawn from Steele, 1978).

Enclosed Ecosystems

To overcome the effects of dispersion, patchiness and the difficulties associated with sampling at sea, scientists have attempted to study the functioning of ecosystems in the laboratory. This has led to development of microcosms - large containers which enclose several elements of the biota and allow observation and manipulation. A review of the types of enclosures used in recent yea is given by Boyd (1981). The largest are those of the Controlled Ecosystems Pollution Experiment (CEPEX) projects, where volumes of $1335m^3$ are enclosed. In the Food web I experiment (Grice *et al.*, 1980) a number of trophic levels were maintained in 2 enclosed systems; they were constructed with sufficient depth to allow zooplankton to migrate in relation to vertical phytoplankton distributions. This experiment was designed to test the hypothesis that the structure of the phytoplankton community determines the type and number links in the food chains and controls the transfer efficiency from primary production to the higher trophic levels (Greve and Parsons, 1977; Steele and Frost, 1977). It develops the suggestion made by Pimm and Lawton (1977) that the number of trophic levels in ecological communities is determined not by ecological energetics but by population dynamics of the ecosystem. May (1979) concluded that, although Pimm and Lawtons idea was broadly supported by certain Lokta-Volterra model studies, the "notion must remain subject to mistrust".

Using the CEPEX enclosures a study was made by Harris *et al.*, (1982) on the primary production, population dynamics, feeding behaviour and seconday production of herbivorous and carnivorous zooplankton. These experiments were conducted over 40d during whic the development of individual cohorts were followed, feeding relationships were identified and transfer efficiencies measured. Th food chain efficiency between primary producers and the herbivores (primarily *Paracalanus* and *Pseudocalanus*) was 6.5% and the transfer efficiency between herbivores and carnivores was estimated at 10%. The validity of these efficiencies would be questioned if applied t the open sea system but that was not the object of the experiment. The detailed insights which are gained into the dynamics and functioning of the enclosed ecosystem are more important.

STUDIES OF WHOLE ECOSYSTEMS

To obtain an accurate estimate of secondary production of an ecosystem is a daunting task but not an impossible one. Initially, a simple model is used to predict annual heterotrophic production a a function of net primary production (which sets constraints of heterotrophic production), consumption, assimilation and growth efficiencies of the species populations, which are organised into food webs or trophic structures. The secondary production is a

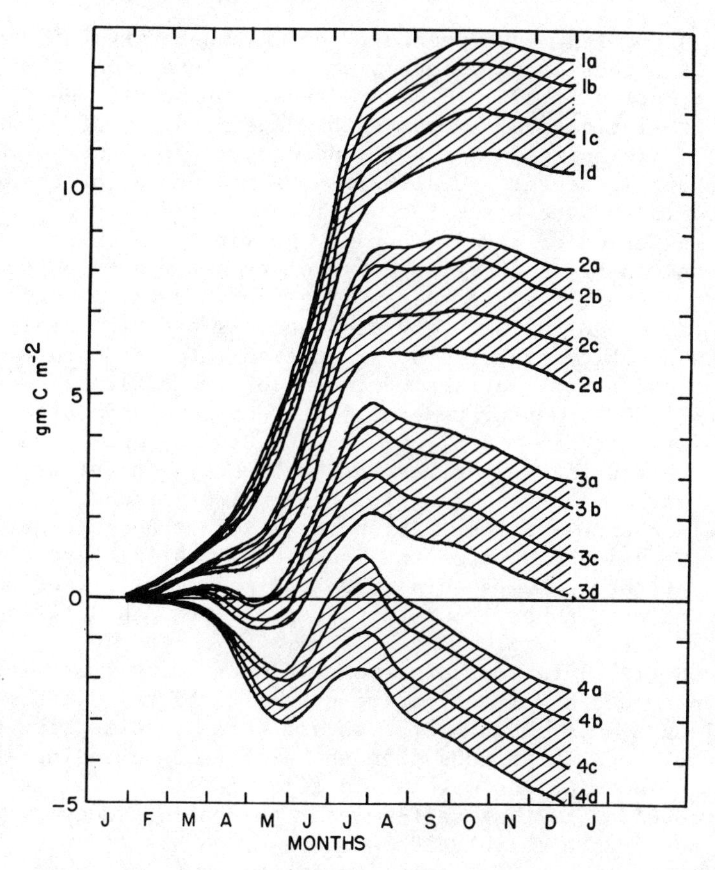

Fig. 11. Cumulative totals of monthly secondary production estmated from field data from Ocean Station P, on the assumption of different grazing schemes, and zooplankton respiratory rates. Curves 1, 2, 3 and 4 show the production estimated on the assumption of zooplankton respiratory rates of 4, 6, 8 and 10% of the body weight, per day, respectively. Curves marked a, and b were estimated assuming nocturnal grazing, each with a different rate of decline of feeding rate. Curves c assume nocturnal grazing at a constant rate equal to the mean of the variable rate. Curve d assumes continuous grazing. (From McAllister, 1970).

logical consequence of combining these factors. Production at the population level may be estimated from changes in population biomass and elimination. The sum of the production of heterotrophic populations across a trophic sequence, from an herbivore (Calanus) to a primary carnivore (Sagitta), and on to a secondary carnivore (Pleurobrachia), is often taken as secondary production.

There are a few measurements of pelagic heterotrophic productivity of either trophic levels or of complete ecosystems (Riley, 1946, 1947; Petipa et al., 1970). More attention has been paid to the study of single species and to the determination of growth and ecological efficiencies, (Marshall and Orr, 1955; Corner, 1961; Conover, 1966; Corner et al., 1967). The production of the planktonic community of the Black Sea was investigated by Petipa et al., (1970) for a period of 4 days in early summer by determining the flow of matter and energy and efficiencies of transfer between trophic levels. To account for all the feeding types in the planktonic community the organisms were divided into six trophic levels. Although reasonable estimates of production were obtained too many assumptions and approximations were made of important rate processes. McAllister (1970) demonstrated the effect on secondary production of assuming different diel herbivore grazing schemes and respiration rates for a given rate of primary production and mortality, (Figure 11). Very marked changes occur in the level or production for relatively small changes in respiration (4 to 10% of body weight). Diel grazing of pelagic herbivores is a subject which still requires a great deal of research, especially as the majority of feeding experiments carried out in the laboratory are of 12 to 24h duration.

Gulland (1970) shows that errors in estimating ecological efficiency introduce a multiplicative error in the calculation of production along a food chain. If fish are three feeding stages away from their phytoplankton food then the difference of using an efficiency of 11 or 14% can make a two fold difference in the estimate of fish production. It is self-evident that each and every variable and rate used in the calculation of production has the potential, if not measured correctly, for creating errors in the estimate of production. However, it can be seen from the examples given that extra care and precision is required in the measurement of rate processes which are applied to each of the trophic levels.

Further study on the temporal changes and latitudinal effects on production processes and more measurements of the processes and rates of different ecosystems are needed. Temporal changes in plankton ecosystems of the North Sea and north-east Atlantic Ocean for the period 1948 to 1972 were observed by Glover et al. (1972). The decline in copepod numbers and zooplankton biomass, together with the delay in the onset of the spring phytoplankton bloom, is shown in Figure 12. The data in this figure represent the most comprehensive set of temporal data collected in the study of long term variability of plankton ecosystems. The latitudinal differences in seasonal amplitudes of algal and zooplankton biomass and production have been shown by Cushing (1959) and Heinrich (1962), Figure 13, and quoted widely in the literature (Parsons et al., 1977; Conover, 1979; Nemoto and Harrison, 1981). However, as Conover (1979) points out, variations from these classical patterns are probably nearer the rule than exception. His point is well illustrated in Figure 14 (from

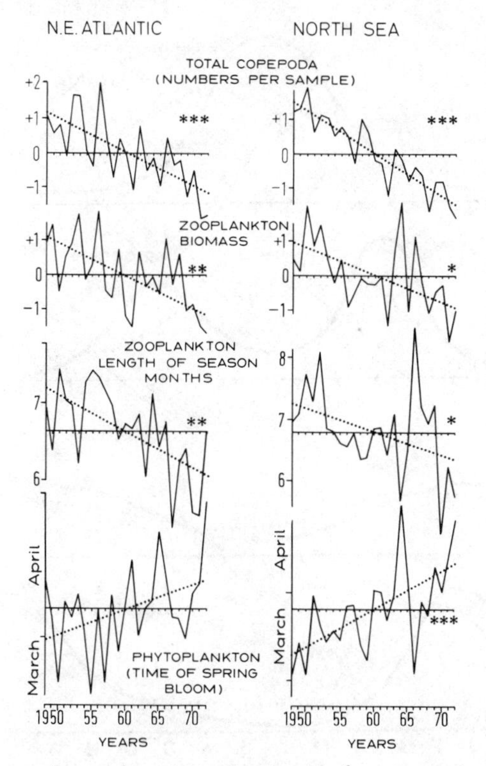

Fig. 12. Fluctuations in the plankton in the north-east Atlantic Ocean (sub-areas C4, C5 and D5 combined, see Figure 14) and the North Sea (sub-areas C1, D1 and D2 combined). The results in the top two pairs of graphs are standardised about a mean of zero, the scales in standard deviation units being shown in the left hand of each pair of graphs. The scales for the bottom two pairs of graphs are plotted about the overall means for 25 years. Calculated trend lines are drawn, the significance of the fit being shown by one, two or three asteristks for P = <5.0%, <1.0% and <0.1% respectively. The line for the bottom left graph just falls short of significance at 5%. (From Glover et al., 1972).

Colebrook, 1979), where the seasonal abundances of phytoplankton and copepods are shown for the northern North Atlantic ocean. Similar regional variabilities in abundance of phytoplankton and zooplankton probably occur in the other three areas given in Figure 12; unfortunately, temporal studies of this nature have not been conducted in these regions. As well as the latitudinal differences associated with the central oceanic areas there are the different ecosystems of neritic or shelf regions, the areas of boundary currents and upwelling regions. Each of these areas is characterised by physical,

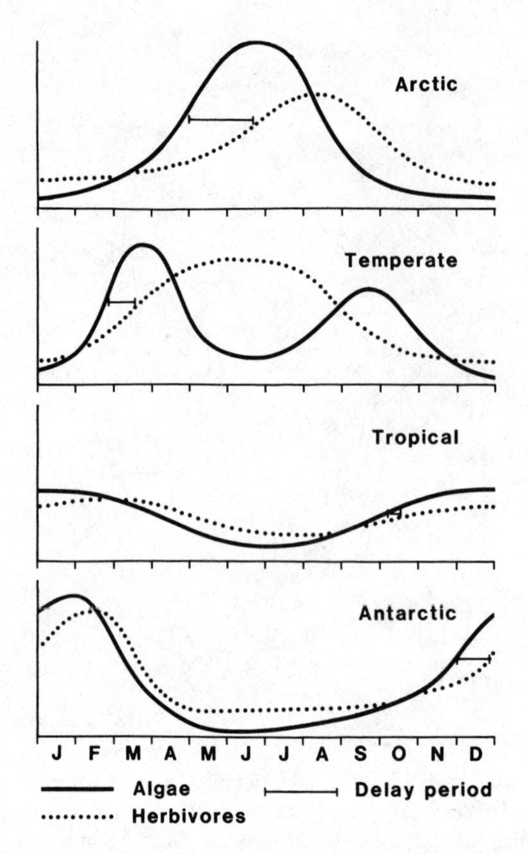

Fig. 13. Diagrammatic representation of the seasonal amplitudes in algal and herbivore production at different latitudes. (Redrawn from Cushing, 1959; Conover, 1979 and Nemoto and Harrison, 1981).

chemical and ecological properties and by specific patterns of speciation and distribution of plankton (Pierrot-Bults and van der Spoel, 1979).

The high latitude ecosystems of the Arctic and Antarctic Oceans are reviewed by Nemoto and Harrison (1981) and include the numerous studies carried out in the Antarctic in recent years. The season, which is short in comparison to the temperate and tropical regions (Figure 13), prohibits some of the zooplankton species from completing one generation within the season. The food chains of the Antarctic ecosystem, and to some extent the Arctic, have three/four trophic levels only, with the top predators being mammals or birds (Figure 15). Most studies of production in the Antarctic have been concerned with the top predators but, since the demise of whaling, more interest has been shown in estimating the standing stock of euphausiids (*Euphausia superba*) (Everson, 1977; El-Sayed, 1978).

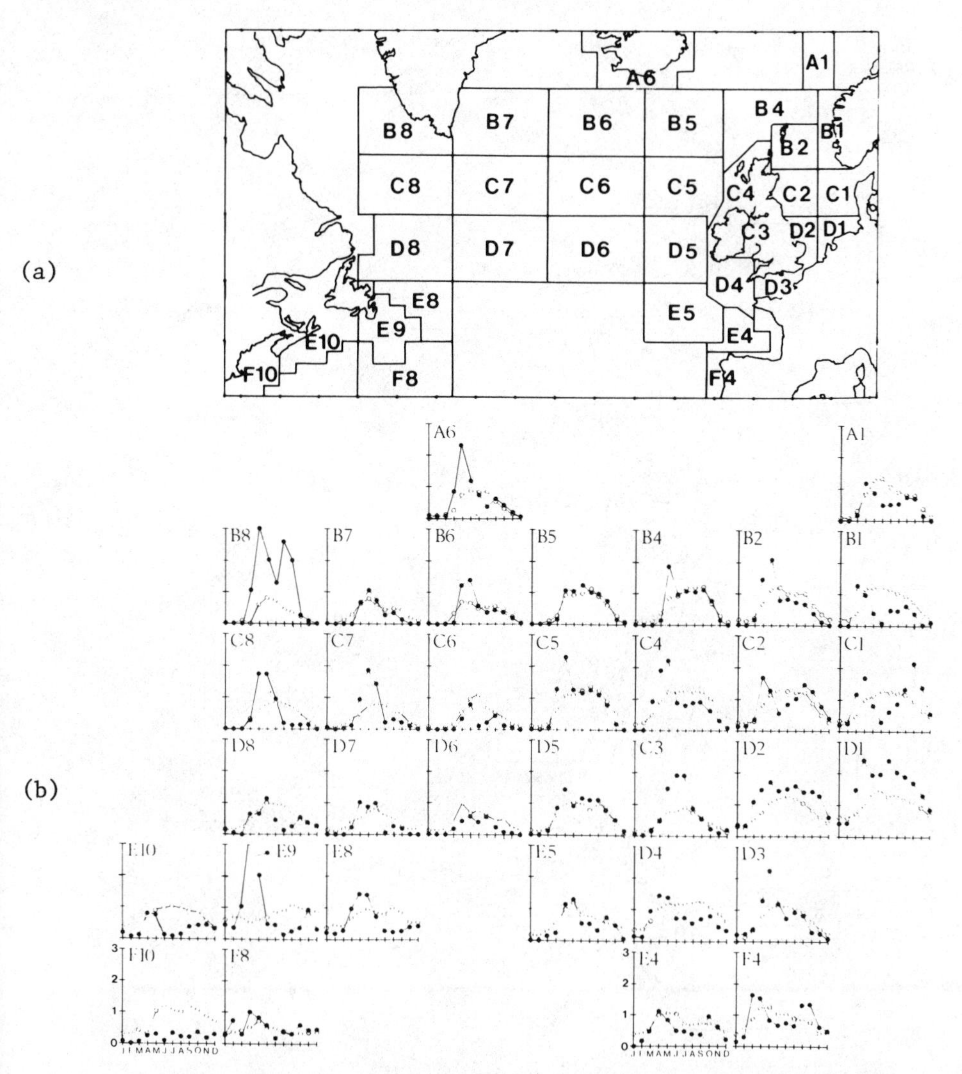

Fig. 14. a) Chart of the North Atlantic Ocean, showing subdivision into areas used in the Continuous Plankton Recorder survey. b) Seasonal variations in abundance of phytoplankton (filled circles), and copepods (open circles) in each of areas shown in Figure 14a. Phytoplankton data are arbitrary units of greenness and copepod data are logarithmic means of numbers. (From Colebrook, 1979).

Similar reviews have been made of low latitude gyral regions (Blackburn, 1981), coastal upwelling ecosystems (Barber and Smith, 1981), equatorial upwelling (Vinogradov, 1981) and shelf sea ecosystems (Walsh, 1981).

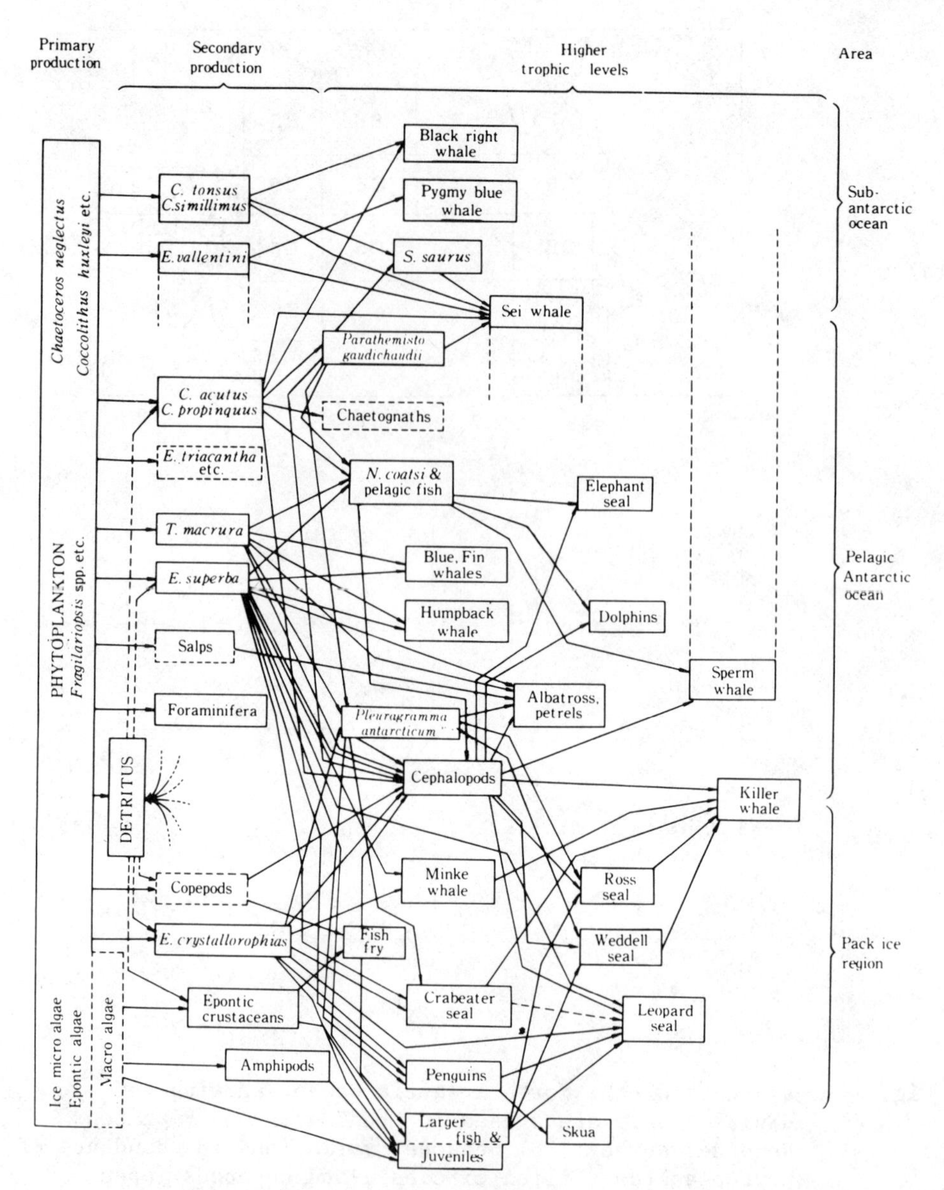

Fig. 15. Main food chains in the Antarctic Ocean (Nemoto and Harrison, 1981). Non swarming plankton are shown in the broken line boxes.

The basic structure of the food webs in the four types of ecosystems shown in Figure 13 probably does not vary greatly latitudinally, but there is a tendency for greater diversity of species away from the poles and away from the land. The strong seasonal signal in the high latitudes diminishes towards the equator, suggesting that the cycles of animals get more synchronous as the climate gets warmer. The lag in the increase of standing stock after the onset of the spring phytoplankton bloom is most marked in the Arctic and temperate waters (Figure 14). It must be remembered that what is usually measured as net zooplankton is frequently caught in a 300 μm mesh net, which poorly samples young copepodites and fails to sample nauplii of herbivorous copepod species such as *Calanus finmarchicus*, *C. helgolandicus*, *C. glacialis* and *C. hyperboreus*. The lag seen in these high latitude ecosystems roughly corresponds to the development time from egg to late copepodite stage, when these numerically abundant species are efficiently sampled. On this basis, the main spawning of these copepods and other zooplankton herbivores is finely tuned to the onset of the phytoplankton bloom. This reconfirms the close relationship suggested by Steeman Nielsen (1957) between the cycles of phytoplankton and zooplankton herbivore production and relies on the basic theme that reproductive success depends on high fecundity being synchronized with phytoplankton blooms. Alternatively, the reproductive success of a herbivore relies on chance "match or mismatch" with its food as suggested for fish larvae by Cushing (1966). In contrast, there are some shelf-sea sites, such as the Southern Bight of the North Sea, where a complete imbalance between the primary production and secondary production is seen; the copepods are completely out of phase with the spring bloom and consume a small percentage of the bloom (Gieskes and Kraay, 1977).

There are other methods of researching the production of individual herbivores, populations or certain size fractions of the zooplankton; one of these utilises radioactive tracers. There are many problems using isotopes to measure the passage of material in food chains, these are given by Edwards and Harris (1955), Conover and Francis (1973), Tranter (1976) and Smith and Horner (1981). Although, with proper labelling techniques and controls to account for a complete tracer analysis, they have been used for studying feeding behaviour of marine zooplankton (Griffiths and Caperon, 1979; Smith *et al*., 1979). A promising method for determining *in situ* zooplankton grazing on natural food concentrations, of autotrophic and heterotrophic particulate matter, has been used by Roman and Rublee (1981). Their technique is to label autotrophic particles with $Na^{14}CO_3$ and heterotrophic particulate matter with (methyl-^{3}H)-thymidine and incubate in a Niskin-type water bottle (Haney, 1971). On completion of the experiment the isotope uptake in various size fractions is measured to assess the grazing impact of the various size groups of zooplankton.

A new technique has been devised by Landry and Hassett (1982) to measure the grazing impact on natural food concentrations by the microzooplankton; an important component of the zooplankton.

The Role of Microheterotrophs in Secondary Production

Approximately 5 to 15% of plant vegetation in terrestrial ecosystems is eaten by herbivores; the remainder goes to the decomposer cycle. Because herbivores consume part of the plant, usually the leaves, leaving the rest of the plant to regenerate, they do not effectively control the biomass of land vegetation. Consequently, the balance between plant and herbivore cannot be simply a matter of available food supply; other limitations, such as control of the herbivore, must exist in plant-dominated terrestrial ecosystems (Crisp, 1975). There is an entirely different mechanism in marine ecosystems in that a large proportion of plant production of the open sea is thought to be eaten solely by the herbivores. The phytoplankton survive because the remaining cells become too sparse to support continual herbivore grazing yet retain sufficient potential to recover after the herbivore population has diminished. How the herbivore zooplankton can get enough to eat from the densities of phytoplankton normally found in the sea, especially in the summer months in temperate regions, is still uncertain (Steele, 1974); although detritus is known to supplement their food supply. There is a growing awareness that microheterotrophs may be fulfilling an important role in the functioning of pelagic ecosystems (Sorokin, 1981; Williams, 1981) and that they are a further source of food for herbivores. Microheterotrophs include microflora (bacteria, fungi and yeasts) and microplankton (2 - 200 μm). The size range covers the two groups, nanoplankton (2 - 20 μm) and microplankton 20 - 200 μm) (Dussart, 1965). Microzooplankton consists of protozoans (zooflagellates, planktonic amoebae, foraminifera, radiolorians and ciliates, including tintinnids) and the young stages (nauplii and eggs) of the larger zooplankton species. From studies in various biotas microheterotrophs appear to be responsible for 20 to 80% of total heterotrophic metabolism and production (Pomeroy and Johannes, 1968; Sieburth, 1977; Sorokin, 1975, 1978). In one of the CEPEX enclosure experiments Williams (1982) found that 70% of all respiration was associated with organisms <10 μm and 50% with organisms <1 μm; bacteria was responsible for approximately 45% of the calculated respiration. Although there are large standing crops of microheterotrophs in tropical oceans (Vinogradov, 1981), it has been suggested that the role of microheterotrophs is greater in temperate waters where the maximum of the phytoplankton standing crop is separated in time from that of the zooplankton biomass (Sorokin, 1981). It is thought that a trophic 'function' of microheterotrophs in temperate waters is to transfer the accumulated organic matter derived from the decay of the phytoplankton bloom into new particulate microheterotroph protein. Their role in this process, I believe, cannot be questioned. The suggestion that, by retaining energy within the system and forming the food for the development of the zooplankton populations, they are effectively bridging the lag between the time of the peak of the phytoplankton bloom and the zooplankton peak, in temperate and northern waters, is more speculative.

Microheterotrophs form part of the whole particle spectrum available to the zooplankton consumers. Foraminifera and radiolarians are known to feed on flagellates and phytoplankton (McKinnon and Hawes, 1961) and foraminifera form part of the diet of copepods (*Pleuromamma robusta*) and molluscs (*Limacina retroversa*). The common shelf-sea copepods, *Paracalanus parvus* and *Pseudocalanus elongatus*), have setule spacings <2 μm, on their feeding appendages (Schnack, 1982), indicating that they possibly filter the smaller flagellates. Tintinnids feed on detritus, bacteria, flagellates but primarily on nanophytoplankton (Johansen, 1976; Burkill, 1982) and are themselves eaten by many zooplankton species, e.g. *Tintinnopsis* by herring larvae (Hardy, 1924). In general, ciliates, especially tintinnids, are important in near shore pelagic food webs and it is suggested (Burkill, 1982) that they may consume, on average, 60% of the annual primary production and may be responsible for the summer decline in nanophytoplankton. The grazing impact of microzooplankton on natural communities of marine phytoplankton in the coastal waters off Washington, USA, has been investigated by Landry and Hassett (1982). Using a 'dilution' technique in their experiments, they concluded that microzooplankton, primarily copepod nauplii and tintinnids, consumed 6 to 24% of the phytoplankton standing crop which was equivalent to 17 to 52% of the daily production. Non-loricate ciliates accounted for 80 to 90% of the microzooplankton numbers but they appeared to contribute little to phytoplankton (diatom) mortality. The authors suggested that their greatest impact was probably on the bacteria and microflagellates.

Zooplankton, which use mucous membranes or food webs to obtain their food, capture and utilise microheterotrophs as well as autotrophs. Pteropod molluscs feed on diatoms, radiolarians and foraminifera (Gilmer, 1974). Appendicularians (*Oikopleura*) retain particles down to 0.1 μm diameter (Jorgensen, 1966) which enable the species to filter bacterioplankton (King *et al*., 1980), although King *et al*. concluded that the grazing by appendicularians had a negligible impact on the bacteria population, even though bacteria were the major food source of the species. Microflagellates also consume bacterioplankton (Haas and Webb, 1979; Fenchel, this volume). The review by Alldredge (this volume) demonstrates the importance of gelatious zooplankton, such as coelenterates, ctenophores, molluscs and tunicates. These gelatinous groups have high daily food rations at high food abundance, high fecundity, short generation times and exceptionally high growth rates, all attributes which enable gelatinous zooplankton to exploit abundant or patchy prey populations.

There is a long-standing discussion of the relative importance of bacterial and autotrophic production in the ocean and shelf-sea ecosystems but bacteria and microflagellates together account for a substantial amount of plant respiration in coastal waters (Williams, 1982). Bacterial production of 5 to 25% of primary production has been measured by Hollibaugh *et al*., (1980) and Fuhrman and Azam (1982) and, assuming a 50% assimilation efficiency, the bacterioplankton would consume 10-50% of total fixed carbon. There are no current estimates of grazing rates on bacteria but if they were in a steady state the production rates would equal grazing rates. Their contribution to the biomass of an equatorial upwelling region is shown by Vinogradov (1981) where at one station (97°W) bacterioplankton reached a value of 2.5 x 10^6 cells ml^{-1} in the upper 50m, about 300 mg m^{-3} and accounted for 30% of phytoplankton biomass; bacterial biomass was greater than the ciliate and flagellate biomass in this upwelling region.

The notion of marine microhetertrophs as 'decomposers' responsible for remineralisation may be misleading since their growth efficiency appears to be greater than that of zooplankton (Williams, 1982). The role of microheterotrophs in recycling organic matter requires careful reassessment; any study or simulation of energy flux in an ecosystem must include a microheterotroph/microbial compartment (Pomeroy, 1974).

Beers and Stewart (1971), working in the eastern tropical Pacific, suggested that the microzooplankton might consume about 70% of primary production in tropical oceans. Bacteria and unicellular heterotrophs were included by Vinogradov and Shuskina (1978) in their coastal upwelling study off Peru, Table 1, Figure 16. They considered that the feeding relationships and trophic exchanges were the most important processes determining the biological structure of the

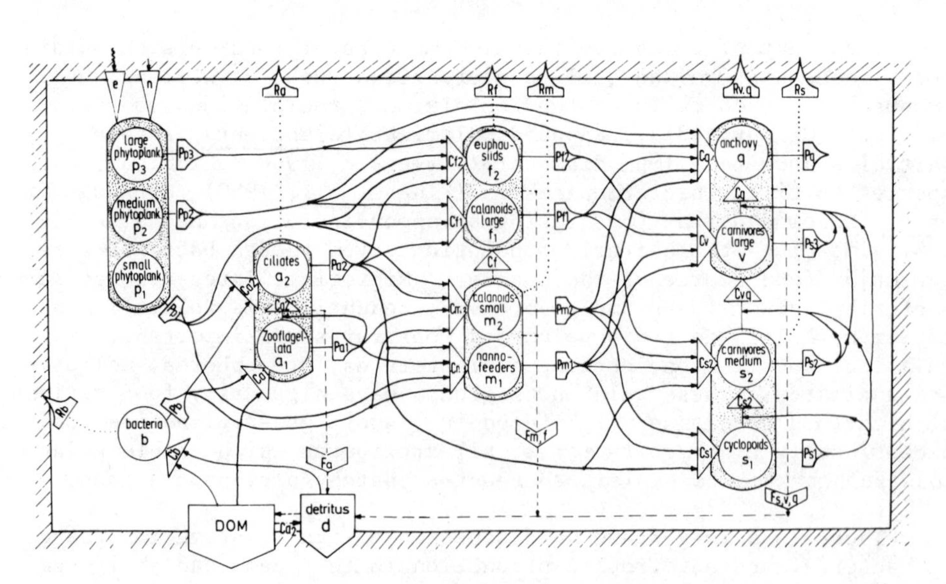

Fig. 16. Scheme of energy transfer through a plankton community in the zone of upwelling. e - solar radiation, n - nutrient. For further explanation see Table 4. (From Vinogradov and Shuskina, 1978).

Table 2. Characteristic elements of plankton communities on Pacific upwelling that are shown in Figure 16. (From Vinogradov and Shuskina, 1978).

Trophic levels and groups	Size(length) L	Predominant weight, w (mcal)	Cal mg wet weight^{-1}	Metablosism $R = aw^b$ a	b	Assimilability U^{-1}
Phytoplankton, p						
Nanno	4-7 µm		1.0			
Small, p_1	8-20 µm		0.7			
Medium, p_2	21-100 µm		0.4			
Large, p_3	>100 µm		0.2			
Bacteria, b	1-5 µm		1.0			1.0
Unicellular heterotrophs, a						
Flagellates, a_1	3-5 µm	$4x10^{-5}$	0.8			0.6
Ciliates, a_2	10-100 µm	$5x10^{-3}$	0.8	0.26	0.76	0.6
Non-carnivorous metazoan plankton						
Fine filters, m						
Meroplankton, m_1	0.1-3.5 mm	2-30	0.7	0.59	0.76	0.6
Appendicularians, m_1	0.1-2.5 mm	1-10	0.1	5.27	0.66	0.6
Doliolids, m_1	1-2.5 mm	4-10	0.01	5.27	0.66	0.6
Small calanoids, m_2	<1 mm	1-5	0.7	0.80	0.73	0.6
Coarse filterers, f						
Medium-sized calanoids, f_1	>1 mm	30-100	0.7	0.80	0.73	0.6
Juvenile euphausiids, f_2	<10 mm	120-2000	0.7	0.42	0.90	0.6
Predatory zooplankton						
Cyclopoids, s_1	0.2-1.5 mm	3-6	0.7	0.44	0.60	0.7
Calanoids, s_2	1.0-4 mm	500-2000	0.7	0.80	0.73	0.7
Small tomopterids, s_2	<3 mm	30-80	0.7	0.59	0.76	0.7
Small coelenterates, s_2	<5 mm	10-30	0.03	6.38	0.58	0.7
Chaetognaths, v	<20 mm	500-1200	0.7	5.30	0.52	0.7
Polychaetes, v	>3 mm	1300-8400	0.7	0.59	0.76	0.7

coastal upwelling ecosystem. An alternative view was given by Margalef (1978) who concluded that the biological character of the ecosystem was determined, qualitatively and quantitatively, by circulation and turbulence; both views are discussed in further detail by Barber and Smith (1981). The point I wish to make from Figure 16 is the inclusion by Vinogradov and Shuskina of compartments for heterotrophs, bacteria and detritus in their conceptual framework and sampling programme. The significance of the energy flux through detritus, micro-organisms to terminal consumers in continental shelf ecosystems was examined by Pomeroy (1979) using a compartmental model of energy flux. If the steps in the model were assumed to be trophic levels with ecological efficiency transfers of 10% then the model demonstrated that not enough energy could be carried to the terminal consumers. Disregarding trophic levels and using data on gross growth efficiencies to estimate the efficiency of transfer of energy, Pomeroy (1979) showed that several pathways could carry sufficient energy flux for the consumers. This modelling exercise showed that either conventional ecological assimilation efficiencies are low by a factor of 2 - 3 or current measurements of photosynthesis are low by a factor of 5 - 10. Alternatively grazers, as well as consuming phytoplankton, were obtaining a large energy flux by consuming micro-organisms which were growing on detritus and utilising dissolved materials.

Microheterotrophs, bacteria and detritus play a greater role in the functioning and passage of energy and material in pelagic ecosystems than was previously realised and are areas where further study should be concentrated.

Acknowledgements

I should like to thank the authors who have granted me approval to reproduce figures and tables from their published work. I wish to acknowledge the copyright owners Blackwell Scientific Publications for permission to reproduce figure 1 and 3 Academic Press Inc. (London) Ltd., for figures 2 and 15, Pergamon Press, Oxford for figure 6 and Plenum Publishing Corporation for figure 9.

REFERENCES

Allen, K. R., 1951, The Horokiwi Stream, Bull.Mar.Dep. N.Z. Fish No.10.

Alldredge, A. L., The quantitative significance of gelatinous zooplankton as planktonic predators, This volume.

Banse, K., and Mosher, S., 1980, Adult body mass and annual production/biomass relationships of field populations, Ecol. Monogr., 50:355.

Barber, R. T.- and Smith, R. L., 1971, Coastal upwelling ecosystems, in: "Analysis of Marine Ecosystems," A. R. Longhurst, ed., Academic Press, London.

Beers, J. R., and Stewart, G. L., 1971, Micro-zooplankton in the plankton communities of the upper waters of the eastern tropical Pacific, Deep-Sea Res., 18:861.

Benke, A. C., 1979, A modification of the Hynes method for estimating secondary production with particular significance for multivoltine populations, Limnol.Oceanogr., 24:168.

Binet, D., 1979, Estimation de la production zooplanktonique sur le plateau continental ivorien, D.Sci.Centre Rech.Oceanogr. Abidjan, 10:81.

Blackburn, M., 1981, Low latitude Gyral regions, in: "Analysis of Marine Ecosystems," A. R. Longhurst, ed., Academic Press, London.

Bougis, P., 1974, Ecologie du plancton marin. II. Le zooplancton, Collect.Ecol., 3, Masson.

Boyd, C. M., 1973, Small scale spatial patterns of marine zooplankton examined by an electronic in situ zooplankton detecting device, Neth.J.Sea Res., 7:103.

Boyd, C. M., 1976, Selection of particle sizes by filter-feeding copepods: a plea for reason, Limnol.Oceanogr., 21:175.

Boyd, C., 1981, Microcosms and experimental planktonic food chains, in: "Analysis of Marine Ecosystems," A. R. Longhurst, ed., Academic Press, London.

Boyd, C. M., and Johnson, G. W., 1969, Studying zooplankton populations with an electronic counting device and the LINC- 8 computer, Trans. Applications of Sea Going Computers Symposium, Mar.Tech.Soc., 83.

Boysen-Jensen, P., 1919, Valuation of the Limfjord 1, Rep.Dan.Biol. Stn., 26:1.

Burkill, P. H., 1982, Ciliates and other microplankton components of a nearshore food-web; standing stocks and production processes, in: "Marine Planktonic Protozoa and Microplankton Ecology," P. Bougis, ed., Annales Inst.Oceanog.

Caswell, H., 1972, On instantaneous and finite birth rates, Limnol. Oceanogr., 17:787.

Clapham, A. R., Lucas, C. E., and Pirie, N. W., 1976, A review of the United Kingdom contribution to the International Biological Programme, Phil.Trans.Roy.Soc.Lond., B 274, No. 934:275.

Colebrook, J. M., 1979, Continuous plankton records: seasonal cycles of phytoplankton and copepods in the North Atlantic Ocean and North Sea, Mar.Biol., 51:23-32.

Conover, R. J., 1966, Assimilation of organic matter by zooplankton, Limnol.Oceanogr., 11:338.

Conover, R. J., 1979, Secondary production as an ecological phenomenon, in: "Zoogeography and Diversity in Plankton," S. van der Spoel and A. C. Pierrot-Bults, eds., Edward Arnold, London.

Conover, R. J., and Lalli, C. M., 1972, Feeding and growth in Clione limacina (Phipps), a pteropod mollusc, J.Exp.Mar.Biol. Ecol., 9:279.

Conover, R. J., and Francis, V., 1973, The use of radioactive isotopes to measure the transfer of materials in aquatic food chains, Mar.Biol., 8:272.

Conover, R. J., and Lalli, C. M., 1974, Feeding and growth in Clione limacina (Phipps), a pteropod mollusc. II. Assimilation, metabolism and growth efficiency, J.Exp.Mar.Biol. Ecol., 16:131.

Conover, R. J., and Huntley, M. E., 1980, General rules of grazing in pelagic ecosystems, in: "Primary Productivity in the Sea," P. G. Falkowski, ed., Environ.Sci.Res., 19, Plenum Press, New York.

Cooke, R. A., Terhune, L. D. B., Ford, J. S., and Bell, W. H., 1970, An opto-electronic plankton sizer, Fish.Res.Bd.Can.Tech.Rep., 170:1.

Corner, E. D. S., 1961, On the nutrition and metabolism of zooplankton. 1. Preliminary observations on the feeding of the marine copepod Calanus helgolandicus (Claus), J.Mar.Biol.Ass. U.K., 41:5.

Corner, E. D. S., Cowey, C. B., and Marshall, S. M., 1967, On the nutrition and metabolism of zooplankton. V. Feeding efficiency of Calanus finmarchicus, J.Mar.Biol.Ass.U.K., 47:259.

Corner, E. D. S., and Davies, A. G., 1971, Plankton as a factor in the nitrogen and phosphorus cycles in the sea, Adv.Mar.Biol., 9:101.

Crisp, D. J., 1975, Secondary productivity in the sea, in: "Productivity of world Ecosystems," Nat.Acad.Sci., Washington, D.C.

Curl, H. Jr., 1962, Standing crops of carbon, nitrogen and phosphorus and transfer between trophic levels in continental shelf waters south of New York, Rapp.P.-v.Réun.Cons.Perm.Int.Explor. Mer., 153:183.

Cushing, D. H., 1959, On the nature of production in the sea, Fishery Invest.(Lond.), Ser. II, 22:1.

Cushing, D. H., 1966, Biological and hydrographic changes in British seas during the last thirty years, Biol.Rev., 41:211.

Cushing, D. H., 1970, Pelagic food chains, in: "Marine Food Chains," J. H. Steele, ed., Oliver and Boyd, Edinburgh.

Dagg, M. J., 1976, Complete carbon and nitrogen budgets for the carnivorous amphipod, Calliopius laeviusculus (Krøyer), Int. Revue Ges.Hydrobiol., 61:297.

Denman, K., and Platt, T., 1976, The variance spectrum of phytoplankton in a turbulent ocean, J.Mar.Res., 34:593.

Denman, K. L., and Mackas, D. L., 1978, Collection and analysis of underway data and related physical measurements, in: "Spatial Pattern in Plankton Communities," J. H. Steele, ed., Plenum Press, New York.

Dickie, L. M., 1972, Food chains and fish production in the Northwest Atlantic, International Commission for the Northwest Atlantic Fisheries, Spec. Publ., 8:202.

Dobben, van, W. H., and Lowe-McConnell, R. H., eds., 1975, "Unifying Concepts in Ecology," Dr. W. Junk, B. V. Publishers, The Hague.

Donaghay, P. L., 1980, Grazing interactions in the marine environment, in: "Evolution and Ecology of Zooplankton Communities," W. C. Kerfoot, ed., University Press, New England.

Donaghay, P. L., and Small, L. F., 1979, Food selection capabilities of the estuarine copepod Acartia clausi, Mar.Biol., 52:137.

Dussart, B. M., 1965, Les différentes catégories de plancton, Hydrobiologia, 26:72.

Edmondson, W. T., 1960, Reproductive rates of rotifers in natural populations, Mem.Ist.Ital.Idrobiol., 12:21.

Edmondson, W. T., 1974, Secondary production, Mitt.Internat.Verin. Limnol., 20:229.

Edmondson, W. T., 1977, Population dynamics and secondary production, Arch.Hydrobiol.Beih.Ergebn.Limnol., 8:56.

Edmondson, W. Y. 1979, Problems of zooplankton dynamics, in: "Biological and Mathematical Aspects in Population dynamics," R. de Bernadi, ed., Mem.Ist.Ital.Idrobiol., Suppl. 37.

Edmondson, E. T., and Winberg, G. G., 1971, "A manual on methods for the assessment of secondary production in fresh waters," (IBP Handbk. 17), Blackwell, Oxford.

Edwards, C., and Harris, E. J., 1955, Do tracers measure fluxes?, Nature, Lond., 175:262.

El-Sayed, S. Z., 1978, Primary productivity and estimates of potential yields of the southern ocean, in: "Polar Research," M. A. McWhinnie, Ed., AAAS Selected Symposium 7, Westview Press, Boulder.

Everson, I., 1977, "The Living Resources of the Southern Ocean. Southern Ocean Fisheries Programme," GLO/SO/77/1, FAO, Rome.

Fasham, M. J. R., 1978, The application of some stochastic processes to the study of plankton patchiness, in: "Spatial Pattern in Plankton Communities," J. H. Steele, ed., Plenum Press, New York and London.

Fenchel, T., 1983, Suspended marine bacteria as a food source, this volume.

Fisher, S. G., and Likens, G. E., 1973, Energy flow in Bear Brook New Hampshire: An integrative approach to stream ecosystem metabolism, Ecol.Mongr., 43:421.

Fuhrman, J. A., and Azam, F., 1982, Thymidine uptake and bacterioplankton production, Mar.Biol., 66:109.

Gamble, J. C., 1978, Copepod grazing during a declining spring phytoplankton bloom in the northern North Sea, Mar.Biol., 49:303.

Gieskes, W. W. C., and van Bennekom, A. J., 1973, Unreliability of the ^{14}C method for estimating primary productivity in eutrophic Dutch coastal waters, Limnol.Oceanogr., 18:494.

Gieskes, W. W. C., and Kraay, G. W., 1977, Primary production and consumption of organic matter in the southern North Sea during the spring bloom of 1975, Neth.J.Sea Res., 11:146.

Gieskes, W. W. C., Kray, G. W., and Baars, M. A., 1979, Current ^{14}C methods for measuring primary production; gross underestimates in oceanic waters, J.Sea Res., 13:58.

Gilmer, R. W., 1974, Some aspects of feeding in the cosomatous pteropod molluscs, J.Exp.Mar.Biol.Ecol., 15:127.

Glover, R. S., Robinson, G. A., and Colebrook, J. M., 1972, Marine biological surveillance, Environment and Change, 2:395.

Greve, W., and Parsons, T. R., 1977, Photosynthesis and fish production; Hypothetical effects of climatic change and pollution, Helgol.Wiss, Meeresunters., 30:666.

Greze, B. S., and Baldina, E. P., 1964, Population dynamics and annual production of *Acartia clausi* Giesbr. and *Centropages kroyeri* Giesbr. in the neritic zone of the Black Sea, Fish, Res.Bd.Can. Trans. Ser. 893.

Greze, V. N., 1978, Production in animal populations, *in*: "Marine Ecology," O. Kinne, ed., Vol. 4, Wiley-InterScience.

Grice, G. D., Harris, R. P., Reeve, M. R., Heinbokel, J. F., and Davis, C. O., 1980, Large scale enclosed water column ecosystems. An overview of Food-web I, the final CEPEX experiment, J.Mar.Biol.Assoc.U.K., 60:401.

Griffiths, F. B., and Caperon, J., 1979, Description and use of an improved method for determining estuarine grazing rates on phytoplakton, Mar.Biol., 54:301.

Gulland, J. A., 1970, Food chain studies and some problems in world fisheries, *in*: "Marine Food Chains," J. H. Steele, ed., Oliver and Boyd, Edinburgh.

Haas, L. W., and Webb, K. L., 1979, Nutritional mode of several non-pigmented microflagellates from the York River Estuary, Virginia, J.Exp.Mar.Biol.Ecol., 39:125.

Haffner, G. D., and Evans, J. H., 1974, Determination of seston-size particle distribution with Coulter counter, models A and B, and the two tube technique, Br.Phycol.J., 9:255.

Hamilton, A. L., 1969, On estimating annual production, Limnol. Oceanogr., 41:771.

Haney, J. F., 1971, An *in situ* method for the measurement of zooplankton grazing rates, Limnol.Oceanogr., 16:970.

Harbison, G. R., and McAlister, V. L., 1980, Fact and artifact in copepod feeding experiments, Limnol.Oceanogr., 25:971.

Hardy, A. C., 1924, The herring in relation to its animate environment. Part 1. The food and feeding of the herring with special reference to the east coast of England, Fishery Invest., Ser II, 7(3):1.

Harris, R. P., 1973, Feeding, growth, reproduction and nitrogen utilization by the harpacticoid copepod *Tigriopus brevicornis*, J.Mar.Biol.Ass.U.K., 53:785.

Harris, R. P., Reeve, M. R., Grice, G. D., Evans, G. T., Gibson, V. R., Beers, J. R., and Sullivan, B. K., 1982, Trophic interactions and production processes in natural zooplankton communities in enclosed water columns, *in*: "Marine Mesocosms," G. D. Grice and M. R. Reeve, eds., Springer-Verlag, New York.

Haury, L. R., McGowan, J. A., and Wiebe, P. H., 1978, Patterns and processes in the time-space scales of plankton distributions, *in*: "Spatial Pattern in Plankton Communities," J. H. Steele, ed., Plenum Press, New York.

Heinle, D. R., 1966, Production of a calanoid copepod Acartia tonsa in the Patuxent River estuary, Chesapeake Sci., 7:59.

Heninrich, A. K., 1962, The life histories of plankton animals and seasonal cycles of plankton communities in the oceans, J.Cons. Int.Explor.Mer., 27:15.

Hollibaugh, J. R., Fuhrman, J. A., and Azam, F., 1980, Radioactively labeling of natural assemblages of bacterioplankton for use in trophic studies, Limnol.Oceanogr., 25:172.

Horwood, J. W., 1978, Observations on spatial heterogeneity of surface chlorophyll in one and two dimensions, J.Mar.Biol.Ass. U.K., 58:487.

Hynes, H. B., 1961, The invertebrate fauna of a Welsh mountain stream, Arch.Hydrobiol., 57:344.

Hynes, H. B., and Coleman, M. J., 1968, A simple method of assessing the annual production of stream benthos, Limnol.Oceanogr., 13:569.

Ivlev, V. S., 1945, The biological productivity of waters, Usp. Sovrem.Biol., 19:98 (J.Fish.Res.Bd.Can., 23:1727, 1966).

Johansen, P. L., 1976, A study of tintinnids and othe Protozoa in eastern Canadian waters with special reference to tintinnid feeding, nitrogen excretion and reproductive rates, PhD. Thesis, Univ. of Dalhousie.

Jørgensen, C. B., 1966, "Biology of Suspension Feeding," Pergamon Press, London.

Juday, C., 1940, The annual energy budget of an inland lake, Ecology, 21:438.

Kachel, V., 1976, Basic principles of electrical sizing of cells and particles and their realisation in the new instrument "Metricell", J.Histochem.Cytochem., 24:211.

Karuhn, R., Davies, R., Kaye, B. H., and Clinch, M. J., 1975, Studies on the Coulter counter. Part 1. Investigation into the effect of orifice geometry and flow direction on the measurement of particle volume, Powder Technol., 11:157.

Keen, R., and Nassar, R., 1981, Confidence intervals for birth and death rates estimated with the egg-ratio technique for natural populations of zooplankton, Limnol.Oceanogr., 26:131.

Ketchum, B. H., 1962, Regeneration of nutrients by zooplankton, Rapp.p-v.Réun.Cons.Int.Explor Mer, 153:142.

King, D. R., Hollibaugh, J. T., and Azum, F., 1980, Predator-prey interaction between larvacean Oikopleura dioica and bacterioplankton in enclosed water columns, Mar.Biol., 56:49.

Krueger, C. C., and Martin, F. B., 1980, Computation of confidence intervals for the size-frequency (Hynes) method of estimating secondary production, Limnol.Oceanogr., 25:773.

Landry, M. R., and Hassett, R. P., 1982, Estimating the grazing impact of marine micro-zooplankton, Mar.Biol., 67:283.

Lasker, R., 1966, Feeding growth, respiration and carbon utilization of a euphausid crustacean, J.Fish Res.Bd.Can., 23:1291.

LeBlond, P. H. and Parsons, T. R., 1978, Reply to comment by W. E. Ricker, Limnol.Oceanogr., 28:380.

Le Borgne, R. P., 1978, Evolution de la production secondaire planctonique en milieu océanique par le méthode des rapports C/N/P, Oceanol.Acta., 1:107.
Le Borgne, R. P., 1981a, Les facteurs de variation de la respiration et de l'excrétion d'azote et de phosphore du zooplancton de l'Atlantique intertropical oriental, 1. Les conditions expérimentales et la temperature, Océanogr.Tropicale, In press.
Le Borgne, R. P.,1981b, Les facteurs de variation de la respiration et de l'excrétion d'azote et de phosphore du zooplancton de l'Atlantique intertripical oriental. 11. Nature des populations zooplanktonique et facteurs du milieu, Oceanogr. Tropicale, In press.
Le Borgne, R. P., 1982, Zooplankton production in the eastern tropical Atlantic Ocean: net growth efficiency and P:B in terms of carbon, nitrogen and phosphorus, Limnol.Oceanogr., 27:681.
Lindeman, R. L., 1942, The tropho-dynamic aspect of ecology, Ecology, 23:399.
Longhurst, A. R., 1981, Significance of spatial variability, in: "Analysis of Marine Ecosystems," A. R. Longhurst, ed., Academic Press, London.
Macfadyen, A., 1964, Energy flow in ecosystems and its exploitations by grazing, in: "Grazing in Terrestrial and Marine Environments," D. Crisp, ed., Blackwell Scientific, Oxford.
Mackas, D., 1976, Horizontal spatial hererogeneity of zooplankton on the Fladen Ground, I.C.E.S., C.M. 1976/L:20, Plankton Committee.
Mackinnon, D. L., and Hawes, R. S. J., 1961, "An Introduction to the Study of Protozoa," Clarendon Press, Oxford.
McAllister, C. D., 1970, Zooplankton rations, phytoplankton mortality and the estimation of marine production, in: "Marine Food Chains," J. H. Steele, ed., Oliver and Boyd, Edinburgh.
Malovitskaya, L. M., 1971, The production of the numerous species of copepods of the Gulf of Guinea, Trudy Atlant- NIRO, 37:401.
Mann, K. H., 1964, The pattern of energy flow in the fish and invertebrate fauna of the River Thames, Verh.Int.Ver.Limnol., 15: 485.
Mann, K. H., 1969, Dynamics of aquatic ecosytems, in: "Advances in Ecological Research," J. B. Cragg, ed., Academic Press, New York.
Margalef, R., 1978, What is an upwelling ecosystem?, in: "Upwelling Ecosystems," R. Boje and T. Tomczak, eds., Springer-Verlag, New York.
Marshall, S. M., and Orr, A. P., 1955, On the biology of Calanus finmarchicus VIII. Food uptake, assimilation, and excretion in adult and stage V Calanus, J.Mar.Biol.Ass.U.K., 34:495.
May, R. M., 1979, The structure and dynamics of ecological communities, in: "Population Dynamics," R. M. Anderson, B. T. Turner and L. R. Taylor, eds., Blackwell Scientific, Oxford.

Menzie, C. A., 1980, A note on the Hynes method of estimating secondary production, Limnol.Oceanogr., 25:770.

Mullin, M. M., 1969, Production of zooplankton in the ocean: The present status and problems, Oceanogr.Mar.Biol.Ann.Rev., 7: 293.

Nemoto, T., and Harrison, G., 1981, High latitude ecosystems, in: "Analysis of Marine Ecosystems," A. R. Longhurst, ed., Academic Press, London.

Nihoul, J. C. T., ed., 1975, "Modelling of marine systems," (Elsevier Oceanogr. Ser. 10.), Elsevier Scientific, Amsterdam.

Odum, H. T., 1957, Trophic structure and productivity of Silver Springs, Florida, Ecol.Monogr., 27:55.

Paloheimo, J. E., 1974, Calculation of instantaneous birth rate, Limnol.Oceanogr., 19:692.

Paloheimo, J. E., Crabtree, S. J., and Taylor, W. D., 1982, Growth model of Daphnia, Can.J.Fish.Aquat.Sci., 39:598.

Parslow, J., Sonntag, N. C., and Matthews, J. B. L., 1979, Technique of systems identification applied to estimating copepod population parameters, J.Plank.Res., 1:137.

Parsons, T. R., 1969, The use of particle size spectra in determining the structure of a plankton community, J.Oceanogr.Soc.Jpn., 25:172.

Parsons, T. R., LeBrasseur, R. J., and Fulton, J. D., 1967, Some observations on the dependence of zooplankton grazing on cell size and concentrations of phytoplankton blooms, J.Oceanogr. Soc.Jpn., 23:10.

Parsons, T. R., Le Brasseur, R. J., Fulton, J. D., and Kennedy, O, D., 1969, Production studies in the Strait of Georgia. Part II. Secondary production under the River Fraser plume, February-May 1967, J.Exp.Mar.Biol.Ecol., 3:39.

Parsons, T. R., and Le Brasseur, R. J., 1970, The availability of food to different trophic levels in the marine food chain, in: "Marine Food Chains," J. H. Steele, ed., Oliver and Boyd, Edinburgh.

Parsons, T. R., Takahashi, M., and Hargrave, B., 1977, "Biological Oceanographic Processes," Pergamon Press, Oxford.

Peterson, B. J., 1980, Aquatic primary productivity problem, Ann.Rev. Ecol.Syst., 11:359.

Petipa, T. S., 1967, in: "Structure and Dynamics of Aquatic Communities and Populations," Akad.Nauk.SSSR.Inst.Biol. Southern Seas, Kiev (English trans).

Petipa, T. S., Pavlova, E. V., and Mironov, G. N., 1970, The food web structure, utilization and transport of energy by trophic levels in the plankton communities, in: "Marine Food Chains," J. H. Steele, eds., Oliver and Boyd, Edinburgh.

Phillipson, J., 1964, A miniature bomb calorimeter for small biological samples, Oikos, 15:130.

Pierrot-Bults, A. C., and van der Spoel, S., 1979, General conclusions, in: "Zoogeography and diversity in plankton," S. van der Spoel and A. C. Pierrot-Bults, eds., Edward Arnold, London.

Pimm, S. L., and Lawton, J. H., 1977, Number of trophic levels in ecological communities, Nature, 268:329.

Platt, T., Mann, K. H., and Ulanowicz, R. E., eds., 1981, "Mathematical models in bilogical oceanography," (Mongr.Oceanogr. Methodol. 7.), Unesco, Paris.

Pomeroy, L. R., 1974, The ocean's food web a changing paradigm, BioSciences, 24:499.

Pomeroy, L. R., 1979, Secondary production mechanisms of continental shelf communities, in: "Ecological processes in coastal and marine ecosystems," R. J. Livingston, ed., Plenum Press, New York.

Pomeroy, L. R., and Johannes, R. E., 1968, Occurrence and respiration of ultra-plankton in the upper 500m of the ocean, Deep-Sea Res., 15:381.

Poulet, S. A., 1973, Grazing of Pseudocalanus on naturally occurring particulate matter, Limnol.Oceanogr., 18:564.

Poulet, S. A., 1974, Seasonal grazing of Pseudocalanus minutus on particles, Mar.Biol., 25:109.

Poulet, S. A., 1976, Feeding of Pseudocalanus minutus on living and non-living particles, Mar.Biol., 34:117.

Poulet, S. A., and Chanut, J. P., 1975, Non-selective feeding of Pseudocalanus minutus, J.Fish.Res.Bd.Can., 32:706.

Pugh, P. R., 1978, The application of particle counting to an understanding of the small-scale distribution of plankton, in: "Spatial Pattern in Plankton Communities," J. H. Steele, ed., Plenum Press, New York.

Rakusa-Suszczewski, S., 1969, The food and feeding habits of Chaetognatha in the seas around the British Isles, Pol.Arch. Hydrobiol., 16:213.

Reichle, D. E., O'Neill, R. V., and Harris, W. F., 1975, Principles of energy and material exchange, in: "Unifying Concepts in Ecology," W. H. van Dobben and L. McConnell, eds., D. W. Junk, B.V. Publishers, The Hague.

Richman, S., Heinle, D. R., and Huff, R., 1977, Grazing by adult estuarine copepods of the Chesapeake Bay, Mar.Biol., 42:69.

Ricker, W. E., 1968, in: "Methods of assessment of fish production in freshwaters," (IBP Handbk. 3), Blackwell, Oxford.

Ricker, W. E., 1978, On computing production, Limnol.Oceanogr., 28:379.

Rigler, F. H., 1975, The concept of energy flow and nutrient flow between trophic levels, in: "Unifying Concepts in Ecology," W. H. van Dobben and L. McConnel, eds., D. W. Junk, B.V. Publishers, The Hague.

Riley, G. A., 1946, Factors controlling phytoplankton populations on Georges Bank, J.Mar.Res., 6:54.

Riley, G. A., 1947, A theoretical analysis of the zooplankton population of Georges Bank, J.Mar.Res., 6:104.

Roger, C., 1973, Recherches sur la situation trophique d'un groupe d'organismes pilagiques (Euphausiacea), 1. Niveau trophiques des espèces, Mar.Biol., 18:312.

Roman, M. R., and Rublee, P. A., 1981, A method to determine in situ zooplankton grazing rates on natural particle assemblages, Mar.Biol., 65:303.

Sage, A. P., and Milsa, J. I., 1971, "Systems Identification," Academic Press, New York.

Schnack, S. B., 1982, The structure of the mouthparts of copepods in Kiel Bay, Meeresforsch., 29:89.

Sheldon, R. W., and Parsons, T. R., 1967a, A continuous size spectrum for particulate matter in the sea, J.Fish.Res.Bd.Can., 24:900.

Sheldon, R. W., and Parsons, T. R., 1967b, A practical manual on the uses of the Coulter Counter in marine science, Coulter Electronics, Inc., Toronto, Ont.

Sheldon, R. W., Prakash, A., and Sutcliffe, Jr., W. H., 1972, The size distribution of particles in the ocean, Limnol.Oceanogr., 17:327.

Shuskina, E. A., and Kisliakov, Iu.Ia., 1975, An estimation of the zooplankton productivity in the equatorial part of the Pacific Ocean in the Peruvian upwelling, in: "Ecosystems of the pelagic zone of the Pacific Ocean," M. E. Vinogradov, ed., v.102. Trudy Instituta Okeanologii (Transl. CUEA Office, Duke Univ., Mar.Lab., Beaufort, N.C.).

Sieburth, J. McN., 1977, Report on biomass and productivity of microorganisms in planktonic ecosystems, Hegol.Wiss.Meeresunters., 30:694.

Slobodkin, L. B., 1962, Energy in animal ecology, Adv.Ecol.Res., 1:69.

Smayda, T. J., 1966, A quantitative analysis of the phytoplankton of the Gulf of Panama. III. General ecological conditions and the plankton dynamics at 8° 46'N, 79° 23'W from November 1954 to May 1957, Inter.Amer.Trop.Tuna Comm.Bull., 11:355.

Smith, D. F., Bulleid, N. C., Campbell, R., Higgins, H. W., Rowe, F., Tranter, D. J., and Tranter, H., 1979, Marine food-web analysis: an experimental study of demersal zooplankton using isotopically labelled prey species, Mar.Biol., 54:45.

Smith, D. F., and Horner, S. M.J., 1981, Tracer kinetic analysis applied to problems in Marine Biology, in: "Physiological bases of phytoplankton ecology," T. Platt, ed., Can.Bull.Fish. Aquat, Sci., 210:346.

Sonntag, N. C., and Parslow, J., 1981, Technique of systems identification applied to estimating copepod production, J.Plank.Res., 3:461.

Steele, J. H., 1974, The Structure of Marine Ecosystems, Harvard Univ. Press, Cambridge, Mass.

Steele, J. H., 1978, Some comments of plankton patches, in: "Spatial pattern in plankton communities," J. H. Steele, ed., Plenum Press, New York.

Steele, J. H., and Frost, B. W., 1977, The structure of plankton communities, Phil.Trans.R.Soc.Lond.B.Biol.Sci., 280:485.

Steele, J. H., and Henderson, E. W., 1979, Spatial patterns in North Sea plankton, Deep-Sea Res., 26:955.

Steeman Nielsen, E., 1957, The balance of phytoplankton and zooplankton in the sea, J.Cons.Int.Explor.Mer., 23:128.

Sorokin, Yu. I., 1975, Heterotrophic microplankton as a component of marine ecosystems, J.Gen.Biol., 36:716 (in Russian).

Sorokin, Yu. I., 1978, Decomposition of organic matter and nutrient regeneration, in: "Marine Ecology," O. Kinne, ed., 4:501, Interscience, London and New York.

Sorokin, Yu. I., 1981, Microheterotrophic organisms, in: "Analysis of Marine Ecosystems," A. R. Longhurst, ed., Academic Press, London.

Sysoeva, T. K., and Degterova, A. A., 1965, The relation between the feeding of cod larvae and pelagic fry and the distribution and abundance of their pricipal food organisms, ICNAF Spec. Publ., 6:411.

Tansley, A., 1935, The use and abuse of vegetational concepts and terms, Ecology, 16:284.

Taylor, B. E., and Slatkin, M., 1981, Estimating birth and death rates of zooplankton, Limnol.Oceanogr., 26:143.

Teal, J. M., 1957, Community metabolism in a temperate cold spring, Ecol.Monogr., 27:283.

Teal, J. M., 1962, Energy flow in the salt marsh ecosystem of Georgia, Ecology, 43:614.

Tranter, D. J., 1976, Herbivore production, in: "The Ecology of the Seas," D. H. Cushing and J. J. Walsh, eds., Blackwell Scientific, Oxford.

Vanderploeg, H. A., 1981, Effect of algal length/apeture length ratio on Coulter analyses of lake seston, Can.J.Fish.Aquat,Sci., 38:912.

van der Spoel, S., and Pierrot-Bults, A. C., eds., 1979, "Zoogeography and Diversity in Plankton," Edward Arnold, London.

Vinogradov, M. E., 1981, Ecosystems of equatorial upwellings, in: "Analysis of Marine Ecosystems," A. R. Longhurst, ed., Academic Press, London.

Vinogradov, M. E., and Shuskina, E. A., 1978, Some development patterns of plankton communities in the upwelling areas of the Pacific Ocean, Mar.Biol., 48:357.

Walsh, J. J., 1981, Shelf-sea ecosystems, in: "Analysis of Marine Ecosystems," A. R. Longhurst, ed., Academic Press, London.

Wiebe, P. H., Hulbert, E. M., Carpenter, E. J., John, A. E., Kapp III, G. P., Boyd, S. H., Ortner, P. B., and Cox, J. L., 1976, Gulf Stream cold core rings: large scale interaction sites for ocean plankton communities, Deep-Sea Res., 23:695.

Williams, P. J. LeB., 1981, Incorporation of microheterotrophic processes into the classical paradigm of the plankton food web, Kieler Meeresforsch., Sonderhaft 5:1.

Williams, P. J., Leb, 1982, Microbial contribution to overall plankton community respiration - studies in enclosures, in: "Marine Mesocosms," G. D. Grice, M. R. Reeve, eds., Springer-Verlag, New York.

Williams, P. J. LeB., Raine, R. C. T., and Bryan, J. R., 1979, Agreement between the ^{14}C and oxygen methods of measuring phytoplankton production of the photosynthetic quotient, Oceanol.Acta., 2:411.
Winberg, G., 1936, Some general problems concerning the productivity of lakes, Zoolog.zhurn., 15:587.
Winberg, G. G., 1971, "Methods for the estimation of production of aquatic animals," Academic Press, London and New York.
Wyatt, T., 1976, Food chains in the sea, in: "The Ecology of the Seas," D. H. Cushing and J. J. Walsh, eds., Blackwell Scientific Publishers, Oxford.
Yablonskaya, E. A., 1962, A study of the seasonal population dynamics of the plankton copepods as a method of their production, Rapp.P.-v.Réun.Cons.Int.Explor.Mer., 153,224.

THE QUANTITATIVE SIGNIFICANCE OF GELATINOUS ZOOPLANKTON AS PELAGIC CONSUMERS

Alice L. Alldredge

Oceanic Biology Group
Department of Biological Sciences and
Marine Science Institute
University of California
Santa Barbara, California 93106 USA

ABSTRACT

Gelatinous zooplankton, including the planktonic coelenterates, ctenophores, molluscs and tunicates are ubiquitous and often abundant members of both neritic and oceanic communities. The quantitative impact of these primary and secondary consumers on their food stocks has been estimated in previous studies by calculating the proportion of the prey standing stock consumed per day. In general, gelatinous zooplankton usually consume less than 10% of their food populations per day although consumption is occasionally greater than 50%. Thus, gelatinous consumers may periodically decimate their food populations. Our understanding of the quantitative significance of gelatinous zooplankton is constrained by the relatively few studies which integrate consumption rates with information on the growth rates and population dynamics of either the zooplankton or their prey. These few studies suggest that gelatinous consumers have the greatest affect on food populations which are already limited by other environmental requirements. Attributes shared by many gelatinous consumers including high ingestion rates at high food abundances, and the potential for high growth rates, high fecundity and short generation time enable many gelatinous consumers to reach high population densities rapidly when food resources increase or when rich patches of food are encountered. Moreover, these attributes may enable some gelatinous zooplankton to overwhelm other, more slowly growing planktonic consumers when food densities are high. Preliminary data on excretion and on the production and sinking rates of fecal material suggest that gelatinous consumers may also contribute significantly to particulate flux and nutrient recycling. Almost all of our knowledge of the quantitative significance of gelatinous zooplankton is based on studies of neritic forms.

INTRODUCTION

The gelatinous zooplankton are a heterogeneous assemblage of fragile, soft-bodied, marine macroplankton which includes the Hydro medusae, Scyphomedusae, Siphonophora, Ctenophora, Heteropoda, pseu thecosomateous Pteropoda, Thaliacea Appendicularia (or Larvacea), and several meroplanktonic larval forms. Although these organisms represent many phyla and span three trophic levels, several shared morphological and physiological adaptations set them apart from ot zooplankton. The bodies of gelatinous zooplankton contain at least 95% water compared to 70-90% for crustaceans, chaetognaths and fis (Curl, 1962; Omori, 1969; Kremer, 1976a). Most have no hard parts and specific gravities near that of seawater, rendering them neutrally buoyant. They are usually transparent and appear nearly invisible in sunlit surface waters. Generally they are large relative to other forms of zooplankton.

Gelatinous zooplankton are ubiquitous and often abundant memb of both neritic and open ocean communities. Although the ecology many taxa is still relatively unknown, there has been a rapid increase in our knowledge of some groups in the last few years. The recent studies suggest that gelatinous zooplankton can be signific consumers in marine pelagic communities. In this review I will examine the quantitative impact of gelatinous zooplankton on the abundance and population dynamics of their food populations. In addition, I will discuss indirect effects of feeding by gelatinous zooplankton, including excretion, egestion and particulate flux, which may have a significant impact on the cycling of nutrients an flow of energy in marine ecosystems.

ABUNDANCE OF GELATINOUS CONSUMERS

Gelatinous zooplankton have been rarely incorporated into ecosystem models because reliable data on the abundance and biomass of these organisms, particularly in the open ocean, is limited (Hamne *et al*., 1975; Harbison *et al*., 1978; Biggs *et al*., 1981). Assessm of the biomass of oceanic gelatinous zooplankton is particularly difficult because these organisms are both fragile and patchy in distribution. Some species of gelatinous zooplankton are particularly delicate and readily break apart when collected in plankton nets. A survey of literature from this century shows that ctenophores are rarely collected in plankton nets in the open ocean, yet we know from other collecting techniques that ctenophores are not always so rare (Harbison *et al*., 1978). For example, conventional net collected samples contained no specimens of the cestid ctenophore, *Velamen*, where simultaneous visual counts by scuba divers yielded 10 to 305 Velamen·m^{-3} (Stretch, 1982). Other forms may fragment in plankton nets. Siphonophores, in particular, are subject to fragmentation and it is often difficult to tell whether the resultant pieces in a plankton tow represent one individual or many. Those species of gelatinous zooplankton which survive net capture often dissolve in preservatives, further complicating estimates of abundance (Harbison *et al*., 1978; Stretch, 1982). Thus, plankton tows generally yield accurate densities for only the most hardy gelatinous forms. However, many species of neritic ctenophores, medusae, calycophoran siphonophores, salps, doliolids and larvaceans can be quantitatively collected in identifiable condition if towing speeds, length of tows, and net characteristics are adjusted to accomodate them.

Many gelatinous zooplankton are present in low abudance relative to planktonic copepods and other small crustaceans. Even if they are not destroyed by the net, the volume sampled by a standard plankton tow is often too small to estimate accurately the population size of gelatinous consumers. Colonial salps, for example, may be present in sufficient numbers in the open ocean to have a grazing impact equal to or greater than the copepods present, yet be absent from plankton tows of brief duration due to their relative rarity and contagious distribution (Harbison and Gilmer, 1976).

Where fragility or patchy distribution limit the use of plankton nets, techniques for measuring densities of large, readily visible gelatinous zooplankton directly *in situ* using trained scuba divers, have yielded some of the first estimates of the abundance of some species of oceanic ctenophores, salps and siphonophores in surface waters (Harbison *et al*., 1978; Biggs *et al*., 1981). Biggs *et al*. (1981) counted animals passing through a 5 by 5m grid towed under a small boat. Although his technique can be criticized for underestimating the abundance of certain gelatinous zooplankton because many small, transparent individuals might be overloooked in a grid this size, his method represents a first step toward direct population assessment of fragile zooplankton in the open ocean. Refinements of this or similar techniques, including submersibles, and their wider application are now needed to increase our information on the biomass and abundance of gelatinous consumers in the open sea.

Gelatinous carnivores are generally more abundant in neritic than in oceanic waters. For example, while oceanic ctenophores rarely exceed 0.1 animals·m^{-3} (Harbison *et al*., 1978; Greze and Bileva, 1980; Biggs *et al*., 1981) swarms of ctenophores are not uncommon in neritic and estuarine areas. *Bolinopsis* has been reported in densities of 400 animals·m^{-3} in the Arctic Sea (Kamshilov, 1960), *Pleurobrachia bachei* occurs in densities up to 20 animals·m^{-3} in the Southern California Bight (Hirota, 1974), and the cestid ctenophore *Velamen parallelum* has occurred in densities of 300 animals·m^{-3} in the Gulf of California (Stretch, 1982). Likewise, siphonophores and medusae may reach densities of 1 or more animals·m^{-3} in semi-enclose bays and seas (Purcell, 1981a; Rogers et al., 1978; Möller, 1980) bu rarely exceed 0.1 animals·m^{-3} in the open ocean (Biggs et al., 1981) Aggregations of *Nanomia bijuga* can occur in densities of 0.3 animals·m^{-3} in the deep scattering layer off California (Barham, 1963).

Gelatinous herbivores are also found in swarm densities near shore. Berner (1967) reported densities of salps between 5 and 500 individuals·m^{-3} on 22 of 26 cruises off the coast of California. Swarms were usually restricted to the upper 100m but extended for hundreds of kilometers. Likewise, doliolids have been reported at densities of 3000 animals·m^{-3} (Deibel, 1980) and appendicularians at 25,000·m^{-3} (Seki, 1973). Some representative abundances and biomasses of a variety of gelatinous species gathered in association with feeding studies appear in Table 1.

IMPACT OF GELATINOUS ZOOPLANKTON ON POPULATIONS OF FOOD ORGANISMS

In recent years investigations of the ecology of gelatinous zooplankton in marine food webs have moved well beyond the descriptive phase. Numerous studies now exist which determine rates of energy transfer between neritic gelatinous consumers and their prey. However, the task of estimating rates of energy transfer to gelatinous consumers may be especially difficult where food web relationships are particularly complex. For example, while adult ctenophore feed on copepods, high population densities of copepods may severely reduce survival rates of larval ctenophores (Greve, 1972; Stanlaw et al., 1981). Likewise, the same gelatinous carnivore may be a pre item for, a competitor with, and a predator on a fish species, depending upon the particular phase of each species' life cycle being considered (Fraser, 1970). Implementing accurate functional definitions of these life phases into mathematical models requires considerable prior knowledge of the biology of each species and its complex interactions within a food web throughout its lifetime. Although such information is known for some gelatinous consumers, particularly neritic ctenophores, even basic life cycle data remain unknown for many oceanic species.

Methods of Estimating Consumption

Recent studies of the ecological importance of gelatinous zooplankton have focused on the quantity of food eaten and the impact of this consumption on prey population dynamics. Several possible approaches for estimating food consumption and its impact on food populations have been applied to gelatinous zooplankton:

1) Gut content - The most direct approach has been to determine the gut contents of freshly collected carnivores, their digestion rates of prey, and the population densities of both predators and prey organisms. The total proportion of the prey population consumed daily can then be calculated. This approach has been applied to the ctenophores *Pleurobrachia* (Anderson, 1974; Harris *et al.*, 1982) and *Bolinopsis* (Harris *et al.*, 1982) and to some species of siphonophores (Purcell, 1981a, b) and medusae (Möller, 1980). Although gut content analysis is tedious, labor intensive, requires that prey leave identifiable remains and may be in error from net feeding if net collection was used, it does emphasize natural diets and requires minimum laboratory maintenance to determine digestion times. Thus, it may be the most versatile approach for fragile, oceanic carnivores not amenable to laboratory culture.
2) Extrapolation of laboratory feeding rates - The most widely used approach extrapolates natural consumption rates from laboratory feeding rates in combination with *in situ* consumer and food abundances, to estimate the quantity of food eaten. This approach has been applied to populations of ctenophores (Miller, 1970; Reeve *et al.*, 1978; Kremer, 1979), appendicularians (King *et al.*, 1980) and doliolids (Deibel, 1980). However, many authors have cautioned that laboratory results may not be applicable to field populations (e.g. Conover, 1968; Hirota, 1974). Salps and doliolids are known to undergo rapid reduction in their natural growth and feeding rates within 5 hours of capture (Heron, 1972; Deibel, 1980) and of the gelatinous zooplankton, only two taxa, the ctenophores (*Beroe*, *Pleurobrachia*, and *Mnemiopsis*; Greve, 1968, 1970; Baker and Reeve, 1974) and the appendicularians (*Oikopleura dioica*, *Fritillaria borealis*, and *Fritillaria pellucida*; Paffenhöfer, 1973, 1976a; Fenaux, 1976) have been cultured through numerous generations in the laboratory. Thus, this method requires careful interpretation of results and may produce only approximations of feeding rates in nature.
3) Metabolic demands - Minimum ingestion rates of gelatinous carnivores have been calculated from measurements of minimum energy demands for the siphonophores *Agalma okeni* (Biggs, 1976) and *Sphaeronectes gracilis* (Purcell, pers. comm.). However, since many coelenterates can shrink in size under severe food limitation (e.g. Hamner and Jenssen, 1974), minimum metabolic demands may not reflect actual ingestion rates in nature.
4) Tracers - Radioactive tracer methodology as applied to gelatinous zooplankton has been used only as an extension of method 2 above. King *et al.* (1980) measured the feeding rate of *Oikopleura dioica* on H^3-labeled marine bacterioplankton. They then extrapolated these laboratory results to natural populations in large, free-floating enclosures (Controlled Ecosystem Population Experiments, CEPEX). Mechanical tracers, such as plastic beads, may be particularly effective tracers for

gelatinous herbivores, which tend to be generalist feeders relying little on complex behavioral patterns for food selection. Alldredge (1981) calculated the impact of appendicularians on natural food assemblages *in situ* by measuring clearanc rates using trace concentrations of plastic beads.

5) Secondary production - The percent of daily herbivore producti consumed by gelatinous carnivores has been calculated from estimates of secondary production for *Pleurobrachia* (Mullin an Evans, 1974; Hirota, 1974) and *Mnemiopsis* (Reeve and Baker, 1975). Hirota (1974) and Reeve and Baker (1975) calculated secondary production of ctenophores from the mean daily biomas of field populations and daily growth rates obtained in the laboratory. Mullin and Evans (1974) measured this production directly by harvesting predators and prey from a quasi-steady state community established in a deep tank.

6) Exclusion experiments - Predation impact of ctenophores has be determined experimentally by comparing similar communities wit and without gelatinous predators. Large floating enclosures such as the CEPEX bags have provided a practical means for suc experimentation (Reeve and Walter, 1976; Harris *et al*., 1982). Although control and experimental ecosystems may not be initially identical in all respects, and not all ctenophores cou be removed, such experiments yield direct assessment of the effect of predation on semi-natural populations of prey withou the complications of extensive assumptions and calculations inherent in less direct methods.

Percent of Food Stock Removed Per Day

The quantitative significance of zooplankton consumption in pelagic communities is generally accepted based on innumerable examples of the strong inverse correlation between abundance of co sumers and depletion of their food (see Reeve and Walter,1978; Kremer, 1979; King *et al*., 1980; Deibel, 1980 for a few examples). Field observations of the increased biomass of predators coinciden with prey population declines have tended to confirm intuitive assumptions that consumers may be a major factor regulating the abun dance and distribution of food populations in marine planktonic communities. However, relatively few studies have attempted to quantify the actual magnitude of this effect.

Most quantitative studies of gelatinous zooplankton have measured the percentage of the food stock removed per day by the c sumers using one or more of the methods just discussed. The liter ture on the quantitative impact of gelatinous zooplankton on food stocks in nature is summarized in Table 1. Gelatinous zooplankton generally remove less than 10% of their food populations each day. Only at high densities do gelatinous zooplankton remove large proportions of their prey populations daily. However, Table 1 clearl illustrates that gelatinous zooplankton can greatly reduce standing stocks of food organisms.

The data in Table 1 illustrate several major limitations and biases in our current understanding of the role of gelatinous zooplankton in planktonic communities. First, relatively few species have been studied. Only a few ctenophores, appendicularians and some siphonophores have been studied in detail. Little quantitative data exist on the impact of gelatinous herbivores such as salps, pyrosomes or pteropods, despite the occasional high abundances of these animals. Likewise little data exist on medusae or heteropods.

Second, detailed studies of the quantitative significance of gelatinous consumers have been limited to neritic forms. Data on the quantitative impact of oceanic gelatinous consumers on food populations has not yet been obtained. No doubt, neritic forms have been emphasized because these organisms are more accesssible to most laboratory based marine scientists and many, partiuclarly neritic ctenophores and appendicularians, are more amenable to laboratory culture and maintenance.

Third, calculations of predation effects rarely take into account variations in feeding through time or space, undoubtedly due to the complexity of measuring these variables. Variations in prey spatial distributions and vertical migration of both consumers and food organisms may all affect the final quantitative impact of gelationous zooplankton as consumers.

Finally, only a few studies have examined the effect of feeding by any zooplankter, gelatinous or otherwise, relative to the population dynamics of either the consumer or the food organisms. An estimate of the percent of the food stock consumed per day has little meaning unless the growth rate of the prey population is also known. Consumption rates of 5% per day may rapidly reduce a prey population which grows slowly or not at all. Likewise, a consumption rate of 50% per day may have little impact on a prey population which doubles every few hours. Meaningful estimates of the quantitative significance of zooplankton consumption thus require considerable additional understanding of plankton population dynamics in nature.

Gelatinous Consumers and Prey Population Dynamics

Several investigators have examined the effect of feeding by gelatinous zooplankton on the population dynamics of their food stocks using mathematical models or calculations based on data from both the laboratory and field. Kremer (1979) estimated growth rates of copepod prey and observed the decline in their standing stock in nature to calculate total prey mortality as the sum of the growth increment and the overall biomass change. She then assessed the

Table 1. Proportion of Food Stock Consumed per Day by Various Gelatinous Consumers. NA = North Atlantic

Consumer	Food	Consumer Density NO.·m^{-3}	Consumer Density Dry weight mg·m^{-3}	Food NO.·m^{-3}
Siphonophora				
Rhizophysa eysenhardti[1]	fish larva	0.91	-	2
Rosacea cymbiformis[1]	crustaceans	0.008-0.1	-	1,50
Muggiaea atlantica[1,2]	copepods	1.4	2.3	9,12
	copepods	0.9	6.0	38
Medusae				
Aurelia aurita[1]	fish larva	0.1-1.0	-	0.1-8.
Chrysaora quincuecirrha[2]	ctenophores	>2	-	-
Ctenophora				
Mnemiopsis leidyi[2]	copepods	13	-	70,00
	copepods	-	-	-
	copepods	30-50[5]	-	100,00
Mnemiopsis mccradyi[2]	copepods	-	0.01-1.6	-
Pleurobrachia[2]	copepods	-	8-189[6]	8,00
Pleurobrachia and Bolinopsis[1]	copepods	-	13-275[6]	-
Tunicata				
Stegasoma magnum[2,3]	natural POM[9]	11-63	-	-
Oikopleura dioica[2,3]	natural POM[9]	205-4587	-	-
	bacteria	-	276[7]	-
Dolioletta gegenbauri[2]	natural POM	>3000	400[7]	-

[1]Gut content method; [2]feeding rates extrapolated to field; [3]tracer method; [4]calculated assuming 1 copepod = 7 mg dry wt.; [5]author, per comm.; [6]calculated assuming carbon = 4% of dry wt. (Reeve et al., 1978); [7]calculated assuming carbon = 8% dry wt. (Madin et al., 1981 [8]mg carbon·m^{-3}; [9]POM = particulate organic matter.

Density Dry weight $mg \cdot m^{-3}$	% Food stock consumed $\cdot day^{-1}$	Location	Source
67	28	Gulf of California	Purcell, 1981a
-	<<1	Gulf of California	Purcell, 1981b
-	0.2	Puget Sound	Purcell, 1981c
2.1	2-4	Southern California Bight	Purcell, pers. comm.
-	2-5	Baltic Sea	Möller, 1980
-	51	Pamlico Estuary, western N.A.	Miller, 1974
490[4]	31	Patuxent Estuary, western N.A.	Bishop, 1969
-	1-48	Pamlico Estuary, western N.A.	Miller, 1970
700[4]	1-30	Narragansett Bay, western N.A.	Kremer, 1979
0.68-90	0.02-10	Biscayne Bay, Florida	1978
48	3-4	CEPEX Bags	Reeve and Walter, 1976
65-247	4-104	CEPEX Bags	Harris *et al.*, 1982
81-132[8]	5-13	Gulf of California	Alldredge, 1981
125[8]	1-37	Gulf of California	Alldredge, 1981
1-5[8]	5.5	CEPEX Bags	King *et al.*, 1980
-	2-100	Georgia Bight	Deibel, 1980

relative contribution of ctenophore predation to the estimated total mortality of the copepod prey population. A survey of zooplankton populations over different years and locations in Narragansett Bay indicated the copepods declined at a rate of about 10% per day from the time of their peak biomass, usually in mid-July, to a population low in mid-August. If copepod populations were growing at a rate of 10-15% per day as estimated from field and laboratory data, then a total mortality of 20-25% was necessary to produce this decline. Since *Mnemiopsis leidyi* cropped 5 to 10% of the standing stock per day, Kremer (1979) concluded that this ctenophore predation was responsible for 20 to 50% of the total copepod mortality. However, Kremer urged caution in generalizing from such computations. Often the sequence of copepod decline did not clearly precede the appearance of substantial ctenophore biomass, suggesting that unknown factors other than ctenophore predation affected the prey populations.

Reeve *et al*. (1978) calculated the potential predation impact o ctenophores in Biscayne Bay. Maximum removal of copepod standing stock by ctenophores was 10% daily, and throughout most of the year it was about 1%. They estimated the rate of change of the copepod populations when the ctenophores were clearing 1 and 10% of the wate column per day by assuming specific rates of production for predator and prey. They then mathematically generated a series of curves relating the change in copepod biomass after 7 days to various combinations of ctenophore and copepod production assuming peak ctenophor standing stock (Figure 1). At one extreme, they assumed that the copepod food supply was limited, permitting no copepod population growth. Under these conditions, copepod prey populations were

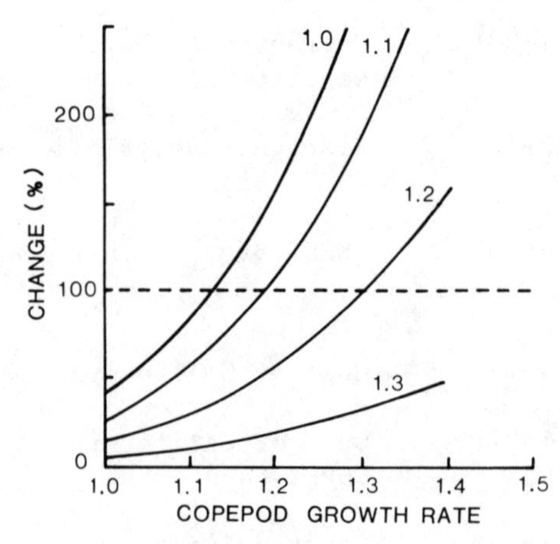

Fig. 1. Computer derived curves for the percentage of initial copepod population remaining after 7 days, assuming different copepod population daily potential increase rates (abscissa) and ctenophore increase rates (values on curves), (from Reeve *et al*., 1978).

reduced to 50% in less than 7 days even when the predator population was not increasing. At a predator population growth rate of 1.3 per day the copepod population declined over 7 days even if the prey population growth rate was as high as 1.4 per day (Figure 1). The results of both Reeve et al. (1978) and Kremer (1979) indicate that field populations of ctenophores can, at their peak densities, severely reduce the abundance of their zooplankton prey. Both Kremer (1976b) and Reeve et al. (1978) performed computations on the fecundity and potential population growth of the ctenophores as well. Both conclude that observed population increases of ctenophores in nature can be readily explained from laboratory derived knowledge of ctenophore biology.

Mathematical models have also been used to calculate the effect of appendicularian grazing on bacterioplankton in large enclosed water columns (CEPEX). King et al. (1980) coupled grazing rates of Oikopleura dioica, measured in the laboratory, with biomass measurements of bacteria and appendicularians in the enclosures to develop a model predicting the impact of grazing on bacterial standing stock. The model assumed a bacterial growth rate necessary to attain the actual bacteria population densities observed in the enclosures if grazing mortality was zero. It then compared this with the growth rate and standing stock of bacteria under appendicularian grazing pressure. Despite high predator abundance, the range of bacterial growth rates needed to overcome the effects of appendicularian grazing in the enclosures was on the order of 1 to 3 doublings per day. Thus King et al. (1980) concluded that appendicularian grazing had a negligible impact on bacterioplankton prey populations, even though bacteria was the major food source for these herbivores.

The data of King et al. (1980) clearly demonstrate the pitfalls of inferring significant predation impact from a strong inverse correlation between the abundances of consumers and their food populations. In the enclosures, rapid population increases and declines of the appendicularian grazers sequentially followed the increase and decline of their food populations. However, the model of King et al. (1980) demonstrates that this classic relationship was clearly fortuitous, and, in fact, the consumer had little impact on its food population. Although high consumer abundances may correlate with declining food abundances in field populations, the studies of Kremer (1979), Reeve et al. (1978) and King et al. (1981) all demonstrate that such correlations do not imply causation.

Mathematical modeling is a useful approach for predicting prey population dynamics in nature. However, the quantitative effect of zooplankton consumption on the population dynamics of food organisms can sometimes be measured directly. Alldredge (1981) calculated the grazing impact of appendicularians while also measuring the biomass and rate of increase (under conditions of no grazing) of food particles in situ. At densities ranging from 205 to 4587 appendicularians·m^{-3}, appendicularian populations cleared 1 to 37% of each cubic meter per day. However, growth of their particulate food was so rapid (maximum rate of increase = 37% per day), that despite high grazing rates, only at maximum population densities did the appendicularians consume more than the daily production of their particulate food.

Harris *et al*. (1982) also measured prey production and predation impact of ctenophores directly in CEPEX enclosures. During the experiment the major herbivorous copepods, *Paracalanus* and *Pseudocalanus*, multiplied rapidly within the enclosures during the first 18 days. A few adult ctenophores within the enclosures began reproducing and by day 25 large numbers of ctenophore larvae were present in the containers. By day 30, the copepod population had been severly reduced. Analysis of gut contents of the ctenophore and chaetognath predators within the enclosures, combined with laboratory observations of their feeding, indicated that predation could account for this drastic decline. However, the production rate and standing stock of the copepod populations had already begun to decline several days prior to significant consumption of prey by gelatinous predators. Harris *et al*. (1982) conclude that predation was only significant when prey biomass was already declining.

Exclusion experiments also permit direct measurement of the effect of predation relative to prey population dynamics. Reeve and Walter (1976) removed 70% of the ctenophore predators (*Pleurobrachia*) from one CEPEX enclosure and compared changes in the zooplankton prey populations with those in an undisturbed enclosure. The copepod population in the enclosure with reduced numbers of ctenophores increased by a factor of 4 compared with the undisturbed enclosure, suggesting that predation had an important regulating effect on population dynamics of copepod prey. However, Reeve and Walter (1976) also observed high mortalities of both predators and prey which could not be related to predation.

These studies all demonstrate that gelatinous consumers have the potential to substantially reduce food stocks in nature. They also suggest that gelatinous zooplankton have the greatest effect on food populations which are already limited by other environmental requirements.

In systems where immigration of new consumers is rare and consumer population size depends primarily on consumer growth and reproduction in response to food abundance, food populations appear to be limited primarily by environmental factors or organisms at trophic levels below them and not by consumer populations at trophic levels above (i.e. phytoplankton food might be limited by light or nutrients, planktonic herbivores might be limited by the abundance of phytoplankton food, etc.). This is a direct result of the different, generally lower, population growth rate of the consumer population relative to its food and of the energy constraints of the trophic pyramid. Immigration of new consumers into patches of food can result in rapid reduction of food stocks because the size of the consumer population is not tied directly to the size of the food stock. There is no reproductive lag time in the ability of the consumer population to respond to increases in food supply.

Secondary Production

Only a few studies exist which estimate secondary productivity of populations of gelatinous predators. Secondary production has been estimated in three populations of ctenophores as the product of mean daily biomass (calculated from sampled biomass in the field and computed mortality rates) and daily growth rate (estimated from laboratory cultures) for each size class of ctenophore. Summation of these products over all the size classes observed yields an estimate of the secondary production of the total population. Mullin and Evans (1974) obtained a secondary production for populations of *Pleurobrachia* raised in a deep tank equal to 12.6% of prey production and 2.6% of primary production. Reeve and Baker (1975) obtained production rates for *Mnemiopsis* of 4.2g ash-free dry weight$\cdot m^2 \cdot d^{-1}$ and a production/biomass ration of 0.12. *Mnemiopsis* production equaled 9 to 19% of prey production in south Florida coastal waters. Hirota (1974) obtained a production rate of 5.4g ash-free dry weight$\cdot m^{-2} \cdot y^{-1}$ for *Pleurobrachia* in a much deeper water column off southern California. The productivity/biomass ratio was 0.02.

INTERACTION WITH OTHER PLANKTONIC CONSUMERS

Feeding by gelatinous zooplankton may affect not only their food organisms, but the ecology of other planktonic consumers as well. Competitive interactions between major taxa of marine zooplankton are very poorly understood. However, both laboratory and field data have suggested that neritic ctenophores and tunicates, in particular, may have higher intrinsic rates of population increase (r) than other planktonic consumers at high food densities (Heron, 1972; Deibel, 1980; Reeve *et al*., 1978). Many species of gelatinous zooplankton share attributes which enable them to reach high population densities rapidly when food resources increase or when rich patches of food are encountered.

First, whereas the ingestion rates of most other zooplankton taxa do not increase but become independent of food density at very high food concentrations, the ingestion rates of the gelatinous herbivores and carnivores studied to date remain proportional to food concentration over extremely wide ranges of food density. Ingestion rates of ctenophores and siphonophores level off and become independent of prey density only at very high densities of several hundred prey$\cdot liter^{-1}$ (Reeve and Walter, 1978; Purcell, 1981c,1982).

Table 2. Maximum Growth Rates of Various Planktonic Taxa.

Species	Temp. (C°)	Maximum coefficient of exponential growth	Source
Ctenophora			
Mnemiopsis mccradyi	26	.78	Reeve and Baker, 1975
Mnemiopsis leidyi	20	.25	Miller, 1970
Bolinopsis infundibulum	16	.2	Greve, 1970; Reeve and Walter, 1976
Pleurobrachia bachei	20	.47	Hirota, 1974
	20	.76	Reeve and Walter, 1976
Pleurobrachia pileus		.2	Greve, 1970; see Reeve and Walter, 1978
Beroe		.4	Greve, 1970; see Reeve and Walter, 1978
Siphonophora			
Muggiaea atlantica	8-10	.2-.35	Purcell, 1982
Agalma okeni	23-29	.35	Biggs, 1976
Mollusca			
Clione limacina	15	.2-.5	Conover and Lalli, 1974
Tunicata			
Oikopleura dioica	13	.57-1.09	Paffenhöfer, 1976a
	11-15	.45-1.20	King, 1981
Dolioletta gegenbauri (*phorozooid*)	20	.69[1]	Deibel, 1982
Thalia democratica	20	.03-.36[1]	Deibel, 1982
	-	>1.0[2]	Heron, 1972
Chaetognatha			
Sagitta hispida	26	.45	Reeve and Baker, 1975
Copepoda			
Calanus helgolandicus	10-15	.41	Paffenhöfer, 1976b
Pseudocalanus elongatus	12.5	.38	Paffenhöfer and Harris 1976
Temora longicornis	12.5	.54	Harris and Paffenhöfer 1976
Rhincalanus nasutus	10-15	.64	Mullin and Brooks, 197[illegible]

[1]Deibel estimates these are minimum rates
[2]Calculated from Figure 17 in Heron, 1972 using carbon vs. length data from Madin et al., 1982, Figure 1.

Likewise, salps, doliolids and appendicularians all filter at a constant rate regardless of food concentration (see Alldredge and Madin, 1982, for review). The medusa, Chrysaora, also feeds at a rate directly proportional to prey density even at high prey density (Clifford and Cargo, 1978). These are the only taxa where feeding rates have been studied in detail and additional information is needed to extend this generalization to other gelatinous zooplankton. However, the functional response curves of these diverse taxa suggest that many gelatinous zooplankton can sustain high ingestion rates at very high food densities. Thus they are adapted to ingesting high daily rations of food when food is abundant.

Secondly, many gelatinous zooplankton are capable of very high growth rates under favorable conditions. Table 2 summarizes the maximum growth rates of a variety of marine plankton. The maximum coefficient of exponential growth of gelatinous carnivores is often above 0.5 and in the case of pelagic tunicates, usually greater than 1. Such values are more characteristic of phytoplankton (see Parsons and Takahashi, 1973) and are higher than reported for zooplankton generally (Table 2).

Gelatinous zooplankton apparently maintain high growth rates through high food consumption and high assimilation when food is abundant. The nutritional ecology of most taxa of gelatinous zooplankton is poorly known. However, available data suggest that the assimilation efficiency of gelatinous carnivores is high, usually greater than 80% (Table 3). This is similar to the assimilation efficiencies of other marine planktonic carnivores (Table 3) and typical of carnivores in general (Welch, 1968). Assimilation efficiencies are not known for most gelatinous herbivores but the appendicularian, Oikopleura dioica, also has an assimilation efficiency greater than 90% (King et al., 1980). This is very high for a herbivore (Table 2).

Although most gelatinous zooplankton assimilate a high proportion of ingested food, data on their gross growth efficiencies (the proportion of ingested food devoted to growth) tends to be both sparse and contradictory. The few siphonophores and medusae studied have gross growth efficiencies greater than 35%. This is similar to the gross growth efficiencies of other planktonic carnivores, including Sagitta and euphausiids (Table 3). Gross growth efficiencies of O. dioica were also high, decreasing with increasing body size and ranging from 37 to above 90% at moderate food densities (King, 1981).

Gross growth efficiencies of neritic ctenophores may be low, generally less than 10% (Table 3), although recent data from Kremer et al. (1980) suggests that ctenophores may also have higher gross growth efficiencies than previously estimated (Table 3). The gelatinous zooplankton represent many diverse taxa from many diverse habitats. Growth efficiencies may be equally as variable. However,

Table 3. Assimilation and Gross Growth Efficiencies of Various Planktonic Taxa.

Taxa	Assimilation efficiency (%)	Gross growth efficiency (%)	Temp. C°	Source
<u>Ctenophora</u>				
Mnemiopsis mccradyi	74	2-6	26	Reeve <u>et al</u>., 1978
	64-100	40	21	Kremer <u>et al</u>., 1980
Pleurobrachia bachei	19-74	3-11	13	Reeve et al., 1978; Hirota, 1974
<u>Medusae</u>				
Cyanea capillata	-	37[1]	12	Fraser, 1969
Aurelia aurita	-	37[1]	12	Fraser, 1969
<u>Siphonophora</u>				
Agalma okeni	-	33-48	23-29	Biggs, 1976
Rhizophysa eysenhardti	98	-	23	Purcell, 1981 and pers. comm.
Sphaeronectes gracilis	-	45-58	14	Purcell, pers. comm.
<u>Tunicata</u>				
Oikopleura dioica	90	37-100	13.5	King, 1981
<u>Mollusca</u>				
Clione limacina	90	49-78	15	Conover and Lalli, 1974
<u>Chaetognatha</u>				
Sagitta hispida	80	36	16-26	Reeve, 1970; Cosper and Reeve, 1975
<u>Crustacea</u>				
Euphausia pacifica	88	30(net)	6-16	Lasker, 1966
Calanus helgolandicus	74-91	20-30	15	Paffenhöfer, 1976b; Corner, 1961
Rhincalanus nasutus	-	30-45	10-15	Mullin and Brooks, 1970
Psuedocalanus elongatus	-	13-18	12.5	Paffenhöfer and Harris, 1976
Temora longicornis	50-98	9-17	12.5	Harris and Paffenhöfer, 1976; Berner, 1962
<u>Carnivorous fish</u>	80	20	-	Welch, 1968

[1]Data based on one specimen.

moderate to high growth efficiency certainly facilitates high growth rates in some species.

Third, certain gelatinous zooplankton have the potential under favorable environmental conditions for short generation times. Some species of Oilopleura and Fritillaria have generation times of 3 to 12 days (Fenaux, 1976; Paffenhöfer, 1976a), while the salp, Thalia democratica, may have a generation time as short as 48 hours (Heron, 1972). Likewise, neritic ctenophores begin producing eggs 2-3 weeks after hatching (Reeve and Walter, 1978). Oceanic salps, pteropods and coelenterates probably have generation times on the order of months when food is sparse.

Fourth, many taxa of gelatinous predators exhibit high fecundity. Planktonic ctenophores are simultaneous hermaphrodites capable of self fertiization (Reeve and Walter, 1978). While most crustacean zooplankton produce only a few hundred eggs per female (see Paffenhöfer and Harris, 1979, for review) species of Pleurobrachia and Mnemiopsis produce from 1100 to 14,000 eggs per female (Hirota, 1974; Greve 1972; Baker and Reeve, 1974; Kremer, 1976b). Likewise doliolids and salps may produce thousands of offspring through asexual reproduction (Alldredge and Madin, 1982). Populations of Thalia democratica, a small neritic salp, may increase 1.6 to 2.5 times in size each day (Heron, 1972). Some medusae are also highly fecund. The leptomedusae Phialidium gregarium produces 3000 eggs per female (Roosen-Runge, 1970). Asexual reproduction of the medusae stage through schizogamy (Stretch and King, 1980) or budding may also increase the reproductive potential of some species.

Currents, upwelling, seasonal changes in nutrients and other environmental variables all result in unpredictable fluctuations and patchiness in the abundance of food available to planktonic consumers. The above four attributes of high daily food rations when food is abundant and the potential for rapid growth, short generation time and high fecundity all enable many gelatinous zooplankton, especially neritic ctenophores and tunicates, to reach high population densities when food resources increase or when rich patches of food are encountered.

These same attributes may enable certain gelatinous zooplankton to overwhelm other, more slowly growing planktonic consumers at high food densities. High growth rate and fecundity enable gelatinous tunicates to exploit food resources more rapidly than other plankton herbivores (Heron, 1972; Deibel, 1980). Other planktonic grazers are rare or absent in swarms of planktonic tunicates (Fraser, 1962; Berner, 1967; Deibel, 1980). Consumption of ultraplankton and bacteria by gelatinous herbivores may also reduce food availability to juvenile stages of copepods and other herbivores, thus reducing recruitment of potential competitors. Salps actually ingest small nauplii and larval stages (Madin, 1974). Moreover, exceptionally high clearance rates of many gelatinous herbivores (see Alldredge and

Madin, 1982, for review) may greatly reduce phytoplankton stocks before populations of other grazers can exploit them. Rapid reproduction may be the most significant factor limiting the abundance of competing herbivores in thaliaceans swarms. In older swarms, both copepods and appendicularians may co-occur with thaliaceans in high densities, suggesting that given enough time and sufficient food, herbivores of longer generation time can also exploit these phytoplankton blooms (Deibel, 1980).

Some gelatinous carnivores may also overwhelm potential competi tors at high food densities. Reeve et al. (1978) discuss competitiv interactions between ctenophores and chaetognaths. If ctenophores u a greater fraction of their food for metabolism than chaetognaths, t at low environmental food levels, chaetognaths apportion a larger percentage of food to growth. Alternatively, at high food abundance the ingestion rate of chaetognaths becomes limiting. The maximum growth coefficient of Sagitta hispida was only 0.45 compared with 0.78 for young Mnemiopsis under the same food conditions. Therefore ctenophore populations would be expected to reach higher densities than chaetognaths and other more slowly growing carnivores at high food abundances (Reeve et al., 1978).

This generalization is complicated by the interaction of larval ctenophores and adult copepods. At high copepod densities, the survival rate of larval ctenophores is reduced significantly (Stanlaw et al., 1981). Thus above a certain copepod density, ctenophores ma become less effective competitors. Moderate levels of prey are know to support ctenophore blooms (Reeve, 1980).

Gelatinous carnivores may also compete with fish for food resources. In the Baltic Sea the medusa, Aurelia aurita, feeds primar ily on copepods, a prey source which also makes up 92% of the food co-occurring fish larvae. Aurelia also preys upon these fish larva removing from 2 to 5% of the standing stock of larvae per day (Möller, 1980). Predation by Bolinopsis on planktonic crustaceans known to affect the nutritional state of co-occurring herring larva Where Bolinopsis is abundant, less crustacean food is available to competing herring larvae, which in turn experience a delay in their summer fattening. Herring larvae ultimately experience lower growt rates and lower overall fat content (Fraser, 1970).

Confer and Blades (1975) suggest that planktonic carnivores ma be a link between smaller prey, which fish don't see, and the fish themselves. Gelatinous carnivores are preyed upon by fish, includi sunfish, blue fin tuna, cod, sardines, butterfly fish, and flying fish (Phillips et al., 1969; Oviatt and Kremer, 1977). Since gelat nous carnivores can rarely be identified in the guts of fish, the quantitative significance of this predation is unknown.

SECONDARY IMPACTS OF FEEDING BY GELATINOUS CONSUMERS

Fecal Flux

The vertical flux of biogenic materials in the ocean appears to be mediated by fecal pellets (Wiebe *et al*., 1979; Bishop *et al*., 1977; Honjo and Roman, 1978; Urrere and Knauer, 1981). Although the importance of gelatinous zooplankton as sources of this fecal material is poorly known, information on fecal production rates by gelatinous herbivores suggest that they may play a significant role in particulate flux to the deep sea. Fecal material produced by thaliaceans and pteropods sink at rates consistently faster than those reported for other zooplankton with maximal sinking rates for *Salpa fusiformis* pellets of 2700 $m \cdot d^{-1}$. This is 3 times faster than the highest rate noted for euphausiid pellets and an order of magnitude higher than those for copepods (Bruland and Silver, 1981).

The importance of salps in the vertical flux of materials is suggested by their common presence in epipelagic communities as well as by their occasional formation of dense population swarms. Such swarms are major processors of primary production in surface waters and as a result produce copious fecal material. Madin (1982) calculated that the fecal production of a population of *Cyclosalpa pinnata* at a station in the central North Atlantic would supply the benthic community with approximately 6% of its daily carbon needs, assuming that all the fecal material reached the bottom intact. Likewise, Wiebe *et al*. (1979) estimated fecal production rates of a population of *Salpa aspera* at swarm densities in the North Atlantic. If all the salp fecal matter produced had reached the sea floor, it would have represented over 100% of the daily energy requirements of the deep sea benthos at that location.

Thus, at high population densities salps may make a significant contribution to particulate flux. Since salp swarms are opportunistic and relatively short lived, such contributions may result in periodically high but unpredictable influxes of surface derived material to the deep sea. Moreover, even when present in numbers several orders of magnitude lower than other herbivores (e.g. copepods), salps may consume a proportion of the phytoplankton standing stock equal to that of all other grazers due to their high clearance rates and large size (Harbison and Gilmer, 1976). Thus, even at low population densities, salps may make a significant contribution to the flux of biogenic materials in the ocean.

Calculations of fecal flux based on fecal production rates and abundance of surface populations of zooplankton necessitate making two assumptions of uncertain validity. First, although gelatinous herbivores produce very rapidly sinking fecal pellets, not all pellets will reach the benthos intact. Many are reconsumed or break apart and sink more slowly (Silver and Bruland, 1981). Second, calculations of fecal production also assume constant feeding rates and food supplies. Ideally, the best approach to estimating the contribution of gelatinous predators to fecal flux might be through the accurate identification of pellet origins in sediment traps (Urrere and Knauer, 1981). Some pellets are distinctive and might be identifiable (Silver and Bruland, 1981). However, current sediment trap methodology is inadequate for measuring flux rates of fecal material produced by large gelatinous consumers in the open sea. The chances that fecal material released by rare or patchily distributed gelatinous consumers might fall directly over the relatively small mouth opening of most sediment traps, seems remote.

Differences in the species composition of gelatinous herbivores inhabiting surface waters can result in significant changes in the chemistry and composition of materials entering the deep sea as well For example, in the California current, salps tend to concentrate ultraplankton, including coccolithophorids and produce fecal materia rich in calcium carbonate. Herbivorous pteropods, on the other hand tend to collect larger phytoplankton, particularly diatoms, and thus produce pellets rich in silica (Silver and Burland, 1981).

Excretion and Nutrient Recycling

The importance of zooplankton excretion to nutrient recycling in the ocean varies with different nutrient regimes. For example, i upwelling systems, excretion by large zooplankton may provide only a small fraction of the nutrient pool (Dagg et al., 1980), while in oligotrophic water 45 to 50% of the ammonium assimilated by phytoplankton may be excreted by zooplankton (Eppley et al., 1973). Data on the excretion of gelatinous zooplankton is scanty and complicated because some investigators have not expressed nitrogen excretion as function of comparable units of zooplankton biomass. Because of their watery composition, gelatinous zooplankton are deceptively large relative to their organic content. Generally, carbon content is less than 10% of dry weight (Ikeda, 1974; Madin et al., 1981) and for some ctenophores it is as low as 1.7% (Kremer, 1976a). Thus excretion rates tend to be large on a per animal basis, but low or similar on an animal weight specific basis to those of other zooplankton (Ikeda, 1974; Biggs, 1977; Kremer, 1977).

Biggs (1977) estimated that ammonium excretion by gelatinous organisms such as salps and siphonophores at densities of 1 animal m^{-3} in the North Pacific Gyre would be sufficient to supply 39-63% o the nitrogen requirements of the phytoplankton and contribute as muc ammonium as all the other zooplankton present in a cubic meter in that area. Such extrapolations only serve to illustrate the potential significane of gelatinous consumers to nutrient recycling in th ocean. Accurate, quantitative data on the abundances, sizes, and size specific excretion rates of gelatinous zooplankton are sorely needed in order to evaluate the role of gelatinous consumers in nutrient recycling. At present, only Kramer (1975) has attempted to integrate excretion data of gelatinous predators with population parameters to calculate their potential impact on nutrient recycling within a pelagic community. She assessed the population biomass and weight specific excretion rates of ctenophores at various temperatures, and the average ambient nutrient levels in the water during periods of ctenophore abundance in Narragansett Bay to calculate the maximum nutrient recycling by the *Mnemiopsis leidyi* population. Daily excretion by this ctenophore was comparable to the nutrients released by all the other zooplankton taxa although it was small relative to the contribution of the benthos (Kremer, 1975).

Additional data are also needed on the nature of the products excreted by gelatinous predators. Kremer (1976a) found that nearly half of the nitrogen excreted by ctenophores is in organic form, and thus may not be immediately available to phytoplankton. Kremer (1977) speculates that because ctenophores have such low organic content, they pump nutrients rapidly through their systems and do not bind these elements into their structure.

These results suggest that gelatinous carnivores stimulate phytoplankton growth in three major ways.

1) They reduce grazing pressure by preying upon herbivorous zooplankton. For example, predation by the medusa, *Phialidium*, reduced herbivorous plankton and allowed a second spring bloom to occur off British Columbia (Huntley and Hobson, 1978).
2) They increase availability of inorganic nutrients in the short-term through excretion, and
3) they produce a long-term, positive feedback on phytoplankton through bacterial regeneration.

Much additional information is needed to assess the quantitative impact of these factors on phytoplankton growth and population dynamics.

CONCLUSION

While most gelatinous zooplankton usually consume less than 10% of their prey populations per day, percent consumption is occasionally greater than 50%. Thus, gelatinous zooplankton may occasionally severely reduce prey populations or affect their population dynamics. The unique attributes of many gelatinous zooplankton including high ingestion rates at high food densities, and the potential for high growth rates, high fecundity and short generation time under favourable conditions, all enable them to be particularly effective consumers when food is abundant or patchy. Moreover, these attributes may enable gelatinous consumers to reach much higher population densities than crustaceans, chaetognaths and other more slowly growing planktonic consumers when food densities are high. Efficient and immediate exploitation of abundant food resources through rapid reproduction by gelatinous consumers may maintain the matter and energy incorporated in these food populations within the pelagic zone. Unconsumed by higher trophic levels, such material might accumulate in the sediments and be lost, at least in the short-term, to the water column (Reeve *et al*., 1978).

However, our understanding of the quantitative impact of gelatinous zooplankton is constrained by the relatively few studies which integrate food consumption rates with information on the nutritional physiology, growth rates and population dynamics of either the consumer or its food. In particular, information on oceanic gelatinous zooplankton is lacking. In addition, preliminary studies on secondary effects of feeding by gelatinous consumers including excretion, nutrient recycling, fecal production and particulate flux suggest that these may also be significant factors in the ecology of marine ecosystems. Our present knowledge of the quantitative impact of gelatinous zooplankton in pelagic communities serves only as a titillating introduction to the possible trophic interactions and ecological significance of this ubiquitous, diverse and poorly understood group of marine consumers.

Acknowledgements

I thank M. R. Reeve, L. P. Madin, J. E. Purcell, B. H. Robison, B. W. Frost and P. Kremer for their constructive criticisms of an earlier version of this paper.

REFERENCES

Alldredge, A. L., 1981, The impact of appendicularian grazing on natural food concentrations *in situ*, *Limnol.Oceanogr.*, 26:247

Alldredge, A. L., and Madin, L. P., 1982, Pelagic tunicates: Unique herbivores in the marine plankton, *Bioscience*, 32:655.

Anderson, E. A., 1974, Trophic interactions among ctenophores and copepods in St. Margaret's Bay, Nova Scotia, Ph.D Dissertation, Dalhousie University.

Baker, L. D., and Reeve, M. R., 1974, Laboratory culture of the lobate ctenophore *Mnemiopsis mccradyi* with notes on feeding and fecundity, *Mar.Biol.*, 26:57.

Barham, E. G., 1963, Siphonophores and the deep scattering layer, *Science*, 140:826.

Berner, A., 1962, Feeding and respiration in the copepod *Temora longicornis* (Müller), *J.Mar.Biol.Ass.U.K.*, 42:625.

Berner, L., 1967, Distributional atlas of Thaliacea in the Californ Current region, *CalCOIF Atlas*, 8:1.

Biggs, D. C., 1976, Nutritional ecology of Agalma okeni, in: "Coelenterate Ecology and Behavior," G. O. Mackie, ed., Plenum Press, New York.

Biggs, D. C., 1977, Respriation and ammonium excretion by open ocean gelatinous zooplankton, Limnol.Oceanogr., 22:108.

Biggs, D. C., Bidigare, R. R., and Smith, D. E., 1981, Population density of gelatinous macrozooplankton: In situ estimation in oceanic surface waters, Biol.Oceanogr., 1:157.

Bishop, J. K. B., Edmond, J. M., Ketten, D. R., Bacon, M. P., and Silker, W. B., 1977, The chemistry, biology and vertical flux of particulate matter from the upper 400m of the equatorial Atlantic Ocean, Deep Sea Res., 24:511.

Bishop, J. W., 1967, Feeding rates of the ctenophore, Mnemiopsis leidyi, Chesapeake Sci., 8:259.

Bruland, K. W., and Silver, M. W., 1981, Sinking rates of fecal pellets from gelatinous zooplankton (salps, pteropods, doliolids), Mar.Biol., 63:295.

Clifford, H. C. , and Cargo, D. G., 1978, Feeding rates of the sea nettle, Chrysaora quinquecirrha, under laboratory conditions, Estuaries, 1:58.

Confer, J. L., and Blades, P. I., 1975, Omnivorous zooplankton and planktivorous fish, Limnol.Oceanogr., 20:571.

Conover, R. J., 1968, Zooplankton - life in a nutritiionally dilute environment, Am.Zool., 8:107.

Conover, R. J., and Lalli, C. M., 1974, Feeding and growth in Clione limacina (Phillips), a pteropod mollusc. II. Assimilation, metabolism, and growth efficiency, J.Exp.Mar.Biol.Ecol., 16:131.

Corner, E. D. S., 1961, On the nutrition and metabolism of zooplankton. I. Prelininary observations on the feeding of the marine copepod, Calanus helgolandicus, J.Mar.Biol.Ass.U.K., 41:5.

Cosper, T. C., and Reeve, M. R., 1975, Digestive efficiency of the chaetognath Sagitta hispida Conant, J.Exp.Mar.Biol.Ecol., 17:33.

Curl, H., 1962, Standing crops of carbon, nitrogen and phosphorus, and transfer between trophic levels in continental shelf waters south of New York, Rapp.Proc.Verb.Cons,Int.Explor.Mer., 153:183.

Dagg, M. T., Cowles, T., Whitledge, T., Smith, S., Howe, S., and Judkins, D., 1980, Grazing and excretion by zooplankton in the Peru upwelling system during April 1977, Deep Sea Res., 27:43.

Deibel, D. R., 1980, Feeding, growth and swarm dynamics of neritic tunicates from the Georgia Bight, Ph.D. Dissertation, University of Georgia.

Deibel, D. R., 1982, Laboratory determined mortality, fecundity and growth rates of Thalia democratica Forskal and Dolioletta gegenbauri Uljanin (Tunicata, Thaliacea), J.Plank.Res., 4:143.

Eppley, R. W., Renger, W. H., Venrick, E. L., and Mullin, M. M., 1973, A study of plankton dynamics and nutrient cycling in the central gyre of the North Pacific Ocean, Limnol.Oceanogr., 18:534.

Fenaux, R., 1976, Cycle vital, croissance et production chez Fritillaria pellucida (Appendicularia), dans la baie de Villefranche-sur-Mer, France, Mar.Biol., 34:229.

Fraser, J. H., 1962, The role of ctenophores and salps in zooplankton production and standing crop, Rapp. et Proc.-Verb.Cons.Int. Explor.Mer., 153:121.

Fraser, J. H., 1969, Experimental feeding of some medusae and chaetognaths, J.Fish.Res.Bd.Can., 26:1743.

Fraser. J. H., 1970, The ecology of the ctenophore Pleurobrachia pileus in Scottish waters, J. du Conseil,33:149.

Greve, W., 1968. The "planktonkreisel", a new device for culturing zooplankton, Mar.Biol., 1:201.

Greve, W., 1970, Cultivation experiments on North Sea ctenophores, Helgo.wiss.Meeresunter., 20:304.

Greve, W., 1972, Okologische Untersuchungen an Pleurobrachia pileus 2. Laboratoriumuntersuchungen, Helgo.wiss,Meeresunter., 23:141.

Greze,V. N., and Bileva, O. K., 1980, Zooplankton and its structure in the pelagic zone of the Caribbean Sea, Soviet J.Mar.Biol., 5:79.

Hamner, W. M., and Jenssen, R. M., 1974, Growth, degrowth, and irreversible cell differentiation in Aurelia aurita, Amer. Zool., 14:833.

Hamner, W. H., Madin, L. P., Alldredge, A. L., Gilmer, R. W., and Hamner, P. P., 1975, Underwater observations of gelatinous zooplankton: Sampling problems, feeding biollogy, and behavior, Limnol.Oceanogr., 20:907.

Harbison, G. R., and Gilmer, R. W., 1976, The feeding rates of the pelagic tunicate Pegea confederata and two other salps, Limnol.Oceanogr., 21:517.

Harbison, G. R., Madin, L. P., and Swanberg, N. R., 1978, On the natural history and distribution of oceanic ctenophores, Deep Sea Res., 25:233.

Harris, R. P., and Paffenhöfer, G. A., 1976, Feeding, growth and reproduction of the marine planktonic copepod Temora longicornis Müller, J.Mar,Biol.Ass.U.K., 56:675.

Harris, R. P., Reeve, M. R., Grice, G. D., Evans, G. T., Gibson, V. R., Beers, J. R., and Sullivan, B. K., 1982, Trophic interactions and production processes in natural zooplankton communities in enclosed water columns, in: "Marine Mesocosms: Biological and Chemical Research in Experimental Ecosystems," G. D. Grice and M. R. Reeve, eds., Springer-Verlag, New York

Heron, A. C., 1972, Population ecology of a colonizing species: The pelagic tunicate Thalia democratica I. Individual growth rates and generation time, Oecologia (Berl.), 10:269.

Hirota, J., 1974, Quantitative natural history of Pleurobrachia bachei in La Jolla Bight, United States National Marine Fisheries Service Fishery Bulletin, 72:295.

Honjo, S., and Roman, M. R., 1978, Marine copepod fecal pellets: Production, preservation and sedimentation, J.Mar.Res., 36:45.

Huntley, M. E., and Hobson, L. A., 1978, Medusae predation and plankton dynamics in a temperate fjord, British Columbia, J.Fish. Res.Bd.Can., 35:257.

Ikeda, T., 1974, Nutritional ecology of marine zooplankton, Memoirs of the Fac.Fish., Hokkaido Univ., 22:1.

Kamshilov, M. M., 1960, Feeding of the ctenophore Beroe cucumis Fabricius, Doklady Akademii Nauk Union of Soviet Socialist Republics, 130:1138.

King, R. K., 1981, The quantitative natural history of Oikopleura dioica (Urochordata: Lavacea) in the laboratory and in enclosed water columns, Ph.D. Dissertation, University of Washington.

King, R. K., Hollibaugh, J. T., and Azam, F., 1980, Predator-prey interactions between the larvacean Oikopleura dioica and bacterioplankton in enclosed water columns, Mar.Biol., 56:49.

Kremer, P. M., 1975, Nitrogen regeneration by the ctenophore Mnemiopsis leidyi, in: "Mineral Cycling in Souteastern Ecosystems," G. G. Howwell, J. B. Gentry and M. M. Smith, eds., United States Energy Research and Development Administration Symposium Series, N.T.I.S. No. CONF-740513.

Kremer, P. M., 1976a, Excretion and body composition of the ctenophore Mnemiopsis leidyi (A. Agassiz): Comparisons and consequences, in: "Proceedings of the 10th European Symposium on Marine Biology," G. Persoone and E. Jaspers, eds., Vol. 2:351, Univera Press, Wetteren, Belgium.

Kremer, P. M., 1976b, Population dynamics and ecological energetics of a pulsed zooplankton predator, the ctenophore Mnemiopsis leidyi, in: "Estuarine Processes," M. L. Wiley, ed., Academic Press, New York.

Kremer, P. M., 1977, Respiration and excretion by the ctenophore Mnemiopsis leidyi, Mar.Biol., 44:43.

Kremer, P. M., 1979, Predation by the ctenophore Mnemiopsis leidyi in Narragansett Bay, Rhode Island, Estuaries, 2:97.

Kremer, P. M., Reeve, M. R., Walter, M. A., and Liedle, S. D., 1980, The effect of food availability on the carbon and nitrogen budgets of the ctenophore Mnemiopsis mccradyi, Abst. 3rd Winter ASLO Meeying, Seattle, Wa.

Lasker, R., 1966, Feeding, growth, respiration, and carbon utilization of a Euphausiid crustacean, J.Fish.Res.Bd.Can., 23:1291.

Madin, L. P., 1974, Field observations on the feeding biology behavior of salps (Tunicata: Thaliacea), Mar.Biol., 25:143.

Madin, L. P., 1982, Production, composition and sedimentation of salp fecal pellets in oceanic waters, Mar.Biol., 67:39.

Madin, L. P., Cetta, C. M., and McAlister, V. L., 1981, Elemental and biochemical composition of salps (Tunicata: Thaliacea), Mar. Biol., 63:217.

Miller, R. J., 1970, Distribution and energetics of an estuarine population of the ctenophore, Mnemiopsis leidyi, Ph.D. Dissertation, North Carolina State University, Raleigh.

Miller, R. J., 1974, Distribution and biomass of an estuarine ctenophore population, Mnemiopsis leidyi (A.Agassiz), Chesapeake Sci., 15:1.

Möller, H., 1980, Scyphomedusae as predators and food competitors of larval fish, Meeresforschung, 28:90.

Mullin, M. M., and Brooks, E. R., 1970, Growth and metabolism of two planktonic marine copepods as influenced by temperature and type of food, in: "Marine Food Chains," J. Steele, ed., Oliver and Boyd, Edinburgh.

Mullin, M. M., and Evans, P. M., 1974, The use of a deep tank in plankton ecology, 2. Efficiency of a planktonic food chain, Limnol.Oceanogr., 19:902.

Omori, M., 1969, Weights and chemical composition of some important oceanic zooplankton in the North Pacific Ocean, Mar.Biol., 3:4.

Oviatt, C. M., and Kremer, P. M., 1977, Predation on the ctenophore Mnemiopsis leidyi by butterflyfish, Peprilus tricanthus, in Narragansett Bay, Phode Island, Chesapeake Sci., 18:236.

Paffenhöfer, G. A., 1973, The cultivation of an appendicularian through numerous generations, Mar.Biol., 22:183.

Paffenhöfer, G. A., 1976a, On the biology of appendicularia of the southeastern North Sea, in: "Proceedings of the 10th European Symposium on Marine Biology," G. Persoone and E. Jaspers, eds., Vol. 2:437, Universa Press, Wetteren, Belgium.

Paffenhöfer, G. A., 1976b, Feeding, growth and food conversion of the marine planktonic copepod Calanus helgolandicus, Limnol. Oceanogr., 21:39.

Paffenhöfer, G. A., and Harris, R. P., 1979, Laboratory culture of marine holozooplankton and its contribution to studies of marine planktonic food webs, Adv.Mar.Biol., 16:211.

Parsons, T. R., and Takahashi, M., 1973, "Biological Oceanographic Processes," Pergamon, Oxford.

Phillips, P. J., Burke, W. D., and Keener, E. J., 1969, Observations on the trophic significance of jellyfish in Mississippi Sound with quantitative data on the associative behavior of small fishes with medusae, Trans.Amer.Fish.Soc., 98:703.

Purcell, J. E., 1981a, Feeding ecology of Rhizophysa eysenhardti, a siphonophore predator of fish larvae, Limnol.Oceanogr., 26: 424.

Purcell, J. E., 1981b, Selective predation and caloric consumption by the siphonophore Rosacea cymbiformis in nature, Mar.Biol., 63:283.

Purcell, J. E., 1981c, Dietary composition and diel feeding patterns of epipelagic siphonophores, Mar.Biol., 65:83.

Purcell, J. E., 1982, Feeding and growth of the siphonophore Muggiaea atlantica, J.Plank.Res., In press.

Reeve, M. R., 1970, The biology of Chaetognatha. I. Quantitative aspects of growth and egg production in Sagitta hispida, in: "Marine Food Chains," J. H. Steele, ed., Oliver and Boyd, Edinburgh.

Reeve, M. R., 1980, Population dynamics of ctenophores in large scale enclosures over several years, in: "Nutrition in the Lower Metazoa," D. C. Smith and Y. Tiggon, eds., Pergamon Press, New York.

Reeve, M. R., and Baker, L. D., 1975, Production of two planktonic carnivores (chaetognath and ctenophore) in south Florida inshore waters, United States National Marine Fisheries Service Fishery Bulletin, 72:238.

Reeve, M. R., and Walter, M. A., 1976, A large scale experiment on the growth and predation potential of ctenophore populations, in: "Coelenterate Ecology and Behavior," G. O. Mackie, ed., Plenum Press, New York.

Reeve, M. R., and Walter, M. A., 1978, Nutritional ecology of ctenophores - A review of recent research, Adv.Mar.Biol., 15:249.

Reeve, M. R., Walter, M. A., and Ikeda, T., 1978, Laboratory studies of ingestion, and food utilization in lobate and tentaculate ctenophores, Limnol.Oceanogr., 23:740.

Rogers, C. A., Biggs, D. C., and Cooper, R. A., 1978, Aggregation of the siphonophore Nanomia cara in the Gulf of Maine: Observations from a submersible, United States National Marine Fisheries Service Fishery Bulletin, 76:281.

Roosen-Runge, E. C., 1970, Life cycle of the hydromedusae Phialidium gregarium (A. Agassiz, 1862) in the laboratory, Biol.Bull., 139:203.

Seki, H., 1973, Red tide of Oikopleura in Saanich Inlet, La Mer (Bull.Soc.Francojap.Oceanogr.), 11:153.

Silver, M. W., and Bruland, K. W., 1981, Differential feeding and fecal pellet compostion of salps and pteropods, and the possible origin of the deep water flora and olive green "cells", Mar,Biol., 62:263.

Stanlaw, K. A., Reeve, M. R., and Walter, M. A., 1981, Growth, food and vulnerability to damage of the ctenophore Mnemiopsis mccradyi in its early life history stages, Limnol.Oceanogr., 26:224.

Stretch, J. J., 1982, Observations on the abundance and feeding behavior of the cestid ctenophore, Velamen parallelum, Bull. Mar.Sci., In press.

Stretch, J. J., and King, J. M., 1980, Direct fission: An undescribed reproductive method in the Hydromedusae, Bull.Mar.Sci. 30:522.

Urrere, M. A., and Knauer, G. A., 1981, Zooplankton fecal pellet production and vertical transport of particulate organic material in the pelagic environment, J.Plank.Res., 3:369.

Welch, H. E., 1968, Relationships between assimilation efficiencies and growth efficiencies for aquatic consumers, Ecology, 49:755.

Wiebe, P. H., Madin, L. P., Haury, L. R., Harbison, G. R., and Philbin, L. M., 1979, Diel vertical migration by Salpa aspera and its potential for large scale particulate organic matter transport to the deep sea, Mar.Biol., 53:249.

FISH PRODUCTION IN OPEN OCEAN ECOSYSTEMS

K. H. Mann

Marine Ecology Laboratory
Bedford Institute of Oceanography
Dartmouth, Nova Scotia
Canada B2Y 4A2

INTRODUCTION

It has been known for several decades that there exists in all the major oceans a sonic scattering layer which during the day is often found between 450 and 750m depth, but which divides at night into components that rise close to the surface and others that stay at depth. The organisms responsible for this sonic reverberation are predominantly fish and crustaceans. Two major symposia have been devoted to the ecology of sound scattering organisms (Farquhar, 1971; Andersen and Zahuranec, 1977) and it is now known that the scattering layers are more or less continuous across the oceans. There are untold millions of fish rising to feed in surface waters at night. This review explores the relationship of these fish populations to those that permanently inhabit the upper layers (tuna, sharks, and their prey species, such as mackerels, jacks and sauries) and those that live permanently at great depth. There appears to be a potential for exploitation of the fish of the sonic scattering layer (at least in the area of greatest concentration), but before this exploitation is seriously contemplated it is important to understand the ecological relations of mesopelagic fishes in open water ecosystems.

It should be noted that by open ocean ecosystems I mean the oceans beyond the margins of the continental shelves. The abundant fish populations of shelf waters are specifically excluded from this discussion. It is also worth noting that, of the total area of the open ocean, 44.5% is between the latitudes 25°N and 25°S, and 75% is between 45°N and 45°S. Hence, the area under consideration is predominantly tropical and subtropical water.

SYSTEM STRUCTURE

The functioning of open ocean fish production systems will be considered from the point of view of their vertical structure (see Marshall, 1979, for details of zonation). The upper, epipelagic zone roughly corresponds with the euphotic zone and extends to a depth of 100 - 200m according to the transparency of the water. In this zone the dominant predators in daytime are the scombroid fishes: tunas (Scombridae), billfish (Istiophoridae) and swordfish (Xiphiidae). Some tunas may feed to a depth of about 400m by day, and swordfish may feed deep at night (Carey and Robison, 1981). The breeding grounds of almost all these fish are in tropical waters, but the various species make migrations of varying extent into temperate waters. For example, bluefin tuna Thunnus thynnus (Figure 1) breed in the Caribbean area, medium sized fish migrate north in summer as far as Cape Cod, but large, older fish may migrate as far as Newfoundland on the coast of North America, or may cross the Atlantic and spend the summer in Norwegian waters. On the other hand, yellowfin tuna Thunnus albacares spend the whole of their life history in tropical or subtropical waters within which they make significant migrations (Figure 2). The ranges of billfish (e.g. striped marlin

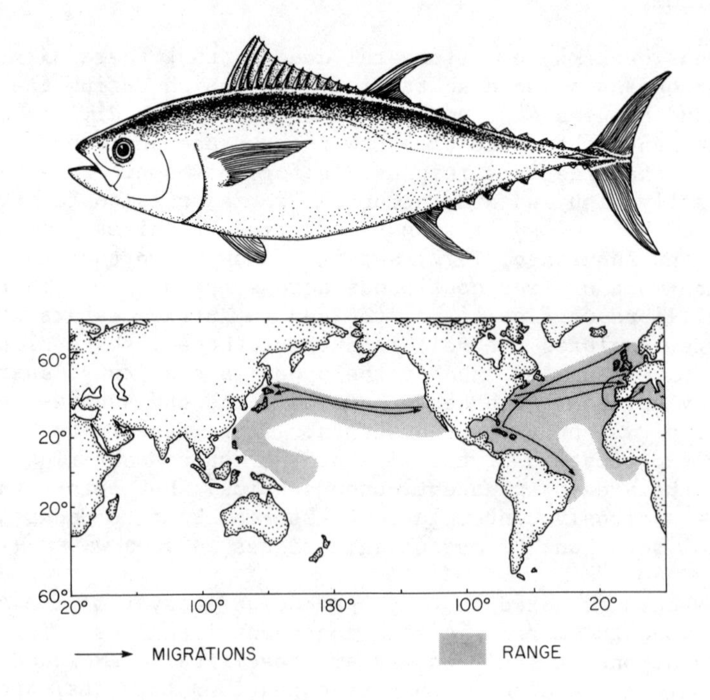

Fig. 1. Bluefin tuna (Thunnus thynnus). These fish may exceed 4m in length and weigh up to almost 1 tonne. The map shows Atlantic distribution and some migration routes.

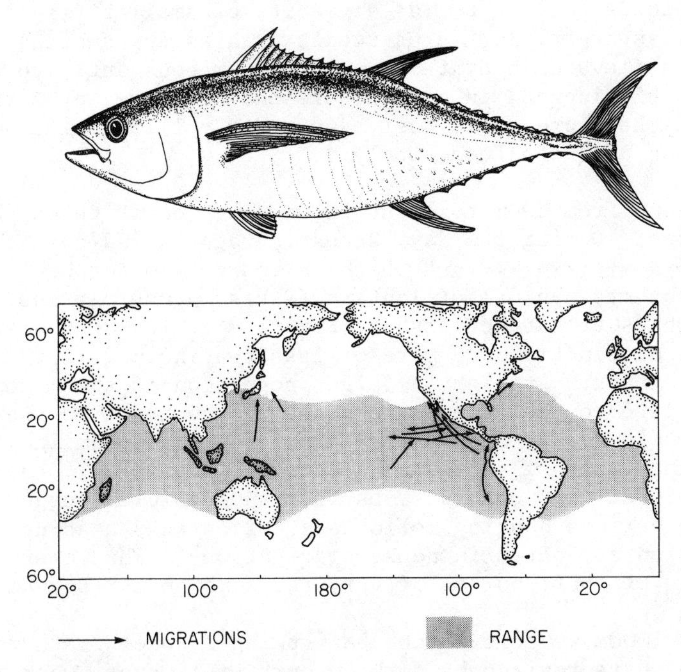

Fig. 2. Yellowfish tuna (Thunnus albacares). Length to 2m and weight to 150 kg. The map shows the Pacific range, and some recorded migrations.

and blue marlin) are less well known, but they are captured over approximately the same range as tunas. Swordfish move further into colder water than the other groups, and greatest catches occur between 30°N and 45°N. The world catch of tunas, bonitos and billfishes in 1979 was 2.4 million tonnes, rather less than 4% of the total world catch of marine fishes (FAO 1980).

In addition to the scombroids, two other major groups of fish, the salmons and the sharks, are present in the epipelagic zone. The salmons are subject to open-ocean fisheries in temperate and subarctic waters, but the world catch is less than one third of the catch of scombroids, and by far the greater part is taken in shelf or estuarine waters. Sharks inhabit much the same oceanic range as scombroids and may be regarded as occupying the same general trophic level, except that large sharks may be predators on scombroids. The world catch of sharks is less than one quarter the catch of scombroids, but this may not be representative of their biomass, since in general they are not highly prized as human food. We need more information on the biomass, production and ecological importance of sharks.

Scombroids feed on a wide diversity of smaller fishes, on squid and to a lesser extent on crustaceans such as euphausiids. They maintain an elevated body temperature, swim constantly to prevent themselves sinking and in general are characterized by a tendency to move in on their prey very fast, and to travel long distances in search of an adequate food supply.

The zone from 200m to about 1000m depth constitutes the mesopelagic zone. During the day, acoustic records indicate the presence of a "deep scattering layer" which is known from net hauls to be dominated by myctophids, or lantern fishes, gonostomatids, and sternoptychids or hatchet fishes (Figure 3). The deep scattering layers are distinctive and more or less continuous all the way across the major oceans. At night a large proportion of the mesopelagic fishes migrate to the epipelagic zone. In contrast to the tunas, these fish often have shapes that are far from streamlined, suggesting that they perform a limited amount of locomotion, other than vertical migration. Observations from submersibles suggest that they spend long periods hanging motionless, often in a head-up or head-down position and often in dense aggregations. When disturbed they are capable of short bursts of very rapid movement (Barham, 1971).

Below 1000m we come to the bathypelagic zone. Bathypelagic fishes are characterized by dark colour, small eyes, weak musculature, but large mouths (Figure 4). They include angler fishes (ceratioids) and dark coloured species of the genus Cyclothone

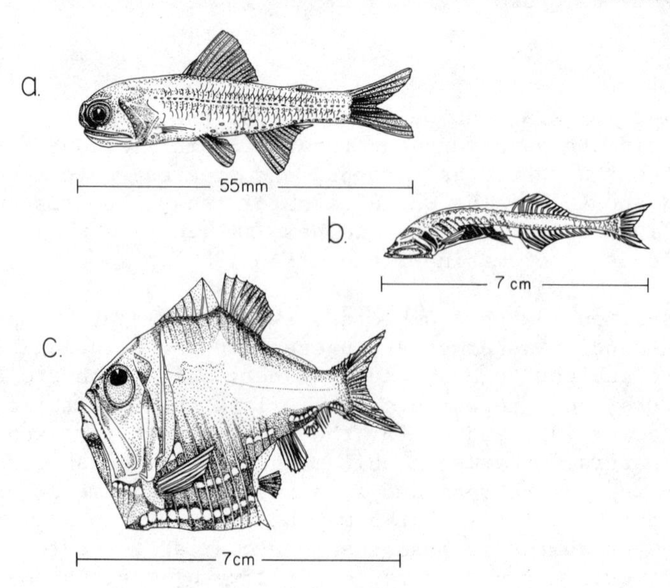

Fig. 3. Mesopelagic fishes: (a) a myctophid or lantern fish, (b) a gonostomatid or anglemouth and (c) sternoptychid, or hatchet fish.

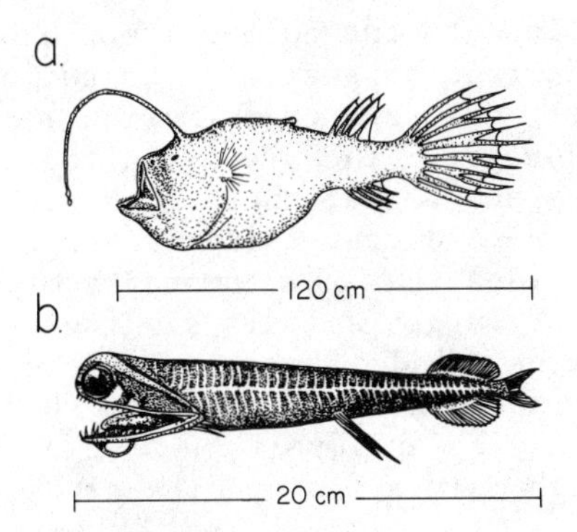

Fig. 4. Bathypelagic fishes: (a) a ceratioid, or angler-fish, (b) a malacostid or loosejaw.

(Marshall 1971a). They appear to be adapted to life in a food-poor environment by minimizing energy expenditure but many have the ability to take prey over a large spectrum of size, whenever it comes within range. Angler fishes have developed elaborate lures to attract prey. They have also maximized reproductive effort in relation to available food by evolving dwarf males which may not feed after metamorphosing from the larval stage. Luminescent organs while well developed in some mesopelagic fishes appear to be a particularly important aspect of communication between individuals in this otherwise dark environment.

Close to the bottom are two distinct communities of bottom-living fish. On the bottom are 'sit and wait' predators (e.g. Bathysaurus and chlorophthalmids) which lack swimbladders and are

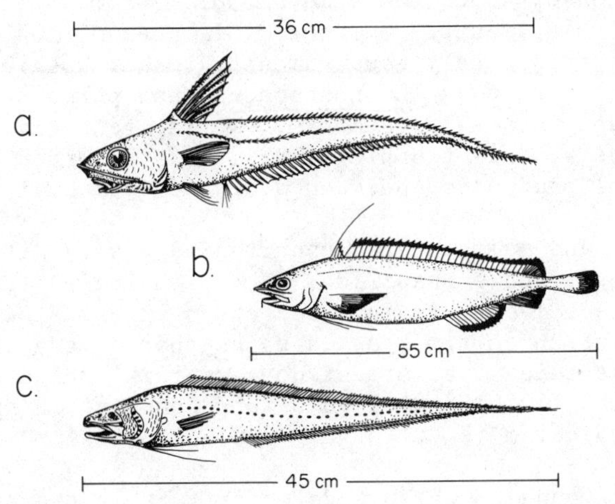

Fig. 5. Deep-sea benthic fishes: (a) a macrourid, or rat tail, (b) a morid, or deep-sea cod and (c) a brotulid.

negatively buoyant. Close to the bottom are the benthopelagic fishes, a group of wide-ranging species with swimbladders, among them rat-tails (Macrouridae), deep-sea cods (Moridae) and brotulids (Marshall, 1971a) (Figure 5). The use of baited cameras (Isaacs and Schwartzlose, 1975) has shown that these fish are adept at detecting carcasses from a distance and converging in surprisingly large numbers. This led to the idea that the community may be supported mainly by the corpses of larger animals sinking from above. There is also a community of large mobile amphipods of the family Lysianassidae, together with shrimps and other decapods (Dahl, 1979; Hessler *et al*., 1978). These may be an important food source for the benthic fish community, although only a few species are so far known to exploit them.

Almost all mesopelagic, bathypelagic and bottom-living fish make use of the epipelagic zone as feeding grounds for their young. Eggs are buoyant, and rise from depth during development. The larvae then feed and grow before starting their downward migrations to the adult living depth.

FISH BIOMASS AND PRODUCTION

Epipelagic Fish

One of the best studied tuna populations is that of the yellowfin tuna (*Thunnus albacares*) in the eastern Pacific ocean between latitudes 40°N and 10°S (Figure 2). Sharp and Francis (1976) assembled data for the construction of an energetics model. They estimated that in the area of exploitation (16.96 million km^2) the biomass is at present of the order of 200,000 tonnes, with a gross growth production of about the same amount. From their data we may calculate that this amounts to about 0.014 kcal m^{-2} of biomass, and production of approximately equal amount. Other species at the same trophic level (bigeye tuna *Thunnus obesus*, and porpoises *Stenella* spp.) are present in smaller numbers. We may tentatively conclude that the biomass at this trophic level is of the order of 0.03 kcal m^{-2}, with annual production between 0.02 and 0.03 kcal m^{-2}.

These fish and mammals are supported by a very diverse fauna of forage fish, squid and euphausiids. The fish include those commonly known as mackerel, jacks, flying fish and sauries. The world catch of these is about three times as large as that of the tuna group, but there are many species that are of no commercial value. The model of Sharp and Francis (1976) indicated a food consumption by the yellowfin of 1.5 -4.25 cal $m^{-2}d^{-1}$, roughly 0.5 - 1.3 kcal $m^{-2}yr^{-1}$.

I have not found comparable data for tuna in the Atlantic ocean. A remarkably clear account of the seasonal distribution of tuna (bluefin, *Thunnus thynnus*; albacore, *T. alalunga*; bigeye, *T. obesus*;

and yellowfin, T. albacares) in the north and south Atlantic is given by Shiohama et al. (1965). The paper is in Japanese, but the maps showing distribution of "hooking rates" in longline fisheries show clearly that populations move with the seasons in the area between 30°N and 30°S, but that few fish are caught in waters close to the equator. Blackburn (1965) showed that tuna distribution is strongly influenced by physical oceanographic features such as surface temperature, surface currents, fronts, and upwelling. Detailed discussion of the ecology of tuna is beyond the scope of this article since, as we shall see, fish production in the open ocean is dominated by mesopelagic species.

Mesopelagic Fish

Since mesopelagic fish form continuous layers across all the major oceans, they are probably the most abundant vertebrates on earth. Two main methods have been used to determine their biomass, namely acoustics and net sampling. Gjøsaeter and Kawaguchi (1980) have reviewed sampling problems and put forward tentative estimates of biomass on a global scale. The largest data sets have been obtained with Isaacs-Kidd midwater trawls and other microtekton nets with mouth openings ranging in area from 1.0 to $10m^2$. In many cases the nets were without closing mechanisms or flowmeters. In some parts of the ocean large commercial pelagic trawls with openings of 100 to $1000m^2$ are in use. These are more efficient at retaining larger fish, less efficient at retaining smaller fish, since they have coarser meshes. Two of the chief sources of error in sampling are net avoidance (when fish sense the approaching net and move out of its path) and escapement (usually through the meshes of the net). Sampling error will depend on the net design and the size and behaviour of each species. Hence, present estimates of abundance and biomass are at best very crude. Acoustic techniques are extremely useful for determining the location of aggregations of mesopelagic fish and following their vertical migrations, but the ability of acoustic gear to determine biomasses within aggregations is at present very limited. It depends on the frequency used and is subject to error when, for example, the air sacs of siphonophores give echoes similar to those from the air bladders of fish. Some of the best estimates have been obtained by locating aggregations by acoustic methods, then sampling them with nets (e.g. Baird and Wilson, 1977).

Table 1 is a condensed version of the data presented by Gjøsaeter and Kawaguchi (1980). Estimates from central gyres are usually of the order of 1 or 2g wet weight m^{-2}, while higher densities, up to 8g m^{-2} have been obtained from arctic and subarctic areas. The small number of acoustic surveys carried out in open ocean areas have tended to give figures similar to those obtained by net sampling. In coastal, shelf-break or upwelling areas, acoustic samples have yielded estimates in the range 10 - 100g m^{-2}. Gjøsaeter

Table 1. Estimated densities of mesopelagic fishes in offshore waters in various parts of the world, obtained by net sampling (data from Gjøsaeter and Kawaguchi, 1980).

Area	Dominant Species	Biomass (g m ^{2})
N.E. Atlantic. N of 60°N	*Benthosema glaciale*	0.1 - 2.0
S of 60°N	*Benthosema glaciale*	0.5 - 2.0
N.W. Atlantic. N of 60°N	*Benthosema glaciale*	5.0 - 8.0
S of 60°N	*Ceratoscopelus maderensis*	0.1 - 1.7
	Benthosema glaciale	
E.C. Atlantic	*Cyclothone spp.*	4.0 - 6.0
	Argyropelecus hemigymnus	
W.C. Atlantic	*Diogenichtys atlanticus*	1.0 - 0.2
S.E. Atlantic		1.0
S.W. Atlantic	*Diaphus dumerili*	3.0
N.E. Pacific		
Subarctic area	*Stenobrachius leucopsarus*	4.5
Transitional area	*Stenobrachius leucopsarus*	
	Diaphus theta	3.6
	Protomyctophum crockeri	
W.C. Pacific		
Equatorial current	*Diaphus garmani*	2.6
Central water	*Diaphus malayanus*	1.0
E.C. Pacific		
Transitional area	*Triphoturus mexicanus*	
	Protomyctophum crockeri	3.6
	Ceratoscopelus townsendi	
Central water	*Ceratoscopelus warmingi*	
	Lampanyctus steinbecki	2.0
	Triphoturus nigrescens	
N.W. Pacific		
Subarctic area	*Stenobrachius nannochir*	
	S. leucopsarus	6.5
	Diaphus theta	
Central water	*Cyclothone atraria*	
	Gonostoma gracile	1.3
	Diogenichthys atlanticus	
	Benthosema suborbitale	
Indian Ocean Eastern	*Diaphus leutkeni*	1.8 - 4.7
	D. splendens	
Western	Numerous spp.	0.5

and Kawaguchi (1980) concluded that, even allowing for net avoidance and escapement, the biomass of mesopelagic fish in the open ocean is always less than 10g m^{-2}. One might tentatively set the figures at 1-2g m^{-2}in the central gyres and 8g m^{-2} in the more productive fringes of those gyres.

Recent years have seen two interesting attempts to produce energy budgets for mesopelagic fish. Childress *et al.* (1980) studied the growth and metabolism of 4 species of mesopelagic fish off Southern California, two of which (*Triphoturus mexicanus* and *Lampanyctus ritteri*) were from offshore waters with a depth of about 2000m. As can be seen from Table 1, these two genera are of widespread occurrence in the Pacific. It was possible to age the fish from their otoliths and hence determine growth rates. The fish are rich in lipids, so that their caloric content was rather high, 1.75 kcal g^{-1} wet weight. Rates of metabolism were measured by bringing fish to the surface in nets, carefully transferring them to containers of water, and placing those that seemed to be in best condition in sealed containers so that their oxygen utilization could be measured. When growth and metabolism were summed, they were very similar for the two species, 4.11 kcal g^{-1} wet weight yr^{-1} for *Triphoturus* and 4.16 for *Lampanyctus*. Of this, growth constituted about 25%.

Childress *et al.* (1980) cross-checked their results by calculating the implied daily ration, which came to 5% of body dry weight per day for 0+ fish, and rather less for older fish. This may be an underestimate. Tseytlin and Gorelova (1978), using direct observations of stomach contents, modelled the feeding process in a common tropical Pacific mesopelagic migrator, *Myctophum nitidulum*. They concluced that fry of 11-20 mm required a daily ration equivalent to 11-16% of their body energy. Considering that the caloric equivalent of a gramme of food is probably lower than that of a gramme of fish, this implies a daily intake equivalent to perhaps 15-20% of the body dry weight. The daily requirement was estimated to be progressively lower for larger and larger fish, coming down to 3% of the body energy for fish of 50-60 mm. In this size group their estimates are not far from those of Childress *et al.* (1980).

Baird and Hopkins (1981) studied a small (about 30 mm) non-migrating species, *Valenciennellus tripunctulus*, which occurs at 200-600m in the Gulf of Mexico and Caribbean waters, and which is common, but not dominant in many tropical and subtropical gyre regions. They could not age the fish, so they took the indirect approach of constructing several energetics models to see which appeared to be viable (Table 2). There was a clear diurnal pattern of feeding with stomachs filling between 1200h and 2200h, so they used the contents of full stomachs as estimates of maximum daily ration. It came to 5.6% of body weight for a 15 mm fish and 3.7% for a 30 mm fish. Using metabolic data from Childress and Nygaard (1973) and various other assumptions inferred from the literature, they showed that on maximum ration it was possible for the fish to grow to 30 mm in 1 year or 35 mm in 1.5 years. When food availability was taken into account they concluded that an annual life history was probably the optimum strategy.

Table 2. Valenciennellus tripunctulatus. Caloric budget assuming various size ranges, rations and periods of growth. Linear growth assumed.

Daily ration	Maximum length (1 max, mm)	Time at 1 max (y)	Energy budget (calories yr^{-1}) Q_c Consumption	Q_g Growth	Q_s Losses	Q_r Respiration	Balance cal.	Balance $\%Q_c$	Conversion efficiency $Q_g/Q_c(\%)$
Maximum	30	0.5	652	245	261	184	- 38	(-) 6	(-)38
	30	1.0	1435	245	575	405	210	15	17
	30	1.5	2220	245	888	625	462	21	11
	30	2.0	3006	245	1202	847	712	24	8
Minimum	30	0.5	314	245	136	184	-224	(-)66	(-)72
	30	1.0	750	245	300	405	-200	(-)27	(-)33
	30	1.5	1161	245	464	625	-173	(-)15	(-)21
	30	2.0	1571	245	629	847	-150	(-)10	(-)16
Maximum	35	1.0	1868	510	747	664	- 53	(-) 3	(-)27
	35	1.5	2891	510	1156	1036	189	6	18
Minimum	35	2.0	1889	510	755	1391	-767	(-)41	(-)27
Maximum	40	2.0	4894	1046	1958	2428	-538	(-)11	(-)21

Triphoturus and Lympanyctus live 5 or more years and from the data of Childress et al. (1980) it can be inferred that they have an annual population production to biomass ratio of about 0.56. Valenciennellus, on the other hand, with an annual life history, can be expected to have a population P/B ratio in the range of 3 - 5 depending on the pattern of mortality (Waters, 1977). However, Valenciennellus was found to be present in very low densities, about, one per $2000m^3$, and an individual biomass of only 0.1 - 0.2g, so its contribution to community production is probably quite small. If we assume that Triphoturus and Lympanyctus are good models for the mesopelagic fish populations, our best estimates would be that in the central gyres the mean biomass is 1.75 - 3.5 kcal m^{-2} and the annual production is about 1 - 2 kcal m^{-2}. Using the data in the models of Childress et al. (1980), the energy requirements for growth and metabolism of such a population would be about 4 - 8 kcal m^{-2} yr and with an assimilation efficiency of 80% the food requirements would be 5 - 10 kcal m^{-2}. In the more productive fringes of the gyres these figures might all be 2 - 3 times larger.

Bathypelagic Fish

There are very few assessments of the biomass of bathypelagic fish. For those in the water column, Marshall (1971b) has stated that the biomass is probably 2 orders of magnitude below that in the mesopelagic. If that is the case, they can almost be ignored in any calculation of fluxes in the total ecosystem. Torres et al. (1979) made careful shipboard measurements of the metabolic rates of fish whose minimum depth of occurrence ranged from near-surface to 1000m (Figure 6). They showed that there was a sharp decline in routine

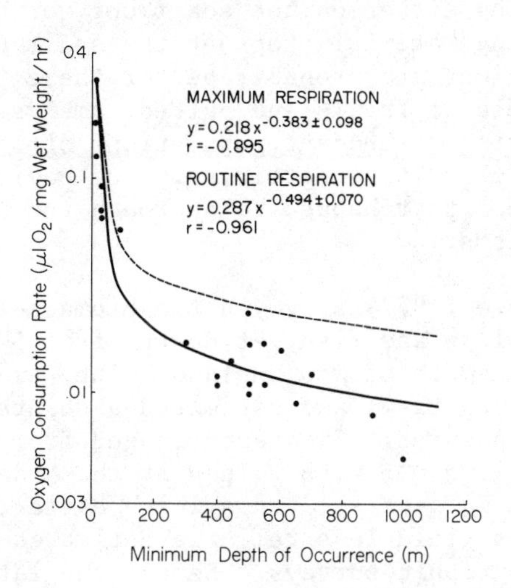

Fig. 6. Metabolic rates of fish whose minimum depth of occurrence range from near-surface to 1000m. After Torres et al. (1979).

respiration from around 0.1 μl O_2·mg wet $wt^{-1}·h^{-1}$ characteristic of fish in the epipelagic zone to less than one tenth that figure for fish penetrating below 500m. They therefore concluded that the bathypelagic fish community was characterized by low metabolic rate, as well as low biomass. An interesting modification of these ideas was provided by Smith and Laver (1981). As was mentioned earlier, gonostomatid fish of the genus *Cyclothone* tend to be one of the major components of the bathypelagic fish fauna, at least down to 2500m. Operating from the submersible ALVIN at a depth of 1300m, Smith and Laver (1981) were able to gently suck three specimens of *Cyclothone acclinidens* into three respiration chambers, using what they called a "slurp gun" technique. Oxygen utilization was monitored for 26h and the fish were found to have daytime levels of respiration very similar to those found by Torres *et al*. (1979), but nighttime levels were 3 - 5 times higher. The authors considered that nocturnal feeding was the most probable explanation of elevated metabolism and speculated on the cues that might serve to synchronize a diurnal rhythm at such great depths. Tidal currents showed no obvious correlation with the rhythm, light was considered not to penetrate to that depth, and the food organisms were not migrators. Although the authors did not discuss them, the possibilities of diurnal cues being received from plankton that migrate to the lighted zone, or from a periodicity in the rain of fecal pellets cannot be ruled out.

Benthic and Benthopelagic Fish

When we reach bottom, whether on the continental slope, continental rise, or on the abyssal plain, there is evidence of a higher biomass of fish living either on the sea floor, or hovering close to it. Since all sinking material stops at the sediment-water interface, feeding conditions are probably better there than in the water column above. The use of trawls and baited cameras has revealed the existence of surprisingly abundant populations of scavenging fish, adapted to making use of large particles: the bodies of dead animals, midwater species that happen to approach the bottom, and other benthopelagic organisms.

Haedrich and Rowe (1977) surveyed the biomasses of demersal fish on the continental slope and rise, at depths from 500m to 2500m in the Western Atlantic south of New England. They examined species composition with a deep trawl and estimated absolute abundances from 3500 photographic exposures. Biomasses ranged from 5.78g m^{-2} wet weight to less than 1.0g m^{-2} with values at the 4 deeper stations al less than 2.0g m^{-2}. In general, traps which attract scavengers from unspecified distances yield less reliable estimates of fish biomass than trawl and photographic surveys. Hence, the latter will be used in the calculations which follow.

Metabolic rates of these deep ocean benthic fishes are known to be low. Smith and Hessler (1974) made *in situ* measurements of the rattail *Coryphenoides* and of the hagfish *Eptatretus* by catching them in baited plexiglass traps operated by a remote controlled vehicles at 1230m and monitored by television. The oxygen consumption of the rattail was two orders of magnitude lower than that of a cod, *Gadus morhua*, of similar weight and measured at a similar temperature. The respiration of the hagfish was significantly lower than that of comparable shallow-water species of mixiniids. The same is almost certainly true for fish in the water column above the bottom. In the study mentioned earlier of the energetics of pelagic fish, by Childress *et al*. (1980), information was obtained on the energy budgets of 5 species of bathypelagic non-migrators. Whereas the metabolism of the mesopelagic migrators was over 4 kcal g^{-1} yr^{-1}, that of the bathypelagic species ranged from 0.78 to 1.29.

Rannou (1975) showed that macrourid fishes from the depths of the western Mediterranean have annual growth rings on their scales and a very slow growth rate. Similar results have been obtained elsewhere (Hureau *et al*., 1979). Seasonal growth is accompanied by seasonal breeding, which is not surprising when one remembers that eggs rise to the surface and the larvae develop in the epipelagic zone. There appears to be a clear selective advantage in releasing eggs at a time of year when conditions at the surface are best for the larvae. Guennon and Rannou (1979) showed that in the Bay of Biscay, at depths between 2000m and 4700m, there was a semi-diurnal feeding activity in macrourids and ophiids. This observation reinforces the interpretation of Smith and Laver (1981) of a diurnal rhythm in respiration in *Cyclothone*. It is not at all clear what provides the cue for diurnal and seasonal rhythms at such great depths, but weak tidal currents and changes in sedimentation rate throughout the year have been suggested.

From the foregoing we may conclude that the fish biomass on or near the sea floor is in the range of 1g m^{-2} wet weight on the abyssal plains to about 5g m^{-2} on the continental slopes, and that growth and metabolism are 1-2 orders of magnitude lower than in the epipelagic zone so that P/B ratios may be very low indeed, indicating a production of the order of 0.1 - 0.5 kcal m^{-2} yr^{-1}.

Biomass and production values for the various zones are summarized later (see Figure 9).

TOTAL SYSTEM FUNCTION

Trophic Relationships

Before we can understand open ocean fish production and the food webs that support it, there are a number of questions to be resolved. For example, what is the relationship between the large epipelagic predators and the mesopelagic fish that migrate into surface waters at night? How important are larger invertebrates such as cephalopods and euphausiids in supporting the production of fish? What is the relative importance of large particles such as dead fish, and small particles such as zooplankton feces, in supporting benthic fish production? Some of these questions have been addressed directly by Roger and Grandperrin (1976). These authors examined the stomach contents of 365 albacore (Thunnus alalunga) and yellowfin (T. albacares) tuna from three regions of the south tropical Pacific Ocean. From these they identified 1399 micronektonic fishes, and the stomachs of 667 micronektonic fishes contained identifiable remains. They found that euphausiids accounted for about 12% in volume of the food of the micronektonic fishes, which in turn comprised about 60% of the diet of tunas. The micronektonic fishes ingested by the tunas were almost solely epipelagic residents. Mesopelagic migrators, which came into the epipelagic zone at night were almost never found in tuna stomachs. It was concluded that these two species of tuna are daytime feeders. The euphausiids found in the stomachs of the epipelagic micronektonic fishes were predominantly epipelagic non-migrators of the genus Stylocheiron. Those migrating to deeper waters during the day were poorly represented. Hence it was concluded that the micronektonic fishes which form the food of tunas are also daytime feeders. All this leads to the conclusion that the food chain tuna - forage fish - euphausiid is based mainly on epipelagic production in the phytoplankton and zooplankton with little exploitation

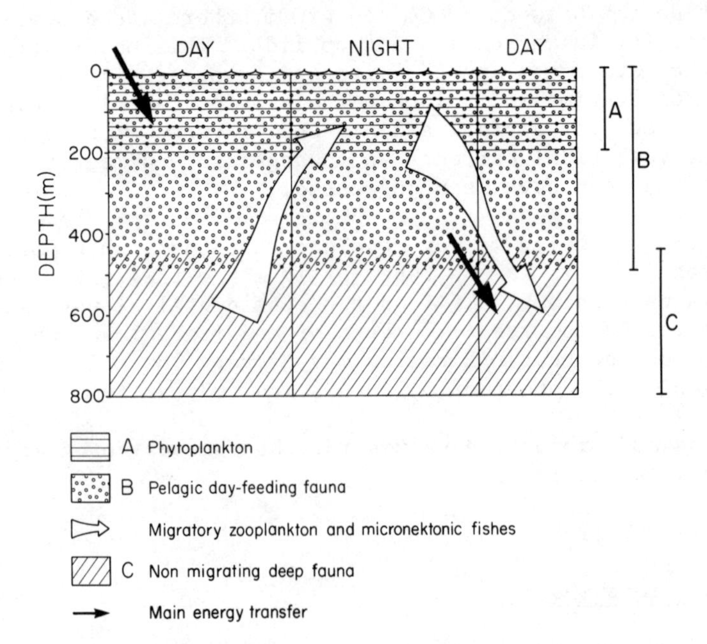

Fig. 7. Diagram summarizing the relationship between migratory and nonmigratory organisms in the south tropical Pacific ocean, by day and by night. Modified from Roger and Grandperrin (1976).

of mesopelagic fish production. There is another branch of the food web involving squid, which form nearly 40% of the tuna diet. These organisms may feed to some extent on the migrating fauna. From this analysis, Roger and Grandperrin (1976) concluded that the mesopelagic migrators are able to take food from the surface layers without themselves contributing very much to the productivity of higher trophic levels in the epipelagic. There is, therefore, a substantial and active net downward transport of energy and materials to support mesopelagic and bathypelagic production (Figure 7).

While the paper by Roger and Grandperrin (1976) represents the most complete study of the place of tuna in the open-ocean food web, it is not necessarily representative of all species and all oceans. Blunt (1960), Sund and Richards (1967) and Mimura (1963) all report representatives of midwater families in the guts of yellowfin tuna. It seems that larger, older fish may range more deeply in pursuit of food. It also seems probable that bigeye tuna (Thunnus obesus) feeds more at night than yellowfin and albacore, and takes more mesopelagic myctophids (Parin, 1970; N. R.Merrett, pers. comm.). Nevertheless, on balance, the literature suggests that mesopelagic fish are not a major component of the diet of the large epipelagic predators.

Olson (in press) has reported on the stomach contents of over 4000 yellowfin tuna from offshore areas of the eastern Pacific. Most fish were associated with dolphins (Stenella spp.) at the time of capture. Small tunas (Auxis sp.) were the most important food items, followed by cephalopods and flying fishes. He estimated that the 200000 tonnes of yellowfin tuna in the eastern Pacific consumed about 5 million tonnes of food, of which about 2 million tonnes were Auxis.

In their review of the feeding ecology of mesopelagic fishes, Hopkins and Baird (1977) concluded that for the three most abundant groups (Myctophidae, Gosnostomatidae and Sternoptychidae) crustaceans are the staple food, with copepods, euphausiids, ostracods, amphipods and decapods constituting more than 70% by volume of the stomach contents of almost all species. Not surprisingly, the mean size of prey taken increased with the growth in size of the fish, but in general mesopelagic fish appear to be opportunistic, ingesting a wide range of particle sizes. Those undergoing strong vertical migrations feed mainly near the surface at night, while non-migrators have less well marked diurnal rhythms of feeding. Some species are selective feeders. For example, Merrett and Roe (1974) found that Valenciennellus tripunctulatus fed extensively on calanoid copepods, Argyropelecus aculeatus on ostracods and Lampanyctus cuprarius on amphipods.

From his intensive and detailed study of the myctophidae of the Pacific ocean near Hawaii, Clarke (1973) concluded that in that area at night the migrating mesopelagic fishes have a nocturnal biomass about 6 times as great as the epipelagic fishes. He estimated the average biomass of micronekton as at least 1.0g m^{-2}, with most myctophids having an annual life cycle and with the micronekton as a whole having a P/B ratio about 2. This led him to conclude that micronekton production was about 0.2 gC m^{-2} yr^{-1}, and that from a primary production of about 50 gC m^{-2} yr^{-1} there could not be more than 2-3 steps in the food web linking micronekton to primary production.

The review by Blackburn (1977) of pelagic animal biomasses in the central Pacific supports the view that mesopelagic fish have 5-10 times more biomass than epipelagic fish in the 0-200m layer at night. Apart from mesopelagic fish, euphausiids and decapod crustaceans are the most abundant kinds of micronekton. Blackburn's (1977) review also shows that the total biomass of micronekton and zooplankton in the water column is at a minimum (<15g m^{-2}) at latitudes about 10°-30° north and south of the equator, is higher (15-25g m^{-2}) in equatorial regions, and is much higher (around 200g m^{-2}) at latitudes 40°-50°N (Figure 8). The ratio of micronekton to zooplankton in the epipelagic zone at night is highest (7-12%) in the oligotrophic waters, but only 1% in the highly productive northern waters. Blackburn (1977) suggested that the transfer efficiency from zooplankton to micronekton might be higher in the unproductive waters than in the more northern areas. This is difficult to under-

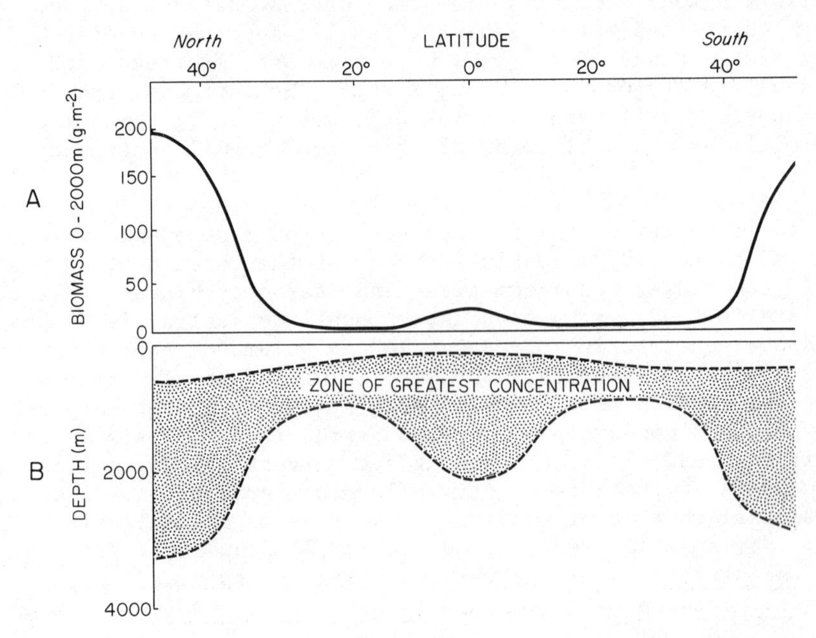

Fig. 8. Distribution of zooplankton biomass, and zone of high concentration, along a Pacific transect running north-south across the equator. From Blackburn (1977) utilizing data of Vinogradov.

stand, and one wonders whether the turnover, and hence production, of zooplankton may be higher in subtropical waters, thus accounting for the relatively higher nektonic biomass.

The question of nutrition for benthic fish populations was mentioned earlier. At least two possible routes exist for conveying food energy from the euphotic zone to the bathypelagic non-migrators, which tend to be adapted to ingesting large particles. Either the organisms on which they prey have in turn preyed upon organisms which made periodic migrations to surface waters, in which case bathypelagic fish are sustained by a "ladder of migration" (Vinogradov, 1953), or there is a food chain in the bathypelagic zone based upon sinking detritus. Harding (1974) examined the gut contents of several hundred copepods taken from depths of 1000-4000m in the northwest Atlantic and concluded that they obtained their nourishment both from vertically migrating zooplankters and from a detritus food web. It is probable that these copepods in turn help support the bathypelagic fish. However, we shall not pursue the matter further, since bathypelagic fish production is very small compared with the production of other groups of fish.

When we come to the benthic boundary layer, there is a considerable increase in fish population density and it is not clear whether these populations are supported mainly by a rain of small particles of detritus, or by relatively large particles such as dead fish or mammals. Rowe and Gardner (1979) made a carbon budget for the slope region of the Western Atlantic by measuring sedimentation rates and comparing them with sediment respiration rates. They found that the primary downward flux of small particles was in the range 2.3 to 6.3g C m^{-2} yr^{-1} and that sediment respiration rates were about one third of the value. From this they concluded that the sediment-dwelling benthic macro- meio- and microfauna processed less than half of the sedimenting material, and that most of the latter was intercepted by a mobile fauna at the sediment surface. From the work of Dahl (1979), Hessler _et al_. (1978) and others we know that this mobile epifauna is usually dominated by amphipods of the family Lysianassidae, and that shrimps and other decapods are also present. It seems probable that the mobile, active benthic fishes such as rattails are supported by three main food sources: large particles on which they scavenge, epifauna on which they prey, and benthic infauna. Almost certainly the benthic infauna is quantitatively the least important food source for these fish; the relative importance of the other two sources is not clear, and may vary from place to place (Marshall and Merrett, 1977).

When Sedberry and Musick (1978) examined the gut contents of demersal deepsea fishes from the lower continental slope and rise, to a depth of 2700m, off Virginaia U.S.A., they found that they had consumed considerable quantitites of mesopelagic and bathypelagic invertebrates. They suggested that these might have migrated to the benthic boundary zone to exploit the concentration of suspended particulate organic matter in the nepheloid layer.

Quantitative Aspects

It is clear both from general biological oceanography and from the results of fish biomass measurements that the central oceanic gyres are the least productive, while their boundary areas, east and west towards the continents and north and south towards poles and equator, are considerable more productive. We may begin by synthesizing available information about the central gyres. Apparently fish production is dominated by the migratory mesopelagic fishes, especially myctophids, gonostomatids and sternoptychids. At night their biomass is 5-10 times that of the epipelagic fishes, but even so it is only of the order of 1-2g m^{-2} (1.75-3.5 kcal m^{-2}). Nevertheless, it is large compared with the mean biomass of large predators (0.03 kcal m^{-2}). Hence, migrating mesopelagic fishes are the major consumers among fish of the epipelagic zone. Their food requirements, as we have seen, are of the order of 5-10 kcal m^{-2} yr^{-1}. When we consider the food web by which they are supported, there is difficulty in balancing the budget on the basis of the current estimates of primary production. Thus Blackburn (1981), reviewing the literature for subtropical gyres assumed phytoplankton production to be 30-40 gC m^2 yr^{-1} with a biomass of micronekton in the upper 1000m of 1g m^2 wet weight. The latter figure is fairly well in agreement with our previous discussions. Blackburn (1981) also agreed with the points made in the foregoing analysis of trophic relations, namely that micronekton feed on a variety of crustaceans, not all of which are herbivores.

Ryther (1969) argued that, since open ocean primary production is performed mainly by nannoplankton, there are three to four trophic levels between phytoplankton and animals no more than 1-2 cm long. If we accept this view, there are likely to be three transformations at 10% efficiency in the food web linking 400 kcal of primary production to the 5-10 kcal required by the mesopelagic fish. Even if we consider the possibility that the ecological efficiency at each level is 20%, we have only 400 x 0.2 x 0.2 x 0.2, i.e. 3.2 kcal available to the mesopelagic fish, with no provision for parallel food chains leading to squid or epipelagic fish, and thence to tuna. The conclusion seems inescapable, that primary production in the subtropical gyres has been underestimated, as has been suggested by Schulenberger and Reid (1981), and others.

Turning to more productive areas peripheral to the gyres, we find that Gjøsaeter and Kawaguchi's (1980) estimates of mesopelagic fish biomass (Table 1) include figures in the range 4.5-8.0g m^{-2} for subarctic waters. There are additional data, not included in Table 1 which indicate biomasses of 3.0-5.0 in equatorial regions.

These trends support the view taken in Blackburn's (1977) review (Figure 8) that micronekton and zooplankton are more abundant at the equator than in the subtropical gyres, and are even more abundant at latitudes higher than 40°N and S.

On the east-west axis, there are even more spectacular increases in biomass of mesopelagic fish as one approaches the continental shelves. Gjøsaeter and Kawaguchi (1980) recorded 10-30g m^{-2} in shelf waters west of Great Britain and Norway, 10-60g m^2 in shelf waters south of Newfoundland, 15-60g m^2 in the Mauritanian upwelling, up to 220g m^{-2} in the Arabian Sea, and a mean of 5.2g m^{-2} in the Kuroshio system area.

It is probably not sensible to attempt any ecosystem budget for these waters peripheral to the gyres. Not only is the productivity of coastal waters extremely variable on account of river runoffs, local upwelling systems, fronts, gyres and so on, but the mesopelagic fish are probably greatly outnumbered from time to time by shoals of mackerel, herring and other pelagic fishes associated with coastal waters. The tuna make highly seasonal migrations through coastal and northern waters, and marine mammals probably have a major impact on epipelagic fish stocks of northern waters. As Vinogradov (1981) has shown, the equatorial regions are characterized by a pattern of upwelling and advection which also makes it difficult to produce a detailed budget for a particular area.

SYNTHESIS

Figure 9 shows a summary of the data reviewed in this paper. It must be emphasized that, with the exception of the biomass data on mesopelagic fish, almost all of the numbers are derived from one or two studies of a small number of species. These studies may or may not be representative of their class, but they appear to be the best we have. This kind of work is still in its infancy.

In a central gyre region the largest biomass of fish is in the mesopelagic migrators. When these rise to the epipelagic zone at night, they greatly outnumber the epipelagic fish, and appear to consume large quantities of epipelagic crustaceans. By their downward migration with full guts they transfer large amounts of euphotic zone production rather rapidly to mesopelagic depths.

The larger bathypelagic fish are opportunistic predators adapted to consuming a wide range of particle sizes. The smaller ones, e.g. *Cyclothone*, seem to depend largely on small crustaceans, which in turn are likely to feed on detritus. This may be an important bathypelagic food web based on sinking detritus. In any case, both biomass and production of bathypelagic fish are believed to be small.

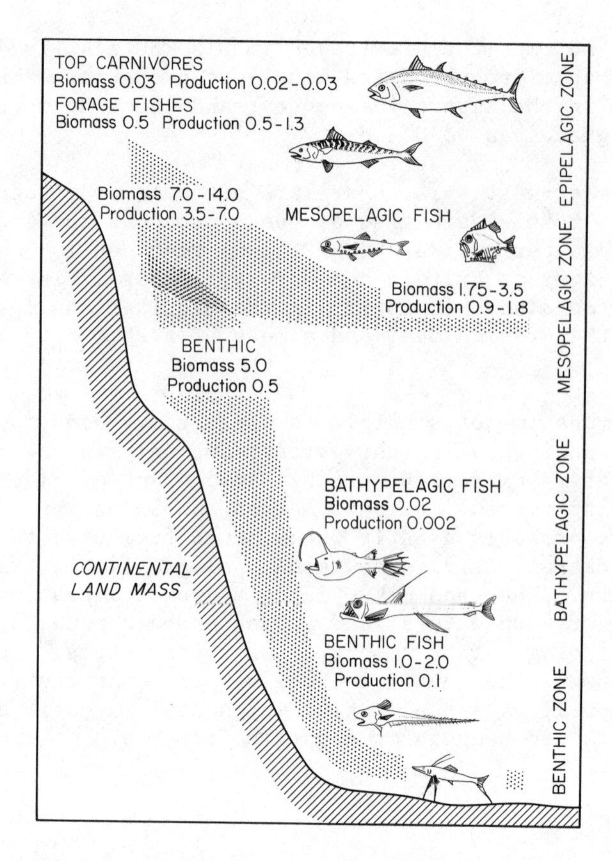

Fig. 9. Diagram summarizing mean values of fish biomass and production in the open ocean. Shading symbolizes decreasing biomass as one moves from the shelf break towards the centre of a gyre.

Biomasses of benthic fishes in central gyre areas are of the order of 1-2 kcal m^{-2}, apparently supported largely by a food chain based on a rain of small particles, but also, to an unknown degree, by scavenging on carcasses that sink to the bottom.

As one moves towards the continental rise and slope, biomasses of both mesopelagic and benthic fishes increase. The same trend is seen when moving into subarctic or subantarctic waters, and to a lesser extent when approaching the equatorial zone. The biomass of top carnivores such as tuna and sharks in the epipelagic zone is estimated to be 2 orders of magnitude less than that of the mesopelagic fish. The top carnivores feed mainly by day and prey on epipelagic fish and squid. It is not known what are the major predators on the mesopelagic fish which are present in the upper layers only at night. Tunas are thought to obtain about two thirds of their food from forage fish and the other one third from invertebrates such as squid and euphausiids. Details of the food web involving large invertebrates is far from clear.

The FAO Technical Paper (Gjøsaeter and Kawaguchi 1980) called "A Review of the World Resources of Mesopelagic Fish" gives rough estimates of the biomass in each area of the world's oceans. Their figures add up to nearly 1000 million tonnes. If 10% of this biomass were harvested annually, the landings would exceed the present world catch of fish. Of course, there are great technical and economic difficulties, when the biomass is distributed over such a large area and volume, compared with the relatively small area of continental shelves from which the present catches are taken. Nevertheless, the documented existence of this large biomass is an open invitation to exploitation, and it is important to have a much clearer understanding of the ecological role of the mesopelagic fish stocks.

Acknowledgements

The author wishes to thank Mr. N. R. Merrett for guidance on literature sources and access to his unpublished data, Drs. A. R. Longhurst, B. T. Hargrave, S. R. Kerr and D. D. Sameoto for critical reading of the manuscript, Mr. R. Pottle for assistance with literature search, and Mrs. Pat Lindley for the illustrations.

REFERENCES

Andersen, N. J., and Zahuranec, B. J., eds., 1977, "Oceanic Sound Scattering Prediction," Plenum Press, New York.

Baird, R. C., and Hopkins, T. L., 1981, Trophodynamics of the fish *Valenciennellus tripunctulatus* III. Energetics, resources and feeding strategy, *Mar.Ecol.Prog.Ser.*, 5:21.

Baird, R. C., and Wilson, D. F., 1977, Sound scattering and oceanic midwater fishes *in*: "Oceanic Sound Scattering Prediction," N. R. Andersen and B. J. Zahuranec, eds., Plenum Press, New York.

Barham, E. G., 1971, *in* "Proceedings of an International Symposium on Biological Sound Scattering in the Ocean," G. B. Farquhar, ed., U.S. Government Printing Office, Washington.

Blackburn, M., 1965, Oceanography and the ecology of tunas, *Oceanogr. Mar.Biol.Ann.Rev.*, 3:299.

Blackburn, M., 1977, Studies on pelagic animal biomasses *in*: "Oceanic Sound Scattering Prediction," N. R. Andersen and B. J. Zahuranec, eds., Plenum Press, New York.

Blackburn, M., 1981, Low latitude gyral regions *in*: "Analysis of Marine Ecosystems," A. R. Longhurst, ed., Academic Press, London.

Blunt, C. E., 1960, Observations on the food-habits of longline - caught bigeye and yellowfin tuna from the tropical eastern Pacific, 1955-56, California Fish and Game, 46:69.

Carey, F. G., and Robison, B. H., 1981, Daily patterns in the activities of swordfish Ziphias gladius, observed by acoustic telemetry, Fish.Bull., 79:277.

Childress, J. J., and Nygaard, M. H., 1973, The chemical composition of midwater fishes as a function of depth of occurrence off Southern California, Deep-Sea Res., 20:1093.

Childress, J. J., Taylor, S. M., Caillet, G. M., and Price, M. H., 1980, Pattterns of growth, energy utilization and reproduction of some meso- and bathypelagic fishes off Southern California, Mar.Biol., 61:27.

Clarke, T. A., 1973, Some aspects of the ecology of lantern fishes (Myctophidae) in the Pacific Ocean near Hawaii, U.S.Fish. Wildl.Serv.Fish.Bull., 71:401.

Dahl, E., 1979, Deep-sea carrion-feeding amphipods: Evolutionary patterns in niche adaptation, Oikos, 33:167.

FAO, 1980, "Yearbook of fishery statistics for 1979," FAO Fisheries Series Vol. 48, F.A.O., Rome.

Farquhar, G. B., 1971, "Proceedings of an International Symposium on Biological Sound Scattering in the Ocean," U.S. Government Printing Offcie, Washington.

Gjøsaeter, J., and Kawaguchi, K., 1980, "A Review of World Resources of Mesopelagic Fish," Technical Paper No. 193, F. A.O., Rome.

Guennon, Y., and Rannou, M., 1979, Semi-diurnal rhythmic activity in deep-sea benthic fishes in the Bay of Biscay, Sarsia, 64:113.

Haedrich, R. L., and Rowe, G. T., 1977, Megafaunal biomass in the deep sea, Nature, 269:141.

Harding, G. C. H., 1974, The food of deep-sea copepods, J.Mar.Biol. Assoc.U.K., 54:141.

Hessler, R. R., Ingram, C. L., Yayanos, A. A., and Burnett, B. B., 1978, Scavenging amphipods from the floor of the Philippine Trench, Deep-Sea Res., 25:1029.

Hopkins, T. L., and Baird, R. C., 1977, Aspects of the feeding ecology of oceanic midwater fishes in: "Oceanic Sound Scattering Prediction," N. R. Andersen and B. J. Zahuranec, eds., Plenum Press, New York.

Hureau, J.-C., Geistdoerfer, P., and Rannou, M., 1979, The ecology of deep-sea benthic fishes, Sarsia, 64:103.

Isaacs, J. D., and Schwartzlose, R. A., 1975, Active animals of the deep-sea floor, Sci.Am., 233:85.

Marshall, N. B., 1971a, "Explorations in the Life of Fishes," Harvard University Press, Cambridge.

Marshall, N. B., 1971b, Animal ecology in: "Deep Oceans," P. J. Herring and M. R. Clarke, eds., Praeger, New York.

Marshall, N. B., 1979, "Developments in Deep-Sea Biology," Blandford Press, Poole, Dorset, U.K.

Marshall, N. B., and Merrett, N. R., 1977, The existence of benthopelagic fauna in the deep sea, in: "A Voyage of Discovery:

George Deacon 70th Anniversary Volume," M. Angel, ed., Pergamon Press, Oxford.

Merrett, N. R., and Row, H. S. J., 1974, Patterns and selectivity in the feeding of certain mesopelagic fishes, Mar.Biol., 28:115.

Mimura, K., 1963, Synopsis of biological data on yellowfin tuna in: "Proceedings of the World Scientific Meeting on the Biology of Tunas and Related Species," H. Rosa, ed., F.A.O., Rome.

Olson, R. J., in press, Feeding and energetics studies of yellowfin tuna. Food for ecological thought. Proceedings of symposium "The definition of tuna and billfish habitats and effects of environmental variations on apparent abundance and vulnerability to fisheries," Nov. 1981, Tenerife, Canary Islands.

Parin, N. V., 1970, "Ichthyofauna of the epipelagic zone." Translated from the Russian, Israel Program for Scientific Translations, Jerusalem.

Rannou, M., 1975, "Reacherches écologiques sur les poissons bathyaux et abyssaux," Thèse Doctorat ès Sciences, Montpellier.

Roger, C., and Grandperrin, R., 1976, Pelagic food webs in the tropical Pacific, Limnol.Oceanogr., 21:731.

Rowe, G. T., and Gardner, W., 1979, Sedimentation rates in the slope water of the northwest Atlantic ocean measured directly with sediment traps, J.Mar.Res., 37:581.

Ryther, J. H., 1969, Photosynthesis and fish production in the sea, Science, 166:72.

Schulenberger, E., and Reid, J. L., 1981, The Pacific shallow oxygen maximum, deep chlorophyll maximum, and primary productivity reconsidered, Deep-Sea Res., 28A:901.

Sedberry, G. R., and Musick, J. A., 1978, Feeding strategies of some demersal fishes of the continental slope and rise off the mid-Atlantic coast of the U.S.A., Mar.Biol., 44:357.

Sharp, D. G., and Francis, R. C., 1976, An energetics model for the exploited yellowfin tuna, Thunna albacores, population in the eastern Pacific ocean, Fish.Bull., 77:36.

Shiohama, T., Myojin, M., and Sakamoto, H., 1965, The catch statistic data for the Japanese tuna long-line fishery in the Atlantic ocean and some simple considerations of it, Rep.Nankai regional Fish.Res.Lab., 21:1.

Smith, K. L., and Hessler, R. R., 1974, Respiration of benthopelagic fishes: in situ measurements at 1230 metres, Science 184:72.

Smith, K. L., and Laver, M. B., 1981, Respiration of the bathypelagic fish Cyclothone acclinidens, Mar.Biol., 61:261.

Sund, P. N., and Richards, W. J., 1967, Preliminary report on the feeding habits of tunas in the Gulf of Guinea, U.S. Fish and Wildlife Service, Special Scientific Report No. 551.

Torres, J. J., Belman, B. W., and Childress, J. J., 1979, Oxygen consumption rates of midwater fishes as a function of depth of occurrence, Deep-Sea Res., 26A:185.

Tseytlin, V. B., and Gorelova, T. A., 1978, Study of the feeding of the lanternfish Myctophum nitidulum (Myctophidae, Pisces), Oceanology, 18:488.

Vinogradov, M. E., 1953, The role of vertical migration of the zooplankton in the feeding of deep sea animals, Priroda Mosk., 6:95.

Vinogradov, M. E., 1981, Ecosystems of equatorial upwellings in: "Analysis of Marine Ecosystems," A. R. Longhurst, ed., Academic Press, London.

Waters. T. F., 1977, Secondary production in inland waters, Adv.Ecol. Res., 10:91.

ECOLOGICAL EFFICIENCY AND ACTIVITY METABOLISM

Gary D. Sharp

Fisheries Resources Officer
Food and Agricultural Organization of the
United Nations
Rome, Italy

INTRODUCTION

CONSIDERATIONS OF WORK DONE AND ECOLOGICAL EFFICIENCY

Quite simply stated the total productivity of an ecosystem is the sum of the productivities of the components. However, this in no obvious way simplifies the aggregation of trophic levels, for simplicity's sake, in order that only some few measurements or estimates might be made and their products summed to yield a general solution of the productivity of a given system. The problems lie in the differences in the efficiencies of the various inter and intra-trophic level components, their distributions in time and space, the typical number of interactions they participate in within the food web, and the life history pathways followed. My comments will be primarily directed at an evaluation of efficiencies of various trophic groups, which will, by their nature, require a series of comparisons and contrasts of distributions, behaviors, and life histories.

To deal realistically with ecosystems one must recognize the inherent variety and plasticity of biological responses to physical-chemical perturbations. It is also useful to realize that the histories of ecosystem evolution are dependent on rates of sequential colonizations and the relative opportunities and efficiencies of the colonists in the milieu which they face. This does not suggest that classical sequential or successional changes are fixed at all, but more that each stage or step will be a one-time opportunity which will successively affect all future states. Fortunately this aspect of ecology is blessed with abundant recent reviews and I will therefore defer comments by recommending that those interested in a) development of species diversity consult Huston (1979); b) food webs consult Odum (1968), Pomeroy (1974) and Platt _et al_., (1981); and c) biological adaptability and ecosystem stability consult Conrad (1972, 1976) and Johnson (1981).

ECOLOGICAL EFFICIENCY

Ecological efficiency is usually estimated in terms of change of biomass compared to amount in weight consumed. This can be calculated in terms of calories, dry weights, wet weights, etc.

The efficiency of any ecosystem component depends upon several conditions:

a) Abundance and concentration of each component's food sources in relation to its own abundance and concentration.
b) Vulnerability of these food sources.
c) Work done to encounter (search out or blunder into) and capture these food sources.
d) Proportions of food sources which are assimilated into useful energy, growth, and/or nutrients.
e) Proportion of assimilated food source available to the component for growth, somatic or reproductive.
f) Complexity of somatic organization of the organism itself.

The efficiency of a food web is the product of the efficiencies of the interacting components, i.e. predator-prey web, which is directly determined by the limits of primary production on one scale, the ambient properties on others, and the feeding specificities, reproductive capacities and requirements, and to varying degrees biomasses of the components.

Any individual component's energetic requirements, i.e. utilization of energy of foodstuffs, can be described in this fashion

$$I = M + G + E$$

where I is energy input in the form of light, proteins, carbohydrates, fats, often monitored in the form of transfer of trace elements, nitrogen, carbon, oxygen, etc., depending upon the organisms or systems being studied.

where M is metabolic energy which may usefully be divided into numerous kinds of metabolic work resulting in internal-biochemical conversion and ultimately, external work, i.e. heat and simple turbulence remaing for a short period after an organism has expended work against gravity or has moved;

where G is growth in terms of biomass, either somatic or reproductive products, and is the only variable which can be either positive or negative; and

where E is excretion, secretion or sloughing of biogenous materials.

The energy requirements for either optimal or maximal energy flow through an organism depend to a great degree on the system characteristics, particularly for poikilotherms. Temperature and available O_2 (also CO_2 or H_2S depending upon the organism) levels determine the respiratory limits of the majority of aquatic organisms. In response to the variations encountered in these variables, diverse adaptations and specializations have evolved to assist or buffer the organisms against stress, i.e. elaborate respiratory surfaces; respiratory pigments; biochemical/genetic and circulatory specializations. The abilities of organisms to persist in uncertain, stressful environments will have been determined by pre-adaptations to environmental gradients and patchy distributions of the individual's respective nutritional requirements. The storage of excess energy in the form of carbohydrates (e.g. glycogen), lipids and proteins for use in metabolic work facilitate the organisms' resilience to varying food availability as well as various stasis requirements, i.e. buoyancy increases with lipid content.

The importance of motility, or mobility and temperature during the various food web components' life history stages is rarely given adequate consideration. In fact the relevance of time and distance scales of biophysical influences have only recently been given their appropriate hierarchic place in the ecosystem framework (Fasham, 1978; Owen, 1981; Sharp, 1981a, b; Simpson *et al.*, 1979; Iles and Sinclair, 1982; Bakun and Parrish, 1981; Parrish, Nelson and Bakun, 1981). Perhaps the most informative place to start an evaluation of efficiencies is in an examination of energy expenditures for mobility with regard to organism size, which often relates to trophic position.

From Rheinheimer's review of marine microorganisms (1980) we find that many marine bacteria are flagellates. The typical marine bacterium has a temperature optimum within the range 18°-22°C, although a full range of optima exist including faculative psychophils with temperature optima as low as 0°C. The bacteria are usually aerobic but are facultatively anaerobic in most cases. There are few dissolved organic substrata which they cannot derive nutrients from and nitrite or sulphate can be utilized by some. One important source of nutrients is the leakage or excretions from primary producers, i.e. phytoplankton.

Bacteria are supreme opportunists and have among the highest potential conversion efficiencies possible, on the order of 50 to 60 percent. Bacteria are found in high densities in coastal zones, and in the sediments. Benthic organisms depend upon them for nutrition, and corals, i.e. polyps, obtain 10-20 percent of their carbon via micro-organisms. Harpacticoid copepods, i.e. *Tisbe holothuriae*, can thrive on bacteria. Copepods and lobster larvae in the zooplankton feed on bacteria (Rheinheimer, 1980). Grazing depression and bloom cycles are readily observed between bacterial and predator populations. Benthic cyanophytes are grazed by rotatoria, nemotodes, crustaceans and insect larvae. Planktonic cyanophytes are eaten by plankton and fish, as supplements, but are of decreasing importance as the predators increase in size.

Of the incident light, approximately 5 percent is used in photosynthesis indicating the first "inefficiency" stage in the production cycle. Of the light utilized in photosynthesis, 97-97 percent is respired away, resulting in 2-3 percent primary productivity. Also, one should not forget Odum's (1957) observations showing that half the net production by algae was decomposed by bacteria, of which only about 30-50 percent is respired away.

Larval fishes and planktonic stages of other organisms graze heavily on unicellular and colonial phytoplankton. Their conversion or production rates vary enormously, depending on species, mobility and such variables as distribution or contagion of appropriate food particles (Vlymen, 1977; Beyer, 1981; Beyer and Laurence, 1981). The quantification of productivity in fish populations is not dealt with simply in the larval through adult stages, as once the post-larvae and juvenile stages become mobile, they become very difficult to sample.

Many phytoplankton organisms are not motile, and therefore their existence is a strict function of the local conditions in which they find themselves, i.e., light, nitrogen and other essential nutrients, local turbulence and system level transport. In contrast, there are motile phytoplankton, i.e., flagellates, and some bacteria which are capable of resisting turbulent dissipation or disruption at low energy levels. All of these species depend to a great extent upon system transport physics for their distribution and encounters with their requirements in order to become productive. In this sense they are very "efficient" as individuals sampling a vast, uncertain volume, as well as in the fact that they do not need to dissipate large proportions of their metabolic energy in swimming, as do many other species at higher trophic levels.

FISHES AND OTHER HIGHER ORGANISMS

As many, or even most, species in the marine environment start life in the plankton and subsequently grow into higher trophic levels, it should be obvious that at smaller stages there is a requirement for both expendability and superior efficiency at small time and distance scales, compared with the larger, later stages. Growth and capacity for persistence are all-consuming objectives among living organisms. The inverse relation between size at reproduction and turnover time is a clear indication of the trade-offs in the ecolocigal continuum between various somatic organizations, behavior, mobility and the opportunistic investments among this suite of options on the persistence and colonization potential of each organism.

After simple fission reproduction and until, in some species, encapsulation of larger embryos or live-bearing is adopted, the evolutionary motivation seems to be one of investing as much energy as necessary in placing fertilized eggs into a "safe" nuturing environment (Alvariño, 1980; Sharp, 1981a, b). The objectives of reproductive strategies and adaptations appeared to be toward optimizing either the numbers of eggs sowed into the immediate environment, or optimizing the selection of the post egg-larval (i.e. limited mobility stages) life history characteristics which would allow the gametes to be placed into a less uncertain, nurturing situation. The closure of this cycle by internal fertilization is an optimal solution for stabilization of the developmental environment of the embryo and also permits the young to enter the ecosystem at larger, apparently more secure sizes. The nurturing of young up through their birth, or beyond in the case of species exhibiting parental care, shifts the emphasis of natural selection from the developmental problems of short time and distance scales, i.e. those pertaining to the distribution of the eggs and larvae, to those of the adults. This would seem to suggest that the energy spent per individual, i.e. adult fish, might well be less in the highly evolved livebearing fishes or those exhibiting parental care. However, numbers of young produced are usually less numerous and would tend to limit total number in a given population, which is in sharp contrast to the situation in the highly fecund pelagic species leading to this state. The differences in capabilities of pelagic larvae and live-born ones certainly sets the stage for very different requirements, as well as very different selection pressures.

The oviparous species' eggs, like phytoplankton, are carried about by turbulence, unless they are spawned on substrates, i.e. herring eggs - (*Clupea* species), or are demersal, i.e. more dense than sea water, and sink to the bottom where turbulence plays a more subtle role in their development. These eggs are from about 0.5 mm to 3.5 mm in size range, with pelagic eggs being about 1 mm in diameter. The larvae hatch at various time, depending primarily on temperature, for each species (Theilacker and Dorsey, 1981). The larvae have from a few days to about ten days after hatching in which they utilize their yolk for further development. After the yolk is used they have from a few days to a week to obtain food. The timing of this hatching in relation to the availability of appropriate food and abundance of predators is critical (Alvariño, 1980).

Parrish, Nelson and Bakun (1981) describe the zoogeographic distinctions in the California current due to local transport characteristics of the various habitats. Where offshore transport dominates, the major fish species do not have epipelagic eggs, however, live bearers (e.g. *Sebastes* and embiotocid species) are abundant and demersal egg producing species dominate these areas. Wherever closed gyral systems occur, pelagic fishes with epipelagic eggs and larvae are found. Other, more migratory pelagic species appear seasonally, entering the coastal system in search of appropriate nursery areas. They spawn, and depart.

There are many differences in relative abilities and modes of swimming in planktonic organisms, from passive drift, to near independence of local physical transport. For example, Vlymen (1970) calculated the proportional energy expenditures of _Labidocera trispinosa_, a calanoid copepod, dur to vertical migratory behavior and short burst accelerations characteristic of avoidance behavior. The relative value of total expenditure of energy due to activity compared to total respiration is on the order of 0.1 to 0.3 percent. These incredible efficiencies are directly related to the high accelerations that copepods exhibit.

Compared with a similar and even more interesting series of studies on larval fish swimming energetics, Vlymen (1974, 1977) showed that _Engraulis mordax_ (California anchovy) larvae have a relatively higher proportional expenditure of their total respiration due to swimming, i.e. 24.6 percent under average activity level conditions for a 1.4 cm larvae. Smaller anchovy larvae are less efficient due to their sporadic swimming behavior, but efficiency increases as they grow larger and settle into their pump-and-glide mode of swimming.

Vlymen's (1977) analysis of food microdistribution, growth and larval behavior for anchovy larvae brings one immediately to the relevance of scale, disruptive turbulence phenomena and survival of larval fishes. (These concepts are further examined in IOC Workshop Report No. 28). There are discontinuities and contagion of particles and organisms in the marine environment (Sheldon and Parsons, 1967; Sheldon _et al_., 1972; Platt and Denman, 1975, 1977: Owen, 1981). These have profound effects on local processes and particularly on local abundance in the primary through tertiary predators some time after the initial bloom in primary production (Sharp, 1981b). The relative energetic efficiency of the more mobile, secondary and tertiary predators which search from one primary bloom and progress to the next is far less than the larval fishes, or even adult neritic fishes.

For example, Sharp and Francis (1967) and Sharp and Dotson (1977) have examined the swimming energetics of two tunas, _Thunnus albacares_ and _Thunnus alalunga_, the yellowfin and albacore tuna, respectively. The yellowfin is a nomadic species living all of its life in the tropical oceans (≥23°C surface temperature) with an individual range radius of about 600 to 800 nautical miles. The albacore, on the other hand, are oceanwide nomads as juveniles, ranging within the surface temperature bounds of the 21° and 15°C isotherms. At maturity they return to the central, tropical ocean areas, well into the depths, for reproduction.

Yellowfin tuna are opportunistic predators which reside in the upper thermocline and mixed layer of the tropical oceans of the world. From rough approximations of their activities (e.g. swimming), growth and stasis metabolism Sharp and Francis (1976) estimated that a cohort of 84 cm (±8cm,~12.1 kg) yellowfin expended about 9.5 percent of their consumption energy on growth; 19 percent on stasis metabolism and about 71.5 of their energy was expended in swimming work.

Recently Olson (1982) has followed up on this study by examining yellowfin tuna stomach contents, their caloric values, and evacuation times.

From recent sonic tracking studies of two yellowfin (Carey and Olson, 1982) I have estimated that an 88 cm yellowfin expends about 95 percent of its daily energy flux on swimming. (See Table 1 for estimated parameters and swimming speeds for the diurnal period). Growth and metabolism account for the remainder. The difference between these two estimates lies in this recently acquired knowledge of the swimming speeds and time rather than the rough estimates utilized in the Sharp and Francis model above. Also, the energy expenditure of the cohort included mortalities and an additional proportion of biomass consumed but not assimilated, so that the similar computations for the recent data would yield 73 percent swimming dissipation of total consumption compared with the 71.5 percent estimated earlier.

From a related approach Sharp and Dotson (1977) utilized fat measurement data and hydrodynamic methods to determine that the flux in fat content in albacore could be accounted for by the energy expended in "directional migration" alone, while any day-to-day activities would be accounted for within the daily diet and energy conversion.

The question of resident versus migrant components has been bothersome for many widely distributed resources. Use of fat content data, the dynamic estimation of fat level and growth rates can be useful in estimating both, where fish might have come from, and what proportion of available fish at any one time might be entering grazing-fishing areas. Fat content values ranging from 18 percent of total body weight to about two percent have been measured in albacore (Dotson, 1978; Sharp and Dotson, 1977) indicating the dynamic nature of this information. Coastal fish of 63 cm average length had lost an average of 404 gm of fat on entry to the fishery compared to fish of the same size sampled about 1000 nautical miles offshore prior to the commencement of the coastal fishery. Two weeks after the initial

Table 1. Observed average activity level and estimate energy expenditures are given for two yellowfin tuna which were tracked over a total of 55.4 hours (Carey and Olson, 1982).

Yellowfin: Length = 88 cm Mass = 11.979 kg Observations every 0.074h

Obs.	$\overline{V}$ cm/s	C_D	% day (hours)	mg O_2 h^{-1} (Dist./h)	mg O_2 day^{-1} (Distance)	cal g^{-1} km^{-1}
63	31	0.071	9.1 (2.18	480 (1.116 km)	1 046 (2.433)	0.121
224	66	0.16	31.7 (7.61)	652 (2.376)	4 962 (16 667)	0.077
190	114	0.01	27.1 (6.50)	3 362 (4.104)	21 853 (26.676)	0.230
179	149	0.01	25.2 (6.05)	7506 (5.364)	45 411 (32.452)	0.392
32	193	0.01	4.7 (1.13)	16 312 (6.948)	18 432 (7.851)	0.658
7	323	0.01	1.1 (0.26)	76 461 (11.628	19 880 (3.023)	1.844
9	391	0.01	1.3 (0.31)	135,632 (14.076)	42 046 (4.364)	2.702
					153 630 mg O_2 = E_s (94.88 kg)	

Change weight/d = 42.8g

E growth = 315.8 mg O_2 (includes specific dynamic action)

E maint = 7003.5 mg O_2

E swim = 153,630 mg O_2

$E_{excretion}$ = .30(E_g + E_m + E_s) = 44,565.6 mg O_2

Estimated daily Oxygen consumption = 205,515 mg O_2

arrival on the coast fish of similar size were sampled again in the coastal fishery and their fat content was back in the range of the offshore material.

The conflicting observations in these last examples can be treated as single, unrelated observations, and indeed they are too often treated as such. However, a series of inter-relations exist which need to be pointed out and discussed in an ecological context if we are ever to understand ecological energetics.

On the other hand we have literally volumes of material which have been recently summarized by Brett and Grove (1979), Brett (1979) and Ricker (1979) on the laboratory and controlled habitat (Aquaculture) studies of growth and physiological energetics. The "average observed" energy budget of fishes shown by Brett and Groves (Figure 18, p. 337, 1979) is summarized in Table 2.

Note that I have put growth and activity metabolism on the same line, as I find that they are both completely inter-related. The activity level determines growth, recognizing that the ambient biophysical milieu varies.

Recent attempts to relate activity metabolism expenditures directly to observed growth rates (Kerr, 1982) are simply inadequate in the light of what is known about the relations between swimming energetics and the non-uniform distributions in time and space of food resources. Whereas a fish with a passive feeding strategy, i.e. one held in the laboratory, responds in a linear fashion to food presentation up to its satiation point, continuously swimming fishes in nature have to balance between energy expended in moving between food patches, their consumption capabilities and the availability-vulnerability of their food resources (Vlymen, 1977; Sharp and Dotson, 1977; Sharp and Francis, 1976; Olson, 1982; Carey and Olson, 1982).

←

Table 1. (cont.)

Tabled are the numbers of observations (Obs): average speed in cm/s ($\overline{V}$); a calculated estimate of the coefficient of total proportional time (% day, hours), respiration rate and distance covered at this rate (gm O_2/h, Dist./h, daily rates at this observed speed, and cost per unit work done (cal g^{-1} km^{-1}). Observations were made every 0.074 hours. The two fish were 87 cm and 89 cm, fork length, respectively and estimates were therefore made for an 88 cm, 11.979 kg yellowfin using the methods of Sharp and Francis (1976). Estimated oxygen consumption due to swimming alone is 75 percent of average daily consumption, and the average ecological efficiency in terms of average growth from total consumption is in the order of 0.15 percent.

Table 2. Average Percent Total Energy Expended of Total Ingested

Gross Energy Ingested	100
Feces	20 loss
Digestible Energy	80
Non-fecal losses	7 loss
Metabloizable Energy	73
Heat Increment	14 loss
Net Available Energy	59
Metabolic Maintenance	7 loss
Heat Increment	30 loss
Growth and Activity	22 percent

Kerr's (1982) model cannot be applied to general populations in its present form since it is only applicable to fishes with constant positive growth and does not account for fish activity energetics which are not directly proportional to food consumed. My interpretation of the correspondence between the gadoids studied in the laboratory and in the field and their apparent growth related activity is that these species are not "Actively Predatory", but are somewhat passive and opportunistic as their diet composition, and relative cannibalism (Daan, 1973, 1978) would suggest. Certainly seasonal thermal and oxygen changes, long and short-term variations in contagion and distribution of prey species and the local density of competitor species determines the activity expenditures, whereas growth simply reflects the relative extent of energy accumulation above and beyond that required for metabolic maintenance and foraging. It is also quite clear that fish will trade off somatic growth under stress situations in order to make more energy available to reproduction, as a general evolutionary survival strategy.

Consideration of rates of respiration, excretion, reproduction and growth in nature as simple constants is folly. To paraphrase Ricker's (1979) summarizing comments regarding attempts to "generalize" growth relations in relation to food consumption, it is unlikely that simple relationships exist between ingestion rates and subsequent growth or production from aquatic species. Therefore, data on growth should be collected and portrayed in context of ambient conditions, temperature being the primary controlling variable, but the spatial distributions and abundances of food items and the work done by the various organisms to obtain them is likely to be the ultimate "density dependent" variable needing careful accounting.

Although many general considerations have been summarized regarding locomotion costs, particulary in fishes (Hoar and Randall,

1978; Webb, 1975), it is still an unusual marine ecological manuscript which includes "work done" as a variable, rather that utilizing literature or laboratory-derived estimates of these extremely variable energetic expenditures. Far more effort should be made to more realistically quantify these variables.

Since it is perhaps now possible to scale the relative efficienies of some members of a basic food web by reassembling the preceding scenaria and other contributions to this book, then perhaps it will also be useful to try to effectively describe the suite of determinant factors leading from the special capabilities and spatial limitations of these components back to the system in which they are united, i.e. a simplified local food web as seen from each organism's own perspective in its habitat. The problems of transport of energy and materials into and out of local situations is one of mobility, or scale of the habitat, of each of the various predators and food components.

The less mobile marine forms, small invertebrates, fish eggs and larvae and their forage are known to carry on their interactions on the meter-day scale (Sharp, 1981a) whereas the more mobile post-transformation fishes explore kilometers in the period of days to weeks. Plankton communities vary tremendously in relative mobility, ranging from hours to months with mobile stages ranging over meters to kilometers, but the basic dispersion of these groups is related to the physical transport processes of the water column occupied. Large, more exploratory species range over oceans on time scales of years. These species often have great "local" impacts, and can add or remove energy and materials which are transported, or dissipated over very great distances. For example baleen whales feeding in the Antarctic, calve in tropical areas, dissipating enormous amounts of energy and materials outside the source system, in excretion, lost energy and reproduction. The distance scales vary with species, but are important.

Marine birds, particularly in guano island areas, remove enormous amounts of material from the sea, and deposit it in a stored form where only rainfall can wash it back into the sea. These rainfalls are only frequent when the upwelling systems are retarded, as, for example, during an El Niño off South America. The rainfall and subsequent run-off fertilizes the area near the islands and MacCall (in press) has pointed out that the plankton growth and subsequent fish attraction might well provide the guano birds with needed food during these infrequent "hard-times" as far as general upwelling productivity is concerned.

Jordán (1967) described the consumption patterns of the guano birds off Peru. The cormorant, *Phalacrocorax bougainvillii*, diet comprises 96 percent anchovy, *Engraulis ringens*. The birds consume an average of 430g per day, and the minimum daily requirement is

200g. Each bird produces about 40-70g of guano per day. The piscivorous bird population, including other marine birds, peaked a about 18,000,000 in 1963. Jordán estimated that in 1962-1963 these birds consumed 7,000-8,000t of anchovy per day, or between 2.5 and 2.9 million t per year. This is equivalent to the present day biomass of anchovy off Peru, as well as equivalent to the catch of ground fish in the North Sea.

As a final example of predator effects in oceanic systems, the studies of Olson (1982) show that for the year 1970, the exploited yellowfin tuna in the eastern Pacific Ocean consumed at a minimum 13,800t of forage per day, comprising about 2,000t each *Vinciguerri* species and cephalopod species, over 2,500t of nomeids and over 5,300t per day of another tuna, i.e. *Auxis* species. The *Auxis* alon represents about two million t a year of major oceanic predator consumed by the exploited yellowfin population. (This is equivalent to the entire tuna harvest of man every year). The *Auxis* biomass is likely on the order of 5 millions t, or more, in the eastern Pacifi and harvesting yellowfin leaves an additional million t in the ecosystem to be eaten, or to eat other species. *Auxis* are certainly among the more voracious and ecologically inefficient of the scombrids, consuming well over 20 percent of their biomass per day, so that we can perhaps expect as much as an additional seventy million metric tons of scombrid food to be consumed by the *Auxis* left to forage, due to the yellowfin harvest (equivalent to man's total harvest from the seas). If they are cannibalistic and/or eat yellowfi tuna larvae or juveniles we might reflect on the general feed back loops into both *Auxis* and *T. albacares* populations due to man's removal of the adult yellowfin tunas.

Probably more important is the fact that the *Auxis* efficiency even less than that of the yellowfin tuna discussed in an earlier section. This suggests that besides lowering the overall size distribution of predators in the eastern Pacific by harvesting the larger yellowfin tuna, man is also probably decreasing the overall production of desirable species like yellowfin tuna by shifting the ecosystem to a more "dissipative", and therefore less ecologically, efficient species such as *Auxis*. Whether *Auxis* (or *Vinciguerria*, nomeids or cephalopods for that matter) can be harvested or somehow made desirable is another question. It is obvious that major resource biomasses exist, particularly after harvesting the high valu tunas.

CONCLUSIONS

Until man's influences are better recognized and understood it seems unlikely that a more ecologically oriented and appropriate utilization of the seas will evolve. Man *is* the ultimate opportunistic predator, and probably the least efficient in the sea, particularly since man takes away only higher level ecosystem components and returns mostly poor quality or even destructive compounds at ever increasing rates.

However, understanding man's influences will certainly require far more knowledge of the biology and ecological efficiency of the major species components in the seas. This will require intense study of activities and distributions of species over the ocean, as well as complete evaluations of predator-prey relations. The energetic costs due to activity, among the several variables needing study, appear to be significant and highly variable among the more mobile organisms. These efficiencies need more study than they have been given before ecosystem relations will be adequately defined and therefore usefully employed in realistic contexts.

Acknowledgements

I am pleased to have been allowed to extend my earlier interests in ecological energetics of oceanic species. It would not have been as satisfying without the knowledge arising from the recent studies of Robert Olson and his colleagues, and the persistant efforts of Francis Carey in his search for insight into the behavior of ocean going nomads through long, arduous at-sea tracking studies.

REFERENCES

Alvarino, A., 1980, The relation between the distribution of zooplankton predators and anchovy larvae, CALCOFI Reports Vol. XXI.

Bakun, A., and Parrish, R. H., 1981, Environmental inputs to fishery population models for eastern boundary currents, in: "Workshop on the effects of environmental variation on the survival of larval pelagic fishes, April 20 to May 5, Lima, Peru," G. D. Sharp, ed., IOC Workshop Report No. 28, Unesco, Paris.

Beyer, J. E., 1981, "Aquatic Ecosystems: and operational research approach," University of Washington Press, Seattle, London.

Beyer, J. E., and Laurence, G. C., 1981, Aspects of stochasticity in modelling growth and mortality of clupeoid fish larvae, in: Early Life History of Fishes: Recent Studies," R. Lasker and K. Sherman, eds., Rapp.P.-v. Réun.Cons.Int.Explor.Mer., 178: 17.

Brett, J. R., 1979, Environmental factors and growth, in: "Fish Physiology, Vol. VIII, Bioenergetics and Growth," W. S. Hoar, D. J. Randall and J. R. Brett, eds., Academic Press, San Francisco, New York, London.

Brett, J. R., and T. D. Groves, 1979, Physiological energetics, in: Fish Physiology, Vol. VIII Bioenergetics and Growth," W. S. Hoar, D. J. Randall and J. R. Brett, eds., Academic Press, San Francisco, New York, London.

Carey, F. G., and Olson, R. J., 1982, Sonic tracking experiments with tuna, in: "The definition of tuna and billfish vulnerability to fisheries," SCRS symposium of the International Commission for the Conservation of Atlantic Tunas, Puerto de la Cruz, Canary Islands, 1981, Collected Volume of Scientific Papers, Vol. XVII.

Conrad, M., 1972, Statistical and hierarchical aspects of biological organization, in: "Towards a Theoretical Biology," C. H. Waddington, ed., Edinburgh University Press, Edinburgh.

Conrad, M., 1976, Biological adaptability: the statistical state model, Bioscience, 26:319.

Daan, N., 1973, A quantitative analysis of the food intake of North Sea cod, Gadus morhua, Neth.J.Sea Res., 6:479.

Daan, N., 1978, Changes in cod stocks and cod fisheries in the North Sea, Rapp.P.-v.Réun.Cons,.Int.Explor.Mer, 172:39.

Dotson, R. C., 1978, Fat deposition and utilization in albacore, in: The Physiological Ecology of Tunas," G. D. Sharp and A. E. Dizon, eds., Academic Press, San Francisco, New York, London.

Fasham, M. J. R., 1978, The statistical and mathematical analysis of plankton patchiness, Oceanogr.Mar.Biol.Ann.Rev., 16:43.

Hoar, W. S., and Randall, D. J., 1978, "Fish Physiology, Vol. VII, Locomotion," Academic Press, San Francisco, New York, London.

Huston, M., 1979, A general hypothesis of species diversity, Amer. Natur., 13:81.

Iles, T. D., and Sinclair, M., 1982, Atlantic herring: stock discreteness and abundance, Science, 215:627.

Johnson, L., 1981, The thermodynamic origin of ecosystems, Can.J. Fish.Aquat.Sci., 38:571.

Jordán, R., 1967, The predation of guano birds on the Peruvian anchovy, Engraulis ringens, Calif.Coop.Ocean.Fish.Invest.Rep., 11:105.

Kerr, S. R., 1982, Estimating the energy budgets of actively predatory fish, Canadian J.Fish.Aq.Sci., 39:371.

MacCall, A. D., in press, Seabird-fishery trophic interactions in eastern Pacific boundary currents: California and Peru, Canadian Wildlife Service Occasional Paper series.

Odum, E. P., 1968, Energy flow in ecosystems: a historical review, Am.Zoologist, 8:217.

Odum, H. T., 1957, Trophic structure and productivity of Silver Springs, Florida, Ecol.Monogr., 25:55.

Olson, R. J., 1982, Feeding and energetic studies of yellowfin tuna: food for ecological thought, in: "The Definition of Tuna and Billfish Vulnerability to Fisheries," SCRS symposium of the International Commission for the Conservation of Atlantic Tunas, Puerto de la Cruz, Canary Islands, 1981, Collected Volume of Scientific Papers, Vol. XVII.

Owen, R., 1981, Patterning of flow and organisms in the larval anchovy environment, in: Intergovermental Oceanographic Commission Workshop Report No. 28, G. D. Sharp, ed., Unesco, Paris.

Parrish, R. H., Nelson, C. S., and Bakun, A., 1981, Transport mechanisms and reproductive success of fishes in the California Current, Biol.Oceanogr., 1:175.

Platt, T., Mann, K. H., and Ullanowicz, eds., 1981, "Mathematical Models in Biological Oceanography," The Unesco Press, Paris.

Platt, T., and Denman, K., 1975, Spectral analysis in ecology, Ann. Rev.Ecol.Syst., 6:189.

Platt, T., and Denman, K., 1977, Organization in the pelagic ecosystem, Helgol.Wiss.Meersunters, 30:575.

Platt, T., and Denman, K., 1978, The structure of pelagic marine ecosystems, Rapp.P.-v.Réun.Cons.Int.Explor.Mer, 173:60.

Pomeroy, L. R., 1974, The oceans' food web, a changing paradigm, Bioscience, 24:499.

Rheinheimer, G., 1980, "Aquatic Microbiology," 2nd ed., John Wiley and Sons, Chichester, New York, Brisbance, Toronto.

Ricker, W. E., 1979, Growth rates and models, in: "Fish Physiology, Vol. VIII, Bioenergetics and Growth," H. S. Hoar, D. J. Randall and J. R. Brett, eds., Academic Press, San Francisco, New York, London.

Sharp, G. D., 1981a, Report and supporting documentation, in: "Workshop on the Effects of Environmental Variation on the Survival of Larval Pelagic Fishes, April 20 to May 5, 1980, Lima, Peru," G. D. Sharp, ed., IOC Workshop Report No. 28, Unesco Paris.

Sharp, G. D., 1981b, Colonization: models of opportunism in the ocean, in: ""Workshop on the Effects of Environmental Variation on the Survival of Larval Pelagic Fishes, April 20 to May 5, 1980, Lima, Peru," G. D. Sharp, ed., IOC Workshop Report no. 28, Unesco, Paris.

Sharp, G. D., and Dotson, R. C., 1977, Energy for migration in albacore, Thunnus alalunga, Fish.Bull., U.S., 75:447.

Sharp, G. D., and Francis, R. C., 1976, An energetics model of the exploited yellowfin tuna, Thunnus albacares, population in the eastern Pacific Ocean, Fish.Bull., U.S., 74:36.

Sheldon, R. W., and Parsons, T. R., 1967, A continuous size spectrum for particulate matter in the sea, J.Fish.Res.Bd.Canada, 24: 909.

Sheldon, R. W., Prakash, A., and Sutcliffe, W. H., 1972, The size distribution of particles in the ocean, Limnol.Oceanogr., 17:327.

Simpson, J. H., Edelston, D. J., Edwards, A., Morris, N.C. G., and Tett, P. B., 1979, The Islay front: physical structure and phytoplankton distribution, estuarine and coastal, Est.Coastal Mar.Sci., 9:713.

Theilacker, G., and Dorsey, K., 1981, Larval fish diversity, a summary of laboratory and filed research, in: "Workshop on the Effects of Environmental Variation on the Survival of Larval Pelagic Fishes, April 20 to May 5, 1980, Lima, Peru," G. D. Sharp, ed., IOC Workshop Report No.28, Unesco, Paris.

Vlymen, W. J., 1970, Energy expenditure of swimming copepods, Limnol.Oceanogr., 15:348.

Vlymen, W. J., 1974, Swimming energetics of the larval anchovy, Engraulis mordax, Fish.Bull., U.S., 72:885.

Vlymen, W. J., 1977, A mathematical model of the relationship betwee larval anchovy (E. mordax) growth, prey microdistribution an larval behaviour, Environ.Biol.Fishes., 2:211.

Webb, P. W., 1975, Hydrodynamics and energetics of fish propulsion, Bull,of the Res.Board of Canada, 190:1.

Whittow, G. C., 1973, Evolution of thermoregulation, in: "Comparativ Physiology of Thermoregulation," G. C. Whittow, ed., Academic Press, San Francisco, New York, London.

DETRITAL ORGANIC FLUXES THROUGH PELAGIC ECOSYSTEMS

M. V. Angel

Institute of Oceanographic Sciences (N.E.R.C.)
Wormley, Godalming, Surrey GU8 5UB, U.K.

INTRODUCTION

Menzel (1974) wrote 'particularly lacking is information on the rates of input, utilisation and decomposition of organic matter to and in the deep sea, and the extent to which these processes influence the chemistry of seawater.' Here we are particularly concerned with the ecological effects and fluxes of detrital organic matter throughout the water column. The steady increase of knowledge of the vertical structure of oceanic midwater communities has gradually increased awareness of the importance of the rain of detrital material in fuelling the bathypelagic and deep benthic communities (e.g. Vinogradov, 1968; Angel and Baker, 1982). Consequently within the last few years the classic approach to qualitative and quantitative analysis of the particulate spectrum by the use of water bottles has been expanded by filter-pump sampling both *in situ* (Bishop and Edmond, 1976; Bishop, Edmond, Ketten, Bacon and Silker, 1977; Bishop, Ketten and Edmond, 1978; Bishop, Collier, Ketten and Edmond, 1980) and with ship-borne systems (e.g. Pugh, 1978), and by sediment traps (Wiebe, Boyd and Winget, 1976; Honjo, 1978; 1980; Spencer *et al.*, 1978; Hinga, Seiburth and Heath, 1979; Müller and Suess, 1979; Brewer *et al.*, 1980; Deuser and Ross, 1980; Wakeham *et al.*, 1980; Fellow, Karl and Knauer, 1981; Deuser, Ross and Anderson, 1981; Knauer and Martin, 1981; Sasaki and Nishizawa, 1981), and direct collection either by divers (Shanks and Trent, 1980) or from submersibles (Silver and Alldredge, 1981) or from captured organisms held temporarily in the laboratory (La Rosa, 1976; Small, Fowler and Ünlü, 1979). The main thrust of many of these sampling programmes has been to resolve questions concerning water column chemistry or sedimentary processes on the ocean floor, consequently not many have been supplemented with biological sampling.

In this paper I want to ask a number of questions. Firstly how far have we advanced since Menzel (1974) in gathering information on the rates of input, utilisation and decomposition of the particulate organic flux? Have all the important sources been identified? What are the factors controlling the recycling of detrital material in the water column. Is the general state of knowledge adequate to start modelling the detrital flux, and where is additional information needed?

SOURCES AND FLUXES

A schematic detrital flow chart through a parcel of water is shown in Figure 1. Sources from the surface photic zone are generally derived from primary production and inputs from atmospheric, river and neritic sources. Menzel (1974) considered that the oceani inputs from via the continental margins are insignificant (cf. Peterson, 1981; Dunbar and Berger, 1981), although Walsh et al. (1981; 1983) on the evidence of C: N ratio and ^{13}C data in sediments have estimated that 1-25% of shelf primary production may be deposited on the upper regions of the continental slope either as phytodetritus or as fecal pellets. Smetacek (this volume) described how, following the depletion of the nutrients in the euphotic zone in the Kiel Bight, half of the organic matter produced during the Spring Bloom sediments out. Within a period of a week 10 g C m^{-2} settles on the sea bed as a loose green carpet consisting of living phytoplankton cells. In the Bornholm Basin of the Baltic the cells sink out at rates cf around 40 m day^{-1}. The confirmation that such a mechanism may be important beyond the shelf-break comes from the observation of a similar floc carpet at depths of 2000m and more, in the Porcupine Sea Bight off S.W. Ireland (Billett, personal communication). So, in areas where there are substantial diatom blooms such as temperate or upwelling zones, the direct sinking out of phytoplankton may be an important, if not the major source of organic input (Smetacek, this volume).

The major source of inorganic particles is via atmospheric sources either as direct fall-out of wind-borne dust or suspended in rain water. Biological inputs are derived from primary producti via dead and moribund organisms, feeding inefficiencies by grazers and carnivores (e.g. the loss of larvacean houses (Alldredge, 1972; 1976; Taguchi, 1982) or the feeding webs of pteropods (Gilmer, 1972) or other invertebrates (Hamner et al., 1975)), or by the external breaking up of prey by feeding carnivores, the production of fecal pellets and crustacean moults, and the conversion of DOM by microorganisms into particulate material. All these biological sources are highly heterogeneous in time and space (Haury, McGowan and Wiebe, 1978), and the consequent patchiness over the whole spectrum of time-space scales has received relatively little attention in sampling programmes (exceptions include Gordon, 1970;1971;

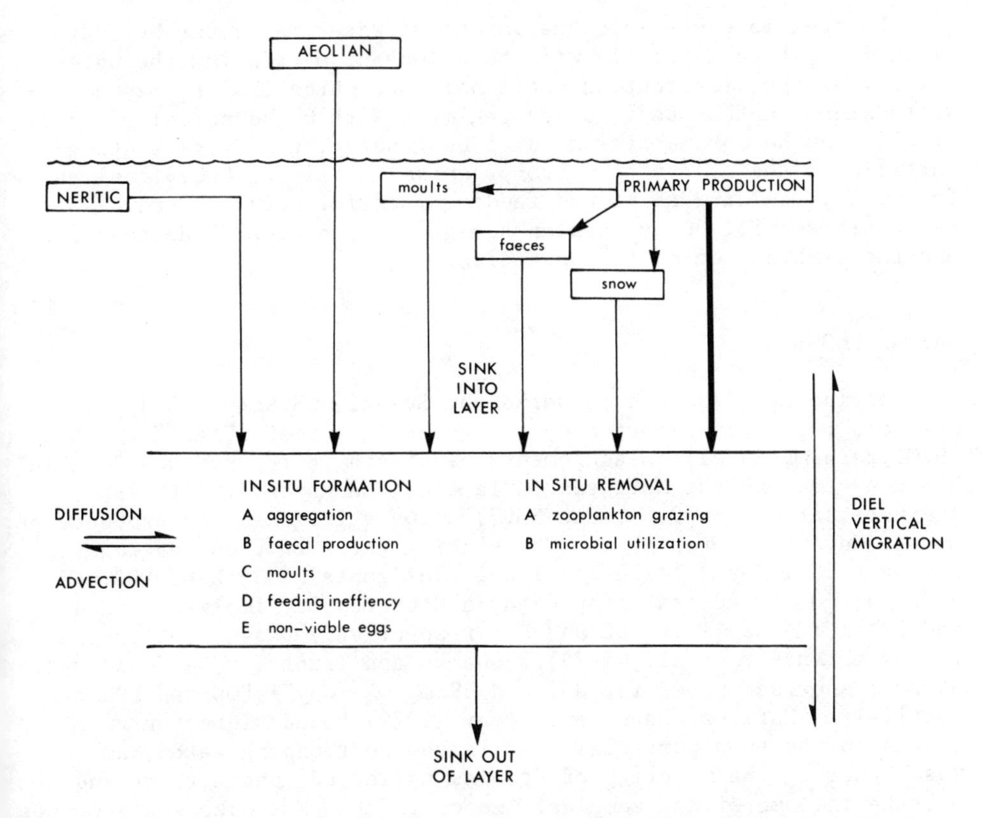

Fig. 1. Schematic flow of detrital material through a parcel of water showing the main sources of particulates and the main *in situ* generation and degeneration of particles.

Wangersky, 1978; Gordon, Wangersky and Sheldon, 1979; Deuser and Ross, 1980; Deuser, Ross and Anderson, 1981; Silver, Shanks and Trent, 1978). Modifications of particulate spectra occur through mechanisms of aggregation, such as the filter feeding by organisms like salps (Pomeroy and Diebel, 1980) and the formation of marine snow (Suzuki and Kato, 1953; Shanks and Trent, 1979; 1980; Alldredge, 1979; Silver and Alldredge, 1981), and also by particle utilization and their mechanical fragmentation by feeding animals. Bubbling has also been suggested as being a potentially important mechanism for the generation of aggregates.

The rate of input of these particles will depend on their sinking velocities which are a function of particle size and density (relative to the medium), and the viscosity of the medium. Active vertical mixing and density discontinuities may slow or even reverse sinking rates under certain conditions. Similarly particles may be accelerated through the stratum by migrating organisms.

Lateral movement into the parcel of water may occur by eddy-diffusion, which generally will be a mechanism reducing the heterogeneity of the distribution patterns. The parcel, also, may be advected into another sedimentary regime. Within the parcel of water itself a number of mechanisms will be generating or regenerating particles - the growth of micro-organisms utilising DOM, fecal pellet formation, moulting rates and feeding inefficiencies. Removal of particulates will be by zooplankton grazing, bacterial degradation, and the sinking out of the particles.

MARINE SNOWS

Marine snow was a term coined by Suzuki and Kato (1953), and has been regularly reported by observers in submersibles (e.g. Costin 1970; Barham, 1979). Measurements by divers in the surface 20 m in Monterey Bay and the Northeast Atlantic (Shanks and Trent, 1980) suggest that 3.6 - 72.6 (mean 30.7) x 10^4 aggregates are produced per m^2 per day in the top 20 m. The sinking rates of these aggregates averaged 68 m day^{-1}. The resultant flux contains 3-5% of POC and 4-22% of PON. Sediment trap data in Monterey Bay implied that 432 and 57 mg m^{-2} day^{-1} of POC and PON respectively passed out of the top 50 m (Knauer _et al._, 1979). Shanks and Trent (1980) found that at 20 m snow accounted for 362 and 59 mg m^{-2} day^{-1} POC and PON respectively. Earlier Shanks and Trent (1979) found higher nutrient levels in the snow particles than in the surrounding water; ammonia was higher in the majority of samples collected, and nitrite and nitrate in some of the samples. Hence in the photic zone aggregates are centres of locally elevated nutrient levels.

Alldredge (1979) studied shallow water snow aggregates both in the Santa Barbara Channel and in the Gulf of California. 20% of the Gulf of California aggregates and 5% of Santa Barbara aggregates were recognisable as being of zooplankton origin. In the Santa Barbara Channel 26-34% of POC, protein, carbohydrate, and lipid occurred in aggregates. In the Gulf of California 4-8% of the POC occurred in aggregates >3 mm, with a 3 mm aggregate containing about 38 μg dry weight of material and having a density of 1.0254 g cm^{-3} (cf _in situ_ sea water density of 1.0250). Although the measured sinking rates for these particles were about 90 m day^{-1}, the small density difference between aggregates and medium means that small localised vertical movements, such as observed in Langmiur Circulations, may aggregate the 'snow' particles and keep them in suspension.

Silver and Alldredge (1981) collected snow particles in deep water using the submersible Alvin. High concentrations of snow (~1/ℓ) were observed at 250-340 m and in 10-20 m layers at 850 m and 1500 m. They identified two main categories of floculent snow, mucus sheets from larvacean houses and smaller flake-like now of unknown origin. Both types contained large numbers of cells and biogenic debris, including some possibly viable cells. 1/3 - 1/2 of the biomass was in the form of nanoplankton with up to 3/4 of the total organic carbon being in the form of phytoplankton. Silver and Alldredge postulated two sources for this material, resistant cells in fecal pellets and cells which had been carried down on aggregates sinking at rates of 50-100 m day^{-1}. The age of cells at 1000 m sinking out on snow would be 10-20 days. In this context it is worth bearing in mind the results of Tabor et al. (1981) in studying the distribution of filterable marine bacteria in the deep sea. They found that significant numbers of colony-forming deep water bacteria pass through a 0.45 μm filter, compared with shallow-water and neritic forms. Concentrations of ATP in the water were proportional to the cell counts. These small-celled bacteria belonged to familiar shallow-water genera, such as Pseudomas, Vibrio and Flavobacterium, in all of which nutrient starvation for around nine weeks causes a dramatic cell shrinkage. Thus shallow-water bacteria may get carried down on aggregates and remain viable.

Silver and Alldredge (1981) could not determine the origin of most marine snow by in situ observations or by low magnification light microscopy. Examination by electron microscopy revealed that larvacean houses constituted part of the matrix of both the floc samples they collected. Characteristically larvaceans occur in the euphotic zone, but giant houses have been observed from submersibles down to depths of 375 m (Barham, 1979). In addition the occurrence of numerous algal cells such as pennate diatoms indicated that much of the snow originated from the surface layers. Descending snow appears to scavenge finer particles (algal cells and small fecal pellets), and so may play an important role in accelerating the descent of finer particulates, particularly of apparently viable, procaryotic and eucaryotic algal cells and cyanobacteria (Silver and Alldredge, 1981). Both the eucaryotic cells and the cyanobacteria appear to be identical to Chlorella-like cells and bacteria observed in the fecal material from salps and pteropods (Silver and Bruland, 1981). The cyanobacteria appeared in photographs to be actively growing within the fecal material, but there was no firm confirmation.

Generally the observations on marine snow suggest that at least some of the aggregates are of biological origin and originate in the photic zone. There is no clear evidence that snow is generated at greater depths, and the rapid decline in abundance of mucus-web feeders with increasing depth implies that the frequency with which snow forms around discarded webs or 'houses' will decrease. In the absence of any clear alternative mechanism of aggregate formation other than bubble aggregation at or close to the sea surface, it is reasonable to assume that biogenic substances also form the nuclei around which most floc develop.

CRUSTACEAN MOULTS

Mauchline (1980) has reviewed the literature on euphausiids. Euphausiid moults range from 3% of the body dry weight in Euphausia superba to 6-10% in Euphausia pacifica and Thysanoessa spp. Sameoto (1976) observed similar values of 6.2%, 7.1% and 5.0% in Meganyctiphanes norvegica, Thysanoessa inermis and T. raschi respectively He calculated that the flux of casts alone could reach 26 mg $m^{-2}day^{-1}$ in the Gulf of St. Lawrence. Shallow water species like E. pacifica moult every 3-5 days. Fowler, Small and Kěckěs (1971) showed an inverse relationship between intermoult periods and temperature over temperature ranges that the animals would normally encounter in the sea.

In marked contrast the life cycle of the mesopelagic mysid Gnathophausia ingens is much more prolonged. Instar times at in situ temperatures range from 166-253 days (Childress and Price, 1978) and maximum longevities were around seven years. These data were compared by Childress and Price (1978) with data for a shallower living mysid Metamysidopsis elongata for which the intermoult period was 4-10 days. Thus it seems likely that the various life-history tactics for shallow and deep-living species change dramatically, with shallower-living animals growing faster, moulting more frequently and squandering more organic material on casts. The data of Childress, Taylor, Caillet and Price (1980) imply that in fishes, metabolic rates decrease and growth efficiencies increase progressively from epipelagic, to mesopelagic migrators to bathypelagic non-migrators. There may be a parallel increase in general efficiency in crustaceans with increasing depth, so the input from casts may be inversely related to depth.

Casts provide a rich source of nitrogen as 46.7% of euphausiid casts are protein and chitin with additional traces of carbohydrate (Mauchline, 1980). They also sink rapidly, Fowler and Small (1972) observed sinking rates of 348-800 m day^{-1} in euphausiids casts. Small and Fowler (1973) extended these observations; freshly shed moults of Meganyctiphanes norvegica from the Mediterranean sink at 1600 m day^{-1}, but this rapidly declines to 400 m day^{-1} after five days as a result of microbial degradation. However, Fowler, Small and Kěckěs (1971) observed that the majority of euphausiids moult at night, so the carapaces will tend to be shed at shallow night time depths where the water is warmer and where re-cycling processes are faster. Lasker (1976) reported 288 m day^{-1} for the casts of Euphausia pacifica. However, casts are relatively uncommon in plankton samples so either they are too fragile to be sampled or they are recycled rapidly. Small and Fowler (1973) observed that, whereas the tightly-packed fecal pellets did not deteriorate for weeks so there was no measurable slowing of their sinking rates over the first few days, casts broke down relatively quickly, so did corpses (Flowler and Small 1972). It is noticeable that few of the sediment trap observations record significant fluxes of crustacean moults or corpses, for example Dunbar and Berger (1981) reported only one zooplankton carapace and several moults in a 1/16 subsample of a 45 day trap experiment (1700 cm^2) in the Santa Barbara basin. Honjo, Manganini and Cole (1982) observed that zooplankton remains were a significant fraction of the combustible part of mesopelagic trap samples, particularly at 400 m. Much of this flux is the result of living animals entering the traps and being killed by the preservative, rather than from a passive sink of moults or corpses. At 1000 m only 20% of the organic flux was of zooplankton remains, and only a small fraction in bathypelagic traps consisting mostly of copepod moults. Fowler, Small, Elder, Ünlü and La Rosa (1979) in their study on the role of fecal pellets in transporting PCB's to the benthos, found intact copepod moults in their particle collection samples from a depth of 100 m in the Mediterranean, in greater abundance in early summer compared with late summer and early autumn.

There are no published data available as to whether animals modify their vertical migrations or distributions at the time they moult. Any such modification could contribute to a change in the pattern of organic flux.

Fecal pellets

Fecal pellets are now widely assumed to be an important form in which organic and inorganic material is transported down through the water column. Assimilation efficiencies of around 70% are often considered reasonable (Dagg *et al.*, 1982) although they vary with food type (Conover, 1966) and lower values are often reported e.g. 30% in *Calanus* (Corner *et al.*, 1980). The increases in fish growth efficiency with depth (Childress *et al.*, 1980) and the evolution of distinctive energy-conserving characteristics in bathypelagic organisms, are probably also associated with greater assimilation efficiency in deep-living organisms. However, if it is assumed that almost all primary production is grazed by herbivores, then at least 30% is almost immediately turned into detrital form. The "taxonomy" of fecal pellets is relatively new (e.g. Martens, 1978; Bruland and Silver 1981) and tends to be based on the particle having a distinct entity and often being enclosed in a peritrophic membrane. Consequently any pellet that is not enclosed in a membrane and which is irregularly shaped may not be readily recognisable as fecal material. Robison and Bailey (1981) observed that the feces of most midwater fishes are amorphous and often filamentous, but in *Melanostigma pammelas* it is invested in a thick mucous layer. Bruland and Silver (1981) showed that the pellets of the doliolid *Dolioletta gegenbaurii* are irregularly-shaped packages that are only distinguishable as fecal material on microscopic examination. Therefore, much of the material caught in sediment traps that is considered to be aggregates of non-fecal origin, may indeed be irregular or disintegrating fecal pellets. In addition, as pointed out above, marine snow particles scavenge small fecal pellets from the water and so accelerate their descent. The numbers of small pellets in these snow particles appear to be higher than would be expected from the aggregate merely clearing its descent path. Langmuir convection cells have been postulated to accumulate zooplankton (Stavn, 1971); consequently they may be a source of the high degree of patchiness observed in the horizontal occurrence of snow, either by aggregating the organisms which produce the detritus, or by aggregating the actual detrital particles.

The integrated flux of fecal pellets is influenced by three major factors, 1) the production rate of pellets which is controlled by the feeding rates and ingestion efficiencies of the source organisms; 2) the downward flux rates which are controlled by the sinking rates of the pellets, their aggregation into larger more rapidly sinking units, and the diel vertical migration of the producer organisms, provided the retention time of food material in the gut is long relative to the time taken for migration; 3) the loss rate which is controlled by the grazing rate of detritivores on the pellets and the microbial breakdown of the pellets resulting in loss of organic matter, disintegration of the pellet, or possibly increased palatability of the pellets to detritivores.

OBSERVED DETRITAL FLUX RATES

Detrital flux rates have mostly been estimated using three techniques a) large water bottle samples, b) *in situ* pump samples and c) sediment traps. The early data on POC, based mainly on water bottle samples, are summarised by Menzel (1974). These data concerned what Fellows, Karl and Knauer (1981) considered to be 'suspended' particles i.e. abundant, low density, fine-grained particles, but because of the small volumes of water processed would not have included 'sediment-trap' particles. In general the organic material associated with these suspended particulates is non-labile (especially below 300 m (Menzel 1974)) and highly patchy in its spatial distribution (Gordon, Wangersky and Sheldon, 1979). Menzel (1974) concluded that the average concentration of bottle POC was 3-10 $\mu g\ \ell^{-1}$, lower (by a factor of 5-50) than the average DOC concentration of 0.5-0.8 mg ℓ^{-1}. Observations of ^{14}C in deep Pacific water POC indicated an age of 2000 years which Menzel (1974) considered was difficult to reconcile with a source of rapidly sinking particles. Similarly the $^{13}C/^{12}C$ ratios of -22 to -24 observed by Williams and Gordon (1970) in the deep N.E. Pacific suggest a very slow rate of fall-out of the particles.

In Figure 2 the possible sources of the suspended POC are summarised. The majority of the sources of DOC are not easily quantified. However, the observation that most suspended POC below

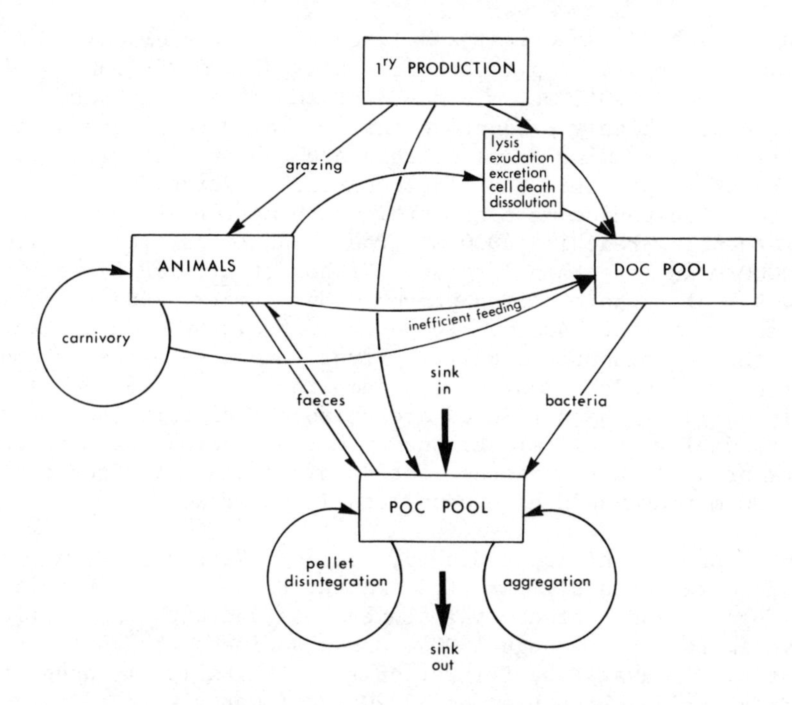

Fig. 2. Summary of the main sources of particulate organic carbon (POC) in the water column. The thicker lines indicate what are probably the greater fluxes (DOC = dissolved organic carbon).

300 m is non-labile is not consistent with most of it being derived from micro-organisms; the most likely alternative source is from sinking particulates and the disintegration of fecal material both of which may have been stripped of labile organic compounds by microbial action at shallower depths.

In situ filter-pump samplers have been used mostly by geochemists. Bishop, Ketten and Edmond (1978) examined vertical profiles of three particulate size fractions, <1 μm, 1-53 μm, and <53 μm along a four station transect off S.W. Africa and one shallow water station close to Cape Town. Poorest agreement between the filter-pump data and large water bottles occurred when particles >53 μm were abundant. These large particles settled out so rapidly in the water bottles that they were missed by the filtration method. C (organic): N ratios in the middle size fraction were about 7 at the productive stations, but 9-11 at the outermost oligotrophic station of the transect. The smallest fraction had lower ratios suggesting the presence of bacteria. Two flux models were derived, one for fecal pellets and foraminiferans assuming they had a spherical shape, the other for discoidal fecal 'matter.' Those data implied organic fluxes of 0.27-0.30, 0.049-0.069 and 0.0033-0.013 mgC m^{-2} day^{-1} at the three deep-water stations moving from the highly productive inshore water to oligotrophic water well offshore (assuming fecal pellets are 24% organic carbon). The flux estimates suggested that large particles which contributed <4% of the total suspended mass of particulate matter at 400 m provided most of the flux reaching 4000 m. These larger particles would reach 4000 m in less than 30 days. Bishop et al. (1978) estimated that 94-99% of organic carbon produced by primary production is recycled in the surface 400 m. Two points should be made now which will be further elaborated below. Firstly no time series of observations were made at any of the sampling depths, so the estimates are only valid if the particles are uniformly distributed horizontally. Secondly, diel vertical migrations by both plankton and micronekton may have produced a significant net downward flux of organic carbon across 400 m that would have been totally undetected.

In a more recent paper Bishop, Collier, Ketten and Edmond (1980) reported on a similar study in the Panama Basin in which a comparison was made between the fluxes observed with their filter pump system passing through 1500 m and a sediment trap set out at 2500 m for 234 days. The filtration times below 100 m ranged from 2½-6 hours during which between 12-20 m^3 of water were filtered. Despite these relatively large volumes of water filtered (large relative to previous filter pump experiments) the flux estimates through 1500 m were 1/40 those observed by sediment trap at 2500 m. So, either the sedimentation is highly variable in time and space, or the main flux was in the form of particles sampled by the sediment trap but not by the filter-pump.

Fecal 'matter' and three categories of fecal pellets showed profiles in which there was a sharp decline in particulate abundance below 200 m. In the deeper water particles >53 μm made up 95% of the total particle volume. The percentage organic content of the fecal material declined with depth from 57-68% in the wind mixed layer to around 40% at 1500 m. The ratio between the organic carbon and the calcium carbonate decreases rapidly from 20 in the near surface water to about 8 at 200 m, suggesting a rapid recycling of the organic carbon. The ratio increased to around 12 at 700 m, possibly due to the generation of particulates by the mesopelagic fauna; although this increase coincided with the strong oxygen minimum so it may have resulted from a relative decrease in the carbonate. Below 700 m the organic carbon content of particulates collected by the filter-pump decreased steadily. Even so, the sediment trap material had a much lower organic carbon content, but whether the oxidation occurred in the water column or in the trap is not clear.

The first observations on fecal pellet flux in deep water were made in the Tongue of the Ocean at a depth of 2050 m (Wiebe, Boyd and Winget, 1976). 650 fecal pellets m^{-2} day^{-1} were collected and the pellets had sinking rates of 50-941 m day^{-1} (mean 159), however, this was still sufficient to provide around 30% of the benthic respiration requirement of 38-42 mg C m^{-2} day^{-1}. At a depth of 5367 m in the Sargasso Sea, fecal pellets made up the bulk of the particulate flux (Honjo, 1978; Spencer et al., 1978). The trap material contained 3-5% organic carbon, giving a flux of about 1.4 mg C m^{-2} day^{-1}, which was about 1% of the surface productivity. Parflux II data (Honjo, 1980) from deep water gave flux values of 0.7-1.7 mg C m^{-2} day^{-1} in the Caribbean and Sargasso regions. At Parflux E ($13\frac{1}{2}^{o}$N,54°W) the organic content of the particles decreased from around 50% at mesopelagic depths (389 m and 988 m) where the measured organic carbon flux represented 4-6% of the primary production, to ≃35% at deep bathypelagic depths (3755 m and 5068 M) where the flux was equivalent to 1% of the primary production.

Rowe and Gardner (1979) observed higher sedimentation rates of 0.13-1.47 g m^{-2} day^{-1} in the slope water of the north-east United States, of which the average organic carbon flux averaged 12.8 mg C m^{-2} day^{-1}. They observed that fecal pellets were larger and were more abundant in the nephloid layer. Wishner (1980) has shown that there is a marked increase in the standing crop of plankton in close proximity to the sea bed which acts as a particle collector and from which biological and physical resuspension occurs to form the nephloid layer. Data collected from a series of slope stations in the Northeast Atlantic from 'Discovery' in 1979 confirm Wishner's observations for both plankton and micronekton (Hargreaves, personal communication). Consequently the flux pattern close to the sea bed will change both because of resuspension, horizontal transport by turbidity flows and the increase in biological activity. Chemical profiles suggest that this influence may extend as much as 1000 m up into the water column (Spencer et al., 1978).

In the Santa Barbara Channel at a depth of 340 m the annual flux was 660 g m^{-2} day^{-1} of which 22 g m^{-2} day^{-1} was carbon representing about 7-22% of the local primary production (Dunbar and Berger, 1981). At least 60% and possibly as much as 90% was in the form of fecal pellet aggregates. Three main types of pellets were tubular (60%), ellipsoid (25%) and cylindrical (15%). Only the ellipsoid pellets were observed to have remains of a membranous cover. Pellet sinking velocities ranged from 71-1694 m day^{-1} and around 120,000 m^{-2} day^{-1} settled out.

Hinga et al. (1979), also sampling in productive waters, observed organic carbon fluxes of 14-30 mg C m^{-2} day^{-1} at 3500 m, sufficient to supply the needs of benthic sediment respiration. At the shallower depths studied (600 m, 1300 m) an additional input was required, not only to satisfy sediment respiration but also the respiration of the larger members of the benthic and bethopelagic communities. The trap material contained 3.8-6.6% of organic carbon whereas the sediment below contained 1.2-1.3%. So about 75% of the carbon is utilised on the bottom.

Müller and Suess (1979) found empirically that the percentage of organic carbon in sediments doubles with each ten-fold increase in sedimentation. However, sedimentation rates varied seasonally (Deuser and Ross, 1980; Deuser, Ross and Anderson, 1981) in traps set off Bermuda at depths of 3200 m between 20-60 mg m^{-2} day^{-1}, where 85-95% of the sediment was of biogenic origin. At a time scale of two months, there was close synchronisation between the increase in surface production and the arrival of all sizes of particles, even the fine particles.

Off Japan, Sasaki and Nishizawa (1981) observed a flux of 418 mg C m^{-2} day^{-1} at 100 m equivalent to about 28% of the mean daily primary production, mostly in the form of long cylindrical pellets (mean dimensions 100 x 30 μm) possibly of euphausiid origin, and oval pellets (mean dimensions 300 x 50 μm) of possible copepod origin. The flux decreased to 300 mg C m^{-2} day^{-1} at 300 m and to only 28 mg C m^{-2} day^{-1} at 500 m. They also observed a straight line relationship with a slope of -3 between the logarithm of number of particles trapped per unit area and logarithm of the maximum particle dimension. If the particles were spheres, this would imply that the volume of the particulate flux is constant across the size spectrum (McCave, 1975).

Individual groups of organisms may at times contribute a large proportion of the flux. Madin (1982), extrapolating from his data on fecal pellet production and sinking rate in salps, estimated that in a swarm of _Salpa aspera_ (Wiebe _et al_., 1979) with a density of 65 salps m^{-3} the salps would have produced 8.5-139 mg C m^{-2} day^{-1}. Combined with the sinking out of dead salps of 3.6 mg C m^{-2} day^{-1}, the salp pellet flux alone would have provided about 180% of the metabolic needs of the local benthic infauna. In another example, Madin (1982) estimated that a sparse population of _Cyclosalpa pinnata_ in the N. Atlantic occurring at a density of 8 salps per 1000 m^3 would produce about 142± 72 μg C m^{-2} day^{-1} equivalent to about 6% of the local benthic requirement.

Fellows, Karl and Knauer (1981) observed the distribution, production and flux of organic carbon at 12 depths ranging from 1-1250 at a station off the California coast with sediment traps which preserved the organic material entering traps. They used ATP to estimate the living microbial biomass (see also Tabor, Ohwada and Caldwell (1981)). In the surface 50 m the microbial biomass contributed 30-40% of POC, but this decreased to 17-24% in the vicinity of the seasonal thermocline (100-200 m) and further steadily decline down to 1250 m where only 1.8% of POC was living (Figure 3). There was an increase in the ATP flux below the depth of the oxygen minimum which rose from a minimum of about 80 μg m^{-2} day^{-1} at 200 m to 300 - 400 μg m^{-2} day^{-1} at 1100-1500 m. This was associated not only with an increased flux in organic carbon and nitrogen (Knauer and Martin, 1981) but also in the total numbers of fecal pellets (Urrère and Knauer, 1981) which was attributed to the repackaging of organic material by feeding mesopelagic organisms (diel vertical migration may also be important, see below). Eppley and Peterson (1979) have also stressed the ecological importance of repackaging and new production in deep water.

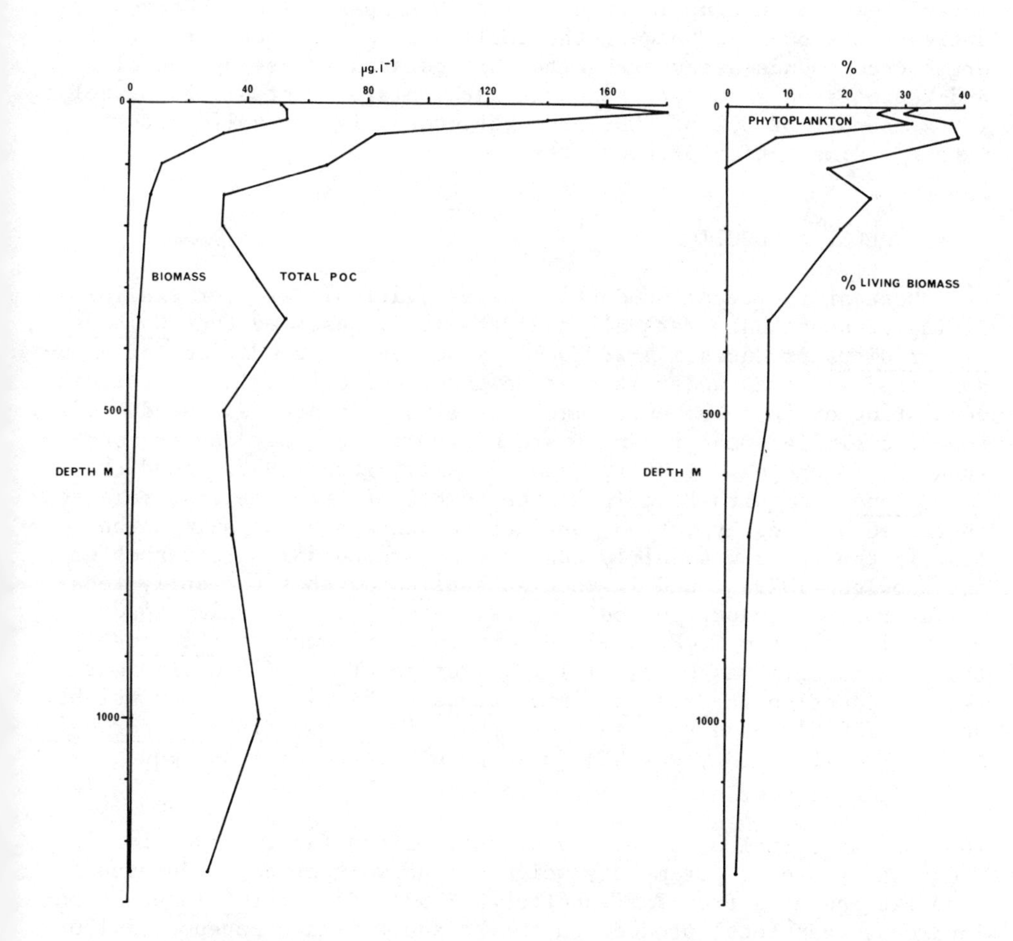

Fig. 3. Vertical profile of 'suspended' POC showing its organic carbon or biomass content, the content of living microbial biomass based on ATP concentrations and the contribution of phytoplankton (after Fellows, Karl and Knauer, 1981).

Other evidence of the changing nature of the particulate flux was given by Wakeham et al. (1980) based on the Parflux E 'profile' of sediment traps. The lipid flux decreased from 15% at 389 m to 2% at 5068 m, many of the compounds involved being structural features in bacteria cells. Free fatty alcohols decreased considerably between 389 m and 988 m, but showed a slight increase deep down, whereas the flux of true sterols, steryl esters and triacylglycerols was maximum at 988 m. Generally the hydrocarbons were of planktonic origin (C17, C19 and some pristane) but traces of longer chain (C25-C31) hydrocarbons were detected which indicate some long-range transport of continental origin. Sterenes were detected at 389 and 988 m and they were more abundant than alkanes in the same carbon range; this suggests that active microbial transformations of organic material was occurring in or on the sinking particles. Fatty-acids decreased by 60-fold between the shallowest and deepest traps. Clearly organic compounds associated with particulates are being recycled and reprocessed at mesopelagic and bathypelagic depths. So Menzel's (1974) report of 80% of POC being refactory, is not valid for the larger sediment-trap particulates.

FECAL PELLET PRODUCTION

Most of the fecal production rates available are for shallow-living crustaceans. Marshall and Orr (1955) observed that *Calanus finmarchicus* produces 144-288 pellets per day in conditions of abundant food and that under starvation ghost pellets would be produced consisting of just the outer membrane with no contents. Gaudy (1974) reported similar mean rates of around 200 pellets per day in four species of copepod. Corner, Head and Kilvington (1972) found that for *Calanus* fed with *Biddulphia* the number of fecal pellets released increased with daily ration, but the percentage of organic carbon lost in the feces was fairly constant at around 66%. Paffenhoffer and Knowles (1979) found in *Eucalanus pileatus* that the daily fecal pellet volume production could be estimated from the animals' dry weight (W mg) from 0.0855W + 1.9, whereas for *Temora turbinatus* the relationship was 0.711W + 1.18. Petipa et al. (1970) estimated fecal production in *Calanus finmarchicus* to be 14.8% of body weight. Heyraud (1979) found that in the euphausiid *Meganyctiphanes norvegica* fecal production (dry weight) could be estimated from the equation:

$$E = 0.35W^{0.42}$$

Thus for an animal of 25 mg dry weight, E, the fecal production is 0.054 mg dry feces per mg dry weight of animal per day. Heyraud also extrapolated from Paffenhoffer's (1971) data for *Calanus helgolandicus*, that fecal production for an adult female copepod of 170 μg is 0.32 μg mg^{-1} day^{-1} whereas for a copepodite of 35 μg it is 0.66 μg $mg^{-1}day^{-1}$. Thus small species tend to have higher production rates than larger species, and even within the same species small individuals produce relatively more feces, weight for weight, than large individuals. However, Small, Fowler and Kěckěs (1973) found that the egestion rage of *Artemia* by *Meganyctiphanes* increased three-fold from 1.8% to reach a plateau 5% of body dry weight as the concentration of available food increased. Furthermore bearing in mind the physiological data for midwater fishes derived by Childress, Taylor, Caillet and Price (1980), it seems likely that deeper-living organisms may assimilate more efficiently so that fecal production rates may decrease with increasing depth.

Certain organisms may prove to have relatively high fecal production rates. Madin (1982) observed in three species of salps that production ranged from 9.9-17.8 μg C (mg C body weight)$^{-1}$ h^{-1}. Taguchi (1982) studying the particulate fluxes in a subtropical bay Kaneohe Bay, Hawaii observed that 42% of the daily production sedimented out mostly in the form of fecal pellets. The predominant type of fecal pellets (on average 44% of total) were those of the appendicularian *Oikopleura longicauda*. Daily production rate was 243 ± 105 pellets, with each animal producing 5.3 ± 3 houses per day each containing 65 ± 32 pellets. Under optimal feeding conditions a *Oikopleura* can produce a pellet every 2.2 minutes. Table 1 summarises some of the published production rates.

VERTICAL DISTRIBUTION OF ORGANISMS IN RELATION TO FECAL PRODUCTION

With primary production restricted to the surface tens of metres the availability of food decreases with increasing depth and the standing crop of organisms declines (Vinogradov, 1968). Wishner (1980) using data from the literature as well as her own data from a plankton net attached to the "Deep Tow" instrument package fished close to the sea bed, found that the regression of the logarithm of standing crop of plankton against the logarithm of depth had similar slopes over a wide geographic range of samples. Angel and Baker (1982) have extended these observation throughout the water column at three stations in the N.E. Atlantic to include both plankton and micronekton (Figure 4). In general the standing-crop of micronekton in the total water column was equivalent to between a third and a half that of the plankton. Even taking into account the reduction in fecal production rate in large animals (Heyraud, 1979), because the larger animals produce larger faster sinking feces, the micronekton is likely to make an important contribution to the flux of detrital carbon down through the water column. Angel and Baker (1982) also show how fishes, decapod crustaceans and mysids provide substantial parts of the deep mesopelagic and shallow bathypelagic communities, all of which produce large and fast falling fecal masses. Yet, neither *in situ* pumps nor sediment traps seem to sample these sorts of particles which are probably > 1 mm in size.

Table 1. Some observed fecal pellet production rates

Species	Feeding Conditions	Fecal Pellets per day	Authors
Acartia tonsa	3 order of magnitude of food concentration	1-6	Reeve and Walter, 1977
Acartia clausi	5 x $10^5 m\ell^{-1}$ natural particles	24.5	Honjo and Roman, 1978
	1.6 x $10^5 m\ell^{-1}$ coccolithophores	91	Honjo and Roman, 1978
	2.0 x $10^4 m\ell^{-1}$ coccolithophores	8	Honjo and Roman, 1978
Temora turbinatus	Natural conditions	10-169	Paffenhofer and Knowles, 1979
Eucalanus pileatus	Natural conditions	55-160	Paffenhofer and Knowles, 1979
Oikopleura longicauda	Natural conditions	243±105	Taguchi, 1982
Cyclosalpa affinis	Excretion of animals collected by SCUBA divers and held in aquaria	9.9*	Madin, 1982
Salpa maxima	Excretion of animals collected by SCUBA divers and held in aquaria	13.0*	Madin, 1982
Pegea confoederata	Excretion of animals collected by SCUBA divers and held in aquaria	17.8*	Madin, 1982
Meganyctiphanes norvegica	25 mg animals fed on *Acartia* in laboratory	5.1^+	Small *et al.*, 1973
M. norvegica	" "	5.3^+	Heyraud, 1979

*μg fecal C mg $C^{-1}h^{-1}$ $^+$mg feces (mg animal dry Wt)$^{-1}$ day^{-1}
(N.B. a *Meganyctiphanes* fecal pellet weighs ≃ 17 μg, Fowler, personal communication).

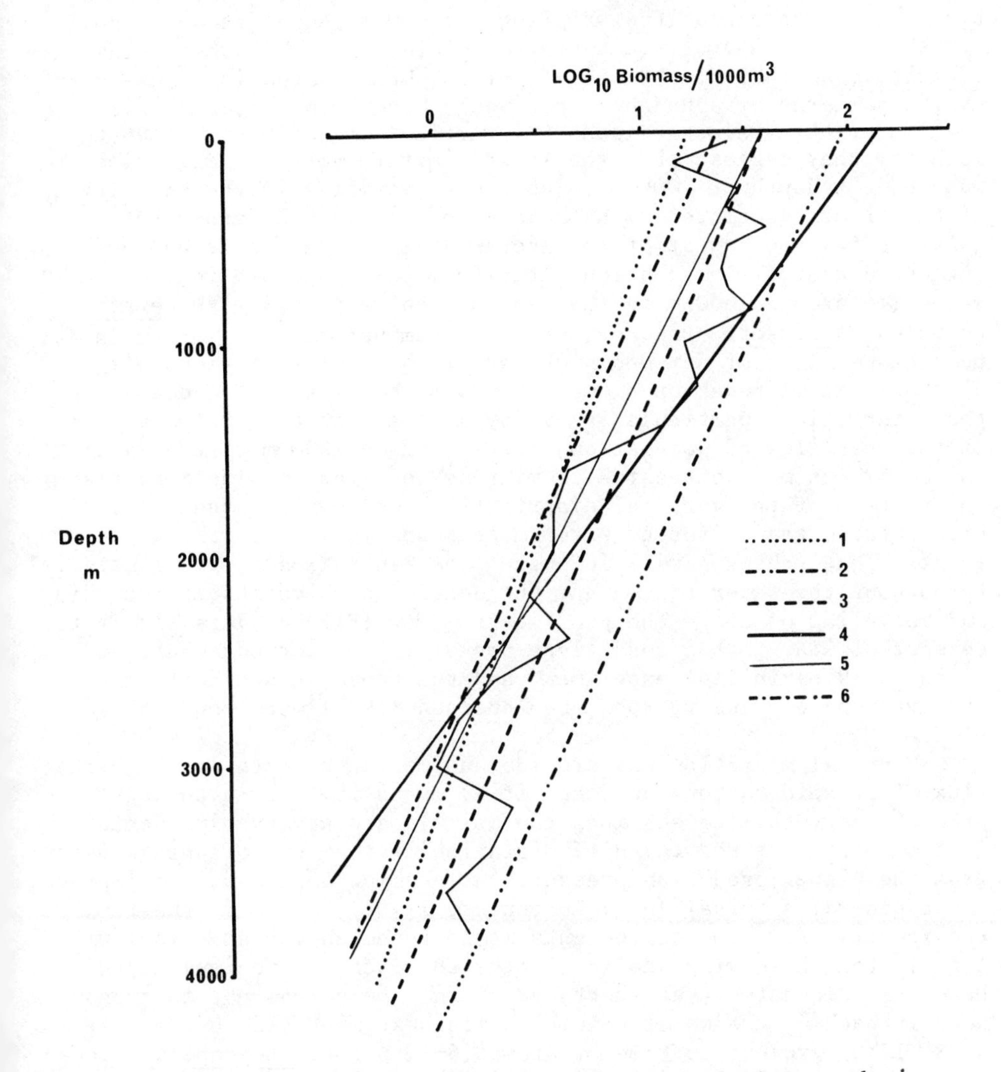

Fig. 4. Regression lines of the logarithm of plankton and micronekton biomass standing crop profiles expressed as displacement volume (mℓs) per 1000 m^3 again depth at three stations in the Northeast Atlantic (from Angel and Baker, 1982). 1. micronekton at 20°N, 21°W; 2. plankton and 3. micronekton at 42°N, 17°W; 4. plankton and 5. micronekton at 49°40'N, 14°W; 6. plankton in the Northeast Atlantic (after Wishner, 1980).

Angel (1979) suggested that the lower limit of diel vertical migration in plankton occurred at around 700 m. At 42°N, 17°W in the N.E. Atlantic repeated sampling at 1000 m (Angel et al., 1982) has shown that, apart from a single species of fish Notoscopelus elongatus, neither planktonic nor micronektonic organisms showed the cyclic patterns of abundance that would have been expected if they were undertaking synchronised diel vertical migrations. So at this locality they suggest that the lower limit to vertical migration was above a depth of 1000 m. Subsequent sampling in the vicinity of the front associated with the edge of the 18°C Sargasso Water (WAW) to the south-west of the Azores at latitudes of around 30°N suggested that diel migration activity may extend down into deeper water (to around 1600 m in the case of the myctophid fish Ceratoscopelus warmingeri, Badcock, personal communication). Accumulative percentage plots of day and night profiles at five stations (Figure 5) show that micronekton biomass tends to be distributed deeper in the water column particularly by day so that the range of movement of the quartiles of population in the surface 1200 m of the water column is far more extensive in micronekton than in plankton; also the pattern of the vertical distributions changes substantially both between the different water masses and in the region of the front. These changes will influence the generation of fecal material throughout the water column and the degree to which its reprocessing and recycling occurs. The profile from WAW (Figure 5) is likely to be most similar to the conditions prevailing of Bermuda where both Honjo's (1978) initial experiment and the repeated sediment trap observations of Deuser, Ross and Andersen (1981) were conducted.

Vertical migration may provide an important route by which the flux of organic carbon can occur if 1) mortality occurs through predation or other causes when the animals are at day-time depths (N.B. One explanation for diel migration is that the organisms escape from the higher predation pressures at shallow depths and so improve their general survival (e.g. Longhurst, 1976)), or 2) if the retention time for food in the guts is long enough for an animal with a full stomach leaving shallow depths to reach its maximum depth before it defecates, yet short enough for the animal not to transport it back up the water column on its next migration cycle. At 42°N, 17°W a retention time of around 6-12 h for Notoscopelus elongatus would ensure its maximum contribution to the transfer of organic carbon from the surface 100 m to depths of 1000 m.

In copepods retention time of $1\frac{1}{2}$ h have been observed in Centropages typicus (Dagg and Gill, 1980) and 1-3 h in a mixture of copepod species from the North Pacific Central Gyre (Hayward, 1980). Even more interesting in the context of vertical migration are the observations of Petipa (1964) that in Calanus Helgolandicus food retention time was only 20 minutes in animals kept at surface temperatures, but at sub-thermocline temperatures of 7° the retention time increased to 3-4 h.

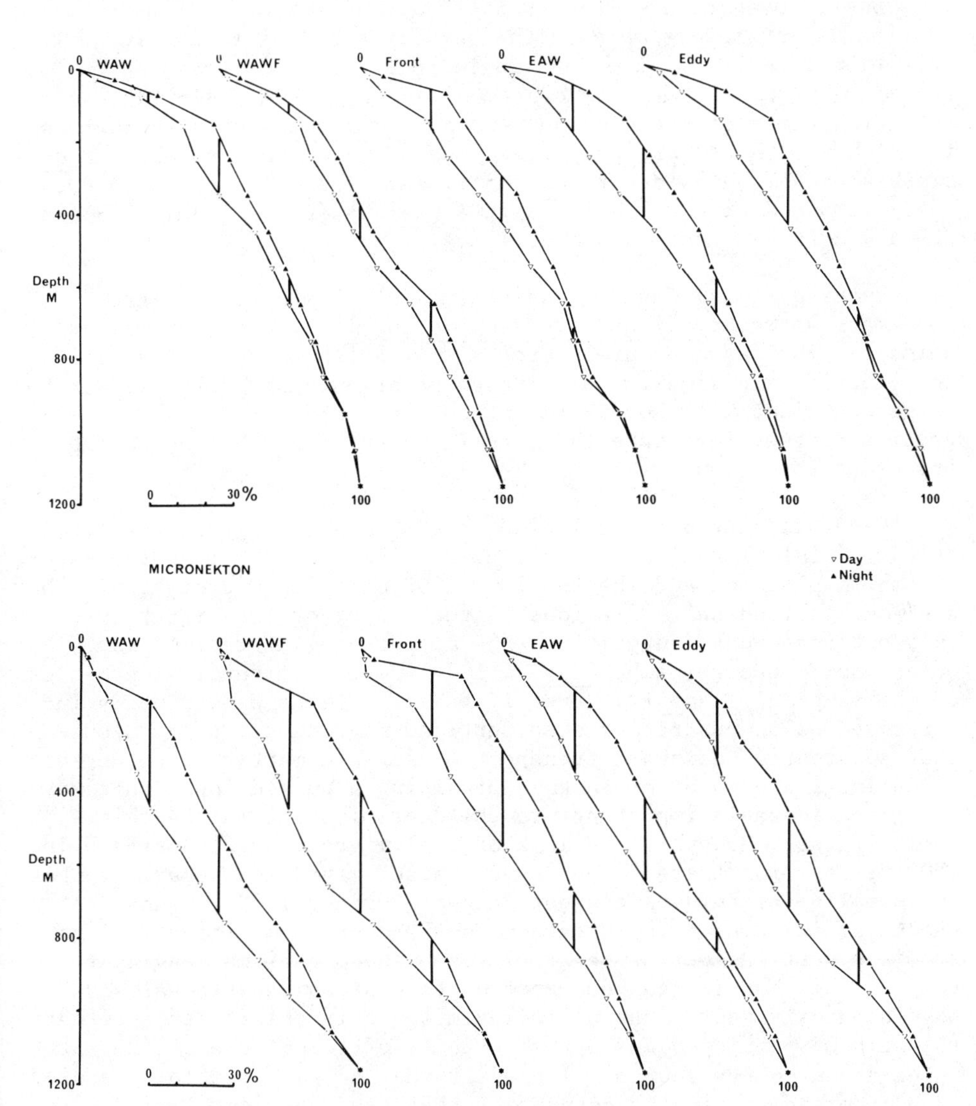

Fig. 5. Accumulative percentages of biomass by day and by night in the top 1200 m of the water column at five stations all within 300 nautical miles of each other in the N.E. Atlantic close to 30°N 30°W. The five stations were in 1) 'Western' Atlantic water with a thick layer of 18°C Sargasso Sea Water (WAW), 2) In Western Atlantic water but in a meander of the front within the East Atlantic water (WAWF), 3) In the Front, 4) In the Eastern Atlantic Water (EAW), 5) In an eddy of Eastern Atlantic Water which had pinched off the meander about a month previously.

Beside temperature, the fullness of the gut may influence retention times, for example in the chaetognath *Sagitta hispida* gut retention times are 1-2 h for single food items but increase to 5 h if 5 or more food items are consumed (Reeve, 1980). Similarly gut retention times of the ctenophore *Mnemiopsis mccradyi* also increase from 1 h to 5 h as more prey items are taken. The ctenophore *Pleurobrachia bachei* has a gut retention time of 1.2 h (Sullivan in Reeve, 1980) compared with 3.2 h in *Sagitta enflata* (Feigenbaum, 1979) and 3.5-4.0 h in *S. elegans* (Reeve, 1980).

Heyraud (1979) found transit times of 30 minutes in large specimens of *Meganyctiphanes norvegica* and 15 minutes in smaller animals. The range of diel vertical migration shows a progressive ontogenetic extension in many groups of organisms; so it may be the large organisms with longer retention times and more extensive migration ranges, that make the more important contribution to the organic carbon flux.

Generally the data collected at IOS suggest that in the N.E. Atlantic diel migration is restricted to the top 700 m of the water column in plankton and the top 1000-1500 m in the micronekton. There are some latitudinal variations in the ranges of the migrations. The only published data suggesting that deeper diel migrations may occasionally occur (Wiebe *et al.*, 1978) were from vertical profiles of the salp *Salpa aspera*. Even if these profiles were wrongly interpreted as demonstrating deep vertical migrations, they do show that substantial downward transport of organic matter may occur beneath large swarms of organisms, providing a pulsed input that may be of considerable importance to the deep midwater and benthic communitites. However, the lack of diel migrations to depths below 1500 m, assuming there is no asynchronised migration behaviour that our sampling techniques can not detect (Pearre, 1979), means that the major input of organic carbon to the deep communities will be from the settling out of particulates. The organisms, having to rely on this source, can be expected to possess behavioural and physiological adaptations which maximise their ability to intercept the rain of particles and optimise their efficient use of the particles as an energy source. I predict that despite the low standing crop of organisms in the deep water that the organisms will be found to have a significant influence on the organic carbon fluxes, and probably many other chemical fluxes through the water column (e.g. ^{210}Po and trace elements, Cherry, Higgo and Fowler, 1978; Cherry and Heyraud, 1981). The total lack of physiological information on deep midwater organisms, as to their feeding rates, retention times and assimilation efficiencies is a serious gap in our knowledge that prevents our proper understanding of the ecological function of these deep communities, how they influence the underlying benthic communities, and how they affect geochemical fluxes.

The potential importance of large food falls on benthic communities has been discussed by Stockton and DeLaca (1982). Such falls may be mostly exploited by the benthic-pelagic scavengers who probably disperse their fecal material over wide areas of the sea bed as they scatter after feeding. However, there may also be plumes of particulate organic material suspended during the feeding of the scavengers and of dissolved organic matter. These plumes will both diffuse and be advected in the water currents. If the effects of the falls are localised, they are likely to give rise to persistent changes in the benthic community structure.

At high latitudes >40°, a further mechanism for vertical transport occurs via the downward seasonal migration performed by many organism prior to overwintering. For example, calanoid copepods such as C. finmarchicus C. plumchrus and C. cristatus overwinter at depths down to 2000m. Angel and Baker 1982 have observed very large populations of scyphozoans around 1000-1500 m in the N.E. Atlantic, and one possible food source for these medusae are the overwintering copepods. Although this flux is only equivalent to the standing crop at the time of the migration and, in terms of the total annual flux, is probably relatively small, it may be associated with a major change in the depth of input of particulates and of the pattern of the flux.

UTILISATION OF FECAL PELLETS BY PLANKTON

The role of detrital material in the functioning of aquatic ecosystems has been well reviewed by Turner and Ferrante (1979), so only a brief summary is needed here. The utilization of fecal pellets as a food source has been inferred from the gut contents (e.g. Harding 1974) which include phytoplankton remains (e.g. Angel, 1972) and chlorophyll pigments or their derivatives, (Nemoto and Saijo, 1968; Nemoto, 1972) in non-migrant bathypelagic organisms. The existence of re-cycling of fecal material has been demonstrated both by modelling the flux of organic carbon (e.g. Paffenhoffer and Knowles, 1979; Hofmann, Klinck and Paffenhoffer, 1981) and by mesocosm experiments (e.g. Davies, Gamble and Steele, 1975). Actual observations of fecal pellet utilization by plankton are surprisingly few (e.g. Paffenhoffer and Strickland, 1970; Petipa, Pavlova and Mironov, 1970; Poulet, 1976, Paffenhoffer and Knowles, 1979), but this probably reflects the technical difficulties of demonstrating the utilization rather than its infrequency. Paffenhoffer and Knowles (1979) emphasise that pellet utilization leads toward complete removal of all available carbon and nitrogen from a previously incompletely assimilated food source, and that this process is likely to be of particular importance in 'nutritionally dilute' environments.

Iseki (1981) found that the salp fecal flux of 10.5 mg C m^{-2} day^{-1} at 200 m had decreased by 36% at 900 m. Given the high sinking rates of salp feces (Madin, 1982; Bruland and Silver, 1981) and the relatively low temperature range (4°C at 900 m to 7°C at 200 m), it seems unlikely that microbial degradation was responsible for this decline. So either a significant amount of disintegration occurred or coprophagy by the midwater plankton were responsible for the decrease in the flux.

MICRO-ORGANISMS AND FECAL MATERIAL

Pomeroy and Deibel (1980) have described the sequence of events occurring during the aging of salp fecal pellets. Fresh pellets contain partially digested phytoplankton and inclusions embedded in an amorphous matrix. After 18-36 h a population of large bacteria develops in the matrix and phytoplankton remains. Between 48-96 h a population of ciliate protozoans develop which consume the bacteria. Up to this time the feces resemble the flocculent organic aggregates containing populations of micro-organisms which have been described from highly productive regions of the ocean. After 96 h few micro-organisms remain and the pellet begins to fragment. At this stage the feces resemble the near sterile flocculent aggregates described by Alldredge (1979).

The rate at which these events occur are related to temperature. Honjo and Roman (1978) found that at 20°C the membrane, enclosing copepod fecal pellets which were produced after feeding with cultured algae, was degraded in 3 h leading to the distinegration of the pellet. Whereas at 5°C the membrane remains intact for 20 days. Pellets produced by laboratory-cultured animals appear to be degraded faster than those produced by animals which have fed in the field. Even so, temperature seems to have an important influence, so that in surface tropical waters not only is the pellet flux likely to be reduced because the producing organisms tend to be smaller than at higher latitudes, but also the bacterial degradation will be faster because of the higher water temperature. This may also have an influence on the geographical distribution of benthic biomass (e.g. Rice, 1978). Similarly bacterial degradation of likely to be less important in pellets generated in deeper cooler water.

There is an acceleration in the rate at which the organic carbon is utilised as the microbial population develops. Turner (1977) found that for copepod fecal pellets at 22°C, 16% of the carbon was lost after one day and 67% by the end of the second day. At 5°C, however, the loss was only 10% even after 14 days.

Roman and Rublee (1981) in discussing their method for determining *in situ* zooplankton grazing rates on particulates, reported that there appeared to be rejection of particles labelled with ^{14}C-thymidine. The thymidine was used to assess the uptake of heterotrophs, and the results suggest that particulates rich in heterotrophs may be unpalatable. Micro-organisms may enrich particles by stripping DOM from the surrounding water and by breaking down non-labile organic compounds into more available forms. In salt marsh conditions the feces produced by the snail Hydrobia are initially rejected by the snail (Koefoed, 1975) but after five days of conditioning the pellets had become palatable. If a similar conditioning process occurs in pelagic pellets, it could modify the pattern of flux, as 'taste' plays a role in copepod feeding (Poulet and Marsot, 1978),

One form of 'conditioning' arises from the observation that midwater fish feces are rich in bacteria (Baguet and Marechal, 1976) which may cause the fecal matter to luminesce (Robison and Bailey, 1981). Luminescent fecal pellets have also been observed to be produced by the mesopelagic ostracod Macrocypridina castanea (Angel, unpublished observation). If luminescence of fecal material is a widespread phenomenon, then it would make the location of rapidly sinking pellets a much simpler problem for organisms, so that their re-cycling rate may prove higher than expected. Furthermore, detritivores living at bathypelagic depths may have special adaptations to eyes which would assist their detection of sinking pellets. One further apparently quaint exploitation of micro-organisms in fecal material is by the mysis larvae of the penaeid shrimp Solenocera atlantidis. The peritrophic membrane does not detach in these larvae and forms a trailing string five times the animal's body length (Youngbluth, 1982). Periodically the string is scraped by the larva and it is suggested that the fecal string is being used as a substrate for growing micro-organisms as a dietary supplement.

FECAL PELLET SINKING RATES

Table 2 summarises some of the available data on fecal pellet sinking rates. The majority of the data refers to either natural pellets or fecal 'material' whose source is unknown, or to easily cultured crustaceans; and just as the degradation rate appears to be faster in pellets produced by laboratory-cultured animals, so too the sinking-rates appear to be slower (Small, Fowler and Ünlü, 1979). Only recently has much attention been paid to the other groups (e.g. Bruland and Silver, 1981). In some of the early work (e.g. Smayda, 1969; 1971) the material was preserved prior to measurements being made. Variations in sinking size occur in relation to diet (Turner, 1977; Bienfang, 1980) and weather conditions; Small et al., (1979) for example, observed that the incorporation of resuspended sediment into fecal material will increase the density of the pellet and so increase its sinking velocity. Krause (1981) in attempting to explain the profiles he observed during the FLEX Experiment in the North Sea, which he belived were inconsistent with natural pellets (mostly Calanus finmarchicus) sinking at all, suggested that anaerobosis by gut bacteria in fecal pellets may generate gas in

Table 2. Some observed fecal pellet sinking rates

Source	Author	Rate m day^{-1}
Natural pellets of unknown origin (? some benthic)	Smayda, 1969	36-376
Sedimented pellets	Wiebe et al., 1976	50-941 (mean 159)
Snow (20 m depth)	Shanks and Trent, 1980	43-95
(1000-1600 m)	Silver and Alldredge, 1981	50-100
Copepods		
Acartia tonsa	Smayda, 1971	74-210*
Pontella meadi	Turner, 1977	15-153*
Acartia tonsa	Honjo and Roman, 1978	80-220*
Calanus finmarchicus	" " " "	180-220*
Temora turbinatus	Paffenhoffer and Knowles, 1979	5-10*
Eucalanus pileatus	" " " "	14-28*
Small copepods (Acartia type)	Small et al., 1979	20-120
Anomalocera patersoni	" " " "	40-150
Euphausiids		
Euphausia pacifica	Osterberg et al., 1963	43*
3 spp.	Fowler and Small, 1972	126-862
Pteropoda		
Corolla spectabilis	Bruland and Silver, 1981	440-1800

Doliolid		
Doliolletta gegenbaurii	Bruland and Silver, 1981	41-208
Salps		
Salpa fusiformis	Bruland and Silver, 1981	450-2700
Pegea socia	" " " "	450-2700
Pegea socia (aggregate)	Madin, 1982	588-1218
P. socia (solitary)	" "	1797-2238
Cyclosalpa pinnata (agg.)	" "	320-950
Salpa maxima (agg. small)	" "	588-1642
Salpa maxima (agg. large)	" "	1210-1987

* Pellets derived from laboratory-fed animals for which the sinking rates are likely to be under-estimates of naturally produced pellets.

sufficient quantities to make the pellets buoyant. Bruland and Silver (1981) show clearly that most published data are consistent with the hypothesis that sinking rates of fecal pellets increase with pellet size (Figure 6). Doliolid feces sink rather more slowly than might be expected possibly because of their more diffuse structure and irregular shapes.

MATHEMATICAL MODELS OF FECAL MATERIAL SINKING RATES

The sinking rate of fecal material is influenced by the density differences between the pellet and the medium and factors influencing the drag. Size and shape of fecal pellets varies between species, but they can also vary between individuals of the same species, or even according to the diet fed to an individual (Marshall and Orr, 1955). The flow of a fluid of density ρ (g cm^{-3}) and viscosity η (poise) past a sphere of diameter d (cm) sinking with a velocity V_d (cm sec^{-1}) is determined by the Reynolds number

$$R_e = \rho d\ V_d/\eta \qquad \text{Equ. 1.}$$

If R_e is small then the flow is laminar and the sinking velocity can be calculated from Stoke's equation,

$$V_d = \frac{g\Delta\rho\ d^2}{18\eta} \qquad \text{Equ. 2.}$$

where g is the acceleration due to gravity and $\Delta\rho$ the density contrast between the particle and seawater. This formula has been used by Paffenhofer and Knowles (1979) and Bishop et al. (1977) to calculate the sinking rate of fecal pellets. Paffenhofer and Knowles (1979) used a $\Delta\rho$ of 0.175 g cm^{-3}, while Bishop et al. (1977) used $\Delta\rho$ of 0.3g cm^{-3}, derived from the data of Smayda (1971) and Fowler andmall (1972).

Stoke's Law is only accurate for Reynold's numbers less than 0.5 Alldredge (1979) has used the Oseen modification of Stoke's equation which is more appropriate for Reynolds numbers in the range 0.5-5. For a sinking spherical particle,

$$\text{Drag force} = C_D\ \pi\rho\ d^2 V^2{}_d/8.$$

$$= \pi g d^3 \Delta\rho/6 \qquad \text{Equ. 3.}$$

The quantity C_D is given by

$$C_D = \frac{24}{R_e}\left(1 + \frac{3}{16} R_e\right), \qquad \text{Equ. 4.}$$

and so by substituting for R_e from equation (1) V_d can be estimated by solving the resulting quadratic equation.

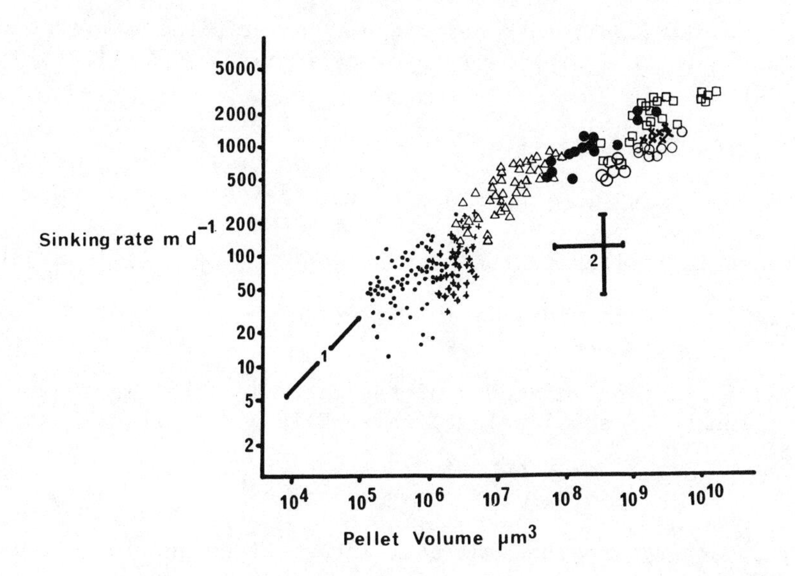

Fig. 6. The relationship between pellet size and sinking rates (modified by Bruland and Silver, 1981, from Small et al. 1979). Salp data from three cruises (○, □, x), the pteropod Corolla spectabilis (●), the doliolid Dolietta gegenbaurii (2), copepod nauplii (1), (Paffenhofer and Knowles, 1979) and small copepods (●), medium copepods (+) and euphausiids (△) (Small et al. 1979).

When the particle is not spherical in shape the situation is more complicated. Bishop et al. (1978) assumed that the fecal matter was approximately disk shaped for which the sinking velocity is given by

$$V_d = \frac{g\Delta\rho \ hd}{10.2 \ \eta} \qquad \text{Equ. 5}$$

where h is the disk thickness and d the diameter of the disk. In the Cape Basin they found an empirical relationship between h and d of

$$h = 0.052d + 0.0045$$

and also found the $\Delta\rho$ decreased exponentially with d.

Komar et al. (1981) analysed the sinking rates of natural copepod and euphausiid fecal pellets which are more cylindrical in shape. The data on copepod pellets (Small et al. 1979) gave Reynolds numbers of between 0.01-0.45. For euphausiid pellets (Fowler and Small, 1972) R_e was in the range 0.22-4.1, but for most pellets R_e was less than 0.5. Shanks and Trent (1980) quoted Reynolds numbers of 0.8-16.0 for snow particles.

Komar et al. (1981) have described a semi-empiracle modification of Stoke's Law for the sinking rate of cylindrical and ellipsoidal fecal pellets.

For cylindrical pellets

$$V_d = 0.790 \frac{1}{\eta} g \, \Delta\rho h^2 \left(\frac{h}{d}\right)^{-1.664} \qquad \text{Equ. 7.}$$

This equation is limited to the region $V_d \rho \eta^{-1} < 2$. For ellipsoidal pellets

$$V_d = \frac{1}{18} \frac{1}{\eta} g \Delta\rho d_n^{\,2} E^{0.380} \qquad \text{Equ. 8.}$$

where d_n is the nominal diameter of the particle (that is the diameter of a sphere of equal volume) while E is a measure of shape defined as

$$E = d_s \cdot \left(\frac{d_s^{\,2} + d_i^{\,2} + d_e^{\,2}}{3} \right)^{-\frac{1}{2}} \qquad \text{Equ. 9.}$$

where d_s, d_i and d_e are the smallest, intermediate and longest axial diameter of the ellipsoid.

The influence of diet in changing sinking rates acts by changing the pellet density (Bienfang, 1982). In enclosure experiments in Saanich Inlet, pellets produced by Calanus feeding predominantly on diatoms were 333 x 61 μm in size, sank at 123 m day^{-1} and had a $\Delta\rho$ of 0.156, whereas when the diet consisted mostly of flagellates the respective values were 398 x 65 μm, 87.9 m day^{-1} and 0.093.

The observations that Reynolds numbers ⩾ 0.5 are found for large euphausiid pellets (Komar et al., 1981) indicate that for larger fecal pellets Stokes Law is inapplicable. This accounts for the high velocities observed for salp and pteropod pellets (Bruland and Silver, 1981) and for fish fecal matter (Robison and Bailey, 1981). Without much better observational data on the sinking rates of these larger fecal particles >2 mm, it is doubtful if an adequate empirical relationship between their sinking rates and size can be deduced in order that their contribution to the total organic flux can be estimated.

MODELLING THE FECAL FLUX

Hofmann, Klinck and Paffenhoffer (1981) have made the only attempt to model the flux of fecal material. They worked in the shelf waters off the south-east coast of the United States where there is a simple plankton community dominated by two species of copepod. The model used laboratory and field parameters measured by Paffenhoffer and Knowles (1979) and included pellet sinking rates, concentrations and production rates, the growth, mortality and grazing rates of the plankton, and an estimate of microbial degradation rates. The plankton was partitioned into three size classes each with its own influence on pellet production, sinking rate and utilisation, and its own growth and mortality. Their model showed that although the nauplii of *Paracalanus* produce 50% of the pellet mass and the adults produce only 13%, the contribution of their pellets to the flux is 4% and 63% respectively. However, only 2% of the average daily production reaches the sea bed at 35 m, the rest being recycled which may account for the low benthic productivity of the shelf. The model estimated that ingestion of fecal pellets contributed <10% of zooplankton growth because of the low concentrations of pellets and their low nitrogen content (1 mm^3 of pellets contain only 3.4 μg N). They concluded that size of plankton significantly affects the flux of fecal material and that, 'the standing stock of pellets gives little information on the rates of pellet production, sinking and consumption, since the pellets present in the water column represent only a small fraction of the pellets produced daily.' If this last conclusion is indeed correct then Krause's (1981) data for the FLEX experiment needs re-evaluation.

In Figure 7 a selection of Krause's results are illustrated. In 1976 the phytoplankton bloom was expressed as a marked peak in the standing crop between 24 April and 6 May in the Fladden Ground in the North Sea. The fecal pellet profiles were measured using water bottle samples which generally showed that maximum numbers occurred above the thermocline in the surface 30 m of the water column; only during a short period in May (e.g. 11/12 see Figure 7) did the maximum abundance of pellets sink into the thermocline. Krause compared the pellet profiles with the profile of the copepodites (i-vi) of *Calanus finmarchicus* which dominated the plankton populations. The numbers of copepods fluctuated considerably (cf. the day and night profile on 23/24 May). Once released the pellets will be dispersed passively whereas the copepods may either modify their horizontal distributions behaviourally or have it modified by the differential shears in the water column acting during their vertical migrations. So the pellet profiles, at the water bottle scale of sampling are more conservative than the organisms and probably reflect the history of the water column rather than the physiological activity of the animals occupying that part of the water column at the time of sampling. Krause believed that his pellet profiles could not be interpreted in terms of the 'experimentally-determined high rates' of pellet sinking. He applied the sinking rates of adult *Calanus*, which are almost certainly high in relation to the pellets produced by copepodites. No attempt was made to compare phytoplankton profiles with pellet profiles to see if the maxima of these profiles coincided or not. Questions such as, is there selective feeding of phytoplankton as compared with pellets, are left unanswered. So it cannot be determined if the apparent changes in pellet production rates which occurred might have resulted from the copepods switching more to pellets as the algal concentrations fall. This would have resulted in an increase in the degree

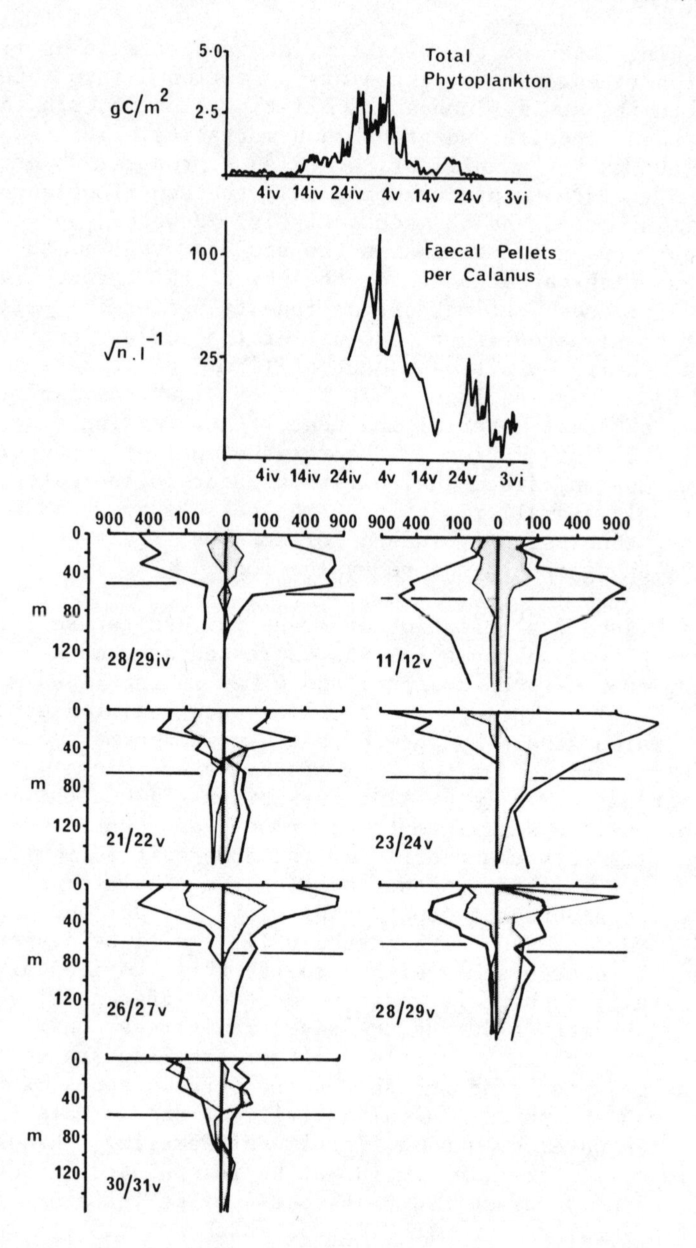

Fig. 7. Events which occurred during the Spring Bloom on the Fladen Ground during FLEX in 1976 (after Krause, 1981), showing th time sequence of the standing crop of phytoplankton (mg C m and the numbers (n) of fecal pellets per Calanus per litre, and seven sets of vertical profiles observed by night (left and by day (right) of fecal pellet concentrations ($\sqrt{n}\ \ell^{-1}$) and of the copepodites (i-vi) of Calanus finmarchicus ($\sqrt{n}\ \ell$ shaded profiles).

and the rate of recycling of detritus and a reduction in the standing cop of pellets that may not have correlated with a decrease in their production.

Another important inference from this model is related to the depth related changes in the sizes of planktonic communities. Angel (1979) showed that daytime near-surface planktonic ostracod communities off Bermuda contained small animals, and that the average size of the ostracods increased down to depths of around 700 m. A cursory glance at plankton samples rapidly confirms that this observation is true for plankton organisms in general. Consequently the amount of recycling of fecal material is likely to be greatest in the near surface layers where the producers are smaller and produce smaller pellets, and to decline in deeper water.

SUMMARY

In the epipelagic zone there is a high concentration of fine slow-sinking particles. In normal oceanic conditions the downward flux through the thermocline is about 30% of the primary production (Table 3). There is normally a considerable amount of recycling within the epipelagic, particularly where high water temperatures in conjunction with the slow sinking rates of the particulates favour a rapid rate of microbial degradation of organic detritus. However, at temperate latitudes (or higher) and in upwelling zones where large blooms of diatoms occur, the final depletion of the nutrients in the photic zone may be followed by a rapid sinking out of viable phytoplankton, providing a relatively vast pulse of rich organic material into the deep ocean, during which as much as half the bloom's primary production may sediment out.

Apart from the occurrence of this pulse, the normal downward flux through 400 m is around 5% of the primary production although this may be a substantial underestimate if diel migrants carry significant quantities down. The average diel vertical migrations of micronekton in mid-gyre oligotrophic waters range over 300-400 m, and ranges of 1000-1500 m are known for individual species. However, too little is known about feeding rates, feeding times and gut retention times of these migrants to estimate the level of flux they may be responsible for.

The lower depth for most planktonic diel migrations appears to be 700-1000 m in the N. Atlantic and 900-1500 m for micronekton. Below these depths the major flux of organic material will be by the particles, and these will be the main source of organic input for the deep bathypelagic communities. The disparity between the data from large volume filtration systems and sediment traps implies that the main flux is in the form of relatively large fecal pellets with fast

Table 3. Summary of published data on the flux rates of organic carbon (mg m^{-2} day^{-1})

Authors	Location	Depth (m)	Method of estimation	Flux	Estimated % of Primary Production
Menzel, 1974	Global	>200	^{14}C and O_2 utilisation	0.0066	
McCave, 1975	S. Pacific	300)		32.8	
		900)	Trap	16.4	
		Bottom)		6.4	
Wiebe *et al.*, 1976	Tongue of Ocean	2050	Trap	5.7	
Honjo, 1978	Sargasso (Parflux)	5367	Trap	1.4	~ 1
Bishop *et al.*, 1977	Tropical Atlantic	400	Pump	30.13	
Bishop *et al.*, 1978	S.E. Atlantic				
	Nearshore	400)	Pump	147.4	
	Middle	400)	"	23.9	
	Offshore	400)	"	2.9	
Hinga *et al.*, 1979	N.W. Atlantic				
	27°42'N. 72°54'W.	660)	Trap	144.0	
	33°30'N, 76°15'W.	1345)	"	29.8	
	38°23'N, 69°45'W	3520)	"	15.4	
Rower and Gardner, 1979	N.W. Atlantic	2156)	Trap	18.3	
	38°50'N. 72°31'W.	2159)	"	16.6	
	(June)	2162)	"	17.5	
	38°28'N. 72°01'W	2788)	"	12.6	
	(August)	2794)	"	10.1	
		2800)	"	9.5	
	38°28'N. 72°02'W	2316)	"	7.0	

		2715)	Trap	9.0	
		2803)	"	9.8	
Shanks and Trent, 1980	Monterey Bay	20	Snow collected by divers	326 (POC)	~ 30
		20	"	59 (PON)	
Iseki, 1981	Japan Sea	200	Trap	10.5 (Salp feces)	
Bishop *et al.*, 1980	Panama Basin	15)	Pump	169.0	~30
		40)		481.0	
		60)		224.0	
		100)		258.0	
		200)		30.7	
		350)		21.0	
		500)		43.3	
		725)		19.4	
		900)		29.5	
		1100)		15.1	
		1300)		0.40	
		1500)		1.09	
		2570	Trap	~40	~3
Honjo, 1980	Central Sargasso	976)	Trap	5.32	~6
	31°32.5'N. 55°55.4'W.	3694)	"	1.99	~1.5
	7-10/77, 110 days	5206)	"	?3.83	
	Tropical Atlantic	389)	"	13.7	
	13°30.2'N. 54°00.0'W.	988)	"	8.6	
	11/77 - 2/78 98 days	3755)	"	4.8	
		5068)	"	4.9	
	North Central Pacific	378)	"	6.8	
	5°21.1'N. 151°28'W.	978)	"	1.2	
	9 - 11/78, 61 days	2778)	"	2.4	
		4280)	"	1.8	
		5582)	"	1.5	

(continued)

Table 3. Continued.

Authors	Location	Depth (m)	Method of estimation	Flux	Estimated % of Primary Production
Deuser and Ross, 1980) Deuser et al., 1981)	Sargasso Sea	3200	Trap	1-3	≃ 1
Fellows et al., 1981	Monterey Bay	60	Suspended	0.128	
			Trap	137.5/355.0	
		210	Suspended	0.024	
			Trap	205	
		385	Suspended	0.135	
			Trap	27.5/345	
		1100	Suspended	0.005	
			Trap	100/110	
		1500	Suspended	0.002	
			Trap	77.5/82.5	
Sasaki and Nishizawa, 1981	Sea of Japan	100	Trap	418	28
		300	"	300	20
		500	"	28	1.9
Dunbar and Berger, 1981	Santa Barbara Channel	340	Trap	22	7-22
Knauer and Martin,	Monterey Bay	35	Trap	161	46
		65	"	78	22
		150	"	41	12
		500	"	12.8	3.7
		750	"	11.9	3.4
		1500	"	37	10.3

sinking rates. The slowness of the flux of fine particles has been confirmed by the early measurements of their ^{14}C content and $^{13}C : ^{12}C$ ratios in the Pacific. The low organic content of these fine particles combined with their low C : N ratios implies that what organic carbon is present is not readily utilisable by micro-organisms which are not attached to these fine particles in high numbers.

Sediment traps suggest that an organic flux of 4-6% of surface primary production occurs at about 1500 m, which decreases to 1-2% at about 4000 m (Table 3), mostly in the form of fecal pellets derived from plankton. However, it is suggested here that, because the sediment traps do not appear to intercept any large fecal material, and because the standing crop of micronekton is equivalent to 40% of the standing crop of plankton, these are underestimates. There seems likely to be an important fraction of the flux that is made up of large particles sinking at rates of 1000 m day^{-1} or more, that the traps used at present appear to fail to intercept. These particles are likely to be rich in organic carbon, to occur at very low concentrations and to have exceeding short residence times in the water column. They will tend to be produced deeper in the water column than are planktonic pellets and in cooler water. Consequently, unless they are heavily seeded with gut microflora, they will not be degraded by micro-organisms to the same extent as plankton fecal pellets.

The decrease in the standing crop of organisms down through the water column will result in the steady reduction in the chance that a sinking particle will be intercepted before it reaches the sea bed. Furthermore, the stability of the slopes of the linear regressions of the standing crop of biomass with depth, together with the close linking of increases in particulate flux with increases in surface production and with the rapidity with which seasonally produced terrigenous inputs such as pollen appear at bathypelagic depths, all serve to confirm that there is an unexpected rapid flux of large particulates down through the water column. The rapidity of this flux especially following the collapse of Spring blooms at temperate latitudes has important implications for how the benthic communities may be linked to seasonal variations in surface production, and for the way in which inorganic elements and compounds become distributed throughout the water column.

Acknowledgements

I would like to thank Dr. M.J.R. Fasham for inviting me to participate in such a stimulating workshop, and Dr. Scott Fowler for his most pertinent and constructive comments on the first draft of the paper.

REFERENCES

Alldredge, A.L., 1972, Abandoned larvacean houses: a unique food source in pelagic environment, Science, 177: 885.
Alldredge, A.L., 1976, Discarded appendicularian houses as sources of food, surface habitats, and particulate organic matter in plankton environments, Limnol. Oceanogr., 21: 14.
Alldredge, A.L., 1979, The chemical composition of macroscopic aggregates in two neritic seas, Limnol. Oceanogr., 24: 855.
Angel, M.V., 1972, Planktonic oceanic ostracods - historical, present and future, Proc. Roy. Soc. Edinburgh (B), 73: 213.
Angel, M.V., 1979, Studies on Atlantic halocyrpid ostracods : their vertical distributions and community structure in the central gyre region along 30°N from off Africa to Bermuda, Prog. Oceanogr 8: 3.
Angel, M.V.and Baker, A. de C., 1982, Vertical distribution of the standing crop of plankton and micronekton at three stations in the Northeast Atlantic, Biol. Oceanogr., 2: 1.
Angel, M.V., Hargreaves, P.M., Kirkpatrick, P. and Domanski, P., 1982, Variability in planktonic and micronektonic populations at 1000 m in the vicinity of 42°N, 17°W in the Northeast Atlantic Biol. Oceanogr. 1: 287.
Baguet, F. and Marechal, G., 1976, Bioluminescence of bathypelagic fish from the Strait of Messina, Comp. Biochem. Physiol., 53C: 75.
Barham, E.G., 1979, Giant larvacean houses : observations from deep submersibles, Science, 205: 1129.
Bienfang, P.K., 1980, Herbivore diet affects fecal pellet settling, Can. J. Fish. Aquat. Sci., 37: 1352.
Bienfang, P.K., 1982, Phytoplankton sinking-rate dynamics in enclosed experimental ecosystems, in: "Marine Mesocosms, Biological and Chemical Research in Experimental Ecosystems", G.D. Grice and M.R. Reeve, eds, Springer-Verlag, New York, Heidelberg and Berlin.
Bishop, J.K.B., and Edmond, J.M., 1976, A new large volume filtratior system for the sampling of oceanic particulate matter, J. Mar. Res., 34: 181.
Bishop, J.K.B., Edmond, J.M., Ketten, D.R., Bacon M.P., and Silker, W.B., 1977, The chemistry, biology and vertical flux of particulate matter from the upper 400 m of the equatorial Atlantic Ocean, Deep-Sea Res., 24: 511.
Bishop, J.K.B., Ketten, D.R., and Edmond, J.M., 1978, The chemistry, biology and vertical flux of particulate matter in the upper 400 of the Cape Basin in the Southeast Atlantic Ocean, Deep-Sea Res., 25: 1121.
Bishop, J.K.B., Collier, R.W., Ketten, D.R., and Edmond, J.M., 1980, The chemistry, biology and vertical flux of particulate matter from the upper 1500 m of the Panama Basin, Deep-Sea Res., 27: 615
Brewer, P.G., Nozaki, Y., Spencer, D.W., and Flear, A.P., 1980, Sediment trap experiments in the deep North Atlantic: isotopic

and elemental fluxes, J. Mar. Res., 38: 703.

Bruland, K.W., and Silver, M.W., 1981, Sinking rates of fecal pellets from gelatinous zooplankton (Salps, Pteropods, Doliolids), Mar. Biol., 63: 295.

Cherry, R.D., and Heyraud, M., 1981, Polonium-210 content of marine shrimp: variation with biological and environmental factors, Mar. Biol., 65: 165.

Cherry, R.D., Higgo, J.J.W., and Fowler, S.W., 1978, Zooplankton fecal pellets and elemental residence times in the ocean, Nature Lond., 274: 246.

Childress, J.J. and Price, M.H., 1978, Growth rate of the bathypelagic crustacean Gnathophausia ingens (Mysidacea: Lophogastridae) 1. Dimenstional growth and population structure, Mar. Biol., 50: 47

Childress, J.J., Taylor, S.M., Caillet, G.M., and Price, M.H., 1980, Patterns of growth, energy utilization and reproduction in some meso- and bathypelagic fishes of Southern California, Mar. Biol., 61: 27.

Collins, N.R., and Williams, R., 1981, Zooplankton of the Bristol Channel and Severn Estuary. The distribution of four copepods in relation to salinity, Mar. Biol., 64: 273.

Conover, R.J., 1966, Assimilation of organic matter by zooplankton, Limnol. Oceanogr., 11: 338.

Corner, E.D.S., Head, R.N., Kilvington, C.G., 1972, On nutrition and metabolism of zooplankton. VIII The grazing of Biddulphia cells by Calanus finmarchicus J. Mar. Biol. Ass. U.K., 52: 847.

Costin, J.M., 1970, Visual observations of suspended-particle distributions at three sites in the Caribbean Sea. J. Geophys. Res., 75: 4144.

Dagg, M.J., and Gill, D.W., 1980, Natural feeding rates of Centropages typicus females in the New York Bight, Limnol. Oceanogr., 25: 597.

Dagg, M.J., Vidal, J., Whitledge, T.E., Iverson, R.L., and Goering, J.J., 1982, The feeding, respiration and excretion of zooplankton in the Bering Sea during a spring bloom, Deep-Sea Res., 29: 45.

Davies, J.M., Gamble, J.C., and Steele, J.H., 1975, Preliminary studies with a large plastic enclosure, in: "Estuarine Research" Volume 1, L.E. Cronin, ed., Academic Press, New York and London.

Deuser, W.G., and Ross, E.H., 1980, Seasonal change in the flux of organic carbon to the deep Sargasso Sea, Nature Lond., 283: 364.

Deuser, W.G., Ross, E.H., and Anderson, R.F., 1981, Seasonality in the supply of sediment to the deep Sargasso Sea and implications for the rapid transfer of matter to the deep ocean, Deep-Sea Res., 28: 495.

Dunbar, R.B., and Berger, W.H., 1981, Fecal pellet flux to modern bottom sediment of Santa Barbara Basin (California) based on sedimentary trapping, Geol. Soc. Amer. Bull., Pt. 1., 92: 212.

Eppley, R.W., and Peterson, B.J., 1979, Particulate organic matter flux and planktonic new production in the deep ocean, Nature Lond., 282: 677.

Feigenbaum, D.L., 1979, Daily ration and specific ration of the Chaetognatha Sagitta elegans, Mar. Biol., 54: 57.
Fellows, D.A., Karl, D.M., and Knauer, G.A., 1981, Large particle fluxes and the vertical transport of living carbon in the upper 1500 m of the northeast Pacific Ocean, Deep-Sea Res., 28: 921.
Fowler, S.W., Small, L.F., and Kĕckĕs, S., 1971, Effects of temperature and size on moulting of euphausiid crustaceans, Mar. Biol., 11: 45.
Fowler, S.W., and Small, L.F., 1972, Sinking rates of euphausiid fecal pellets, Limnol. Oceanogr., 17: 293.
Fowler, S.W., Small, L.F., Elder, D.L., Ünlü Y., and La Rosa J., 1979, The role of zooplankton fecal pellets in transporting PCBs from the upper mixed layer to the benthos, IVes Journées Étud. Pollutions: 289.
Gaudy, R., 1974, Feeding four species of pelagic copepods under experimental conditions, Mar. Biol., 25: 125.
Gilmer, R.G., 1972, Free floating mucus webs : A novel feeding adaptation for the open ocean, Science, 176: 1238.
Gordon, D.C., 1970, Some studies on the distribution and composition of particulate organic carbon in the North Atalntic Ocean, Deep-Sea Res., 17: 233.
Gordon, D.C., 1971, Distribution of particulate organic carbon and nitrogen at an oceanic station in the central Pacific, Deep-Sea Res., 18: 1127.
Gordon, D.C., Wangersky, P.J., and Sheldon, R.W., 1979, Detailed observations on the distribution and composition of particulate organic material at two stations in the Sargasso Sea, Deep-Sea Res., 26: 1083.
Hamner, W.M., Madin, L.P., Alldredge, A.L., Gilmer, R.W., and Hamner, P.P., 1975, Underwater observations of gelatinous zooplankton: sampling problems, feeding biology and behaviour, Limnol. Oceangr., 20: 907.
Harding, G.C.H., 1974, The food of deep-sea copepods, J. Mar. Biol. Ass. U.K., 54: 141.
Haury, L.R., McGowan, J.A., and Wiebe, P.H., 1978, Patterns and processes in the time-space scales of plankton distributions, in: "Spatial Pattern in Plankton Communities", J.H. Steele, eds., Plenum Press, New York.
Hayward, T.L., 1980, Temporal and spatial feeding pattern of copepods from the North Pacific Central Gyre, Mar. Biol., 58: 295.
Heyraud, M., 1979, Food ingestion and digestive transit times in the euphausiid Meganyctiphanes norvegica as a function of animal size, J. Plankt. Res., 1: 301.
Hinga, K.R., Sieburth, J. McN. and Heath, G.R., 1979, The supply and use of organic material at the deep-sea floor, J. Mar. Res., 37: 557.
Hoffman, E.E., Klinck, J.M., and Paffenhoffer, G-A., 1981, Concentrations and vertical fluxes of zooplankton fecal pellets on a continental shelf, Mar. Biol., 61: 327.

Honjo, S., 1978, Sedimentation of materials in the Sargasso Sea at at 5,367 m deep station, J. Mar. Res., 36: 469.
Honjo, S., 1980, Material fluxes and modes of sedimentation in the mesopelagic and bathypelagic zones, J. Mar. Res., 38: 53.
Honjo, S., Manganini, S.J. and Cole, J.J., 1982, Sedimentation of biogenic matter in the deep ocean, Deep-Sea Res., 29: 609.
Honjo, S. and Roman, M.R., 1978, Marine copepod fecal pellets: production, preservation and sedimentation, J. Mar. Res., 36: 45.
Iseki, K., 1981, Particulate organic matter transport to the deep-sea by salp fecal pellets, Mar. Ecol. Prog. Ser., 5: 55.
Krause, M., 1981, Vertical distribution of fecal pellets during FLEX '76, Helgolanders Meersunter., 34: 313.
Knauer, G.A. and Martin, J.H., 1981, Primary production and carbon-nitrogen fluxes in the upper 1500 m of the northeast Pacific, Limnol. Oceanogr., 26: 181.
Knauer, G.A., Martin, J.H. and Bruland, K.W., 1979, Fluxes of particulate carbon, nitrogen and phosphorus in the upper water column of the northeast Pacific, Deep-Sea Res., 26: 97.
Koefoed, L.H., 1975, The feeding biology of Hydrobia ventricosa (Montagu). 1. The assimilation of different components of the food, J. Exp. Mar. Biol. Ecol., 19: 233.
Komar, P.D., Morse, A.P., Small, L.F., and Fowler, S.W., 1981, An analysis of sinking rates of natural copepod and euphausiid fecal pellets, Limnol. Oceanogr., 26: 172.
La Rosa, J., 1976, A simple system for recovering zooplankton faecal pellets in quantity, Deep-Sea Res., 23: 995.
Lasker, R., 1976, Feeding, growth, respiration and carbon utilization of a euphausiid shrimp, J. Fish. Res. Bd Can., 23: 1291.
Longhurst, A.R., 1976, Vertical Migration, in: "The Ecology of the Sea", D.H. Cushing and J.J. Walsh, eds., Blackwell Scientific Publications, Oxford.
McCave, I.N., 1975, Vertical flux of particles in the ocean, Deep-Sea, Res., 22: 491.
Madin, L.P., 1982, Production, composition and sedimentation of salp fecal pellets in oceanic waters, Mar. Biol., 66: 39.
Marshall, S.M., and Orr, A.P., 1955, On the biology of Calanus finmarchicus, VII Food uptake, assimilation and excretion in adult and stage V Calanus, J. Mar. Biol. Ass. U.K., 34: 495.
Martens, P., 1978, Fecal pellets, Fich. Ident. Zooplankt., 162: 1.
Mauchline, J., 1972, The biology of bathypelagic organisms especially Crustacea, Deep-Sea Res., 19: 753.
Mauchline, J., 1980, The biology of mysids and euphausiids, Advances Mar. Biol., 18: 1.
Menzel, D.W., 1974, Primary productivity, dissolved and particulate organic matter and the sites of oxidation of organic matters, in: "The Sea", 5, E.D. Goldberg, ed., John Wiley, New York.
Muller, P.J., and Suess, E., 1979, Productivity, sedimentation rate and sedimentary organic matter in the oceans - 1. Organic carbon preservation, Deep-Sea Res., 26: 1347.

Nemoto, T., 1972, Chlorophyll pigments in the stomach and gut of some macrozooplankton species, in: "Biological Oceanography of the Northern North Pacific Ocean", A.Y. Takenouti et al., eds., Idemitsu Shoten, Tokyo.

Nemoto, T., and Saijo, Y., 1968, Traces of chlorophyll pigments in stomachs of deep sea zooplankton, J. Oceanogr. Soc. Japan, 24: 310.

Osterberg, C., Carey, A.G. and Curl, H., 1963, Acceleration of sinking rates of radionuclides in the ocean, Nature Lond., 200: 1276.

Paffenhoffer, G.-A., 1971, Grazing and ingestion rates of nauplii, copepodids and adults of the marine planktonic copepod Calanus helgolandicus, Mar. Biol., 11: 286.

Paffenhoffer, G.-A., and Knowles, S.C., 1979, Ecological implications of fecal pellet size, production and consumption by copepods, J. Mar. Res., 37: 35.

Paffenhoffer, G.-A., and Strickland, J.D.H., 1970, A note on feeding of Calanus helgolandicus on detritus, Mar. Biol., 5: 97.

Pearre, S., 1979, Problems of detection and interpretation of vertical migration, J. Plankt. Res., 1: 29.

Peterson, B.J., 1981, Perspectives on the importance of the oceanic particulate flux in the global carbon cycles, Ocean Sci. Eng., 6: 71.

Petipa, T.S., 1964, Diurnal rhythm of the consumption and accumulatic of fat in Calanus helgolandicus (Claus) in the Black Sea, Dokl. Acad. Sci. USSR, 156: 361.

Petipa, T.S., Pavlova, E.V., and Mironov, G.N., 1970, The food web structure, utilization and transport of energy by trophic levels in plankton communities, in: "Marine Food Chains", J.H. Steele, ed., Oliver and Boyd, Edinburgh.

Pomeroy, L.R. and Deibel, D., 1980, Aggregation of organic matter by pelagic tunicates, Limnol. Oceanogr., 25: 643.

Poulet, S.A., 1976, Feeding of Pseudocalanus minutus on living and non-living particles, Mar. Biol., 34: 117.

Poulet, S.A., and Marsot, P., 1978, Chemosensory grazing by marine calanoid copepods, Science N.Y., 200: 1403.

Pugh, P.R., 1978, The application of particle counting to an understanding of the small-scale distribution of plankton, in: "Spatial Pattern in Plankton Communities", J.H. Steele, ed., Plenum Press, London and New York.

Reeve, M.R., 1980, Comparative and experimental studies on the feeding of chaetognaths and ctenophores, J. Plankt. Res., 2: 381.

Reeve, M.R., and Walter, M.A., 1977, Observations on the existence of lower threshold and upper critical food concentrations for the copepod Acartia tonsa Dana, J. Exp. Mar. Biol. Ecol., 29: 211.

Rice, A.L., 1978, Radioactive waste disposal and deep-sea biology, Oceanologica Acta, 1: 483.

Robison, B.H., and Bailey, T.G., 1981, Sinking rates and dissolution of midwater fish fecal matter, Mar. Biol., 65: 135.

Roman, M.R. and Rublee, P.A., 1981, A method to determine in situ zooplankton grazing rates on natural particle assemblages, Mar. Biol., 65: 303.

Rowe, G.T. and Gardner, W.D., 1979, Sedimentation rates in the slope water of the northwest Atlantic Ocean measured directly with sediment traps, J. Mar. Res., 37: 581.

Sameoto, D.D., 1976, Respiration rates, energy budgets and molting frequencies of three species of euphausiids found in the Gulf of St. Lawrence, J. Fish. Res. Bd Can., 33: 2568.

Sasaki, H. and Nishizawa, S., 1981, Vertical flux profiles of particulate material in the sea off Sanriku, Mar. Ecol. Prog. Ser., 6: 191.

Schrader, H. -J., 1971, Fecal pellets: Role on sedimentation of pelagic diatoms, Science, N.Y., 174: 55.

Shanks, A.L. and Trent, J.D., 1979, Marine Snow: Microscale nutrient patches, Limnol. Oceanogr., 24: 850.

Shanks, A.L. and Trent, J.D., 1980, Marine Snow: sinking rates and potential role in vertical flux, Deep-Sea Res., 27: 137.

Silver, M.W. and Alldredge, A.L., 1981, Bathypelagic marine snow: deep-sea algal and detrital community, J. Mar. Res., 39: 501.

Silver, M.W. and Bruland K.W., 1981, Differential feeding and fecal pellet composition of salps and pteropods, and the possible origin of the deep-water flora and olive-green 'cells', Mar. Biol., 62: 263.

Silver, M.W., Shanks, A.L. and Trent, J.D., 1978, Marine Snow: Microplankton habitat and source of small-scale patchiness in pelagic populations, Science, N.Y., 201: 371.

Small, L.F. and Fowler, S.W., 1973, Turnover and vertical transport of the zinc by the euphausiid Meganyctiphanes norvegica in the Ligurian Sea, Mar. Biol., 18: 284.

Small, L.F., Fowler, S.W. and Kěckěs, S., 1973, Flux of zinc through a macroplanktonic crustacean, in: "Radioactive contamination of the marine environment" IEAE, Vienna.

Small, L.F., Fowler, S.W. and Ünlü, Y., 1979, Sinking rates of natural copepod fecal pellets, Mar. Biol., 51: 233.

Smayda, T.J., 1969, Some measurements on the sinking rate of fecal pellets, Limnol. Oceanogr., 14: 621.

Smayda, T.J., 1971, Normal and accelerated sinking of phytoplankton in the sea, Mar. Geol. 11: 105.

Spencer, D.W., Brewer, P.G., Fleer, A., Honjo, S., Krishnaswani, S., and Nozaki, Y., 1978, Chemical fluxes from a sediment trap experiment in the deep Sargasso Sea, J. Mar. Res., 36: 493.

Stavn, R.H., 1971, The horizontal-vertical distribution hypothesis; Langmuir circulations and Daphnia distributions, Limnol. Oceanogr., 16: 453.

Stockton, W.L. and DeLaca, T.E., 1982, Food falls in the deep sea: occurrence, quality and significance, Deep-Sea Res., 29: 157.

Suzuki, N. and Kato, K., 1953, Studies on suspended marine snow in the sea Pt. 1. Sources of marine snow, Bull. Fac. Fish. Hokkaido Univ., 4: 132.

Tabor, R.S., Ohwada, K. and Caldwell, R.R., 1981, Filterable marine bacteria found in the deep sea: Distribution, taxonomy and response to starvation, Microb. Ecol., 7: 67.

Taguchi, S., 1982, Seasonal study of fecal pellets and discarded houses of Appendicularia in a subtropical Inlet, Kaneohe Bay Hawaii, Est. Coast. Shelf Sci., 14: 533.

Turner, J.T., 1977, Sinking rates of fecal pellets from the marine copepod Pontella meadii, Mar. Biol., 40: 249.

Turner, J.T. and Ferrante, J.G., 1979, Zooplankton fecal pellets in aquatic ecosystems, Bioscience, 29: 670.

Urrère, M.A. and Knauer, G.A., 1981, Zooplankton fecal pellet fluxes and vertical transport of particulate organic matter in the pelagic environment, J. Plankt. Res., 3: 359.

Vinogradov, M.E., 1968, "Vertical Distribution of the Oceanic Plankton", Nauka, Moscow (Translation: Israel Program for Scientific Translation, Jerusalem, 1970).

Wakeham, S.G., Farrington, J.W., Gagosian, R.B., Lee, C., DeBaer, H., Nigrelli, G.E., Tripp, B.W., Smith, S.O., and Frew, N.W., 1980, Organic matter fluxes from sediment traps in the equatorial Atlantic Ocean, Nature, Lond., 286: 798.

Walsh, J.J., Premuzic, E.T. and Whitledge, T.E., 1981, Fate of nutrient enrichment on continental shelves as indicated by the C/N content of the bottom sediment, in: "Ecohydrodynamics". J.C.J.Nihoul, ed., Elsevier, Amsterdam.

Walsh, J.J., 1983, Death in the sea: enigmatic loss processes, Prog. Oceanogr., 12: 1.

Wangersky, P.J., 1978, The distribution of particulate organic carbon in the oceans: ecological implications, Int. Rev. Hydrobiol., 63: 567.

Wiebe, P.H., Boyd, S.H. and Winget, C., 1976, Particulate matter sinking to the deep-sea floor at 2000 m in the Tongue of the Ocean, Bahamas, with a description of a new sedimentation trap, J. Mar. Res., 34: 341.

Wiebe, P.H., Madin, L.P., Haury, L.R., Harbison, G.R. and Philbin, L.M., 1978, Diel vertical migration by Salpa aspera: potential for large-scale particulate organic matter transport to the deep-sea, Mar. Biol., 53: 249.

Williams, P.M. and Gordon, L.I., 1970, Carbon-13: Carbon-12 ratios in dissolved and particulate organic matter in the sea, Deep-Sea Res., 17: 19.

Wishner, K.F., 1980, The biomass of the deep-sea benthopelagic plankton, Deep-Sea Res., 27: 203.

Youngbluth, M.J., 1982, Utilization of a fecal mass as formed by the pelagic mysis larva of the penaeid shrimp Solenocera atlantidis, Mar. Biol., 66: 47.

THE SUPPLY OF FOOD TO THE BENTHOS

Victor Smetacek

Institut für Meereskunde, Düsternbrooker Weg 20
2300 KIEL 1, Fed. Rep. of Germany

INTRODUCTION

The food supply of the bulk of the marine benthos is provided by the downward transport of organic matter produced by phytoplankton in the surface layer. The importance for the benthos of this vertical transfer of matter and energy from the pelagic to the benthic system has been recognised for a long time, although only recently has effort been devoted towards measuring this vertical flux. In the past, there has been a widespread tendency amongst aquatic ecologists to regard sedimentation out of the pelagic community as an elimination of waste products and hence of little direct consequence to the functioning of the pelagic system. However, results obtained during the last decade from direct monitoring of the export of particulate matter out of the pelagic system with sediment traps now necessitate a revisal of earlier concepts. At present, attention is still centered primarily on measuring rates of sedimentation and examining the nature of this material rather than on elucidating the mechanisms governing the export of biogenous material from the surface layer. Thus, the ecological processes regulating magnitude and quality of the benthic food supply are still not well documented, and most of the available information is of an indirect nature.

The lack of a satisfactory understanding of these processes has been accentuated by the failure of compartmental flow models such as those of Steele (1974) and Mills (1980) in budgeting energy flow through marine ecosystems (Mann, 1981). The main problem encountered by most of these budgetary models has been an apparent food shortage; this has led to a critical reappraisal of the data on primary production (Gieskes, this volume) as well as the widely accepted 10%

transfer efficiencies between trophic levels suggested by Slobodkin (1961) (Mills et al., in press). Although it is well known that these models are merely hypothetical because only a few of the flux rates have actually been measured, they nevertheless reflect explicitly not only the state of our knowledge but also reveal to a great extent the intuitive thinking of the ecological community as a whole.

To my mind, this problem of a food shortage can only be satisfactorily resolved by approaching it both on the conceptual or holistic level as well as by acquiring more detailed and dependable data on the complexities of interaction within and between the various subsystems. On the holistic level, more regard should be paid to the physical boundaries between systems and the magnitude and temporal scales of imports and exports passing through these boundaries. Thus, a pycnocline is a boundary for dissolved substances but not necessarily for sinking particles or swimming organisms. Based on this simple fact, Margalef (1978) developed a conceptual model of the major environmental factors determining the nutrient load of the surface layer and hence the structure of the phytoplankton communities as well as their temporal evolution.

Margalef's (1978) straightforward conceptual model also provides a useful framework for a better understanding of the sedimentation process in the sea because it relates qualitative with quantitative aspects of pelagic systems and this accounts for the patterns of development that are characteristic of the annual cycles of various regions. In a given geographical region, temporal succession in the various components of the pelagic system will be reflected in the rate of sedimentation and the composition of the sinking particles. Marked seasonal variation in the quantity and quality of the vertical particle flux has been found in all localities where the annual cycle of sedimentation has been studied so far. The majority of these studies has been carried out in coastal environments (Zeitzschel, 1965; Steele and Baird, 1972; Webster et al., 1975; Hargrave and Taguchi, 1978; Smetacek 1980a; Hargrave, 1980; Kuparinen et al., in press; Peinert et al., 1982), but Honjo (in press) also obtained a distinct annual cycle in sedimentation monitored over monthly intervals in the Panama Basin. He found an identical pattern in material collected by traps suspended at 890 m, 2590 m and 3560 m depths that reflected the annual production cycle of the plankton in the surface layer. These data and those obtained by many other investigators from deep sea environments (see Angel, this volume, for a review) indicate that sinking particles, once they have left the surface layer, reach the sea bed comparatively rapidly, thus refuting the older view of a slow and steady drizzle of small particles, long supposed to be the major source of sustenance of the deep sea benthos.

In this article, I shall attempt to incorporate the sedimentation process in the conceptual framework presented by Margalef (1978). Because of the restrictions of space, I shall not deal with the food supply of benthos living within the euphotic zone as it can be independent of sedimentation. Further, other sources of energy such as terrestrial input or hydrothermal vents will not be dealt with, as also the important but quantitatively little known transport of macroalgae and seagrasses from the coast to the deep. The following presentation is both an attempt at sorting the available, often conflicting data, but also at providing an *a priori* model for the planning of future investigations of this process. As such, many of the conclusions reached below are of a speculative nature and are intended to stimulate and provoke rather than to convince and placate.

THE CONCEPTUAL FRAMEWORK

Depending on the source of the nutrient supply, it is possible to differentiate pelagic communities into "bloom" and "steady state" types (Yentsch *et al*., 1977). In the former type, primary production is based on 'new' nutrients, introduced into the surface layer by physical transport processes, whereas in the latter, nutrients are provided by the remineralisation of heterotrophs within the surface layer, i.e. production is based on 'regenerated' nutrients. Thus, 'new production' always leads to an increase in the total biomass of the surface layer, whereas 'regenerated production' tends to maintain a constant, or a decreasing level of biomass. As Margalef (1978) has pointed out, the bloom type community dominated by diatoms matures into a regenerating community dominated by flagellates within the same water body, this progression being driven by the loss of essential elements bound in particles sinking through the pycnocline, i.e. maturation of the pelagic system is accompanied by diminishing biomass. Apart from some spatially restricted or minor mechanisms (terrestrial run-off, pore water input, atmospheric discharge etc.), renewal of the lost resources and hence an increase in biomass can only occur following upward transport of water from below the mixed layer. It is apparent that the total amount of organic matter that can be exported from the surface layer and eventually to the benthos living below it is determined by the amount of new production in the system and not by the total annual primary production which also includes regenerated production.

The concept of new and regenerated nutrients was introduced by Dugdale and Goering (1967) in a study of an upwelling system where it was possible to differentiate the two in terms of nitrate and ammonia respectively. Although this is a useful working definition for offshore areas, it is not applicable to coastal regions where new nutrients can well include ammonia. Thus, in regions with an oxygen deficit below the pycnocline, remineralised nitrogen will accumulate as ammonia; when this subpycnocline water is mixed upwards, new nutrients will be introduced to the surface layer but in the form of ammonia rather than nitrate. It is obvious that new and regenerated nutrients can only be defined in relation to one another and based on spatial and temporal scales. Thus, a precise and general definition cannot be given, particularly for localities such as estuaries that are characterised by large fluctuations in comparatively short spatial and temporal scales. Kemp et al. (1982) have addressed this problem and provided a conceptual framework for the better understanding of estuarine dynamics. They differentiate five interfaces or spatial boundaries as being of importance in such environments. In this presentation however, I shall confine myself to a consideration of only two boundaries, viz. the sediment/water interface and the pycnocline, i.e. I shall deal only with coastal and open sea areas and neglect the far more complex estuaries.

In the former environments, regenerated nutrients include not only nutrients remineralised by pelagic heterotrophs within the surface layer but also those released by benthic organisms, particularly suspension-feeding macrobenthos such as mussels, living within the surface layer and hence in direct interaction with the phytoplankton. The importance of these benthic organisms in fuelling phytoplankton production has been demonstrated by Kautsky and Wallentinus (1980).

New nutrients, on the other hand, are taken to include all those nutrients introduced to the surface layer by physical transport processes. Most of the 'new' nutrients in marine environments are those added to subsurface water, either by benthic or pelagic remineralisation or by mobilisation from the sediment. In this connection, the surface layer is meant to include the uppermost mixed layer that is normally, but not necessarily, bounded below by the seasonal pycnocline.

In the tropics and sub-tropics, new production only occurs under upwelling conditions; in temperate and some arctic regions, nutrient renewal during the autumnal turnover and winter mixing is characteristic. Thus, diatom blooms based on new production and culminating in large biomass stocks are typical for upwelling regions and for the spring and sometimes the autumn in higher latitudes. Flagellate communities based on regenerated nutrients and with low biomass are typical for stratified water, i.e. throughout most of the tropics and sub-tropics and in the summer of higher latitudes (Parsons, 1979). The regions of high primary production and large benthic biomass below the euphotic zone are also those in which new production occurs at some time of the year. In a given system, e.g. a temperate shelf with stable summer stratification, the bulk of the annual primary production figure might well be provided by regenerated summer production, not because of high production rates in this season but simply because of its long duration relative to the highly productive but short bloom phases. Invoking higher phytoplankton production rates will thus not help in balancing pelagic/benthic budgets if the production is merely due to a higher turnover rate within a regenerating system and not to import of new nutrients to the system. The latter is effected by vertical transport of water and it is thus the hydrography of a region, together with the nutrient load of the water, that determines the food supply of the benthos.

PLANKTON GROWTH AND SEDIMENTATION IN KIEL BIGHT

In this section, a brief account of the interdependance between the annual cycle of the pelagic system and that of sedimentation is given, observed over several years in Kiel Bight - a coastal shallow-water system with an average depth of 17 m. Kiel Bight is used as an example, because this aspect in particular has been studied here in greater detail than elsewhere (Smetacek, 1980a,b; Peinert _et al_., 1982; Smetacek _et al_., in press). In following sections, I have attempted to generalise from the Kiel Bight observations, undoubtedly a risky undertaking considering the many special circumstances of the Kiel Bight system. However, as the generalisations are based more on the conceptual framework presented above than on the observed sequence of events in Kiel Bight, I feel that they are justified at this stage of our knowledge.

In Kiel Bight, the plankton growth season is initiated by the advent of calm weather in March/April which results in development of the spring phytoplankton bloom. Diatoms dominate the population and, as growth is rapid, nutrients accumulated in the water column over the winter are depleted within 2 weeks after bloom initiation. Organic carbon production by this bloom is in the order of 20 gCm^{-2}, estimated from NO_3^-, NH_4^+ and PO_4^{3-} uptake (using the Redfield ratios) as well as from direct ^{14}C-measurements. The response of zooplankton to this rapid increase in the food supply is slow and even the herbivorous protozooplankton that dominate zooplankton biomass in this period, manage to graze only a small part of the population. Following nutrient depletion, sedimentation of the bloom occurs and more than half of the organic matter produced by the bloom is deposited, largely as resting spores and living vegetative cells that form a loose green carpet on the sea bed. The process of sedimentation lasts for about a week with a 3 - 4 day period where sedimentation rates are in the order of 1 - 2 $gCm^{-2}d^{-1}$. In all, approximately 10 gCm^{-2} reach the benthos during this week (Figure 1). This flux has been measured directly over several years with sediment traps and calculated indirectly from water column measurements (Peinert _et al_., 1982), both results being in reasonable agreement.

As a large portion of the sedimented matter is in the form of living phytoplankton cells, and hence of high nutritional quality, benthic response to this input is rapid. This was demonstrated by

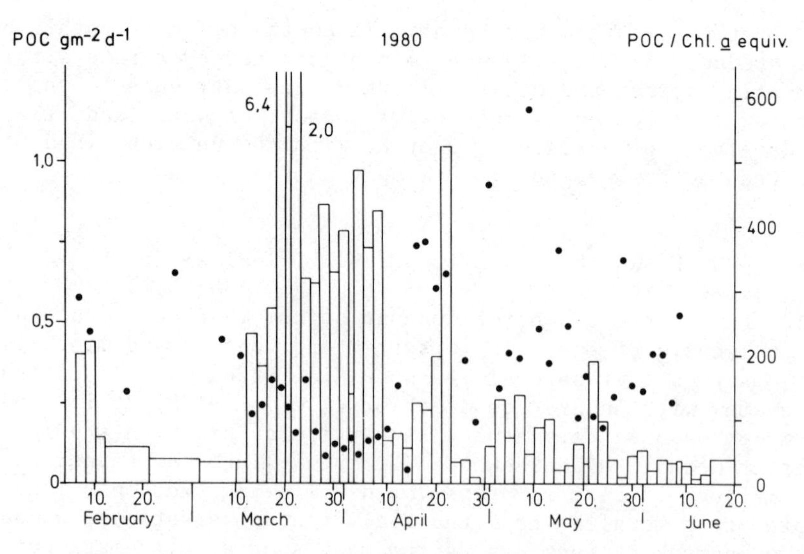

Fig. 1. Sedimentation rates of particulate organic carbon (POC) monitored at 2-day intervals with a multisample sediment trap, deployed 2 m above bottom in a 20 m water column of the Hausgarten, Kiel Bight during 1980. The dots are ratios POC/chlorophyll a equivalent (= chlorophyll a + pheopigments) of the collected material. The bulk of POC collected till 23 March was resuspended sediment, the 6,4 $gCm^{-2}d^{-1}$ collected on 18-20 March is the highest value ever recorded in Kiel Bight. Bloom sedimentation occurred in the period 20 March - 8 April. (Redrawn from PEINERT et al., 1982).

Graf et al. (1982) who monitored benthic activity during late winter and spring concomitantly with water column and sediment trap measurements. Their results, in comparison to food input rates measured with sediment traps, are presented in Figures 1 and 2. The increase in benthic activity immediately following bloom input was clearly apparent from all relevant parameters measured. Direct observation by divers indicated that the carpet of phytoplankton was no longer visible after a week and heat production data indicate that organic matter equivalent to input by the bloom was "burned" within 2 - 3 weeks. However, as activity did not decline significantly after this event although sedimentation rates did (0.2-0.3 $gCm^{-2}d^{-1}$), the benthos was evidently using organic material already present in the sediment. This "spring rain" of phytoplankton apparently has a triggering effect on the benthos as it leads to an increase in biomass of both microbenthos and, somewhat later, meiobenthos and to an elevation in the level of activity of the benthic system.

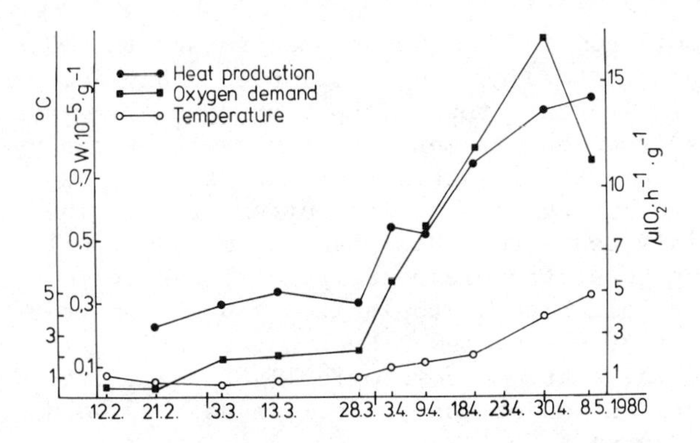

Fig. 2. Benthic activity parameters and in situ temperatures measured at the sediment surface below the trap. Heat production (measured with a Pamatmat calorimeter) and potential oxygen demand (based on ETS measurements from the top 1 cm sediment layer) increase significantly following bloom input. (From GRAF et al., 1982).

The influence of this bloom input on 2 species of macrobenthos with differing feeding modes was also interesting as it reflects the different strategies followed by these organisms. *Macoma balthica* - a filtering clam - increased its carbohydrate and lipid content significantly whereas the selective particle feeder *Nephthys ciliata* showed no such direct response (Graf *et al.*, 1982). Ankar (1980) also observed maximum growth of *Macoma* to occur in the spring in the Baltic proper.

During late spring and early summer, primary production levels fluctuate, with peaks as high as 0.5 $gCm^{-2}d^{-1}$ (v.Bodungen, 1975; Peinert *et al.*, 1982)although phytoplankton biomass levels tend to be low. As this is the period of the copepod maximum, much of the primary production is based on regenerated nutrients, although a significant nutrient contribution also comes from the sediments in the period before establishment of the seasonal thermocline. Sedimentation rates during this period are the lowest of the year (<0.1 $gCm^{-2}d^{-1}$). Apparently, the pelagic community conserves matter within the surface layer in this phase. Occasional inputs of new nutrients from the sediments as well as from deeper layers in the course of the summer are at least partly retained within the pelagic system as there is a net biomass increase over the summer. Bacterial activity is high (Bölter, in press) and the ratio phytoplankton : zooplankton biomass low (<4:1). The community is highly diverse and cycling of matter efficient as demonstrated by the high levels of primary production (0.5 - 1.0 $gCm^{-2}d^{-1}$) and low rates of sedimentation ($<<0.2\ gCm^{-2}d^{-1}$). Occasional bursts in sedimentation rates during the summer ($>0.3\ gCm^{-2}d^{-1}$) are presumably the result of an imbalance in the system brought about by unusually large inputs of new nutrients from the sediments at the depth of the pycnocline via flushing of interstitial water (Smetacek et al., 1976)and from restricted upward mixing of stagnant bottom water. These new nutrients are rapidly converted into plant biomass which is in excess of the prevailing heterotrophic capacity of the pelagic system, the "surplus" material, both phytoplankton and phytodetritus, evidently sedimenting out of the system. Such occasional peaks in summer sedimentation rates are meteorologically forced, i.e. due to wind mixing and oscillation of the pycnocline as they do not occur with regularity every year.

For reasons not yet clearly identified, the summer community breaks down in late August and metazooplankton stocks decline rapidly. Upward mixing of nutrients accumulated in deeper layers over the summer by increasing wind speeds in autumn gives rise to a massive autumn bloom of ceratia, frequently followed by a large diatom bloom. The Ceratium bloom attains roughly the same biomass as the spring bloom though the autumn diatom bloom is somewhat smaller. Both these blooms also sediment out ($\approx 10\ gCm^{-2}$) and there is thus another significant input of fresh organic matter to the sediments during October/November. The processes triggering sedimentation of these blooms are unclear, nutrient depletion being certainly not responsible as in the case of the spring bloom. Perhaps a decline in light input has a similar effect on sinking rates of the phytoplankton as does nutrient depletion.

The seasonal pattern of sediment trapping is clearly seen from Figure 3 where results of a year's monitoring with a multisample sediment trap (Zeitzschel et al., 1978) suspended at 2 m above the bottom in a 20 m water column are depicted. The winter peaks are due to resuspension of surface sediments and the spring (March) and autumn (October/November) bloom peaks are conspicuous because of the high chlorophyll content of the material. The low values from April through September are interrupted by two peaks, in July and August respectively. The phytoplankton origin of this material is indicated by its relatively high chlorophyll content.

The benthic response to this annual pattern in food supply in Kiel Bight is interrupted by a period of oxygen depletion in late summer and again in autumn that recurs every year. These semi-anoxic phases accompanied by sedimentary nutrient release, particularly phosphate, are a result of stagnation of the bottom water (v.Bodungen, 1975), a situation characteristic of many other enclosed bodies of water.

One can draw the following conclusions from the above account of the processes regulating input of organic matter to the benthos in a shallow coastal system :

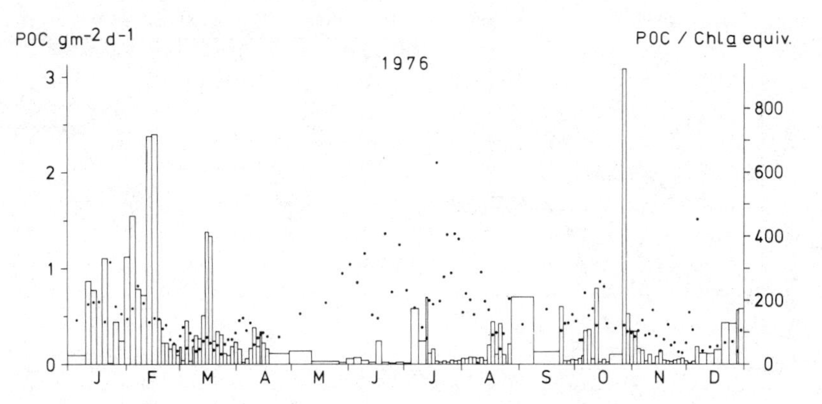

Fig. 3. Sedimentation rates of particulate organic carbon (POC) and POC/chlorophyll a equiv. of the material collected in 1976, from the same depth and site as in Figure 1. (Redrawn from SMETACEK, 1980a).

1. Sedimentation rates are highest during periods of new production (unstable water column) and lowest during regenerated production (stratified water column).

2. Accordingly, the bulk of the material (about 2/3) settling out of the system in the course of an annual cycle is in the form of phytoplankton cells or phytodetritus resulting from cell mortality and independent of zooplankton grazing.

3. Zooplankton fecal pellets are largely retained within the pelagic system and hence their contribution to the total particle flux is negligible; possibly, the most important contribution by zooplankton to the particle flux is in the form of carcasses (see later discussion).

SEDIMENTATION OF PLANKTON AND DETRITUS IN OTHER REGIONS

Phytoplankton and phytodetritus

In deeper, more open areas not significantly influenced by tidal mixing or fronts, such as in the Baltic proper or in large areas of the North Sea, a similar seasonal pattern in input to the benthos might be expected. Phytoplankton cells from the sedimenting spring bloom reach depths well above 50 m with considerable rapidity and apparently at sinking speeds greater than the observed maximum rates under controlled laboratory conditions (v. Bodungen *et al.*,

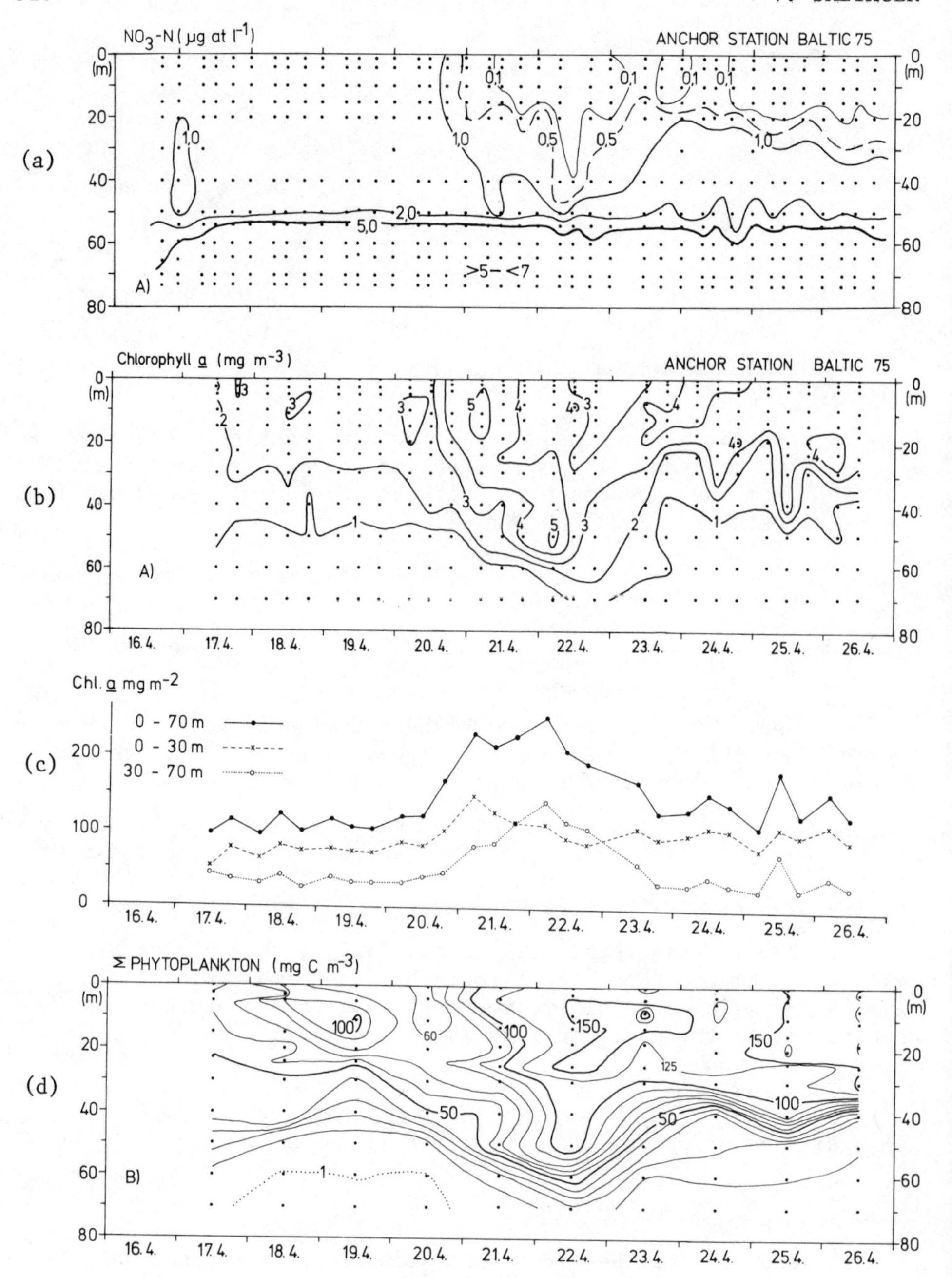

Fig. 4. Data collected from the water column during the spring bloom in the Bornholm Basin (Baltic Sea) during April 1975.

a. Nutrient concentration (NO_3^-). Depletion in surface layers following 20 April is apparent.

1981). The mechanisms effecting these rapid sinking rates were observed in operation during an intensive study of the spring bloom in the 80 m Bornholm Basin of the Baltic proper (Smetacek et al., 1978; v. Bodungen et al., 1981). The growth phase of the bloom was much more prolonged than in Kiel Bight because of intermittent vertical mixing to depths in excess of 20 m. Phytoplankton sedimentation appeared to occur in events governed by the weather (Figure 4). On the first calm sunny day following a stormy period (20 April), biomass in the upper 10 m increased 2 fold accompanied by concomitant nutrient depletion in this layer. By the next day (21 April), biomass in the entire water column had doubled and a portion of the population, apparently the nutrient depleted cells from the surface layer, started to sink downwards in bulk and had reached the pycnocline at 50 m depth. By the third day the sinking population had passed through the halocline - a density difference of 4 δt - and settled out on the sea bed. On 24 April, the situation in the entire water column had reverted to that of the period preceding the sedimentation event and remained so till the investigation was terminated (26 April). Presumably, more such events occurred in the following days.

The sinking rate of diatom cells is a function of their physiological state (Smayda, 1970; Titman and Kilham, 1976) and nutrient depleted cells sink much more rapidly than vigorously growing cells of the same species. The difference is approximately four-fold and the reported maximum sinking rate of senescent cells of small species such as *Skeletonema costatum* are in the order of 10 m d^{-1} (Smayda and Boleyn, 1966). The unusually high sinking rate of *S. costatum* cells indicated by the above field observations (40 m d^{-1}) can possibly be attributed to aggregate formation during the course

←

Fig. 4. (cont.)

b. Chlorophyll *a* concentrations. Chlorophyll build-up at the surface on 21 April and the downward movement of the bulk on the 22 and 23 April is clearly visible.

c. Integrated values for chlorophyll *a* in the 0-70 m, 0-30 m, and 30-70 m water columns. The sedimentation event of 22 and 23 April is apparent.

d. Phytoplankton biomass, derived from cell counts and carbon conversion factors ; *Skeletonema costatum* was the dominant species.

The permanent halocline was located at ca. 50 m depth, the seasonal thermocline was established in May. (From v. Bodungen et al., 1981).

of descent. As at least portions of the populations appear to sink out in concert, their temporary accumulation at density discontinuities will certainly favour aggregate formation. Such aggregates have a faster sinking rate than the individual particles of which they are composed (McCave, 1975). This mechanism would explain the occurrence of accelerated sinking rates in nature suggested by Smayda (1970).

Sedimentation of the spring diatom bloom in coastal areas is well documented (Platt and Subba Rao, 1970; Skjoldal and Lännergren, 1978; Raymont, 1980; Kuparinen et al., in press); it is widely believed that this is due to the absence of overwintering populations of zooplankton in such areas. However, little consideration has been given to the question whether this also occurs in regions accessible to oceanic zooplankton.

Direct monitoring of the vertical particle flux with sediment traps has rarely been carried out in open shelf environments. During the spring bloom in the northern North Sea studied by FLEX 76, where a large population of Calanus developed, sediment traps were put out by workers from Aberdeen and Kiel. Both groups collected considerable amounts of phytoplankton in their traps deployed well below the developing thermocline (40 m). Quantitative evaluation of these samples however, proved difficult because of the large quantities of copepods that apparently swam into and died in the traps (Davies, pers. comm.). Indirect evidence is now emerging, voiced by Walsh (1981), that during periods of new production, loss rates of organic matter from the pelagic zone can be substantial. Thus, Coachman and Walsh (1981) and Dagg et al. (1982), in studies of the Bering Sea spring phytoplankton bloom, also suggested that a fairly large portion of plant biomass sedimented directly to the benthic system in spite of the presence of sizable overwintering zooplankton populations. In upwelling regions as well, a substantial portion of the organic matter produced by the initial diatom bloom also apparently settles out of surface layers as shown by the rapidity with which the total concentration of biogenous material in the surface layer declines away from the centre of upwelling (Margalef, 1978). Eppley et al. (1979) also suggested that the loss rate of organic matter is higher during periods of new production than when regenerated production prevails. The decisive factor here will be the size of the grazer population.

Fransz and Gieskes (in press) have recently shown that, in contrast to the prevailing view on the subject, there was an imbalance in the temporal development patterns of the phytoplankton and zooplankton during the spring bloom in the southern and central North Sea and that a large portion of the spring bloom was not grazed by the zooplankton but settled out directly to the sea floor (Gieskes, pers. comm.).

Much less is known of sedimentation during and following dinoflagellate blooms and the presence of a relationship between the sinking rate and physiological state in this group such as that present in diatoms has, to my knowledge, not been reported so far. However, that dinoflagellate populations can also settle out of the water column *en masse*, sometimes with devastating consequences for the benthos has been reported from red tide blooms. Many of the red tide organisms (e.g. *Gonyaulax* spp.) form resting spores that collect on the sediment surface in large numbers (Provasoli, 1979). One might speculate that the toxic forms are able to increase their survival chances by deterring grazing much as many land plants do. Similarly, some herbivores will develop immunity to the particular poison or to the particular species by not digesting its cell wall, which might explain the selective effect of the dinoflagellate toxins on marine animals. Thus, the toxins of *Gymnodinium breve* are reportedly lethal for cowries and horseshoe crabs but apparently not for the other molluscs and arthropods living in the area (Tufts, 1979). Vertebrates appear to be most susceptible to these plant poisons, which is again similar to the situation on land.

Sedimentation of an immense bloom of *Ceratium tripos* in the New York Bight in June 1976 resulted in oxygen depletion at the sediment surface and mass mortality of the benthos over an area of 13,000 km^2 (Mahoney and Steimle, 1979). This bloom resulted from an overwintering autumn population that built up its enormous biomass during spring and early summer (Malone, 1978). Such spectacular events are unusual, but because of the devastation they cause, they will be of considerable importance in shaping the structure of the benthic communities. The composition of the sediment will in all likelihood be influenced more by occasional blooms of this magnitude than the regular annual input.

Much less information is available on the sedimentation of flagellates although it is well known that coccolithophorid skeletons are an important constituent of the sediments in many regions. Reynolds and Wiseman (1982) found in enclosure experiments carried out in a lake that only a very small percentage of nanoflagellates sank out of the 10 m water column in contrast to diatoms where up to 100% of the population was observed to sink out. The nanoflagellates were apparently largely grazed down (Reynolds *et al*., 1982).

Sedimentation of Zooplankton and Derivates

From the above discussion, it would appear that the role played by zooplankton in the sedimentation process is secondary when compared to that of the phytoplankton especially under conditions of new production. This is in contrast to the prevalent views on the subject that are based on examination of sediment trap material (Angel, this volume). In three recent studies, carried out on the relationship between copepods and the fecal pellets produced by them, it has been shown that the fate of the majority of fecal pellets is retention and utilization within the trophogenic layer rather than sedimentation to the sea floor. In two of the environments, (SE continental shelf of the USA, Paffenhöfer and Knowles, 1979 and Kiel Bight, Smetacek, 1980b), the dominant zooplankters were small copepods In the third environment, the North Sea, large copepods - Calanus - formed the bulk of the zooplankton (Krause, 1981). Coprophagy by the copepods (Paffenhofer and Knowles, 1979) together with bacterial breakdowns of the pellets, were considered to be the major mechanisms of fecal pellet retention in the water column.

The smaller copepods generally dominate zooplankton biomass during the summer months when the pelagic system is characterized by regeneratated production. These copepods are too small to migrate to richer food sources and yet have generation times of several weeks. It is obvious that if all the fecal pellets produced by a growing cohort of copepods were to sink out of the mixed layer, the population would starve before it reached maturity (Krause, 1981). An increase in trophic efficiency would not suffice as a survival strategy and coprophagy would be a more effective mechanism within such a system as it would drastically reduce leaking of essential elements.

If a regenerating, and hence self-sufficient, pelagic population could be considered an ecosystem, then retention of essential elements within it would be an important property of such a system. Support for this assumption is provided by the variety of feeding modes present in summer copepod communities in contrast to those dominating diatom blooms : whereas the large Calanus spp. (temperate and arctic) and Calanoides spp. (e.g. NW African upwelling) are typically herbivores, the smaller summer copepods apparently have a more miscellaneous diet as reflected by their more versatile mouth parts (e.g. Temora, Acartia, Centropages, Oithona, etc., Schnack, 1982). This is also shown by their ^{13}C content which is an indication of the "trophic distance" of a given heterotroph from the primary producers. Thus, Mills et al., (in press) reported that large copepods and euphausiids were apparently only one trophic step away from the phytoplankton, whereas the small copepods were significantly further removed. This trophic "distance" of the smaller copepods from the phytoplankton could be easily explained if fecal material and adhering bacteria were indeed an important part of the diet of these small copepods.

Under conditions of new production, the growth rate of the phytoplankton exceeds grazing rate of the zooplankton and there is an abundance of food. Under such conditions, copepods not only increase their pellet production rate (Marshall and Orr, 1972) but assimilator efficiency also declines, as evidenced by the large numbers of undigested algal cells in the pellets (Turner and Ferrante, 1979). It is possible that coprophagy is of lesser importance when there is an abundance of food allowing a greater proportion of the fecal pellets to sink out of the surface layer. The implication is that the ratio pellets produced/pellets sedimented will be variable and dependent on conditions within the surface layer. This ratio will presumably be inversely related to the ratio phytoplankton/copepod biomass. The presence of such a relationship would make ecological sense for the mesozooplankton that share their environment with their plant food and would serve to explain the discrepancies between results obtained from sediment traps from different localities and seasons (Smetacek, 1980b).

Migratory organisms would have no obvious need to conserve matter within the surface layer as they are able to change their feeding grounds. Fish fecal matter, particularly of nektonic swarms would presumably settle out rapidly. Staresinic et al.(in press) collected large quantities of anchoveta fecal pellets in sediment traps deployed by them in the Peru upwelling region. This also applies to organisms such as salps and Antarctic krill, of the macrozooplankton that occur in distinct swarms that are able to change their feeding grounds by vertical migration. Iseki (1981) found salp fecel pellets to be an important constituent of the particles collected by sediment traps in the Pacific ocean. Sediment traps deployed in the Antarctic occasionally collected large quantities of krill feces (SFB 95, Kiel University, unpublished data). However, large salp swarms have been reported to persist in the same region for over 2 months (Alldredge, this volume). Maintenance of such a large swarm over such a long period will necessitate some form of recycling of fecal material, although we can at present only specullate at the possible mechanisms involved. As a rule, however, large swarms of macrozooplankton tend to concentrate at the shelf break where the water depth is sufficient to permit extensive vertical migrations. Dagg et al. (1982) have shown that much of the biomass produced by the spring bloom in the outer reaches of the Bering Sea shelf was indeed grazed by zooplankton, in contrast to the situation on the mid-shelf and coastal region. Possibly, the major form of input to the benthos on the outer shelf is in the form of macrozooplankton fecal pellets, whereas towards the coast, the contribution of phytoplankton to the benthic food supply increases.

Another source of food for the benthos that has received little attention so far is dead zooplankton. This is because it is widely assumed that predation rather than natural mortality, i.e. death due to old age, is the major factor controlling the size of zooplankton stocks. As in the case of the phytoplankton/herbivore relationship, discussed above, one can advance the same argument with regard to balance between herbivores and carnivores to support the contention that prey and predators do not necessarily have to be balanced. This point can be illustrated with an extreme example: the common cladoceran *Penilia avirostris* which occurs in the coastal regions of the tropical and warm temperate zones. It has benthic eggs that hatch synchronously releasing a large cohort into the pelagic system. The animals reproduce parthenogenetically and build up a large population with great rapidity, generally dominating the zooplankton biomass during such "blooms". When the bloom is over, the organisms form resistent resting eggs that are transported to the bottom in the carcass of the parent (Wickstead, 1963). Because of the large biomass of the population, such an event is bound to be of significance to the benthos.

Natural mortality may also be widespread with species that occur in distinct cohorts as in the case of overwintering euphausiid or copepod populations. Presumably, many of the adult gelatinous macro- and megazooplankton also die a natural death after reproduction. Assessing this source of input to the benthic system is indeed difficult partly because of its patchy nature. In shelf and coastal areas where zooplankton stocks are large, the organisms tend to swim into and die in sediment traps, which makes it difficult to assess natural mortalit rates from trap material. Smetacek et al. (1978)showed that almost the entire overwintering zooplankton stock in the southern Baltic (0.4 gCm^{-2}) succumbed to old age and was collected in a trap located below the pycnocline, i.e. in oxygen depleted water not normally inhabited by them. Although this observation might indeed be typical only for the Baltic because of the paucity of carnivores at this time of year, it would also serve to explain the comparatively steep decline often observed in large copepod stocks following reproduction. Zooplankton carcasses sink faster than phytoplankton and fecal pellets (Smayda, 1970) and they are likely to be even more fancied by the benthos than the preceding "shower" of diatoms and phytodetritus.

Another well-known example of pelagic input to the sea bed is in the form of nekton carcasses. The presence of large, active scavengers in the deep sea, comparable to vultures in terrestrial systems, shows that this form of food supply is of ecological significance. To what extent these carcasses are the result of old age, sickness or merely "messy feeding" by nektonic carnivores is unknown. The role of these large parcels of organic matter ("food falls") as a benthic food source has been recently reviewed and discussed by Stockton and DeLaca (1982).

Nutritional Value of Sedimenting Particles

Data on the chemical composition of sedimenting matter has largely been restricted to its elemental composition. Thus, matter of marine origin collected in sediment traps generally has C/N ratios below 10 and is hence considered to be of direct nutritional value, i.e. without the agency of bacteria, to benthic animals. However, little is known of the digestibility of such material. Smetacek (1980a) and Pollehne (1981) have shown that C/N and C/P ratios of sedimenting matter varied seasonally and were lowest during spring and autumn bloom sedimentation and highest during summer stratification. Sometimes, summer material showed even higher C/N and C/P ratios than present in the surface sediment. C/chlorophyll ratios followed an identical pattern indicating that food quality was highest during new production and poorest during regenerated production. These observations indicate that regenerating systems conserve N and P relative to C. The particles sinking out of such systems are truly wastes, i.e. they are composed of refractory material with low essential element content. Smetacek *et al.* (1980) found in tank experiments with flagellate populations that settling particles always had a higher C/N and C/chlorophyll ratio than particles that remained in suspension. These 'waste' particles evidently had higher sinking rates. On reaching the sea bed, such particles will have to be enriched by bacterial growth (Johannes and Satomi, 1966), involving uptake of inorganic nutrients from the environment and loss of energy, before they can be utilized by benthic animals. Thus, benthos living below a regenerating pelagic system receive not only small quantities of food, but also food of poor nutritional quality. Evidently, the seasonal cycle in new and regenerated production in the pelagic system is of vital importance to the benthos both in terms of quantity and quality of the food supply.

Regional Considerations

In a given region, the magnitude of new production and the seasonal cycle of the pelagic community will determine the temporal cycle in quantity and quality of the food sinking out of surface layers. In some coastal areas such as in Bedford Basin, high sedimentation rates have been recorded in summer (Hargrave, 1980). Seasonal cycles in sediment trapping recorded from various other coastal localities (Steele and Baird, 1972; Ansell, 1974; Webster *et al.*, 1975; Platt, 1979) also differ considerably from that in Kiel Bight (see Parsons *et al.*, 1977 for a review). However, in many of these localities, additional factors such as resuspension, terrestrial input and input from macroalgal beds play a far greater role than in Kiel Bight. It may be that the basic pattern in pelagic input to the benthos exhibited by Kiel Bight will be found in other temperate enclosed bodies of water as well. In regions where tidal mixing is important, the situation is likely to be more complicated, depending on whether new nutrients are introduced to surface layers in the course of a tidal cycle or not. Further, sediment resuspension and redeposition will serve to complicate the picture.

In shelf regions, the pattern of sedimentation will depend on latitude and on the hydrography. The higher production of frontal regions is maintained by input of new nutrients. As dinoflagellates play an important role in such areas (Holligan, 1979) the seasonal cycle in food input to the benthos will be quite different here although sedimentation rates are bound to be considerably higher under a front than in adjoining regions where regenerated production dominates. Some shelf systems such as Georges Bank do not develop stable summer stratification as do most others. Because of its unstable water column and its location in an area with high nutrient concentrations below the seasonal pycnocline, the contribution of new to total production over the bank is greater than in other adjoining shelf areas. Sissenwine et al. (in press) attribute its higher fishery yield to this factor. In such a system, it is likely that downward mixing of phytoplankton besides sedimentation will play an important role in providing food to the benthos.

It is a well-known fact that benthic biomass decreases from coastal and shelf regions to the deep sea (Rowe, 1971; Parsons et al., 1977). One explanation generally advanced is that sinking particles must penetrate through a longer gauntlet of pelagic heterotrophs during their descent in deeper water columns and are hence less likely to reach the sea floor. However, as most of the productive regions are located over shallow depths, a comparison with deep productive regions such as the equatorial divergence would be necessary to confirm the above explanation. Hargrave (1975), in a comparative study of lakes and marine bays, found an inverse correlation between sedimentation and depth of the mixed layer which he attributed to more effective remineralisation within a deeper mixed layer. However, as new nutrients are added more easily and hence more frequently to shallow mixed layers than to deeper ones, the percentage of new to total production will be greater in the former (eutrophic) as compared to the latter (oligotrophic) type. Thus, although pelagic remineralisation is indeed more efficient in the latter environment, the cause for this is not the longer water column, but rather the lower input rates of new nutrients and their effect on the phytoplankton community.

That sedimentation of organic matter is higher under areas with significant new production is clearly demonstrated by the organic carbon content of the sediment on a global basis (Müller and Suess, 1979). These authors and Müller and Mangini (1979) have shown that the higher the sedimentation rate, the higher the percentage of organic carbon that is burried. Walsh (1981) suggested that the cont nental shelves are a major global site of burial of organic carbon, which if true, would indicate that the benthos is receiving food well in excess of its demand in these areas. This is hard to believe, if budgetary models of the pelagic system (Steele, 1974; Mills, 1980) are considered where even the total annual primary production is insufficient to fuel the apparent demand of the combined pelagic and benthic heterotrophs. Either we are missing sedimentation events on short time scales where occasional bursts in new production result in corresponding bursts in sedimentation, or events occurring once in several years such as the Ceratium bloom in the New York Bight are decisive. In any case, more information on the relationship between pelagic system structure and sedimentation is urgently called for before such fundamental issues can be resolved.

BENTHIC RESPONSE TO THE FOOD SUPPLY

It is well-known that benthic biomass and activity is closely correlated with the food supply (e.g. Parsons *et al.*, 1977) however, detailed information on the response of the benthos to its food supply is scanty. Standing stocks of benthos are largest in regions with significant new production, the length of the growth season, i.e. latitude is not the decisive factor. Mills (1975), in a critical evaluation of the contribution of benthic ecology to biological oceanography has compared benthic populations from different regions and has pointed out that interaction within the benthic community can exert considerable influence on its structural characteristics, e.g. large epifauna vs. small infauna. He assumes, however, that food input is "a rather sparse but constant rain" of various particles of pelagic origin. The most recent evidence shows that this assumption is false, even for the deep sea where the overlying water exhibits some degree of seasonality (Honjo, in press). I believe that the seasonal pattern in the food supply is the single most important factor determining benthic community structure as it sets the rhythm to which the various organism types respond in their individual ways. However, the postulate cannot be verified at present.

Høpner Petersen and Curtis (1980) and Høpner Petersen (in press) have compared various environments from the Arctic to the tropics with one another in terms of the contribution of pelagic and demersal fish to the total catch. They show that at high latitudes, demersal fish form the main portion of the catch in contrast to the tropics where pelagic fish are much more important. The authors suggest that, as the northern ecosystems are young, the pelagic food web is inefficient and a proportionately larger portion of primary production is diverted to the benthos. However, it is also likely that because of the shorter growth season in high latitudes, pelagic food is available for only a short period of time, and these environments are therefore not conducive to development of large populations of pelagic fish. Stated in other terms, the higher the proportion of new to total production in a system the greater the relative amount of total energy fixed in the system that is diverted to the benthos. A closer comparison between high latitude spring blooms and low latitude upwelling blooms in terms of pelagic/benthic partitioning of energy will be necessary to resolve this question.

Within a geographical region, benthic biomass is rarely evenly distributed and, depending on the spatial scale regarded, the distributional patterns can be in the form of zones or merely patches. On the SE shelf of the USA, Hanson *et al.* (1981)found benthic biomass and activity to be higher at the shelf break than in the mid-shelf. The authors attribute this to nutrient input via intrusions of deep Gulf Stream waters at the shelf break, i.e. the contribution of new to total production is higher here. Paffenhöfer and Knowles (1979) suggest that the paucity of benthos in the mid-shelf of the same region is due to coprophagy by the copepod population that effectively reduces exports to the benthos. Turning their argument around, one might conclude that in regions with high benthic biomass such as the North Sea, the copepods either do not practise coprophagy or there are sources other than copepod fecal pellets, i.e. phytoplankton and zooplankton carcasses, providing food to the benthos.

Benthic zonation can also be brought about by physical factors of the environment. Thus, food settling out of surface layers can be concentrated in certain localities by currents interacting with bottom topography. A particularly striking example from the southern North Sea has been presented by Creutzberg *et al.* (in press). Their results are presented in Figures 5 and 6. They found benthic biomass to increase from South to North, as the mud content of the sediment increased. At the transition zone of sand to mud, benthic biomass was highest. In this zone the tidal current velocity drops to a critical value at which deposition of silt can take place. Thus, organic matter emanating from the regions where deposition cannot occur is concentrated in this zone giving rise to a benthic biomass 4 times the size of the sand community. The authors also found a spatial succession in the dominant feeding types within the zone of high benthic biomass. Going northwards, surface deposit and suspension feeders (*Angulus fabula*) occurred in the region of fine sand, to be succeeded first by subsurface feeders (*Nucula* and *Echinocardium*) and then by primary deposit feeders (*Abra alba*). Thereafter, suspension feeders became dominant (*Turritella* followed by *Chaetopteris*). The authors suggest that this distributional pattern reflects environmental properties pertaining to resuspension, deposition and sediment stability. However, new production in the water column immediately overlying this zone is likely to be higher than elsewhere in the region as it is the site of a front (Pingree *et al.*, 1978). Whether this is a significant contribution to the food supply relative to the concentration effect suggested by the authors is worth looking into.

METHODOLOGY

There has been a general reluctance amongst ecologists to study the sedimentation process on an equal footing with other major pelagic processes. In the following, I shall therefore briefly outline some of the approaches that have been used in the past to study various aspects of the sedimentation process in the hope that more marine ecologists will include them in their routine sampling programme.

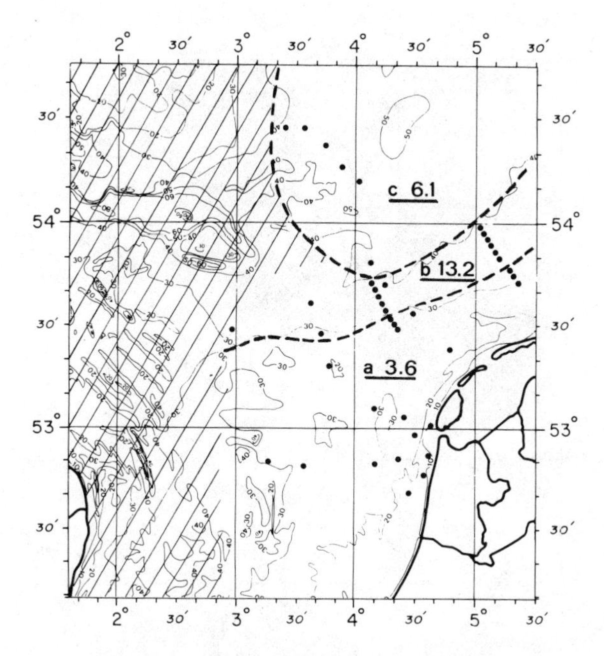

Fig. 5. Depth contours and mean biomass of the benthic macrofauna (ash free dry weight, g m^{-2}) from the southern North Sea between the English and Dutch coasts.
a. Sandy sediments, b. Transition zone from sand to mud, c. Muddy sediments. Dots are sample stations (from Creutzberg et al., in press).

Sinking Rate Measurements

Measurements of the sinking rate of particles by direct observation have been carried out on various marine particles, e.g. phytoplankton cells, zooplankton fecal pellets and carcasses. Such direct measurements have to be carried out under artificial conditions - i.e. in the absence of natural turbulence patterns - and although they yield important baseline data, the actual sinking behaviour of particles in the field can be quite different. Some laboratory and ship-board methods have been described by Smayda and Boleyn (1965), Eppley et al. (1967), Turner (1977) and Bienfang (1979). Smayda (1970) and Angel (this volume) have compiled available data on sinking rates of phytoplankton and fecal pellets respectively measured by such methods.

This simple approach, i.e. monitoring of sinking particles within an experimental vessel, has been expanded for in situ use by Lännergren (1979) who used a horizontally partitioned plexiglass cylinder suspended in the sea. Changes in the sinking rate of a sping diatom population, enclosed periodically in this water column

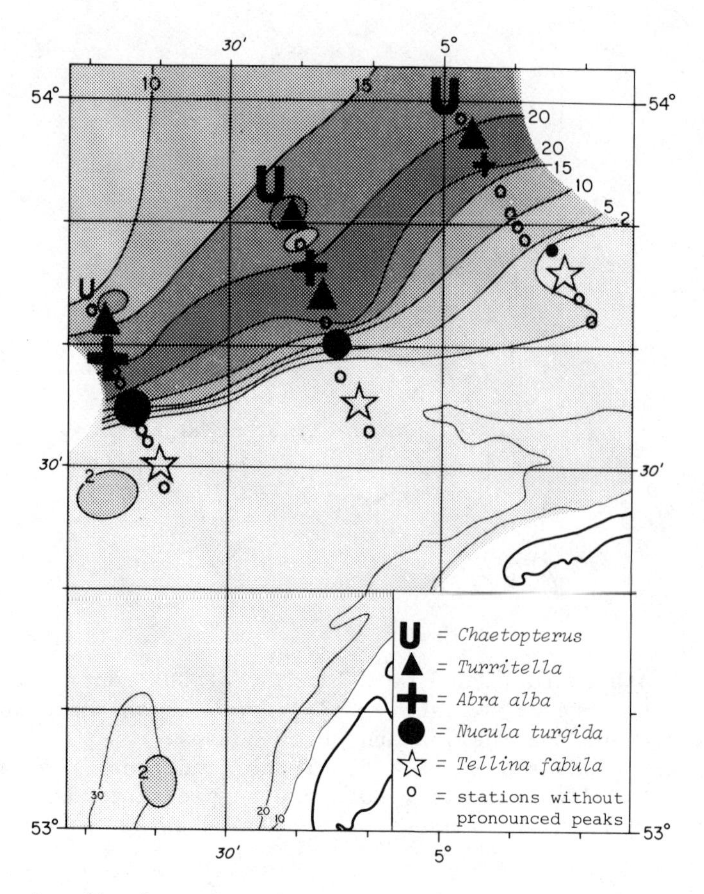

Fig. 6. Distribution of peaks of certain macrofauna species along three transects in relation to mud content from the transition zone (Figure 5 b.). Mud content is given in % and size of symbols denotes relative abundance (from Creutzberg et al., in press).

were followed at short intervals. The author demonstrated that at the peak of the bloom, the diatom cells were actually positively buoyant. The results from such experiments, if combined with field observations, provide much useful information and help greatly in interpreting the field data. The use of mesocosms has also helped in elucidating sinking behaviour of natural plankton populations (Bienfang, 1982; Reynolds and Wiseman, 1982; Reynolds et al., 1982).

Sediment Traps

The use of sediment traps - i.e. containers placed in the path of downward moving particles - to directly assess the vertical flux of material via passive sinking has become widespread during the last decade. A wide variety of containers, generally cylinder or funnel-shaped, have been employed, and experimental studies have shown that the amount and possibly the composition or particles collected by the various types of traps is dependent on many environmental factors other than the actual sedimentation of particles occurring during the period of deployment. Several discussions of the methodology as well as reviews of the various types in usage have appeared recently (e.g. Hargrave and Burns, 1979; Gardner, 1980a; 1980b; Blomqvist and Håkanson, 1981), therefore, in the following, only some of the possible sources of bias inherent to this methodology will be dealt with.

As sediment traps are rigid objects placed in a turbulent environment, their presence alone will significantly change the turbulence pattern in the immediate vicinity and hence influence the sedimentation regime. This influence will be greatest in slow-sinking small particles and it has been shown that depending on trap configuration and current speed, estimation of vertical flux of these particles can be less or even more than the actual flux. Large particles with high sinking rates will be trapped more realistically but, being rarer by nature than the small particles, the area of the collection surface will decide here how representative a given sample is. As the range of particle sizes likely to contribute significantly to the vertical flux will vary both spatially and temporally by as much as several orders of magnitude, the size of the sediment trap used as well as its configuration can seriously bias the information thus obtained. One way to lessen distortion of the turbulence pattern around a sediment trap is to let it drift freely with its environment (Staresinic et al., 1978).

A problem inherent to results from sediment traps, particularly in coastal environments or when the trap is in the proximity of the nepheloid layer of the deep sea, is the fact that the trap collects both primary settling material, i.e. particles settling out for the first time, as well as resettling material, i.e. particles resuspended from the sediment surface by water movement. The problems involved in differentiating between these two types of material have been discussed by Smetacek (1980a). In coastal localities subject to frequent bottom-near turbulence, only indirect clues can be used and it is probably utterly futile attempting to estimate primary settling rates using sediment traps in highly energetic coastal waters such as those strongly influenced by tides.

As particles collected by a trap generally clump together at the bottom, it is difficult to deduce from this material their original shape at the time of entering the trap, i.e. whether particles were sinking individually or in aggregate form. Further, concentration of particles into clumps significantly changes the biological environment and hence the break-down rate of particles.

Although use of preservatives prevents organism activity, it also affects the chemical composition of the particles (Smetacek and Hendrickson, 1979). Results obtained from sediment traps have thus to be treated with caution, just as in the case of all other methods for measuring rates that are in use in marine ecology.

It is important to bear the space and time scales of data provided by sediment traps in mind, particularly when incorporating the same into holistic models. On a m^2 basis, they provide more useful information on export out of the pelagic system than import to the benthos, because of near-bottom redistribution of the food particles. Logistical problems involved with sediment trap usage rather than the quality of information derived from them is certainly the major disadvantage associated with this methodology. So far, sediment traps have been deployed mainly in inshore areas or in the deep sea for any length of time, presumably because of the hazard posed to moored sediment traps by fishing activity; their routine use on continental shelves will be necessary before these important ecosystems can be successfully budgeted.

Direct Field Observation

Direct sampling of vertical gradients in particle composition and quantity is another means of monitoring the vertical flux. In shallow areas with large spatial and temporal fluctuations in particle concentration and composition this method is of only restricted use. In deeper regions with less surface patchiness, the vertical gradients will provide information on the vertical flux particularly if repeated in the same water column over short periods. The use of large-volume water bottles or pumps together with large-mouthed, fine-meshed nets will aid in representatively sampling the larger particles and *in situ* fluorometers and transmissometers can provide useful additional data.

Resuspension of Sediment Particles

Resuspension of sediment particle and near-bottom transport to other areas has been the domain of geologists for a long time. Only recently have benthologists employed some of the available methods e.g. flumes, to study the effect of this potential food supply on the benthos. Taghon *et al.* (1980) showed in a flume experiment how the feeding behaviour of spionid polychaetes varied in relation to water velocity. At low velocities the animals practised deposit feeding but lifted their tentacles from the sediment surface and switched to suspension feeding when higher current velocities increased the horizontal flux of particles in the overlying water layer. Such direct observation of feeding by benthic organisms will help greatly in classifying benthic communities according to their relative proportions of food acquisition strategies. This knowledge, when combined with field observations of feeding type distribution such as those of Creutzberg et al. (in press), will considerably illuminate the picture of pelagic/benthic coupling that is now in the process of emerging.

CONCLUDING REMARKS

A purpose of this presentation has been to draw attention, particularly of pelagic ecologists, to the general importance of the sedimentation process in the sea. I am convinced that a closer consideration of this process within the context of general plankton ecology will bring to light previously neglected aspects of systems functioning; these would provide new perspectives for resolving some contemporary controversies such as those pertaining to the fate of phytoplankton cells and the relative roles of bacterioplankton, protozooplankton and metazooplankton in pelagic remineralisation (see Williams 1981 for a review). The comprehensive studies on loss rates of phytoplankton carried out by Jewson et al. (1981) and Reynolds et al., (1982) in lacustrine environments provide good examples for the type of data that are rare for marine ecosystems. Obtaining such data from the sea is, of course, a much more difficult task.

One of the main problems is differentiation of new and regenerated production which can be resolved by recording NO_3-uptake rates (Dugdale and Goering, 1967) or by measuring net diel production (net growth of the plankton community plus sedimentation) within a given water body. Sedimentation can best be monitored in such daily budget studies by deployment of free-floating sediment traps (Staresinic et al., 1978) which can simultaneously serve as drogues for marking the water mass under study. Such short-term budgetary analyses of various types of pelagic community will provide criteria for classification of different structural types based on their dynamic features. Such criteria might well be the 'macroscopic properties' characteristic of particular systems as suggested by Mann, (1981). Such a classification scheme for the Kiel Bight pelagic system has been presented by Smetacek et al. (in press).

In this paper, the case has been made that the benthic food supply is a function of the controlling element (generally nitrogen) budget of the area. More explicitly stated, the hydrography of a given region together with its latitude, i.e. its patterns of external and radiant energy input respectively, using Margalef's (1978) terms, will determine the quantity and quality of food supply to its benthos. With the possible exception of the oligotrophic oceanic gyres, this annual rhythm in the food supply is likely to be a fundamental factor in shaping the benthic community structure just as its quantity sets the upper limit for the total benthic biomass.

Acknowledgements

I am grateful to Michael Kemp and an anonymous reviewer for useful suggestions and valuable comments on the first draft, as also to my colleagues of the Sonderforschungsbereich 95. This is contribution No. 393 of the Joint Research Programme (SFB 95), Kiel University, funded by the German Research Council (DFG).

REFERENCES

Alldredge, A., 1983, The quantitative significance of gelatinous organisms as zooplankton predators, this volume.

Angel, M.V., 1983, Detrital organic fluxes through pelagic ecosystems, this volume.

Ankar, S., 1980, Growth and production of Macoma balthica (L.) in a northern Baltic soft bottom, Ophelia, Suppl. 1: 31.

Ansell, A.D., 1974, Sedimentation of organic detritus in Lochs Etive and Creran, Argyll, Scotland, Mar. Biol., 27: 263.

Bienfang, P.K., 1979, A new phytoplankton sinking rate method suitable for field use, Deep-Sea Res., 26/6A: 719.

Bienfang, P.K., 1982, Phytoplankton sinking-rate dynamics in enclosed experimental ecosystems, in: "Marine Mesocosms", G. D. Grice and M.R. Reeve, eds., Springer-Verlag, New York.

Blomqvist, S., and Håkanson, L., 1981, A review on sediment traps in aquatic environments, Arch. Hydrobiol., 91: 101.

Bodungen, B. von, 1975, Der Jahresgang der Nährsalze und der Primärproduktion des Planktons in der Kieler Bucht unter Berücksichtigung der Hydrographie, Ph. D. Thesis, Univ. Kiel, Kiel.

Bodungen, B. von, Bröckel, K. von, Smetacek, V., and Zeitzschel, B., 1981, Growth and sedimentation of the phytoplankton spring bloom in the Bornholm Sea (Baltic Sea), Kieler Meeresforsch., Sonderh. 5: 49.

Bölter, M., in press, Seasonal variation in microbial biomass production and related carbon flux in Kiel Fjord, Mar. Ecol.

Coachman, L.K., and Walsh, J.J., 1981, A diffusion model of cross-shelf exchange of nutrients in the southeastern Bering Sea, Deep-Sea Res., 28A: 819.

Creutzberg, F., Wapenaar, P., Duineveld, G., and Lopez, N.L., in press, Distribution and density of the benthic fauna in the southern North Sea in relation to bottom characteristic and hydrographic conditions, Rapp. Proc. - Verb. Cons. Int. Explor. Mer.

Dagg, M.J., Vidal, J.J., Whitledge, T.E., Iverson, R.L., and Goering, J.J., 1982, The feeding, respiration, and excretion of zooplankton in the Bering Sea during a spring bloom, Deep-Sea Res., 29: 45.

Dugdale, R.C., and Goering, J.J., 1967, Uptake of new and regenerated forms of nitrogen in primary productivity, Limnol. Oceanogr., 12: 196.

Eppley, R.W., Holmes, R.W., and Strickland, J.D.H., 1967, Sinking rates of marine phytoplankton measured with a fluorometer, J. Exp. Mar. Biol. Ecol., 1: 191.

Eppley, R.W., Renger, E.H., and Harrison, W.G., 1979, Nitrate and phytoplankton production in southern California coastal waters, Limnol. Oceanogr., 24: 483.

Fransz, H.G., and Gieskes, W.W.C., in press, The imbalance of phytoplankton production and copepod production in the North Sea, Rapp. Proc.-Verb. Cons. Int. Explor. Mer.

Gardner, W.D., 1980a, Field assessment of sediment traps, J. Mar. Res., 38: 41.

Gardner, W.D., 1980b, Sediment trap dynamics and calibration: a laboratory evaluation, J. Mar. Res., 38: 17.

Gieskes, W.W.C.,1983, The state-of-the-art of primary production measurements, this volume.

Graf, G., Bengtsson, W., Diesner, U., Schulz, R., and Theede, H., 1982, Benthic response to sedimentation of a spring phytoplankton bloom: process and budget, Mar. Biol. 67: 201.

Hanson, R.B., Tenore, K.R., Bishop, S., Chamberlain, C., Pamatmat, M.M., and Tietjen, J., 1981, Benthic enrichment in the Georgia Bight related to Gulf Stream intrusions and estuarine outwelling, J. Mar. Res., 39: 417.

Hargrave, B.T., 1975, The importance of total and mixed-layer depth in the supply of organic material to bottom communities, Symp. Biol. Hung., 15: 157.

Hargrave, B.T., 1980, Factors affecting the flux of organic matter to sediments in a marine bay in: "Marine Benthic Dynamics," K.R. Tenore and B.C. Coull, eds., University of South Carolina Press, Columbia, S.C.

Hargrave, B.T., and Taguchi, S., 1978, Origin of deposited material sedimented in a marine bay, J. Fish. Res. Board Can., 35: 1604.

Hargrave, B.T., and Burns, N.M., 1979, Assessment of sediment trap collection efficiency, Limnol. Oceanogr., 24: 1124.

Holligan, P.M., 1979, Dinoflagellate blooms associated with tidal fronts around the British Isles, in: "Toxic dinoflagellate blooms," D.L. Seliger, ed., Elsevier, North Holland.

Honjo, S., in press, Seasonality and interaction of biogenic and lithogenic particulate flux at Panama Basin, Science.

Hϕpner Petersen, G., in press, Energy flow in comparable aquatic ecosystems from different climatic zones, Rapp. Proc.-Verb. Cons. Int. Explor. Mer.

Hϕpner Petersen, G., and Curtis, M.A., 1980, Differences in energy flow through major components of subarctic, temperate and tropical shelf ecosystems, Dana, 1: 53.

Iseki, K., 1981, Particulate organic matter transport to the deep sea by salp fecal pellets, Mar. Ecol. Progr. Ser., 5:55.

Jewson, D.H., Rippey, B.H., and Gilmore, W.K., 1981, Loss rates from sedimentation, parasitism and grazing during the growth, nutrient limitation, and dormancy of a diatom crop, Limnol. Oceanogr., 26: 1045.

Johannes, R.E., and Satomi, M., 1966, Composition and nutritive value of fecal pellets of a marine crustacean, Limnol. Oceanogr. 11: 191.
Kautsky, N. and Wallentinus, I., 1980, Nutrient release from a Baltic Mytilus-red algal community and its role in benthic and pelagic productivity, Ophelia, Suppl. 1: 17.
Kemp, W.M., Wetzel, R.L., Boynton, W.R., D'Ella, C.F., and Stevenson, J.C., 1982, Nitrogen cycling and estuarine interfaces: some current concepts and research directions, in: "Estuarine Comparisons," V.S. Kennedy, ed., Academic Press, New York.
Krause, M., 1981, Vertical distribution of fecal pellets during FLEX '76, Helgoländer Meeresunters., 34: 313.
Kuparinen, J., Leppanen, J.-M., Sarvala, J., Sundberg, A., and Virtanen, A., in press, Production and utilization of organic matter in a Baltic ecosystem off Tvärminne, SW coast of Finnland, Rapp. Proc.-Verb. Cons. Int. explor. Mer.
Lännergren C., 1979, Buoyancy of natural populations of marine phytoplankton, Mar. Biol., 54: 1.
Mahoney, J.B., and Steimle Jr., F.W., 1979, A mass mortality of marine animals associated with a bloom of Ceratium tripos in the New York Bight, in: "Toxic dinoflagellate blooms," D.L. Taylor, and H.H. Seliger, eds., Elsevier, North Holland.
Malone, T.C., 1978, The 1976 Ceratium tripos bloom in the New York Bight: causes and consequences, NOAA Tech. Rep. NMFS Circ. 410.
Mann, K.H., 1981, The classes of models in biological oceanography in: "Mathematical Models in Biological Oceanography," T. Platt, K.H. Mann, and R.E. Ulanowicz, eds., The Unesco Press, Paris.
Margalef, R., 1978, Life-forms of phytoplankton as survival alternatives in an unstable environment, Oceanol. Acta, 1: 493.
Marshall, S.M., and Orr, A.P., 1972, "The Biology of a Marine Copepod,", Oliver and Boyd, London.
McCave, I.N., 1975, Vertical flux of particles in the ocean, Deep Sea Res., 22: 491.
Mills, E.L., 1975, Benthic organisms and the structure of marine ecosystems, J. Fish. Res. Board Can., 32: 1657.
Mills, E.L., 1980, The structure and dynamics of shelf and slope ecosystems off the North East coast of North America, in: "Marine Benthic Dynamics," K.R. Tenore and B.C. Coull, eds., University of South Carolina Press, Columbia, S.C.
Mills, E.L., Pittman, K., and Tan, F.C., in press, Food web structure on the Scotian shelf, Eastern Canada: A study using ^{13}C as a food-chain tracer, Rapp. Proc. -Verb. Cons. Int. Explor.Mer.
Müller, P.J., and Suess, E., 1979, Productivity, sedimentation rate and sedimentary organic matter in the oceans. I. Organic carbon preservation, Deep-Sea Res., 26A 1347.
Müller, P.J., and Mangini, 1979, Productivity, sedimentation rate and sedimentary organic matter in the oceans. III. Organic carbon decomposition rates, Earth Planet. Sci. Lett., 51: 94.

Paffenhöfer, G.A. and Knowles, S.C., 1979, Ecological implications of fecal pellet size, production and consumption by copepods, J. Mar. Res., 37: 35.

Parsons, T.R., 1979, Some ecological, experimental and evolutionary aspects of the upwelling ecosystem, S. Afr. J. Sci. 75: 536.

Parsons, T.R., Takahashi, M., and Hargrave, B.T., 1977, "Biological Oceanographic Processes," 2nd ed., Pergamon, Oxford.

Peinert, R., Saure, A., Stegmann, P., Stienen, C., Haardt, H., and Smetacek, V., 1982, Dynamics of primary production and sedimentation in a coastal ecosystem, Neth. J. Sea Res. 16:276.

Pingree, R.D., Holligan, P.M. and Mardell, G.T., 1978, The effects of vertical stability on phytoplankton distribution in the summer on the northwest European shelf, Deep-Sea Res., 25: 1011.

Platt, H.M., 1979, Sedimentation and the distribution of organic matter in a sub-Antarctic marine Bay, Estuar. Coast. Mar. Sci. 9: 51.

Platt, T. and Subba Rao, D.V., 1970, Primary production measurements on a natural plankton bloom, J. Fish. Res. Bd. Canada, 27: 887.

Platt, T., Mann, K.H., and Ulanowicz, R.E., eds., 1981, "Mathematical models in Biological Oceanography," The Unesco Press, Paris.

Provasoli, L., 1979, Recent progress, an overview, in: "Toxic dinoflagellate blooms," D.L. Taylor and H.H. Seliger, eds., Elsevier, North Holland.

Pollehne, F., 1981, Die Sedimentation organischer Substanz, Remineralisation und Nährsalzrückführung in einem Flachwasserökosystem, Ph.D.Thesis, Kiel Univ., Kiel.

Raymont, J.E.G., 1980, "Plankton and productivity in the oceans. I. Phytoplankton," Pergamon, Oxford.

Reynolds, C.S., Thompson, J.M., Ferguson, A.J.D., Wiseman, S.W., 1982, Loss processes in the population dynamics of phytoplankton maintained in closed systems, J. Plankton Res., 4: 561.

Reynolds, C.S., and Wiseman, S.W., 1982, Sinking losses of phytoplankton in closed limnetic systems, J. Plankton Res., 4: 489.

Rowe, G.T., 1971, Benthic biomass and surface productivity, in: "Fertility of the Sea," J.D. Costlow, ed., Gordon and Breach, London.

Schnack, S.B., 1982, The structure of the mouth parts of copepods in Kiel Bay, Kieler Meeresforsch., 29: 89.

Sissenwine, M.P., Cohen, E.B., and Grosslein, M.D., in press, Structure of the Georges Bank ecosystem, Rapp. Proc. -Verb. Cons. Int. Explor. Mer.

Skjoldal, H.R., and Lännergren, C., 1978, The spring phytoplankton bloom in Lindaspollene, a land-locked Norwegian Fjord. II. Biomass and activity of net- and nanoplankton, Mar. Biol., 47: 313.

Slobodkin, L.B., 1961, "Growth and Regulation of Animal Populations," Holt, Rinehart and Winston, New York.

Smayda, T.J., 1970, The suspension and sinking of phytoplankton in the sea, Oceanogr. Mar. Biol. Ann. Rev., 8: 353.

Smayda, T.J., and Boleyn, B.J., 1965, Experimental observations on the flotation of marine diatoms. I. Thalassiosira cf. nana, Thalassiosira rotula and Nitzschia seriata, Limnol. Oceanogr., 10: 499.

Smayda, T.J., and Boleyn, B.J., 1966, Experimental observations on the flotation of marine diatoms. II. Skeletonema costatum and Rhizosolenia setigera, Limnol. Oceanogr., 11: 18.

Smetacek, V., 1980a, Annual cycle of sedimentation in relation to plankton ecology in western Kiel Bight, Ophelia, Suppl 1: 65.

Smetacek, V., 1980b, Zooplankton standing stock, copepod fecal pellets and particulate detritus in Kiel Bight, Estuar. Coast Mar. Sci., 11: 477.

Smetacek, V., Bodungen, B. von, Bröckel, K. von, and Zeitzschel, B., 1976, The plankton tower. II. Release of nutrients from sediments due to changes in the density of bottom water, Mar. Biol., 34: 373

Smetacek, V., Bröckel, K. von, Zeitzschel, B. and Zenk, W., 1978, Sedimentation of particulate matter during a phytoplankton spring bloom in relation to the hydrographical regime, Mar. Biol., 47: 211.

Smetacek, V. and Hendrikson, P., 1979, Composition of particulate organic matter in Kiel Bight in relation to phytoplankton succession, Oceanol. Acta, 2: 287.

Smetacek, V., Bodungen, B. von, Knoppers, B., Neubert, H., Pollehne, F. and Zeitzschel, B., 1980, Shipboard experiments on the effect of vertical mixing on natural plankton populations in the Central Baltic Sea, Ophelia, Suppl. 1: 77.

Smetacek, V., Bodungen B. von, Knoppers, B., Peinert, R., Pollehne, F., Stegmann, P. and Zeitzschel, B., in press, Seasonal patterns in the structure of an inshore pelagic system, Rapp. Proc. -Verb. Cons. Int. Explor. Mer.

Staresinic, N., Rowe, G.T., Shaughnessey, D., Williams III, A.J., 1978, Measurement of the vertical flux of particulate organic matter with a free-drifting sediment trap, Limnol. Oceanogr., 23: 559.

Staresinic, N., Hovey Clifford, C. and Hulburt, E.M., in press, Role of the Southern Anchovy, Engraulis ringens, in the downward transport of particulate matter in the Peru coastal upwelling, in: "Coastal Upwelling: Its sediment record," E. Suess and J. Thiede, eds., Plenum Press, New York.

Steele, J.H. and Baird, I.E., 1972, Sedimentation of organic matter in a Scottish sea loch, Mem. Ist. Ital. Idrobiol., 29 (Suppl.) 73.

Steele,J.H., 1974, "The structure of Marine Ecosystems," Harvard Univ. Press, Cambridge, Mass.

Stockton, W.L., and DeLaca, T.E., 1982, Food falls in the deep sea: occurrence, quality, and significance, Deep-Sea Res., 29: 157.

Taghon, G.L., Nowell, A.R.M., and Jumars P.A., 1980, Induction of suspension feeding in spionid polychaetes by high particulate fluxes, Science, 210: 562.

Titman, D., and Kilham P., 1976, Sinking in freshwater phytoplankton: some ecological implications of cell nutrient status and physical mixing processes, Limnol. Oceanogr. 21: 409.

Tufts, N.R., 1979, Molluscan transvectors of paralytic shellfish poisoning, in: "Toxic dinoflagellate blooms," D.L. Taylor and H.H. Seliger, eds., Elsevier, North Holland.

Turner, J.T., 1977, Sinking rates of fecal pellets from the marine copepod Pontella meadii, Mar. Biol., 40; 249.

Turner, J.T., and Ferrante, J.G., 1979, Zooplankton fecal pellets in aquatic ecosystems, BioScience, 29: 670.

Walsh, J.J. 1981, Shalf-sea ecosystems, in: "Analysis of marine ecosystems," A.R. Longhurst, ed., Academic Press, London.

Webster, T.J.M., Paranjape, M.A., and Mann, K.H., 1975, Sedimentation of organic matter in St. Margaret's Bay, Nova Scotia, J. Fish. Res. Bd. Can., 32: 1399.

Wickstead, J.H., 1963, "An Introduction to the Study of Tropical Plankton," Hutchinson, London.

Williams, P.J.leB., 1981, Incorporation of microheterotrophic processes into the classical paradigm of the planktonic food web, Kieler Meeresforsch., Sonderh. 5: 1.

Yentsch, C.M., Yentsch, Ch. S., and Strube, L.R., 1977, Variations in ammonium enhancement, an indication of nitrogen deficiency in New England coastal phytoplankton populations, J. Mar. Res., 35: 537.

Zeitzschel, B., 1965, Zur Sedimentation von Seston, eine produktionsbiologische Untersuchung von Sinkstoffen und Sedimenten der westlichen und mittleren Ostsee, Kieler Meeresforsch., 21: 55.

Zeitzschel, B., Diekmann, P., and Uhlmann, L., 1978, A new multisample sediment trap, Mar. Biol., 45: 285.

COUPLING THE SUB-SYSTEMS - THE BALTIC SEA AS A CASE STUDY

B.-O. Jansson, W. Wilmot and F. Wulff

Institute of Marine Ecology, The Askö Laboratory
University of Stockholm, Box 6801
S-113 86 Stockholm, Sweden

INTRODUCTION

Even if we adopt two commonly used criteria for defining a subsystem, namely, stronger interactions within the system than over the boundaries (Webster, 1979) and physically recognized boundaries (Rowe, 1961) we can differentiate two types of subsystems, the geographical, for example the coastal region, and the ecological such as the pelagic zone. This does not exclude the possibility of treating the coastal region as an ecological system but usually the region would include several clear-cut ecosystems more or less highly connected. Being a higher organizational level the regions have a slower dynamic response and determine the climate for the enclosed ecosystems. The urgent need for better knowledge of the processes and flows of the biological systems (Platt et al., 1981) has been exemplified by stressing the importance of primary production, respiration, excretion feeding, etc. This is certainly of importance but there is an equal lack of understanding of the degree of connection between the phytal, pelagic and softbottom subsystems or between the coastal and offshore areas. We should therefore not concentrate our efforts on one organizational level alone but through continuously climbing up and down the hierarchical ladder bring our knowledge towards an increasing understanding of the total system's behaviour.

The Baltic Sea is one of the most intensively studied marine systems in the world. We will not give another review (see e.g. Jansson, 1972, 1978; Magaard and Rheinheimer, 1974; Voipio, 1981) but rather give some illustrations of the importance of studying couplings between subsystems in order to understand the behaviour of the total system.

CHARACTERIZATION OF THE TOTAL BALTIC SYSTEM

A short description of its main characteristics should stress the temperate climate with a heavy annual solar pulse, the dominance of rocky shorelines, often greatly increased through archipelagoes, the stratified brackish water and an organism assembly of comparatively few euryhaline marine and freshwater species. Figure 1 tries to summarize the main structures of the system. The estuarine region where the salinity still fluctuates due to land runoff variations is merged with the coastal zone for simplicity. The diffuse border between the coast and the offshore regions is set somewhat seaward of the point at which the primary halocline strikes the bottom. The role of the physical processes as for the transfer of biological matter is indicated by using the complex concept "coastal dynamics" as a forcing function driving both horizontal and vertical transports. The connection to the North Sea is only indicated and does not show the actual effects on nutrient and oxygen levels nor the transports of organisms. The various connections between regions and subsystems and their physical and biological characteristics will be described in greater detail below but this rough model sets the scene for our play.

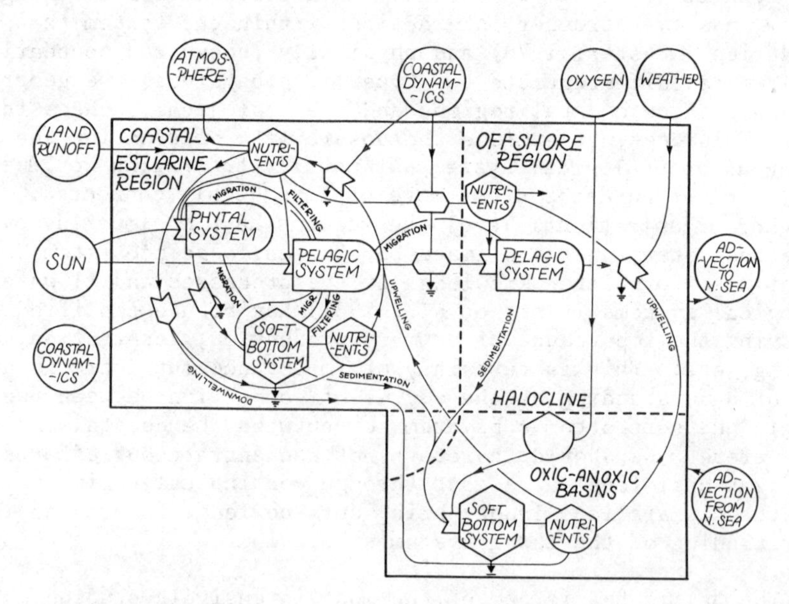

Fig. 1. Conceptual model of the total Baltic system, showing the main connections between the subsystems and regions. For clarity the estuarine and coastal region have been merged and the thermocline in the surface layer omitted. (Jansson, 1983).

Bathymetry

The Baltic Sea is an inland shelf-sea with a meridional extent of 1500 km and zonal extent of 650 km. The drainage basin is 4.3 times the surface area of roughly 400 000 km^2. The Baltic consists of a series of deep basins separated by sills. Three large gulfs, the Gulf of Bothnia, the Gulf of Finland and the Gulf of Riga extend far inland from the Baltic Proper (Figure 2). The Gulf of Bothnia is further separated into a shallow northern Bay (Bothnian Bay) and a deeper Sea (Bothnian Sea) connected to the Baltic through the Åland Sea. The Baltic has a volume of roughly 20000 km^3 and has an average depth of 60 meters with a maximum of 459 meters and a sill depth of 18 meters.

Water Balance

The river flow of 479 km^3 $year^{-1}$ represents some 40% of the total water balance (Falkenmark and Mikulski, 1974). The inflow of groundwater has been estimated to be only a per cent of the water balance. Evaporation (183 km^3 $year^{-1}$) and precipitation (183 km^3 $year^{-1}$) each contribute 15% of the total but roughly balance each other over a year. Salt water inflow through the Danish Sounds driven by density differences contributes nearly twice as much (737 km^3 $year^{-1}$) to the water balance as runoff. Outflow of the brackish surface water roughly balances runoff and inflow over longer time scales (1216 km^3 $year^{-1}$). Water level variations due to water exchange with the North Sea are on the order of tens of centimeters over several days resulting in transports of possibly 10 km^3 day^{-1}. Of course, the long term average is zero.

Salinity, Temperature and Density

The dominant feature of the Baltic proper is a permanent halocline separating a surface brackish layer, 6-7 $^{o}/_{oo}$, from deep water, 8-12 $^{o}/_{oo}$. The halocline deepens from 30 meters near the entrance to the basin to 70 meters in the deepest parts. The halocline is maintained by inflow of North Sea water underneath an outflowing layer of brackish water formed from mixing of runoff with deep water. The highly intermittant inflowing water has a mean salinity of 17.5$^{o}/_{oo}$ and has been observed to be as much as 22 $^{o}/_{oo}$. In the summer a thermocline is formed at ca 30 meters with temperatures of 15 - 20°C in the surface layer and 4 - 5°C below the halocline. In the winter mixing down to the halocline reduces the surface temperatures near to zero (Figure 3A). This water, 0 - 3°C, forms a layer between the the summer thermocline and the halocline (Figure 3B). The sigma-t distribution in the summer (Figure 4) shows the additional stratification due to the thermocline absent in the winter but still reflects primarily the salinity distribution.

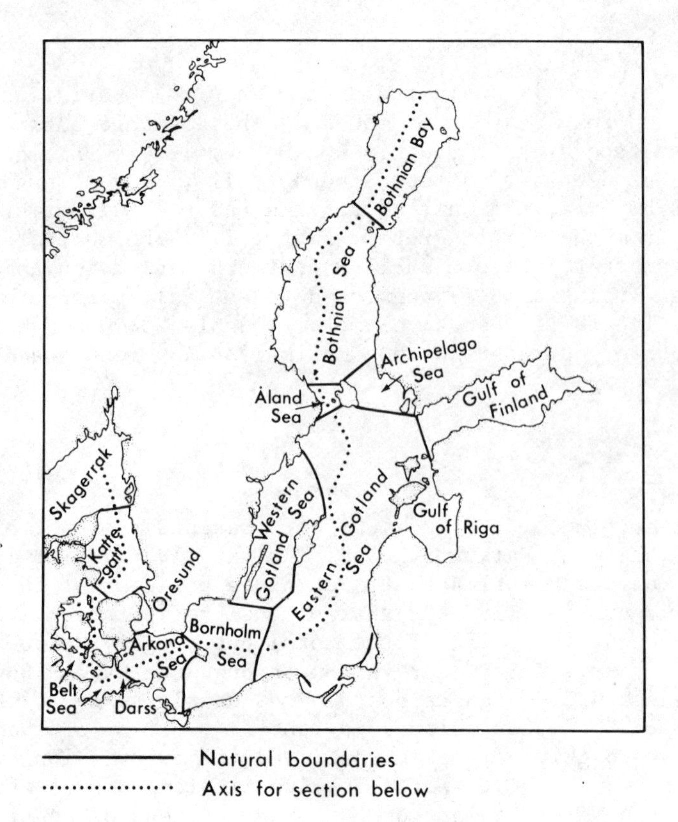

Fig. 2. Map of the Baltic with regional names. (Jansson, 1978).

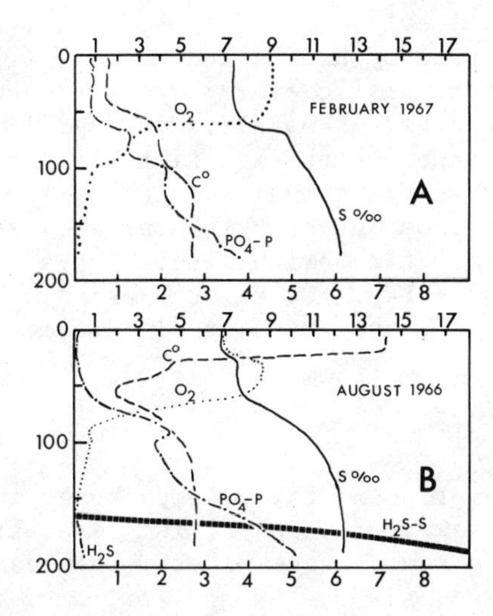

Fig. 3. Representative vertical profiles of temperature (C), salinity (‰), oxygen (ml/l), and H_2S (ml/l) according to the upper scale and phosphate (μg-at/l) and H_2S (μg-at/l) according to the lower scale during the winter (A) and summer (B) in the Baltic Sea. (Fonselius, 1969).

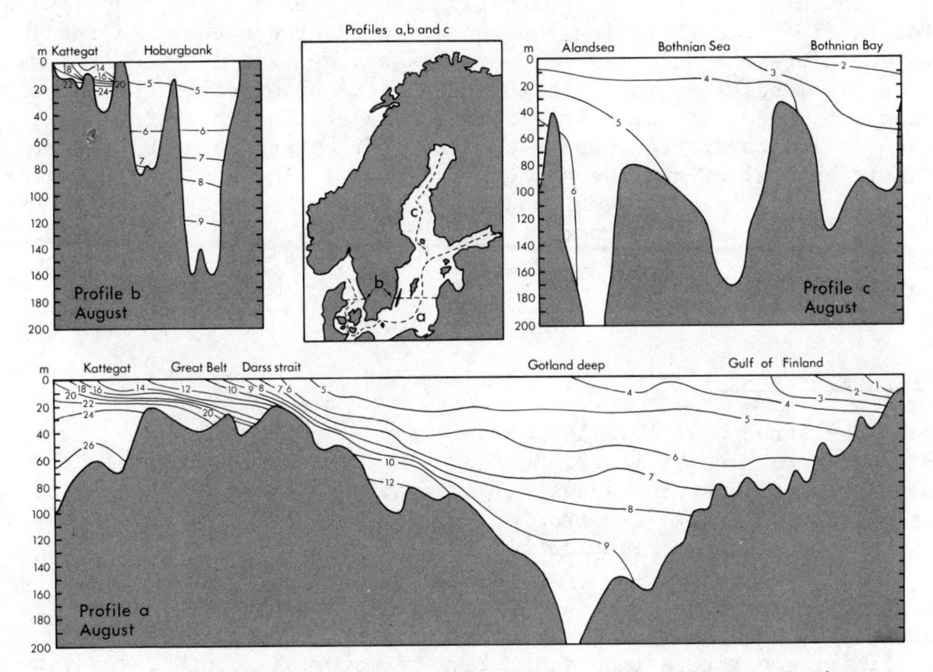

Fig. 4. Sections of sigma-t in the Baltic (Bock, 1971).

The Baltic is clearly a fjord type estuary according to the classification proposed by Hansen and Rattray (1966) with strong stratification characteristic over most of its extent except the Gulf of Bothnia. Further, the Baltic is a sill controlled fjord since the bottom reaches up to significantly influence the density and inflow of water in the transition zone. The water of the Bothnian Sea originates from the brackish surface layer of the Baltic Proper and exhibits little variability. The salinity decreases northward into the Bothnian Bay. Over large areas of the Gulf of Bothnia the weak halocline is erased by unstable convection in the fall.

Mean Circulation

The mean circulation in the Baltic Proper is weak and clearly related to the excess water supply. The velocities are of the order of a few cm sec^{-1} in the surface water. The mean circulation is cyclonic in all three main basins and reflected in the monthly mean temperature charts. Due to weak mean winds the mean circulation is maintained by density differences between water in the North Sea and runoff.

Characteristics of ecological subsystems

The difficulties in demarcating ecosystems are not made less by the task of characterizing the connections between them. Although we could present examples of couplings between small-scale systems like algal belts we shall chiefly restrict ourselves to the three main subsystems of the Baltic Sea: the vegetation covered bottoms here called the phytal subsystem, the soft bottoms devoid of macroscopic vegetation and the pelagic subsystem. The quality and quantity of the fluxes between subsystems are set by the properties of respective systems. A characterization of the major Baltic systems is therefore necessary before going into an analysis of the different interactions.

The phytal subsystem

The structural elements of most phytal systems can roughly be divided into robust, large, more or less branched, perennial producers and filamentous annual algae growing on hard substrates or as epiphytes. The phycial structure of the phytal system is an important habitat character but difficult to quantify.

The primary production shows a clear annual periodicity: the annuals succeed each other in bursts of organic production and when declining constitute important flows of material to other subsystems. The spring bloom of sessile diatoms, starting already below the ice cover, is followed by brown (Pilayella littoralis), green (Cladophora glomerata) and red (Ceramium tenuicorne) blooms. The latter also extend over the deepest hard bottoms where it constitutes an important structural element and substrate for microscopic epiflora and fauna.

The perennial algae are dominated by Fucus vesiculosus, an important habitat and substrate for both microscopic and macroscopic organisms down to at least 8 meters depth (figure 5), where it is replaced by other brown and red algae often forming loose-lying mats entangled in the dense Mytilus beds which cover the deeper hard bottoms.

The sunlit shallow sediment bottoms are inhabited by a vegetation of phanerogames (Zostera marina, Ruppia spiralis, Potamogenton spp.), more important as structures than as producers of organic material except in the northernmost basins which are totally dominated by euryhaline freshwater plants and animals with low biomasses (Kautsky et al., 1981).

The phytal system provides good examples of the ecosystem attributes stated by Reichle et al. (1975) as decisive for securing the persistence of a system: formation of an energy base, a carbon reservoir, element recycling and rate regulation. The combination of small/fast and big/slow producer components is examplified by the annual algae which rapidly reacts to changes in insolation with a production of up to 30 mg C (g dry wt)$^{-1}$ day^{-1}, compared to perennial algae with values of ca 4 mg C (g dry wt)$^{-1}$ day^{-1} (Wallentinus, 1978). This gives a turnover rate of roughly 20 and 130 days for the respective groups. The carbon base is secured mainly by the large biomasses of Fucus.

Jansson and Kautsky (1977) calculated a net macrophyte production of 750 ton C for the 160 km^2 primary research area of the Askö laboratory in the northern Baltic proper. Elmgren (in press) used 75 g C m^{-2} year^{-1} as an estimate of total benthic primary production including the microalgae.

The fauna of the phytal system, reaching biomasses in the northern Baltic proper close to 500 g dry wt m^{-2}, is dominated by filter feeders, mainly the mussel Mytilus edulis (Figure 5). The constant low salinity excludes the main predators of the mussels and the hard bottoms, especially where vegetations is sparse, are covered with dense mats of mussels of all sizes. The total filtering capacity of the mussels is imposing. The total 10,000 tons of mussels (dry wt including shells) in the 160 km^2 investigation area could filter a water volume equivalent to the total volume of the area in three months (Kautsky, 1981b). Extremely clear water just above the mussel beds is conspicuous to the diver.

Far less abundant are the invertebrate omnivores and carnivores, dominated by Amphipods.

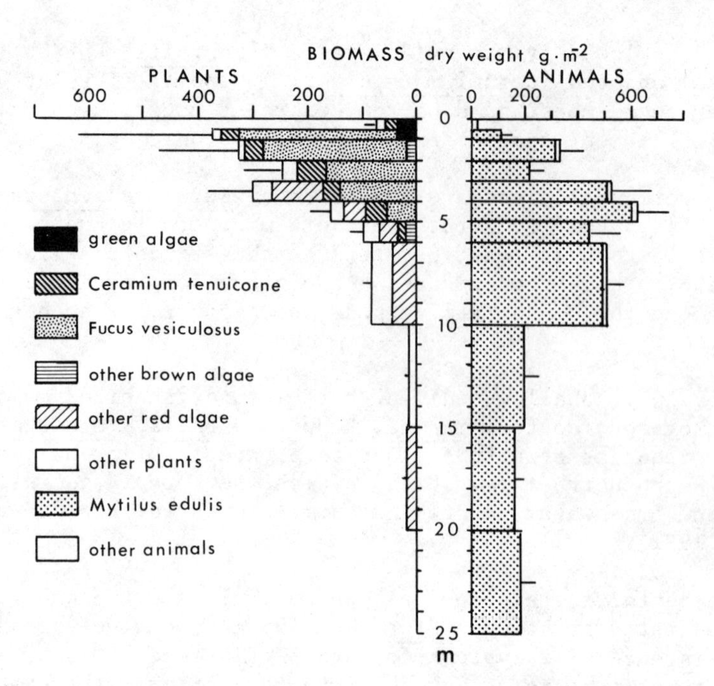

Fig. 5. Biomass and vertical distribution of producers and consumers of the phytal subsystem in the Askö-Landsort area (160 km^2). (Jansson and Kautsky, 1977).

The top carnivores, the fish, show a mixture of freshwater species like pike, perch, bream, roach and ruff, and saltwater species like sand goby, black goby, father lasher and eel-pout. Line-transect quantifications of demersal fish in the Askö-Landsort area show a mean value of 4 g wet wt m^{-2} composed of some 15 species (Jansson, Aneer and Nellbring, pers comm.).

To extract from this qualitative and quantitative diversity, measures of the trophic status of whole phytal system, community metabolism techniques (Odum and Hoskin 1958) have been applied (Jansson, 1974; Guterstam 1979; Jansson and Wulff, 1977, and Kautsky and Wallentinus, 1980). As the Fucus covered bottoms in many respects constitute the dominating phytal community structure, annual studies of their metabolism, made by Guterstam (1979) for the Askö-area, represent the changes in subsystem properties (Figure 6A). The system is clearly autotrophic during the cold periods with a maximum P/R-ratio around 2 in February (Figure 6 B). The growth of the producers, measured on single Fucus plants, is highest in July (Figure 6A) and lowest in November-December.

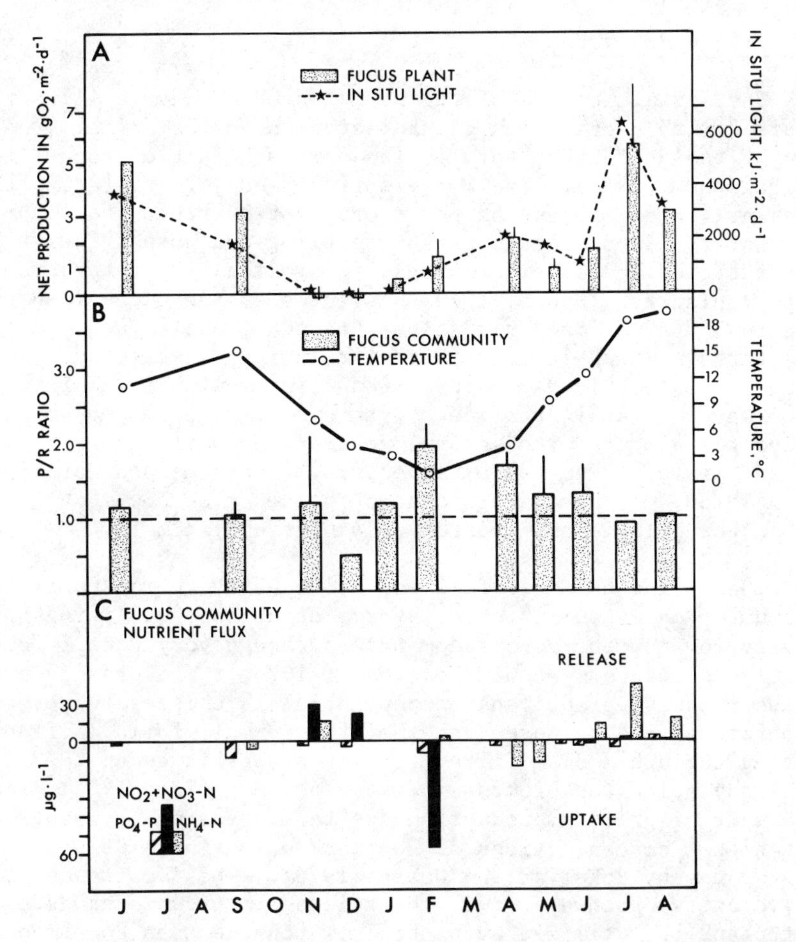

Fig. 6. Annual dynamics of a Fucus community measured in plastic enclosures with plants devoid of macrofauna (A) and intact communities (B and C). Explanations in the text. (Based on Guterstam, 1979).

The nutrient kinetics is roughly presented in Figure 6C, where the concentration changes in the plastic enclosures after the separate 24-hour experimental periods are plotted. Due to the absence of transports to and from outside, only trends can be read from these experiments. Nitrate-nitrite are released during late autumn due to the annual maximum in consumer biomass. In early spring the accumulated nutrients in the water and the low recycling from the consumers is depicted in a strong uptake in February. During the summer, strong growth and a high nitrogen demand is apparently more than covered through the respiration by the highly active consumers since ammonia appears in excess.

The pelagic subsystem

The dynamic nature of the plankton community is illustrated in Figure 7 with data from a coastal station in the primary research area of the Askö Laboratory in the northern Baltic proper (Larsson and Hagström, 1982). The heavy spring bloom, initiated by the increasing light, exhausting the storage of nutrients built up during the winter, is to a large extent settling out when the water column is stratified and nutrient supply is limited. Long-term monitoring of phytoplankton (Hobro, 1979) has revealed the general succession pattern but also large variations between years. The spring bloom may start in March in years with little ice and calm sunny weather or as late as early May after a long winter with thick ice. The bloom may be over in two weeks with one very sharp peak or be prolonged for a month with several smaller peaks if interrupted by wind induced mixing or unstable convection. The time and duration of the spring bloom may have a pronounced influence on how much is utilized within the pelagic or exported to other subsystems.

The spring bloom is followed by an early summer nutrient limited minimum. During summer, small forms capable of utilizing nutrients at very low concentrations dominate although very conspicuous blooms of large bluegreens occur (Horstmann, 1975). The grazing pressure of the increasing zooplankton population is compensated by the higher temperatures and turnover rates of these small forms. Picoplankton, passing through a 3 μm filter, may then constitute up to 25% of the total phytoplankton biomass (Larsson and Hagström, 1982). An enhancement of primary production due to increasing nutrient recycling with higher concentrations of zooplankton has been demonstrated for this system by McKellar and Hobro (1976). The importance of alternative energy pathways, i.e. the release of organic exudates by the phytoplankton, utilized by bacteria which are then consumed by ciliates and macrozooplankton (Williams, 1981) has been demonstrated for the Askö area (Larsson and Hagström, 1979 and 1982) as well as for the Kiel Bight (see review by Rheinheimer, 1981). About 16% of the annual primary production (30% of the production during summer stratification) is released as exudates contributing to 50% of the bacterial energy requirements in the Askö area.

The overwintering zooplankton population is very small and very little of the spring bloom, which may constitute more than 30% of the annual production, is grazed (Hobro et al. in press). However, bacteria and ciliates with much higher turnover rates than the larger zooplankton can respond faster to the rapid changes in organic production. With a bacterial food chain, a larger proportion of the highly pulsed primary production can be channeled to higher trophic levels than would be possible if only a phytoplankton-zooplankton chain was present. The increase of bacteria and ciliates after the spring bloom and their subsequent decrease with increasing zooplankton as illustrated in Figure 7, support this hypothesis.

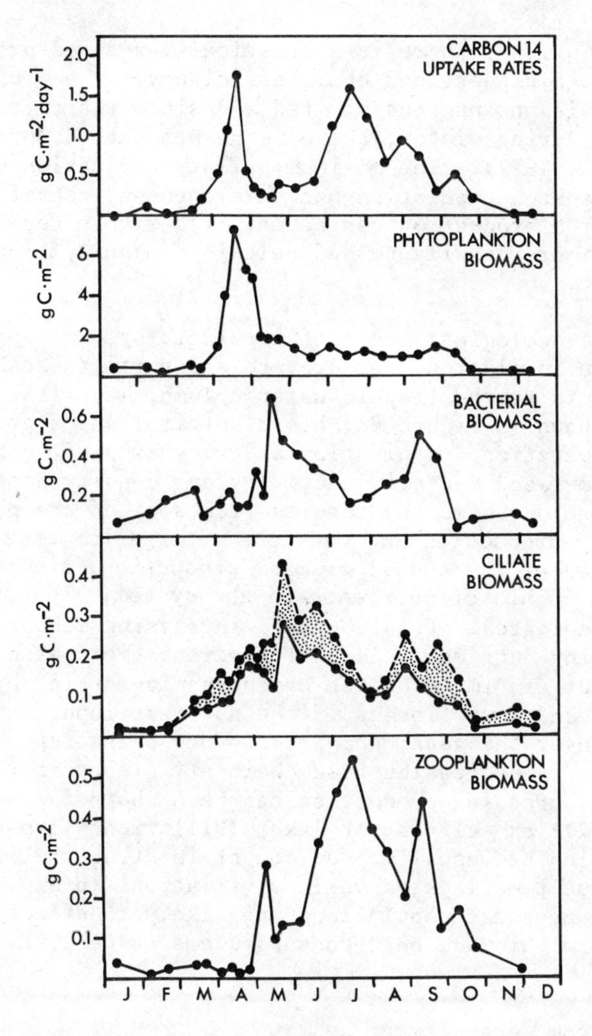

Fig. 7. Annual dynamics of the pelagic system in the Askö-Landsort area. The surves represent integrated values for 0 to 20 m (zooplankton 0 to 35 m) water column. Shaded area represents the autotrophic Mesodinium rubrum. (Hagström and Larsson, in press).

The salinity gradient in the Baltic affects the species composition but there are also clear gradients in productivity due to differences in nutrient inputs and hydrography of the different basins. Annual phytoplankton and zooplankton production estimated with comparable methods (Ackefors et al., 1978) is 191 and 20 g C m^{-2} in the southern Baltic proper whereas the same calculations for the nothernmost part of the Bothnian Bay only reach 13 and 4 g C m^{-2}, respectively. Thus, there is a drastically reduced productivity in the north but a higher food chain efficiency. The production in the Bothnian Bay is phosphorous limited and since there is no nutrient accumulation during winter, there is no pronounced spring bloom (Wulff et al., 1977). The maximum primary production occurs in summer when a zooplankton population has developed and therefore, a larger fraction of the production can be utilized within the system, compared to the more heavily pulsed pelagic community found in the Baltic proper.

A simple ecological model of a pelagic system, forced by seasonal variations in insolation, temperature and vertical stability driving nutrient inputs to the trophic water column, may illustrate this. The model, shown in Figure 8A, has a detrital and a grazing food chain, sedimentation, respiration and recycling of nutrients dependant on consumer and bacterial activity and density dependant mortality rates. Figure 8B shows the results from some of the pertubations. The secondary production and the loss of organic matter, expressed as percentages of the annual primary production, are plotted against total annual inputs of nutrients to the system. The model shows a decreasing ecological efficiency and increasing losses of organics with increasing nutrient loads. The pertubations also show the important effect of timing of the hydrodynamic forcing by giving higher efficiencies and lower losses if the nutrient inputs are evenly distributed over the year, compared to one pulse input prior to the spring bloom. A decreasing food chain efficiency in the pelagic system with increasing production has been shown for marine areas (Cushing, 1971) as well as for lakes (Hillbricht-Ilkowska, 1977). The explanation has usually been sought in structural (size) changes that occur but time lags as well as variations in amplitude between producer and consumer populations are likely to affect the transfer efficiencies within and between subsystems as well.

The soft bottom subsystem

The overwhelming part of the Baltic bottoms are composed of soft sediments, ranging from gravel and sand in the north and south to mud in the central and deeper parts. The uniform macroscopic structure compared to the phytal benthic system is to some extent compensated for by an internal heterogeneity created by horizontal and vertical gradients of physical and chemical parameters. The successively decreasing oxygen level in the bottom water since the 1950's has now affected ca 25% of the total area of the Baltic Proper (Andersin et al., 1978).

This detritus feeder dominated system shows clear changes in the extended salinity gradient. In the south the hydrographical noise due to proximity to the North Sea causes irregular changes in the bottom communities from invasions of marine larvae and some-

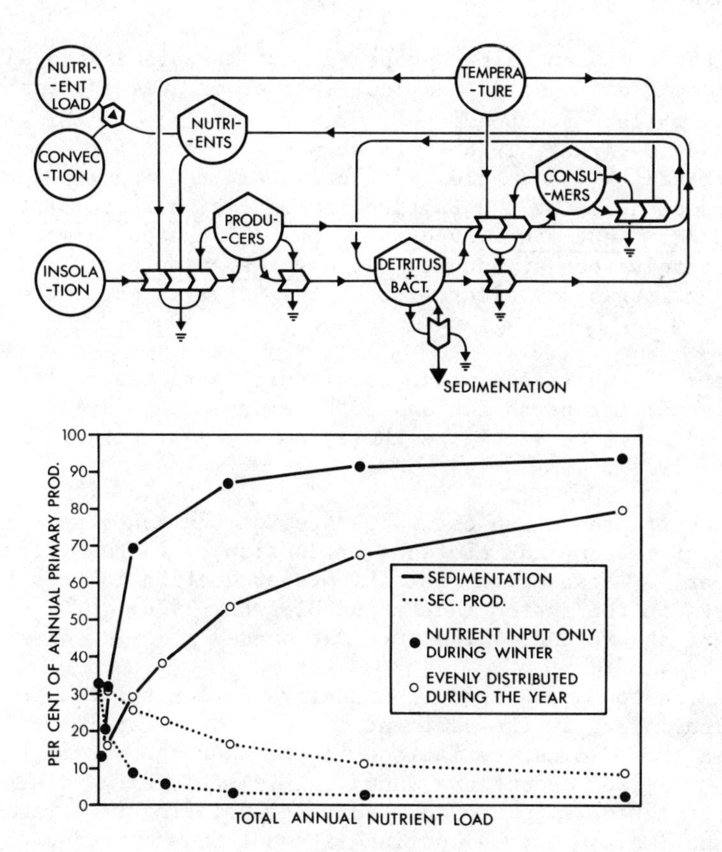

Fig. 8. Conceptual model of the pelagic subsystem, forced by temperature, nutrient flow and insolation. The grazing and detrital food-chains control the recycling of nutrients. The effect of pulsing or steady nutrient inputs during the year are shown in the lower diagram.

times even adult invertebrates and fish. In the Baltic proper the number of taxa per grab sample attains 20 compared to 0-2 in the extreme north (Elmgren, 1978). Also the biomass shows a clear N-S gradient with ca 600 g m^{-2} wet weight in the Kiel Bay (Arntz, 1971), 150 g in the Baltic proper and less than 1 g in the extreme north (Elmgren, 1978). This heavy drop is caused by changes in community structure due to the decreasing salinity and, as will be discussed later, by changes in the overall productivity and couplings to other subsystems.

In the northern Baltic proper 90% of the biomass is distributed over 6 macrofauna species and four species of demersal fish (Elmgren, in press) while corresponding figures for the Kiel Bight in the south are 20 and 10 (Arntz, 1978). Large parts of the Baltic mud bottoms are dominated by two species of the deposit feeding amphipod *Pontoporeia* accompanied by the Baltic clam *Macoma balthica*, the omnivorous isopod *Saduria entomon*, the carnivorous polychaete *Harmotoe sarsi* and the carnivorous sipunculid *Halicryptus spinulosus*. This combination of organisms clearly shows the detritus based character of this system.

The meiofauna has a weaker latitudinal gradient. The 2 g wet wt m^{-2} in the far north increases three- to fourfold in the central Baltic and seems to stabilize there, not increasing further south (Elmgren, 1978).

The best production estimates are for the Askö-Landsort area where total macro- and meiofauna production is in the order of 7 g C m^{-2} $year^{-1}$. Most of the detritivore production is used by the carnivores in the system (Ankar and Elmgren, 1976). The dominating demersal fish are cod, flounder, four-horned sculpin and sand goby.

The shallow areas show an annual variation related to food and oxygen conditions in the sediment. A mud bottom at 10 meters in the Askö area showed clear variations in the abundance of microflora, micro-, meio- and macrofauna (Ankar, 1979) primarily related to the depth of the redoxcline in the sediment, pointing to a minimum of macrofauna (mainly *Macoma*) during late autumn and a tendfold maximum during summer (Figure 9).

Respiration measurements on sediment cores from the depth range 20 - 150 meters of sandy and highly organic mud, showed a range of 100 - 800 mg O_2 m^{-2} day^{-1} with a mean of 400 (Jansson and Hagström, 1978). For a system in steady state this would correspond to the burning of 0.15 g C m^{-2} day^{-1}. On a yearly basis that would give ca. 50 g C m^{-2}.

Estimates of the nutrient release/uptake of Baltic softbottoms are very scarce. Bell-jar studies of shallow mud bottoms have shown release during anoxic conditions of ca 5 mg PO_4 m^{-2} day^{-1} (Schippel *et al.*, 1973) and a release of total nitrogen of 0.25 g N m^{-2} $year^{-1}$ (recalculated from Engwall, 1978). On the basis of field surveys in the southern Baltic proper, Holm (1978) calculated the mobile phosphorus pool to be 5-10 g m^{-2} with a release of 18 mg m^{-2} day^{-1} from bottoms below the halocline. The sediments of the Gotland basin, anoxic since the 1950's, contain no pool of potentially mobile phosphorus.

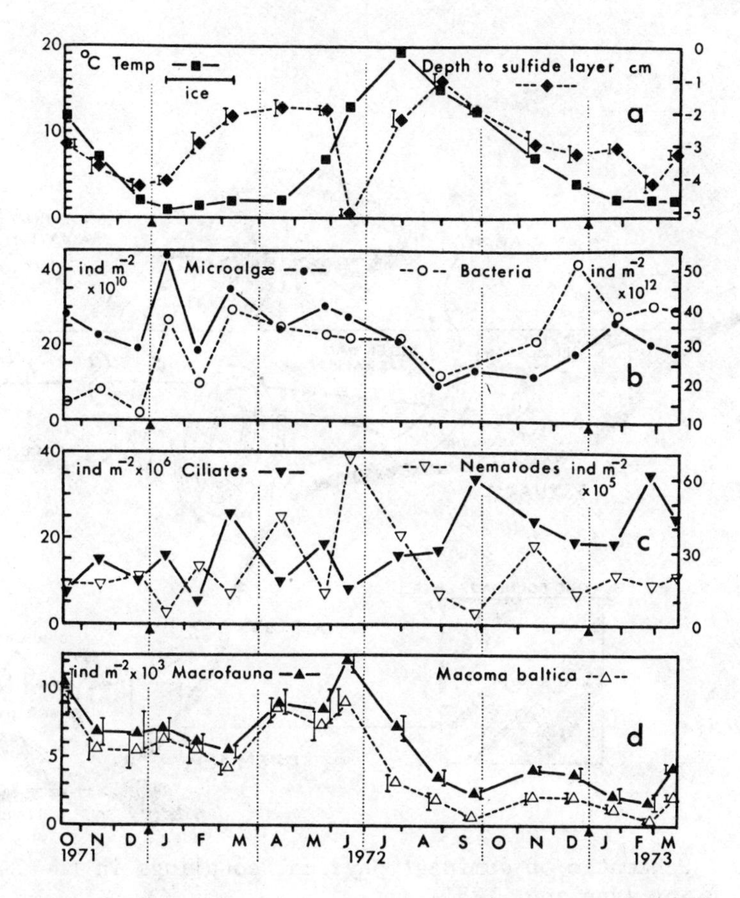

Fig. 9. Annual variations in a softbottom subsystem at 10 m depth, Askö. (Anker, 1979).

CONNECTIONS BETWEEN REGIONS AND SUBSYSTEMS

Physical Couplings

The movement of water in the marine ecosystem is the dominant coupling of organisms and inorganics. Motion on scales for molecular processes such as friction and diffusion (millimeter scales) up to the size of the Baltic (1000 kilometers) contribute to the redistribution of organic matter and nutrients. It is often convenient to classify motion on the basis of time and space scale properties. The most elementary distinction is between mean motion and fluctuating motion (Reynolds, 1895). Inherent in this division is the specification of the time period used to define the mean. Variations with a time scale longer than the averaging period are interpreted as time dependency of the mean motion. Fluctuations of shorter time

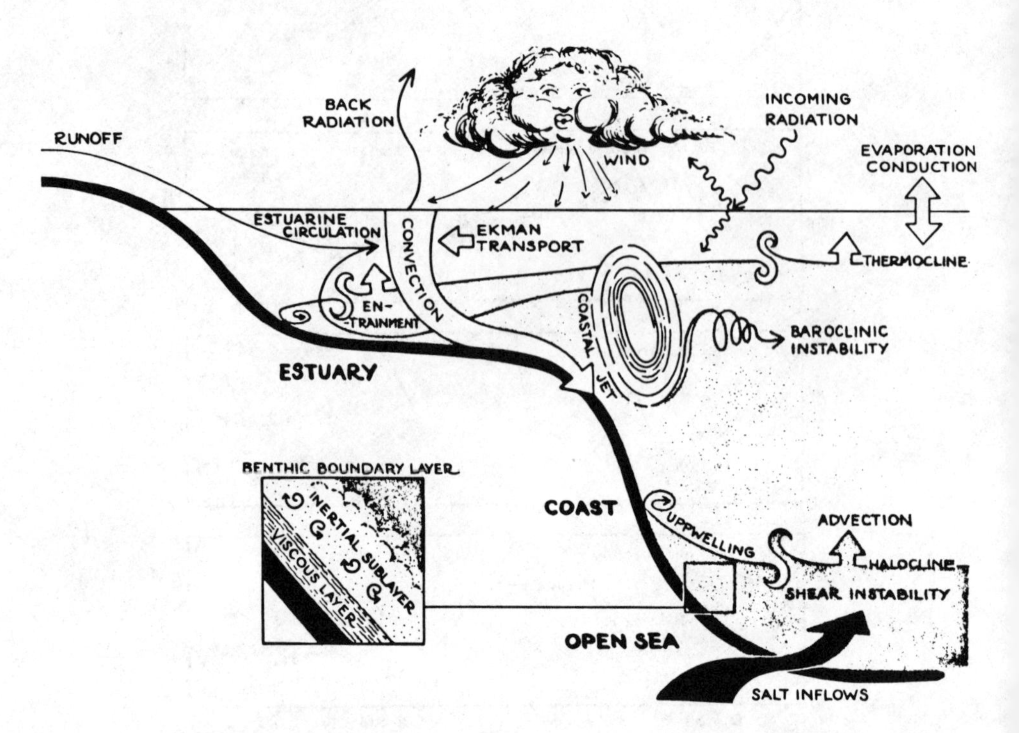

Fig. 10. Schematic of dominant physical couplings in the Baltic Sea (Jansson, 1981)

scale do not contribute to the mean, but are rather regular oscillations (waves) or erratic vascillations (turbulence) about the mean. Although the classification including the averaging period is arbitrary it is often useful because linearity applies to many different processes. Even non-linear processes are only so over a limited area of the permissible time-space domain.

In order to isolate dominant coupling processes in the Baltic it is necessary to consider the nature of driving forces as well as scales (Figure 10).

Mean estuarine circulation

The motion with the longest time scale (40 years) and largest space scale (1000 km) is the mean estuarine circulation driven by the density difference between sea water outside the Baltic and the mean runoff of river water to the surface layer around the Baltic (Ekman, 1893). Runoff flows out at the surface carrying salt which has been mixed upward from heavier oceanic water flowing into the Baltic at depth.

A time dependent salt water inflow also exists with a much shorter time scale relating to the dominant forcing: large quantities of North Sea water are pushed by storms (days) through the Danish Sounds, accelerating into deep basins in the Baltic (months) where it remains trapped (years) after the storm has passed. The probability of such a storm induced inflow increases as the deep salty water is slowly mixed upward into the brackish water outflow reducing the density of the deep Baltic. Such a forcing is rather nonlinear (inflow of North Sea water followed by outflow of Baltic water result in a huge net gain of salt) and is dependent on other processes contributing to vertical transports (Wilmot, 1974).

Characteristic estuarine circulation occurs in coastal estuaries of the Baltic as well. Since these estuaries are much smaller than the Baltic the time scale of water renewal is much shorter (weeks). Another mechanism exists in these not strongly sill controlled areas to accelerate flushing to a time scale of days. Coastally trapped variations in the pycnocline depth cause flushing of the estuaries in a matter of days (Toll *et al.*, 1982).

Coastal Upwelling Dynamics

Much of what we know about coastal upwelling in enclosed seas is a result of the International Field Year of the Great Lakes in 1972 (Csanady, 1978; Simons, 1980; Bennet, 1974). Observations of coastally trapped thermocline fluctuations (Walin, 1972a) and a boundary layer analysis of the dynamics (Walin, 1972b) gave impetus to the idea that coastal upwelling could be important in the Baltic. Wind blowing parallel to the coast results in currents flowing in the same direction as the wind near both coasts and against the wind in the deep central region. The wind must move a larger mass of water in the deep portions of the Baltic than in the shallow coastal region. The coastal water acquires greater velocities than in the interior. The ends of the basin force a return flow of the coastal water in the interior. The earth's rotation plays a role since the Baltic is so large. In the surface layer a component of velocity to the right of the wind across the Baltic results from a combination of the earth's rotation and friction (Ekman transport). A return flow is caused by the piling up of water at the side of the Baltic.

On the side of the Baltic where the surface water is piled up a depression of the thermocline (downwelling) occurs. On the opposite side shere surface water has been withdrawn from the coast the thermocline causes an additional velocity alongshore (coastal jet). The halocline (especially in the absence of the thermocline) responds in a similar manner. Such coastally trapped motions contribute to onshore-offshore transport of nutrients and organic matter as well as upwelling-downwelling. Material from coastal estuaries flushed by large pynococline excursions are quickly dispersed along the coast by the coastal jet.

Offshore only a very weak residual mean counter clockwise circulation has been observed. However, transient wind driven currents interacting with depth variations generate mesoscale eddies causing upwelling which can be persistent for much longer periods (weeks) than the forcing (days).

Mesoscale Eddies

Dynamical investigations have shown that the growth of the eddies can be described adequately using linear theory (Gill *et al.*, 1974). A small pertubation on a basic state in which the density surfaces are level and the bottom is flat can be expressed as a planetary wave with time scales long compared to a pendulum day. In reality, of course, the isopycnals are not horizontal and neither is the sea bottom. If the disturbance has a scale small compared with that on which isopycnals and topography vary then the perturbation still has the Rossby wave form except that the vertical structure depends on the vertical density gradient and the bottom slope. The major difference introduced by sloping isopycnals is that there is a source of energy available to drive unstable perturbations, the so called baroclinic instability. Near the Gulf Stream the growth rate of eddies is days because of the large slope of density surfaces. In the Baltic it is expected that the growth rate is comparable due to the enormous density gradients. Eddies grow spontaneously in baroclinicly unstable regions and can extract energy from the mean potential energy field resulting in preferred scales and vertical structure. The growth rate calculated typically shows a short wave and a long wave maximum. The long wave disturbances which are preferred extend over the whole depth and are sensitive to the bottom slope. Model simulations of mesoscale eddies (Holland and Lin, 1973) have revealed that the linear theory holds well for the growth of disturbances but that an equilibrium state approached is dependent on a balance between non-linear effects and linear propagation balance. The effect of an eddy field is that most of the energy in scales smaller than the eddy is kinetic whereas in scales larger it is primarily in potential energy. Eddy fields seem to contribute extensively to the vertical transport of nutrients in the interior of the Baltic (Kahru *et al.*, 1981).

Unstable Convection

If the vertical gradient of density decreases enough any small motion will cause the water column to mix rapidly. This is called dynamic instability. A further decrease of stratification will result in static instability. When the vertical density gradient divided by the density, i.e. the stability, is negative it is expected that overturning will occur. Such overturning results in a rapid (hours) redistribution of nutrients and organic matter in the water column.

Turbulent Mixing

Although it is impossible to separate mixing from processes generating mixing (wave breaking, upwelling, eddies, boundary drag, shear instabilities, etc.) it is often convenient to view turbulence as a phenomena in itself. It can be characterized as having randomness, diffusiveness (i.e. decreases gradients of material), being dominated by inertial effects, three dimensional and rotational, dissipative (i.e. continuous loss of energy to molecular friction), and a property of the flow and not the fluid. The kinetic energy of turbulence must be maintained by some instability at large scales and dissipated at small scales. The "cascade" of turbulent energy from large scales to small scales is controlled by various dynamic balances. The time change of turbulent kinetic energy is due to work by pressure, viscous transport, advective transport, buoyancy work (changes in potential energy) and generation and dissipation. In order to evaluate the dispersive transport properties of turbulence it is parameterized in a form analogous to molecular diffusion. Although such a practice is known to be valid under only unrealistically restrictive conditions it continues to be the only avenue available for description of large scale redistribution of material by turbulence.

Such parameterization relates a correlation of fluctuating current and material concentration to the mean gradient of concentration and the concentration itself (Lumley, 1975). The proportionality constant is called the eddy (not mesoscale eddy, but turbulent eddy) diffusion. The parameter can be related to bulk parameters and scales characterizing sources and sinks of turbulent kinetic energy. Often a well mixed layer at the surface reaching down to the thermocline (or halocline if the thermocline is absent) is observed. When input of energy to this layer exceeds dissipation it eats its way into deeper quiescent water bringing up nutrients. The downward velocity of the well mixed layer is called the entrainment velocity and has been related to the eddy diffusion using energy arguments (Kullenberg, 1976).

Transports

The function of the processes discussed so far are the transport of material. Physical transport of material can result from mean motion (advection) or random exchange of water masses of different properties (turbulent diffusion). In both cases material is moved around altering the distribution of water parcels carrying material. Molecular processes of diffusion enter with the essential task of moving material from one parcel to another.

In an effort to develop a budget for water, salt, nutrients and oxygen, Walin (1977) presented a description of material transports in a density stratified estuary. If one imagines a situation in which the deep water of the Baltic tries to upwell without mixing i.e. exchanging water parcels with the surface layer, the only result is that the halocline lifts upward. Without mixing advection through the halocline cannot occur. In the presence of a mixing flux of salinity, F, through the isohaline, s, an advective flux, G*s, occurs. Furthermore, $G = -\partial F/\partial s$, which means that a relation exists between the mixing flux of salt and advection of salt through s. Knowing the amount of water transported into the deep waters below the halocline and the change in volume below the halocline, the advective flux and the diffusive flux though the halocline can be calculated. The total upward transport of salt through the moving halocline is the sum of the two, H. If the distribution of some material, C, dependent only on salinity is known (assuming that it is relatively uniform along isohalines) then the transport (advective, diffusive and total) of C is known:

$$H_C = G{*}C + F{*}\partial C/\partial s$$

Budgets of nutrients and organic matter can be made using this description if the assumptions are satisfied, as they are in the Baltic Sea.

EXAMPLES OF COUPLINGS

Water Transports

Coastal Upwelling

Coastal upwelling has long been recognized as a major mechanism for the vertical transport of nutrients, especially in the presence of an appreciable vertical stratification as in the Baltic. Data of the surface temperature (Figure 11) reveal colder water patches in regions of upwelling which have been shown to be nutrient enriched (Figure 12).

Data taken in the Baltic conclusively show a coastal jet and strongly imply a Kelvin-topographic wave representation of pycnocline

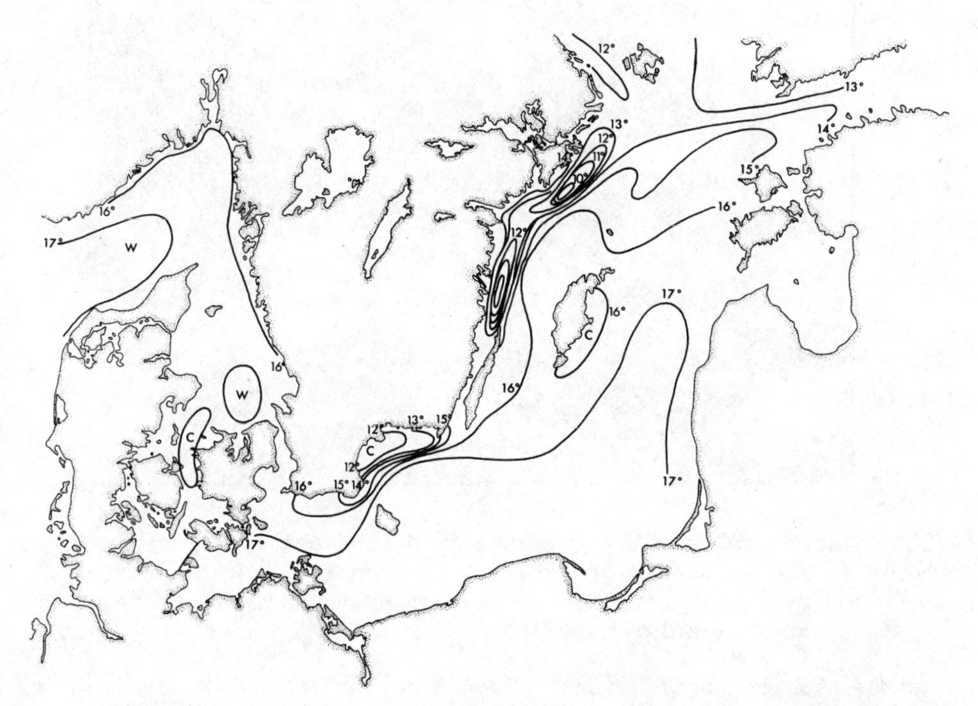

Fig. 11. Surface temperatures showing three recurrent nuclei of cold water after a period of dominating SW-winds, indicating coastal upwelling. (Jansson, 1978 based on values from the Swedish Meteorological and Hydrological Institute).

motions as predicted by theory (Shaffer, 1975). A model of the Baltic coastline dynamics has been developed for use in estimating upwelling/downwelling transports as well as onshore/offshore transports of organic matter and nutrients. The motion of the permanent halocline maintained by the time dependent estuarine circulation is an added feature in the Baltic. A section has been chosen between Landsort and the island Gotska Sandön for comparison with data taken by Shaffer (1979b). The model formulation follows Bennett's (1974) scaling for a vertical cross section with a rigid lid and no net volume transport normal to the section. The coastal response is a transient one to the wind (days). A model simulation was made using the model response to a NE wind without bottom friction effects during a period when the thermocline was absent. The longshore transport simulated (Figure 13A) corresponds to the velocity observed (Figure 13B). The surface (A) and deep currents (B) are in opposite directions near Gotland in both simulations and observations. The deep current (B) is absent and the surface jet (A) is broader near

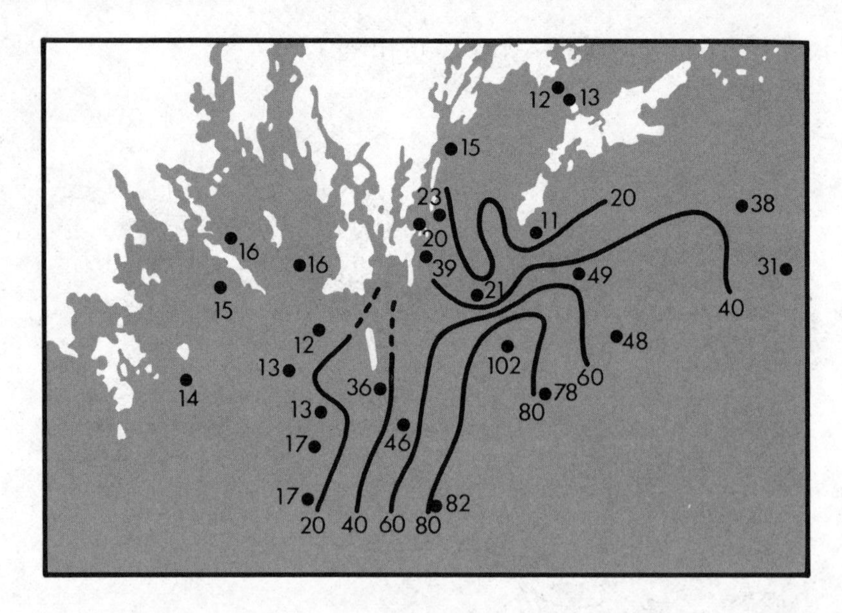

Fig. 12. Coastal upwelling in the Askö-Landsort cold water nucleus. The isopleths of total phosphorus (μg/l) show an area of high concentrations, contrasting to the surrounding poor areas (Karlgren, 1978).

Askö as in the observations. The halocline depths (Figure 13C) agree with the data (Figure 13D) in that there is an elevation close to Gotland larger than a depression near Askö. The simulation depicts the response after one day's wind. The magnitudes of velocity and elevation agree well if the wind has blown several days (as stated by Shaffer).

Offshore Upwelling

Satellite observations of the surface layer of the Baltic have revealed an abundance of mesoscale (10 km - 100 km) eddies as evidenced in a photograph of the blue green algae *Nodularia spumigena* in an area east of Gotland recorded by LANDSAT2 (Figure 14).

The eddy kinetic energy in the Baltic gives rise to thinning of the layer between the thermocline and halocline and an increased vertical mixing (Kahru *et al.*, 1981). Associated with thinning of the intermediate layer by 30% (Figure 15A) is a doubling or tripling of chlorophyll implying an increased upward supply of nutrients (Figure 15B). The abundance and persistence of such eddies indicates that they may be the major mechanism controlling upward nutrient flux in the Baltic.

The relative importance of coastal and offshore upwelling is still unclear and there are no comparable data available to show any

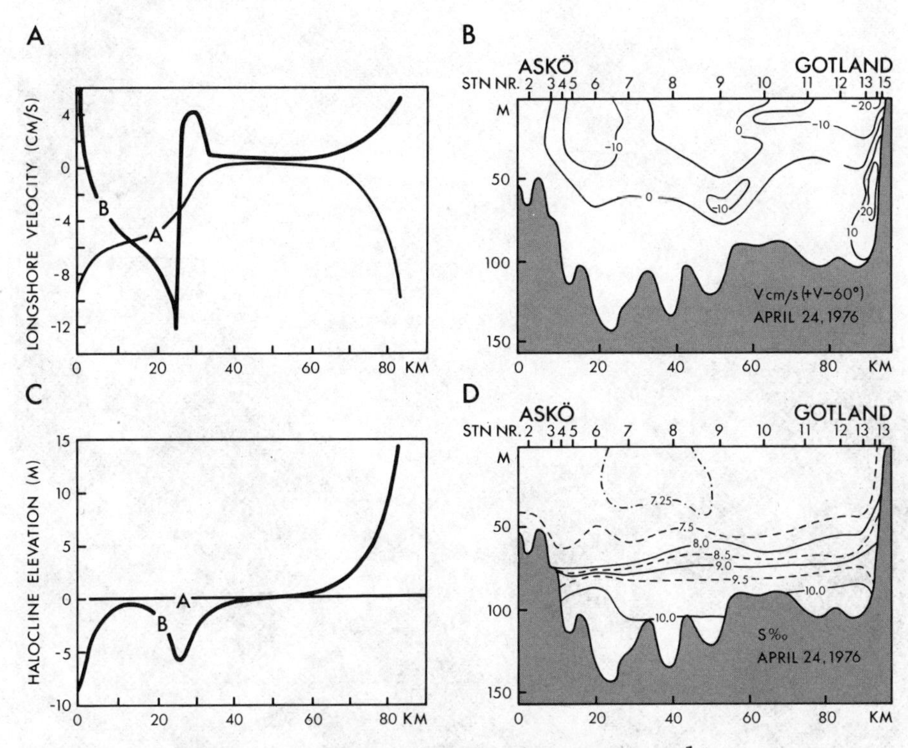

Fig. 13. (A) Simulated longshore current (cm sec^{-1}) in a section between Landsort and Gotska Sandön.
(B) Synoptic section of longshore velocity between Askö and Gotland. (Schaffer, 1979b).
(C) Simulated halocline elevation (m) in a section between Landsort and Gotska Sandön.
(D) Synoptic section of salinity between Askö and Gotland. (Schaffer, 1979b).

differences in the primary productivity. The importance of the site for upwelling is significant especially during summer stratification. Offshore systems would be able to maintain a higher primary productivity and a higher food chain efficiency if offshore nutrient upwelling occur (as illustrated with the simple model, Figure 8). Otherwise, offshore systems have to be maintained by recycling and "seeding" of organic matter produced in the coastal upwelling areas in order to compensate for sedimentation losses.

Estuarine Circulation

The deep waters of the Baltic have become increasingly anoxic over the last three decades. The question has been raised whether

Fig. 14. Satellite registration (Landsat-2) of eastern central Baltic, showing the bloom of the blue-green algae *Nodularia spumigena*. The algal masses are good indicators of the surface currents, here showing mesoscale eddies (Nyqvist, pers comm).

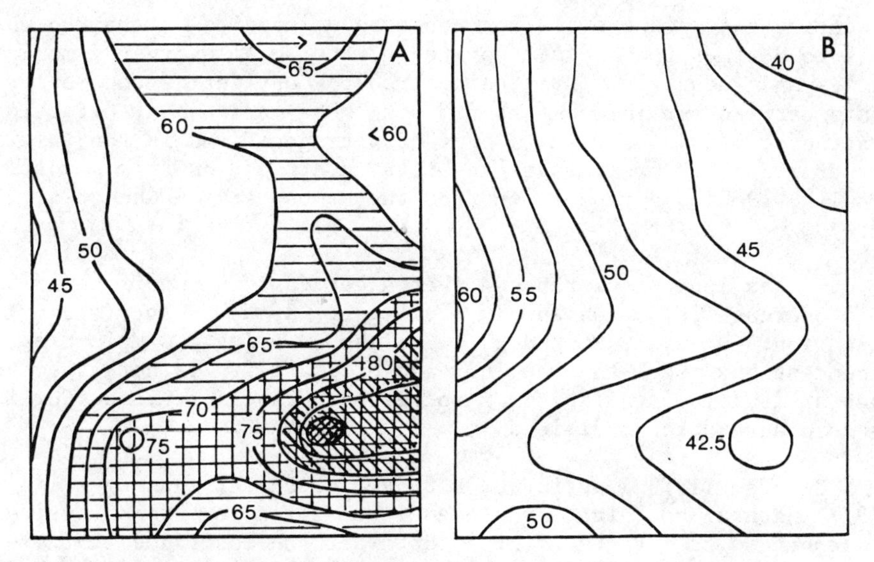

Fig. 15. Offshore upwelling due to mesoscale eddies.
(A) Map of the total chlorophyll (integral of the Chl a concentration between 0 - 60 m) in mg m^{-2} in a 25 km square area east of Gotland in the Baltic Sea. (From Kahru et al., 1981).
(B) Map of the intermediate layer thickness of m in the same area.

this is a result of changes in the mean estuarine flow due to a decreasing runoff during the century, or changes in time dependent supply of oxygen to the deep Baltic or increased oxygen consumption. In order to investigate the first possibility a numerical model of the phosphorous budget was used to simulate the effects of changing loading on the oxygen climate of the deep Baltic (Sjöberg et al., 1972). Assuming phosphorous to be the limiting nutrient, they identified a phosphorous feedback loop which could accelerate growth of anoxic conditions: increased phosphorous in the surface layer caused increased production of sedimenting organic matter to the bottom causing increased oxygen consumption leading to release of phosphate frome the sediment which when upwelled would start the cycle again. An amplification factor of 3 was found for an increased loading. It was later found (Karlgren, 1978) that nitrogen and not phosphorous was the limiting nutrient in the pelagic zone of the Baltic Proper. Recently, it has been recognized that a mechanism exists to promote the feedback loop even though nitrogen is limiting: phosphorous dependent blue green algal nitrogen fixation during July-August. Atmospheric nitrogen can be considered as a practically unlimited source and it is thus no wonder that nitrogen fixation by the abundant blue-green algae is responsible for a nitrogen input to the Baltic which is 2 to 3 times the amount supplied by the Swedish rivers, municipalities and industries (Brattberg, 1980). Considering the measurements carried out in the Arkona basin (Hübel and Hübel, cited in Brattberg, 1980) the total amount of nitrogen fixed in the Baltic proper might very well equal the total contribution from runoff (Pawlak, 1980) and deposition from the atmosphere (Rodhe *et al*., 1980).

It has long been proposed that a decrease in runoff contributed to a decrease in oxygen supply (Fonselius, 1969) by increasing stratification. However, Welander (1974) showed using a simple time dependent box model that an increase of stability is not expected in such a situation, ruling out a decrease in runoff as cause for increased anaerobic conditions.

Shaffer (1979a) performed a budget calculation using Walin's (1977) natural coordinate representation for advective/diffusive transports with a rather limited data set and concluded that biological consumption caused by increased manmade loading was the cause of anaerobic bottom areas.

Although the cause for the anoxic conditions is unclear the effects are striking even outside the deep basins. Total phosphorous as well as phosphate have increased not only in the deep basins but also in the mixed winter surface layer of the Baltic Proper (Nehring, 1979). Although there are no long term series of primary production measurements available, Cederwall and Elmgren (1980) have shown a four-fold increase of the macrobenthic biomass above the halocline in the Central Baltic, compared to samples taken in the 1920's. This is a clear example of the strong coupling in the form of a feedback-loop between the pelagic and softbottom subsystems.

Unstable Convection

In a numerical model analysis of a spring phytoplankton bloom Sjöberg and Wilmot (1977) found that an observed deepening of biomass and replenishment of nutrients could not be explained by wind induced mixing. It was concluded that unstable convection must have caused the redistribution. During the spring the water is warmed up from below the temperature of maximum density, 2.4°C at about 7 $^{o}/_{oo}$, tending to promote formation of dense water at the surface. Further numerical experiments with a mixed layer model for stable and unstable mixing (Smeda and Wilmot, 1978) confirmed that conditions sufficient for unstable convection existed during the exponential phase of the bloom.

This mechanism was isolated as possibly very important for increasing the total production by renewing nutrients and exporting organic matter from the coastal pelagic zone to deep offshore benthic regions. Because the water is shallower in the coastal areas warming is more effective and convection occurs there first.

Turbulent Mixing

Kullenberg (1974) extensively studied parameterization of the eddy diffusion and entrainment velocity in the Baltic using dye experiments in the pycnocline. He estimates that open sea vertical transports can account for the upward flux of salt and nutrients required. Recent refinements in calculating the eddy diffusion in the Baltic include numerical simulation of a closed form of the turbulent energy equation (Omstedt and Sahlberg, 1982). Comparison with observed temperature distribution is encouraging but it is still too soon to evaluate how much better the estimates are than earlier numerical experiments with bulk parameter mixing (Svensson and Wilmot, 1978). The next step is to attempt to separate all pathways of energy from the wind and buoyancy to mixing and make relative comparisons of each generating process.

Zooplankton as couplings between benthic-pelagic, coastal-offshore Systems

Besides regional differences, there are distinct differences in species composition of the zooplankton community between coastal and offshore areas. In the Baltic proper, the copepods *Temora longicornis* and *Pseudocalanus minutus elongatus*, which usually dominate the zooplankton biomass in offshore regions, overwinter as late copepodite stages and *Pseudocalanus* has even been found in high concentrations below the halocline at oxygen concentrations less than 2 ml l^{-1} (Hernroth and Ackefors, 1979). The adults of these cold stenotherm species never occur above the thermocline during summer which prevents their distribution into shallower coastal regions.

Rotifers of the genus *Synchaeta*, cladocerans and the copepod *Acartia bifilosa*, dominants in the coastal regions, overwinter as resting diapausing eggs in the bottom sediments. Figure 16 (redrawn from Kankaala, in press) shows the seasonal variation of the cladoceran *Bosmina longispina maritima* in the water column and the number of resting eggs in the sediment at a coastal station in the northern Baltic. There is a continuous decrease of eggs during the whole summer rather than just a drastic decrease prior to the massive occurence of *Bosmina* in the water column in August. A continuous recruitment to the small pelagic population with a high mortality ensures that there is always an initial population present, able to grow rapidly by parthenogenetic reproduction when conditions are favorable. If all eggs were hatched simultaneously, the risk of a

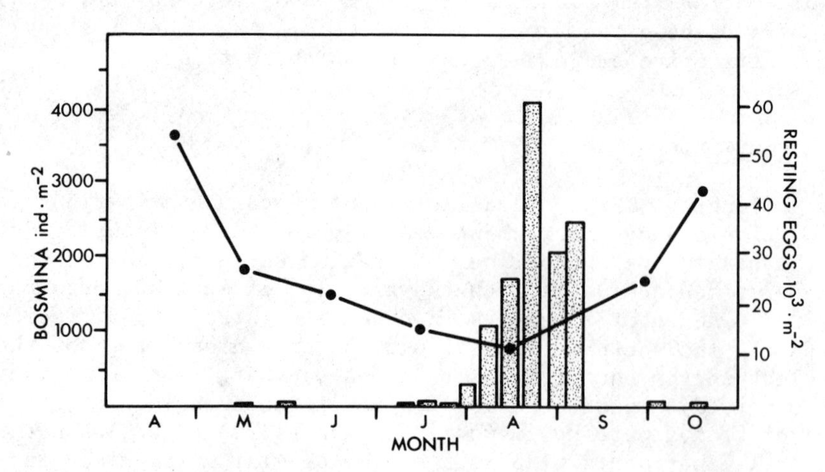

Fig. 16. The seasonal variation of individuals of the cladoceran Bosmina longispina maritima (bars) and number of resting eggs in the sediment (line). (Redrawn from Kankaala, in press).

mismatch in this highly unpredictable environment would be high. The anoxic conditions in the deeper parts of the Baltic prevent the hatching of eggs deposited by offshore populations and the recruitment must occur from eggs deposited in oxic coastal sediments.

Figure 17, redrawn from Hernroth (1977), shows the occurrence nauplii of A bifilosa and P m. elongatus at one coastal and one station 25 km offshore in the Askö-Landsort area, monitored simultaneously during the spring of 1976. The 40 meter deep coastal station shows twice as many Acartia mauplii recruited from benthic eggs during the maximum in March compared to the 160 m deep offshore station, where the sediment is anoxic. Adult Pseudocalanus were almost absent in the coastal zone which is reflected in the very few nauplii found there. In the offshore area the overwintering late copepodite stages of Pseudocalanus have developed into reproducing adults which is shown by the increasing numbers of nauplii found.

Cladocerans, rotifers as well as the naupliar stages of most Baltic copepods are found in the upper water mass subjected to the same advective transports. A mixing of the populations initiated in coastal and offshore areas during spring may then occur during summer. Bosmina, capable of rapid parthenogenetic reproduction during favourable conditions, may often be the most abundant species in late summer in offshore areas all over the Baltic proper (Hernroth and Ackefors, 1979).

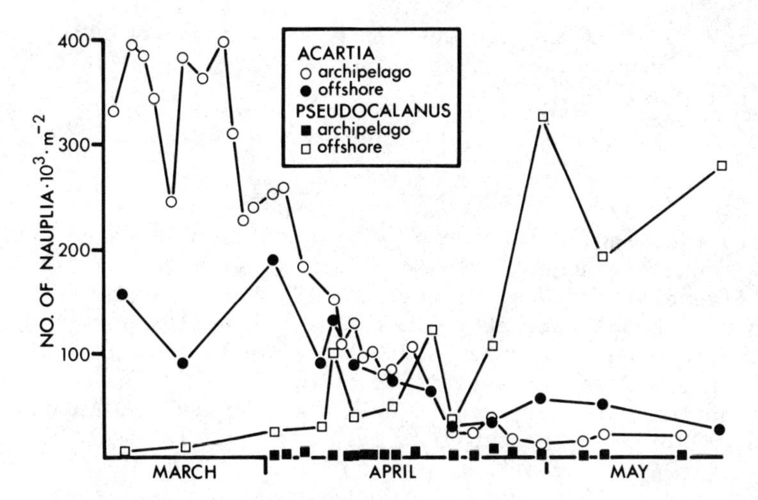

Fig. 17. Zooplankton couplings between coastal and offshore areas due to differences in overwintering strategies of the dominant copepods. Further explanation in the text. (Based on data from Hernroth, 1977).

Migration

Availability of food, space and water quality are responsible for the migration of aquatic organisms. At temperate latitudes these factors show annual variations which do not run parallel for the different subsystems. There are often time lags: the peak of the phytal spring bloom is usually earlier than the pelagic one and the maximum of the soft bottom biomass comes later than for the pelagic zooplankton. Systems which react fast will create a dynamic heterogeneity for the total system as answers to sudden increases in production potentialities like nutrient pulses, and insolation patterns. As not only the availability of food but also the concentration of it has a strong selective value, migration patterns have evolved, often with temperature or day length as triggers, which enable the organism to utilize food when and where it is in excess. Fish are well known examples of this, acting as links between subsystems and probably acting as stabilizers for the total system. This choice of habitat has a wider energetic background. The feeding aspect does not only concern the adult stages but often also the next generation as the spawning grounds coincide with suitable foraging areas for the future fry. The temperature of the location also bears a strong correlation to the physiological state of the animals, starving for example is more easily endured at low temperatures. Although there is a common knowledge that fish do migrate for example between shallow and deeper areas of the sea there are, at least for the Baltic, few quantitative data.

Couplings between phytal and soft bottom subsystems

Figure 18 includes a map of Öresundsgrepen, an area west of Åland, at the entrance to the Bothnian Sea from the Baltic proper. Of the five stations Nos 1-3 were established close to the mainland in the phytal zone at 2-8 meters depth while Nos 4 and 5 represent more open continuous softbottoms of 14 and 15 meters depth respectively. Fishing was carried out with gillnets from surface down to bottom for every second week (No.4 every week) during 1975-79 (Neuman, 1982). As the smallest mesh size was 17 mm, smaller fish like sticklebacks and gobies were not caught. Of the 27 recorded species the distribution of the most common demersal species are shown in Figure 18. The three freshwater species, roach (*Rutilus rutilus*), perch (*Perca fluviatilis*)and ruff (*Acerina cernua*) show the same annual distribution pattern. As the trapping of the animals is a function not only of their presence but also of their activity, the data have to be interpreted with caution, however. For all three species there is an immigration during April towards shallower areas, though partly masked in the diagram as stations 1-3 are pooled. In May-June they spawn, in late summer returning to the outer stations. Emigrating roach were trapped mainly in the pelagic zone in August-September before returning to the bottom waters, apparently exploiting the pelagic subsystem more than the other two species.

The fourhorn sculpin (*Myoxocephalus quadricornis*) shows another pattern. This fish is a glacial relict which during summer prefers temperatures below 9-10°C. In late autumn the population migrates to shallow areas to dig their nests and spawn. They remain in the phytal zone through March whereafter the adults migrate to deeper areas and the fry start their pelagic life (Westin, 1970).

The energetic background for these migrations is not clear for all the species. While the roach practically does not feed during winter the fourhorn sculpin feeds all the year round and the stomach contents clearly reveals the actual habitat; the soft bottom amphipod *Pontoporeia affinis*, another glacial relict, dominating the summer menu: its own roe, phytal sand gobies and gammarids being the main items during winter. The locomotory activity patterns measured in the laboratory agree well with the field behavior. The fish is night-active during summer and day-active in winter with a phase shift in spring and autumn (Westin, 1971). The perch shows a decrease in activity in temperatures below 10°C and feeding during winter is low (Aneer and Westin, pers comm.). Thus the soft bottom subsystem acts more as a refuge for perch, roach and ruff away from the lower temperature in shallower areas than as a food resource. The fourhorned

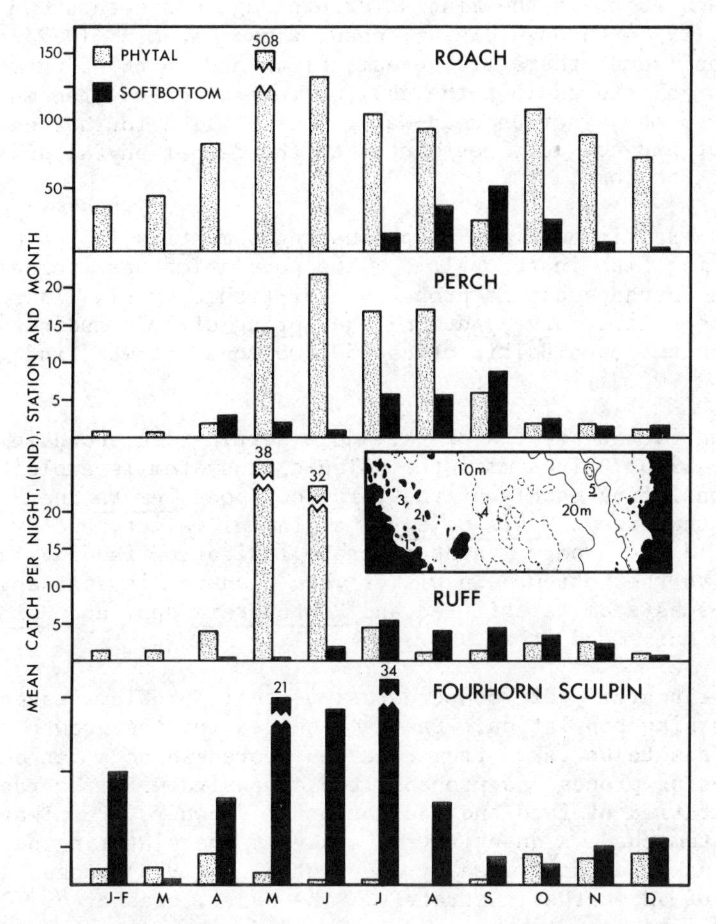

Fig. 18. Seasonal migration of the dominating demersal fish in Öresundsgrepen, South Bothnian Sea (map) as determined by gill net fishing. Stations 1, 2 and 3 are pooled to represent the phytal subsystem, 4 and 5 to give the softbottom subsystem. (based on Neuman, 1982).

sculpin, however, is apparently dependent on winter-feeding. Hansson (1980) recorded no summer migration to deeper water for the population in Lulea archipelago (Bothnian Bay) where the deeper water shows extremely low macrofauna biomasses. This brings the sculpin into potential competition with roach during summer which is in practice avoided by the former occupying the inner, the latter the outer shallow parts of the archipelago.

A downward migration towards the autumn from the green algal belt, via the Fucus-belt to deep sandy bottoms by fish, mainly perch, roach and ruff, in the Finnish archipelago has been shown by Bagge et al. (1975) through diving census work. Neuman (1977) classified those and some others like bream, bleak and roach as warm-water species able to exploit the shallow vegetation-covered mud bottoms during summer, whereas cold-water species like fourhorned sculpin, eel-pout and cod were restricted to the deeper phytal system or to the soft bottoms.

The Baltic herring, Clupea harengus membras is the economically dominating fish in the Baltic. The population has increased during the last decades but is probably over-fished by now (Thurow, 1980). Eco-integration surveys during the spring of 1976 indicated a total stock in the open Baltic of ca 450,000 tons wet wt (Lindqvist et al., 1977).

The feeding habits of the adult herring are shown in Figure 19 (Aneer, 1975). In summer the pelagic subsystem is exploited, the food consisting mainly of the large copepods Temora and Pseudocalanus. Toward the autumn Mysis relicta, a glacial relict, and Mysis mixta, both cold-stenotherm mysids dominate indicating feeding taking place closer to the bottom. In winter when plankton is scarce the soft-bottom subsystem is utilized and Pontoporeia spp. and Harmotoe sarsi are the main food items.

The scarcity of food creates energetic problems especially for the spawning population. Energy required for the growth of the gonads has to be taken from reserves stored as body fat during autumn The feeding process is probably too expensive with regards to the low concentration of food and the potential spawners are always found wit empty stomachs. A quantitative study of overwintering pelagic fish in the Askö-Landsort area by Aneer et al., (1978) suggests the evolution of an energetic strategy. The survey area of ca 100 km^2 was divided into an outer exposed area and an inner fairly sheltered area with several islands. The herring found in the outer area were feeding juveniles with a mean biomass of 0.5 g wet wt m^{-2}. The sheltered archipelago area contained only non-feeding adults forming large schools, splitting up or merging now and then during the period December - January. The total biomass for the area was estimated at about 12,000 tons wet wt, equivalent to one quarter of the total yearly Swedish herring catch ! This gives a mean biomass of 7.8 kg m^{-2}. A scanning survey the following year along the Swedish coast in the northern Baltic proper showed similar overwintering concentrations at three localities of the same topography.

This mechanism would be energy saving in at least three differen ways:

1. The schooling behaviour decreases the disturbances from predators,
2. The choice of sheltered but ventilated areas a/ decreases the stress from waves and currents but secures good oxygen

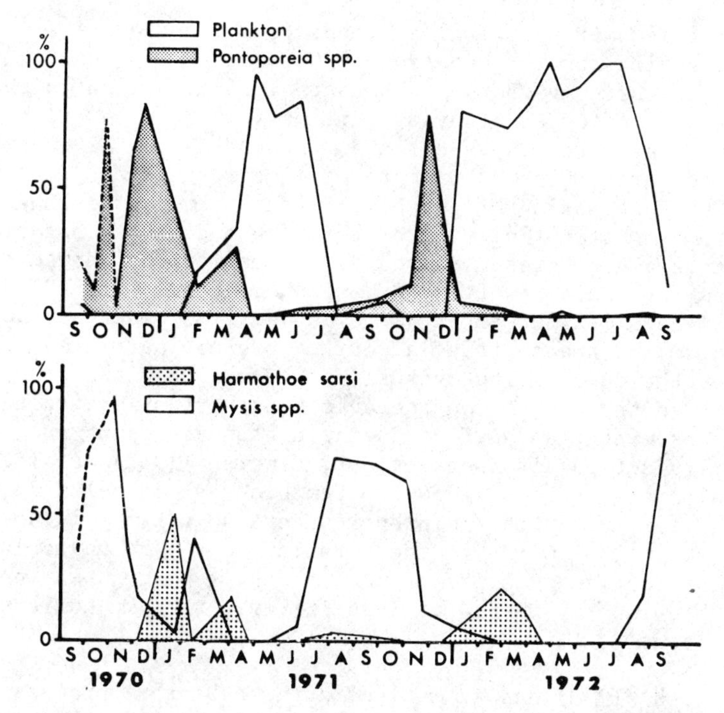

Fig. 19. Annual variation in the food of the Baltic herring plotted as percent of dominating food items at each sampling occasion. (Aneer, 1975).

conditions, and b/ concentrates the fish to the coldest water.

3. the population is close to optimal spawning grounds.

Spawning starts in May and continues intermittently during two months. The phytal subsystem is utilised as spawning grounds and most of the roe is deposited on filamentous brown algae like Pilayella. Hatching takes place after approximately two weeks and the postlarvae are caught in the pelagic subsystem from August (Aneer, pers. comm.).

Sedmentation

The measurements of sedimentation made so far in the Baltic, reviewed by Elmgren (1982), indicate that about 30% of the phytoplankton settles out annually. However, the data are only approximate due to the use of less than optimal sedimentation traps and the difficulties of compensating for resuspension of bottom sediment during storms. Macroalgal debris may form a substantial contribution to the shallow softbottoms. After each burst of primary production the senescent algae are freed from the substrate. Depending on the water circulation they are collected in more or less restricted areas into dense mats. When consisting of perennial algae they can be long lived, on softbottoms constituting shelter for fish and phytal organisms. For filamentous algae Persson (personal communication) showed closed to total remineralisation after 60 days while Schmidt (1980) found a total remineralisation after 40 to 100 days for macroscopic plants. A mechanism for transporting deposited organic particles offshore exists in the downwelling events in the area when the outward bottom currents are capable of considerable transport of particles (Shaffer, 1975). Assuming that half of the allochtonous inputs and benthic primary production also settles out, Elmgren (1982) estimates that the total annual organic input to benthic systems below 25 meters is 50-60 g C m^{-2} in the Baltic proper, about 40 g C m^{-2} in the Bothnian Sea and about 20 g C m^{-2} in the Bothnian Bay. The phytoplankton primary production constitutes more than 90% of organic input except in the Bothnian Bay where the large rivers, rich in humic substances provide about half.

Elmgren (1978) has summarised data on benthic biomasses as well as on phytoplankton primary productivity from different regions of the Baltic. These data show an increase in the macrofauna biomass by a factor of over 100 going from north to south. However, the corresponding increase in phytoplankton primary productivity is less than ten times. Elmgren *et al.* (in press) sought an explanation in the absence of filter feeders (the marine bivalves *Macoma baltica* and *Mytilus edulis*) in the Bothnian Bay, leading to longer and less efficient food chains in the benthic community. An alternative or complementary explanation could be the higher food chain efficiency within the pelagic community in the Bothnian Bay and thus less production available for other subsystems.

Calculations on the energy demands of the benthic systems, based on biomass and energy budget calculations for macro- and meiofauna and of oxygen consumption of the sediment show a fair agreement with the estimates on inputs, considering the large uncertainities inherent in most of the data. These calculations (Elmgren, 1982) give a total respiration corresponding to about 50 g C m^{-2} for the Baltic proper and the Bothnia Sea. The fauna is responsible for about 50% of the respiration in the Bothnian Sea but only to 25% in the Baltic proper due to the large anoxic areas devoid of metazoans.

The large seasonal variation in the pelagic community is reflected in the magnitude and quality of the organic matter that finally settles on the bottoms. About 40-60% of the large diatoms sinks out, almost undecomposed, after the peak of the spring bloom

(Hobro et al., in press). A much smaller proportion of the primary production settles out in the summer when small phytoplankton forms dominate and the grazing pressure of zooplankton is increased. The organic carbon content of the sedimentated material is lower than in the spring (U. Larsson, pers comm.) suggesting a lower nutritive value. A low organic content is also typical for the large amounts of matter caught in the traps during fall, originating from resuspension of bottom sediments during storms.

This highly variable energy input has a pronounced influence on the benthic communities. Elmgren (1978) compared data from two consecutive but highly different years showing the seasonal variations of primary production and biomass (Hobro and Nyqvist, pers. comm.) and the energy budget of the populations of Pontoporeia affinis and P. femorata in the Askö area. For Pontoporeia the variations in production (P), biomass (B) and elimination (E) are calculated from a cohort growth analysis on data collected by Cederwall (1977). Respiration (R) is calculated from the temperature-weight relationships on Pontoporeia by Johnson and Brinkhurst (1971). These data, Figure 20, show that after an initial low biomass of the amphipods in the winter of 1972, there is a threefold increase in three months after the release of juveniles in March-April. This rapid increase that occurs in spite of a heavy mortality of juveniles and spent females, is synchronized to the sedimentation of the spring bloom of phytoplankton. The Pontoporeia populations show a decreased growth rate and assimilation is almost balanced by respiration and elimination during summer when sedimentation is low. The production is however high enough to maintain the total biomass until the next spring. The spring bloom of 1973 is less intensive than in the previous year and this is reflected in the lower production of Pontoporeia that barely compensates the mortality of juveniles and spent females. The biomass is maintained during summer but the production is not sufficient in the fall to compensate for the heavy mortality associated with the mating period and the size of the population declines drastically. The annual mean biomass as well as the total eliminated biomass (mortality) is almost identical for the two years but the production of Pontoporeia is 2.7 times higher the first year. Elmgren (1978) therefore draws the conclusion that food, i.e. input from the pelagic subsystem, and not predation limits the populations.

Filter feeding

Filter feeding is an additional coupling to sedimentation, especially in the more shallow areas of the Baltic proper where Mytilus constitute the bulk of the biomass. As these mussels can filter out 1 μm particles with a 50% efficiency increasing to 100% for particles larger than 4 μm (Möhlenberg and Riisgaard, 1978), particles that do not easily sediment due to their small size are removed from the water.

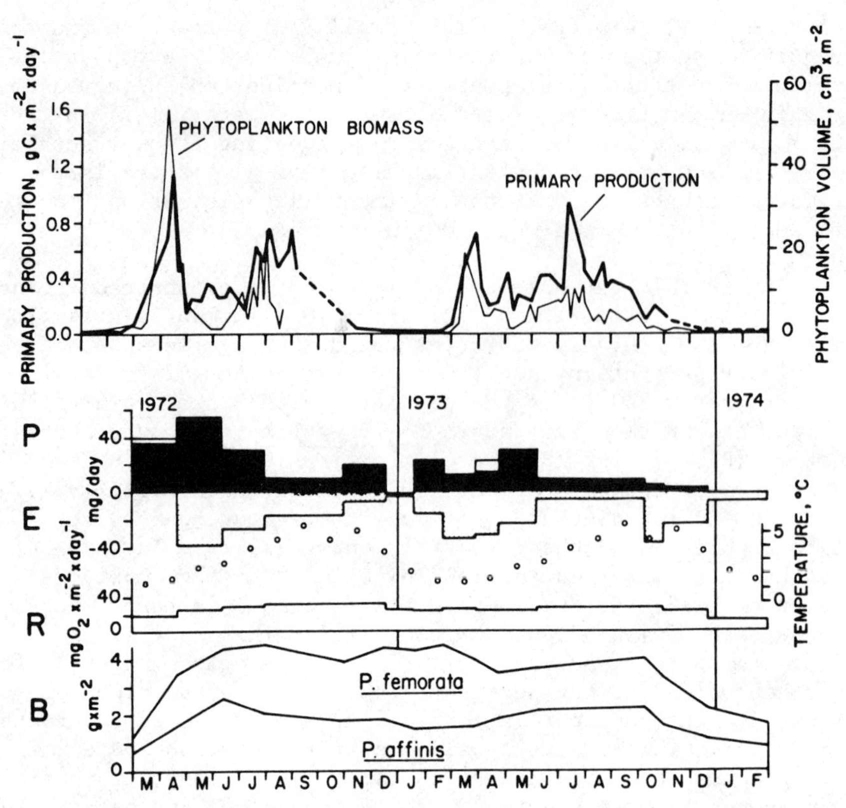

Fig. 20. Connection between pelagic and softbottom subsystems. The growth of the Pontoporeia populations (Amphipoda) are dependent on the strength of the phytoplankton spring bloom. P = production, E = elimination, R = respiration and B = biomass. (Elmgren, 1978, based on data from several authors)

The importance of filter feeding on a larger scale has been explored by Kautsky (1981b) who from detailed biomass and energy budget studies calculated that one third (56 g C m^{-2}) of the phytoplankton primary production in the Askö area could be filtered off by the resident Mytilus population. The animals respire 32 g C m^{-2} and release 20 g C m^{-2} as feces, the remaining net production of 4 g C m^{-2} is almost entirely released as gonads. The somatic growth of this Baltic population is very small and almost entirely respired during periods with scarce food. The reproduction is closely correlated to the spring maximum of phytoplankton as illustrated in Figure 21. The drastic drop in meat content is due to the spawning in early summer. As in the example of Pontoporeia production, there is a correlation between the size of the mussel gonad output and the variable size of the spring bloom for the consequitive two years shown in the figure.

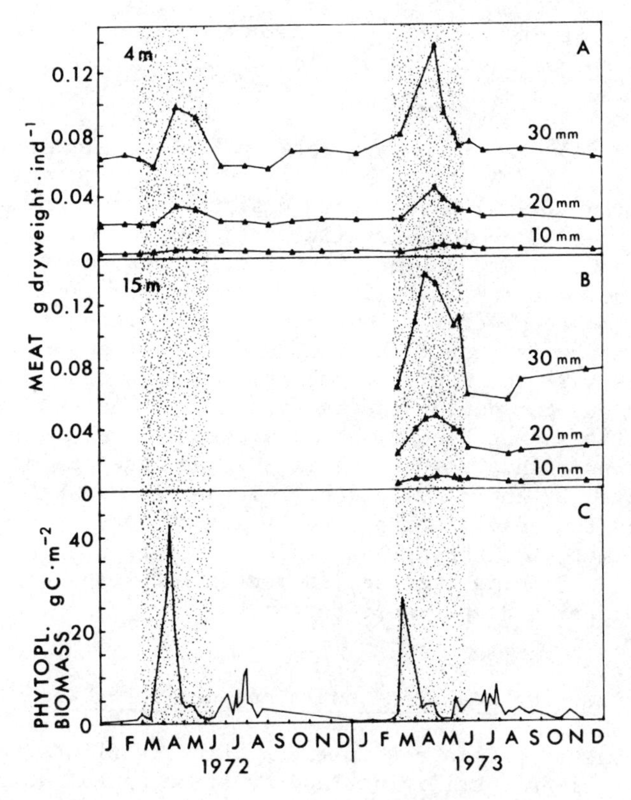

Fig. 21. Couplings between pelagic and hardbottom subsystems. Different size-classes of filter feeding Mytilus edulis at two depths (A and B) react by increasing meat weight to the spring bloom of phytoplankton (C). Hatched period indicates reproductive period of the mussel. (Kausky, 1982b)

The average gonad output of 4 g C m^{-2} is a substantial input of energy from the benthic to the pelagic subsystem. It corresponds to half of the annual net production of zooplankton, calculated (S. Johansson, pers. comm.) for the same area.

The 56 g C m^{-2} $year^{-1}$ filtered off by the mussels can be compared to the sedimentation of 50 g C m^{-2} $year^{-1}$, calculated from sedimentation traps (U. Larsson, pers. comm.). However it is not advisable to sum up these figures to get an estimate of the total energy transport from pelagic to benthic subsystems. We do not know to what extent the food filtered off by the mussels is taken from sedimentated and later resuspended particles. Nor do we have figures of input of organic matter from offshore areas. Direct simultaneous measurements of filtration rates, food composition and water transports are needed. The numbers should now be seen as indicators of the potential importance of these different flows in a coastal marine system.

Another important coupling between benthic and pelagic subsystems facilitated by the mussels is nutrient regeneration. Kautsky and Wallentinus (1980) have, from in situ measurements of nutrient excretion rates of enclosed Mytilus-red algal communities, estimated the total annual output of nutrients in the Askö area. With a total Mytilus standing stock of 10,200 tons dry weight including shells in the 160 km^2 research area, about 339 tons of nitrogen and 104 tons of phosporus are excreted. Kautsky and Wallentinus show that this excretion is well in excess of the demands of benthic algae which may explain why the deeper red and brown algal belts can continue to grow after the spring bloom maximum when the water column has been depleted of nutrients. This nutrient input might be more important than indicated by annual values of release and uptake. A substantial part of this remineralisation takes place on bottoms above the thermocline in summer, thus being available in the trophogenic layer for both benthic and pelagic algae, in contrast to the remineralisation that take place in deeper soft bottoms.

Seabirds connecting sea, land and atmospheric regimes

The large archipelagoes in the Baltic constitute favourable areas for seabirds. Utilizing land as breeding space, atmosphere for fast transportation they exploit the marine subsystems when feeding. Only the diving duck population during winter is capable of consuming in the order of 16,000 ton C of bottom fauna, mainly Mytilus (Nilsson, 1980). The eider duck population alone probably consumes as much during the rest of the year (Elmgren, in press). The consumption of fish of the seabirds around Gotland, the biggest island in the Baltic, equals the total Baltic fishery around that island (Jansson and Zuchetto, 1978).

A system in a system - the metabolism of an archipelago sound

The different characteristic entities of the biological subsystems can be illustrated by measures of metabolic rates for whole communities. This is exemplified in Figure 22, with data obtained during an intensive study of subsystems within an archipelago sound around midsummer (Jansson and Wulff, 1977). The P/R ratios describing the degree of self-maintenance show that, although the production per unit area is high in the benthic algal communities, their ratios are close to one due to the respiration of the large standing stock of consumers in these systems. The Ruppia and soft bottom communities are true heterotrophic communities with ratios less than

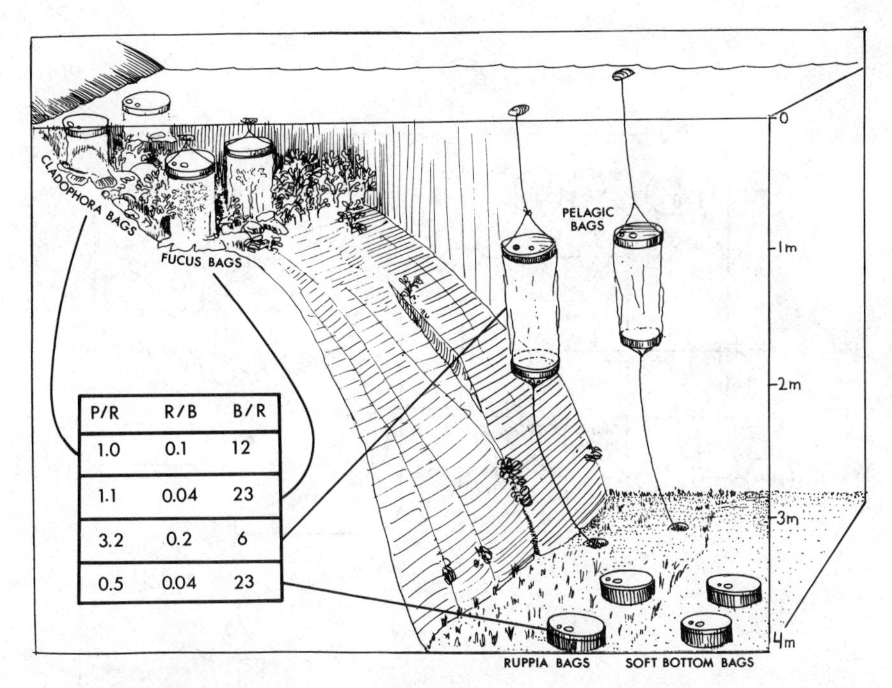

Fig. 22. Subsystem properties of a shallow sound as measured by community metabolism. Explanations in the text. (Jansson and Wulff, 1977).

one. Only the pelagic is clearly autotrophic showing a production more than twice the respiration.

The R/B ratios can be considered as measure of "maintenance to structure" (Odum, 1967). The highest maintenance costs per structured unit is to be found in the areas of the greatest physical noise, i.e. in the pelagic and upper littoral, the site for the Cladophora-belts. The frequent need of repairing damaged tissues or replacing whole individuals has favored the selection of fast-growing organisms in these regions. The lowest costs of keeping an ordered structure is found in the Fucus system. This also means that the cost of replacement, the B/R ratio or the turnover time, is longest here, 23-74 days compared to the 1.5 days for the pelagic system.

The total model of the sound, based on energy budgets of the subsystems and their areal contributions is drawn in Figure 23 in order to show the connections between the different parts. A rough compartmentalization of the consumers into filter feeders, browsers plus omnivores and carnivores is used. If values for gross community production and respiration are summed up for the whole sound a P/R ration of 1 is found which may indicate a high degree of self-maintainance. But of the total production 77% is due to plants within

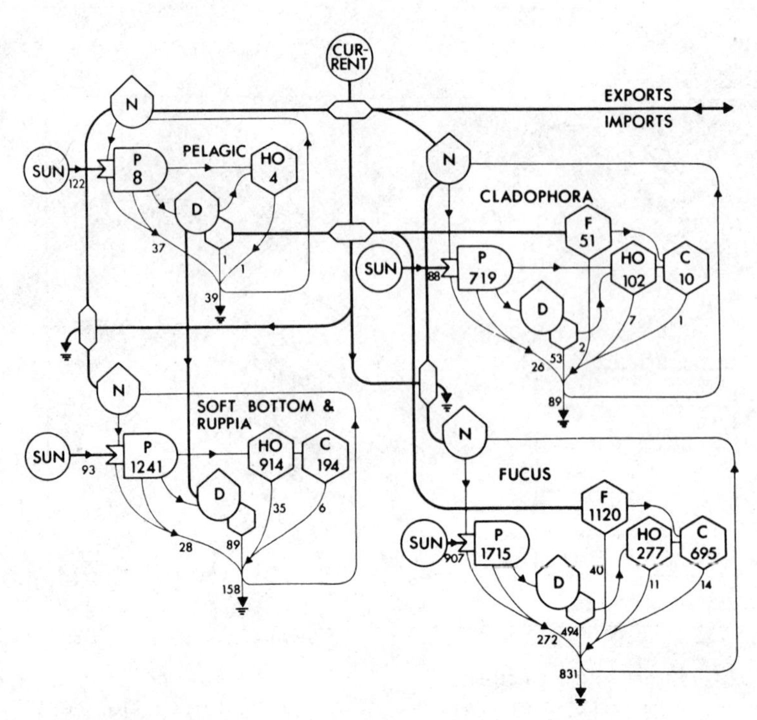

Fig. 23. An energetic model of the sound in Fig. 22. Values of storages are in MegaJoules respectively MegaJoules per day for the total sound of 14,000 m^2. P = producers, F = filter feeders, HO = herbivores and omnivores, C = carnivores, D = detritus and bacteria and N = nutrients. (Jansson and Wulff, 1977).

the brown algal belt where Fucus vesiculosus constitutes over 99% of the total plant biomass. This means that a large proportion of the production is stored up as algal tissues which are not readily eaten by herbivores. About 35% of the total animal biomass in the sound is due to filter feeders, depending on the suspended organic matter as a food source. The soft bottom and Ruppia communities are dependent on organic storages, supplied by sedimentation of seston or benthic algae. Only a minor part of the primary production within the system is directly available to the consumers and not sufficient to meet their energy demands. A continuous transport of seston through the sound is necessary to maintain the consumer biomasses found.

It is tempting to compare all the energy flow measurements made for a larger system like the Askö archipelago area and use the agreement or disagreement in the total energy budget as a means to validate these calculations. As an example, we may consider our data inaccurate if we find that estimates of food consumption, production and respiration cannot be met by the total primary production for the area. Such conclusions are not warranted since we still do not have any measurement of the magnitude of couplings to other areas, via especially the physical transports.

LITERATURE CITED

Ackefors, H., Hernroth, L., Lindahl, O., and Wulff, F., 1978, Ecological production studies of the phytoplankton and zooplankton in the Gulf of Bothnia, Finn. Mar. Res., 244: 116.

Andersin, A. -B., Lassig, J., Parkonen, L., and Sandler, H., 1978, The decline of macrofauna in the deeper parts of the Baltic and the Gulf of Finland, Kieler Meeresforsch. Sonderh., 4: 23.

Aneer, G. ,1975, Composition of food of Baltic herring (Clupea harengus var. membras L.), fourhorn sculpin (Myoxocephalus quadricornis L.), and ellpout (Zoarces viviparus L.) from deep soft bottom trawling in the Askö-Landsort area during two consecutive years. Merentutkimuslaitoksen Julk. /Havsforskningsinst. Skr. 239: 146.

Aneer, G., Lindqvist, A., and Westin, L., 1978, Winter concentrations of Baltic herring (Clupea harengus var membras L.), Contr. Askö Lab. Univ. Stockholm, 21: 1.

Aneer, G., 1979, On the ecology of the Baltic herring- studies on spawning areas, larval stages, locomotory activity patterns, respiration, together with estimates of production and energy budgets, Ph. D Thesis, Dep. Zool. Univ. Stockholm.

Aneer, G., 1980, Estimates of feeding pressure on pelagic and benthic organisms by Baltic herring (Clupea harengus v. membras L.), Ophelia, Suppl. 1: 265.

Ankar, S., 1979, Annual dynamics of a Northern Baltic soft bottom, In: "Cyclic phenomena in marine plants and animals," E. Naylor and R.G. Hartnoll, eds., Pergamon Press, Oxford.

Ankar, S., and Elmgren, R., 1976, The benthic macro- and meiofauna of the Askö-Landsort area - A stratified random sampling survey, Contr. Askö. Lab. Univ. Stockholm, 11: 1.

Arntz, W.E., 1971, Biomasse und Produktion des Makrobenthos in den tieferen Teil der Kieler Bucht im Jahr 1968, Kieler Meeresforsch, 27: 36.

Arntz, W.E., 1978, The "upper part" of the benthic food web: the role of macrobenthos in the Western Baltic, Rapp. P. -V Reun. Cons. int. Explor. Mer., 173: 85.

Bagge, P., Ilus, E., and Motzelius, F., 1975, Line census of fish made by the SCUBA diving method in the archipelago of Loviisa (Gulf of Finland), Merentutkimuslait. Julk/Havsforskningsinst. Skr., 240: 57.

Bennet, J.R., 1974, On the dynamics of wind-driven lake currents, J. Phys. Oceanogr., 4: 400.

Bock, K. -H., 1971, Monatskarten des Dichte des Wassers in der Ostsee dargestellt für verschiedene Tiefenhorizonte. Ergänzungsh, Dtsch. Hydrogr. Z., Reihe B: 13.

Brattberg, G., 1980, Kvävefixering i marin miljö - Östersjön, In: "Processer i kvävets kretslopp," T. Rosswall, eds., SNV PM 1213: 95.

Cederwall, H., 1977, Annual macrofauna production of a soft bottom in the northern Baltic proper, In: "Biology of bentic organisms, B.F. Keegan, P.O'Ceidigh and P.J.S. Boaden, eds., Pergamon Press London.

Cederwall, H., and Elmgren, R., 1980, Biomass increase of benthic macrofauna demonstrates eutrophication of the Baltic Sea, Ophelia, Suppl. 1: 287.

Csanady, G.T., 1978, The coastal jet conceptual model in the dynamics of shallow seas, In: "The Seas," Vol. 6., E. D. Goldberg, I.N. McCave, J.J. O'Brien and J.H. Steele, eds., John Wiley & Sons, New York.

Cushing, D.H., 1971, A comparison of production in temperate seas and the upwelling areas, Trans. Roy. Soc. S. Afr. 40: 17.

Ekman, F.L., 1893, Den svenska hydrografiska expeditionen år 1877, I.K. Sven Vetenskapsakad Handl., 25: 1.

Elmgren, R., 1978, Structure and dynamics of Baltic benthic communities, with particular reference to the relationship between macro- and meiofauna, Kieler Meeresforsch. Sonderh., 4: 1.

Elmgren, R., 1982, Ecological and trophic dynamics in the enclosed, brackish Baltic Sea, ICES Symposium on Biological Productivity if Continental Shelves in the Temperate Zone of the North Atlantic, Kiel, March 1982, No 27: 1.

Elmgren, R., Rosenberg, R., Andersin, A. -B., Evans, S., Kangas, P., Lassig, J., Leppakoski, E., and Varmo, R., in press, Benthic macro- and meiofauna in the Gulf of Bothnia (Northern Baltic) Pr. morsk. Inst. ryb. Gdyni.

Engwall, A. -G., 1978, Ammonium release at the sediment water interface. In situ studies of a Baltic sediment during negative redox turnover, Contr. Microbial Geochemistry, Dept. of Geology Univ. Stockholm, 2: 1.

Falkenmark, M., and Mikulski, Z., 1974, "Hydrology of the Baltic sea International Hydrological Decade Project No. 1, Stockholm-Warzawa.

Fonselius, S.H., 1969, Hydrography of the Baltic deep basins III. Fishery Board of Sweden Ser. Hydrogr. Rep., 23: 1.

Gill, A.E., Green, J.S.A. and Simmons, A.J., 1974, Energy partition in the large scale ocean circulation and the production of mid-ocean eddies, Deep-Sea Res., 21: 499.

Guterstam, B., Wallentinus, I., and Iturriaga, R., 1978, In situ primary production of Fucus vesiculosus and Cladophora glomerat Kieler Meeresforsch. Sonderh., 4: 257.

Guterstam, B., 1979m In situ-Untersuchungen uber Sauerstoffumsatz und Energiefluss in Fucus-gemeinschaften der Ostsee, Ph.D. Thesis, Christian-Albrechts-Universitat, Kiel.

Hagström, A., and Larsson, U., in press, Diel and seasonal variation in growth rates of pelagic bacteria, In: "Heterotrophic Activity in the Sea," J. Hobbie and P.J.LeB. Williams, eds., Plenum Press.

Hansen, D., and Rattray, M., 1966, New dimensions in estuary classification, Limnol. Oceanogr., 11: 319.

Hansson, S., 1980, Distribution of food as a possible factor regulating the vertical distribution of fourhorn sculpin (Myoxocephalus quadricornis L.) in the Bothnian Bay, Ophelia, Suppl. 1: 277.

Hernroth, L., 1977, Zooplankton dynamics of a spring bloom in the northern Baltic proper, (Mimeo), Dept. of Zoology, Univ. Stockholm.

Hernroth, L. and Ackefors, H., 1979, The zooplankton of the Baltic proper, Report, Fish. Bd. Sweden, Inst. Mar. Res., 2: 1.

Hillbricht-Ilkowska, A., 1977, Trophic relations and energy flow in pelagic plankton, Pol. ecol. Stud., 3: 3.

Hobro, R., 1979, Annual phytoplankton successions in a coastal area in the northern Baltic, In: "Cyclic phenomena in marine plants and animals," E. Naylor and R.G. Hartnoll, eds., Oxford and New York, Pergamon Press.

Hobro, R., Larsson, U., and Wulff, F., in press, Dynamics of a phytoplankton spring bloom in a coastal area of the northern Baltic, Prace morsk. Inst. Ryb. Gdyni.

Holland, W. and Lin, L., 1973, On the generation of mesoscale eddies and their contribution to the oceanic general circulation, J. Phys. Oceanogr., 5: 642.

Holm, N., 1978, Phosphorus exchange through the sediment-water interface. Mechanism studies of dynamics processes in the Baltic Sea, Contrib. in Microbial Geochemistry, Dep. Geology, Univ. Stockholm, 3: 1.

Horstmann, U., 1975, Eutrophication and mass production of blue-green algae in the Baltic, Merentutkimuslait. Julk/ Havsforskningsinst. Skr., 239: 83.

Jansson, A.M., 1974, Community structure, modelling and simulation of the Cladophora ecosystem in the Baltic Sea, Contr. Askö Lab. Univ. Stockholm, 5: 1130.

Jansson, B. -O., 1972, Ecosystem approach to the Baltic problem, Bulletins from the Ecological Committee NFR 16: 1.

Jansson A.M., and Zuchetto, J., 1978, Energy, economy and ecological relationships for Gotland, Sweden. A regional systems study, Ecol. Bull. Stockholm, 28: 1.

Jansson, A. -M. and Kautsky, N., 1977, Quantitative survey of hard bottom communities in a Baltic archipelago, In: "Biology of bethic organisms," B.F. Keegan, P. O'Ceidigh and P.J.S. Boaden, eds., Pergamon Press, London and New York.

Jansson, B. -O., 1978, The Baltic - a systems analysis of a semi-enclosed sea, In: "Advances in Oceanography," H. Charnock and Sir G. Deacon, eds., Plenum Publ, New York.

Jansson, B. -O., 1983, Baltic Sea ecosystem analysis : critical areas for future research, Limnologica, In press.

Jansson, B. -O. and Wulff, F., 1977, Ecosystems analysis of a shallow sound in the northern Baltic - A joint study by the Askö group, Contr. Askö Lab. Univ. Stockholm, 18: 1.

Jansson, B. -O., and Hagström, A., 1978, Sea-bed respiration: core samples for oxygen consumption measurements, XI Conference of the Baltic Oceanographers, Paper Nr 62, 806.

Johnson, M.G. and Brinkhurst, R.O., 1971, Production of benthic macroinvertebrates of Bay of Quinte and Lake Ontario, J. Fish Res Bd Canada, 28: 1699.

Kankaala, P., in press, Resting eggs, seasonal dynamics, and production of Bosmina longispina maritima (P.E. Muller) (Cladocera) in the northern Baltic proper, J. Plankton Res.

Kahru, M., Aitsam, A., and Elken, J., 1981, Coarse-scale spatial structure of phytoplankton standing crop in relation to hydrography in the open Baltic Sea, Marine Ecology Progress Series, 5: 311.

Karlgren, L., 1978, Närsalternas roll i kustvatten, In: "Diagnos Östersjon," A. Åkerblom, ed., Statens Naturvardsverk, Solna, Sweden.

Kautsky, H., Widbom, B., and Wulff, F., 1981, Vegetation, macrofauna and benthic meiofauna in the phytal zone of the archipelago of Lulea - Bothnian Bay, Ophelia, 20: 53.

Kautsky, N., 1981a, On the trophic role of the blue mussel (Mytilus edulis L.) in a Baltic coastal ecosystem and the fate of the organic matter produced by the mussels, Kieler Meeresforsch. Sonderh., 5: 454.

Kautsky, N., 1981b, On the role of the blue mussel, Mytilus edulis L. in the Baltic ecosystem, Unpubl. Ph. D. Thesis summary, Dep. Zoology, Univ. Stockholm.

Kautsky, N., 1982a, Growth and size structure in a Baltic Mytilus edulis L. population, Mar. Biol. 68: 117.

Kautsky, N., 1982b, Quantitative studies on the gonad cycle, fecundity, reproductive output and recruitment in a Baltic Mytilus edulis L. population, Mar. Biol., 68: 143.

Kautsky, N., and Wallentinus, I., 1980, Nutrient release from a Baltic Mytilus-red algal community and its role in benthic and pelagic productivity, Ophelia, Suppl. 1: 17.

Kullenberg, G., 1974, An experimental and theoretical investigation of the turbulent diffusion in the upper layer of the sea, Rep 25, Inst. Phys. Oceanogr. Univ. of Copenhagen, Denmark.

Kullenberg, G., 1976, Note on the entrainment velocity in natural stratified vertical shear flow, The 10th Conference of Baltic Oceanogr., Gothenburg, Sweden.

Larsson, U., and Hagström, A., 1979, Phytoplankton exudate release as an energy source for the growth of pelagic bacteria. Mar. Biol., 52: 199.

Larsson, U., and Hågström, A., 1982, Fractionated phytoplankton primary production, exudate release and bacterial production in a Baltic eutrophication gradient, Mar. Biol., 67: 57.

Lindqvist, A., Hagström, O., Håkansson, N., and Kollberg, S., 1977,

Preliminary results from echo integrations in the Baltic, 1976 and 1977, ICES C.M. 1977/P : 13:1.

Lumley, J., 1975, Modelling turbulent flux of passive scalar quantities in inhomogeneous flows, J. Phys. of Fluids, 18: 619.

Magaard, L., and Rheinheimer, G., 1974, "Meereskunde der Ostsee," Springer Verlag, Berlin.

McKellar, H. and Hobro, R., 1976, Phytoplankton Zooplankton relationships in 100 liter plastic bags, Contr. Askö. Lab. Univ. Stock holm, 13: 183.

Möhlenberg, F. and Riisgard, H.U., 1978, Efficiency of particle retention in 13 species of suspension feeding bivalves, Ophelia, 17: 234.

Nehring, D., 1979, Relationships between salinity and increasing nutrient concentration in the mixed winter surface layer of the Baltic from 1969 to 1978, ICES C.M. 1979/ C: 24_1.

Neumann, E., 1977, Activity and distribution of benthic fish in some Baltic archipelagos with special reference to temperature, Ambio Spec. Rep., 5: 47.

Neumann, E., 1982, Species composition and seasonal migrations of the coastal fish fauna in the southern Bothnian Sea. In: "Coastal Research in the Gulf of Bothnia," K. Muller, ed., Dr. W. Junk Publ., The Hague.

Nilsson, L., 1980, Wintering diving duck populations and available food resources in the Baltic, Wildfowl, 31: 131.

Odum, H.T., 1967, Biological circuits and the marine systems of Texas, In: "Pollution and marine ecology," T.A. Olsson and F.J. Burgess, eds., Interscience Publ., New York.

Odum, H.T. and Hoskins, C.M., 1958, Comparative studies on the metabolism of marine waters, Publ. Inst. mar. Sci. Univ. Texas, 5: 16.

Omstedt, A. and Sahlberg, T., 1982, Vertical mixing restratification in the Bay of Bothnia during cooling, SMHI Report RHO32, Norrköping, Sweden.

Pawlak, J., 1980, Land-based inputs of some major pollutants to the Baltic Sea, Ambio 9: 163.

Platt, T., Mann, K.H., and Ulanowicz, R.E., eds., 1981, "Mathematical models in biological oceanography," The UNESCO Press, Paris.

Reichle, D.E., O'Neill, R.V., and Harris, W.F., 1975, Principles of energy and material exchange in ecosystems, In: "Unifying concept in ecology," W.N. van Dobben and A.H. LoweMcConnell, eds., Junk Publ., the Hague.

Reynolds, J., 1895, On the dynamical theory of incompressible viscous fluids and the determination of the criteria, Phil. Trans. of the Royal Soc. of London, Series A, 186.

Rheinheimer, G., 1981, Investigations on the role of bacteria in the food web of the Western Baltic, Kieler Meeresforsch., Sonderh., 5: 284.

Rodhe, H., Söderlund, R., and Ekstedt, J., 1980, Deposition of airborne pollutants on the Baltic, Ambio 9: 168.

Rowe, J.S., 1961, The level of integration concept and ecology. Ecology 42: 420.

Schippel, F., Hallberg, R.O., and Oden, S., 1973, Phosphate exchange at the sediment water interface, Oikos, Suppl 15: 64.
Schmidt, C., 1980, Some aspects of marine algae decomposition, Ophelia, Suupl. 1: 257.
Shaffer, G., 1975, Baltic coastal dynamics project - the fall downwelling regime off Askö, Contr. Askö Lab. Univ. Stockholm, 7: 1.
Shaffer, G., 1979a, On the phosphorus and oxygen dynamics of the Baltic Sea, Contr. Askö Lab. Univ. Stockholm, 25: 1.
Shaffer, G., 1979b, Conservation calculations in natural coordinates, J. Phys. Oceanogr., 9: 847.
Simons, T.J., 1980, Circulation models of lakes and inland seas, Dept. of Fish. and Oceans Bull., 203, Ottawa, Canada.
Sjöberg, S., Wåhlstrom, P., and Wulff, F., 1972, Computer simulations of hydrochemical and biological processes in the Baltic, Contr. Askö Lab. Univ. Stockholm, 1: 1.
Sjöberg, S. and Wilmot, W., 1977, Systems analysis of a spring phytoplankton bloom in the Baltic, Contr. Askö Lab. Univ. Stockholm, 20: 1.
Smeda, M., and Wilmot, W., 1978, A mixed layer model of wind and density driven mixing during a phytoplankton bloom, Report DM-24, Dept. of Meteorol. Univ. of Stockholm, Sweden.
Svensson, J. and Wilmot, W., 1978, A numerical model of the circulation in Öresund, SMHI Report RHO-15, Norrköping, Sweden, 1.
Thurow, F., 1980, The state of fish stocks in the Baltic, Ambio, 3-4: 153.
Toll, T., Wilmot, W., and Kjerve, B., 1982, Nutrient transports in a Swedish estuary, Manuscript.
Voipio, A., ed., 1981, "The Baltic Sea," Elsevier, Amersterdam.
Walin, G., 1972a, Some observations of temperature fluctuations in the coastal region of the Baltic, Tellus 24: 187.
Walin, G., 1972b, On the hydrographic response to transient meteorological disturbances, Tellus, 24: 169.
Walin, G., 1977, A theoretical framework for the description of estuaries, Tellus, 29: 128.
Wallentinus, I., 1978, Productivity studies on Baltic macro-algae, Botanica Marina, 21: 365.
Wallentinus, I., 1979, Environmental influences on benthic macrovegetation in the Trosa-Askö area, northern Baltic proper II, Contr. Askö Lab. Univ. Stockholm, 25: 1.
Webster, J.R., 1975a, Hierarchial organization of ecosystems. In: "Theoretical systems ecology," E. Halfon, ed., Acad. Press.
Welander, P., 1974, A two-layer exchange in an estuary with special reference to the Baltic Sea, J. Phys. Oceanogr., 4: 542.
Westin, L., 1970, The food ecology and the annual food cycle in the Baltic population of fourhorn sculpin, Myoxocephalus quadricornis (L.), Pisces. Rep. Inst. Freshwater Res. Drottningholm., 50: 168.
Westin, L., 1971, Locomotory activity patterns of fourhorn sculpin, Myoxocephalus quadricornis (L.) (Pisces). Rep. Inst. Freshwater Res. Drottningholm, 51: 184.

Williams, P.J. LeB., 1981, Incorporation of microheterotrophic processes into the classical paradigon of the planktonic food web, Kieler Meeresforsch., Sonderh. 5:1.

Wilmot, W., 1974, A numerical model of the gravitational circulation in the Baltic. ICES Special Meeting on Models of Water Circulation in the Baltic, Copenhagen, Denmark.

Wulff, F., Flygh, C., Foberg, M., Hansson, S., Johansson, S., Kautsky, H., Klintberg, T., Samberg, H., Skärlund, K., Sörlin, T., and Widbom, B., 1977, Ekologiska undersökningar i Luleå Skärgård 1976, Slutrapp. Statens Naturvardsverk, Kontr. 5860401-8 (In Swedish).

ON THE RELATION OF PRIMARY PRODUCTION TO GRAZING DURING THE FLADEN GROUND EXPERIMENT 1976 (FLEX'76)

G. Radach[1], J. Berg[1], B. Heinemann[1] and M. Krause[2]

[1]Institut für Meereskunde, Universität Hamburg
Heimhuder Str. 71, D-2000 Hamburg 13
[2]Sonderforschungsbereich 94, Meeresforschung Hamburg
Universität Hamburg, Bundesstr. 55, D-2000 Hamburg 13

INTRODUCTION

To study the flow of matter in marine ecosystems means firstly to find out the most important paths of transport to higher trophic levels of the food web, and secondly to quantify the transports. It is critical for these systems, whether primary production is channelled upwards in the food web only through herbivorous zooplankton, or whether part of the particulate primary production goes directly to other consumers by sinking down to the benthos.

Steemann Nielsen (1957, p. 182) states that it is *a priori* possible for the sea to assume either more or less stationary conditions in which for instance the zooplankton population matches the production of phytoplankton or variable conditions in which periods of flourishing phytoplankton alternate with periods in which the standing crop of phytoplankton is eaten right down by the herbivorous zooplankton. He continues that since about 1930 the latter concept has been prevalent, but accepts this point of view only 'for regions in which conditions for phytoplankton production change suddenly from bad to good', as is true for high latitudes in spring.

For the North Sea Cushing and Vucetic (1963) found that in spring 'grazing mortality is the most effective controlling agent on algal production'. They investigated the grazing capacity of a Calanus patch, observed for 66 days, from 19 March to 25 May 1954. Grazing was found to amount to 64% of total mortality of algae, with non-grazing mortality composed mainly of sinking and diffusion.

The same line of thinking is adopted by Steele (1974, p.19) in setting up his model of the North Sea food web, by assuming that all particulate primary production goes through the herbivores. On the other hand, Fransz and Gieskes (1982) conclude from observations in the North Sea 'that the delay between the copepod production and the primary production makes it impossible that all phytoplankton is directly eaten by the copepods' during the phytoplankton spring bloom They report 'that there is little correspondence between primary production and the production of copepods'. Only during summer they found a high grazing pressure on phytoplankton, thus maintaining a balance between phytoplankton and zooplankton.

For the Fladen Ground Experiment 1976 (FLEX '76) Williams and Lindley (1980) suggest that 'the depletion of nutrients limited the size of the spring bloom but that it was the grazing pressure exerted by Calanus finmarchicus which was responsible for the control and depletion of the phytoplankton'.

In this paper primary production and grazing by herbivores are calculated from data of FLEX '76, applying as many different theoretical formulations as possible, to investigate whether the phytoplankton bloom was limited by nutrient depletion or by zooplankton grazing. The results obtained are used for validating a simmulation model for the spring bloom during FLEX'76 (Radach, 1982, 1983).

DATA

The data used in this investigation were obtained at the Central Station (58° 55' N., 0° 32' E.) of the Fladen Ground Experiment 1976 in the northern North Sea. In this experiment the interrelations between the spring plankton bloom and the development of the seasonal thermocline were studied. Details about the experiment are given by Ramster (1977) and Lenz, Ramster and Weidemann (1980).

Most of the data sets used here have been presented elsewhere before. Phytoplankton was counted and converted to carbon values by Gillbricht (Gassmann and Gillbricht, 1982- chlorophyll was analysed by spectral photometry by Weber (Radach, Trahms and Weber, 1980). Zooplankton was counted from 101 water samples by Krause (Krause and Radach, 1980). Calanus finmarchicus accounted for 80% of the biomass. The other species entering the calculation of zooplankton biomass were Pseudocalanus elongatus, Microcalanus pusillus, Paracalanus parvus, Oithona similis, Metridia lucens, Acartia clausi and Oithona plumifera stages I to VI copepodites. The nutrients phosphate, nitrate and ammonia were determined by Eberlein et al. (1980).

All these concentrations and standing stocks were measured 6-hourly in 10 depths (3, 10, 20, 30, 40, 50, 60, 75 or 80, 100, 150 m or bottom) from 25 March to 5 June 1976, with one gap for several data sets (16-21 May). A few characteristic state variable of the

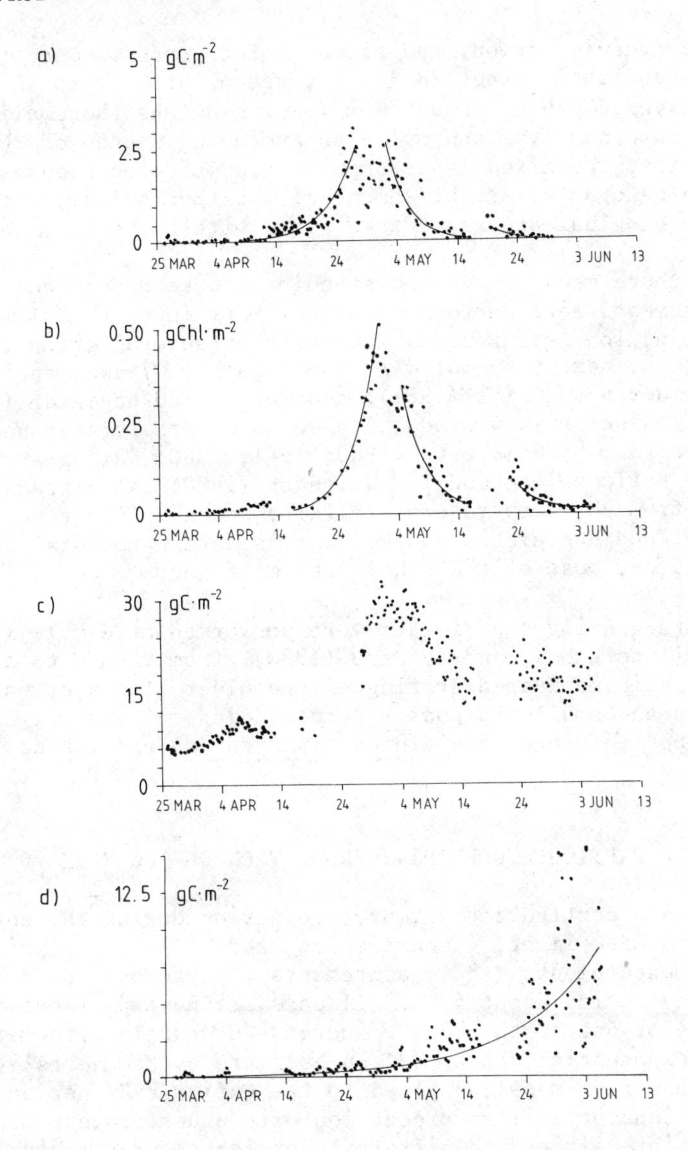

Fig. 1. Characteristic state variables of the biological development during FLEX '76 integrated over depth: (a) phytoplankton carbon from cell counts (0-50 m); originator: Gillbricht (unpubl. data): (b) chlorophyll (0-50 m), after Radach, Trahms and Weber (1980); (c) particulate organic carbon (0-100 m); originator: Hickel (unpubl. data); (d) biomass of all copepods (0-100 m): originator: Krause.

biological development are shown in Figure 1 as time series of vertically integrated quantities : phytoplankton carbon, chlorophyll, particulate organic carbon, and biomass of copepods. Once the thermocline is established (after 28 April), the natural depth scale is the mixed layer depth of about 50 m (excluding the thermocline of about 20 m thickness). To provide an averaging depth for the entire FLEX time, the mixed layer depth scale was also used as averaging scale before the establishment of the thermocline, because the important biological processes take place within the 50 m layer.

Beside these measurements of standing stocks and concentrations, flux measurements were performed for the most important fluxes. Primary production measurements were done by several groups, using the ^{14}C method, namely by Baird (unpubl. data), Gieskes and Kraay (1980), Mommaerts and Nijs (1981), and Weigel and Hagmeier (unpubl. data). All these primary production measurements were presented in a unified form by Mommaerts (1980; 1981; 1982), using a modification of the Vollenweider model. Weichart (1980) calculated primary production from pH, temperature, salinity and alkalinity measurements. While Weigel and Hagmeier's and Weichart's data were obtained at the Central Station, most of the other data were not.

Shipboard grazing experiments were performed by Gamble (1978) from 24 April to 2 May, and by Daro (1980), from 22 May to 5 June, 1976. Gamble investigated grazing of the older stages of *Calanus finmarchicus* and small popepods. Daro studied the grazing of different copepodite stages of *Calanus finmarchicus* and its daily variation.

ESTIMATION OF PARTICULATE PRIMARY PRODUCTION DURING FLEX'76

To estimate particulate primary production during the entire Fladen Ground Experiment, 4 sources are used: (1) direct primary production measurements (^{14}C measurements), performed for a few hours per day during about 45 of all days of the experiment, (2) the decrease of nutrients and the increase of particulate matter during the exponential phase, (3) an estimate by Mommaerts' version of the Vollenweider model, applied to the entire FLEX period, and (4) calculations of primary production with other process models using light intensities and different nutrients, prescribing the input for these models *a priori*.

Later we attempt to calculate primary production with a set of constants assumed to be valid for the full period, and we check whether such a procedure (as also used in model simulations) may successfully reproduce primary production.

Estimates from primary production measurements

The data by Weigel and Hagmier and by Weichart, extrapolated

Table 1a. Estimates of gross particulate primary production from measurements, adding 20% to Weichart's (1980) values to account for respiration and further adding Weigel and Hagmeier's starting value of 6.19 gC m^{-2}, to make his values comparable.

	particulate primary production until T (number of days in the year 1976)				
measurements by	109 19 April	121 1 May	136 16 May	154 3 June	109-121
Weichart (1980)	6.19	34.57	no data	no data	28.4
Weigel and Hagmeier	6.19	15.12	27.67	39.18	8.93
Mommaerts (1981)	8.86	22.04	40.33	64.41	13.18

over time and depth, both from the Central Station, are compiled in Table 1a. During the exponential growth phase of FLEX '76, from 19 to 30 April, particulate primary production integrated over depth and time amounts to 8.93 gC m^{-2} (gross) after Weigel and Hagmeier and to >22.7 gC m^{-2} (net) after Weichart (Table 1a, Figure 2).

Estimates from nutrient consumption and particulate matter

Rough estimates of particulate production can be also calculated from phosphate and nitrate decrease during the exponential growth phase (25 March to 30 April), using appropriate C:P and C:N ratios. The same was attempted for particulate nitrogen. As there were no direct measurements of these ratios from algae during FLEX, we have to take values from literature. Thus the values of particulate production range from 20 to 31 gC m^{-2} for small element ratios and 8 to 15 gC m^{-2} for large ratios, as given in Table 1b.

Estimate by Mommaerts' version of the Vollenweider model

All the production data indicated in the previous section except those by Weichart (1980) have been presented by Mommaerts (1980) in a unified form via a modification of the Vollenweider model. Mommaerts fitted each primary production profile to the Vollenweider model, measured biomass in chlorophyll and calculated primary production for each day with a separate set of constants. For the exponential phase Mommaerts gives 13.2 gC m^{-2} (Figure 2, Table 1a). Mommaerts' estimate follows Weigel and Hagmeier's data closely because they are his main data base.

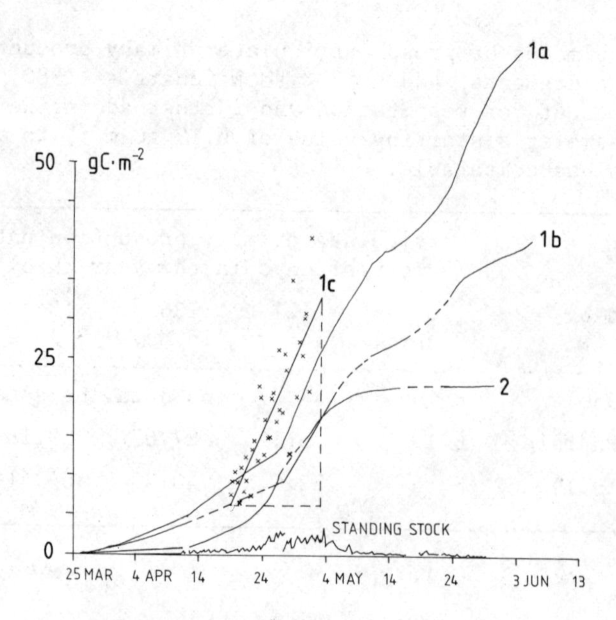

Fig. 2. Various estimates of primary production (gC m^{-2}) during FLEX, cumulated over time and depth, as calculated by Mommaerts (1981) (1a); as extrapolated from the data by Weigel and Hagmeier (1976, unpubl. manuscript) (1b); as shown by data (x and regression line) of Weichart (1980), (1c) (the broken lines of the triangle show that during 19 April to 3 May the cumulated particulate primary production value by Weichart is more than twice that of Weigel and Hagmeier), and as calculated with the primary production process model (2) after formulae (1) and (8) (r_p = 1.5 day^{-1}). For comparison the standing stock (from cell counts by Gillbricht) is also shown. Gaps in the measurements were filled by interpolation (broken lines in curves 1a, 1b).

CALCULATION OF PRIMARY PRODUCTION USING PROCESS MODELS

The estimates from nutrient consumption support both of the different primary production measurements, Weigel and Hagmeier's as well as Weichart's. In order to decide between the measurements, further estimates of primary production were calculated by applying commonly used process models to the data. Primary production is assumed to be proportional to the phytoplankton standing stock, $P(z,t)$. The proportionality factor is the growth rate, consisting of the optimum growth rate r_p modified by limitation factors, $I(t,z)$ for light $I(t,z)$ and /or r_N $(N(t,z))$ for nutrients $r_N(t,z)$. The production term is specified as

Table 1b. Estimates of particulate primary production from changes of nutrients and particulate matter during the exponential growth phase, using element ratios of (1) 0.6944 m mol P $(gC)^{-1}$ (Weichart, 1980), (2) 1.764 m mol P $(gC)^{-1}$ (Parsons et al., 1977), (3) 11.90 m mol N $(gC)^{-1}$ (Kremer and Nixon, 1978), (4) 23.810 m mol N $(gC)^{-1}$ (Parsons et al., 1977).

Type of estimate	particulate primary production ($gC\ m^{-2}$), using different element ratios. 'small ratios'	'large ratios'
phosphate decrease of 21.7 m mol P m^{-2} (Weichart, 1980)	31.3 [1] gross	12.3 [2] gross
particulate phosporus increase of 14 m mol P m^{-2} (Eberlein et al., 1980)	20.2 [1] net	7.9 [2] net
nitrate decrease of 4.9 gN m^{-2} (Eberlein et al., 1980)	29.4 [3] gross	14.7 [4] gross
particulate nitrogen increase of 4.7 gN m^{-2} (Eberlein et al., 1980)	28.0 [3] net	14.0 [4] net

$$\langle \text{primary production} \rangle = r_p \cdot \min(r_I, r_N) \cdot P(t,z). \qquad \text{Equ. 1.}$$

The function r_I is a defined light-response function (e.g. of Blackman (1905) or Steele (1962), see next section), and r_N is a Michaelis-Menten formula

$$r_N = \frac{N(t,z)}{N(t,z) + k_s} \qquad \text{Equ. 2.}$$

where k_s is the half-saturation constant. Phosphate, nitrate, ammonia, and nitrate plus ammonia were studied as possible limiting nutrients.

Equation (1) was calculated from measured light intensities, I(t,z), data of the different nutrients, N(t,z), and phytoplankton standing stock data, P(t,z), to yield (after integration over time and depth) values to be compared to primary production measurements, namely

$$\int_{T_o}^{T}\int_{o}^{50m} \langle \text{prim.prod.} \rangle dz dt = r_p \int_{T_o}^{T}\int_{o}^{50m} \min (r_I, r_N) P dz dt \qquad \text{Equ.3.}$$

known from measurements — not known — based on theory, fed with measurement

This equation serves to check the consistency between production measurements (on the left) and the theoretical representation of particulate primary production (on the right) by claculating both sides independently. The constant optimum growth rate r_p could be used to match the two expressions. We would expect realistic values of r_p to be in the range between 0.5 and 2.0 day^{-1}, based on values reported by Eppley (1972) and Schöne (1977), as discussed by Radach (1982b).

The light response functions

The following light response functions were applied to try to obtain a consistent description of the relations between the data, according to equation (3):

Blackman's function (1905):

$$r_I(t,z) = \begin{cases} I/I_1, & \text{for } I<I_1 \\ 1, & \text{for } I \geq I_1 \end{cases} : \qquad \text{Equ.4.}$$

or Steele's function (1962):

$$r_I(t,z) = \frac{I}{I_2} \exp (1 - I/I_2) \qquad \text{Equ.5.}$$

with $I_2 = e \cdot I_1$ (The factor e = 2.71828 originates from the mathematical procedure of equating the slopes of formulae (4) and (5) at light intensity I=0.);
or no light limitation down to a local compensation depth where $I=I_o$:

$$r_I(t,z) = \begin{cases} 1, & \text{for } I>I_o \\ 0, & \text{for } I \leq I_o \end{cases} \qquad \text{Equ.6.}$$

or light limitation, which is determined by the surface light intensity, the surface value of the limitation function being continued down to the seasonal thermocline:

$$r_I(t,z) = r_I(I(t,z=0)), \qquad \text{Equ.7.}$$

using Blackman's or Steele's function;
or no light limitation down to the thermocline, if it is daytime,

$I(t,z=0) > 0,$

$$r_I(t,z) = \begin{cases} 1, \text{ if } I(t,z=0) > 0 \\ 0, \text{ else} \end{cases} \qquad \text{Equ. 8.}$$

The constants

The constants necessary for this approach have been determined a priori. They were discussed in Radach (1982b). From literature and FLEX data the optimum growth rate ranges between 0.3 and 2.0 day^{-1}. For Chaetoceros species which formed the main bloom event, we may take 0.9 day^{-1}. The optimum light intensity can be calculated to be in the range $48 \cdot 10^{18} < I_{opt} < 123 \cdot 10^{18}$ quanta m^{-2} s^{-1} for Blackman's functional. Steele's functional has to be used with this optimum light intensity multiplied by e=2.71828 to keep the results comparable with respect to the slope of the light production relation. The half-saturation constants to be used in the Michaelis-Menten type expressions for the nutrient limitation factors were found in the literature in the range 0.06 - 0.36 μmol P l^{-1} for phosphate and 0.2 - 0.8 μmol N l^{-1} for nitrate and nitrate plus ammonia (here called favourable or unfavourable conditions).

Nutrient limitation of the bloom

The limitation of the phytoplankton spring bloom is investigated evaluating the reduction in growth predicted by Michaelis-Menten equations for the nutrients phosphate, nitrate and ammonia plus nitrate. The mean concentrations of phosphate and nitrate in the upper 0-30 m sank below the assumed half-saturation values (0.12 m mol P m^{-3} and 0.4 m mol N m^{-3}) on 30 April for phosphate and on 27 May for nitrate (Figure 3a). The phosphate limitation function attains mean values less than 0.5 in the upper layer of 0-30 m from 30 April to 11 May (Figure 3b). At a few depths, there occur local minima less than 0.2. For unfavourable conditions (i.e. the largest half-saturation constants chosen), phosphate would be limiting from the surface to 50-60 m, from 28 April. For favourable conditions (i.e. the smallest half-saturation constants chosen), between 1-11 May, limitation is restricted to the top 20 m. The nutrient concentrations in the 0-30 m layer will not cause mean values of the limitation function below 0.5 (Figure 3b).

Inorganic nitrogen and even nitrate alone take over strongest limitation only on one day during the main bloom period (7 May). During the latter part of the experiment, from 27 May on, nitrogen takes over the role as the limiting nutrient (Figure 3b), if judged from Michaelis-Menten kinetics. Only for unfavourable conditions limitation occurs on 6 and 7 May and continues from 24 May onwards.

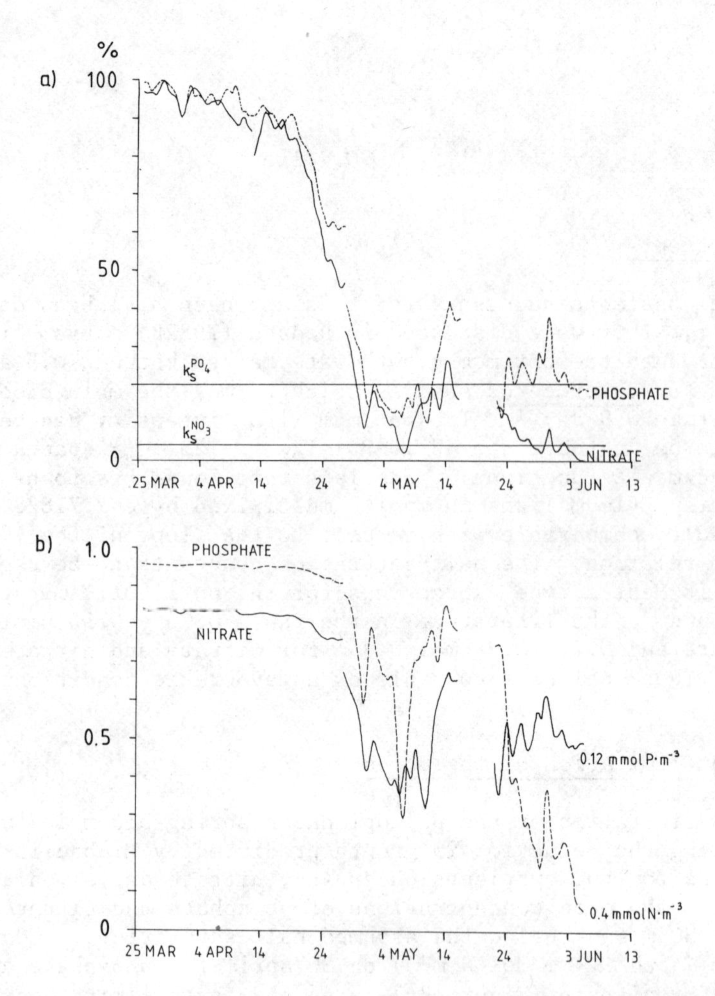

Fig. 3. Phosphate and nitrate and their limitation effects during FLEX'76: (a) means of phosphate and nitrate concentrations in the upper 30 m as percent of their maximum values (smoothed by running weighted averages); (b) mean of the values of the Michaelis-Menton function at the nutrient concentrations in the upper 30 m for phosphate (half saturation constant k_s = 0.12 m mol P m^{-3}) and nitrate (half saturation constant k_s = 0.40 m mol N m^{-3}).

Results of primary production process modelling

Calculations using formula (3) with different sets of constants for the optimum light intensity and half-saturation constants have been performed, using the different nutrients. The results of these attempts are summarized in Table 2.

The double integral on the right side of (3) attains values between 0.97 and 9.42 for the exponential phase, depending on the choice of constants. Comparing these values to the extrapolated measurements (left side of (3)), growth rates r_p as indicated in the last two columns of Table 2 would be necessary to make measurements and calculations consistent with each other. Only the fully unlimited response of algae down to about 50 m (equations (7) and (9)) yields acceptable growth rates r_p in conjunction with Weigel and Hagmeier's data. All light response functions, which limit growth within the euphotic zone or the mixed layer, imply larger growth rates.

There are, of course, several errors involved in the calculations, which should be mentioned. The standing stock data are the result of several processes acting, including herbivorous grazing which is nearly avoided in ^{14}C bottles. To calculate, from standing stock profiles, the same primary production as measured in the bottles, the phytoplankton standing stock data in (3) would have to be corrected for the amount grazed during the ^{14}C experiments. This proves impossible because the phytoplankton carbon profiles show extreme variability over one day and a daily trend in the stock development cannot be delineated. The use of chlorophyll instead of carbon does not help much because it shows no distinct daily variations in the upper layers, again necessitating the use of daily mean profiles in calculating (3).

Further problems would occur in converting chlorophyll to carbon. Assuming a C:Chl ratio of 20 for vigorously growing algae (Parsons et al., 1977), the peak concentration of 0.55 gChl m^{-2} (Figure 1b; see Radach et al., 1980) would correspond to 11 gC m^{-2}. This is high compared to the value of 2.5 gC m^{-2} as determined by cell counts (Figure 1a; Gieskes and Kraay, 1980; Gassmann and Gillbricht, 1982). Carbon from cell counts would yield a C:Chl ratio of 4.6 during the exponential phase of the bloom (Figure 4). This value seems rather low but was determined independently by Gieskes and Kraay (1980). This gives room to speculate that either primary production is overestimated (which is unlikely) or the standing stock determinations from cell counts are underestimated by a factor of about 4. The available data leave this question open.

In addition to these problems, some uncertainty originates from the knowledge of the half-saturation constants and the optimum light constants used, as shown by Table 2.

Gross production during FLEX amounted to 64 gC m^{-2} (for 69 days, in the mean 0.928 gC m^{-2} day^{-1}) following Mommaerts (1981). During the exponential growth phase (19 April - 1 May) the estimates range from 7.9 (net) to 31.3 gC m^{-2} (gross). Weigel and Hagmeier's and Mommaerts' estimates agree with estimates from particulate matter and nutrients using conversion factors of 1.764 m mol P $(gC)^{-1}$ and

Table 2. Evaluation of the primary production integral (3), right hand side (see text), for different types of the production - light response

Type of Calculation (Figures in brackets refer to text equation numbers)	Optimum light intensity (10^{18} quanta/ sec.m^2)	half-saturation contant for	
		PO_4	NO_3+NH_4
		(mg-at/m^3)	
1) Steele's light function, (5):			
favourable cond.	130	0.06	0.2
mean cond.	204	0.12	0.4
unfavour. cond.	334	0.24	0.8
2) Blackman's light function, (4):			
favourable cond.	48	0.06	0.2
mean cond.	75	0.12	0.4
unfavour. cond.	123	0.24	0.8
3) full growth response for light $\geqslant 10^{18}$ quanta/m^2sec, (6):			
favourable cond.	-	0.06	0.2
mean cond.	-	0.12	0.4
4) light limitation by continuing surface limitation value down to 50 m (mixed layer), (7):			
Steele function	204	0.12	0.4
Blackman function	75	0.12	0.4
5) full response down to 50 m, if light at surface >0, (8):	-	0.06	0.2
	-	0.12	0.4
	-	0.36	0.8

value of integral (3) until day T, with growth rate 1 (T number of day of 1976)				estimate for growth rate r_p^o (day^{-1}): prim. prod. after	
109.	121.	149.75	109.-121.	Weichart (22.7)	W. & H. (8.93)
0.39	2.14	3.36	1.75	13.0	5.10
0.30	1.68	2.53	1.38	16.5	6.47
0.21	1.18	1.72	0.97	23.3	9.21
0.444	2.46	3.75	2.01	11.28	4.44
0.331	1.82	2.71	1.49	15.20	5.99
0.230	1.26	1.81	1.03	22.04	8.67
1.36	5.94	8.04	4.58	4.96	1.95
1.24	5.18	6.67	3.94	5.76	2.27
1.21	7.12	10.77	5.91	3.84	1.51
1.33	7.62	11.41	6.30	3.61	1.42
2.17	11.59	17.56	9.42	2.41	0.95
1.98	10.09	14.56	8.11	2.80	1.10
1.69	8.08	11.10	6.39	3.55	1.40

W. & H. = Weigel and Hagmeier

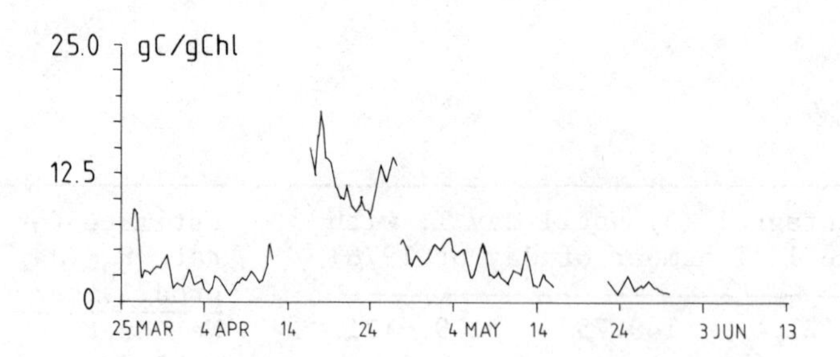

Fig. 4. Ratio of the carbon to chlorophyll content of the upper 30 m of the water column (compare Figure 1a,b for the 0-50 m layer).

23.81 m mol N $(gC)^{-1}$ as given by Parsons et al. (1977) for vigorously growing phytoplankton. Weichart's estimate of net production agrees with estimates from particulate matter and nutrients using conversion factors of 0.6944 m mol P $(gC)^{-1}$ derived by Weichart from the incorporation of CO_2 and phosphate along $\Delta C:\Delta P = 120$ (by atoms) and 11.9 m mol N $(gC)^{-1}$ as reported by Kremer and Nixon (1978). The gross production would be higher by about 20%, yielding about 28.4 gC m^{-2} during the exponential phase. The process modelling yields gross particulate production values of the order of 9 to 12 gC m^{-2} for an optimum reproduction rate of 1.5 day^{-1} during the exponential growth phase. This suggests that Weigel and Hagmeier's data should be relied on for further consideration.

Summarizing, we suggest that light limitation is no important factor in the mixed layer during the exponential phase of FLEX. We suppose that turbulent mixing in spring is strong enough that the algal cells will get sufficient light energy within a few hours on their way through the uppermost, illuminated layers so that production is independent of depth as long as they are above the thermocline. The situation may be very different in subtropical or tropical regions, where there may be very little turbulent mixing in the upper layers.

GRAZING CAPACITY OF HERBIVOROUS ZOOPLANKTON

During FLEX grazing experiments were performed by Gamble (1978) from 24 April to 2 May '76, and by Daro (1980) from 22 May to 5 June 1976. Estimates of herbivorous grazing capacity have to be based upon these measurements and the additional assumptions that (i) these results may be extrapolated to the entire FLEX time, and that (ii) *Calanus finmarchicus*, making up 80% of the total biomass, yields also about 80% of total grazing. An estimate for the amount of grazing by the other animals than Calanus finmarchicus will be attempted below.

The assumptions (i) and (ii) as well as further assumptions to be made for any estimate cannot be checked. They are only approximately valid. Conclusions must therefore be based on as many attempts of estimating the grazing capacity as possible. Four formulae are used to estimate the grazing capacity. First the formulae are explained, then the results are given. The constants necessary for the computations are given in Table 3.

Estimate of grazing capacity from biomass

Certain percentages of biomass were assumed as ingestion rations per day for the j-th copepodite stage of Calanus finmarchicus. The biomass was calculated from the numbers of animals per unit volume, $n_j(t,z)$ ($ind \cdot m^{-3}$), (Krause and Radach, 1980) by applying stage specific biomasses per individual, b_j ($gC\ ind^{-1}$). The percentages p_j (day^{-1}) of intake per day in terms of body weight are taken from Daro (1980). The estimate of the local grazing rate $g(t,z)$ ($gC\ m^{-3}\ day^{-1}$) is then given by

$$g(t,z) = \sum_{j=1}^{6} p_j\ b_j\ n_j(t,z) \qquad \text{Equ. 9}$$

Implicitly we assumed that (i) the biomass of an individual of a specific stage is constant during FLEX, and (ii) food intake is a constant percentage of this biomass.

Estimate of grazing capacity from filtration rates

This estimate takes notice of the available food concentration, $F(t,z)$ ($gC\ m^{-3}$), as given by particulate organic carbon (POC). Food intake is estimated via filtered water volumes by assuming that (i) the filtration rate for each state, f_j ($m^3 ind^{-1} day^{-1}$), is constant, i.e. independent of food concentration during FLEX, and (ii) that all food in the filtered water volume is taken out by the animal. Then the local grazing rate is calculated

$$g(t,z) = \sum_{j=1}^{6} F(t,z)\ f_j\ n_j(t,z) \qquad \text{Equ. 10}$$

The numerical values of the filtration rates were derived from Paffenhöfer (1971) and Gamble (1978). Gamble distinguishes between Calanus finmarchicus stages IV to VI and small copepods. As the rates for C. helgolandicus, stages V and VI given by Paffenhöfer are about 1.5 times those given by Gamble for C. finmarchicus, the

Table 3. Constants used for the various estimates of grazing capacity, equations (9) to (16).

Constants	*C. finmarchicus* stage I	II	III	IV	V+VI
for equ. (9)					
biomass/individual:					
b_j (μgC ind. $^{-1}$)	1.22	2.57	9.05	18.09	64.34
uptake in % of biomass:					
p_j (% day^{-1})	148	148	115	83	31
for equ. (10)					
filtration rate f_j (ml day^{-1} ind.$^{-1}$)	11.67	26.67	48.33	73.33	100.0
for equ. (11), (12)					
optimum grazing:					
rates r_j (mgC day^{-1} ind. $^{-1}$)	0.003	0.003	0.01	0.014	0.021
threshold values:					
ε_j (gC m^{-3})	0.04	0.04	0.05	0.06	0.065
half-saturation values					
β_j (gC m^{-3})	0.145	0.145	0.110	0.125	0.130
for equ. (13) - (16)					
threshold values:					
s_D^1 (mgC m^{-3})	80	80	70	60	60
s_D^2 (mgC m^{-3})	250	250	150	170	250
s_N^1 (mgC m^{-3})	80	80	70	60	60
s_N^2 (mgC m^{-3})	286	286	223	140	160
maximum intake rates (mgC day^{-1} ind.$^{-1}$):					
I_D^{max}	0.00168	0.00168	0.003	0.006	0.0144
I_N^{max}	0.006	0.006	0.006	0.0151	0.042

rates for stages I to IV of C. helgolandicus were reduced by 1/3 to be applied to C. finmarchicus. There are only a few measurements of the size spectrum of POC during FLEX (Gamble, 1978), and we do not know which part of POC has been utilized by herbivorous zooplankton. Thus we take 100% as an upper limit.

Estimate of grazing capacity with food limited intake.

The relation between food concentration and food intake can be represented by a hyperbola shaped (dimensionless) relation $h_j(t,z)$, introducing a stage dependent threshold value ε_j ($gC\ m^{-3}$) and a half-saturation constant β_j ($gC\ m^{-3}$)

$$h_j(t,z) = \frac{P(t,z) + \varepsilon_j}{P(t,z) + \varepsilon_j - \beta_j} \qquad \text{Equ. 11}$$

The formula and the constants were derived from Gamble's grazing experiments (1978). The local grazing rate can then be estimated as

$$g(t,z) = \sum_{j=1}^{6} r_j \; h_j(t,z) \; n_j(t,z) \qquad \text{Equ. 12}$$

where r_j is the maximum grazing rate of stage j.

This estimate implies that (i) food intake for all individuals of a specific stage is the same, (ii) factors like particle size and composition of food are negligible, and (iii) the relations for $h_j(t,z)$ is valid for the entire FLEX time.

Estimate of grazing capacity with day/night rhythm in grazing

Grazing activity depends on the time of the day, as shown for FLEX by Daro (1980). Her results were used to construct actual grazing rates: Observed maximum (mid night) and minimum (noon) intake curves were interpolated by a cosine function

$$I^j(t') = I_D^j + (I_N^j - I_D^j)\,(\cos(2\pi t') + 1) \qquad \text{Equ. 13}$$

where t' is daytime (as fraction of 24 h), $I_D{}^j$ and $I_N{}^j$ are intake rates ($gC\ ind^{-1}\ day^{-1}$) at noon and midnight for stage j. The rates $I_D{}^j$ and $I_N{}^j$ are dependant on food availability and stage specific thesholds $s_x{}^1$, $s_x{}^2$ (x=N or D),

$$I_x^j = \begin{cases} 0, & \text{if } P(t,z) < s_x^1 \\ I_x(P(t,z)), & \text{if } s_x^1 \leqslant P(t,z) < s_x^2 \\ I_x^{max} & \text{else} \end{cases} \qquad \text{Equ. 14}$$

where

$$I_x(P(t,z)) = I_x^{max} \frac{P(t,z) - s_x^1}{s_x^2 - s_x^1} \qquad \text{Equ. 15}$$

All occurring constants are stage dependent. The local grazing rate is then

$$g(t,z) = \sum_{j=1}^{6} I^j(t') \, n_j(t,z) \qquad \text{Equ. 16}$$

Implicitly the assumptions were made that (i) food intake by Calanus finmarchicus is at its maximum at midnight and at its minimum at noon, (ii) the dependence of intake from food concentration can be described by the ramp function (14) at any time of the day, and (iii) the instantaneous food intake functions can be interpolated by a cosine function between minimum and maximum intake.

Calculation of the estimates of grazing capacity

With the formulae (9) to (16) for the local grazing rate estimates, the total grazing capacity, G(T) (gC m^{-2}), from the begin of FLEX, T_o, to time T within the upper 100 m of the water column are calculated as

$$G(T) = \int_{T_o}^{T} \int_{o}^{100m} g(t,z) \, dz \, dt \qquad \text{Equ. 17}$$

with T_o = 26 March 1976, 6.00 h and last T = 5 June 1976, 12.00 h.

Figure 5 a-d shows the grazing capacities per day, related to zooplankton biomass, for the 4 methods. For the biomass orientated approach this ratio is larger than 1 in the earlier stages of the experiment, because of the great number of young stages; from mid-April on mean percentages between 40 and 90% dominate (Figure 5a). The filtration rate approach yields the opposite picture, which does not seem trustworthy until mid-April because it is unlikely that the young stages should have an intake of about 40% of their body weight, while the older stages show an intake of more than 150% of their body weight (Figure 5b). The same appearance is given by the POC limited approach of intake (Figure 5c). Clearly, the POC concentration is determining the results in Figure 5b and c. Finally, Figure 5d shows the grazing/biomass ratio for the day/night-rhythm approach, which seems to underestimate the first phase (of low biomass), but becomes very similar to the body weight approach from mid-April on, with a mean intake of approximately 50% of biomass, decreasing from 80 to 30% within about a month.

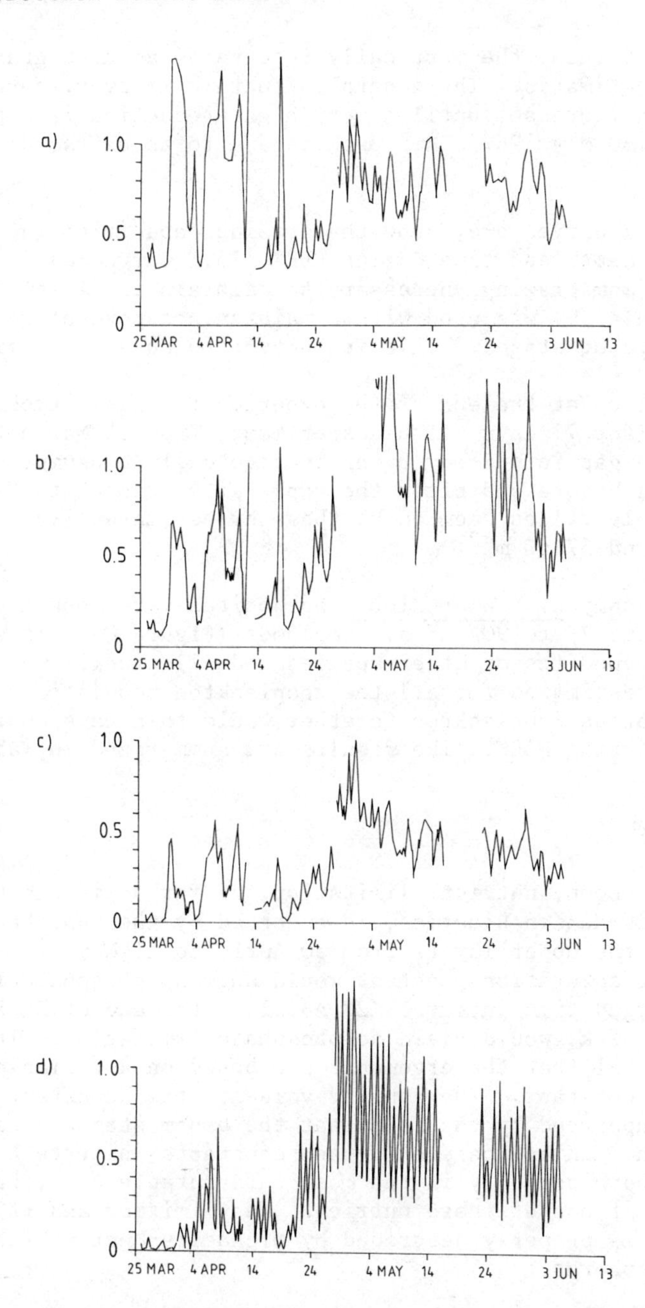

Fig. 5. Grazing capacities (gC m^{-2}) per day, as a fraction of the C. finmarchicus biomass (gC m^{-2}) within the upper 100 m of the water column for (a) the biomass approach, formula (9), (b) the filtration approach, formula (10), (c) the POC limited approach, formula (11), and (d) the day/night rhythm approach, formula (16). Values larger than 1.5 are indicated by crosses.

Figure 6 shows the vertically integrated amounts grazed per day for the 4 estimates. The general structure is everywhere the same. Consumption increases until a very high production rate per day appears at 30 May '76. This peak is due to an extraordinary high biomass.

Figure 7 curves a-e, show the grazing capacities (gC m^{-2}) cumulated over depth and time during FLEX '76. Curve (e) is an estimate of the minimum grazing, necessary to maintain the life of the animals, where formula (9) was used with a maintenance percentage of 10% irrespective of stages. This is thought to be a lower limit.

The values at the end of the experiment range between 35 and 77 gC m^{-2} (for 71 days of the experiment, from 25 March to 5 June), filling the gap in mid-May by an average daily consumption derived from values before and after the gap. With respect to Figure 5 the most reliable values seem to be those by estimates (9) and (16), namely 47 and 67 gC m^{-2}.

So far only C. finmarchicus copepodites were considered, which make up about 75 to 90% of all copepods (Figure 8). If we add 20% for the copepodites of other species and all nauplii we may get reasonable estimates for all the zooplankton populations together. All herbivorous zooplankton together would then have consumed 59 to 84 gC m^{-2} during FLEX. The results are summarized in Table 4.

CONCLUSIONS

As was shown, nutrient limitation, as far as it can be specified by Michaelis-Menten kinetics, is enforced by the phosphate concentration in the upper layers from 30 April to 11 May. A two times larger half-saturation constant would suggest phosphate limitation in an enlarged time interval (28 April to the end of FLEX). Half that value of k_s would yield no phosphate limitation. It is therefore essential that the arguments are based on a reliable half-saturation constant. The applied value of this constant for phosphate is supported by the fact that the bloom starts decaying from 30 April on. In any case of chosen constants, nitrate has been limiting the bloom only in the most unfavourable case, from 24 May on. Thus, if one of these nutrients was limiting and if nutrient limitation is properly described by Michaelis-Menten kinetics, then it was phosphorus.

The comparison of daily production and daily grazing capacity (Figure 9) shows that grazing was equal to production around 8 May, when both are about 0.8 gC m^{-2} day^{-1}. Even assuming 20% higher grazing capacity would fix this point somewhere between 5 and 11 May. But production stagnated already from about 29 April, when phosphate was supposed to be limiting. Production amounts up to 23.25 gC m^{-2}

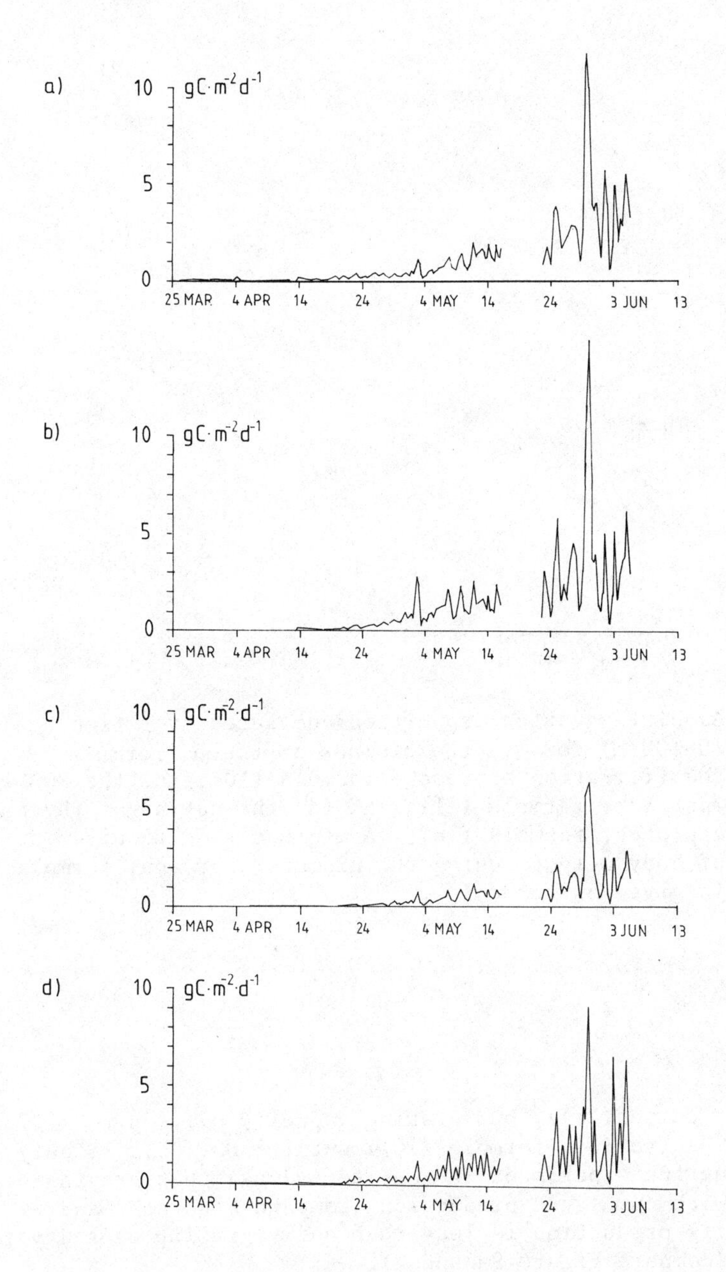

Fig. 6. Absolute values of grazing capacities per day within the upper 100 m of the water column for (a) the biomass approach, formula (9), (b) the filtration approach, formula (10), (c) the POC limited approach, formula (11), and (d) the day/night rhythm approach, formula (16).

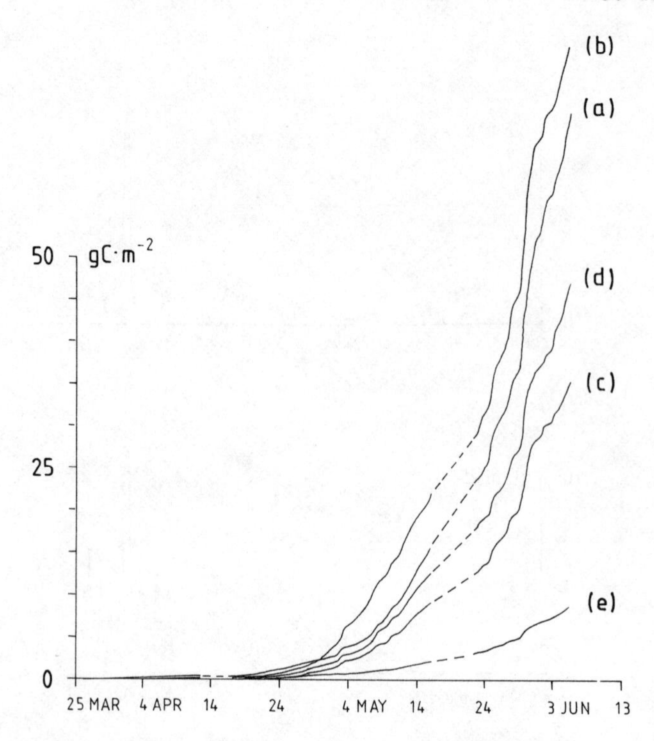

Fig. 7. Absolute grazing capacities cumulated over time and depth (0-100 m) for (a) the biomass approach, formula (9), (b) the filtration approach, formula (10), (c) the POC limited approach, formula (11), and (d) the day/night rhythm approach, formula (16). A minimum food requirement of 10% of body weight, using the biomass approach, formula (9), is given by curve (e).

(Figure 2, curve (1b)), but grazing capacity only up to 5.27 gC m (Figure 7, curve (3), formula (16)) until 8 May. It is only very late during the experiment (20 May to 4 June) that grazing capacity can cope with cumulated production (compare Figures 2 and 7), although daily production is less than daily grazing capacity from 8 May on (compare Figure 9a and b).

Until about 28 May herbivorous zooplankton could satisfy its demand, if the primary produced material would remain in the upper 100 m of the water column (which is studied here), possibly converted into non-living particulate material, and if zooplankton

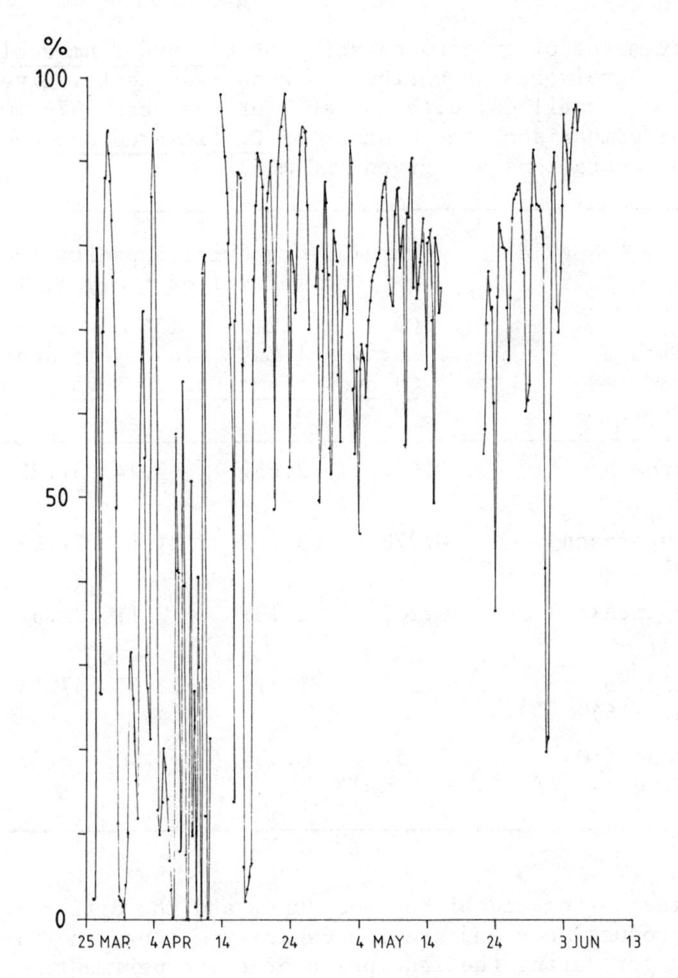

Fig. 8. Biomass of C. finmarchicus as percent of the biomass of all copepodites of all copepod species (without nauplii) within the upper 100 m of the water column.

would graze on this material. There is, of course, the possibility that detritus feeding by copepods has not to be evoked at all, if the suggestion by Williams and Lindley (1980) is accepted that half of the Calanus population is remaining in deep water for weeks without any feeding. Thus, grazing capacity would be less than given by

Table 4. Estimates of grazing capacity of Calanus finmarchicus and all herbivores by various methods (see text), given in gC m^{-2} until day with number T of the year 1976 (0-100 m). For comparison, the biomass of C. finmarchicus and POC concentrations are given (gC m^{-2}).

type of estimate	109 19 April	121 1 May	136 16 May	156.5 5 June	156.5 5 June all copepods
	cumulated grazing capacity (gC m^{-2}) until day T of the year 1976 — Calanus finmarchicus I-VI				
biomass approach equ. (9)	0.617	2.637	15.74	67.22	84.03
filtration approach equ. (10)	0.278	3.103	21.89	77.09	96.36
POC limited intake equ. (12)	0.105	1.131	9.11	35.31	44.14
day/night rhythm approach, equ. (16)	0.185	1.810	12.62	47.13	58.91
minimum demand (10% of biomass), equ. (9)	0.157	0.534	2.20	8.83	11.04

our estimates during end of May and June, and the differences of cumulated production P and grazing capacity G shown in Figure 10 would be larger during the last phase of the experiment.

The differences P-G (curves 1 and 4 in Figure 10) should be comparable to non-living particulate material (calculated as particulate organic carbon minus phytoplankton carbon minus zooplankton carbon), the latter of which is shown in Figure 10 denoted as "detritus". Detritus shows two maxima at about the same time as the difference P-G. The surplus of P-G compared to detritus must be interpreted as being lost material by processes like sinking and regeneration. A detailed analysis may possibly provide the shares of the different loss processes, but this is beyond the scope of this paper.

Our conclusion is that the phytoplankton bloom during FLEX was phosphate limited. Its decay was caused by the phosphate depletion and not by herbivorous grazing. Before zooplankton grazing could overcome primary production, so much material was already produced

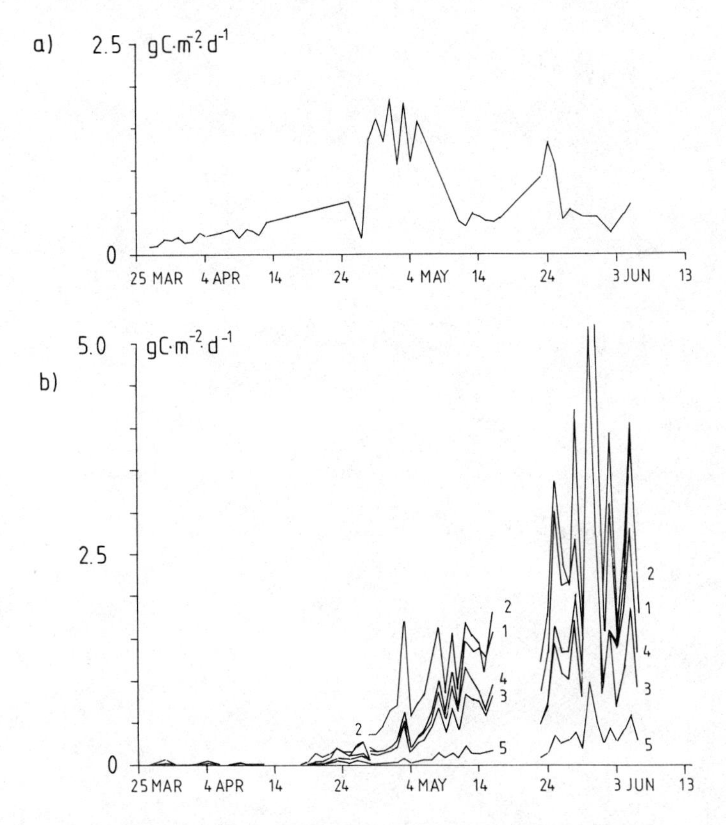

Fig. 9. Daily particulate primary production (0-50 m), represented by Weigel and Hagmeier's data (a), compared to daily grazing capacity (0-100 m) (b) during FLEX '76. Grazing capacity is shown for the biomass approach (1), the filtration approach (2), the POC limited approach (3), and the day/night rhythm approach (4), and for minimum intake (5).

that zooplankton would not have to starve, if satisfied with detrital material and assuming that this material remained in the mixed layer, or if only half of the Calanus population was feeding from mid-May on.

Acknowledgements

Data analysis of the type presented here depends on the availability of the many different data. The authors are therefore in-

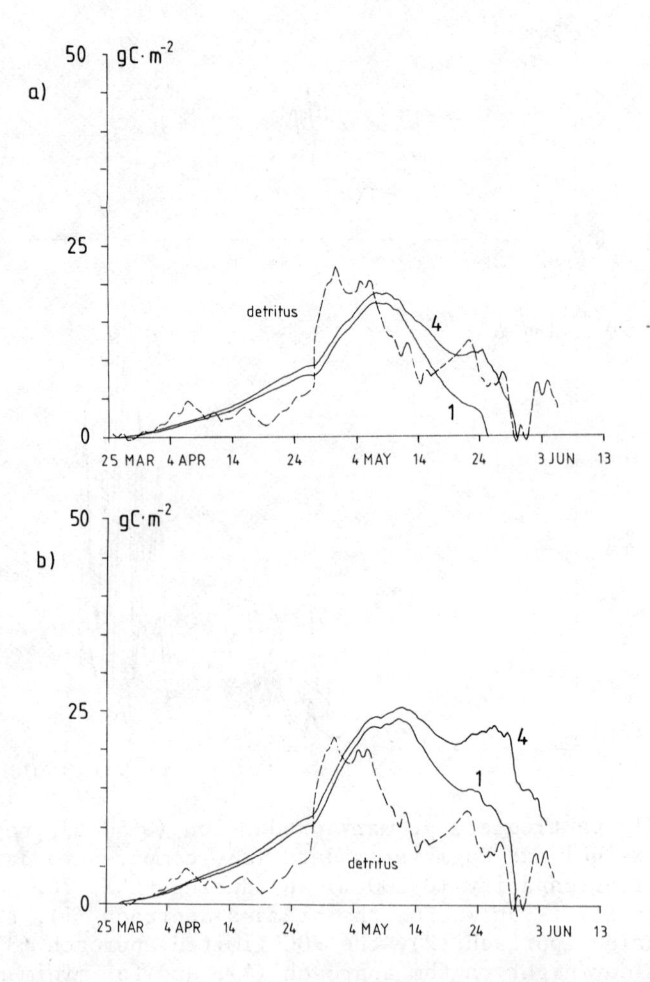

Fig. 10. Difference of cumulated production P and grazing capacity G calculated as (a) Weigel and Hagmeier's cumulated production (curve (1b) of Figure 2) minus cumulated grazing capacity calculated by the biomass approach (curve 1) and by the day/night rhythm approach (curve 4, cp. curves (a) and (d) of Figure 7), and (b) as (a) but using Mommaerts' cumulated production. In addition, detritus is shown in both graphs (0-100 m).

debted to the many data originators, who kindly provided data, namely Drs. U. Brockmann, K. Eberlein, M. Gillbricht, K. -D. Hammer, E. Hagmeier, W. Hickel, N.K. Höjerslev, G. Kattner, J.P. Mommaerts, J. Trahms, A. Weber, G. Weichart, P. Weigel.

We are grateful to Drs. D. Cushing, H.G. Fransz, B. Frost, M. Gillbricht, E. Hagmeier, W. Hickel and R. Williams for valuable comments. Thanks are due to Ms. R. Krautwald for drawing figures. This research was sponsored by the Deutsche Forschungsgemeinschaft through the "Sonderforschungsbereich 94 - Meeresforschung Hamburg". This is JONSDAP '76 contribution no. 91.

REFERENCES

Blackman, F.F., 1905, Optima and limiting factors, Ann. Bot. London. 19: 281.

Cushing D.H., and Vucetic, T., 1963, Studies on a Calanus patch, III. The quantity of food eaten by Calanus finmarchicus, J. Mar. Bio. Ass. U.K., 43:349.

Daro, M.H., 1980, Field study of the diel feeding of a population of Calanus finmarchicus at the end of a phytoplankton bloom. FLEX '76 22 May - 5 June, "Meteor" Forsch. -Ergebnisse, Series A. 22:123.

Eberlein, K., Kattner, G., Brockman, U. and Hammer, K.D., 1980, Nitrogen and phosphorus in different water layers at the central station during FLEX '76, "Meteor" Forsch. -Ergebnisse, Series A, 22: 87.

Eppley, R.W., 1972, Temperature and phytoplankton growth in the sea, Fish. Bull., 70: 1063.

Fransz, H.G., and Gieskes, W.W.C., 1982, The unbalance of phytoplankton production and copepod production in the North Sea, ICES Symp. on Biol. Prod. of Cont. Shelves, paper no. 22, Kiel, F.R.G., 2-5 March 1982, 19pp.

Gamble, J.C., 1978, Copepod grazing during a declining spring phytoplankton bloom in the northern North Sea, Mar. Biol., 49: 303.

Gassmann, G., and Gillbricht, M., 1982, Correlations between phytoplankton, organic detritus and carbon during the Fladen Ground Experiment FLEX'76, Helgoländer Meeresunters, 35: 253.

Gieskes, W.W.C., and Kraay, G.W., 1980, Primary production and phytoplankton pigment measurements in the northern North Sea during FLEX'76, "Meteor" Forsch. -Ergebnisse, Series A, 22: 105.

Krause, M. and Radach, G., 1980, On the succession of development stages of herbivorous zooplankton in the norther North Sea during FLEX '76. 1. First statements about the main groups of the zooplankton community, "Meteor" Forsch. -Ergebnisse, Series A, 22: 133.

Krause, M. and Radach, G., 1983, On the vertical distribution and daily migration of herbivorous zooplankton in the norther North Sea during FLEX '76. In preparation.

Kremer, J.N. and Nixon, S.W., 1978, "A coastal marine ecosystem. Simulation and analysis," Ecol. Stud. 24, Springer-Verlag, Berlin, Heidelberg and New York.

Lenz, W., Ramster, J., and Weidemann, H., 1980, First steps in the realization of the Joint North Sea Data Acquisition Project for 1976 (JONSDAP '76), "Meteor" Forsch. -Ergebnisse, Series A, 22: 3.

Mommaerts, J.P., 1980, Seasonal variations of the parameters of the photosynthesis-light relationship during the Fladen Ground Experiment 1976, Proc. Final ICES/JONSIS Workshop on JONSDAP '76, Liege, Belgium, 29 April - 2 May 1980, ICES C.M. 1980/C. 3: 31.

Mommaerts, J.P., 1982, The calculation of particulate primary production in a stratified body of water, using a modification of the Vollenweider model formula, "Meteor" Forsch. -Ergebnisse, Series D, 34: 1.

Mommaerts, J.P., 1981, Atlas of particulate primary production results from the central station during the FLEX '76 campaign as calculated with a model of photosynthesis-light relationship, Management Unit of the North Sea Model, Ministry of Public Health, I.H.E., Brussels, Belgium.

Mommaerts, J.P., and Nijs, J., 1981, MECHELEN primary production data from the Flagen Ground (North Sea) during the spring bloom 1976 (FLEX'76), Publ. by Management Unit of the North Sea Model, Ministr of Public Health, I.H.E., Brussels, Belgium.

Paffenhöfer, G.A., 1971, Grazing and ingestion rates of nauplii, copepodids and adults of the marine planktonic copepod Calanus helgolandicus, Mar. Biol., 11: 286.

Parsons, T.R., Takahashi, M., and Hargrave, B., 1977, "Biological oceanographic processes," Pergamon Press, Toronto, 2nd ed.

Platt, T., Denman, K.L., and Jassby, A.D., 1977, Modeling the productivity of phytoplankton, In: "The Sea, Ideas and observations of progress in the study of the sea," Vol. 6, Goldberg, E.D., McCave, I.N., O'Brien, J.J. and Steele, J.H. (eds.), John Wiley and Sons, New York.

Radach, G., 1980, Preliminary simulations of the phytoplankton and phosphate dynamics during FLEX '76 with a simple two-component model, "Meteor" Forsch. -Ergebnisse, Eries A, 22: 151.

Radach, G., 1982, Dynamical interactions between the lower trophic levels of the marine food web in relation to the physical environment during the Fladen Ground Experiment, Netherl. J. Sea Res. 16: 231.

Radach, G., 1983, Simulations of phytoplankton dynamics and their inter actions with other system components during FLEX'76, In:'North Sea Dynamics', J.Sündermann and W.Lenz, eds., Springer Verlag Berlin, Heidelberg : 584.

Radach, G. and Maier-Reimer, E., 1975, The vertical structure of phytoplankton growth dynamics - a mathematical model. Mem. Soc. Roy. des Sciences de Liege, Series 6,7: 113.

Radach, G., Trahms, J. and Weber, A., 1980, The chlorophyll development at the Central Station during FLEX '76. Two data sets Proc. Final ICES/JONSIS Workshop on JONSDAP '76, Liege, Belgium, 29 April - 2 May, 1980, ICES C.M. 1980/C. 3: 3.

Ramster, J.W., 1977, Development of cooperative research in the North Sea. The origins, planning and philosophy of JONSDAP '76, Mar. Pol., 1: 318.

Schöne, H.,1977, "Die Vermehrungsrate mariner Planktondiatomeen als Parameter in der ökosystemanalyse", RTWH Aachen, Habil. Schrift.

Steele, J.H., 1962, Environmental control of photosynthesis in the sea, Limn. Oceanogr., 7: 137.
Steele, J.H., 1974, "The structure of marine ecosystems," Harvard University Press, Cambridge.
Steeman Nielsen, E., 1957, The balance between phytoplankton and zooplankton in the sea, J. Cons. perm. int. Explor. Mer., 23: 178.
Weichart, G., 1980, Chemical changes and primary production in the Fladen Ground area (North Sea) during the first phase of a spring phytoplankton bloom, "Meteor" Forsch. -Ergebnisse, Series A, 22: 79.
Weigel, P. and Hagmeier, E., 1978, Primary production and the timing of phytoplankton spring bloom at 58 55'N, 0 32' E (North Sea) during the Fladen Ground Experiment (FLEX '76), Unpubl. manuscript.
Williams, R. and Lindley, J.A., 1980, Plankton of the Fladen Ground during FLEX '76. III. Vertical distribution, population dynamics and production of Calanus finmarchicus (Crustacea: Copepoda), Mar. Biol., 60: 47.

PART 2

WORKING GROUP REPORTS

STUDIES ON MARINE AUTOTROPHS: RECOMMENDATIONS FOR THE 1980s

L. Legendre[1], Y. Collos[2], M. Elbrächter[3], M.J.R. Fasham[4], W.W.C. Gieskes[5], A. Herbland[6], P.M. Holligan[7], R. Margalef[8], M.J. Perry[9], T. Platt[10], E. Sakshaug[11], and D.F. Smith[12].

1. GIROQ, Dép. biologie, Université Laval, Québec Canada G1K 7P4

2. Laboratoire d'Océanographie, Campus de Luminy Case 902, 13288 Marseille Cedex 9, France

3. Biologisches Anstalt Helgoland, Litoralstation 2282 List, F.D.R.

4. Institute of Oceanographic Sciences, Wormley Godalming, Surrey GU8 5UB, U.K.

5. Netherlands Institute for Sea Research, PO Box 59 Texel, The Netherlands.

6. Antenne Orstom, Centre Océanologique de Bretagne BP 337, 29273 Brest Cedex, France

7. Marine Biological Association of the U.K., The Laboratory, Citadel Hill, Plymouth, Devon PL1 2PB, U.K.

8. Departamento de Ecologia, Facultad de Biologia Universidad de Barcelona, Gran Via 585, Barcelona 7 Spain

9. School of Oceanography, WB-10 University of Washington Seattle, WA 98195, U.S.A.

10. Marine Ecology Laboratory, Bedford Institute of Oceanography, PO Box 1006, Dartmouth, N.S. Canada B2Y 4A2

11. Institute for Marine Biochemistry, University of Trondheim, 7034 Trondheim-NTH, Norway

12. CSIRO Division of Fisheries and Oceanography
PO Box 20, North Beach, Western Australia 6020
Australia

RESEARCH DESIGN

There was agreement among members of the working group that, in the next 5-10 years, the process-oriented focus of field studies should be maintained. This orientation is viewed as a third phase in the historical evolution of modern biological oceanography, in which the regional studies of the 1940s and 1950s identified the important classes of biological oceanographic processes, and which were followed by a period during the late 1960s and 1970s when intensive laboratory studies of the physiology, biochemistry and behaviour of single species were emphasized. More recently, there has been a tendency for the scientists to go back into the field and to apply the conceptual understanding and the methods developed in the laboratory to natural systems. The purpose of the process-oriented field studies over the next 5-10 years is to understand better the mechanisms by which both environmental and biological forcing act upon autotrophs to regulate their production rates, biomass and species composition. The scales in environmental forcing functions among different regions in the ocean should be carefully considered in the context of understanding the coupling with autotrophic processes.

(1) To obtain measurements of phytoplankton growth rates, we should sample the measured variable with high frequency (for example every 30 min) for a period of at least one physical (for example 24 h) or biological (for example one mean doubling time) cycle of the process of interest. In order to compare one measurement to another, a record of the environmental variables should be obtained at the same frequency as the samples for growth rate estimates. It is also advisable that nutrients and light be measured at least one cycle preceding the period over which the growth rate estimate was obtained. It should be stressed that, if productivity data (i.e. paired values of phytoplankton growth rates and time) are to be compatible with time-series data denoting other state variables of a model, then all the variable measurements should be done on the same system at the same time (although not necessarily at the same frequency, however).

(2) The choice of sampling areas should allow processes to be studied in the best conditions possible, aiming at optimizing the cost-benefit of the research. These might include, for example, boundary areas - such as fronts, thermoclines, sea grass beds, kelp beds, coral reefs (LIMER 1975 Expedition Team, 1976) etc - where sharp physical gradients are identified and where dynamic physical equilibria are established which force autotrophic processes. On the other hand biologically complex, but relatively uniform systems with high turnover rates, such as some tropical seas (McCarthy and

Goldman, 1979), provide typical conditions to study specific processes. It was also suggested that regions with single species blooms (Elbrächter and Boje, 1978) could be used to study some rate processes without the added complexity brought out by multi-specificity.

(3) Future field experiments in the marine environment should be designed to include a control and the possibility of experimenting with one (or perhaps several) variable(s) at a time.

(4) Oceanographers with practical experience of field conditions recommend and encourage that laboratory studies on primary producers should be focussed on processes relevant to those occurring in the natural environment. Subjects explored in marine laboratories should, for the most part, be derived from field problems (Dugdale, 1977, Kalff and Knoechel, 1978; Sharp et al., 1980).

(5) Despite the emphasis on rate processes studies, regional surveys are still necessary, especially as new measurement techniques and new physical information become available (Pingree et al., 1982). These surveys give essential indications on precisely where and when to study the various processes, and they provide the framework for interpreting process studies. Often, both process studies and regional surveys can be conducted on the same cruise. There is also a need to obtain a comprehensive modern data set on the seasonal changes of phytoplankton, nutrients, herbivores and hydrographic structure in the open ocean with which to critically test recent models of these processes (Kiefer and Kremer, 1981).

(6) Environmental variables can force autotrophic rate processes, but the identify of the involved species may drive future developments. It follows that species succession should be studied and that critical experiments designed both in the laboratory and in the field, in order to understand it (Elbrächter, 1977; Aubert et al., 1981). This new information could also be incorporated in models either by subdividing the primary production compartment in a series of sub-compartments, or by changing the values of the parameters according to the steps of the succession. As pointed out by Prof. Margalef: "organisms are not only carriers or molecules, but they are subject to natural selection".

IT IS RECOMMENDED THAT

Joint programs be organised to study identified processes in terms of the forcing functions and mechanisms acting upon autotrophs; these programs could bring together specialists, mastering different experimental and analytical techniques, to study simultaneously a series of sites located along gradients in some forcing variable(s).

TECHNIQUES

(1) Estimation of gross and net fixation rates by primary producers is still problematical (Peterson, 1980). The question should continue to be investigated, for example by using series of short-term tracer assimilations together with parallel series of light and dark O_2 incubations (using for instance the new O_2 probes* or modern high-precision versions of the Winkler method that involve photometric determination of the titration endpoint). A proper conceptual background for such an approach should be established to derive the desired quantities.

(2) The incubation of samples for rate measurements should not be prolonged beyond the minimum time necessary to satisfy the requirements of the technique employed. For example, in isotope tracer techniques this minimum time can often be established by compartmental analysis (Lancelot, 1979). Prolongation of incubation increases the inevitable, but unknown, bias that accrues from containment of the samples (Venrick _et al._, 1977; Gieskes _et al._, 1979; Glibert and Goldman, 1981). Long-term phenomena are better studied by series of short-term incubations.

(3) New techniques to measure rate processes _in situ_ should be developed since bringing a water sample to the surface (changes in pressure, light and temperature shocks, etc.) and handling it in various ways (shaking the Niskin sampler, pouring water in glass bottles, etc.) may seriously damage many of the phytoplankton cells or at least modify their physiological responses (Doty, 1955; Strickland, 1965; Copin-Amieil, 1974). Samplers with _in situ_ ^{14}C inoculation already exist (Gundersen, 1973) and should be developed for wider use.

(4) Full advantage should be taken of the increasingly precise O_2 and pH electrodes to measure _in situ_ changes of these variables, including from automatic stations equipped with telemetry (Smith, 1982). In addition, drifting buoys with such sensors could be used for Lagrangian measurements of these properties in well-defined water masses.

(5) The question of whether the growth of phytoplankton in the oligotrophic ocean is nutrient-limited cannot be resolved by the analytical techniques available now to determine concentrations of nitrogen compounds in seawater (Jackson, 1980; Glibert and Goldman, 1981). There is an immediate requirement for a method to measure ammonia in nanomolar concentrations in seawater.

(6) For direct measurements of autotrophic "biomass", to which to apply fixation rates, the plant cells must be separated from

*Orion Research, 840 Memorial Drive, Cambridge, MA 02139, U.S.A.

detritus and other particles. An efficient cell sorter is therefore needed that would (a) short photoautotrophic cells in the size range 0.4 - 200 μm, and (b) process reasonably high volumes of water in a short time period. Such a method would also provide valuable data for natural populations on ratios of cell constituents - e.g. carbon to chlorophyll, or carbon to nigrogen: see for instance Eppley (1972), Sakshaug (1977) and Sakshaug and Holm-Hansen (1977) -, especially for conditions of non-balanced growth when the relative uptake rates of different nutrients may be very variable in the short term (Collos and Slawyk, 1980). Labelling different cell constituents, to be separated for instance by HPLC, is another way of estimating the carbon to chlorophyll ratio (Redalje and Laws, 1981).

(7) Biomedical research instruments, such as automatic counters of particles by size class, which discriminate between fluorescent and non-fluorescent particles, should be adapted for phytoplankton research. Simultaneously, improvements are needed in preservation techniques in order to preserve both the shape and the fluorescent characteristics of phytoplankton samples collected at sea.

(8) Close collaboration should be established with optical oceanographers to enable biological oceanographers (a) to take full advantage of on-going technological advances in that field and (b) to promote the development of new optical measurements relevant to primary production. Such optical equipment includes a compact, robust, sea-going red sensitive transmissometer capable of detecting *in vivo* chlorophyll concentrations down to 0.1 mg m^{-3}.

(9) Colour scanner satellite images should be made readily available to biological oceanographers so that sampling strategies at sea can be routinely based, if possible in real time, on synoptic information about the bio-optical properties of surface waters (Yentsch and Phinney, 1982).

Because of the wide range of variation in experimental manipulation (Peterson, 1980) in the application of the so-called "standard technique" for the measurement of C fixation by autotrophs (the ^{14}C method) and of chlorophyll *a*.

IT IS RECOMMENDED THAT

A workshop be held, initially in a shore-based laboratory and subsequently during a cruise, to ascertain which of the variants of the commonly used methods for measuring primary production and pigments are of trivial significance for the results and which have wider implications. The ultimate aims would be to find a consensus on the optimal procedure. This suggestion could be acted upon within 6-18 months. Dr Platt will look into the feasibility of implementation.

PARAMETERISATION

(1) New models of the production-irradiance relationship are being developed from consideration of events occurring within the photosynthetic unit (PSU). These models suggest that primary production studies be conducted in terms of the PSU and quanta rather than chlorophyll and energy (Fasham and Platt, ms.).

(2) Various parameterisations are used in models to describe the carbon-to-chlorophyll ratio. In fact, this ratio varies by about one order of magnitude as a function of species composition, light (Falkowski, 1980), temperature (Eppley, 1972; Sakshaug, 1977), nutrient deficiency (Sakshaug, 1977; Sakshaug and Holm-Hansen, 1977). This ratio is used to estimate carbon concentration from chlorophyll and in some circumstances this ratio is a critical parameter in phytoplankton models (Steele, 1974; Fasham *et al*., 1982). The only satisfactory way to obtain reliable measurements of carbon is to directly measure carbon on microalgal samples uncontaminated by detritus or other organisms (see above). Reliable carbon-to-chlorophyll measurements are needed in order to understand these changes, so that all biological oceanographers are encouraged to record such data whenever they are able to sample micro-algae populations uncontaminated by either detritus or microzooplankton (Sakshaug *et al*., 1981).

(3) Physiological parameters and conversion factors formerly treated as constants are now known to be highly variable in response to fluctuations in their physical environment (see for instance: Fortier and Legendre, 1979; Gallegos *et al*., 1980; Falkowski, 1980; Vincent, 1980; Savidge, 1981; Auclair *et al*., 1982, Walsh and Legendre 1982). Physiological parameters indicating nutrient deficiency are discussed, among others, by Jensen and Sakshaug (1973), Sakshaug and Myklestad (1973), Myklestad (1977), Sakshaug (1977), and Sakshaug and Holm-Hansen (1977). It is also known that environmental fluctuations happen on a wide variety of time and space scales. It is therefore important that research in phytoplankton physiology be related explicitly to its physical context. This can be done through a statement of the interval, in frequency or wavenumber space, to which the results will be relevant.

(4) It is considered that the parameterisation of vertical diffusion is fundamental to the development of models in marine phytoplankton models in various different ways, so that comparative studies are needed to assess their significance. In a few critical cases, comparisons should be conducted between Lagrangian and Eulerian models. Also, simultaneous measurements of vertical distribution of phytoplankton should be made together with measurements of vertical movements. In time series of observations, vertical profiles of eddy

diffusion can be computed from successive temperature profiles, combined with meteorological measurements (air temperature, solar ratiation and wind). See for instance Jassby and Powell (1975).

(5) Given that "balanced growth" is rarely encountered in natural situations (Sakshaug and Myklestad, 1973), description of autotrophic fluxes in terms of carbon are not immediately translatable into nitrogen or phosphorus (Collos and Slawyk, 1980). Biochemical and physiological research aimed at facilitating this conversion is a requirement of some urgency. In addition, the macro-molecular composition of autotrophs is of importance in considering transfers to the next trophic levels (Smith and Morris, 1980; Stoecker et al., 1981).

(6) Satellite measurements of see surface chlorophyll (water colour), temperature and wind stress (scatterometer) can perhaps be used to estimate general levels of phytoplankton production on large spatial scales and over longer temporal scales than are accessible by ships. Field data are needed to estimate the proportion of chlorophyll per square metre that is detectable by remote sensing, and both laboratory and field data are needed to more fully develop and test models for conversion of chlorophyll concentrations into instantaneous rates of production. Preliminary data suggest this is feasible for specific regions (Smith, Eppley and Baker, 1982) and that the proportions of chlorophyll and also production per square metre that is accessible to satellites are very constant for an oceanic region and a season (Platt and Herman, in press).

(7) Models that incorporate spatial organisation should deserve more consideration in the furure (Riley et al., 1949; Dubois, 1975; Margalef, 1978). Also, before designing a model, the characteristic spatial scales of the system should be taken into account (satellite images, batfish surveys, etc.).

(8) Theoretical considerations suggest that primary production could be a direct function of the input of energy into the ocean. Therefore, correlating surface heat exchanges (satellites) with large scale primary production patterns might provide general information on oceanic primary production (Margalef and Estrada, 1980).

(9) Two distinctly different measures of phytoplankton abundance are required to adequately describe the population rates of increase and loss. One is an estimate of population biomass, the other an estimate of the number of particles comprising the population. Two models of bacterial growth, relating the rate of change of cell biomass and cell number to changes in limiting nutrient concentration, were published in close succession (Helmstetter, 1969; Williams, 1971). These might profitably be investigated for future modelling of phytoplankton populations.

REFERENCES

Aubert, M., Gauthier, M., Aubert, J., and Bernhard, P., 1981, Les systèmes d'information des microorganismes marins. Leur rôle dans l'équilibre biologique, Rev. int. Océanogr. méd., 231 p.

Auclair, J.C., Demers, S., Fréchette, M., Legendre, L., and Trump, C.L., 1982. High frequency endogenous periodicities of chlorophyll synthesis in estuarine phytoplankton, Limnol. Oceanogr., 27: 348.

Collos, Y., and Slawyk, G., 1980, Nitrogen uptake and assimilation by marine phytoplankton, in: "Primary productivity in the sea," P.G. Falkowski, ed., Plenum Press, New York.

Copin-Amieil, C., 1974, Contribution à l'étude chimique des particules en suspension dans l'eau de mer, Thèse Doct. ès Sciences, Univ. Paris VI.

Doty, M.S., 1955, Current status of carbon-14 method of assaying productivity of the ocean, Rep. Congr. atom. Energy Comm. U.S. Contr. At (04-3)-15.

Dubois, D.M., 1975, A model of patchiness for prey-predator plankton populations, Ecol. Model., 1: 67.

Dugdale, R.C., 1977, Modeling, in: "The sea," Vol. 6, E.D. Goldberg, I.N. McCave, J.J. O'Brien, and J.H. Steele, eds., Wiley-Interscience, New York.

Elbrächter, M., 1977, On population dynamics in multi-species cultures of diatoms and dinoflagellates, Helgolander wiss. Meeresunters, 30: 192.

Elbrächter, M., and Boje, R., 1978, On the ecological significance of Thalassiosira partheneia in the northwest African upwelling area, in: "Upwelling Ecosystems," R. Boje and M. Tomczak, eds, Springer, Berlin.

Eppley, R.W., 1972, Temperature and phytoplankton growth in the sea, Fish Bull. U.S., 70: 1063.

Falkowski, P.G., 1980, Light and shade adaptation in marine phytoplankton, in: "Primary productivity in the sea," P.G. Falkowski ed., Plenum Press, New York.

Fasham, M.J.R. Holligan, P.M. and Pugh, P.R., In press, The spatial and temporal development of the spring phytoplankton bloom in the Celtic Sea, April 1979, Prog. Oceanogr.

Fasham, M.J.R., and Platt, T. ms., Photosynthetic response of phytoplankton to light: a physiological model, submitted for publication.

Fortier, L., and Legendre, L., 1979, Le contrôle de la variabilité à court terme du phytoplancton estuarien: stabilité verticale et profondeur critique, J. Fish. Res. Board Can., 36: 1325.

Gallegos, C.L., Hornberger, G.M., and Kelly, M.G., 1980, Photosynthesis-light relationships of a mixed culture of phytoplankton in fluctuating light, Limnol. Oceanogr., 25: 1082.

Gieskes, W.W.C., Kraay, G.W., and Baars, M.A., 1979, Current ^{14}C methods for measuring primary production: gross underestimates in oceanic waters, Neth. J. Sea Res., 13: 58.

Glibert, P.M., and Goldman, J.C., 1981, Rapid ammonium uptake by marine phytoplankton, Mar. Biol. Letters, 2: 25.

Gundersen, K., 1973, In situ determination of primary production by means of a new incubator ISIS, Helgol. wiss. Meeresunters., 24: 465.

Helmstetter, C.E., 1969, Regulation of chromosome replication and cell division in Escherichia coli, in: "The cell cycle," G.M. Padilla, G.L. Whiston, and I.L. Cameron, eds., Academic Press, New York and London.

Jackson, G.A., 1980, Phytoplankton growth and zooplankton grazing in oligotrophic oceans, Nature, 284: 439.

Jassby, A., and Powell, T., 1975, Vertical patterns of eddy diffusion during stratification in Castle Lake, California, Limnol. Oceanogr., 20: 530.

Jensen, A., and Sakshaug, E., 1973, Studies on the phytoplankton ecology of the Trondheimfjord. II. Chloroplast pigments in relation to abundance and physiological state of the phytoplankton, J. exp. mar. Biol.Ecol., 11: 137

Kalff, J., and Knoechel, R., 1978, Phytoplankton and their dynamics in oligotrophic and eutrophic lakes, Ann. Rev. Ecol. Syst., 9: 475.

Kiefer, D.A., and Kremer, J.N., 1981, Origins of vertical patterns of phytoplankton and nutrients in the temperate, open ocean: a stratigraphic hypothesis, Deep-Sea Res., 28: 1087.

Lancelot, C., 1979, Gross excretion rates of natural marine phytoplankton and heterotrophic uptake of excreted products in the southern North Sea, as determined by short-term kinetics, Mar. Ecol. Prog. Ser., 1: 179.

LIMER 1975 Expedition Team, 1976, Metabolic processes of coral reef communities at Lizard Island, Queensland, Search (Syd.), 7: 463.

Margalef, R., 1973, Life-forms of phytoplankton as survival alternatives in an unstable environment, Oceanol. Acta, 1: 493.

Margalef, R., and Estrada, M., 1980, Les áreas oceánicas más productivas, Investigatión y Ciencia (Spanish Edition of Scientific American), Oct. 1980: 8.

McCarthy, J.J., and Goldman, J.C., 1979, Nitrogenous nutrition of marine phytoplankton in nutrient-depleted waters, Science, 203: 670.

Myklestad, S., 1977, Production of carbohydrates by marine planktonic diatoms. II. Influence of the N/P ratio in the growth medium on the assimilation ratio, growth rate, and production of cellular and extracellular carbohydrates by Chaetoceros affinis var. willei (Gran) Hustedt and Skeletonema costatum (Grev.) Cleve, J. exp. mar. Biol. Ecol., 29: 161.

Peterson, B.J., 1980, Aquatic primary productivity and the ^{14}C-CO_2 method: a history of the productivity problem, Ann. Rev. Ecol. Syst., 11: 359

Pingree, R.D., Mardell, G.T., Holligan, P.M., Griffiths, D.K., Smithers, J., 1982, Celtic Sea and Armorican Current structure and the vertical distributions of temperature and chlorophyll, Cont. Shelf Res., 1: 99.

Platt, T., and Herman, A.W., In Press, Remote sensing of phytoplankton in the sea: surface layer chlorophyll as an estimate of water-column chlorcphyll and primary production, Int. J. Remote Sensing.

Redalje, D.G., and Laws, E.A., 1981, A new method for estimating phytoplankton growth rates and carbon biomass, Mar. Biol., 62: 73.

Riley, G.A., Stommel, H., and Bumpus, D.F., 1949, Quantitative ecology of the plankton of the western North Atlantic, Bull. Bingham oceanogr. Coll., 12: 1.

Sakshaug, E., 1977, Limiting nutrients and maximum growth rates for diatoms in Narragansett Bay, J. exp. mar. Biol. Ecol. 28: 109.

Sakshaug, E., and Holm-Hansen, O., 1977, Chemical composition of Skeletonema costatum (Grev.) Cleve and Pavlova (Monochrysis) lutheri (Droop) Green as a function of nitrate-, phosphate-, and iron- limited growth, J. exp. mar. Biol. Ecol., 29: 1.

Sakshaug, E., and Myklestad, S., 1973, Studies on the phytoplankton ecology of the Trondheimfjord. III. Dynamics of phytoplankton blooms in relation to environmental factors, bioassay experiments and parameters for the physiological state of the population, J. exp. mar. Biol. Ecol., 11: 157.

Sakshaug, E., Myklestad, S., Andersen, K., Hegseth, E.N., and Jørgensen, L., 1981, Phytoplankton off the Møre coast in 1975-1979: distribution, species composition, chemical composition and conditions for growth, in: "Proceedings of the Symposium on the Norwegian Coastal Current," Geito, Sept. 1980, Univ. Bergen.

Savidge, G., 1981, Studies of the effects of small-scale turbulence on phytoplankton, J. mar. biol. Ass. U.K., 61: 477.

Sharp, J.H., Underhill, P.A., and Frake, A.C., 1980, Carbon budgets in batch and continuous cultures: how can we understand natural physiology of marine phytoplankton ?, J. Plankton Res., 2: 213.

Smith, A.E., and Morris, I., 1980, Pathways of carbon assimilation in phytoplankton from the Antarctic Ocean, Limnol. Oceanogr., 25: 865.

Smith, D.F., 1982, Observations and quantitative analysis of curvilinear regions of time-varying oxygen concentrations with an oxygen electrode and a minicomputer, J. exp. mar. Bio. Ecol., 64: 117.

Smith, R.C., Eppley, R.W., and Baker, K.S., 1982, Correlation of primary production as measured aboard ship in Southern California coastal waters and as estimated from satellite chlorophyll images, Mar. Biol., 66: 281.

Steele, J.H., 1974, "The structure of marine ecosytems," Harvard Univ. Press, London.

Stoecker, D., Guillard, R.R.D., and Kavee, R.M., 1981, Selective predation by Favella ehrenbergii (Tintinnida) on and among dinoflagellates, Biol. Bull., 160: 136.

Strickland, J.D.H., 1965, Production of organic matter in the primary stages of marine food chain, in: "Chemical oceanography," J.P. Riley and G. Dkirrow, eds., Academic Press, London.

Venrick, E.L., Beers, J.R., and Heinbokel, J.F., 1977, Possible consequences of containing microplankton for physiological rate measurements, J. exp. mar. Biol. Ecol., 26: 55.

Vincent, W.F., 1980, Mechanisms of rapid photosynthetic adaption in natural phytoplankton communities. II. Changes in photochemical capacity as measured by DCMU-enhanced chlorophyll fluorescence, J. Phycol., 16: 568.

Walsh, P., and Legendre, L., 1982, Effets des fluctuations rapides de la lumière sur la photosynthèse du phytoplancton, J. Plankton Res., 4: 313

Williams, F.M., 1971, Dynamics of microbial populations, in: "Systems analysis and simulation in ecology," Vol. 1, B.C. Patten, ed., Academic Press, New York.

Yentsch, C.S., and Phinney, D.A., 1982, The use of attenuation of light by particulate matter for the estimation of phytoplankton chlorophyll with reference to the coastal zone scanner, J. Plankt. Res., 4: 93.

RESPIRATION

T.T. Packard[1], C. Joiris[2], P. Lasserre[3], H.J. Minas[4]
M. Pamatmat[5], A.R. Skjoldal[6], R.E. Ulanowicz[7],
J.H. Vosjan[8], R.M. Warwick[9], and P.J. Le B. Williams[10]

1. Bigelow Laboratory for Ocean Sciences, West Boothbay Harbour, Maine 04575, U.S.A.

2. University of Brussels, Fac. Wetenschappen, Lab. voor Ekologie en Systematik, Pleinlaan 2, 1050 Brussel Belgium

3. Station Biologique d'Arcachon, Universite de Bordeaux 2 rue du Professeur-Jolyet, 3312 Arcachon, France

4. Laboratoire d'Oceanographie, Faculte des Sciences de Luminy, Case 902, 13288 Marseille Cedex 2, France

5. Triburon Center for Environmental Studies, San Francisco State University, PO Box 855 Tiburon Calif 94920, U.S.A.

6. Institute of Marine Research, PO Box 1870, N5011 Nordnes, Bergen, Norway

7. Chesapeake Biological Laboratory, University of Maryland, Box 36, Solomons, Maryland 20688, U.S.A.

8. Netherlands Institute for Sea Research, PO Box 59 Texel, The Netherlands

9. Institute for Marine Environmental Research, Prospect Place, The Hoe, Plymouth, Devon PL1 2PB, U.K.

10. Department of Oceanography, The University, Southampton Hampshire, SO9 5NH, U.K.

The respiration working group conducted nine discussions. Each was led by a working group member after presenting his recent ideas and results. The discussions focused on topics that facilitated achieving the following objectives:

1. to develop a fundamental definition of respiration that is accurate and descriptive at all levels of biological organization.

2. to review the limitations and strengths of current methods of measuring respiration.

3. to identify areas for the potential application of new technology.

4. to identify "user interest" in respiration measurements (i.e., the modeling community) and how dimensions, time-scale, and space -scale effects the "usefulness" of a respiration measurement.

5. to determine the feasibility of developing a unifying model of respiration that can be applied to different organisms, communities, and ecosystems.

6. to assess the compatibility of the results of respiration studies with the results of other process studies.

7. to summarize the state of knowledge of respiration, to identify the limitations of that knowledge, and to recommend research for the next decade.

Progress was made on items one, two, four and five, but little progress was made on the remainder. In the following report is: (1) a summary of each member's informal contribution; and (2) the chairman's impressions of the progress made in achieving the above objectives.

SUMMARY OF INFORMAL CONTRIBUTIONS

In the first discussion, Ulanowicz explained how modelers view respiration. He presented a simple conceptual model in which (1) respiration represents the difference between the total input of a component in an ecosystem and the sum of the export out of the ecosystem and the throughput to the next compartment (within the ecosystem) and (2) respiration is guaranteed to be positive by the second law of thermodynamics (Ulanowicz and Kemp, 1979). Throughout his presentation he stressed the quest of modelers for fundamental relationships that would mathematically describe biological processes and would also be consistent with thermodynamics.

Later, Lasserre explained how both microcalorimetry (Wagensberg _et al_., 1978) and the cartesian diver (Zeuthen, 1943; Price and

Warwick, 1980) could be used to determine respiration rates in small planktonic and benthonic organisms. Both methods appear to offer absolute direct measurements of respiration at low levels but they are beset by the effect of starvation and isolation and by the inconvenience of the lengthy incubation time.

Skjoldal suggested that in certain situations the indirect biochemical methods for respiration determinations may be preferred. He presented an analysis of the errors associated with the ETS method and showed that in assessing the respiration rate of mixed zooplankton populations the ETS method is accurate to ±30% of the true rate. Furthermore, he showed by statistical analysis that the error is reduced when species diversity is high and suggested that calibration in the field would also reduce the error.

Joiris presented the results of a Belgian study of the ecology of the southern bight of the North Sea (Joiris _et al_., 1982). The results clearly showed the importance of the microbial heterotrophs in the water column in utilizing the primary productivity. These organisms appear much more important than the zooplankton in respiration and remineralization. Joiris also showed that the local primary productivity could not sustain the local respiratory demands of the planktonic community and argued that either the ^{14}C-method for productivity underestimated true prodictivity or advection served to supply organic matter to the region of study.

Williams presented more evidence that bacteria and other microheterotrophs contribute greatly to total community respiration of the water column. He argued that in the water column, bacterial biomass, growth rates and respiration are much higher than previously thought. In agreement with Joiris, he suggested microheterotrophs may be more important in cycling organic matter and inorganic nutrient salts than the zooplankton. He also briefly described his new microprocessor-controlled Winkler titration system for respiration and productivity measurements (Williams and Jenkinson, 1982) and presented data showing that microbial growth in and on the incubation chamber does not interefere with the measurement.

Minas explained how one can use hydrographic and nutrient data in mixing models to calculate cumulative respiration rates in the oceanic water column below the euphotic zone. Using Mediterranean data, T-S diagrams, O_2-S diagrams and nutrient-salinity diagrams, he demonstrated the effect of deep-ocean respiration on the distribution of oxygen, nitrate and phosphate.

Pamatmat discussed the literature on sediment respiration in relation to new data and concluded that present fractionation techniques of community respiration, involving the separation of infauna from the sediment, yield questionable results. Heat-flow measurements indicate the coupled metabolism of microbes and meiofauna, making it extremely difficult to separate them without disturbing the metabolic activity of both. He suggested that the fractionation approach might be improved if we understood the physiological ecology of the component species in the community and paid more attention to their natural life habits. He argued that calorimetry of intact sediment measures benthic energy flow, which includes undisturbed community respiration. He proposed that the accuracy of the sum of fractionated benthic respiration can be estimated by such measurement (Pamatmat, 1982a and b). Oxygen uptake ignores anaerobic metabolism, which varies in relation to total metabolism in different benthic communities.

Vosjan reminded the group of the diversity of anaerobic types of respiration and their contribution to the carbon flux through benthic communities. He pointed out that most investigators of benthic catabolism have focused on sulfate reduction and that we should learn more about other types of respiration such as, fermentation, denitrification, etc. He pointed out that new microelectrodes offer great promise in facilitating future studies and argued that the indirect biochemical methods (ATP, ETS, etc) are still very useful in survey work where temporal and spatial coverage is required.

Warwick made six points in his discussion of respiration in the microbenthos. (1) In contrast to the plankton, the benthos have a size-frequency particle distribution that is not uniform. There are three peaks corresponding to the bacteria, microfauna, and macrofauna with two minima at 8 and 500 μ between the three peaks. (2) The relative importance of the three peaks has never been determined at one site. The relative importance of microfauna to total catabolism is highly variable and cannot be explained. (3) There are synergistic interactions between the three benthic components such that macrofaunal and meiofaunal presence stimulates respiration and remineralization in the bacterial component. (4) Recent evidence that the size dependence of macro- and meio- fauna respiration cannot be described by the same allometric equation, suggests that modelers will predict erroneous respiration rates if they persist in using a common allometric equation for all sizes of organisms. (5) Anaerobiosis in the meiofauna may explain much of the imbalance that often occurs in benthic respiration budgets. (6) Evidence that Q_{10} for meiofauna respiration varies from 2 to 4 suggest that modelers using a constant Q_{10} throughout the size spectrum will generate erroneous respiration rates.

PROGRESS

After the above series of presentations and discussions the group was able to conclude with the following statements and recommendations.

On the Definition of Respiration

Respiration is an energy yielding process in living systems that degrades organic matter beyond the point of immediate biological utility. The energy released during the degradation is used by the living system to achieve the goals of its survival strategy. The type of organic matter degraded, the degree to which it is oxidized, and the amount of useable energy extracted from the process varies with both the biological system and its environment.

Fermentation is a special case of respiration in which the terminal electron acceptor is an organic compound. Denitrification and sulfate reduction are special cases of respiration in which nitrate, nitrite, nitrous oxide and sulfate serve as electron acceptors in the process. Carbon dioxide production occurs during most but not all types of respiration. Heat production occurs in all cells as organic matter is degraded and as mechanical work is performed. Heat production can be enhanced when respiration is uncoupled from oxidative phosphorylation (ATP production) and it can be reduced when respiration is coupled to extensive growth and biosynthesis. However, under steady state situations, heat production accurately reflects the energy released during respiration.

On Reporting Respiration Data

Measurements of respiration are often converted from their primary units to more conventional ones such as $\mu\ell\ O_2\ h^{-1}$ per mg dry weight. This conversion requires a conversion factor which contributes an error. The working group concluded that the primary measurement as well as the conversion factors should always be reported. If joule sec^{-1} were measured in a calorimeter then these units should be reported. If millielectron equivalents were measured with an ETS assay then these units should be reported. Only after this primary measurement is reported should conversion factors be employed to report the respiration in units of oxygen, carbon or calories. Furthermore, to enable measurements to be used by as many scientists as possible, reports of respiration in marine organisms, populations or ecosystems should not only give the biomass, volume or area-specific rate (i.e. $\mu\ell\ O_2\ h^{-1}\ m^{-3}$); but also the volume and areal distribution of the biomass and the depth over which the rates have been integrated. Also, since respiration rates may have a diel wave function they should only be reported for the time unit over which the primary measurement was made.

On the Calibration Problem

As with any other analytical measurement, respiration measurements are only as good as their calibration. Unlike chemical or biochemical measurements, physiological measurements require two levels of calibration. The first and most obvious is the degree of transducer response to changes in reactants or products of respiration (i.e., CO_2, O_2, heat, etc.). The second level of calibration is more subtle; it requires definition of the relationship between the *in situ* respiration rate and the apparent one as measured. This requirement arises because of the physiological changes that often occur when an organism is removed from its natural environment and is maintained in a respiration chamber. Because of these changes the measured respiration may not be representative of the *in situ* respiration. Since there is no way to measure the *in situ* respiration without some disturbance to the organism a true calibration between the measured and *in situ* rates cannot be made. Nevertheless, investigators can attempt evaluation of the *in situ* rate for a few measurements by running time course experiments. The working group recommends that such experiments be conducted in future experimental programs.

On Recommending Methods

In making recommendations of methods to be used in respiration research, so much depends on the objectives of the researcher. On one hand, geochemists may be interested in the respiration rate in the deep-sea and its effect on the chemistry of seawater over hundreds of years. On the other hand, a benthic ecologist may be interested in the diel respiratory cycle of a certain species of protozoa. In between these two extremes are the plankton ecologists who may be interested in the respiration of entire plankton communities in a water column, or the benthic ecologists who may be interested in community metabolism of the benthos. Needless to say, there is no one method that can satisfy such a spectrum of scientific users. For ecological and physiological studies with individual microorganisms the most promising methods appear to be the cartesian diver techniques of Seuthen (1943) as recently employed by Klekowski (1977); Klekowski *et al.* (1980), and the microcalorimetry techniques (Wagensberg *et al.*, 1978; Castell et al., 1981) as described by Lasserre (this *volume*).

For studies of community respiration in the benthos, direct calorimetry as recently described by Pamatamat (1982 a and b; in press) appears to be the best method. It reflects all types of respiratory metabolism, whether it be aerobiosis, denitrification, sulfate reduction or fermentation. The major problem is the long incubation time (6-8 hr) each measurement requires. Thus only 3-4 samples a day can be run per calorimeter. This precludes the method from being used in surveys in which temporal and spatial distribution must be resolved. Also, calorimetry does reflect the heat production of extracellular chemical reactions that, at times, may be significant and may lead to an over-estimation of respiration. For water column respiration, the best method has a similar limitation. The direct determination of oxygen changes by the new microprocessor-controlled Winkler titration of Williams and Jenkinson (1982) is currently the best way to measure water column respiration in oceanic surface waters. However, it has limitations similar to those of the ^{14}C-method for productivity that limit its use for temporal and spatial distribution work. In both the benthos and in the plankton, the use of a metabolic index, such as ETS activity (Christensen and Packard, 1977; Olańczuk-Neyman and Vosjan, 1977; Skjoldal and Lännergren, 1978; Packard, 1979), in combination with calorimetry and/or Winkler measurements can make such temporal and spatial respiration surveys feasible.

For geochemical work one can use the helium-dating method (Jenkins, 1980), the advection-diffusion model (Craig, 1971) and the ETS method (Packard et al., 1977 and in press). The He dating method should only be used above 1000 m and the Craig method should only be used below 1000 m. The ETS method can be used throughout the water column. In all three methods, assumptions and constants are used that detract from their accuracy.

On Formulating a "Unified Field Theory" for Respiration

The development of a unifying mathematical description of respiration is an objective rarely attempted, but it should be encouraged because such a description would represent a significant thermodynamic hypothesis and would greatly improve our ability to use respiration in ecosystems modeling. Achieving such an objective would represent a "quantum jump" in our understanding of the respiratory process. From the diversity of the above discussions it is easy to understand why so little progress has been made towards this objective. The allometric equation (Bertalanffy, 1964) was offered as a unifying model many years ago and in spite of its many limitations is still accepted and widely used by theoreticians and modelers. One line of research tries to explain the allometric equation on the basis of energy transfers across membrane surfaces (Schmidt-Nielsen, 1970). Another line of research ignores the allometric equation and its dependence on size and instead focuses on the chemical basis of the physiological process (Atkinson, 1968 and 1977; Packard, 1971; Jacobus et al., 1982). Its objective is to develop equations based on enzyme kinetics that will describe respiration. To date neither approach has succeeded in developing the equivalent of the unified field theory or even the equivalent of the perfect gas law for respiration.

Acknowledgements

This report was supported by NATO and by ONR Contract N000 14-76-C-0271 to T.T. Packard. It is contribution number 82021 from the Bigelow Laboratory for Ocean Sciences.

REFERENCES

Atkinson, D.E., 1968, Regulation of enzyme function, Ann. Rev. Micro Biol., 23: 47.

Atkinson, D.E. 1977, "Cellular energy metabolism and its regulation" Academic Press, New York.

Bertalanffy, L.V. 1964, Basic concepts in quantitative biology of metabolism, Helgol. Wiss, Meeresunters, 9: 5

Castell, C., Wagensberg, J., Tejero, A., and Vallespinós, F., 1981, Identificatión de las fases metabólicas en termogramas de cultivos bacterianos, Inv. Pesq., 45:291

Christensen, J.P., and Packard, T.T., 1977, Sediment metabolism from the northwest African upwelling system, Deep-Sea Res., 24:331.

Craig, H., 1971, The deep metabolism: oxygen consumption in abyssal ocean water, J. Geophys. Res., 76:5078

Jacobus, W.E., Moreadith, R.W., and Vandegaer, K.M., 1982, Mitochondrial respiratory control, J. Biol. Chem., 257:2397

Jenkins, W.J., 1980, Tritium and ^{3}He in the Sargasso Sea, J. Mar. Res., 38:533.

Joiris, C., Billen, G., Lancelot, C., Daro, K.H. Mommaerts, J.P., Hecq, J.H., Bertels, A., Bossicart, M., and Nijs, J., 1982, A budget of carbon cycling in the Belgian coastal zone: Relative roles of zooplankton, bacterioplankton and benthos in the utilization of primary production, Neth. J. Sea Res., 16:260.

Klekowski, R.Z., 1977, Microrespirometer for shipboard measurements of metabolic rate of microzooplankton, Pol. Arch. Hydrobiol., 24-Suppl.:455.

Klekowski, T.Z., Kukina, I.V., Tumanseva, N.I. 1977, Respiration in the microplankton of the equatorial upwellings in the eastern Pacific Ocean, Pol. Arch. Hydrobiol., 24-Suppl.:467

Laybourn-Parry, J., Baldock, B., and Kingmill-Robinson, C., 1980, Respiratory studies on two small freshwater amoebae, Microb. Ecol., 6:20-9216.

Olánczuk-Neyman, K.M., and Vosjan, J.H., 1977, Measuring respiratory electron-transport system activity in marine sediment, Neth. J. Sea Res., 11:1

Packard, T.T., 1971, The measurement of respiratory electron transport activity in marine phytoplankton, J. Mar. Res., 29:235

Packard, T.T., 1979, Respiration and respiratory electron transport activity in plankton from the Northwest African upwelling area, J. Mar. Res., 37:711.

Packard, T.T., Minas, H.J., Owens, T., and Devol, A., 1977, Deep-sea metabolism in the eastern tropical north Pacific Ocean, in: "Oceanic Sound Scattering Predictions" N.R. Andersen and B.J. Zahuranec, eds., Plenum, New York.

Packard, T.T., Garfield, P.C., and Codispoti, L.A., In press, Oxygen consumption and denitrification below the Peruvian upwelling, in: "Coastal Upwelling: It's Sediment Record," E. Suess, ed., Plenum, New York.

Pamatmat, M.M., 1982a, Heat production by sediment: Ecological significance, Science, 215:395
Pamatmat, M.M., 1982b, Direct calorimetry in benthos and geochemical research, in: "The Dynamic Environment of the Ocean Floor," K.S. Fanning and F. Manheim, eds., Lexington Books, Toronto.
Pamatmat, M.M., In press, Simultaneous direct and indirect calorimetry, in: "Handbook on polarographic oxygen sensors: aquatic and physiological applications," E. Graiger, H. Forstner, eds., Springer, Berlin.
Price, R., and Warwick, R.M., 1980, The effects of temperature on the respiration rate of meiofauna, Oecologia (Berl.), 44:145
Schmidt-Nielsen, K., 1970, Energy metabolism, body size and problems of scaling, Fed. Proc., 29-1524.
Skjoldal, H.R., and Lännergren, C., 1978, The spring phytoplankton bloom in Lindaspollene, a land-locked Norwegian fjord. II. Biomass and activity of net and nanoplankton, Mar. Biol., 47:313.
Ulanowicz, R.E., and Kemp, W.M., 1979, Towards canonical trophic aggregations, Amer. Naturalist, 114:871.
Wagensberg, J., Castell, C., Torra, V., Rodellar, J., and Vallespinós 1978, Estudio microcalorimétrico del metabolismo de bacterias marinas: detección de procesos rítmicos, Inv. Presq., 42:179.
Williams, P.J. LeB., and Jenkinson, J.W., 1982, A transportable microprocessor-controlled precise Winkler titration suitable for field station and shipboard use, Limnol. Oceanogr., 27:576.
Zeuthen, E., 1943, A cartesian diver micro-respirometer with a gas volume of 0.1 µl, Compt.-rend. Lab. Carlsberg Sér. Chim. 24:479.

NUTRIENT CYCLING IN ESTUARINE AND COASTAL MARINE ECOSYSTEMS

H.Postma[1], W.M. Kemp[2], J.M. Colebrook[3], J. Horwood[4], I.R. Joint[3], R. Lampitt[5], S.W. Nixon[6], M.E.Q. Pilson[6] and F. Wulff[7]

1. Netherlands Institute for Sea Research
   P O Box 59, Texal, Netherlands

2. University of Maryland,
   Horn Point Environmental Laboratories
   Box 775, Cambridge, Mass. 21613, U.S.A.

3, Institute for Marine Environmental Research
   Prospect Place, The Hoe, Plymouth, Devon PL1 3DH, U.K.

4. Ministry of Agriculture, Fisheries and Food
   Fisheries Laboratory
   Lowestoft, Suffolk NR33 OHT, U.K.

5. Institute of Oceanographic Sciences
   Wormley, Godalming, Surrey GU8 5UB, U.K.

6. University of Rhode Island
   Graduate School of Oceanography
   Kingston, Rhode Island 02881, U.S.A.

7. Asko Laboratory
   Box 58, S150-13 Trosa, Sweden

INTRODUCTION

Although secondary production of marine and estuarine ecosystems is the result of energy and carbon flux from lower trophic levels, the rate of primary production may be controlled by the availability of disssolved inorganic (and perhaps organic) nutrients. In this report we briefly review the status of knowledge and relevant questions pertaining to nutrient flux and transformation processes in shallow marine systems. Our emphasis is on nitrogen although we

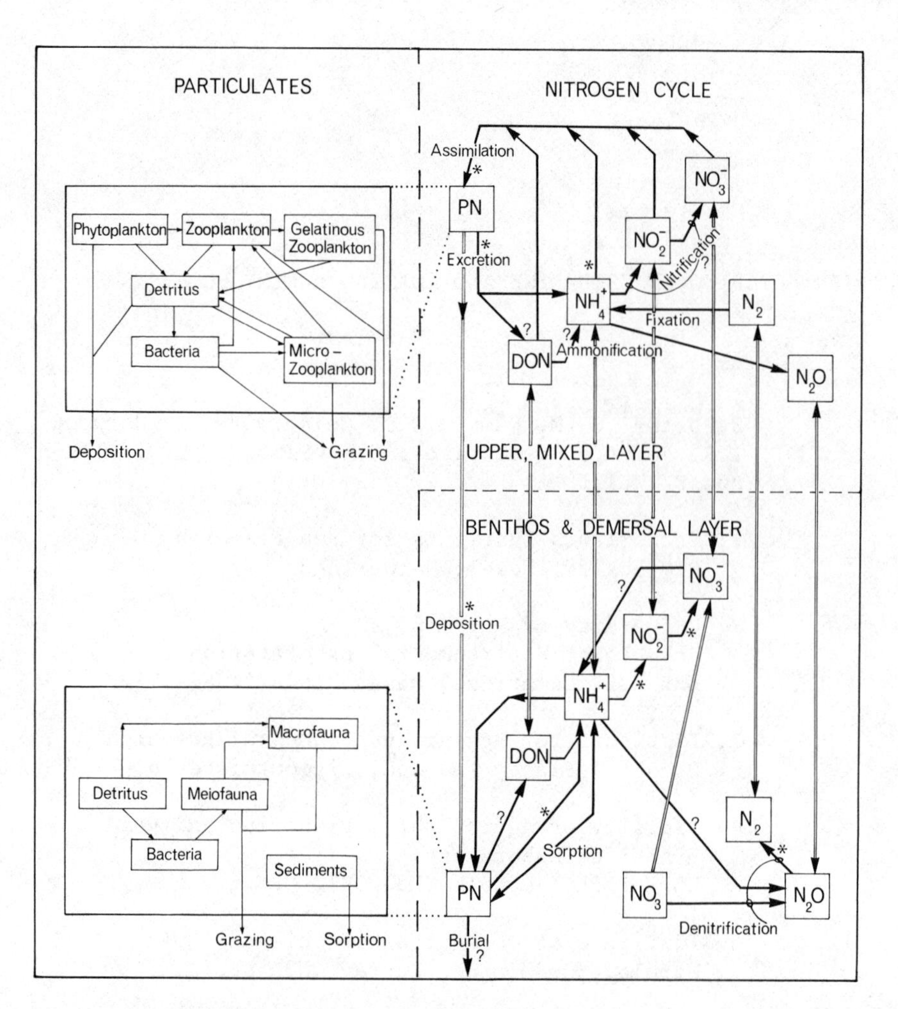

Fig. 1. Schematic diagram of the nitrogen (N) cycle for estuarine and coastal marine ecosystems. Various N species are indicated and labelled in boxes, where PN = particulate N and DON = dissolved organic N. Components of PN in upper layer are phytoplankton, larger zooplankton, micro-zooplankton, bacteria, gelatinous Zooplankton, and detritus, while additional components in the sediment are macroinvertebrates, meiofauna and sediments. Straight solid lines with arrows connecting boxes indicate chemical/biological/physical transformations, while open lines represent advective or diffusive transport. Fluxes which are known to be important in magnitude relative to assimilation are indicated by asterisk, and those about which little is known are designated by a question-mark.

conclude that further research is also needed for phosphorus and silicon. We begin by presenting a conceptual schematic for the nitrogen cycle in these systems (Figure 1). The diagram is organized by vertically separating characteristics of the upper mixed layer from those of the demersal and benthic systems. Even though particulate nitrogen is divided into various living and abiotic compartments, fluxes among those particulate components are not emphasized here. Concentrations and fluxes are discussed one pathway at a time.

NITROGEN CYCLING FLUXES

NH_4 Regeneration

Ammonium is regenerated from organic compounds by microbial decomposition of organic matter and by animal excretion. Regenerated NH_4^+ appears to contribute a substantial portion of the phytoplankton requirements and recent measurements in estuarine waters suggest that water column and benthic processes both usually supply more than 30% of the assimilatory demand (Harrison, 1980; Boynton, et al. 1980; Zeitzschel, 1980; Nixon, 1981). It is presumed that excretion contributes the largest part of NH_4^+ regeneration in the water column, while decomposition of organic matter is most important in the sediments. However, the relative contributions of these pathways awaits further experimentation (Harrison, 1980; Nixon, 1981). There are at present few direct measurements of plankton regeneration using the ^{15}N-dilution technique. These data suggest the importance of this regeneration; however, further information is needed on the contribution of various size fractions to this NH_4 regeneration (Harrison, 1978; Glibert, 1982), as well as the role of fecal matter. Eventually we need to test the relevance of allometric concepts and laboratory rate measurements for field conditions by comparing in situ observations to predictions based on laboratory derived relations for size distribution, temperature and physiological state of organisms (Harrison, 1980). The role of gelatinous consumers deserves further attention, as does the potential significance of plankton vertical migrations and NH_4^+ (or DON) excretion in terms of nitrogen dynamics (Kremer, 1975).

Numerous direct measurements are now available for NH_4^+ flux from estuarine and coastal sediments, and these rates, which range from 50-800 $\mu mol\ m^{-2}h^{-1}$ have been summarized in recent review articles (Boynton et al., 1980; Zeitzschel, 1980; Nixon, 1981). Strong correlations exist between benthic regeneration and primary production (Nixon, 1981). Aerobic decomposition provides the major mechanism for release, particularly the rapid break-down of labile, recently deposited organic matter (Kemp et al., 1982). Decomposition via sulfate reduction occurs deeper in the sediment column (>10cm) and provides an additional (but probably secondary) source of NH_4^+ (Vanderborght et al., 1977; Aller and Yingst, 1980). Benthic regeneration is a function of temperature; however the effects of other factors (such as sediment grain size and physical circulation) on these rates have not been well studied. It has been demonstrated that the activity of fauna (meio- and macro-) enhanced the rate of NH_4^+ release from sediments (Aller, 1978), but whether the mechanism is burrowing and irrigation, direct excretion, or some other factor awaits further research.

Excretion of dissolved organic nitrogen (DON)

The behaviour of DON in estuaries and coastal waters is enigmatic at our present state of knowledge (Boynton et al., 1980; Nixon, 1981; Stanley and Hobbie, 1981). We know very little of the specific organic compounds included in this broad category of nitrogen species. Some of these compounds such as urea and amino acids are readily available for microbial or algal assimilation (Hobbie et al., 1968; Jørgensen, 1982), while others are more resistant to biochemical processes. Dissolved organic nitrogen often seems, at least in some estuaries, to be conservative in time and space (Nixon and Pilson, 1983). However, significant net fluxes of DON between sediments and water have been reported for various coastal areas (Nixon, 1981; Jørgensen, 1982). Because the concentrations of DON are often high (equalling or exeeding those of inorganic forms) the matter is important and deserves more study (Nixon, 1981; Nixon and Pilson, 1983). There are few direct measurements of DON excretion by zooplankton (Schell, 1974), and (to our knowledge) none have been made in situ. Whereas numerous observations have shown the release of organic carbon by phytoplankton to range from 1-50% (Williams and Yentsch, 1976; Lancelot, 1979), fewer data are available for DON excretion. The often observed substantial excretion of organic carbon suggests the possible significance of algal excretion of DON. Improved methods are needed for analysis of DON so that a biologically active portion can be distinguished from the total, and so that specific compounds can be measured conveniently and accurately.

NITRIFICATION, DENITRIFICATION AND N_2-FIXATION

There are few direct measurements of NH_4^+ oxidation in estuarine and coastal waters although spatial-temporal patterns of NO_2^- and NO_3^- strongly suggest the importance of this pathway. Nitrification has been measured in coastal sediments using N-serve, ^{14}C and ^{15}N techniques, all of which indicate rates of the order of 30-100 μmol $m^{-2}h^{-1}$ (Billen, 1975; Henriksen et al., 1981; Hansen et al., 1981). This process is sharply limited by redox conditions and will not occur below about + 100Mv (Billen, 1975). Although nitrification is influenced by temperature, this influence may be less than that for ammonification (Nixon, 1981; Hansen et al., 1981). Periodically high (>3μM) concentrations of NO_2^- have been reported and may reflect NH_4^+ oxidation at the pycnocline of stratified water columns, but this needs to be tested (Kemp et al., 1982). More direct measurements in coastal waters are needed, as are experiments to explain this occurrence of incomplete nitrification to NO_2^-.

Denitrification appears to be tightly coupled to nitrification in sediments and net fluxes of NO_3^- from sediments are generally small (Henriksen et al., 1981; Seitzinger et al., 1980). However under conditions of very high NO_3^- in the overlying water (>30μM), significant fluxes of NO_3^- may enter the sediments into dissimilatory NO_3^- reduction (Sørensen, 1978; Nishio et al., 1982). The few existing measurements of denitrification indicate that it represents a significant loss of nitrogen from the biologically available pool. Rates of 20-200 μg-atoms $m^{-2}h^{-1}$ have been reported (Seitzinger et al., 1980; Nishio et al., 1982; Billen, 1978; Kaplan et al., 1979; Oren and Blackburn, 1979), and these represent typically 30-50% of the flux of N from the sediments (Boynton et al., 1980; Seitzinger et al., 1980; Billen, 1978), but the importance of organic substrate, redox and temperature have not been adequately studied. It has been suggested that low ratios of dissolved inorganic N : P (relative to the Redfield model) characteristic of coastal waters may be, in part, attributable to high rates of denitrification (Nixon, 1981; Seitzinter et al., 1980).

Nitrogen fixation has been widely measured in estuaries and coastal waters using acetylene reduction techniques, with fewer direct ^{15}N experiments to corroborate these data. Rates reported tend to be relatively low in estuarine water, sediments and sea grasses (Lipschultz et al., 1979). Somewhat higher rates have been observed in emergent marsh grass systems, but all of these are low compared with those common in lakes (Lean et al., 1978). On the other hand, the epiphytes of tropical seagrasses (*Thalassia testudinum*) produce N_2-fixation rates which are capable of supplying about 30-100% of the assimilatory demand (Patriquin and Knowles, 1972; Capone and Taylor, 1977). In contrast to lakes there have been no satisfactory evidence provided yet to explain the generally low rates of N_2-fixation occuring in coastal ecosystems (Nixon, 1981).

Nitrogen assimilation and sedimentation

Nitrogen is generally accepted to be the macro-nutrient least available (most limiting) for plankton production in coastal waters. The uptake of dissolved nitrogen species has been investigated under limiting conditions. Evidence suggests that phytoplankton may exhibit a distinct order of preference for nitrogen compounds, i.e. NH_4^+, urea, NO_2^-, NO_3^- (McCarthy et al., 1977). The reduced form NH_4^+, is energetically more efficient as a nitrogen source, and concentrations above 1-2μM tend to inhibit assimilation of other N species (McCarthy et al 1977). Typical concentrations of NH_4^+ and NO_3^- in coastal waters range from <1 to 10 and <2 to 25μM for both (Nixon, 1981; Nixon and Pilson, 1983). There is controversy, however, about K_S values reported for inorganic nitrogen uptake by neritic phytoplankton in oligotrophic waters, which may be generally overestimated (McCarthy and Goldman, 1979). Recent reports of correlations between N-loading and plankton peak biomass (and production) for several estuaries and for one estuary over several years indicate some kind of N limitation (Boynton et al., 1982).

In shallow coastal systems a considerable fraction (10-15%) of net phytoplankton production is deposited at the sediment surface (Taguchi and Hargrave, 1978; Smetacek, 1980; Kemp and Boynton, 1981). Factors involved include herbivory, vertical circulation, and sinking rates of the plankton, all of which must influence the rate of deposition. In coastal waters, river-borne suspended matter and off-shore material is also deposited (input) to the sediment. These rates of sedimentation are the combined result of autochthonous and allochthonous produced particulates (Nixon, 1981), and where fine-grain particulates are deposited this material is about 0.1-0.6% nitrogen (Nixon and Pilson, 1983; Boynton and Kemp, 1981). More information is needed concerning the composition of deposited material and the yearly diagenic processes by which it is transformed (Klump and Martens, 1981). Qualitatively, it is clear that major climatic events (e.g. hurricanes) exert greater effects on deposition rates, but quantitative information is needed (Schubel and Hirschberg, 1978). Similarly, few data are available for deposition rates for neritic coastal regions. There is a considerable need for a methodology sensitive enough to measure net deposition in short time intervals, of the order of weeks.

Nutrient assimilation by macrophytes can be significantly different from that by phytoplankton. Macrophytes have a different physiology, such as an ability to grow on stored nutrients for long periods (Mann et al., 1980). In addition, they do not have the same nutrient preferences (Jackson, 1977). Rooted sea-grasses can assimilate nutrients from sediments and could possibly serve as nutrient pumps (McRoy et al., 1972; Iizumi and Hattori, 1982). The presence of macrophytes can significantly change sediment patterns (Burrell and Schubel, 1977). Massive stands of macrophytes are not only sources of organic material, but they can also increase sedimentation rates by damping water motion. This, in effect, enhances the input of nutrients (as particulates) to these macrophyte systems.

NUTRIENTS OTHER THAN NITROGEN

In recent years, with the growing emphasis on nitrogen and the development of ^{15}N methodologies, there have been relatively fewer studies involving phosphorus and silicon. Nevertheless, bioassay experiments often report positive, significant response to enrichment with P and Si (although usually less than for N) (Thayer, 1974). There is reason to expand research involving P and Si, and to examine multiple nutrient interactions.

Of particular importance relative to phosphorus, is the role of flocculation, sorption, precipitation and early diagenesis in determining transformation and availability of this element for primary production in estuaries (Bricker and Troup, 1975; Krom and

Berner, 1980; 1981). There has been much discussion recently as to the role of Si availability influencing the balance between diatoms and flagellates (Parsons et al., 1978); however, more work is needed to test these notions.

PHYSICAL CONSIDERATIONS

Physical features, such as transport, mass balance and the effects of various forces need increased attention. Far too little is known about the mechanisms and rates of advective and diffusive transport. Physical circulation structures, such as frontal boundaries and zones where water masses are mixed, can transport and mix nutrients and biota and are often of great importance. Nevertheless, physical transport terms may be hard to obtain with high precision, and new approaches, including the use of biological and chemical tracers, may be useful.

Attempts to make mass balances or budgets of various materials have demonstrated the difficulties in obtaining a complete picture of the sources, pathways and sinks for any substance (Nixon, 1981; Nixon and Pilson, 1983). Even so, to understand nutrient cycling such budgets should be attempted, including measurements of all the inputs and outputs.

In defining problems of nutrient uptake, regeneration and transport, careful consideration should be given to the definition of appropriate scales of time and space. The coupling of biological and physical variables must be related to their characteristic frequencies, and these can be investigated both statistically (e.g. by spectral analysis) and mechanistacally (using controlled experiments at all scales)(Therriault and Platt, 1981; Walsh, 1976; Kemp and Mitsch, 1979; Lewis and Platt, 1982). More field data from temporal and spatial series are needed to infer frequency couplings, and sampling intervals must take cognizance of natural periodicities. Physical forces exist at characteristic time-scales, and all of these may be relevant to issues of nutrient cycling. In coastal waters the physical forces associated with waves, tides, winds and solar cycles are manifest at many time scales. Understanding characteristic time-lags between contemporaneous data sets may help to infer causal mechanisms.

"NEW" VERSUS "REGENERATED" SOURCES OF NUTRIENTS

To understand the potential for transfer of energy from net primary production to secondary production, it is essential to distinguish between sources of "new" and recently "regenerated" nutrients. The distinction between these two categories of nutrients is, once again, a matter of temporal and spatial scale (Kemp et al., 1982). On geological time scales, perhaps, all nutrients might be considered regenerated. While nutrients welling up from Ekman circulation are new to that environment, they may be part of a larger regenerative system of deep oceanic waters. Time and space scales, however, can be defined appropriate to scientific questions being asked. Clearly, if small scale mechanisms of nutrient regeneration within the water column do not lead to increased net production from the plankton community, they are of little relevance to fisheries production. Such mechanisms may be of importance to maintaining the "fly-wheel" of gross production, but inputs of new nutrients are needed to sustain a net yield (Nixon and Pilson, 1983).

REFERENCES

Aller, R.C., 1978, The effects of animal-sediment interactions on geochemical processes near the sediment-water interface, in: "Estuarine interactions," M.Wiley, ed., Academic Press, New York.

Aller, R.C., and Yingst, J.Y., 1980, Relationship between microbial distributions and the anaerobic decompositions of organic matter in surface sediments of Long Island Sound, USA., Mar. Biol., 56:29.

Billen, G., 1975, Nitrification in the Scheldt Estuary (Belgium and the Netherlands), Est. Coast. Mar. Sci., 3:79.

Billen, G., 1978, A budget of nigrogen recycling in North Sea sediments off the Belgian Coast, Est. Coastal Mar. Sci., 7:127.

Boynton, W.R., and Kemp, W.M., 1981, The significance of nutrient and carbon fluxes across the sediment-water interface along major estuarine gradients, Rep to USEPA. Univ.Maryland Centr. Envir. Est. Studies, Ref. No. 81-253, Cambridge, MD., USA.

Boynton, W.R., Kemp, W.M., and Keefe, C.W., 1982, a comparative analysis of nutrients and other factors influencing estuarine phytoplankton production, in: "Estuarine comparisons," V.S. Kennedy, ed., Academic Press, New York.

Boynton, W.R., Kemp, W.M., and Osborne, G.C., 1980, Nutrient fluxes across the sediment-water interface in the turbid zone of a coastal plain estuary, in: "Estuarine perspectives" V.S. Kennedy, ed., Academic Press, New York.

Bricker, O.P., and Troup, B.N., 1975, Sediment-water exchange in Chesapeake Bay, in: "Estuarine research," Vol. 1, L.E. Cronin, ed. Academic Press, New York.

Burrell, D.C., and Schubel, J.R., 1977, Seagrass ecosystem oceanography, in: "Seagrass ecosystems," C.P. McRoyand and C. Helfferich, eds., Marcel Dekker, New York.

Capone, D.G., and Taylor, B.F., 1977, Nitrogen fixation (acotylene reduction) in the phyllosphere of Thalassia festudinum, Mar. Biol., 40:19.

Glibert, P.M., 1982, Regional studies of daily, seasonal and size fraction variability in ammonium remineralization, Mar. Biol. 70:209.

Hansen, J.I., Kenriksen, K., and Blackburn, T.H., 1981, Seasonal

distribution of nitrifying bacteria and rates of nitrification in coastal marine sediments, Microb. Ecol., 7:297.
Harrison, W.G., 1978, Experimental measurements of nigrogen remineralization in coastal waters, Limnol. Oceanogr., 23:684.
Harrison, W.G., 1980, Nutrient regeneration and primary production in the sea, in: "Primary productivity in the sea," P.G.Falkowski, ed., Plenum Press, New York.
Henriksen, K., Hansen, J.I. and Blackburn, T.H., 1981, Rates of nitrification, distribution of nitrifying bacteria and nitrate fluxes in different types of sediments from Danish waters, Mar. Biol., 61-299.
Hobbie, J.E., Crawford, C.C., and Webb, K.L., 1968, Amino acid flux in an estuary, Science, 159:1463.
Iizumi, H., and Hattori, A., 1982, Growth and organic production of eelgrass (Zostera marina L.) in temperate waters of the pacific coast of Japan. III. The kinetics of nitrogen uptake, Aquat. Bot., 12:245.
Jackson, G.A., 1977, Nutrients and production of giant kelp, macrocystis pyrifera, off southern California, Limnol. Oceanogr., 22:979.
Jørgensen, N.O.G., 1982, Heterotrophic assimilation and occurrence of dissolved free amino acids in a shallow estuary, Mar. Ecol. Progr. Ser., 8:145.
Kaplan, W., Valiela, I., and Teal, J.M., 1979, Denitrification in a salt marsh ecosystem, Limnol. Oceanogr., 24:726.
Kemp, W.M.,and Boynton, W.R., 1981, External and internal factors regulating metabolic rates of an estuarine benthic community, Oecologia, 51:19.
Kemp, W.M., and Mitsch, W.J., 1979, Turbulence and phytoplankton diversity: A general model of the paradox of plankton, Ecol. Model., 7:201.
Kemp. W.M., Wetzel, R.L., Boynton, W.R., D'Elia, C., and Stevenson, J.C., 1982, Nitrogen cycling at estuarine interfaces, in: "Estuarine comparisons," V.S. Kennedy, ed., Academic Press, New York.
Klump, J.V., and Martens, C.S., 1981, Biochemical cycling in an organic rich coastal marine basin. II. Nutrient sediment-water exchange processes, Geochim. Cosmochim. Act., 45:101.
Kremer, P., 1975, Nitrogen regeneration by the ctenophore, Mnemiopsis leidyi, in: "Mineral cycling in southeastern ecosystems," F.G. Howell, J.B. Gentry and M.H. Smith, eds., ERDA Symp. Serv., (CONF-740513), NTIS, Springfield, VA.
Krom, M.D., and Berner, R.A., 1980, Adsorption of phosphate in anoxic marine sediments, Limnol, Oceanogr., 25:797.
Krom, K.D., and Berner, R.A., 1981, The diagenesis of phorphorus in a nearshore marine sediment, Geochim, Cosmochim.Acta, 45:207.
Lancelot, C., 1979, Gross excretion rates of natural marine phytoplankton and hetrotrophic uptake of excreted products in the Southern North Sea, as determined by short-term kinetics, Mar. Ecol. Progr. Scr., 1:179.

Lean, D.R.S., Liao, C.F.-H., Murphy, T.P., and Painter, D.S., 1978. The importance of nitrogen fixation in lakes, Ecol. Bull. (Stockholm), 26:41.

Lewis, M.R., and Platt, T., 1982, Scales of variability in estuarine ecosystems, in: "Estuarine comparisons," V.S. Kennedy, ed., Academic Press, New York.

Lipschultz, F., Cunningham, J.J., and Stevenson, J.C., 1979, Nitrogen fixation associated with four species of submerged angiosperms in the central Shesapeake Bay, Est. Coast. Mar. Sci., 9:813.

McCarthy, J.J., and Goldman, J.C., 1979, Nitrogenous nutrition of marine phytoplankton in nutrient depeleted waters, Science, 203:670.

McCarthy, J.J., Taylor, W.R., Taft, J.L., 1977, Nitrogenous nutrition of the plankton on the Chesapeake Bay. I. Nutrient availability and phytoplankton preferences, Limnol. Oceanogr., 22:996.

McRoy, C.P., Barsdate, R.J., and Nebert, M., 1972, Phosphrus cycling in an eeelgrass (Zostera marina L.) ecosystem, Limnol. Oceanogr. 17:58.

Mann, K.H., Chapman, A.R.O., and Gagne, J.A., 1980, Productivity of seaweeds: The potential and the reality, in: "Primary productivity in the sea," P.G. Falkowski, ed., Plenum Press, New York.

Nishio, T., Kioke, I., and Hattori, A., 1982, Denitrification, nitrate reduction, and oxygen consumption in coastal and estuarine sediments, Appl. Environ. Microb., 43:648.

Nixon, S.W., 1981, Remineralization and nutrient cycling in coastal marine ecosystems, in: "Estuaries and nutrients," B.J. Neilson and L.E. Cronin, eds., Humana Press, New York.

Nixon, S.W., and Pilson, M.E.Q., 1983, Nitrogen in estuarine and coastal marine ecosystems, in: "Nitrogen in the marine environment," E.J. Carpenter and D.G. Capine, eds., Academic Press, New York, In press.

Oren, A., and Blackburn, T.H., 1979, Estimation of sediment denitrification rates at in situ nitrate concentrations, Appl. Environ. Microb., 37:174.

Parsons, T.R., Harrison, P.J., and Waters, R., 1978, An experimental simulation of changes in diatom and flagellate blooms, J. Exp. Mar. Biol. Ecol., 32:285.

Patriquin, D.G., and Knowles, R., 1972, Nitrogen fixation in the rhizophere of marine angiosperms, Mar. Biol., 16:49.

Schell, D.M., 1974, Uptake and regeneration of free amino acids in marine waters of the Southeast Alaska, Limnol, Oceanogr., 19:260.

Schubel, J.R., and Hirschberg, D.J., 1978, Estuarine graveyards, climatic change and the importance of the estuarine environment, in: "Primary productivity in the sea," P.G. Falkowski, ed., Plenum Press, New York.

Seitzinger, S., Nixon, S., Pilson, M.E.Q., and Burke, S., 1980, Denitrification and N_2O production in near-shore marine sediments, Geochim. Cosmochim. Acta, 44:1853.

Smetacek, V., 1980, Annual cycle of sedimentation in relation to plankton ecology in western Kiel Bight, Ophelia (Suppl.), 1:65.

Sørenson, J., 1978, Capacity for denitrification and reduction of nitrate to ammonia in a coastal marine sediment, Appl. Environ. Microb., 36:301.

Stanley, D.W. and Hobbie, J.E., 1981, Nitrogen recycling in a North Carolina coastal river, Limnol. Oceanogr., 26:30.

Taguchi, S., and Hargrave, B.T., 1978, Loss rates of suspended material sedimented in a marine bay, J. Fish, Res. Bd. Can., 35:1614.

Thayer, G.W., 1974, Identity and regulation of nutrients limiting phytoplankton production in the shallow esturies near Beaufort, N.C. Oecologia, 14:75.

Therriault, J.-C., and Platt, T., 1981, Environmental control of phytoplankton patchiness, Can. J. Fish. Aquat. Sci., 38-638.

Vanderborght, J.P., Wollast, R., and Billen, G., 1977, Kinetic models of diagenesis in disturbed sediments. Part II. Nitrogen diagenesis, Limnol. Oceanogr., 22:794.

Walsh, J.J., 1976, Herbivory as a factor in patterns of nutrient utilization in the sea, Limnol. Oceanogr., 21:1.

Williams, P.J. LeB., and Yentsch, C.S., 1976, An examination of photosynthetic production, excretion of photosynthetic products, and hetrotrophic utilization of dissolved organic compounds with special reference to results from a coastal subtropiical sea, Mar. Biol., 35:31.

Zeitzsche, B., 1980, Sediment-water interactions in nutrient dynamics, in: "Marine benthic dynamics," K.R. Tenore and B.C. Coull, eds., Univ. S. Carolina Press, Columbia.

EXCRETION AND MINERALISATION PROCESSES IN THE OPEN SEA

E. Corner[1], U. Brockmann[2], J.C. Goldman[3], G.A. Jackson[4], R.P. LeBorgne[5], M. Lewis[6] and A. Yayanos[4]

1. Marine Biological Association of U.K., The Laboratory Citadel Hill, Plymouth, Devon, PL1 2PB, U.K.

2. Universität Hamburg, Institut für Organische Chemie u. Biochemie, Martin Luther King Platz 6 2 Hamburg 12, F.D.R.

3. Woods Hole Oceanographic Institution, Woods Hole Mass. 02543, U.S.A.

4. Scripps Institution of Oceanography, La Jolla California 92093, U.S.A.

5. Centre Oceanologique de Bretagne, B.P.337 29273 Brest Cedex, France

6. Marine Ecology Laboratory, Bedford Institute of Oceanography, PO Box 1006, Dartmouth, N.S. Canada B2Y 4A2

INTRODUCTION

A key feature of the sea area discussed was that the vast majority of the organic material was in the dissolved form (Cauwet, 1981). A substantial fraction of this has still not been adequately characterized chemically (Moppe and Degens, 1979). We occasionally touched on the question of what was dissolved and what was not, but had no clear answer. Estimates seemed to depend on the pore-size of the filters used and it should be noted that some bacteria will not be collected by filters with a pore-size of 0.4μm (Sharp, 1973).

We considered processes likely to build up this dissolved organic material and processes involved in breaking it down. One of the inputs was probably horizontal transport from other sea areas.

Another was release from living organisms, for example the exudates from plant cells and the excretion products of animals.

THE PLANT CONTRIBUTION

This would depend on species, the physiological state of the plant and the stage of growth (Brockmann _et al._, 1982; Myklestad, 1977). It seemed that a large fraction of the material had not been fully characterized chemically (Williams, 1975) and we were not clear about how much was dissolved and how much was particulate. We concluded that filters of much smaller pore-size should be used in this kind of work (0.1-0.2μm). Another improvement might be to study exudates in chemostats as well as batch cultures: values ranged from 10-50% of the carbon fixed (Mague _et al._, 1980). We were unable to put a precise figure on the size of the plant contribution.

THE ANIMAL CONTRIBUTION

Excretion by animals also contributes to the pool of dissolved organic material (Conover, 1978). As many of the zooplankton - especially in oligotrophic areas - are very small (<5μm), rates of soluble release would be very rapid because body size is an important factor in excretion as well as respiration. Organic forms of nitrogen such as urea (Smith, 1978; Smith and Whitledge, 1977) and amino acids (Conover, 1978; Corner and Davies, 1971) could account for 20-50% of the total excreted depending on species, stage and previous feeding history. Some of the excreted phosphorus would also be organic - but had not been characterized chemically (Butler _et al._, 1970).

Measurements of excretion rates using very small zooplankton - which would be mixed with phytoplankton - called for the use of the ^{15}N-dilution technique (Gilbert _et al._, 1982). Incorporation of a competitive inhibitor (e.g. methylamine) to block ammonia metabolism might help to measure release rates more accurately. The ^{32}P method used with freshwater zooplankton (Peters and Rigler, 1973) might also be applied in a marine context. A further improvement in methodology was the _in vitro_ technique in which it has been demonstrated that a clear relationship exists between glutamic dehydrogenase activity and ammonia release (Packard, personal communication).

Zooplankton might also contribute to the dissolved organic pool by spilling the contents of food particles during capture (Wangersky, 1978; Whittle, 1977). A sizable contribution to the DOM pool might also occur in some sea areas due to excretion by other marine animals, e.g. schools of fish (Whitledge and Packard, 1971).

The N:P ratio of the soluble excretion products of zooplankton, together with those found for fecal pellets, the animals and the diets have been successfully used to estimate assimilation and food conversion efficiencies as well as P/B values for whole populations (Le Borgne, 1982). Perhaps this approach merits wider application.

LOSSES OF ORGANIC MATERIAL FROM THE ECOSYSTEM

Dissolved organic material could be reduced by photo-oxidation near the surface. It might also be converted into particulate form by adsorption on inert material such as clay particles (Chave and Suess, 1970; Carter and Mitterer, 1978) and by the action of rising bubbles (Riley, 1970). We were unable to quantify these processes. Degradation of dissolved organic material would mostly occur through bacterial activity (Sepers, 1977; Williams and Yentsch, 1976). Much of the dissolved material is refractory, but the quantities are so immense that even a very slow rate of turnover requires a continual transfer from the euphotic zone to the deeper waters.

Particulate organic material would include planktonic organisms and detritus. The living organisms would involve plants, that remove organic compounds such as ammonia and inorganic phosphate as nutrients, and animals that would regenerate these nutrients in their excretion products. In some sea areas excretion by zooplankton was regarded as providing a substantial fraction of the nutrient requirements of plants (Corner, 1973).

One form of particulate material considered in some detail by the Group was the fecal pellet production by animals. Such pellets, along with other particles, would fall out of the euphotic zone and remove material from the system (Osterburg _et al._, 1963). It was noted that many literature values existed for fecal losses by zooplankton animals both large and small; these values varied from 5 to 40% of the ingested food, depending on diet (Petipa _et al._, 1975).

The "rain" of particles has been measured just below the euphotic zone by using sediment traps (Wiebe _et al._, 1976). Fecal pellets were an important contribution to this but over 90% of these appeared to be removed in the euphotic zone (Bishop _et al._, 1977). One mechanism is coprophagy (Hofman _et al._, 1981). Another is mineralization or microbial degradation (Madin, 1982). It seemed probable that the organic material in these particles would be used by bacteria, which would subsequently be taken as a food by microheterotrophs responsible for the excretion of inorganic compounds. Bacteria might have a more direct role in remineralization if their assimilation efficienty is low. However, if they assimilate with high efficiency they could provide a substantial food supply for bacterivores such as ciliates and flagellates. These, together with the bacteria, would then be responsible for the remineralizing process. Relative rates of mineralization of particulate material might be assessed from measurements of changes in organic composition. For example, there were indications that the lipid fraction of the organic content decreases rapidly with depth, implying that lipid material is mineralized faster (Wakeham et al., 1980). More work on the chemical compositions of fecal pellets and other aggregates such as "marine snow" (Silver et al., 1978) seems desirable.

NITRIFICATION, DENITRIFICATION, NITROGEN FIXATION

All measurements of nitrification in the sea with few exceptions have been carried out in coastal waters. In the most detailed work (off Southern California), nitrification rates were inhibited by light (Olson 1981 a and b) and the highest rates in the vertical plane were associated with the primary nitrite maximum below the euphotic zone (Ward et al., 1982). Sensitive fluorometric techniques have recently been developed for the specific determination of nitrifying bacteria (Ward and Perry, 1980); these bacteria appeared to be distributed throughout the water column. Measured rates of nitrification range from 0.4 to 7×10^{-8} mol ℓ^{-1}day^{-1} (specific rates = 3×10^{-14} to 2×10^{-12} mol. cell^{-1}). Areal rates of nitrification should be approximately equivalent to the flux of reduced N out of the euphotic zone since the concentration of DON(or NH_4^+) remains relatively constant below the euphotic zone. Oxidation of reduced nitrogen to NO_3^- provides energy used for carbon reduction; the quantitative significance of this autotrophy remains speculative, particularly in the oligotrophic areas.

Few direct measurements of denitrification have been made in the open sea, although some estimates based on elemental ratios have been published (Joint and Morris, 1982; Knowles, 1982). Therefore no statement can be made of the significance of this potential loss of nitrogen from the marine ecosystem.

Rates of nitrogen fixation in the sea are generally considered to be low with reported rates of 1×10^{-6} to 10^{-3}mol N m^{-2} day^{-1}. Free living *Richelia* or *Oscillatoria* or symbioses involving *Richelia* are organisms identified in this process. Recent observations involving a symbiosis between *Rhizosolenia* (diatom) and a nitrogen-fixing bacterium have been reported, with the demonstration that improper sampling affects the fixation rates (Alldredge and Silver, 1982). Data in the latter study imply fixation rates of the order of 10^{-3}mol N m^{-2} day^{-1}, an amount roughly equal to the flux of NO_3^- into the euphotic zone from deep water.

REFERENCES

Alldredge, A.L., and Silver, M.S., 1982, Abundance and production rates of floating diatom mats (*Rhizosolenia castracahei* and *R.imbricata* var. *shrubsolei*) in the eastern Pacific Ocean, *Mar. Biol.*, 66:83.

Bishop, J.K.B., Edmond, J.N., Darlene, R.K., Bacon, M.P., and Silken, W.N., 1977, The chemistry, biology and vertical flux of particulate matter from the upper 400m of the equatorial Atlantic Ocean, Deep Sea Res., 24:511

Brockmann, U., Ittekkot, V., Kattner, G., Eberlein, K., and Hammer, K.D., 1982, Release of dissolved organic substances in the course of phytoplankton blooms, North Sea Dynamics, In Press.

Butler, E.I., Corner, E.D.S., and Marshall. S.M., 1970, On the nutrition and metabolism of zooplankton. VII. Seasonal survey of nitrogen and phosphorus excretion by Calanus in the Clyde Sea-Area, J. Mar. Biol.Ass. U.K., 50: 525

Carter, P.W., and Mitterer, R.M., 1978, Amino acid composition of organic compounds associated with carbonate and non-carbonate sediments Geochim. Cosmochim. Acta, 42: 1231

Cauwet, G., 1981, Non-living particulate matter, in: "Marine Organic Chemistry," E.K. Duursma and R. Dawson, eds., Elsevier, Amsterdam.

Chave, K.E., and Suess, E., 1970, Calcium carbonate saturation in seawater: effects of dissolved organic matter, Limnol. Oceanogr., 15: 633.

Conover, R.J., 1978, Transformation of organic matter, in: "Marine Ecology," vol.4, O.Kinne, ed., John Wiley & Sons, Chichester.

Corner, E.D.S., 1973, Phosphorus in marine zooplankton. Wat.Res. 7: 93.

Corner, E.D.S., and Davies, A.G., 1971, Plankton as a factor in the nitrogen and phosphorus cycles in the sea, Adv. Mar. Biol., 9: 101.

Glibert, P.M., Lipschultz, F., McCarthy, J.J., and Altabet, M.A., 1982 Isotope dilution models of uptake and remineralization of ammonium by marine plankton, Limnol. Oceanogr., In Press

Hofman, E.E., Klinck, J.M., and Paffenhöfer, G.-A, 1981, Concentrations and vertical fluxes of zooplankton faecal pellets on a Continental shelf, Mar. Biol., 61: 327

Joint, I.R., and Morris, R.J., 1982, The role of bacteria in the turnover of organic matter in the sea, Oceanogr. Mar. Biol. Ann. Rev., 20: 65

Knowles, R., 1982, Denitrification, Microbiol. Rev., 46: 43

LeBorgne, L., 1982, Zooplankton production in the eastern tropical Atlantic Ocean: net growth efficiency and P:B in terms of carbon, nitrogen and phosphorus, Limnol. Oceanogr., 27: 681.

Madin, L.P., 1982, Production, composition and sedimentation of salp fecal pellets in oceanic waters, Mar. Biol., 67: 39

Mague, T.J., Friberg, E., Highes, D.J., and Morris, I., 1980, Extracellular release of carbon by marine phytoplankton: a physiological approach, Limnol. Oceanogr., 25: 262

Moppe, K., and Degens, E.T., 1979, Organic carbon in the Ocean: Nature and cycling, in: "The Global Carbon Cycle," B. Bolin, E.T. Gegens, S. Kempe, and P. Ketner, eds., John Wiley & Sons, Chichester.

Myklestad, S., 1977, Production of carbohydrates by marine planktonic diatoms. II. Influence of the N/P ratio in the growth medium on the assimilation ratio, growth rate, and production of cellular and extracellular carbohydrates by Chaetoceros affinis var. willei (Gran) Hustedt and Skeletonema costatum (Grev.) Cleve, J. Exp. Mar. Biol. Ecol., 29: 161

Olson, R.J. 1981a, ^{15}N tracer studies of the primary nitrate maximum, J. Mar. Res., 39: 203.

Olson, R.J., 1981b, Differential photoinhibition of marine nitrifying bacteria: a possible mechanism for the formulation of the primary nitrate maximum, J. Mar. Res., 39: 227.

Osterburg, C., Carey, A.G., and Curl, H., 1963, Acceleration of sinking rates of radionuclides in the ocean, Nature (Lond.), 200: 1276.

Peters, R.H., and Rigler, F.H., 1973, Phosphorus release by Daphnia, Limnol. Oceanogr., 18 821

Petipa, T.S., Monakov, A.V., Pavlyutin, A.P., and Sorokin, Yu.I., 1975, The significance of detritus and humus in the nutrition and energy balance of the copepod Undinula darwini Scott. Biologicheskaya produkfivnost yuzhnykh morei, Canadian Fisheries and Marine Service, Translation Series No. 3368.

Riley, G.A., 1970, Particulate organic matter in sea water, Adv. Mar. Biol., 8: 1

Sepers, A.B.J., 1977, The utilization of dissolved organic compounds in acquatic environments, Hydrobiol., 52: 39

Sharp, J.H., 1973, Size classes of organic carbon in seawater, Limnol. Oceanogr., 18: 441.

Silver, M.S., Shanks, A.L., and Trent, J.D., 1978, Marine snow: microplankton habitat and source of small-scale patchiness in pelagic populations, Science (N.Y.), 201: 371

Smith, S.L., 1978, Nutrient regeneration by zooplankton during a red tide off Peru, with notes on biomass and species composition of zooplankton, Mar. Biol., 49: 125

Smith, S.L., and Whitledge, T.E., 1977, The role of zooplankton in the regeneration of nitrogen in a coastal upwelling system off northwest Africa, Deep-Sea Res., 24: 49.

Wakeham, S.G., Farrington, J.W., Gagosian, R.B., Lee, C., DeBarr, H., Nigrelli, G.E., Tripp, B.W., Smith, S.O., and Frew, N.M., 1980, Organic matter fluxes from sediment traps in the equatorial Atlantic Ocean, Nature (Lond.) 286, 798.

Wangersky, P.J., 1978, Production of dissolved organic matter, in: "Marine Ecology," vol. 4, O. Kinne, ed., John Wiley & Sons, Chichester.

Ward, B.B., and Perry, M.J., 1980, Immunofluorescent assay for the marine ammonium-oxidising bacterium Nitrosococcus oceanus, Appl. Envir. Microbiol., 39: 913

Ward, B.B, Olson, R.J., and Perry, M.J., 1982, Microbial nitrification rates in the primary nitrite maximum off Southern California, Deep-Sea Res., 29A: 247.

Whitledge, T.E., and Packard, T.T., 1971, Nutrient excretion by anchovies and zooplankton in Pacific upwelling regions, Invest. Pesq., 35: 243.

Whittle, K.J., 1977, Marine organisms and their contribution to organic matter in the ocean, Mar. Chem., 5: 381.

Wiebe, P.H., Boyd, S.H., and Winget, C., 1976, Particulate matter sinking to the deep-sea floor at 2000m in the Tongue of the Ocean, Bahamas, with a description of a new sedimentation trap, J. Mar. Res., 34: 341

Williams, P.J. LeB., 1975, Biological and chemical aspects of dissolved organic material in sea water, in: "Chemical Oceanography, II (2nd edition)," J.P. Riley and G. Skirrow, eds., Academic Press, London.

Williams, P.J. LeB., and Yentsch, C.S., 1976, An examination of photosynthetic production, excretion of photosynthetic products and heterotrophic utilization of dissolved organic compounds with reference to results from a coastal subtropical sea, Mar. Biol., 35: 31

HERBIVORY

K.H. Mann[1], B. Frost[2], L. Guglielmo[3], M. Huntley[4],
B.O. Jansson[5], A.R. Longhurst[6], G. Radach[7] and R. Williams[8]

1. Marine Ecology Laboratory
   Bedford Institute of Oceanography
   P.O. Box 1006, Dartmouth
   Nova Scotia, Canada, B2Y 4A2

2. Department of Oceanography
   University of Washington
   Seattle, Washington 98195
   U.S.A.

3. Universita di Messina
   Ist. di Idrobiologia e Pescicoltura
   Via del Verdi 75
   Messina, Italy

4. Institute for Marine Resources,
   Scripps Institution of Oceanography
   La Jolla, California 92093
   U.S.A.

5. Askö Laboratory
   Box 58
   S150-13 Trosa
   Sweden

6. Bedford Institute of Oceanography
   PO Box 1006, Dartmouth
   Nova Scotia, Canada B2Y 4A2

7. Universität Hamburg
   Institut für Meereskunde
   Heimhuder Str. 71
   2 Hamburg 13, F.D.R.

8. Institute for Marine Environmental Research
   Prospect Place
   The Hoe, Plymouth
   Devon, PL1, 2PB, U.K.

INTRODUCTION

Definition

The working group adopted a simple definition of herbivores:

"Herbivores are simply those animals whose feeding mechanisms and life history strategies favour the ingestion of living autotrophic cells".

In arriving at this definition it was noted that few animals subsist solely on living plant material. Filter feeding animals, plankton and benthos alike, select their food on criteria of particle size and particle quality, and often have appropriate rejection mechanisms. Studies of herbivory must include both the mechanisms by which cells are culled from natural autotrophic populations and subsequent utilization of their protoplasm for nutrition and growth of herbivores. Hence, major sections of the working group report are devoted to measurements of grazing rate and estimation of herbivore production.

Size Classes

In the past, most attention has been given to copepods and their grazing on diatoms. Recent studies have shown that important fluxes between autotrophs and herbivores may occur among both very small size classes of organisms, such as bacteria and ciliates, and through large planktonic herbivores such as gelatinous zooplankton, larger crustacea and clupeid fish.

For example, it is now understood that not only small algae in the size range 2 - 200 μm, but also cyanobacteria in the range 0.5-1.0 μm may contribute importantly to primary production in the ocean (Li *et al.*, submitted). There are also situations in which total chlorophyll in the ocean exceeds that contained in filterable cells, and this has led to the hypothesis that there may be 'phantom' populations of autotrophs not susceptible to sampling by present techniques. The presence of such cells should be confirmed or denied.

We recommend that more attention be given to fluxes involving very small organisms, such as autotrophic and heterotrophic bacteria and ciliates, and to those involving large filter feeders such as gelatinous zooplankton, larger crustacea, and some clupeid fish.

Spatial Relationships

In the past, plankton samples that have not described spatial relationships between producers and consumers with sufficient precision have disclosed apparent paradoxes: for instance, in vertically integrated samples there is frequently a mismatch between the demands of consumers for food particles above a critical theshold abundance, and the food particles apparently available. Such paradoxes have, in some cases, been resolved by the application of sampling techniques with finer spatial discrimination, especially in the vertical plane: perhaps the best known example is the case of the larval anchovy populations off California and their interactions with extensive but very narrow layers of food particles above theshold concentration (Lasker, 1975).

Another puzzling case with general implications for herbivory is supplied by recent studies of the vertical profiles of plant biomass and chlorophyll which commonly, under a variety of oceanographic situations contain a subsurface maximum which is often a very prominent feature of the profile. As Cullen (1982) has shown, it is possible to explain the formation and maintenance of such subsurface maxima without recourse to vertically differentiated grazing pressure by herbivores: yet is now becoming clear (e.g. Longhurst, 1981) there is also a highly ordered differential vertical distribution of zooplankton over the same space scale. It is only occasionally, perhaps especially when grazers aggregate slightly above the chlorophyll maximum, apparently at the production maximum (Longhurst, 1976; Herman, Sameoto and Longhurst, 1981; Cox, Willason and Harding, 1982), that grazing pressure is clearly seen to result in reduced abundances of plants (Herman and Sameoto, 1982).

Such cases should also be interpreted in the light of the recent understanding that marine planktonic ecosystems are not necessarily in trophic balance, and that on continental shelves (as opposed to open ocean ecosystems) as little as 20-30% of plant production may be consumed by herbivores where it is produced, the remainder being available for export, burial and subsequent remineralization by microbial activity (Walsh, 1981; Joiris, et al., 1982; Longhurst, submitted). Under such circumstances the spatial relations of producers and herbivores may not be as meaningful, and may easily lead to false interpretation, as would be the case in an open ocean ecosystem where production and consumption are usually more approximately in balance within the same body of water.

Fluxes in Whole Communities

In order to understand marine ecosystems we need to know the total fluxes from primary production to herbivores, and the total production by herbivores. These can be attempted in two ways: by determining the fluxes for each species or group of species, and summing them to get the total flux, or by developing methods that integrate the flux during measurement. In places where most of the flux is through a relatively small number of species, the species-by-species approach may be feasible. In other cases, an integrative approach may be more useful.

MEASUREMENT OF GRAZING RATE

Because the consumers of phytoplankton, "herbivores," span a very broad range of size, from protozoans to large salps and even clupeid fishes, a variety of methods have been used to measure grazing rates in natural assemblages. No method is likely, by itself, to provide a complete characterization of grazing, that is, information on rates of grazing on various sizes and types of phytoplankton as well as identification of the consumers responsible for the grazing. Therefore we encourage the continued development and application of several methods, and suggest that any serious attempt to measure total grazing in a natural system will require the use of some combination of the methods. The methodology of grazing in pelagic ecosystems was recently reviewed by Frost (1980). Here we discuss those methods, together with some very recently developed techniques, which seem to offer the greatest promise for reliably assessing grazing.

Incubation Techniques

These techniques are used to estimate grazing by measuring the feeding rates of specific grazers, or mixtures of grazers, when incubated in jars containing natural assemblages of phytoplankton. Grazing may be calculated from the rate of disappearance of particles or chlorophyll during the incubation period. From estimates of abundance of the grazers in the field, the grazing rate may be calculated in terms of the fraction of phytoplankton standing stock or the fraction of phytoplankton production consumed by grazers. Specific applications of this approach have differed in the manipulation of both the phytoplankton and the grazers. In one approach individual large grazers, such as copepods, are captured, sorted by pipette into jars containing natural phytoplankton, and incubated for a day or less; recent examples are Gamble (1978) and Dagg *et al.* (1980). In a variant of this approach, Daro (1980) labelled natural phytoplankton assemblages with ^{14}C, then measured the grazing rate of copepods by their incorporation of ^{14}C. These investigations generally have not considered all potentially significant grazer species, and sometimes not even all developmental stages of the presumed dominant grazer. This is a serious shortcoming, as noted by Paffenhöfer (1982). An exception is the study by Griffiths and Caperon (1979). Special sampling and incubation techniques may be required for potentially very important, but delicate, gelatinous grazers, such as salps and larvaceans (Harbison and McAlister, 1979; King *et al.*, 1980; Alldredge, 1981).

A complete analysis of grazing by the incubation technique should include investigation of differential grazing on particle sizes. This may be most conveniently done with electronic particle counters (noting many caveats: Harbison and McAlister, 1980), but can also be accomplished by microscopical counting of cells (e.g. Huntley, 1981) or size-fractionation of phytoplankton with screens. Incubation periods should be designed to detect possible diel periodicity in feeding rates (e.g., Daro, 1980; Dagg and Grill, 1980).

Sorting of individual grazers is either prohibitively time-consuming or totally impractical for very small grazers, such as microzooplankton. Incubation techniques have therefore been developed to estimate the grazing of the entire assemblages. Capriulo and Carpenter (1980) concentrated microzooplankton by filtration (35 µm screen), then estimated grazing from the disappearance of chlorophyll or phytoplankton cells. Controls permitted the descrimination of grazing from other potential losses of phytoplankton during incubation. Grazing was expressed as the fraction of phytoplankton standing stock removed per day. Landry and Hassett (1982) describe a serial dilution technique to estimate both the phytoplankton growth rate and the grazing rate by microzooplankton. The method assumes that dilution has no effect on growth rate of the phytoplankton and that ingestion rate of grazers is proportional to phytoplankton concentration. With this technique addition of nutrients may be required to prevent nutrient depletion during incubation. As both studies on microzooplankton indicated that very small animals can at times exert substantial grazing mortality on the phytoplankton, it is important that grazing studies address the microzooplankton.

In Situ Feeding Rates of Grazers

There is always uncertainty as to whether rates determined in incubation experiments are representative of the rates at which grazers feed when undisturbed in their natural environment. Attempts have been made to estimate feeding rates of grazers from relatively easily measured biochemical parameters which are products or correlates of the grazing process. Two methods are discussed here.

Mackas and Bohrer (1976) inferred that variations in the amount of chlorophyll and its derivatives in the gut contents of copepods were indicative of variations in the feeding intensity of the grazers. A grazing rate can be obtained if it is known how long it takes for the passage of material through the gut. The method was used recently to estimate the feeding rate of certain species of grazers (Dagg and Grill, 1980; Boyd *et al.*, 1980). The method depends on the (currently untested) assumption that gut passage time of food does not vary with feeding rate. The advantage of the method, in addition to the fact that incubation of grazers is not required, is that the procedure for measuring pigments in gut contents is sensitive and can be made rapidly on single animals; therefore, many samples can be analyzed. The main drawback of the method is that it is selective for those larger grazers which can be easily collected and sorted for analysis; very small grazers, such as those among the microzooplankton, cannot at present be analyzed by this technique.

A second method, still under development, depends on the relationship between feeding rate and activity of digestive enzymes specific for substrates which are unique to phytoplankton. Cox (1981) and Cox and Willason (1981) showed that copepods and euphausiids feeding on phytoplankton had high activities of the digestive enzyme laminarinase, but that activities fell to minimal levels when the animals fed carnivorously or were starved. A quantitative relationship between activity of digestive enzymes and food concentrations appears to exist for several species of copepods (Mayzaud and Poulet, 1978; Hirche, 1981). Nevertheless, critical laboratory experiments are yet to be reported wherein activities of digestive enzymes and feeding rates are measured simultaneously. This method has the potential of providing an estimate of total grazing rate if there is a consistent relationship between activity of digestive enzymes and feeding rates of grazers.

Pigment Budget Approach

Lorenzen (1967) suggested that phaeophorbide in natural waters must be produced by grazers and therefore must be indicative in some way of their feeding activity. Shuman and Lorenzen (1975) showed that the chlorophyll _a_ contained in diatoms ingested by a species of copepods was largely converted to phaeophorbide _a_ during passage through the copepod's gut, and was subsequently released in fecal pellets. Generalizing from this result, the production rate of phaeophorbide in natural waters must be quantitatively related to the grazing rate on the phytoplankton. The technique has recently been explored by Welschmeyer (1982).

To estimate the production rate of phaeophorbide it is necessary to sample both the fraction contained in small suspended particles and the fraction in large, rapidly sinking fecal pellets. The former is sampled with water bottles, the latter with sediment traps positioned just below the euphotic zone. Assuming steady state, the production rate of suspended phaeophorbide particles must equal their destruction rate by photodegradation. Production rate of phaeophorbide in large fecal pellets is, of course, obtained from the rate of accumulation in sediment traps.

The advantage of the method is that it potentially provides a measure of total grazing rate on the phytoplankton. In addition, it may allow evaluation of the relative role of large and small consumers as grazers of phytoplankton. It provides no information on size-specific patterns of grazing, however. To obtain an estimate of grazing as the fraction of primary production consumed, it is further necessary to measure the production rate of chlorophyll by the phytoplankton. A potential drawback of the method is that phaeophorbide may be produced by processes other than grazing. Moreover, if fecal material is reingested by grazers within the euphotic zone (Hofman et al., 1981), the phaeophorbide may be broken down and no longer be available as a tracer of grazing activity. Further evaluation of the pigment budget approach is recommended.

ESTIMATES OF PRODUCTION BY HERBIVORES

The majority of methods that have been used to estimate the production of marine plankton herbivores have proven, for various practical or theoretical reasons, to be somewhat unsatisfactory. More than a decade ago, Mullin (1969) reasoned that there were few estimates because there was "no reliable, simple *in situ* method for measuring even the approximate production of zooplankton populations." Although Mullin's statement is still largely true today, we wish to note several methods which remain, or have recently become, worthy of attention.

C : N : P Ratio Method

This method was suggested by work of Ketchum (1962), improved by Butler et al. (1969) and Corner and Davies (1971), and has recently been elaborated and used extensively by LeBorgne (1978;1982). The method requires estimates of the net growth efficiency (K_1) and of metabolic losses; these are acquired from measurements of the N:P ratios of herbivores and their food, and of the excretion rates of N or P by the herbivores. In practical terms, the method requires bottle incubations of herbivores for the purpose of measuring both their excretion rates and the C:N:P ratio of the feces. Incubations can be conducted on shipboard, and N:P ratios of prey and predators are determined from frozen samples. From the calculations, one obtains an estimate of the instantaneous "gross" production (i.e., production is taken to include the production of molts and eggs). Further details of the method are given by LeBorgne (1978).

The advantages of this technique are that (a) it can be applied to mixed zooplankton populations; (b) it can be applied to carnivores as well as herbivores; (c) only one rate measurement needs to be made for each size class of zooplankton; (d) it can be used in tropical communities, where production has proven notoriously difficult to estimate; and (e) it is easy to apply.

The C:N:P method does suffer from several possible sources of error, namely (a) the definition of "food", which in the case of herbivores is assumed to consist of all particulate matter in a certain size class (1 - 50 μm in LeBorgne, 1982); (b) the assumption that assimilation efficiencies of N and P are equal; and (c) the difficulties inherent in measuring an absolute excretion rate which is not influenced either by stress due to capture of the plankton, or by their starvation in the experimental chamber. LeBorgne (1978; 1982) discusses advantages and disadvantages of the technique in more detail.

Weight-dependent Method

This method, described in Conover and Huntley (1980), is a modified carbon-budget approach based upon the von Bertalanffy equation (1938), and is derived from a set of empirical relationships between herbivore body weight, respiration (Ikeda, 1974) and filtering rate (Huntley, 1980). The method requires measurements of the size-frequency distributions of herbivores and their food, as well as temperature; calculations convert these static quantities to dynamic estimates of instantaneous "gross" production (in the same sense as the C:N:P: ratio method).

The advantages of the weight-dependent technique are similar to those of the C:N:P ratio method, in that (a) it can be applied to mixed herbivore populations; (b) no rate measurements are required; (c) it can be used in communities of planktonic herbivores from boreal regions to the tropics; and (d) it is easy to apply. The necessary measurements can be made either with rudimentary sampling equipment or with more sophisticated, automated sampling systems such as Batfish (Herman and Dauphinee, 1980). The latter approach offers the possibility of continuous, real-time mapping over large scales, and has been applied in this manner by C.M. Boyd (unpubl. data). Details of the method are described in Conover and Huntley (1980) and in Huntley (1980).

The weight-dependent method is subject to several sources of error, in that (a) it assumes that the consumption of particulate matter depends only on its size and abundance; and (b) it assumes that assimilation efficiency is constant. An attempt has been made recently to calibrate the method by comparison to simultaneous measurements of production using cohort analysis (M. Huntley and E.R. Brooks, unpubl. data).

Production- Respiration Relationships

McNeill and Lawton (1970) showed that annual production of short-lived poikilotherms (which would include most planktonic herbivores) was related to annual respiration. A more recent analysis of a larger body of data (Humphreys, 1979) failed to improve upon the wide confidence intervals obtained by McNeill and Lawton.

The principal advantage of this approach to estimating production is that it requires the measurement of only one variable, the annual respiration. However, therein also lies its major disadvantage, in that practical application demands an estimate, rather than a measurement, of respiration.

One application of respiration estimates might be to determine a minimum production rate. If all the potential grazers in a natural assemblage can be identified, and their metabolic activity estimated, it is possible to calculate the minimal grazing rate required to meet the metabolic demand. We find little to recommend such an approach at this time because of the large uncertainty in estimates of production rate or metabolic rate.

P/B Ratio Method

The principal advantage of the P/B (production/biomass) method is its simplicity of application. Provided one knows, from laboratory measurements, the value of P/B for a given population and, from field measurements, the biomass of the population then a simple multiplication provides an estimate of the production.

The user of this method should be cautioned that P/B ratios can change as the age structure and the biomass of the community changes, and that it may also vary as a function of temperature, rate of predatory loss and food supply. Noting that P/B may change over a relatively short period of time, Kamshilov (1953) introduced the term "specific production" to designate a daily P/B ratio. Winberg (1968) also recommended that P/B to calculated over the shortest possible time intervals. Such restrictions may not need to be considered as rigorously in oligotrophic, tropical oceans where temperature, food supply and the age structure of herbivore communities are less variable.

There is a disturbing amount of disagreement on regional variation in P/B; Parsons and Takahashi (1973) suggest that P/B is lower in tropical ecosystems, Conover (1968) predicts a lower P/B in boreal ecosystems, and Shushkina (1971) finds no significant differences in P/B the world over. The utility of this approach is severely hampered by the almost complete lack of reliable production measurements, from which valid P/B ratios could be obtained for herbivorous plankton (the situation is a little better for benthic populations). While we find little to recommend the technique in its current form, we believe that the P/B method will ultimately become valuable, and we encourage continued efforts to evaluate P/B ratios.

None of the four methods discussed above is entirely satisfactory. A great deal of refinement and contined research is required before we can produce accurate estimates of the production of planktonic herbivores.

The most acceptable technique for estimating herbivore production is cohort analysis, since growth is measured directly from a time-series of herbivore abundance and increases in body weight with age. However, the primary assumption of this method - that one must repeatedly sample the same discrete population - can rarely be satisfied in the open ocean. Furthermore, one must also make the assumption that the population either (a) has a constant age structure, or (b) consists of discernible cohorts. Given such rigorous assumptions, it is hardly surprising that the successful use of cohort analysis has been restricted to plankton populations in lagoons or coastal embayments (e.g. Landry, 1978; Heinle, 1966; Uye, 1982; Durbin and Durbin, 1981).

Despite the unsuitability of cohort analysis for application to open ocean communities of planktonic herbivores, we strongly recommend it as the method of choice for standardizing any technique to be applied to pelagic communities.

CURRENT KNOWLEDGE OF GELATINOUS ZOOPLANKTON

The working group met for a time with Dr. Alice Alldredge, who provided the following information, much of which is in press.

Larvaceans (Appendicularians)

There are good estimates of biomass and feeding rates for these herbivores from neritic waters and tropical and temperate oceanic regions. Larvaceans are able to filter particles from 0.1 μm upwards and are able to replace their filtering net (or houses) every 4 h.

Salps

These are mostly herbivorous and are able to filter large volumes of water. The only work done recently on filtering efficiency was that of Harbison and McAlister (1979), using coulter counter techniques. There is a concensus that if animals are to be useful for experimental work on feeding or metabolism they must be collected by SCUBA divers. Salps are able to filter particles of 4 μm or greater with 100% efficiency, and particles of 1 μm are taken with a relatively high efficiency.

Estimates of population density and biomass in oceanic waters are poor.

Pteropoda (Mollusca)

These are mostly herbivorous. Biomass estimates are poor, and there is no quantitative information on feeding rates. The study of the feeding of Clione limacina on Limacina retroversa (Conover and Lalli, 1972) shows how a carnivorous pteropod preys on an herbivorous filter-feeding pteropod, but quantifies the feeding rate of only the carnivore.

ECOLOGICAL EFFICIENCY OF COPEPODS

In the course of discussing the overlap between the working group on herbivory and that on detritivory, our working group noted the observations of Paffenhöfer and Knowles(1979) that copepods may ingest faecal pellets. This raises the possibility that the same population of copepods could feed as herbivores in the euphotic zone at night, descend to deeper water by day and then feed on their own sinking faecal pellets. If the same plant material is ingested more than once, this may serve to increase the efficiency of utilization of primary production by copepods. As was noted by Steele (1974) it is apparently necessary for copepods to utilize phytoplankton with an efficiency close to 20% if the energy budget of the North Sea is to be balanced.

REFERENCES

Alldredge, A.L., 1981, The impact of appendicularian grazing on natural food concentrations in situ, Limnol. Oceanogr., 26: 247.

Boyd, C.M., Smith, S.L., and Cowles, T.J., 1980, Grazing patterns of copepods in the upwelling system off Peru, Limnol. Oceanogr., 25: 583.

Butler, E.I., Corner, E.D.S., and Marshall, S.M., 1969, On the nutrition and metabolism of zooplankton. VI. Feeding efficiency of Calanus in terms of nitrogen and phosphorus, J. Mar. Biol. Ass. U.K., 49: 977.

Capriulo, G.M. and Carpenter, E.J., 1980, Grazing by 35 to 202 μm microzooplankton in Long Island Sound, Mar. Biol., 56: 319.

Conover, R.J., 1968, Zooplankton - life in a nutritionally dilute environment, Amer. Zool. 8: 107.

Conover, R.J., and Lalli, C.M., 1972, Feeding and growth in Clione limacina, a pteropod mollusc, J. Exp. Mar. Biol. Ecol., 9: 279.

Conover, R.J., and Huntley, M.E., 1980, General rules of grazing in pelagic ecosystems, in: "Primary Productivity in the Sea," P. Falkowski, Ed., Plenum Press, New York.

Corner, E.D.S., and Davies, A.G., 1971, Plankton as a factor in the nitrogen and phosphorus cycles in the sea, Adv. Mar. Biol., 9: 101.

Cox, J.L., 1981, Laminarinase induction in marine zooplankton and its variability in zooplankton samples, J. Plankton Res., 3: 345.

Cox, J.L., and Willason, S.W., 1981, Laminarinase induction in Calanus pacificus, Mar. Biol. Letters 2: 307.
Cox, J.L., Willison, S.W., Harding, L., in press, Consequences of distributional heterogeneity of Calanus pacificus grazing, Bull Mar. Sci.
Cullen J.J., 1982, The deep chlorophyll maximum: comparing vertical profiles of chlorophyll a, Can. J. Fish Aquat. Sci., 39: 791.
Dagg, M.J., and Grill, D.W., 1980, Natural feeding rate of Centropages typicus females in the New York Bight, Limnol. Oceanogr., 25: 597.
Dagg, M., Cowles, T., Whitledge, T., Smith, S., Howe, S., and Judkins, D., 1980, Grazing and excretion by zooplankton in the Peru upwelling system during April 1977, Deep-Sea. Res., 27: 43.
Daro,M.H., 1980, Field study of the diel feeding of a population of Calanus finmarchicus at the end of a phytoplankton bloom FLEX '76 22 May - 5 June, "Meteor" Forsch. -Ergebn., Ser. A, 22: 123.
Durbin, A.G., and Durbin, E.G., 1981, Standing stock and estimated production rates of phytoplankton and zooplankton in Narragansett Bay, Rhode Island, Estuaries, 4: 24.
Frost, B.W., 1980, Grazing, in: "The Physiological Ecology of Phytoplankton," I. Morris, ed., Blackwell Scientific.
Gamble, J.C., 1978, Copepod grazing during a declining spring phytoplankton bloom in the northern North Sea, Mar. Biol., 49: 303.
Griffiths, F.B., and Caperon, J., 1979, Description and use of an improved method for determining estuarine zooplankton grazing rates on phytoplankton, Mar. Biol., 54: 301.
Harbison, G.R., and McAlister, V.L., 1979, The filter-feeding rates and particle retention efficiencies of three species of Cyclosalpa (Tunicata, Thaliacea), Limnol. Oceanogr., 24: 875.
Harbison, G.R., and McAlister, V.L., 1980, Fact and artifact in copepods feeding experiments, Limnol. Oceanogr., 35: 971.
Heinle, D.R., 1966, Production of the calanoid copepod Acartia tonsa in the Patuxent River estuary, Chesapeake Sci., 7: 59.
Herman, A., and Dauphinee, T.M., 1980, Continuous and rapid profiling of zooplankton with an electronic counter mounted on a 'Batfish' vehicle, Deep-Sea Res., 27: 79.
Herman, A. and Sameoto, D.D., 1982, Copepod distributions on the Peru shelf at 9°S during November 1977, Bol. Instit. Mar. Peru (Volumen Extraordinario): 228.
Herman, A., Sameoto, D.D., and Longhurst, A.R., 1981, Vertical and horizontal distribution patterns of copepods near the shelf break south of Nova Scotia, Can. J. Fish Aquat. Sci., 38: 1065.
Hirche, H.J., 1981, Digestive enzymes of copepodids and adults of Calanus finmarchicus and C. helgolandicus in relation to particulate matter, Kieler Meeresforsch., Sonderh., 5: 174.
Hofmann, E.E., Klinck, J.M., and Paffenhöfer, 1981, Concentrations and vertical fluxes of fecal pellets on a continental shelf, Mar. Biol., 62: 327.
Humphreys, W.F., 1979, Production and respiration in animal populations, J. Animal Ecology, 48: 427.

Huntley, M.E., 1980, Developing and testing a new method for estimating the production of marine zooplankton. Ph. D. thesis, Dalhousie Univ., Halifax.

Huntley, M.E., 1981, Nonselective, nonsaturated feeding by three calanoid copepod species in the Labrador Sea, Limnol. Oceanogr., 26: 831.

Ikeda, T., 1974, Nutritional ecology of marine zooplankton, Mem. Fac. Fish. Hokkaido Univ., 22: 1.

Joiris, C., Billen, G., Lancelot, C., Daro, M.H., Mommaerts, J.P., Hecq., J.H., Bertels, A., Bossicart, M., and Nijs, J., 1982, A budget of carbon cycling in the Belgian coastal zone: relative roles of zooplankton, bacterioplankton and benthos in the utilisation of primary production, Neth. J. Sea. Res., 16: 260.

Kamshilov, M.M., 1953, Production of Calanus finmarchicus (Gunnerus) in the littoral zone of Eastern Murman, Trudy Murmansk Morsk. Biol. Sta.Vol. 4.

Ketchum, B.H., 1962, Regeneration of nutrients by zooplankton, Rapp. P. -v. Réun Cons. Int. Explor. Mer., 153, 142.

King, K.R., Hollibaugh, J.T., and Azam, F., 1980, Predator-prey interactions between the larvacean Oikopleura dioica and bacterioplankton in enclosed water columns, Mar. Biol., 56: 49.

Landry, M.R., 1978, Population dynamics and production of a planktonic marine copepod, Acartia clausi, in a small temperate lagoon on San Juan Island, Washington, Int. Rev. ges. Hydrobiol. Hydrogr. 63: 77.

Landry, M.R., and Hassett, R.P., 1982, Estimating the grazing impact of marine micro-zooplankton, Mar. Biol., 67: 283.

Lasker, R., 1975, Field criteria for survival of anchovy larvae: the relation between the inshore chlorophyll layers and successful first feeding, U.S. Fish. Bull., 71: 453.

LeBorgne, R., 1978, Evaluation de la production secondaire planctonique en milieu océanique par la méthode des rapports C:N:P, Oceanol. Acta., 1: 107.

LeBorgne, R., 1982, Zooplankton production in the eastern tropical Atlantic Ocean: net growth efficiency and P:B in terms of carbon, nitrogen and phosphorus, Limnol. Oceanogr., 27: 681.

Li, W., Subba Rao, D.V., Harrison, W.G., Smith J.C., Cullen, J.J., Irwin, B., and Platt, T., 1983 Autotrophic picoplankton in the tropical ocean, Science, 219: 292

Longhurst, A.R., 1976, Vertical migration, in: "The Ecology of the Seas" ,D.H. Cushing and J.J. Walsh, eds., Blackwell, Oxford.

Longhurst, A.R., 1981, Significance of spatial variability, in: "Analysis of Marine Ecosystems," A.R. Longhurst, ed., Academic Press, London.

Longhurst, A.R., In Press, Benthic-pelagic coupling and export of organic carbon from a tropical Atlantic continental shelf— Sierra Leone. Estuarine, Coastal and Shelf Science.

Lorenzen, C.J., 1967, Vertical distribution of chlorophyll and phaeopigments: Baja California, Deep-Sea Res., 14: 735.

Mackas, D., and Bohrer, R., 1976, Fluorescence analysis of zooplankton gut contents and an investigation of diel feeding patterns, J. Exp. Mar. Biol. Ecol., 25: 77.
Mayzaud, P., and Poulet, S., 1978, The importance of the time factor in the response of zooplankton to varying concentrations of naturally occurring particulate matter, Limnol. Oceanogr., 23: 1144.
McNeill, S., and Lawton, J.H., 1970, Annual production and respiration in animal populations, Nature (London), 225: 472.
Mullin, M.M., 1969, Production of zooplankton in the ocean: the present status and problems, Oceanogr. Mar. Biol. Ann. Rev., 7: 293.
Paffenhöfer, G. -A., 1982, Grazing by copepods in the Peru upwelling, Deep-Sea Res., 29: 145.
Paffenhöfer, G. -A., and Knowles, S.C., 1979, Ecological implications of fecal pellet size, production and consumption by copepods, J. Mar. Res., 37: 35.
Parsons, T.R., and Takahashi, M., 1973, "Biological Oceanographic Processes," Pergamon Press, New York.
Shuman, F.R., and Lorenzen, C.J., 1975, Quantitative degradation of chlorophyll by a marine herbivore, Limnol. Oceanogr., 20: 580.
Shushkina, E.A., 1971, Estimation of the intensity of tropical zooplankton production, in: "Funktsionirovanie Pelagischeskikh Soobchetsv Tropischeskikh Raionov Okeana," Nauka, Moscow. (in Russian).
Steele, J.H., 1974, "The Structure of Marine Ecosystems," Harvard University Press, Cambridge, Mass.
Uye, S., 1982, Population dynamics and production of Acartia clausi Giesbrecht (Copepoda: Calanoida) in inlet waters, J. Exp. Mar. Biol. Ecol., 57: 55.
Walsh, J., 1981, Shelf-sea ecosystems in: "Analysis of Marine Eco-Systems" A.R. Longhurst, ed., Academic Press, London.
Welschmeyer, N.A., 1982, The dynamics of phytoplankton pigments: Implications for zooplankton grazing and phytoplankton growth. Ph. D. Dissertation, University of Washington, Seattle.
Winberg, G.G., 1968, (ed.), "Methods for Determination of Production in Aquatic Animals," Vysheishaya shkola, Minsk.
von Bertalanffy, L., 1938, A quantitative theory of organic growth, Hum. Biol., 10: 181.

DETRITIVORY

P.A. Jumars (Chairman)[1], R.C. Newell (Rapporteur)[2], M.V. Angel[3], S.W. Fowler[4], S.A. Poulet[5], G.T. Rowe[6], and V. Smetacek[7]

1. Department of the Navy, Office of Naval Research, Code 480, Arlington, Virginia 22217, U.S.A.
2. Institute for Marine Environmental Research, Prospect Place, The Hoe, PLYMOUTH, Devon PL1 3DH, U.K.
3. Institute of Oceanographic Sciences, Wormley Godalming, Surrey GU8 5UB, U.K.
4. International Laboratory of Marine Radioactivity Musée Océanographique, Monaco-Ville, MONACO
5. Station Biologique, Place Georges Teissier, 29211 Roscoff, FRANCE
6. Building 318, Brookhaven National Laboratory, Upton New York 11973, U.S.A
7. Institut für Meereskunde, Dusternbrooker Weg 20, D2300 Kiel 1, F.D.R.

DEFINITION

We choose to define detritivory operationally as the relatively frequent ingestion of particulate food whose bulk generally is not composed of animal, plant or bacterial protoplasm. Our definition does not preclude the possibility that a major fraction of a detritivore's requirements for specific nutrients may be met from bacterial, plant or animal constituents (e.g. Anderson, 1976). We consider it largely an open and exciting question whether particular detritivores or the majority of detritivores under our rather inclusive definition meet their metabolic needs by (1) digesting and assimilating non-living detritus directly, (2) employing bacteria in external "gardening" or internal "rumination" to digest detritus, (3) digesting the normal bacterial component of detritus, (4) digesting plant (e.g. benthic diatom or fragmented macroalgal) protoplasm, (5) digesting the living or recently dead faunal components of detritus, or (6) utilizing some combination of these food sources. Within this potential continuum, are there "adaptive peaks" that allow natural functional grouping of detritivores for further investigation and generalization?

Rapid progress can now be expected in the study of detritivory. With a few exceptions (e.g. those noted below with respect to the measurement of the rate of detrital supply), major advances in understanding the detritus feeding process appear to have been blocked by the lack of simple, if only provisional parameterization of the detritus feeding process and its major components. One example of the difficulty of working on detritivory without an explicit rate-process model is the fact that many more data have been published on the assimilation efficiencies of detritivores than on their rates of detrital throughput. Without the measurement of both parameters for the same species fed on the same detrital food source, it is impossible to calculate so basic a quantity as a rate of material or energy gain.

PARAMETERISATION

The need for an explicit rate-process parameterization of marine detritivory has been met only recently (Levinton and Lopez, 1977; Taghon et al., 1978; Taghon, 1981). To illustrate the power of such models in formulating well-poised questions that promise new insights into detritivory, we will extend (Figure.1), the simplified graphical procedure of Sibly (1981). Because we have perhaps over-simplified to make our point, one should consult Sibly (1981) and the other references cited above before carrying out any explicit tests of the detritivore model. We use, for example, the simplifying assumption (often true for detritivores) that the volume or mass of absorption through the gut wall is insignificant relative to the volume or mass of the throughput. Sibly (1981) gives an explicit correction to use when the approximation that ingested volume or mass equals egested volume or mass is a poor one.

Let G be the net gain (grams or energetic equivalent - not a rate) of a limiting nutrient per gram of food ingested. Retention time (t) in the gut begins with ingestion $(t = 0)$. One can argue, without the need to specify an exact functional form for G(t), that it must first be decreasing while mechanical (e.g. masticatory) and chemical (e.g. enzymatic) energy is expended in initial breakdown of the food. This net loss should be followed by a period of relatively

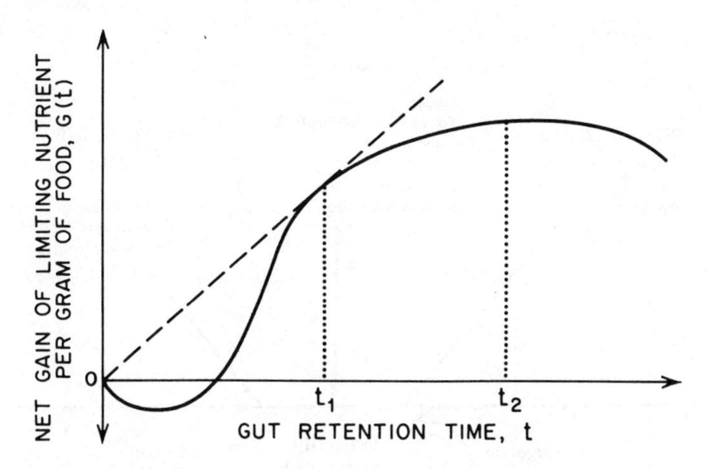

Fig. 1. Net gain of limiting nutrient per gram of food versus gut retention time (modified from Sibly, 1981). Time t_1 and t_2 respectively, optimal gut retention times for non-food-limited cases. (see text).

rapid absorption, with eventually diminishing returns as time proceeds. To these three regions of Sibly's, we would add a final downturn or net loss of energy as the basal catabolism of carrying the weight of gut contents exceeds any gross gain from further retention.

That several carefully selected parameters should be measured in concert to provide new information on detritivory is shown by so simple a question as "How fast should a detritivore feed in order to maximize its average net rate of gain G(t)?" If food is available in unlimited supply (the case treated by Taghon et al., 1978, and Taghon, 1981), as may be true for detritivores more frequently than for members of other feeding guilds, the retention time (t_1) providing maximal net rate of gain is found easily and graphically by drawing the tangent to G(t) that also passes through the origin (figure.1). If food supply is limiting (the case treated by Levinton and Lopez, 1977), however, then food should be held so long (t_2) as any net gain occurs. Without the explicit model, the desirability of measuring rate of supply of detritus and detritivore gut retention time together might not be obvious.

DETRITAL SUPPLY

Although we thus must recommend consideration of detrital supply and utilization in concert, we will continue our report and recommendations by discussing, in turn, supply and the sequential steps involved in detrital utilization. Considering all the factors involved in determining gross and net rates of detrital supply (figure 2), input and export of detrital material to and from a system can

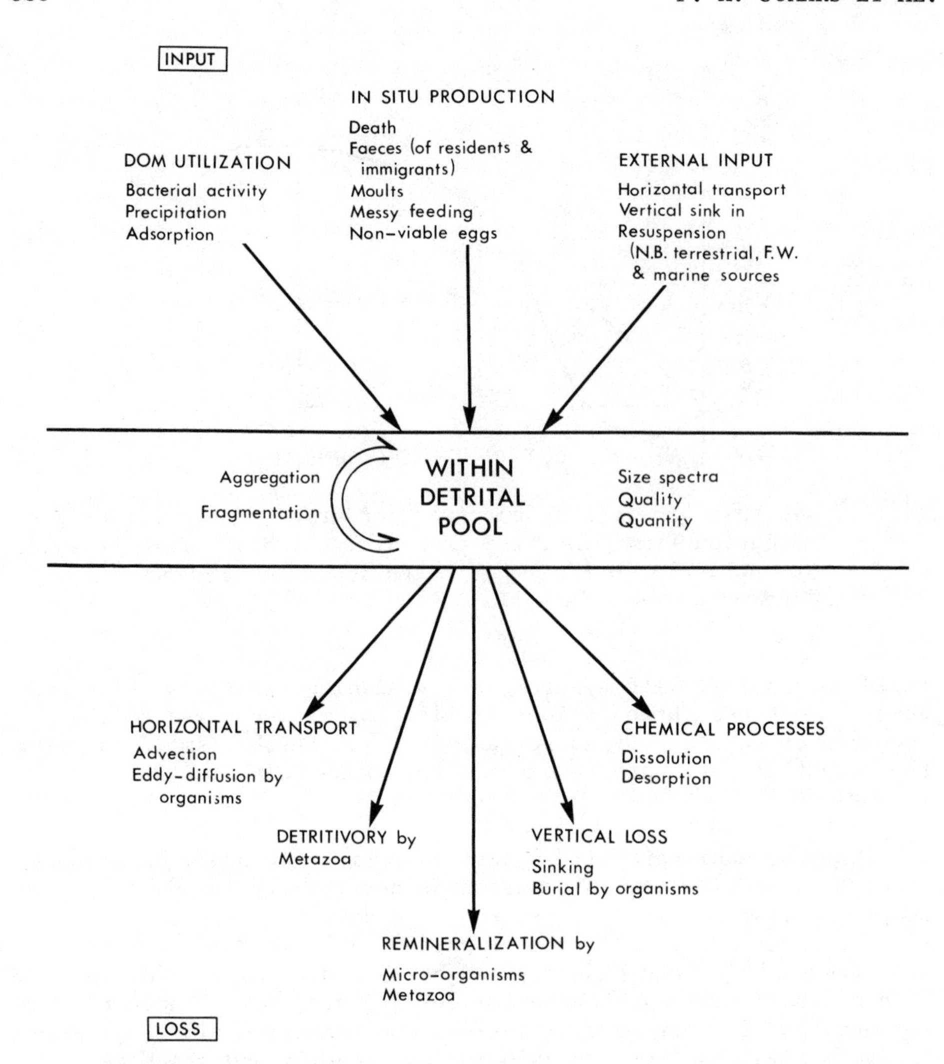

Fig. 2. Outline of detrital sources and sinks that require description and quantification. Spatial and temporal scales of measurement must be selected for the problem at hand, and predictability of detrital pool size and composition will have a major impact on detritivory.

rarely be treated as a constant. Even the deep sea is no exception to this generalization. For some purposes, the time-averaged value of detrital flux will be useful parameter. However, most detritivores are likely to have adaptations in their behavioural repertoires, feeding activities and life cycles in response to variations in detrital availability. Quantities, chemical qualities and size spectra of detrital particles may fluctuate in predictable cycles (e.g. related to ripple-patterns of sand, tidal cycles of resuspension or seasonal variations in production). The more predictable the cycles, the more likely are organisms to be tuned to them. Conversely, large, irregular inputs may pass through the system without causing major functional or numerical responses in the organisms, or they may swamp the processing ability of the community entirely and cause major perturbations in its function (e.g. an episode of anoxia). In such circumstances, a "mean" estimate as provided, for example, by a long-term sediment trap deployment, may be totally misleading.

Where fluxes are large, sampling on temporal and spatial scales appropriate to individual and community responses may not be a major problem, even with extant sampling techniques. Where the fluxes are small, however, such as in many parts of the deep sea, present methodology may not be adequate to discriminate functionally relevant short-term or small-scale spatial variations adequately, especially where analysis of size spectra or chemical composition of detritus is concerned. Once again this problem becomes explicit by attention to the model of Figure. 1; the spatial scale of interest is the foraging area of a single individual and the temporal scale of attention is the gut retention period.

A second technical problem is the estimation of detrital fluxes where fluid motions cause appreciable (gross or net) particle fluxes either vertically upward or horizontally. Here severe bias can be expected in collections by conventional sediment traps. Prominent examples of such regions are surface mixed layers or bottom boundary layers during their respective episodes of active turbulent mixing. To provide even crude estimates of detrital supply rates in these situations, the expert advice of physicists concerned with fluid dynamics and sediment transport will be required; their approach generally is to couple suspended load measurements with models of particle and fluid motion in order to yield the desired vector quantities of net or gross particulate fluxes.

While biologists may benefit directly from measurements made in other disciplines, they run the risk of receiving data on biologically inappropriate parameters or on unsuitable temporal and spatial scales. For example, stratigraphers employing radio-isotopic measures of sedimentation obtain a net rate and generally average over periods well in excess of organism life-spans, and geochemists studying diagenesis generally do not measure the chemical components of detritus that biologists require. Interdisciplinary coordination clearly would benefit all concerned. Besides the obvious benefit to biologists, stratigraphers stand to gain insights into the component processes of bioturbation, and geochemists stand to increase their grasp of biochemical transformations involved in diagenesis.

DETRITAL UTILIZATION

For the study of detritivory itself, rates of detrital supply should be scaled against detrital feeding rates, and estimates of food quality should be scaled against deposit feeder nutritional requirements and assimilative capabilities. Nutritive quality of detritus thus must be determined both by the chemical composition of the detrital supply and by the ability of detritus feeders to utilize the material during its period of gut passage. With few exceptions, little is known of the biochemistry or even bulk organic chemistry of detritus and its associated microbiota or of the extent to which digestive enzymes of consumers can degrade detrital components. Both kinds of data are required to specify more precisely and accurately the shape of the curve in Figure. 1. More subtly, increasing food quality or digestive ability in the food-limited case means a higher plateau in the curve, while in the non-food-limited case it means a more rapid initial rate of assimilation.

It seems likely that in some nearshore systems where freshly produced and fragmented plant debris is available for consumption, detritivores may obtain their daily ration from the detritus itself. For example, the style enzymes of some bivalves have been shown to be fully capable of digesting algal fragments from the kelp beds in which they live (Seiderer et al., 1982; Stuart et al., 1982; Newell and Field, 1983). Insufficient information is available on the gut retention time and digestive physiology of most other detritivores, however, to assess whether the detrital carbon supply meets the consumer's nutritional needs. Nitrogen requirements and, above all, trace requirements for such substances as vitamins and essential fatty acids are also largely unknown and represent areas which require detailed investigation. There is especially great difficulty in obtaining realistic estimates of food quality from hydrolysates of detrital material unless the details of the biochemistry of the intact components and their ease of degradation by consumer organisms is known.

Although some information is available on the carbon and nitrogen content of detrital sources in certain nearshore systems, the chemistry of decomposition, including the formation of humic substances, is in general poorly understood in marine systems. More information is required on the significance of detritus not merely as a possible carbon resource, but as a source of protein and trace nutrients as well. Sufficient information now exists to show that bulk carbon and nitrogen contents alone are not sufficient to characterize food quality.

Provision of food of known quality and composition at a known rate is no guarantee that the detritivore will ingest a representative sample. Several claims of chemical transformation are based on the tenuous assumption that a sample of detritus from the immediate vicinity of a detritivore can be compared with its fecal composition as an estimate of digestive ability. That assumption is true only if (1) the animal is entirely non-selective in its choice of particles for ingestion and (2) the supply of detritus in its vicinity is in steady state. There is a shortage of critical data on the selective abilities of detritivores; note the strong effect that selection could have on the shape of the curve in Figure. 1 if the selected components proved to be highly digestible. An issue which demands attention is the degree to which particle selection is accomplished by purely mechanical means versus via chemosensory evaluation of individual food items. Some kinds of detritivores may process particles at a much greater rate (number of particles per unit time) than do members of other feeding guilds. At what feeding rates does it become uneconomical to reject unwanted items? Particle rejection criteria and costs in detritivores are virtually unknown.

Items selected for ingestion can be processed at varying rates. Details of digestion, absorption and assimilation may be relatively difficult to obtain in detritivores because such a small percentage (by total mass or volume) of ingested material is digested or absorbed. On the other hand, the frequent ability to roughly equate ingestion and defecation rates in detritivores eases some measurement problems inherent in the study of other trophic groups. Various tracer techniques employing both stable isotopes and natural and artificial radionuclides offer great promise in answering questions about detritus origin, age, digestion and assimilation. For example, examining radionuclide ratios in detritus and associated organisms can give useful information on the detrital component of the organism's diet as well as insight into the decompositional characteristics of the released detrital particles (e.g. fecal pellets and molts). Of particular interest is the recent availability of labelled inert plastic and glass microspheres which allow direct measurement of gut transit times under various ecological conditions.

In summary, a great deal can now be learned about detritivory from a few simple measurements, as we will highlight by presenting a few simple questions. Do most detritus feeders have short gut retention times, digesting only the most labile detrital components, or do they have long throughput times, refelcting the relatively refractory nature of detritus as compared with other foods? Will both digestive strategies work on the same food resource? One potential solution to maintaining a fixed rate of energy gain on food of lower quality is to increase the volume of material carried (processing more grams of food per time in the parameterization of Figure. 1), as has been found in some ruminants (Sibly, 1981, Table 5.5). Will a similar inverse relation be found between gut volume and food quality within marine detritivores? Due to the energetics of flight, birds that forage on food of low bulk value are under severe constraints in terms of the gut volume that they can carry - so severe that the gut volume (length) actually changes over periods of a few weeks in response to changes in food quality (Sibly, 1981, Table 5.2). Will actively swimming or burrowing detritivores be similarly, if less severely, constrained in the marine realm?

In addressing these questions and others which by now will have occurred to the reader, we strongly recommend that the alternatives be made explicit by the use of figures (like that of Figure. 1, but showing the alternatives) or equations. Developing this habit will save work both by underscoring those parameters that must be measured very carefully and by pinpointing unnecessary measurements. Given the ease of posing important questions of detritivory in an easily understood graphic fashion and the existence of a wide spectrum of available methods for addressing them, we look forward enthusiastically to rapid progress in the study of this ubiquitous process.

SUMMARY

Understanding of marine detritivory and its component processes is in a primitive state. Until recently, progress appears to have been limited by the lack of simple, if only provisional, parameterizations. Now that such models are available, we recommend that they be used explicitly in designing experiments to investigate (sensu Pielou, 1981) marine detritivory. Given such use, we are confident of rapid progress.

Major gaps in available information are made apparent by our own provisional model, leading to the following recommendations:

1. detrital supply rates and detritivore utilization rates should be measured simultaneously on appropriate scales (e.g. within the foraging area of a detritivore over the time scale of its gut retention period);

2. conventional sediment trapping should be avoided in attempting to measure detrital supply in situations of substantial horizontal or vertically upward particulate fluxes;

3. interdisciplinary co-ordination should be sought in measuring detrital chemistry and supply rates;

4. food quality of detritus should be defined relative to detritivore nutritional requirements, necessitating new studies both of detrital chemistry and of detritivore digestive and assimilatory capacities;

5. nutritional requirements for both major and minor food components require study among detritivores until the limiting nutrients can be specified; and,

6. details of particle selection in detritivores should be elucidated, with particular attention to mechanical versus chemosensory selection and rejection mechanisms and costs.

REFERENCES

Anderson, N.H., 1976, Carnivory by an aquatic detritivore, Clistoronia magnifica (Trichoptera: Limnephilidae), Ecology, 57: 1081.

Levinton, J.S. and Lopez, G.R., 1977, A model of renewable resources and limitations of deposit-feeding benthic populations, Oecologia, 31: 177.

Newell, R.C. and Field, J.G., 1983, The contribution of bacteria and detritus to carbon and nitrogen flow in a benthic community, Mar. Biol. Letters, 4: 23.

Pielou, E.C. 1981, The usefulness of ecological models: a stock-taking, Quart. Rev. Biol., 56: 17.

Seiderer, L.J., Newell, R.C. and Cook, P.A., 1982, Quantitative significance of style enzymes from two marine mussels (Choromytilus meridionalis Krauss and Perna perna Linnaeus) in relation to diet, Mar. Biol. Letters, 3:257.

Sibly, R.M., 1981, Strategies of digestion and defaecation, in: "Physiological Ecology: An Evolutionary Approach to Resource Use," C.R. Townsend and P. Calow, eds., Blackwell Scientific Publications, Oxford.

Stuart, V., Field, J.G. and Newell, R.C., 1982, Evidence for the absorption of kelp detritus by the ribbed mussel, Aulacomya ater (Molina) using a new ^{51}Cr-labelled microsphere technique, Mar. Ecol. Prog. Ser., 9: 263.

Taghon, G.L., 1981, Beyond selection: optimal ingestion rate as a function of food value, Am. Nat., 118: 202.

Taghon, G.L., Self, R.F.L. and Jumars, P.A., 1978, Predicting particle selection by deposit-feeders: a model and its implications, Limnol. Oceanogr., 23: 752.

CARNIVORY

J.B.L. Matthews (Chairman and rapporteur)[1],
A.L. Alldredge[2], R.L. Haedrich[3], J. Paloheimo[4],
L. Saldanha[5], G.D. Sharp[6] and E. Ursin[7]

1. Institute of Marine Biology, University of Bergen
5065 Blomsterdalen, NORWAY

2. Marine Sciences Institute, University of California
Santa Barbara, California 93106, U.S.A.

3. Memorial University of Newfoundland, St John's
Newfoundland, CANADA A1B 3X9

4. Department of Zoology, University of Toronto, Ontario
CANADA M5S 1A1

5. Lab. de Zoologia (Museu Bocage), Faculdade de
Ciencias, Rua Escola Politecnica, 1200 Lisbon
PORTUGAL

6. FAO, Via delle Terme di Caracalla, 00100 Rome, ITALY

7. Danish Institute for Fisheries and Marine Research
Charlottenlund Castle, DK2920 Charlottenlund
DENMARK

DEFINITION

The group chose to define carnivory as:
"The acquisition of animal food resulting in the immediate death of the food organism".

It was considered that the flux of organic material to the benthos is such an important part of many biological systems that benthopelagic and benthic carnivory should not be excluded. On the other hand the definition excludes parasitism and the partial consumption of the food organism which does not result in the death of the prey, e.g. demersal fish "grazing" on the siphons of bivalves or hyperiid amphipods mibbling at gelatinous members of the plankton. Energy and materials which flow along these pathways may be insignificant in the overall budget of an ecosystem, but to particular populations such habits are certainly crucial and may have far-reaching effects, e.g. in converting a food source of high water and inorganic content into a form more palatable for higher carnivores or in returning organic material from the sea bed to the water column.

MODELS OF CARNIVORY

Models including carnivory may have specific uses and be applied for the purposes of management, as in the analysis of the effects of species interactions, fishing predation etc. on yield, or they may be used as a scientific tool in the elucidation of basic problems in biological oceanography. The group discussed models mainly from the latter viewpoint as part of the attempt to identify existing gaps between data supply and modelling needs.

Early models, including the Lotka-Volterra model, embodied not only the functional relationship between the prey abundance and predator's harvest, but the dynamics of both prey and predator relationships. To use the somewhat awkward but commonly used terminology (Solomon, 1949), they were mathematical representations of the numerical relationship between prey and predator populations. There has been some elaboration of the basic Lotka-Volterra models (Schaefer, 1954) since then, and much theoretical work on community dynamics is still based on them (May, 1973).

In more recent work, specifically in simulation models, the functional relationship is often fashioned according to Michaelis-Menten or Holling equations (Holling, 1966a; Paloheimo, 1971), as opposed to the random encounter models with no satiation assumed in the earlier models. Moreover, the growth and sometimes the mortality of both the prey and the predator population are often based on detailed energetic-type models dependent on the available rations and metabolic requirements of the species (Steele and Frost, 1977). More than one prey species are often included in the model with each prey species being assigned a different preference depending on the size, vulnerability etc.

Ursin (1979; in press) has reviewed recent models. Paloheimo (1979) and Andersen (in press) have described the estimation of preference.

The working model shown in Figure 1 was used by the group as the conceptual basis for discussing the fluxes inherent in carnivory and considering the available and foreseeable methods of measuring

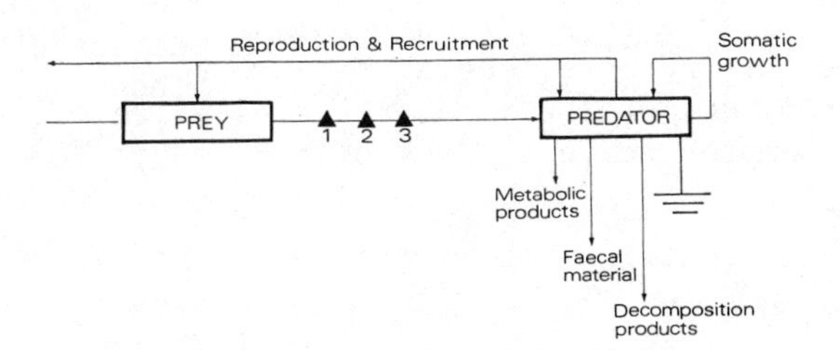

Fig. 1. Conceptual model of fluxes involved in carnivory.

them. The determining factors that affect the flux from prey to predator are:

1. Opportunity, e.g. distribution, abundance, and detectability of prey,

2. Deployment, e.g. mobility of predator, searching strategy, and preference of predator,

3. Requirements, e.g. energetic cost of feeding, defence, maintenance or change of distribution, maximum and minimum rations.

The factors which operate on the main flux in the model are themselves compound determinants. This fact, and our opinion that the trophic level concept is not useful in understanding structured food webs involving carnivores, dictated that the group's discussions developed along reductionist lines.

The model is centred very firmly on a chosen population of carnivores for conceptual reasons. Carnivores in marine ecosystems are a diverse group which obtain their food in a wide variety of ways and which rarely belong to a single trophic level. Perhaps the only common factor is the mobility of their prey. Many carnivores have to expend a great deal of energy in capturing their prey (Sharp and Francis, 1976), energy which is "lost" to the ecosystem, though the food once obtained may be easily assimilable and highly nutritious. The approach taken by the group is similar to that adopted by Holling and co-workers in their attempts to analyze processes such as predation by examining as realistically as possible the basic mechanisms of the processes (Holling and Buckingham, 1976).

The group identified the following rates as relevant in models incorporating carnivory: encounter rate, attack rate, feeding rate, utilization rate, digestion rate, growth rates, metabolic rates. This list includes rates directly involved in carnivorous interaction and more general physiological rates applicable at all trophic levels. They are listed together, not only because a firm distinction between carnivorous and more general rates hardly exists, but also to serve as a reminder of the main processes whose rates need to be measured in comprehensive studies of carnivory.

CLASSIFICATION OF CARNIVORES

The group suggests that a double classification of predators may be useful in further analysis of their role in marine systems:

By Size Category

The size spectrum is useful in the development of ecological theory (Sheldon *et al*., 1972; van Valen, 1973) though apparent differences in the spectral pattern between different communities (Thiel, 1975; Polloni *et al*., 1979) may limit its use in this respect. The group supports its continued application. Linear dimensions are usually applied, very often some expression of body length. Standardization of the dimension is desirable and the group suggests "average diameter" (proportional to the cube root of body weight if it cannot be obtained any other way). The log scale seems to be empirically good for scaling. Grouping is necessary and it is hoped that a theoretical basis for this will develop.

By Functional Grouping

There are a number of ways in which carnivores can be classified according to behavioural aspects of food capture, which are important in the partitioning of energy and the efficiency of populations in the trophic transfer of energy. One of the more obvious schemes contrasts ambushers, such as netters, bait trappers and pouncers, with hunters. The latter group may be divided into migrators and wanderers or alternatively into schoolers and solitaries.

The hunters, scombroid fish for example, can be expected to expend a greater proportion of their energy consumption in obtaining their food than do the ambushers (Ware, 1975; Sharp, this volume). *Pleurobrachia* with its tentacles extended waits passively for its prey while *Sagitta* must make a dart to capture its victim once it has come within sensing distance (Horridge and Boulton, 1967) as does the fish *Synodus* (Saldanha, pers. comm.). On the other hand movement is costly for a gelatinous organism, and a "netter" may need to expend considerable energy and material in repairing body damage, e.g. to tentacles.

Notwithstanding the fact that systems of classification tend to emphasize divisions rather than continuities, realistic modelling of carnivory must take account of behaviour initially in a descriptive manner such as that outlined above, until quantification becomes possible and a theoretical base is developed.

TECHNIQUES FOR STUDYING CARNIVORY

The group made no attempt to review or evaluate the various methods which have been or could be applied to the measurements of "carnivorous processes". Nevertheless, since most reviewers have been concerned with particular classes of method and are only incidentally involved with carnivory, it seems useful to list the various methodological approaches, with a few key or recent references, as an aid in searching for methods suitable for particular problems.

Population Dynamics

With the focus on the population and with good data, estimates can be made of growth, recruitment and mortality. When properly developed, methods of cohort analysis can facilitate the measurement of process rates, such as food requirements and production available to the consumer, from field measurements (Edmondson and Winberg, 1971; Parslow *et al.*, 1979; R. Williams, this volume).

Laboratory and Mesocosm Experiments

The experimental approach tends to emphasize the individual or population level through its emphasis on control and the limitation of manipulable variables. Reality is sacrificed and there always remains a certain imponderable "laboratory factor". Mesocosms are an attempt to find an intermediate "between beakers and bays" (Strickland, 1967), measuring nature under control. Their usefulness and applications can be assessed from a collective presentation of recent results (Grice and Reeve, 1982). Despite the criticisms, experiments remain the only way to measure directly many of the processes inherent in carnivory (Pamatmat, this volume).

Pathway Indicators: Gut Analysis, Immunology, Enzymology, Parasitology

Food webs are complicated in almost any community and ecological realism seems to demand that they be described in sufficient detail for the main pathways to be identified prior to quantification. The most direct method is by gut analysis but problems are manifold: some prey items are quickly digested without trace, masticatory differences between predators may alter the recognizability of the same prey, after capture nothing or everything may be found in the gut (Lancroft and Robison, 1980), and, unless a feeding rhythm can be detected, it is difficult to expand instantaneous observations into expressions of feeding rate without recourse to laboratory investigations. Gut analysis is usually a tedious and unpleasant task.

It may be safer to look for indicators of food pathways within the body of the consumer or in its conditioned response. The enzymes that an organism possesses to digest its food can elucidate its diet and immunological techniques can be used to determine the age at which the enzymes start to be produced, i.e. when a particular diet can be utilized (Doyle and Jamieson, 1978; Alliot et al., 1980; Praët et al., 1980). Biochemical markers may be indicators of the food source (Sargent, 1976; Boon, 1978; Volkman et al., 1980). In some cases parasites with alternate hosts can be proof of a dietary link (Campbell et al., 1980).

Morphological and Physiological Interpretation

Despite all the changing fashions in biological oceanographic research, interpretation of diet is still very largely based on the structures used to obtain the food (Arashkevich, 1969). Organisms can be classified as suspension feeders, deposit feeders and raptors. It is important to realize that this classification alone does not identify the carnivores in a system, so that further evaluation is necessary: a small raptor may be grabbing relatively large phytoplankton cells; a filter-feeding baleen whale is a carnivore. Care must be exercised in making such interpretations.

The physiological, nutritional or biometric condition of an organism can be a measure of the food supply available to it. Changes in condition seen in relation to consumption of food can be used by means of regression methods to estimate the efficiency of food utilization. Such methods are most useful when the estimator of "condition" is easily measured, e.g. body length/weight or depth/length ratios or fat content (Sharp and Dotson, 1977; Dotson, 1978). The more precise measurements, e.g. C/N ratio, tend to be destructive and therefore useless in experiments on individuals.

It is relevant also to mention papers by Childress et al. (1980) and by Ware (1980) under this heading.

Behavioural Studies

Modern techniques in the laboratory (Koehl and Strickler, 1981) and in the field (Grassle et al., 1975; Hamner et al., 1975;

Alldredge, this volume) have added new power to the approach of the observational biologist. A mass of data is accumulating to tell us more about what eats what, and in some studies it is possible to quantify predation studied in this way, which has always been a mainstay of terrestrial ecology.

HOLISTIC APPROACHES

In addition to these methods, which are directed towards better understanding of populations and their interactions, there are some promising holistic methods.

Community Size Distribution

On the principle that in the sea the prey is usually smaller than the predator and the theory that there is an optimal size difference, the size spectrum of a community can give an indication of its trophic structure (Silvert and Platt, 1980; cf. Wilson, 1973). There is some empirical information to support this assertion (Parsons and LeBrasseur, 1970; Haedrich *et al.*, 1980). Theory has been further developed by Platt and Denman (1977) to make it possible to calculate the flux rate along an observed size spectrum. This approach is doubtless the most holistic at present and has been greatly advanced - promoted in fact - by the development of particle-counting techniques (Mackas and Boyd, 1979; Mackas *et al.*, 1981). At the larger end of the scale, where particle-counting techniques so far are inapplicable, there seems to be so much irregularity, even in whole communities, that the group has reservations about its general application. The information conveyed by a particle size spectrum would be greatly enhanced if the particles could be classified according to their functional type.

P/B Ratios

Metabolic relationships with body size have long been suggested. On the basic reasoning that smaller organisms have a shorter generation time than larger ones an inverse relationship has been postulated to exist between average body size and production. Banse and Mosher (1978) have reviewed the evidence for this at the population level and have found support for the idea with some limitation on the organisms and environments to which the rule should be applied. P. Williams (this volume) has related metabolic activity to surface area. Although there is evidence that the small juveniles of a big species are biologically very different from the adults of a small species, this approach may be useful in holistic studies.

Community Processes

Excretion is a tell-tale sign of community activity which it seems possible to measure at a fairly broad level (see report of Working Group on herbivory, this volume). Herbivores being supposedly the only organisms with plant pigments in their guts (a statement repeated here with some reservation), estimates of total herbivore production by means of phaeopigments can be subtracted from total zooplankton production to obtain carnivorous production.

Tracer Elements, Bioaccumulation

Caesium, vanadium, and polonium are among elements which have been postulated, not just as indicators of the origins of food eaten but also as indicators of the organism's position on the trophic ladder - the phenomenon of bioaccumulation well known from pollution studies (Mearns *et al.*, 1981; Fowler, 1982). After an initial selection against ^{13}C in favour of ^{12}C in photosynthetic fixation there seems to be a selective retention of the heavier isotope at each subsequent trophic step. The stable isotope ratio is thus an indicator of trophic level (Mills *et al.*, in press).

This list gives promise of unifying methods becoming available but the group is convinced that they can only be of practical use if there can be sound reasons for considering and treating the carnivores in a comprehensive manner. Two theoretical possibilities suggest themselves in this context:

1. Where there is information on dietary components it may be possible to apply indices of average trophic level (Ulanowicz, this volume).

2. A useful conceptual simplification may be that of Isaacs' unstructured food web (Isaacs 1973; 1977), but applied purely to the carnivores.

Recent studies, however, indicate the need for caution in the application of generalities (Mearns *et al.*, 1981).

SUMMARY

The working group offers the following conclusions and recommendations:

1. We should like to emphasize that the usefulness of obtaining biomass data still remains. There are important components of marine ecosystems for which biomass data are still very scanty. Properly analyzed, with information on size distribution,

species composition or functional grouping, biomass data will continue to be essential information in ecosystem and management studies. Oceanic sampling, especially for large organisms, is a long way from being perfect. We recognize the need for increased effort in the field of invertebrate predation in the open ocean and of carnivorous transfer between the pelagic and benthic regions.

2. We envisage only limited possibilities of developing routine methods for measuring carnivorous processes, at least until there is better ecological and biological understanding of systems. Predation studies are an essential part of ecosystem studies and should be so conducted as to contribute to a gradual accumulation of information on food webs and trophic transfer of energy and material.

3. The biological complexity of carnivory necessitates closer integration of ecological research with other fields, notably physiology and ethology.

4. There are compelling biological reasons for the development of community models from the point of view of the populations that together make up the community. It may be that there is a hierarchical structure in which ecosystems possess properties unique to that level of organization, but by and large populations are "unaware" of them and are not there primarily as links to effect the transfer of energy and materials to the top carnivore or on to the decomposers.

5. We find a surprising amount of terminological confusion in the ecological literature, often of words like metabolism which in the appropriate specialist branch of biology have precise meanings. We strongly recommend that sloppy and imprecise use of scientific terms be strictly eliminated.

Finally, we quote from a paper by Holling (1966b) and ask whether this is not still a relevant exercise in marine ecology in the 1980's: "It would therefore be an entertaining and potentially profitable undertaking to analyze the various processes that affect animal numbers - e.g. predation, competition and parasitism - by initially ignoring the limitations imposed by the restrictions of traditional mathematical models and emphasizing the need for a realistic explanation. Once the analysis is complete it might then be possible to express the explanation in a precise mathematical form without sacrificing reality. In this way the process itself would dictate the form of the model rather than some arbitararily chosen mathematical language".

REFERENCES

Alldredge, A.L., The quantitative significance of gelatinous zooplankton as planktonic indicators, This volume.

Alliot, E., Pastoureaud, A., and Trellu, J., 1980, Evolution des activités enzymatiques dans le tractus digestif au cours de la vie larvaire de la Sole. Variations des proteinogrammes et des zymogrammes, Biochem. Syst. Ecol., 8: 441.

Andersen, K.P., In press, An interpretation of the stomach contents of fish in relation to prey abundance, Dana, 2.

Arashkevich, Ye. G., 1969, The food and feeding of copepods in the north-western Pacific, Oceanology, 9: 695.

Banse, K., and Mosher, S., 1980, Adult body mass and annual production/biomass relationships of field populations, Ecol. Monogr., 50: 355.

Boon, J., 1978, "Molecular biogeochemistry of lipids in four natural environments," Delft Univ. Press, Delft.

Campbell, R.A., Haedrich, R.L., and Munroe, T.A., 1980, Parasitism and ecological relationships among deep-sea benthic fishes, Marine Biology, 57: 301.

Childress, J.J., Taylor, S.M., Cailliet, G.M. and Price, M.H., 1980, Patterns of growth, energy utilization, and reproduction in some meso- and bathypelagic fishes off southern California, Mar. Biol., 61: 27.

Dotson, R.C., 1978, Fat deposition and utilization in albacore, in: G.D. Sharp and A.E. Dizon, eds., "The Physiological Ecology of Tunas," Academic Press, New York.

Doyle, C.M., and Jamieson, J.D., 1978, Development of secretagogue response in rat pancreatic acinar cells, Developmental Biol., 65: 11.

Edmondson, W.T., and Winberg, G.G., 1971, A manual on methods for assessment of secondary productivity in fresh waters, IBP Handbook No. 17, Blackwell, Oxford.

Fowler, S.W., 1982, Biological transfer and transport processes. in: G. Kullenberg, ed. "Pollutant Transfer and Transport in the Sea," C.R.C. Press, Cleveland, U.S.A.

Grassle, J.F., Sanders, H.L., Hessler, R.R., Rowe, G.T., and McLellan, T., 1975, Pattern and zonation: a study of the bathyal megafauna using the research submersible Alvin, Deep-Sea Res., 22: 457.

Grice, G.D., and Reeve, M.R., eds., 1982, "Marine Mesocosms," Springer-Verlag, New York.

Haedrich, R.L., Rowe, G.T., and Polloni, P.T., 1980, The megabenthic fauna in the deep sea south of New England, Mar. Biol., 57: 165.

Hamner, W.M., Madin, L.P., Alldredge, A.L., Gilmer, R.W., and Hamner, P.P., 1975, Underwater observations of gelatinous zooplankton: sampling problems, feeding biology, and behaviour, Limnol. Oceanogr., 20: 907.

Holling, C.S., 1966a, The strategy of building models of complex ecological systems. in: K.E.F. Watt, ed. "Systems Analysis in Ecology," Academic Press, New York.

Holling, C.S., 1966b, The functional response of invertebrate predators to prey density, Mem. entomol. Soc. Canada, No. 48, 5.

Holling, C.S., and Buckingham, S., 1976, A behavioural model of predator-prey functional responses, Behavioral Science, 21: 183.

Horridge, G.A., and Boulton, P.S., 1967, Prey detection by Chaetognatha via a vibrational sense, Proc. Roy. Soc. London. Ser. B, 168: 413.

Isaacs, J.D., 1973, Potential trophic biomasses and trace substance concentrations in unstructured marine food webs, Marine Biology, 22: 97.

Isaacs, J.D., 1977, The life of the open sea, Nature London, 267: 778.

Koehl, M.A.R., and Strickler, J.R., 1981, Copepod feeding currents: food capture at low Reynolds number, Limnol. Oceanogr. 26: 1062.

Lancroft, T.M., and Robison, B.N., 1980, Evidence of post-capture ingestion by midwater fishes in trawl nets, Fish. Bull. nat. mar. Fish. Serv., 77: 713.

Mackas, D.L., and Boyd, C.M., 1979, Spectral analysis of zooplankton spatial heterogeneity, Science, 204: 62.

Mackas, D.L., Curran, T.A., and Sloan, D., 1981, An electronic zooplankton counting and sizing system, Oceans Magazine, 12: 783.

May, R.M., 1973, "Stability and complexity in model ecosystems," Princeton University Press, Princeton.

Mearns, A.J., Young, D.R., Olson, R.J. and Schafer, H.A., 1981, Trophic structure and the cesium-potassium ratio in pelagic ecosystems, CalCOFI Rep., 22: 99.

Mills, E.L., Pittman, K., and Tan, F.C., In press, Food-web structure on the Scotian shelf, Eastern Canada. A study using ^{13}C as a food-chain tracer, Rapp. R.-v. Réun. Cons. int. Explor. Mer.

Paloheimo, J.E., 1971, On the theory of search, Biometrika, 58: 62.

Paloheimo, J.E., 1979, Indices of food type preference by a predator, J. Fish. Res. Bd Can., 36: 470.

Pamatmat, M., Measuring the metabolism of the benthic ecosystem, This volume.

Parslow, J., Sonntag, N.C., and Matthews, J.B.L., 1979, Technique of systems identification applied to estimating copepod population parameters, J. Plankton Res., 1: 137.

Parsons, T.R., and LeBrasseur, R.J., 1970, The availability of food to different trophic levels in the marine food chain, in: "Marine Food Chains," J.H. Steele, ed., Oliver and Boyd, Edinburgh.

Platt, T., and Denman, K., 1977, Organization in the pelagic ecosystem, Helgol. wiss. Meeresunters., 30: 575.

Polloni, P., Haedrich, R.L., Rowe, G.T., and Clifford, C.H., 1979, The size-depth relationship in deep ocean animals, Int. Revue ges. Hydrobiol., 64: 39.

Praët, M. van, and Geistdoerfer, P., 1980, Etudes des zymogrammes des tissus digestifs des poissons et invertébrés abyssaux, Comptes Rendus Acad. Sci., Paris, 290 (D): 1083.

Sargent, J.R., 1976, Marine wax esters, Sci. Progr. Oxford, 65: 637.

Schaefer, M.B., 1954, Some aspects of the dynamics of populations important to the management of commercial marine fisheries, Bull. inter-Amer. trop. Tuna Comm., 1: 26.
Sharp, G.D., Ecological efficiency and activity metabolism, This volume.
Sharp, G.D., and Dotson, R.C., 1977, Energy for migration in albacore, Thunnus alalunga, Fish. Bull. nat. mar. Fish. Serv., 75: 447.
Sharp, G.D., and Francis, R.C., 1976, An energetics model for the exploited yellowfin tuna, Thunnus albacares, population in the eastern Pacific Ocean, Fish. Bull. nat. mar. Fish. Serv., 74: 36.
Sheldon, R.W., Prakash, A., and Sutcliffe, W.H., 1972, The size distribution of particles in the oceans, Limnol. Oceanogr., 17: 327.
Silvert, W., and Platt, T., 1980, Dynamic energy-flow model of the particle size distribution in pelagic ecosystems, in: "Evolution and Ecology of Zooplankton Communities", W.C. Kerfoot, ed., University Press of New England, Hanover, New Hampshire.
Solomon, M.E., 1949, The natural control of animal populations, J. Animal Ecol., 18: 1.
Steele, J.H., and Frost, B.W., 1977, The structure of plankton communities, Phil. Trans. Roy. Soc. London, B 280: 485.
Strickland, J.D.H., 1967, Between beakers and bays. New Scientist, London, 33: 276.
Thiel, H., 1975, The size structure of the deep-sea benthos, Int. Revue ges. Hydrobiol., 60: 575.
Ulanowicz, R.E., Community measures of marine food webs and their possible applications, This volume.
Ursin, E., 1979, Population dynamics and fish behaviour, Invest. Pesquera, 43: 171.
Ursin, E., In press, Multispecies fish stock and yield assessment in I.C.E.S., Can. spec. Publ. Fish. aquat. Sci.
Van Valen, L., 1973, Body size and numbers of plants and animals, Evolution, 27: 27.
Volkman, J.K., Corner, E.D.S., and Eglinton, G., 1980, Transformations of biolipids in the marine food web and in the underlying bottom sediments, Colloque international de C.N.R.S., Marseille, 1979, 185.
Ware, D.M., 1975, Growth, metabolism, and optimal swimming speed of a pelagic fish, J. Fish. Res. Bd Can., 32: 33.
Ware, D.M., 1980, Bioenergetics of stock and recruitment, Can. J. Fish. aquat. Sci., 37: 1012.
Williams, P. Le B., Bacterial production in the marine food chain; the emperor's new suit of clothes?, This volume.
Williams, R.C., An overview of secondary production in pelagic ecosystems, This volume.
Wilson, D.S., 1973, Food size selection among copepods, Ecology, 54: 909.

THE ROLE OF FREE BACTERIA AND BACTIVORY

J.S. Gray[1], J.G. Field (rapporteur)[2], F. Azam[3],
T. Fenchel[4], L.-A. Meyer-Reil[5], F. Thingstad[6]

1. Institutt for Marinbiologi og Limnologi
Universitetet i Oslo, Postboks 1064 Blindern, Oslo 3
NORWAY

2. Zoology Department, University of Cape Town
Rondebosch 7700, SOUTH AFRICA

3. Institute of Marine Resources, Scripps Institution
of Oceanography, La Jolla, California 92093, U.S.A.

4. Institute of Ecology and Genetics, University of
Aarhus, DK 8000 Aarhus-C, DENMARK

5. Institut für Meereskunde, Universität Kiel
Dusternbrooker Weg 20, D2300 Kiel 1, F.D.R.

6. Institute for Microbiology, University of Bergen
Bergen, NORWAY

INTRODUCTION

It is now recognized that bacteria and microheterotrophs play an important part in marine ecosystems (see Hobbie et al., 1972; Sieburth et al., 1977; Sieburth, 1979, 1983; Sorokin, 1979, 1981). This chapter is concerned with bacteria that live freely in the water column or on sand grains, as opposed to those on detritus particles, because detritus and detritivory are covered in another chapter of this volume. For the purposes of this chapter, bactivory is simply defined as feeding on such free bacteria.

There is a general relationship between size of prey and predator which implies that optimal bactivores should not be much larger than bacteria (Fenchel, this volume). Some macrofauna are able to utilize bacteria but with the exception of sponges (Reiswig, 1974, 1975) it is doubtful that they can rely solely on bacteria as a food source. This report therefore concentrates on microflagellates as the most important marine bactivores (Fenchel, 1982d) and is not confined to bactivory but also considers bacterial population dynamics and its role in marine ecological systems. Consideration will be given to the following aspects of bacteria and bactivory: in the water column, in the sediments, and in symbiotic relationships.

METHODS

Estimating Standing Stock:

The traditional methods of enumerating marine bacteria were based on plate counts, serial dilutions, or phase contrast microscopy. However, such methods gave estimates of at best 10% of actual numbers and have generally been discarded for estimating bacterial biomass. Two general approaches are presently in use, one microscopic and the other chemical.

A variety of microscopic staining methods ahve been used, the most promising are: Acridine orange direct counts (A.O.D.C.) (Francisco *et al.*, 1973; Hobbie, Daley and Jasper, 1977), DAPI (Porter and Feig, 1981), and Hoechst 33258 (Paul, 1982). The above, when combined with volume estimates by scanning and transmission electron microscopy (SEM and TEM) which need correction for shrinkage and geometry (Watson *et al.*, 1977; Fuhrman, 1981), give the best current estimates of standing stock. These techniques are being improved by new developments such as the use of image analyzers (Krambeck *et al.*, 1981).

The following chemical methods of estimating standing stock have been used: Muramic acid (Moriarty, 1977; Fazio *et al.*, 1978), ATP (see reviews by Sorokin and Lyutsarev, 1978; Karl, 1980), and lipo-polysaccharides (LPS) (Watson *et al.*, 1977). These indirect chemical methods are not recommended for estimating standing stocks of natural populations of bacteria because of variable conversion factors between the cell component measured and bacterial biomass.

Bacterial Production

Bacterial production needs careful definition to be compatible with other trophic processes: we define it as the increase in biomass of bacteria, whether or not all the biomass survives throughout the period of measurement. All the methods for estimating bacterial production have been introduced recently and are therefore in need of further investigation and of intercalibration.

The following criteria may be applied to assess the suitability of techniques:

1. the method should be specific to bacteria

2. the conversion factor should not be dependent on the growth rate

3. experimental manipulation should not change the growth rate

4. the method should be sensitive enough to allow short incubations (hours).

At this stage we recommend the following techniques:

1. frequency of dividing cells (FDC) (Hagström et al., 1979)

2. tritiated thymidine incorporation (TTI) (Fuhrman and Azam, 1980, 1982).

Both of the above have drawbacks, FDC is very tedious to use and difficult to calibrate satisfactorily, while the TTI leads to underestimates because of the conservative nature of the assumptions. To relate production to the active fraction of bacterial assemblages, tritiated thymidine auto-radiography mey be used in combination with epifluorescence microscopy (Fuhrman and Azam, 1980, 1982).

We recommend that before different techniques are more widely applied in different geographic areas, an inter-calibration workshop be held.

Bactivory

The experimental methods for estimating rates of feeding on a free bacteria are dealt with fully by Fenchel (1982a, b, c, d; this volume) and are not repeated here.

PROPERTIES OF MARINE BACTERIAL POPULATIONS

Standing Stocks

There is little information on carbon:phosphorous:nitrogen (C:P:N) ratios in natural bacterial populations, although cultures

Table 1. Bacterial standing stock estimates based on acridine orange direct counts and electron microscopy, in different environments, assuming a conversion factor of 10% from live mass to carbon equivalent (summarized from Meyer-Reil, 1982; Van Es and Meyer-Reil, 1982).

Environment	Numbers	Standing stock
	($x10^8$ l^{-1})	(μg C l^{-1})
Estuaries	50	200
Coastal	10 - 50	5 - 200
Offshore	0.5 - 10	1 -5
Deep Waters	0.1	?
	(cells g^{-1}dry weight)	(μg C g^{-1}dry wt)
Sediments	10^8 - 10^{11}	1 - 1000

obtained under varying nutrient conditions suggest C:N ratios of 3.5-4:1. Indeed, the biochemical composition generally is poorly known. This is an important gap in calculating nutrient cycles.

Table I gives standing stock estimates in different environments based A.O.D.C. and electron microscopy, summarised from Van Es and Meyer-Reil (1982) and Meyer-Reil (1983). Generally, bacterial numbers in the water column and sediments are of the order 10^9 cells l^{-1}, and 10^9 cells g^{-1} sediment, respectively. Bacterial standing stock appears to increase in proportion to primary production (Linley et al., 1983).

Production Rates

Because all methods have been developed recently, only a limited number of estimates are presently available. These suggest for coastal waters 2-250 μg C l^{-1}day^{-1}, with bacterioplankton production being 5-30% of primary production for both coastal and offshore waters (summarized in the review of Van Es and Meyer-Reil, 1982). It has been generally believed that a large proportion of bacteria in the sea are dormant. Recent results by Fuhrman and Azam (1982) suggest that at least 50% of bacteria in Californian waters are active.

Carbon Conversion Efficiencies

Pure culture experiments have been reviewed by Koch (1971), Calow (1977), Payne and Wiebe (1978), and Williams (this volume), while microcosm experiments are reviewed by Newell in this volume.

The values for carbon conversion efficiency range from below 10% in microcosms containing non-enriched natural sea water which may be limited by nitrogen and/or phosphorus, to 60% or even 80% in N- or P- rich situations. It therefore appears that carbon conversion efficiency by bacteria may be positively related to nitrogen (or other nutrient) availability.

Primary production and bacterial data from off the Californian coast, combined with an assumed conversion efficiency of 50% suggest that the bacterioplankton consume 10-50% of the total fixed carbon (Fuhrman and Azam, 1982).

Respiration

All the evidence suggests that respiration of bacterial populations is linearly related to growth rate, the exact relationship depending upon the carbon conversion efficiency. The proportion of carbon respired is (1 - proportion of carbon assimilated), assuming that a negligible amount of carbon is excreted. Thus present estimates range from 40-90% of absorbed carbon being respired in the water column, but 5-10% has been reported for sediments (Meyer-Reil *et al.*, 1978). However, the latter figures need re-evaluation in the light of recent findings (King and Klug, 1982).

Excretion

There is little data on bacterial excretion in the sea, but some qualitative evidence suggests that molecules of small size may be excreted (Iturriaga and Zsolnay, 1981; Novitsky and Kepkay, 1981). In the sediments anaerobic metabolism is common and is associated with communities of bacteria of different physiological types, giving off and utilizing different chemical excretions in a complementary manner (Fenchel and Blackburn, 1979).

THE PROCESS OF BACTIVORY : FUNCTIONAL RESPONSES

Bactivory in the Water Column

The two factors must important to bactivory are the particle size and density of bacteria in the sea. Physical and physiological constraints favour small organisms as bactivores, because of their large surface to volume ratio which increases the probability of contact with small particles (see Fenchel, this volume for details). The small organisms which are probably most important in this regard are microflagellates (Fenchel, 1982d). An important exception to this is bactivory by Oikopleura which have giant filters (Flood, 1978; King *et al*., 1980) and which Fenchel (this volume) therefore refers to as the baleen whales of the microbial plankton. There is evidence that free-living bacteria in the water column can be utilized to some extent by certain larger animals such as sponges (Reiswig, 1974, 1975), salps, and bivalves (Jorgensen, 1966; Stuart *et al*., 1982; Wright *et al*., 1982), although bacteria are at the lower size limit for their efficient utilization on an extensive scale by macrofauna many orders of magnitude larger than the bacteria themselves. While bacteria may be an important source of vital food substances for some of these animals, it is doubtful whether macrofauna play a significant role in controlling bacterial populations on a large scale.

The lower limit of bacterial density for bactivory is hard to quantify experimentally because the very slow balanced growth of flagellates is difficult to study. Flagellate populations can probably be sustained by bacterial populations of about 2-3 x 10^5 cells per ml. Bacterial densities in the sea are probably controlled by phagocytosis, and predator-prey oscillations have been observed with bacterial densities reaching about 10 million cells per ml. (Fenchel, 1982d).

In Sediments

Ciliates are important bactivores in sediments where bacteria are denser than in the water column and show similar functional responses based on size and density (Fenchel, 1980). Meiofauna include bacteria on their diet and also feed on flagellates when there is capillarity. The bacteria are obtained mainly by epistrate feeding but there is little quantitative information on epistrate feeding. An unusual aspect of bactivory in sediments is the phenomenon of bacteria being gardened by nematodes (Riemann and Schrage, 1978).

SCALES IN TIME AND SPACE

Spatial Scales

There is little information on the macro spatial scale, but Fuhrman and Azam (1980) have shown that off California on a scale of tens of kilometres, there is a good correlation between chlorophyll concentration and bacterial abundance and growth. On the micro-scale, natural selection appears to favour motility amongst water-column bacteria (Koop, 1982), as opposed to those in sediments, although there is little quantitative information. While motile bacteria can move to favourable locations for obtaining dissolved organic matter (DOM), by doing so they may accumulate in densities sufficient to favour bactivory. There is evidence that bacteria shows kinesis in a field of 10-100 μm from algal cells, close enough to take advantage of DOM (Azam and Ammerman, this volume). Under laboratory conditions, bacteria were observed to remain at distances of the order of 10 μm from algal cells, possibly being repelled by antibiotics produced by algae. They attach mainly to moribund algae.

Wangersky (1977) speculated that the bacteria that attached to particles are likely to sink faster than free (motile) ones. Particle-associated bacteria may thus tend to sink through the thermocline (Kirchman and Mitchell, 1982). Fuhrman and Azam (1982) found that free and attached bacteria above the thermocline were adapted to warm temperatures by showing optimal activity at 17°C. However, bacteria obtained from 400 m depth varied in their thermal optima. The free bacteria showed optimal activity at 7°C, implying long residence at depth whilst bacteria attached to particles showed optimal activity at 17°C, like the surface bacteria, suggesting a more recent origin from above the thermocline.

In sediments bacteria have been found at depths down to 11 m (Meyer-Reil, unpublished data). On a scale of microns, there are variations in the composition and density of bacteria on sand grains (Weise and Rheinheimer, 1978), which influence the behaviour of epistrate bactivores (Gray, 1966).

Time Scales

On a seasonal scale we require information on patterns of DOM release, but bacteria do show seasonal patterns of abundance. presumably in response to DOM released (Meyer-Reil _et al._, 1979, Larsson and Hagström 1979, 1982) and one infers that bactivory in turn varies seasonally. On the time scale of a phytoplankton bloom (days), it appears that DOM is released by certain species of phytoplankton as nutrients are depleted at the end of the bloom. The fraction of labile photosynthate released is probably in inverse proportion to the concentration of the limiting nutrient, as shown for nitrogen in Figure 1 (Joiris _et al._, 1982; Sakshaug, pers. comm.). DOM release (in the form of glucan) is also evident on a diel scale, being synthesized and excess released during daylight, and being utilised for protein synthesis that continues through the dark hours (Barlow, 1982; Sakshaug pers. comm.). These cycles in DOM release obviously have consequences for bacterial growth.

Further evidence of inter-related diel rhythms is that the specific growth rate of water column bacteria increases some 30-40% during daylight and decreases at night (Sieburth _et al._, 1977; Azam, 1983,). This is supported by free amino-acid and DOC

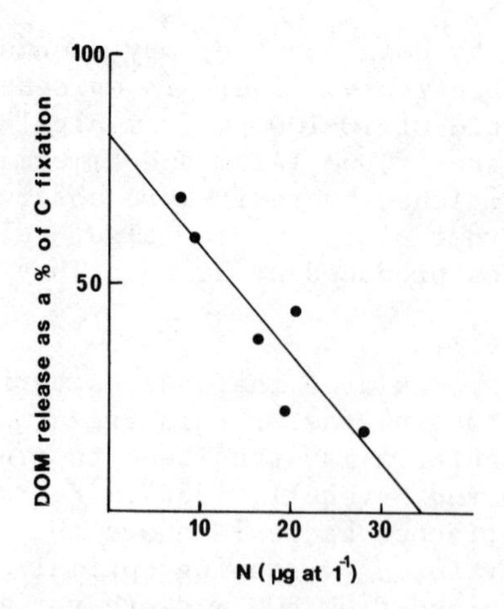

Fig. 1. Release of dissolved organic carbon expressed as a percentage of carbon fixed in photosynthesis, plotted against nitrogen concentration (from Joiris et al., 1982). Similar relationships have been found for several phytoplankton species, with varying slopes. (Reproduced from Joiris et al, 1982, by permission of Netherlands Institute for Sea Research, Texel).

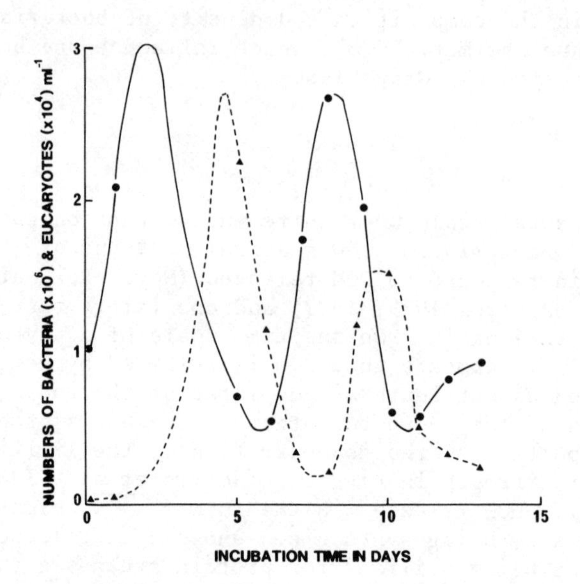

Fig. 2. Oscillations in the density of bacteria——and small (3-10 μm) eucaryotic----organisms after the addition of crude oil to a 10 l sample of natural seawater. The population of eucaryotic organisms was totally dominated by small flagellates. Incubation at 15°C in darkness (Thingstad, personal communication).

measurements showing a similar diel pattern in both water column and sediments (Meyer-Reil _et al_., 1978). Experimental evidence shows that bactivorous flagellates show cycles of the order of 1-2 weeks (Fenchel, this volume; Newell, this volume). Figure 2 shows the results of laboratory experiments in which crude oil provided an artificial and increased carbon source for bacteria, carbon which, under natural conditions, is produced by the release of DOM from living and moribund phytoplankton cells. This evidence, together with mesocosm experiments conducted in 20 m high columns, 1 m in diameter in the Lindaspollen in Norway, showed that as the phytoplankton blooms declined, bacterial populations built up, and these declined when flagellates became abundant (Thingstad, unpublished data). Thus all the evidence to date suggests a remarkably similar pattern, with flagellate bactivory controlling water column bacteria in predator-prey oscillations which lag some 3-4 days behind one another.

THE ROLE OF BACTIVORY IN MARINE FOOD WEBS

There is little quantitative information on the flows of energy and materials in the sediments via bactivory. However, in coastal waters it is likely that bacterial processes in the sediments are related to those in the water column above (Smetacek, this volume). Standing stocks of bacteria in the sediments have been estimated (Meyer-Reil _et al_., 1978; Meyer-Reil, 1982) and oxygen uptake and heat production have been measured in situ and in microcosms to estimate community metabolism (Pamatmat _et al_., 1981; Hargrave and Phillips, 1981). A combination of techniques is needed to estimate rates of bactivory in the sediments. A promising approach, which has not yet been applied to sediments, would be to estimate bacterial growth using techniques such as the frequency of dividing cells or tritiated thymidine incorporation, coupled with experimental estimates of predation rates on bacteria by direct counts of bacteria and of predator abundance, as done on water column microbes by Fenchel (1982b, d; this volume). In anaerobic sediments bacterial numbers are also large and generally decrease with depth. Bactivory probably plays a minor role in controlling anaerobic bacteria since few eukaryotes do not have mitochondria and can tolerate these conditions. Ciliates are the main anaerobic bactivores (Fenchel and Blackburn, 1979). The factors which limit the abundance of anaerobic bacteria are poorly understood.

Symbiotic bacteria have recently come into prominence again with the discovery of chemo-synthesis in the symbionts of Vestimentifera and bivalves of deep-sea hot vents. This phenomenon is probably widespread at all aerobic-anaerobic interfaces. The rates of material and energy transfer between symbiont and host have not been quantified.

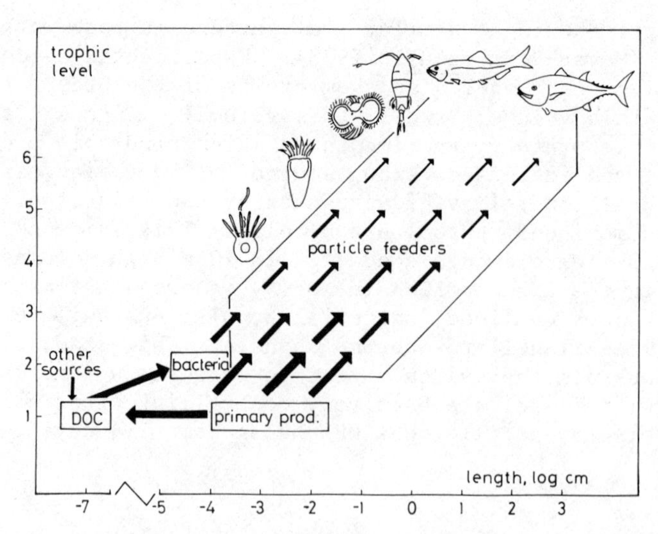

Fig. 3. A semi-qualitative model of planktonic food chains, showing the "microbial loop" of dissolved organic carbon (DOC) bacteria, and heterotrophic flagellates, ciliates and other microzooplankton. It is here assumed that some 25% of primary production is channelled through D.O.C. to the microbial loop. It is also assumed that the most efficient size ratio of predator to prey is 10:1, hence the slope of the arrows relating trophic level to log body length is 1:1. The food chain base represents a size rang 3 orders of magnitude (smallest bacteria~0.2 μm, largest diatoms~200 μm), therefore any trophic level will span a size range of 10^3 and conversely each size class of organisms (>~10^{-2}cm or 100 μm) will represent at least three trophic levels.

In the water column, bacteria utilise DOM which is mainly of phtoplankton origin since it has been estimated that 5-50% of carbon fixed is released as DOM (see Larsson and Hagström, 1979), and there is a close correlation between bacteria and primary production (Fuhrman and Azam, 1982; Van Es and Meyer-Reil, 1982; Linley _et al._, 1983). The chemical composition of DOM needs further investigation. The difficulties experienced in quantifying DOM release are almost certainly caused by rapid uptake of its labile components by bacteria.

Figure 3 presents a modification of the Sheldon _et al_. (1972) particle size model of a water column food chain to illustrate the role of bacteria and other microbes in the water column. DOC released by phytoplankton, and to a much smaller extent by animals, is utilised by bacteria. When sufficient DOC is present for their growth, bacteria are kept below a density of about 10^7 cells per ml, primarily by heterotrophic flagellates which have been found to reach densities of some 3000 per ml (Fenchel, this volume). Flagellates in turn are preyed upon by other micro-zooplankton so that energy and materials diverted from phytoplankton through bacteria are returned as particles accessible to the classical pelagic food chain. We term this the "microbial loop". The food chain thus includes heterotrophic- and cyano-bacteria which are both preyed upon by heterotrophic flagellates. Heterotrophic and autotrophic flagellates are included in the diet of slightly larger microzooplankton which in turn (with larger phytoplankton) form the diet of macrozooplankton.

An important consequence of the "microbial loop" described above, stems from the ability of bacteria to absorb mineral nutrients from the sea. Their small size and large surface-to-volume ratio allow absorption of nutrients at very low concentrations, giving them a competitive advantage over larger phytoplankton cells as documented in Cepex bag and laboratory experiments (Vaccaro *et al*., 1977; Parsons *et al*., 1981). Bactivory, by its strong control of bacterial density, may play an important role in influencing the result of competition for nutrients between phytoplankton and bacteria (Thingstad, unpublished data; Azam, 1982). This needs further study. Bacteria sequester minerals efficiently and are consumed by flagellates with a carbon assimilaton efficiency of some 60% (Fenchel, 1982b), the remaining 40% of carbon being egested as feces. Furthermore, heterotrophic flagellates and other microzooplankton excrete nitrogen and since their C/N ratios are similar to those of their food (3.5 - 4.0, Fenchel and Blackburn, 1979), an amount of nitrogen must also be excreted, corresponding to some 25% of respired carbon. Although bacteria also excrete minerals and respire carbon, they compete efficiently to regain mineral nutrients and the net mineral release will depend upon the supply of carbon and minerals. When DOC is available and mineral supply limited, it therefore remains for the heterotrophic flagellates and other microzooplankton to play the major role in remineralization in the sea. This is contrary to the classical view of bacteria as remineralizers (see also Mann, 1982).

The role of the "microbial loop", including microzooplankton, is likely to be particularly important in rapid recycling of nutrients above the thermocline because the free bacteria, flagellates and other microzooplankton sink slowly and are less likely to fall through the thermocline than the larger phytoplankton cells and bacteria attached to particles. The evidence of Fuhrman and Azam (1982), cited above, that attached bacteria at depths well below the thermocline are not thermally adapted to low temperatures, supports this hypothesis.

SUMMARY AND RECOMMENDATIONS

1. Bacterial biomass is a significant fraction of the total biomass in the sea.

2. Bacterial production accounts for 10-50% of photosynthetically fixed carbon.

3. The "microbial loop" from bacteria-flagellates-microzooplankton is quantitatively comparable to the phytoplankton-herbibore route in the flux of energy and materials.

4. The "microbial loop" is important for rapid, large scale remineralization in the mixed layer.

5. The "microbial loop" is closely coupled to phytoplankton production, and the rate of bactivory by flagellates may control the outcome of competition for limiting nutrients between phytoplankton and bacteria.

Future research directions should include field work to get more estimates of standing stock and production, and experimental micro- and mesocosms to investigate the rates of processes mentioned in 3,4 and 5 above. Laboratory experiments on cultures will also be required to investigate rates of bactivory and physiological processes. Mathematical process models should be an integral part of this research so that there is continual iteration between conceptual development and gathering appropriate data.

REFERENCES

Azam, F., 1983, Measurement of growth of bacteria in the sea and the regulation of growth by environmental conditions, In: "Heterotrophy in the Sea," J. Hobbie and P.J. Le B Williams, eds., Plenum Press, New York.

Azam, F., and Ammerman, J.W., Cycling of organic matter by bacterioplankton in pelagic marine ecosystem : microenvironmental considerations, This volume.

Azam, F., and Hodson, R.E., 1977, Size distribution and activity of marine microheterotrophs, Limnol. Oceanogr., 22: 492.

Barlow, R.G., 1982, Phytoplankton ecology in the southern Benguela Current: 2. Carbon assimilation patterns, J. exp. mar. Biol. Ecol., 63: 229.

Calow, P., 1977, Conversion efficiencies in heterotrophic organisms, Biol. Rev. 52: 385.

Fazio, S.D., Mayberry, W.R. and White, D.C., 1978, Muramic acid assay in sediments, Mar. Biol., 48: 185.

Fenchel, T., 1980, Suspension feeding in ciliated protozoa: feeding rates and their ecological significance, Microb. Ecol., 6: 13.

Fenchel, T., 1982a, Ecology of heterotrophic microflagellates. I. Some important forms and their functional morphology, Mar. Ecol. Prog. Ser., 8: 211.

Fenchel, T., 1982b, Ecology of heterotrophic microflagellates, II. Bio-energetics and growth, Mar. Ecol. Prog. Ser., 8: 225.

Fenchel, T., 1982c, Ecology of heterotrophic microflagellates, III. Adaptations to heterogeneous environments, Mar. Ecol. Prog. Ser., 9: 35.

Fenchel, T., 1982d, Ecology of heterotrophic microglagellates. IV. Quantitative occurrence and importance as consumers of bacteria, Mar. Ecol. Prog. Ser., 9:35.

Fenchel, T., Suspended marine bacteria as a food source, This volume.

Fenchel, T. and Blackburn, T.H., 1979, "Bacteria and mineral cycling," Academic Press, London.

Flood, P.R., 1978, Filter characteristics of appendicularian food catching nets, Experimentia, 34: 173.

Francisco, D.E., Mah, R.A. and Rabin, A.C., 1973, Acridine orange epifluorescence technique for counting bacteria, Trans. Amer. Micros. Soc., 92: 416.

Fuhrman, J.A. 1981, Influence of,method on the apparent size distribution of bacterioplankton cells: epifluorescence microscopy compared to scanning electron microscopy, Mar. Ecol. Prog. Ser., 5: 103.

Fuhrman, J.A. and Azam, F., 1980, Bacterioplankton secondary production estimates for coastal waters of British Columbia, Antartica and California, Appl. environ. Microbiol., 39: 1085.

Fuhrman, J.A. and Azam, F., 1982, Thymidine incorporation as a measure of heterotrophic bacterioplankton production in marine surface waters: evaluation and field results, Mar. Biol. 66: 109.

Gray, J.S., 1966, Factors controlling the localizations of populations in Protodrilus symbioticus Giard, J. anim. Ecol., 35: 435.

Hagström, A., Larsson, U., Hörstedt, P. and Normark, S., 1979, Frequency of dividing cells, a new approach to the determination of bacterial growth rates in aquatic environments, Appl. environ. Microbiol., 37: 805.

Hargrave, B.T. and Phillips, G.A., 1981, Annual in situ carbon dioxide and oxygen flux across a subtidal marine sediment, Estuar. Coastl Shelf Sci., 12 725.

Hobbie, J.E., Daley, R.J. and Jasper, S., 1977, Use of Nuclepore filters for counting bacteria by fluorescence microscopy, Appl. environ. Microbiol., 33: 1225.

Hobbie, J.E., Holm-Hansen, O., Packard, T.T., Pomeroy, L.R., Sheldon, R.W., Thomas, J.P. and Wiebe, W.J., 1972, A Study of the distribution and activity of microorganisms in ocean water, Linmnol. Oceanogr., 17: 544.

Itturiaga, R. and Zsolnay, 1981, Transformation of some dissolved organic components by a natural heterotrophic population, Mar. Biol., 62: 125.

Joiris, C., Billen, G., Lancelot, C., Daro, M.H., Momaerts, J.P., Bertels, A., Bossicarta, M., Nijs, J., and Hecq, J.H., 1982, A budget of carbon cycling in the Belgian coastal zone. Relative roles of zooplankton, bacterioplankton and benthos in the utilization of primary production, Neth. J. Sea, Res., 16: 260.

Jørgensen, C.B., 1976, "Biology of suspension feeding," Pergamon, Oxford, U.K.

Karl, D.M., 1980, Cellular nucleotide measurements and applications in microbial ecology, Microb. Rev., 44: 739.

King, G.M., and Klug, M.J., 1982, Glucose metabolism in sediments of a eutrophic lake : tracer analysis of uptake and product formation, Appl. Environ. Microbiol., 44: 1308.

King, K.R., Hollibaugh, J.T. and Azam, F., 1980, Predator-prey interactions between the larvacean Oilopleura dioica and bacterioplankton in enclosed water columns, Mar. Biol., 56: 49.

Kirchman, D., and Mitchell, R., 1982, Contribution of particulate-bound bacteria to total microheterotrophic activity in five ponds and two marshes, Appl. environ. Microbiol½. 43: 200.

Koch, A.L., 1971, The adaptive responses of Escherichia coli to a feast and famine existence, Adv. Microbial Physiol., 6: 147.

Koop, K., 1982, Fluxes of material associated with the decomposition kelp on exposed sandy beaches and adjacent habitats, Ph. D. thesis, Univ. Cape Town.

Koop, K., Newell, R.C., and Lucas, M.I., 1982, Microbial regeneration of nutrients from the decomposition of macrophyte debris on the shore, Mar. Ecol. Prog, Ser., 9: 91.

Krambeck, C., Krambeck, H.-J. and Overbeck, J., 1981, Microcomputer-assisted biomass determination of plankton bacteria on scanning electron micrographs, Appl. environ. Microbiol., 42: 142.

Larsson, U. and Hagström, A., 1979, Phytoplankton exudate release as an energy source for the growth of pelagic bacteria, Mar. Biol., 52: 199.

Larsson, U. and Hagström, A., 1982, Fractionated phytoplankton primary production, exudate release, and bacterial production in a Baltic eutrophication gradient, Mar. Biol. 67: 57.

Linley, E.A.S. and Newell, R.C., 1981, Microheterotrophic communities associated with the degradation of kelp debris, Kieler Meeresforsch., 5: 345.

Linley, E.A.S. and Field, J.G., 1981, The nature and ecological significance of bacterial aggregation in a nearshore upwelling ecosystem, Estuar. Cstl Shelf Sci., 14: 1.

Linley, E.A.S. Newell, R.C. and Lucas, M.I., 1983, Quantitative relationships between phytoplankton, bacteria and heterotrophic microflagellates in shelf waters, Mar. Ecol. Prog. Ser., In press.

Mann, K.H., 1982, "The Ecology of coastal waters: A systems approach," Blackwell, Oxford.

Meyer-Reil, L.-A., 1979, Bacterial growth rates and biomass production, In: "Microbial ecology of a brackish water environment," G. Rheinheimer, ed., Springer-Verlag, Berlin.

Meyer-Reil, L.-A., 1983, Bacterial biomass and heterotrophic activity in sediments and overlying waters, In: "Heterotrophy in the sea," J.E. Hobbie, and P.J. Williams, eds., Plenun Press, New York.

Meyer-Reil, L.-A., Dawson, R., Liebezeit, G. and Tiedge, H., 1978, Fluctuations and interactions of bacterial activity in sandy beach sediments and overlying waters, Mar. Biol., 48: 161.

Meyer-Reil, L.-A., Bölter, M., Liebezeit, G., Schramm, W., 1979, Short-term variations in microbiological and chemical parameters, Mar. Ecol. Prog. Ser., 1: 1.

Moriarty, D.J.W., 1977, Improved method using muramic acid to estimate biomass of bacteria in sediments, Oecologia, 26: 317.

Newell, R.C., Lucas, M.I., and Linley, E.A.S., 1981, Rate of degradation and efficiency of conversion of phytoplankton debris by marine micro-organisms, Mar. Ecol. Prog. Ser., 6: 123.

Newell, R.C., The biological role of detritus in the marine environment, This volume.

Novitsky, J.A. and Kepkay, P.E., 1981, Patterns of microbial heterotrophy through changing environments in a marine sediment, Mar. Ecol. Prog.Ser., 4: 1.

Pamatmat, M.M., Graf, G., Bengtsson, W. and Novak, C.S., 1981, Heat production, ATP concentration and electron transport activity of marine sediments, Mar. Ecol. Prog. Ser., 4: 135.

Parsons, T.R., Albright, L.J., Whitney, F., Wong, C.S. and Williams, P.J., 1981, The effect of glucose on the productivity of of seawater - an experimental approach using controlled aquatic ecosystems, Mar. Envir. Res., 4: 229.

Payne, W.T., and Wiebe, W.J., 1978, Growth yield and efficiency in chemosynthetic microorganisms, Ann. Rev. Microbiol., 32: 115.

Paul, J.H., 1982, Use of Hoechst Dyes 33258 and 33342 for enumeration of attached and planktonic bacteria, Appl. env. Microbiol., 43: 939.

Porter, K.G. and Feig, Y.S., 1981, The use of DAPI for identifying and counting aquatic microflora, Limnol. Oceanogr., 25: 943.

Riemann, F. and Schrage, M., 1978, The mucus-trap hypothesis on feeding of aquatic nematodes and implications for biodegradation and sediment texture, Oecologia (Berl.), 34: 75.

Reiswig, H.M., 1974, Water transport, respiration and energetics of three tropical marine sponges, J. exp. mar. Biol. Ecol., 14: 231.

Reiswig, H.M., 1975, The aquiferous systems of three marine demospongiae, J. Morphol., 145: 493.

Sheldon, R.W., Prakash, A. and Sutcliffe, W.H., 1972, The size distribution of particles in the ocean, Limnol. Oceanogr., 17: 327.
Sieburth, J.McN., 1979, "Sea microbes", Oxford Univ. Press, N.Y.
Sieburth, J.McN., 1983, Grazing of bacteria by protozooplankton in pelagic marine waters, In: "Heterotrophy in the sea," J.E. Hobbie and P.J. le B. Williams, eds., Plenum Press, New York.
Sieburth, J.McN., Johnson, K.M., Burney, C.H. and Lavoie, D.M., 1977, Estimation of in situ rates of heterotrophy using changes in dissolved organic matter and growth rates of picoplankton in diffusion culture, Helgoländer wiss. Meeresunters., 30: 565.
Smetacek, V., The supply of food to the benthos, This volume.
Sorokin, Y.I., 1979, Zooflagellates as a component of eutrophic and oligotrotrophic communities of the Pacific Ocean, Okeanologiya, 3: 476.
Sorokin, Y.I., 1981, Microheterotrophic organisms in marine ecosystems, In: "Analysis of marine ecosystems," A.R. Longhurst, ed., Academic Press, London.
Sorokin, Y.I. and Lyursarev, S.V., 1978, A comparative evaluation of two methods of determining the biomass of planktonic microflagellates, Oceanol. Acad. Sci. U.S.S.R., 18: 232.
Stuart, V., Field, J.G. and Newell, R.C., 1982, Evidence for absorption of kelp detritus by the ribbed mussel Aulacomya ater using a new ^{14}Cr-labelled microsphere technique, Mar. Ecol. Prog. Ser., 9: 263.
Vaccaro, R.F., Azam, F. and Hodson, R.E., 1977, Response of natural marine bacterial populations to copper: controlled ecosystems pollution experiment. Bull. Mar. Sci., 27: 17.
Van Es, F.B., and Meyer-Reil, L.A., 1982, Biomass and metabolic, activity of heterotrophic bacteria, Adv. Microbial Ecol., 6: 111.
Wangersky, P.J., 1977, The role of particulate matter in the productivity of surface waters, Helgol. wiss. Meeresunters., 30: 546.
Watson, S.W., Novitsky, J.J., Quinby, H.L. and Valois, F.W., 1977, Determination of bacterial number and biomass in the marine environment. Appl. environ. Microbiol., 33: 940.
Weise, W. and Rheinheimer, G., 1978, Scanning electron microscopy and epifluorescence investigation of bacterial colonization of marine sand sediments, Microb. Ecol., 4: 175.
Wiebe, W.J. and Pomeroy, L.R., 1972, Microorganisms and their association with aggregates and detritus in the sea: a microscopic study, In: "Detritus and its role in aquatic ecosystems", U. Melchiom-Santolini and J.W. Hopton, eds., Mem. Ist. Ital. Idrobiol., 24: 325.
Williams, P.J. Le B., 1981, Incorporation of microheterotrophic processes into the classical paradigm of the planktonic food web, Kieler Meeresforsch, 5: 1.

Williams, P.J. le B., Bacterial production in the marine food chain: the emperor's new suit of clothes, This volume.

Wright, R.T., Coffin, R.B., Ersing, C.P. and Pearson, D., 1982, Field and laboratory measurements of bivalve filtration of natural marine bacterioplankton, Limnol. Oceanogr. 27: 91.

PARTICIPANTS

A. Alldredge, U.S.A.
M.V. Angel, U.K.
F. Azam, U.S.A.
U. Brockmann, Germany
J.M. Colebrook, U.K.
Y. Collos, France
E.D.S. Corner, U.K.
M. Elbrächter, Germany
M.J.R. Fasham, U.K.
T. Fenchel, Denmark
J. Field, South Africa
S. Fowler, Monaco
B. Frost, U.S.A.
W.W.C. Gieskes, Netherlands
J. Goldman, U.S.A.
J.S. Gray, Norway
L. Guglielmo, Italy
R.L. Haedrich, Canada
A. Herbland, France
P.M. Holligan, U.K.
J. Horwood, U.K.
M. Huntley, U.S.A
G. Jackson, U.S.A.
B.O. Jansson, Sweden
I.R. Joint, U.K.
C. Joiris, Belgium
P. Jumars, U.S.A.
W.M. Kemp, U.S.A.
R.S. Lampitt, U.K.
P. Lasserre, France
P. Le Borgne, France
L. Legendre, Canada
M. Lewis, Canada
A.R. Longhurst, Canada
K.H. Mann, Canada
R. Margalef, Spain
J.B.L. Matthews, Norway
L.-A. Meyer-Reil, Germany
H. Minas, France
R. Newell, U.K.
S.W. Nixon, U.S.A.
T. Packard, U.S.A.
J. Paloheimo, Canada
M.M. Pamatmat, U.S.A.
M.J. Perry, U.S.A.
M.E.Q. Pilson, U.S.A.
T. Platt, Canada
H. Postma, Netherlands
S. Poulet, France
G. Radach, Germany
G. Rowe, U.S.A.
L. Saldanha, Portugal
E. Sakshaug, Norway
G.D. Sharp, Italy
H.R. Skjoldal, Norway
V. Smetacek, Germany
D.F. Smith, Australia
F. Thingstad, Norway
R.E. Ulanowicz, U.S.A.
E. Ursin, Denmark
J.H. Vosjan, Netherlands
R. Warwick, U.K.
P.J. Le B. Williams, U.K.
R. Williams, U.K.
F. Wulff, Sweden
A.A. Yayanos, U.S.A.

AND MATERIALS IN
NATO ADVANCED RESEARCH INSTITUTE
SEMINAIRE DE RECHERCHES AVANCEES OTAN
ET
MATIERE
LES ECOSYSTEMES
MAI

Front row (left to right):
Y. Collos, J. Goldman, A. Alldredge, S. Fowler, R. Haedrich, F.Azam,
E. Sakshaug, H. Minas, V. Smetacek, R. Ulanowicz, J. Field.

Second row:
L. Meyer-Reil, -, -, Mme. B. Lasserre, M. Pilson, R. Newell,
M. Huntley, M. Pamatmat, L. Saldanha, Mme. Minas, G. Radach,
J. Mathews, M.J. Perry, L. Legendre, I. Joint, G. Rowe.

Third row:
A. Yayanos, E. Corner, P. Le Borgne, A. Herbland, A. Longhurst,
P. Lasserre, H. Postma, S. Nixon, F. Wulff, D. Smith, M. Fasham,
K. Mann, J. Paloheimo, C. Joiris, P. Williams, J. Gray, J. Vosjan.

Fourth row:
U. Brockmann, ?, F. Thingstad, B. Frost, G. Sharp, L. Guglielmo,
T. Platt, R. Margalef, J. Castell, E. Ursin.

Back row:
P. Holligan, M. Angel, M. Colebrook, H. Skjoldal, R. Warwick,
R. Williams, J. Horwood, G. Jackson, P. Jumars, M. Kemp, R. Lampitt,
M. Lewis, Mme Meyer-Reil, W. Gieskes, B.-O. Jansson, ?.

INDEX